Chiral Building Blocks in Asymmetric Synthesis

Chiral Building Blocks in Asymmetric Synthesis

Synthesis and Applications

Edited by

Elżbieta Wojaczyńska
Jacek Wojaczyński

WILEY-VCH

Editors

Prof. Elżbieta Wojaczyńska
Wrocław University of Science and Technology
Faculty of Chemistry
Wybrzeże Wyspiańskiego 17
50 370 Wrocław
Poland

Dr. Jacek Wojaczyński
University of Wrocław
Faculty of Chemistry
14 F. Joliot-Curie St.
50 383 Wrocław
Poland

Cover Image: Courtesy of Prof. E. Wojaczyńska

Library of Congress Card No.:
applied for

British Library Cataloguing-in-Publication Data
A catalogue record for this book is available from the British Library.

Bibliographic information published by the Deutsche Nationalbibliothek
The Deutsche Nationalbibliothek lists this publication in the Deutsche Nationalbibliografie; detailed bibliographic data are available on the Internet at <http://dnb.d-nb.de>.

Print ISBN: 978-3-527-34946-3
ePDF ISBN: 978-3-527-83418-1
ePub ISBN: 978-3-527-83419-8
oBook ISBN: 978-3-527-83420-4

Typesetting Straive, Chennai, India
Printing and Binding CPI Group (UK) Ltd, Croydon, CR0 4YY

C115417_260722

Contents

Preface

The concept of chiral building blocks has emerged with the first total syntheses of natural products. Their complex architecture with a defined stereochemistry could be conveniently afforded with the use of enantiopure, easily available reactants: amino acids, carbohydrates, terpenes, simple alkaloids – compounds from a collection called a "chiral pool." Their introduction at the early stage opens the route to stereoselective creation of subsequent stereogenic centers due to the possible chiral induction. The main drawbacks of this approach lie in the necessity of using stoichiometric amounts of chiral starting materials and in the fact that they are typically found in one enantiomeric form – limiting the accessible configuration of the target molecule.

Over the years, with the development of asymmetric synthesis, the attractiveness of building blocks has faded, as methods based on chiral, reusable auxiliaries, and catalysts (metal complexes, organic molecules, and enzymes) have gained popularity. Chirality multiplication resulting from the use of substoichiometric ("catalytic," to use a common, but imprecise word) doses of such inducers, typically available as both optical antipodes, has captured the imagination of synthetic chemists. Their significance was also manifested by two Nobel Prizes awarded in 2001 to William S. Knowles, Ryōji Noyori, and K. Barry Sharpless for their work on stereoselective catalysis, and 20 years later to Benjamin List and David McMillan for the development of asymmetric organocatalysis.

It does not mean, however, that chiral pool concept has been completely forgotten or neglected. Perhaps pushed to the background by new discoveries and attracting less academic interest, it has always been of practical importance for the industry, in particular, for pharmaceutical companies exploiting all efficient and cheap routes for production of chiral drugs. For many preparations, the use of starting enantiomerically pure material is still the method of choice. And nowadays the chiral pool has been significantly enriched, to include not only compounds isolated from natural sources, but also simple synthetic building blocks, now available through efficient synthesis (sometimes combined with separation of enantiomers) or – more and more often – using biotechnological methods. They deliver both "natural," traditional building blocks (amino- and hydroxyacids, terpenes, certain alkaloids) but also their less abundant enantiomers or diastereomers, and add new types of functionalities, e.g. amines or epoxides. A blend of approaches becomes common: reagents from the chiral pool often serve as the primary source of optical activity of many nonracemic compounds applied in catalytic processes, but, in turn, some of them are produced in asymmetric synthesis or with the use of enzymes. We hope that this book illustrates the current status of chiral building blocks in chemistry.

The book does not pretend to be comprehensive; instead of showing a full landscape, rather versatility and diversity of chiral building blocks are presented, from simple to complex ones, representing various types of chirality, originating from the presence of a stereogenic center (carbon atom or heteroatom), axis, or plane to helical and topological chirality. The authors – specialists in the field, some more experienced, and some at earlier stage of their careers – focus on the preparation of synthetic building blocks, and application of them as well as of natural chiral compounds in the construction of complex chiral scaffolds: drugs, agrochemicals, but also ligands and catalysts for organic synthesis.

We start with the smallest cyclic chiral building blocks, namely cyclopropene derivatives, whose stereoselective preparation is presented in Chapter 1. Two subsequent chapters focus on various ring systems containing heteroatoms: their preparation, and application as chiral reagents and auxiliaries. Again, there is no clear line between building blocks and auxiliaries: up to the moment when the chiral element remains in the product, it can be treated as a building block which may be eventually removed, but sometimes not entirely.

The overlap of chiral building block and chiral ligand/catalyst concept is clearly visible in Chapter 4, devoted to derivatives of 1,2-diaminocyclohexane (DACH), and Chapter 16 which concentrates on 1,1,-bi-2-naphthol (BINOL) and similar compounds – their preparation and applications in asymmetric transformations.

Diketopiperazines, described in Chapter 5, serve as a bridge between synthetic cyclic compounds and – as the simplest cyclic dipeptides – natural chiral building blocks. A series of chapters that follows focus on typical chiral pool members: amino acids (Chapter 6), carbohydrates (Chapter 7), terpenes (monoterpenes in Chapter 8, and examples of diterpenes in Chapter 9), and alkaloids (Chapter 10). In case of steroids (Chapter 11), the stress is on their synthesis from the appropriate building blocks, but representative transformations are also shown.

Chapters 12, 13, and 14 describe applications in asymmetric synthesis of chiral phosphorus, sulfur, and selenium-containing compounds, respectively. The presence of stereogenic P or S atoms in the structure opens the route to efficient transfer of chirality.

Preparation of axially chiral allenes and their use for generation of new stereogenic elements in the product of their transformation is a subject of Chapter 15. After the already-mentioned BINOL representing planar or axial chirality, chiral acenes are described as an example of helical building blocks (Chapter 17). Chiral assemblies based on a low-symmetry isomer of a regular porphyrin are subsequently presented. The book is concluded with a chapter devoted to preparation of molecules exhibiting topological chirality: catenanes, rotaxanes, and molecular knots further used in the construction of more complex chiral architectures.

We believe that the content of the book will meet the demand of all readers interested in the synthesis and applications of chiral compounds who can find here the inspiration for their further research.

The chapters contain a significant amount of the gathered and thoroughly reviewed material. The effort of all the contributors is highly appreciated. We would like to express also our gratitude to Ms. Katherine Wong and Dr. Elke Maase from Wiley for their patience and coordinating the work.

Wrocław University of Science and Technology
University of Wrocław
March 2022

Elżbieta Wojaczyńska
Jacek Wojaczyński

Foreword

Chirality is a "Signature of Life" and beyond any doubt one of the most fascinating and fundamental aspects of the chemical sciences. The great importance of controlling molecular chirality with high precision goes far beyond its key role in medicinal chemistry to encompass among others natural product synthesis, asymmetric catalysis, materials design, supramolecular systems, and molecular machines. At the basis of the design and synthesis of molecules and materials, where homochirality is an essential structural feature, are usually the chiral building blocks available. Frequently, the access to and versatility of chiral building blocks dictate the synthesis strategy to be chosen. Traditionally we have been highly dependent on chiral natural products and their derivatives denoted as the "natural chiral pool." In the past decades, due to the amazing progress in asymmetric synthesis and catalysis, we have seen a multitude of novel chiral buildings blocks rapidly emerging.

In this highly timely volume on chiral building blocks in asymmetric synthesis the authors not only provide an overview of the most important chiral compounds available to the synthetic chemist but also show the richness and versatility of their structures. It goes without saying that sugars, terpenes, amino acids, and alkaloids and their derivatives will continue to occupy central stage as they are a superb gift from mother nature. They are often indispensable as chiral synthons, to make the chiral ligands for transition metal catalysts, are crucial in asymmetric organocatalysis, and serve as chiral auxiliaries in asymmetric transformations. But the availability of novel chiral building blocks, with the benefit of easy access to both enantiomers, enhanced functionality or stereo-elements based on, e.g. sulfur and phosphorus, has drastically expanded our synthetic repertoire.

Despite all the fascinating problems and discoveries in stereochemistry it should be remembered that the synthetic chemist needs chiral building blocks, and their essential role also in the future of our discipline can hardly be overestimated. This book also vividly reminds me of my first strides as a young student in stereochemistry making Binol via asymmetric oxidative coupling. How could I have realized that the chiral Binol unit is decades later one of the key chiral buildings blocks for numerous chiral ligands and catalysts in modern asymmetric synthesis. Having a scholarly and state-of-the-art overview of the variety of chiral building blocks currently available and detailed insight into their synthetic versatility is to be applauded and will be highly beneficial to the practitioner of modern synthetic chemistry. Controlling chirality, especially with increasing molecular complexity, is often still a bit of an adventure and I hope this volume will be a great source of inspiration and provide the stereochemical insights and show opportunities for the next steps on your journey into asymmetric synthesis.

University of Groeningen Ben L. Feringa

1

Enantioselective Synthesis of Cyclopropenes

Virginie Carreras and Thierry Ollevier

Université Laval, Département de chimie, 1045 avenue de la Médecine, Québec, QC G1V 0A6, Canada

1.1 Introduction

Cyclopropene is the smallest unsaturated cyclic hydrocarbon. Its preparation is relatively tricky because it is an unstable compound that has a high ring strain. This strain energy of cyclopropene (228 kJ mol^{-1}) is basically double of the one of cyclopropane (115 kJ mol^{-1}), mainly because of the high angular strain present in the former [1, 2]. However, cyclopropenes containing one or two substituents on the aliphatic position are relatively stable and easily accessible compounds.

1.1.1 Synthesis of Cyclopropenes

The synthesis of the first cyclopropene was carried out by Demyanov and Doyarenko in 1922 via the thermal decomposition at high temperature (320°C) of trimethylcyclopropylammonium hydroxide under a carbon dioxide atmosphere. After the Hofmann elimination step, cyclopropene was obtained with low yield [3, 4]. An improved procedure was reported starting from allyl chloride and sodium bis(trimethylsilyl)amide that allowed to isolate cyclopropene in a better yield (40%) than in the previously published procedure [5]. An enantioenriched cyclopropene was prepared for the first time by Breslow and Douek via a resolution using cinchonine [6].

Cyclopropenes possessing various substitutions can be prepared: (i) by [2+1] cycloaddition between an alkyne and a free carbene or a metal carbene, most often resulting from the decomposition of diazo compounds; by cycle contraction, by retro-Diels–Alder reaction; from the nucleophilic attack on a previously synthesized cyclopropene; by isomerization; by rearrangement initiated by photochemistry; and finally, via elimination reactions induced by strong bases [7].

Diazo compounds have also been used in the development of asymmetric versions of cyclopropenations of alkynes. The various catalytic asymmetric methods are reviewed in this chapter.

Chiral Building Blocks in Asymmetric Synthesis: Synthesis and Applications, First Edition.
Edited by Elżbieta Wojaczyńska and Jacek Wojaczyński.

1.1.2 Reactivity of Cyclopropenes

Cyclopropenes can be involved in a large variety of chemical transformations, whose driving force is the release of the ring strain. A few reviews highlighting the synthetic potential of cyclopropenes have been published [8–14].

The reactivity of cyclopropenes has been illustrated in various synthetically useful transformations. Releasing the strain energy enables cyclopropenes to undergo reactions that would be more challenging in other alkenes, e.g. hydrofunctionalization and cycloaddition reactions.

The catalytic stereoselective functionalization of cyclopropenes has been reported throughout various synthetic transformations (Scheme 1.1), i.e. carbocupration [15], carbozincation [16, 17], carbomagnesiation [18], Fe- [19] and Pd-catalyzed carbozincation [20], hydroboration [21], hydroformylation [22], hydroacylation [23], hydroalkylation [24], hydroalkynylation [25], and hydrosilylation [26, 27]. Also, chiral cyclopropylamines have been obtained via highly enantioselective Cu-catalyzed three-component cyclopropene alkenylamination [28]. Using these methods, a large diversity of enantioenriched cyclopropanes can be accessed. Getting high selectivities, i.e. ring-retention vs. ring-opening processes, is often a challenge. In this regard, Dong and coworkers elegantly demonstrated that the choice of a bisphosphine ligand could orient the hydrothiolation of cyclopropenes toward the formation of cyclopropyl sulfides or allylic sulfides [29]. The Pd-catalyzed selective alkynylallylation of the C—C σ bond of tetrasubstituted cyclopropenes has also been disclosed [30].

Cyclopropenes undergo a ring-opening process to generate rhodium carbenes due to the ring strain. This approach has been used in various synthetic transformations, e.g. the synthesis of dicarbonyl-functionalized 1,3-dienes by the reaction of enaminones with cyclopropenes in the presence of a rhodium catalyst [31].

Cyclopropenes have been used in copper-catalyzed [3+2] cycloaddition reactions as dipolarophiles [32]. Donor–acceptor cyclopropenes have been used in [4+3]-cycloaddition reactions with benzopyrylium salts [33], and in Favorskii-type ring opening [34].

Cyclopropenes have also been used in radical chemistry. Waser has reported the radical azidation of cyclopropenes leading to alkenyl nitriles and polycyclic aromatic compounds [35]. A visible-light-promoted addition of α-bromoacetophenones onto the cyclopropene π-system in the presence of the *fac*-Ir(ppy)$_3$ has been described by Landais and coworkers [36].

The synthesis of difluoro- and trifluoromethylated derivatives of cyclopropenes has been disclosed using diazo compounds [37–39]. Continuous-flow methods have been reported for the difluorocyclopropenation of alkynes using trimethylsilyl trifluoromethane ($TMSCF_3$) [40] and for the photochemical cyclopropenation of alkynes using trifluoromethyl diazirines [41].

Cyclopropenes have appeared to be key intermediates in various total syntheses. The intramolecular Pauson–Khand reaction of a cyclopropene with an alkyne has been used in the synthesis of (−)-pentalenene and (−)-spirochensilide A [42, 43].

The polymerization of cyclopropenes, in particular the well-developed ring-opening metathesis polymerization, takes advantage of their high strain

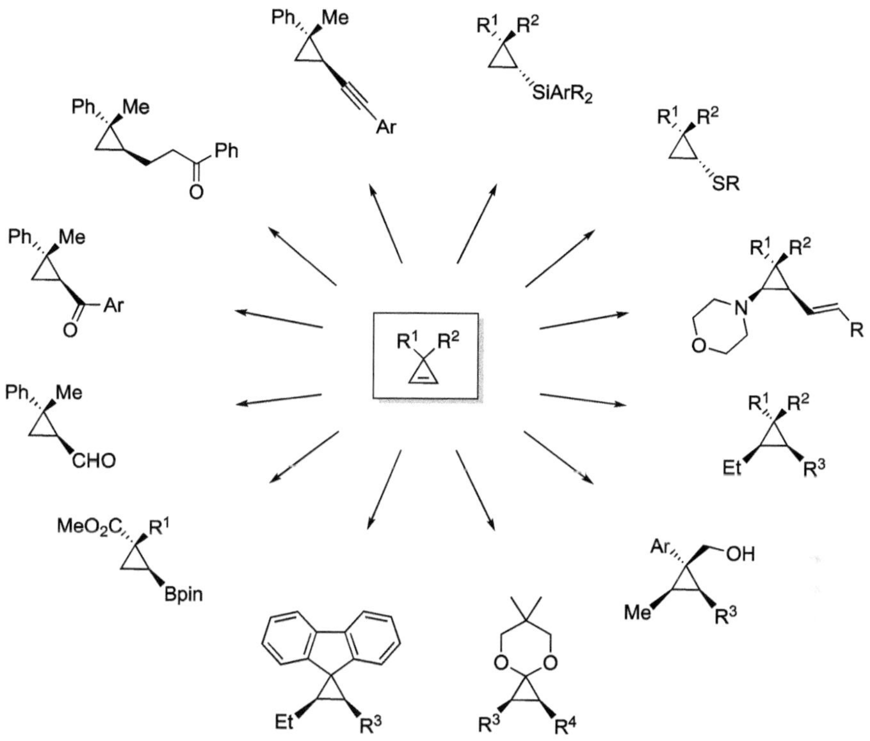

Scheme 1.1 Catalytic stereoselective functionalization of cyclopropenes.

energy [44]. Recent uses of cyclopropene units have recently emerged in the study of biological systems. Cyclopropenes can react quickly in tetrazine and photoclick ligation reactions [43].

1.2 Metal-Catalyzed Enantioselective Syntheses of Cyclopropenes

The [2+1] cycloaddition between an alkyne and a metal carbene is the main enantioselective approach to prepare cyclopropenes. The metal carbene usually results from the decomposition of a diazo compound using a chiral metal catalyst.

A practical non-enantioselective method was first published by Teyssié and coworkers using a Rh-catalyzed [2+1] cycloaddition of methyl diazoacetate with alkynes [45].

1.2.1 Rhodium Catalysis

Asymmetric versions of these transformations came out with pioneering work of Doyle and coworkers and Müller with the use of chiral complexes based on rhodium [46–61]. Doyle and coworkers focused their work on the cyclopropenation of terminal alkylated alkynes with the use of chiral dirhodium(II) tetrakis[methyl 2-oxopyrrolidine-5(*R*)-carboxylate] Rh_2(5*R*-MEPY)$_4$. This method was applied to various diazoacetates and was even extended to chiral diazo possessing auxiliaries derived from menthol (Scheme 1.2) [46, 47]. Using a significant excess of alkyne (20 equiv), cyclopropenes were obtained in high enantioselectivities.

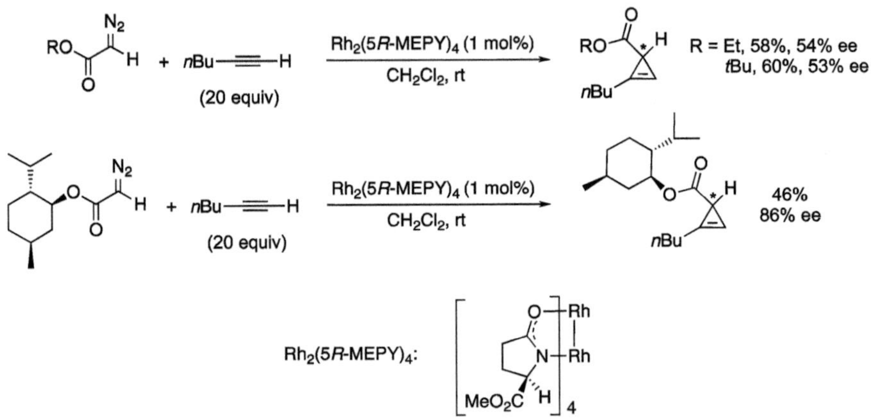

Scheme 1.2 Asymmetric cyclopropenation of terminal alkynes with Rh_2(5*R*-MEPY)$_4$ catalyst.

Müller used Rh_2(5*R*-MEPY)$_4$ for the cyclopropenation of propargylamines bearing two carboxyl or sulfonyl protecting groups using ethyl diazoacetate [48, 49]. The cycloaddition reaction proceeds smoothly at room temperature in CH_2Cl_2 with a slow addition of the diazo compounds via a syringe pump to 10 equiv of the alkyne. This method employing 3–7% of rhodium affords high yields and excellent enantioselectivities in the range of 90–97% ee (Scheme 1.3). Further derivatizations via selective deprotection of the TEOC derivatives (*N*,*N*-di-(2-trimethylsilyl ethoxycarbonyl)) could be illustrated by the synthesis of γ-aminobutyric acid (GABA) analogues containing the cyclopropene ring.

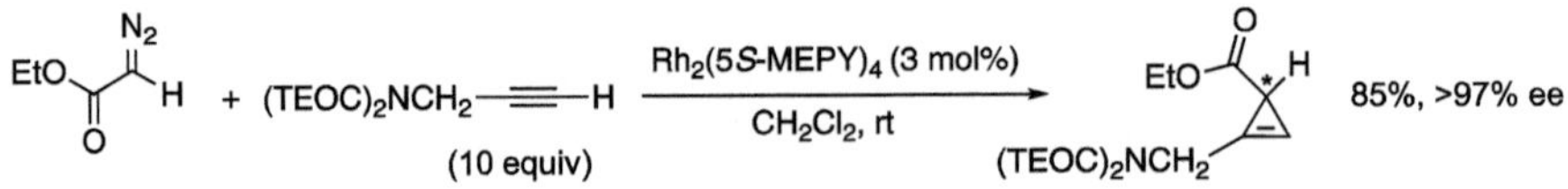

Scheme 1.3 Asymmetric cyclopropenation of terminal alkynes with Rh_2(5*S*-MEPY)$_4$.

These transformations were initially conducted with diazoesters, i.e. ethyl diazoacetate. As shown in Scheme 1.4, Davies' work with dirhodium tetrakis (*S*)-*N*-(dodecylbenzenesulfonyl) prolinate Rh_2(*S*-DOSP)$_4$ demonstrated that the catalyst was efficient for this transformation to get high enantioselectivities of the cyclopropenes via the asymmetric cyclopropenation of terminal alkynes using aryl or vinyl diazoacetates [62, 63]. A large excess of aromatic alkynes is used vs. the aliphatic ones, showing an improved reactivity of the latter. Computational studies

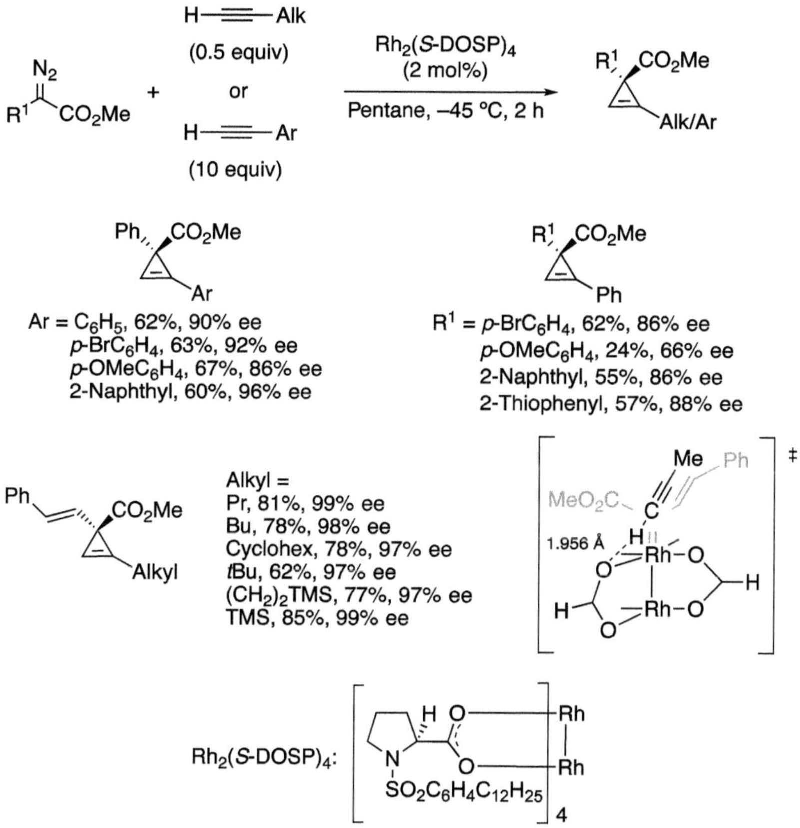

Scheme 1.4 Asymmetric cyclopropenation of terminal alkynes with $Rh_2(S\text{-DOSP})_4$.

demonstrated that the chiral induction of the process is governed by the specific orientation of the alkyne as it approaches the metal carbene. Specific orientation occurs due to the presence of a binding interaction between the alkyne hydrogen and the carboxylate ligand on the dirhodium catalyst.

A closely related observation was made by Hashimoto and coworkers with dirhodium carboxylate catalyst [$Rh_2(S\text{-tbpttl})_4$] emerging as a catalyst of choice for enantioselective cyclopropenation reactions of terminal alkynes with various alkyl-diazoacetates, in which high levels of asymmetric induction (up to 99% ee) as well as high chemoselectivities have been achieved [64]. In addition to the binding interaction between the alkyne hydrogen and the carboxylate ligand, the alkene formation through a 1,2-hydride shift was highlighted.

Corey and coworkers developed a new chiral rhodium catalyst $Rh_2(OAc)_x(DPTI)_y$ (DPTI: diphenyltriflylimidazolidinone) capable of affording excellent yields and enantioselectivities of various C1- and C3-substituted cyclopropenes [50–53]. This route involved mild reaction conditions with the use of 0.05–0.5 mol% of catalyst (Scheme 1.5). A rationale for understanding the chirality induction was provided with the help of X-ray structural data. A mechanistic model in which one of the ligand bridges is broken in the intermediate Rh–carbene complex was disclosed. The synthetic results led to conclusions regarding kinetically and thermodynamically favored pathways for the synthesis of mixed acetate–DPTI complexes. Experimental ^{13}C KIEs were investigated by Singleton and compared to calculated ones to establish the mechanism of the cyclopropenation reaction [65]. The plane geometry of the carbenoid carbon is oriented perpendicular to donating proximal Rh–N bonds. The alkyne approach *anti* to the Rh—N bond remains unobstructed. The chirality is determined by the orientation of the carboalkoxy group on the carbenoid which is affected by the proximal phenyl group from the DPTI ligand.

The use of mixed ligands on paddlewheel complexes was also tackled by Fox and coworkers. They showed that this route offers a versatile handle for diversifying catalyst structure and reactivity for cyclopropenation of terminal alkynes and cyclopropanation reactions [66].

In the same category of donor–acceptor diazo compounds, Koenigs' work with the use of trifluoromethyl diazo compounds further demonstrates the potential of a rhodium chiral catalyst for the enantioselective synthesis of cyclopropenes [38]. Up to 97% yield with 98% ee of trifluoromethylcyclopropenes using aliphatic terminal alkynes but also aromatic alkynes have been obtained (Scheme 1.6). The method was also tested on oligo-alkynes, thus making it possible to generate a subclass of rare racemic oligo-cyclopropenes with excellent yields.

Charette and coworkers developed a simple and highly stereoselective method of cyclopropenation of alkynes catalyzed by rhodium(II) with diacceptor-type diazo compounds [67]. This route represents a very efficient method to access cyano-phosphonate cyclopropenes with high yields and enantioselectivities (Scheme 1.7). Mild conditions and a normal addition of diazo compounds by avoiding the use of a syringe pump afforded excellent results with an easy and practical protocol. Given the reactivity of the cyano-carbene intermediates generated *in situ*, the scope of the substrates was also extended to the use of substituted allenes. This study highlighted the first enantioselective method for the synthesis of cyclopropane alkylidene diacceptors. Scheme 1.6 illustrates the results obtained for the cyclopropenation of alkynes.

Early work to incorporate a fluorine motif on the chiral cyclopropene unit was reported by Marek and coworkers (Scheme 1.8) [37]. This motif is incorporated through the use of difluorodiazoethane, obtained via diazotization of the corresponding amine. This study offers a wide range of enantioenriched difluoromethylated cyclopropenes (40 examples, up to 99% yield, 97% ee) within a short reaction time. Disubstituted alkynes have been mainly employed with an excess of the difluorinated diazo compound. Indeed, the use of terminal alkyne did not give them satisfactory results (no enantioselectivity obtained or no conversion). This cyclopropenation reaction is carried out at low temperature with a reaction time of 40 minutes allowing for total conversions. The synthetic utility of the difluorinated cyclopropenes obtained was demonstrated with subsequent applications, such as cross-coupling reactions, hydrogenation, Diels–Alder cyclization, and Pauson–Khand reaction.

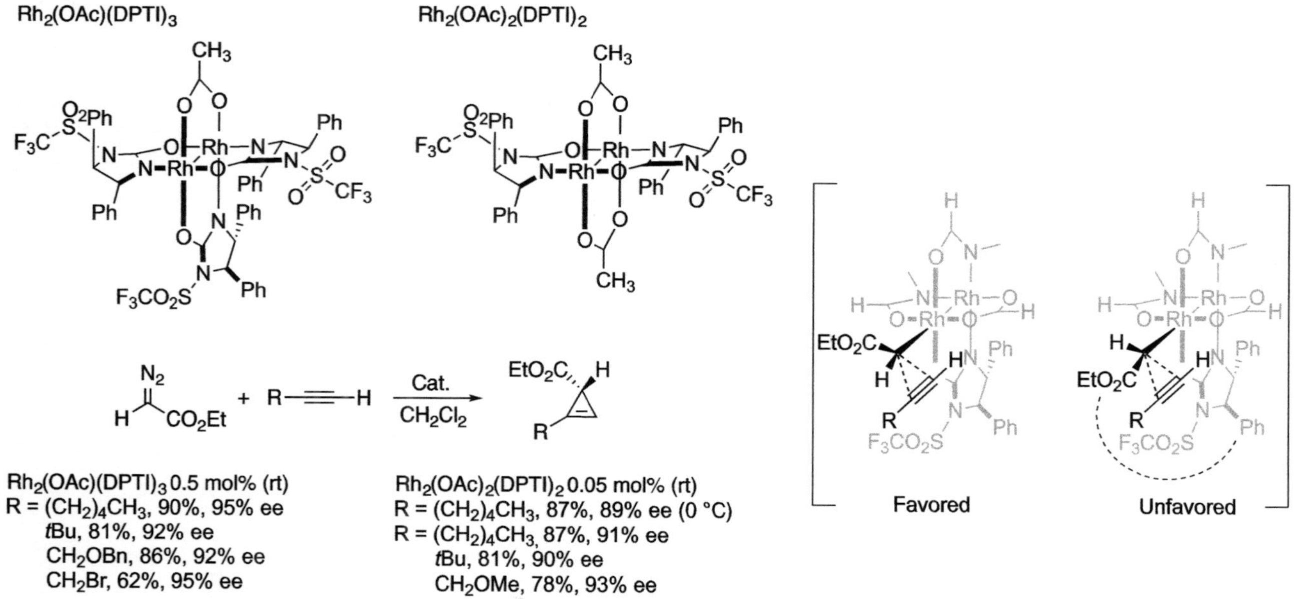

Scheme 1.5 Enantioselective [2+1]-cycloaddition reactions of ethyl diazoacetate and terminal acetylenes using mixed-ligand Rh(II) complexes.

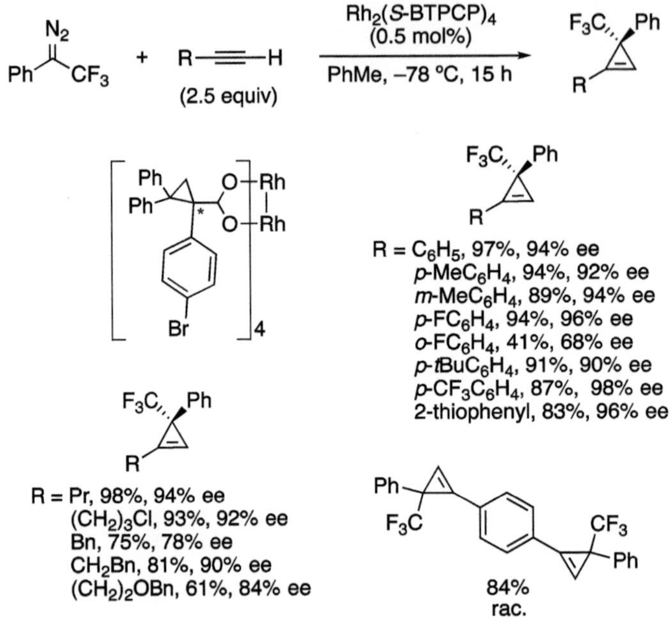

Scheme 1.6 Asymmetric cyclopropenation of terminal alkynes using *Rh*$_2$(*S*-BTPCP)$_4$.

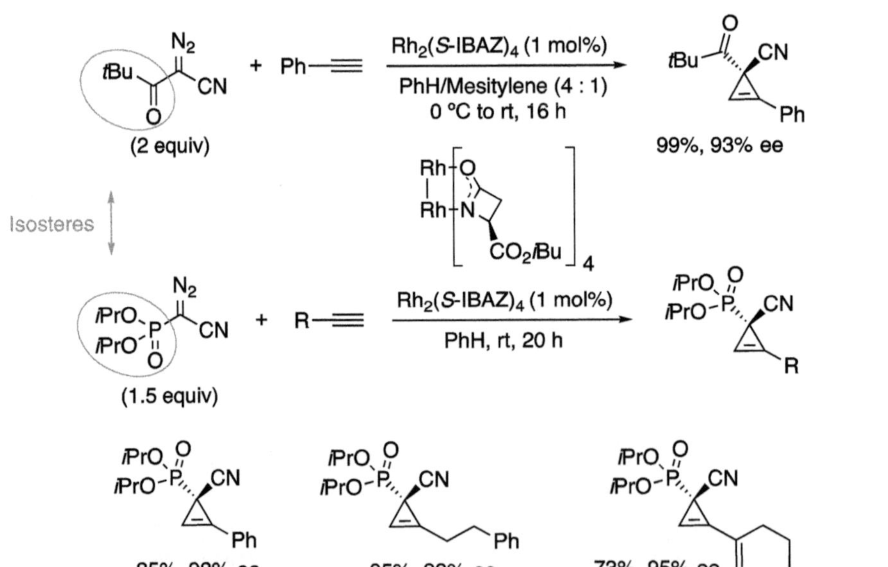

Scheme 1.7 Rh$_2$(*S*-IBAZ)4-catalyzed cyclopropenation of terminal alkynes with diacceptor diazo compounds.

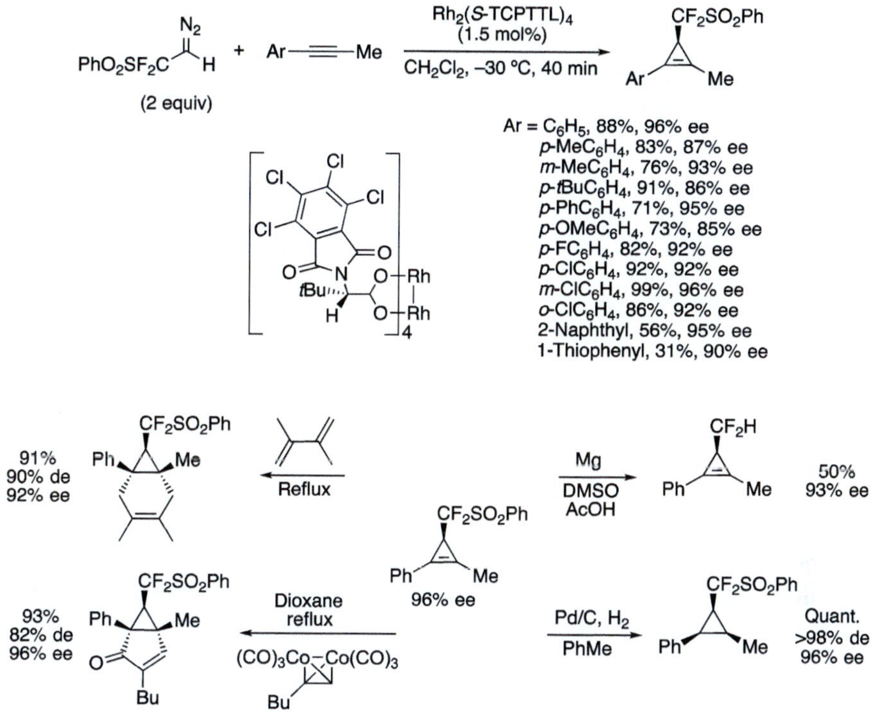

Scheme 1.8 First enantioselective synthesis of fluorinated cyclopropenes and their applications. Source: Zhang et al. [37]/John Wiley & Sons.

1.2.2 Copper Catalysis

Unlike rhodium catalysis, copper catalysis is underused. Indeed, only one example in the literature illustrates this cyclopropenation of alkynes with diazo compounds (Scheme 1.9) [56, 68]. Two intramolecular cyclopropenations allowing the formation of macrocycles were carried out using Cu^I and bis-oxazoline as chiral ligand. Poor results compared to rhodium were generally observed except for the synthesis of a macrocycle where a yield of 85% and 79% ee was obtained.

1.2.3 Iridium Catalysis

Katsuki and coworkers developed a synthesis of a chiral iridium (Ir(salen)) complex suitable to efficiently catalyze the cyclopropenation reaction of alkynes with two families of diazo compounds [69]. This cyclopropenation method can be carried out using donor–acceptor-type diazo compounds, but also using acceptor–acceptor-type diazo compounds (Scheme 1.10). The latter are less reactive toward metal-catalyzed decomposition. However, once decomposition by a metal takes place, the resulting metal carbenes are highly electrophilic and their reactions are less selective. This is why, in order to counterbalance this reactivity, a large excess of alkyne was used. The reactions provided highly enantioenriched cyclopropenes in excellent yields. Alkylated alkynes and terminal aryls were employed to show the selectivity of

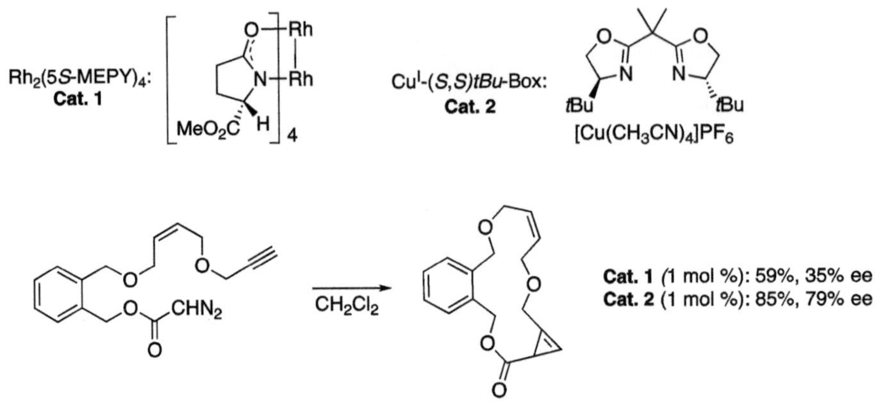

Scheme 1.9 Macrocyclization via cyclopropenation of alkynes using copper catalysis.

this transformation into a cyclopropene. In general, the yields are higher for the cyclopropenation of arylated alkynes than for alkylated ones. Moreover, the levels of enantioselectivities remain unchanged between the various families of alkynes, being more dependent on the nature of the electron withdrawing group (EWG) present on the diazo substrate.

1.2.4 Cobalt–Chiral Porphyrin Catalysis

Cobalt(II) complex with a chiral D_2-symmetrical porphyrin ligand was found to be a highly efficient chiral catalyst for the enantioselective cyclopropenation of alkynes with diazo compounds bearing acceptor–acceptor groups, such as cyanodiazoacetamides and cyanodiazoacetates (Scheme 1.11) [70]. Metal catalysts based on a Co^{II}–porphyrin were shown by Zhang and coworkers to be very effective in asymmetric cyclopropanation of a wide range of olefinic substrates with diazo compounds [71]. This cobalt(II)–porphyrin complex also catalyzes the cyclopropenation reaction of a large variety of terminal aromatic and conjugated alkynes possessing various steric and electronic properties, providing tri-substituted cyclopropenes with high yields and excellent enantioselectivities (up to 99% ee). Only terminal alkynes were used. In addition to the mild reaction conditions, this catalytic transformation exhibits a high degree of tolerance for the presence of functional groups around the alkyne. The consecutive diastereoselective addition of thiols on generated cyclopropenes allowed the synthesis of several cyclopropanes retaining 98% ee.

1.2.5 Gold and Silver Catalysis

Davies and coworker developed a gold catalyst ((*S*)-xylylBINAP$(AuCl)_2$ dimer), which, once activated by silver hexafluoroantimonate ($AgSbF_6$), enables highly enantioselective cyclopropenation reactions of internal alkynes with aryldiazoacetates [72]. Enantioselective cyclopropenation of terminal alkynes with Rh^{II},

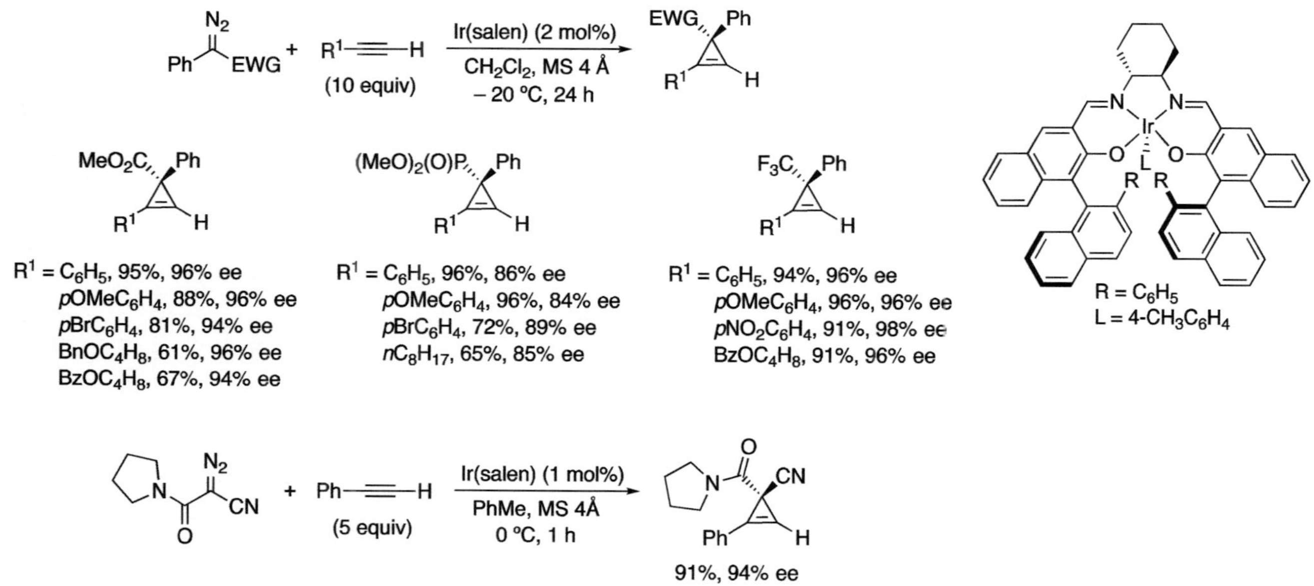

Scheme 1.10 Iridium-catalyzed enantioselective cyclopropenation of terminal alkynes.

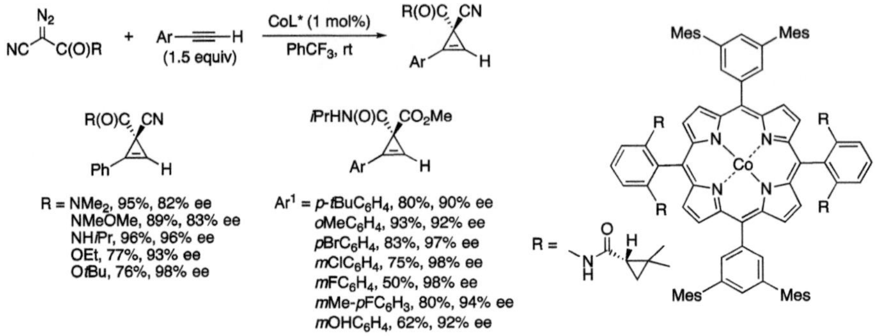

Scheme 1.11 Enantioselective cyclopropenation with a cobalt–porphyrin complex. Source: Cui et al. [70]/American Chemical Society.

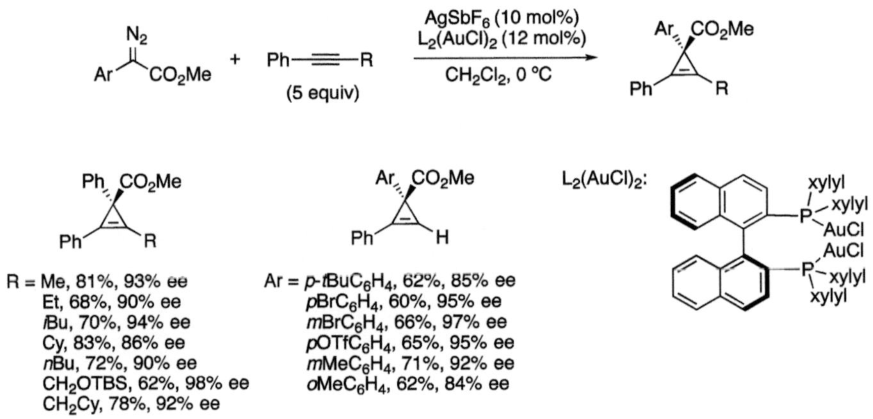

Scheme 1.12 Enantioselective cyclopropenation reaction catalyzed by a gold–silver chiral complex.

Co^{II}, and Ir^{III} catalysts is well established as previously seen. Despite this progress, extending asymmetric cyclopropenation reactions to disubstituted alkynes remains a challenge. Reactivity of gold carbenes and silver carbenes appears to be similar, where gold carbenes have a very different reactivity profile from that of rhodium carbenes. Indeed, these carbenic species are much less sensitive to steric effects, thus allowing an easier approach to disubstituted alkynes. Different disubstituted alkynes in slight excess have been used with various aryl diazoacetates under mild reaction conditions (Scheme 1.12). Yields around 80% have been obtained with high enantioselectivities. The joint use of $AgSbF_6$ is essential for the gold catalyst to initiate the cyclopropenation reaction. However, the exact structure of the active catalyst has not been determined. Hypotheses have nevertheless been proposed, such as mass spectrometry data of $L_2Au_2AgCl_2$ catalyst, suggesting the possibility of the presence of a mixed gold–silver complex. Indeed, it is well known that combining a silver salt with a gold pre-catalyst leads to a more complex structure than a simple phenomenon of ligand exchange by halide removal.

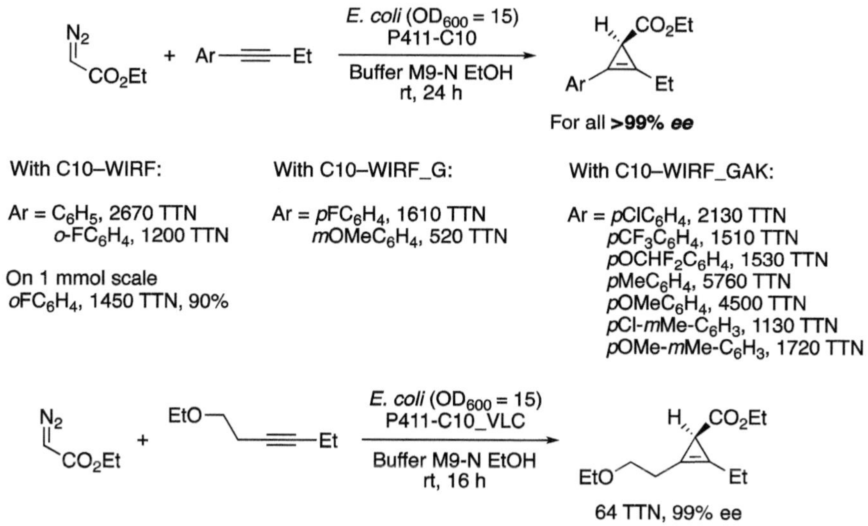

Scheme 1.13 Directed evolution catalysis for enantioselective synthesis of cyclopropenes.

1.2.6 Biocatalysis

Arnold demonstrated the directed evolution catalysis for the enantioselective synthesis of cyclopropenes. She developed a biocatalytic system based on cytochrome P411 in the form of *Escherichia coli* whole-cell catalysts able to provide access to a range of cyclopropenes by transferring carbene from ethyl diazoacetate with disubstituted alkynes in equimolar mixture [73, 74]. This evolutionary biocatalytic system provides high total turnover number (TTN) and cyclopropenes with unprecedented stereoselectivities (>99% ee for all, Scheme 1.13). This enzyme platform is also adaptable for the production of cyclopropenes on large scale (1 mmol), with even higher yields, which is notorious in the field of biocatalysis. Enantioselective cyclopropenation of internal aliphatic alkynes has also been shown possible, but catalytically much less efficient [74].

1.3 Other Synthetic Routes and Derivatizations of Enantioenriched Cyclopropenes

Another approach that could be envisioned for the preparation of chiral cyclopropenes involves the enantioselective desymmetrization of previously generated cyclopropenes [32, 75]. Scheme 1.14 illustrates an easy and stereoselective synthesis of an enriched cyclopropene. Subsequent acyl transfer with azide and Curtius rearrangement provides protected α-amino acid derivatives, which are shown to be stable to harshly acidic and basic reaction conditions. Starting achiral cyclopropenes are generated via the cyclopropenation of the corresponding terminal alkyne with diazomalonate.

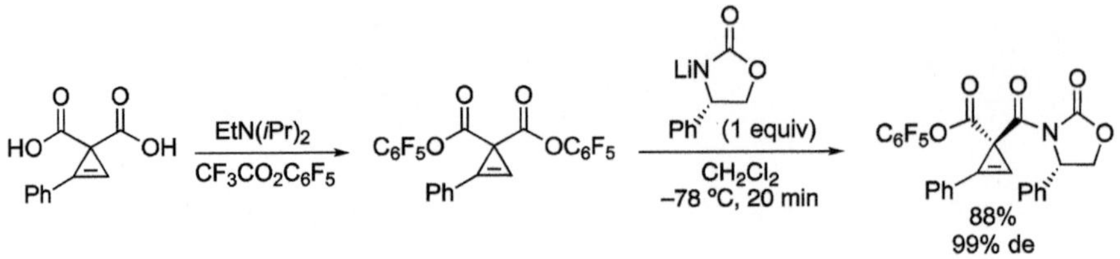

Scheme 1.14 Enantioselective desymmetrization of bis-pentafluorophenylesters.

Marek and coworkers have described the preparation of enantiomerically pure cyclopropenyl carbinols by kinetic resolution via Sharpless epoxidation of racemic cyclopropene alcohols [76]. Only one enantiomer is epoxidized leading to a hypothetical unstable chiral 2-oxabicyclobutane. The remaining non-epoxidized enantiomer is obtained with high enantiomeric excess.

Fox and coworker reported on the first enantioselective synthesis of (−)-pentalenene via the Pauson–Khand cycloaddition involving enantioenriched cyclopropenes (Scheme 1.15) [42]. This enantioselective synthesis has been described with an overall yield of 9% from the known diyne. The key to success for this synthesis was the use of enantiopure cyclopropene for the Pauson–Khand step leading to the desired quaternary center using 0.6 equiv $Co_2(CO)_8$ and TMTU (tetramethylthiourea). This cyclopropene was synthesized via the cyclopropenation of the corresponding terminal alkyne with ethyl diazoacetate using a chiral dirhodium complex.

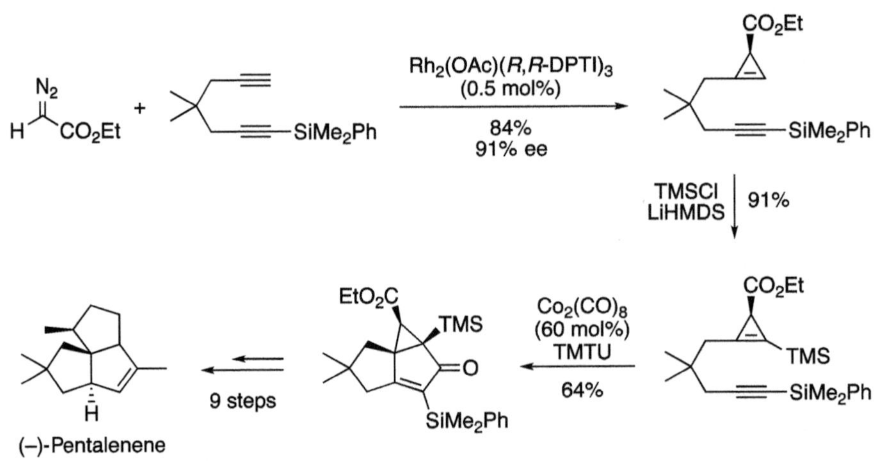

Scheme 1.15 Synthesis of (−)-pentalenene via Pauson–Khand cycloaddition involving enantioenriched cyclopropenes. Source: Pallerla and Fox [42]/American Chemical Society.

Starting with a camphene compound has also been used for the diastereoselective synthesis of cyclopropene derivatives [77].

The preparation of enantiomerically enriched 1,2-disubstituted cyclopropenes remains challenging. A convenient way to access these kinds of cyclopropenes is a post-functionalization of enantioenriched mono-substituted cyclopropenes (Scheme 1.16). These derivatizations include palladium catalysis or the trapping of nucleophilic chiral cyclopropenyl metal species with electrophiles [51, 78, 79].

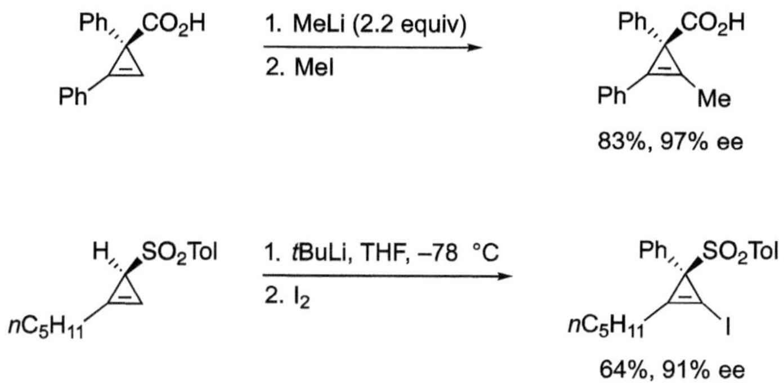

Scheme 1.16 Post-functionalization of enantioenriched mono-substituted cyclopropenes.

1.4 Summary and Prospect

The synthesis of enantioenriched cyclopropenes has been reviewed with the focus on the cyclopropenation of alkynes involving metal-catalyzed decomposition of diazo compounds. Various chiral catalysts are known for the enantioselective cyclopropenation of terminal alkynes. Tremendous work has been done with the use of chiral dirhodium paddlewheels catalysts. Synthetic methods involving other metals have been highlighted. Few reports involve the cyclopropenation of disubstituted alkynes, one of them being the use of enzymatic directed evolution. Enantioselective desymmetrization is one alternative in the synthesis of chiral cyclopropenes, as well as further functionalizations of enantioenriched cyclopropenes.

An interest for the design of new reactions is emerging from the atypical three-membered unit of cyclopropenes. The high angular strain and relative stability of cyclopropenes containing one or two substituents on the aliphatic position will be undoubtedly exploited in the development of new stereoselective reactions.

References

1 Bach, R.D. and Dmitrenko, O. (2004). Strain energy of small ring hydrocarbons. Influence of C-H bond dissociation energies. *J. Am. Chem. Soc.* 126: 4444–4452.

2 Wiberg, K.B. (1986). The concept of strain in organic chemistry. *Angew. Chem. Int. Ed. Engl.* 25: 312–322.

3 Demyanov, N.Y. and Doyarenko, M.N. (1922). Cyclopropene. *Bull. Acad. Sci. Russ.* 16: 297–320.

4 Demyanov, N.Y. and Doyarenko, M.N. (1929). Cyclopropene and some of its derivatives. *Bull. Acad. Sci. Russ.* 23: 653–665.

5 Binger, P., Wedemann, P., Goddard, R., and Brinker, U.H. (1996). Cyclopropene: a new simple synthesis and Diels–Alder reactions with cyclopentadiene and 1,3-diphenylisobenzofuran. *J. Org. Chem.* 61: 6462–6464.

6 Breslow, R. and Douek, M. (1968). Antiaromatic effects in cyanocyclopropenyl anions. *J. Am. Chem. Soc.* 90: 2698–2699.

7 Zhu, Z.-B., Wei, Y., and Shi, M. (2011). Recent developments of cyclopropene chemistry. *Chem. Soc. Rev.* 40: 5534–5563.

8 Vicente, R. (2021). C-C bond cleavages of cyclopropenes: operating for selective ring-opening reactions. *Chem. Rev.* 121: 162–226.

9 Li, P., Zhang, X., and Shi, M. (2020). Recent developments in cyclopropene chemistry. *Chem. Commun.* 56: 5457–5471.

10 Raiguru, P.B., Nayak, S., Mishra, R.D. et al. (2020). Synthetic applications of cyclopropene and cyclopropenone: recent progress and developments. *Asian J. Org. Chem.* 9: 1088–1132.

11 Vicente, R. (2016). Recent progresses towards the strengthening of cyclopropene chemistry. *Synthesis* 48: 2343–2360.

12 Marek, I., Simaan, S., and Masarwa, A. (2007). Enantiomerically enriched cyclopropene derivatives: versatile building blocks in asymmetric synthesis. *Angew. Chem. Int. Ed.* 46: 7364–7376.

13 Rubin, M., Rubina, M., and Gevorgyan, V. (2006). Recent advances in cyclopropene chemistry. *Synthesis* 2006: 1221–1245.

14 Fox, J.M. and Yan, N. (2005). Metal-mediated and -catalyzed nucleophilic additions to cyclopropenes. *Curr. Org. Chem.* 9: 719–732.

15 Isaka, M. and Nakamura, E. (1990). Carbocupration of cyclopropene. Asymmetric synthesis of quaternary carbon centers. *J. Am. Chem. Soc.* 112: 7428–7430.

16 Kubota, K., Isaka, M., Nakamura, M., and Nakamura, E. (1993). Ligand control in the stereoselective allylzincation of cyclopropenes. *J. Am. Chem. Soc.* 115: 5867–5868.

17 Müller, I. and Marek, D.S. (2015). Asymmetric copper-catalyzed carbozincation of cyclopropenes en route to the formation of diastereo- and enantiomerically enriched polysubstituted cyclopropanes. *J. Am. Chem. Soc.* 137: 15414–15417.

18 Liu, J.M. and Fox, X. (2005). Enantioselective, facially selective carbomagnesation of cyclopropenes. *J. Am. Chem. Soc.* 128: 5600–5601.

19 Nakamura, E. and Yoshikai, N. (2010). Low-valent iron-catalyzed C-C bond formation-addition, substitution, and C-H bond activation. *J. Org. Chem.* 75: 6061–6067.

20 Krämer, K., Leong, P., and Lautens, M. (2011). Enantioselective palladium-catalyzed carbozincation of cyclopropenes. *Org. Lett.* 13: 819–821.

21 Edwards, A., Rubina, M., and Rubin, M. (2018). Directed RhI-catalyzed asymmetric hydroboration of prochiral 1-arylcycloprop-2-ene-1-carboxylic acid derivatives. *Chem. Eur. J.* 24: 1394–1403.

22 Sherill, W.M. and Rubin, M. (2008). Rhodium-catalyzed hydroformylation of cyclopropenes. *J. Am. Chem. Soc.* 130: 13804–13809.

23 Phan, D.H.T., Kou, K.G.M., and Dong, V.M. (2010). Enantioselective desymmetrization of cyclopropenes by hydroacylation. *J. Am. Chem. Soc.* 132: 16354–16355.

24 Huang, W. and Meng, F. (2021). Cobalt-catalyzed diastereo- and enantioselective hydroalkylation of cyclopropenes with cobalt homoenolates. *Angew. Chem. Int. Ed.* 60: 2694–2698.

25 Dian, L., Marek, I., and Acquarica, I. (2020). Pd-catalyzed enantioselective hydroalkynylation of cyclopropenes. *ACS Catal.* 10: 1289–1293.

26 Dian, L. and Marek, I. (2020). Cobalt-catalyzed diastereoselective and enantioselective hydrosilylation of achiral cyclopropenes. *Org. Lett.* 22: 4914–4918.

27 Zhang, L. and Oestreich, M. (2019). Copper-catalyzed enantio-and diastereoselective addition of silicon nucleophiles to 3,3-disubstituted cyclopropenes. *Chem. Eur. J.* 25: 14304–14307.

28 Zhang, Y., Li, Y., Zhou, W. et al. (2020). Assembly of polysubstituted chiral cyclopropylamines via highly enantioselective Cu-catalyzed three-component cyclopropene alkenylamination. *Chem. Commun.* 56: 12250–12253.

29 Nie, S., Lu, A., Kuker, E.L., and Dong, V.M. (2021). Enantioselective hydrothiolation: diverging cyclopropenes through ligand control. *J. Am. Chem. Soc.* 143: 6176–6184.

30 Jiang, Z., Niu, S.-L., Zeng, Q. et al. (2021). Selective alkynylallylation of the C-C sigma bond of cyclopropenes. *Angew. Chem. Int. Ed.* 60: 297–303.

31 Jiang, C., Wu, J., Han, J. et al. (2021). An expedient synthesis of highly functionalized 1,3-dienes by employing cyclopropenes as C_4 units. *Chem. Commun.* 57: 5710–5713.

32 Yuan, Y., Zheng, Z.-J., Ye, F. et al. (2018). Highly efficient desymmetrization of cyclopropenes to azabicyclo[3.1.0]hexanes with five continuous stereogenic centers by copper-catalyzed [3 + 2] cycloadditions. *Org. Chem. Front.* 5: 2759–2764.

33 Zheng, H., Wang, R., Wang, K. et al. (2021). Formal [4 + 4]-, [4 + 3]-, and [4 + 2]-cycloaddition reactions of donor–acceptor cyclobutenes, cyclopropenes and siloxyalkynes induced by Bronsted acid catalysis. *Chem. Sci.* 12: 4819–4824.

34 Marichev, K.O., Wang, K., Greco, N. et al. (2021). Strain-induced nucleophilic ring opening of donor–acceptor cyclopropenes for synthesis of monosubstituted succinic acid derivatives. *Chem. Eur. J.* 27: 340–347.

35 Muriel, B. and Waser, J. (2021). Azide radical initiated ring opening of cyclopropenes leading to alkenyl nitriles and polycyclic aromatic compounds. *Angew. Chem. Int. Ed.* 60: 4075–4079.

36 Dange, N.S., Jatoi, A.H., Robert, F., and Landais, Y. (2017). Visible-light-mediated addition of phenacyl bromides onto cyclopropenes. *Org. Lett.* 19: 3652–3655.

37 Zhang, Z.-Q., Zheng, M.-M., Xue, X.-S. et al. (2019). Catalytic enantioselective cyclopropenation of internal alkynes: access to difluoromethylated three-membered carbocycles. *Angew. Chem. Int. Ed.* 58: 18191–18196.

38 Tran, U.P.N., Hommelsheim, R., Yang, Z. et al. (2020). Catalytic synthesis of trifluoromethyl cyclopropenes and oligo-cyclopropenes. *Chem. Eur. J.* 26: 1254–1257.

39 Barroso, R., Jiménez, A., Pérez-Aguilar, C.M. et al. (2016). Synthesis of 1,3-diaryl-3-trifluoromethylcyclopropenes by transition-metal-free reaction of 2,2,2-trifluoroacetophenone tosylhydrazones with alkynes: the effect of the trifluoromethyl group. *Chem. Commun.* 52: 3677–3680.

40 Rullière, P., Cyr, P., and Charette, A.B. (2016). Difluorocarbene addition to alkenes and alkynes in continuous flow. *Org. Lett.* 18: 1988–1991.

41 Tanbouza, N., Carreras, V., and Ollevier, T. (2021). Photochemical cyclopropenation of alkynes with diazirines as carbene precursors in continuous flow. *Org. Lett.* 23: 5420–5424.

42 Pallerla, M.K. and Fox, J.M. (2007). Enantioselective synthesis of (–)-Pentalenene. *Org. Lett.* 9: 5625–5628.

43 Liang, X.-T., Chen, J.-H., and Yang, Z. (2020). Asymmetric total synthesis of (–)-Spirochensilide A. *J. Am. Chem. Soc.* 142: 8116–8121.

44 Elling, B.R., Su, J.K., and Xia, Y. (2021). Polymerization of cyclopropenes: taming the strain for the synthesis of controlled and sequence-regulated polymers. *Acc. Chem. Res.* 54: 356–365.

45 Petiniot, N., Anciaux, A.J., Noels, A.F. et al. (1978). Rhodium catalyzed cyclopropenation of acetylenes. *Tetrahedron Lett.* 19: 1239–1242.

46 Protopopova, M.N., Doyle, M.P., Mueller, P., and Ene, D. (1992). High enantioselectivity for intermolecular cyclopropenation of alkynes by diazo esters catalyzed by chiral dirhodium(II) carboxamides. *J. Am. Chem. Soc.* 114: 2755–2757.

47 Doyle, M.P., Protopopova, M.N., Brandes, B.D. et al. (1993). Diastereoselectivity enhancement in cyclopropanation and cyclopropenation reactions of chiral diazoacetate esters catalyzed by chiral dirhodium(II) carboxamides. *Synlett* 1993: 151–153.

48 Müller, P. and Imogaï, H. (1998). Rh(II)-catalyzed asymmetric cyclopropenation of propargyl derivatives; synthesis of cyclopropene- and cis-cyclopropaneamino acids. *Tetrahedron: Asymmetry* 9: 4419–4428.

49 Imogaï, H., Bernardinelli, G., Graenicher, C. et al. (1998). cis-Disubstituted cyclopropanes via asymmetric catalytic cyclopropenation. Synthesis of cyclopropyl dehydro amino acids and of dictyopterene C'. *Helv. Chim. Acta* 81: 1754–1764.

50 Lou, Y., Horikawa, M., Kloster, R.A. et al. (2004). A new chiral Rh(II) catalyst for enantioselective [2 + 1]-cycloaddition. Mechanistic implications and applications. *J. Am. Chem. Soc.* 126: 8916–8918.

51 Weatherhead-Kloster, R.A. and Corey, E.J. (2006). Theory-guided discovery of unique chemical transformations of cyclopropenes. *Org. Lett.* 8: 171–174.

52 Lou, Y., Remarchuk, T.P., and Corey, E.J. (2005). Catalysis of enantioselective [2 + 1]-cycloaddition reactions of ethyl diazoacetate and terminal acetylenes using mixed-ligand complexes of the series $Rh_2(RCO_2)_n(L^*_{4-n})$. Stereochemical heuristics for ligand exchange and catalyst synthesis. *J. Am. Chem. Soc.* 127: 14223–14230.

53 Remarchuk, T. and Corey, E.J. (2018). Stereodivergent synthesis of novel chiral C_2-symmetric bis-trifluoromethyl-2-oxazolidinones. *Tetrahedron Lett.* 59: 2256–2259.

54 Liao, L.-a., Zhang, F., Yan, N. et al. (2004). An efficient and general method for resolving cyclopropene carboxylic acids. *Tetrahedron* 60: 1803–1816.

55 Liao, L.-a., Zhang, F., Dmitrenko, O. et al. (2004). A reactivity/affinity switch for parallel kinetic resolution: alpha-amino acid quasienantiomers and the resolution of cyclopropene carboxylic acids. *J. Am. Chem. Soc.* 126: 4490–4491.

56 Doyle, M.P. and Hu, W. (2001). Macrocycle formation from catalytic metal carbene transformations. *Synlett* 2001: 1364–1370.

57 Doyle, M.P., Forbes, D.C., Vasbinder, M.M., and Peterson, C.S. (1998). Enantiocontrol in the generation and diastereoselective reactions of catalytically generated oxonium and iodonium ylides. Metal-stabilized ylides as reaction intermediates. *J. Am. Chem. Soc.* 120: 7653–7654.

58 Doyle, M.P. (1998). New catalysts and methods for highly enantioselective metal carbene reactions. *Pure Appl. Chem.* 70: 1123–1128.

59 Doyle, M.P., Protopopova, M., Müller, P. et al. (1994). Effective uses of dirhodium(II) tetrakis[methyl 2-oxopyrrolidine-5(R or S)-carboxylate] for highly enantioselective intermolecular cyclopropenation reactions. *J. Am. Chem. Soc.* 116: 8492–8498.

60 Doyle, M.P. (1994). Asymmetric syntheses with catalytic enantioselective metal carbene transformations. *Izv. Akad. Nauk, Ser. Khim.* 1994: 1879–1892.

61 Müller, P., Grass, S., Shahi, S.P., and Bernardinelli, G. (2004). Rh(II)-catalyzed asymmetric carbene transfer with ethyl 3,3,3-trifluoro-2-diazopropionate. *Tetrahedron* 60: 4755–4763.

62 Briones, J.F., Hansen, J., Hardcastle, K.I. et al. (2010). Highly enantioselective $Rh_2(S\text{-}DOSP)_4$-catalyzed cyclopropenation of alkynes with styryldiazoacetates. *J. Am. Chem. Soc.* 132: 17211–17215.

63 Briones, J.F. and Davies, H.M.L. (2011). $Rh_2(S\text{-}PTAD)_4$-catalyzed asymmetric cyclopropenation of aryl alkynes. *Tetrahedron* 67: 4313–4317.

64 Goto, T., Takeda, K., Shimada, N. et al. (2011). Highly enantioselective cyclopropenation reaction of 1-alkynes with alpha-alkyl-alpha-diazoesters catalyzed by dirhodium(II) carboxylates. *Angew. Chem. Int. Ed.* 50: 6803–6808.

65 Nowlan, D.T. III and Singleton, D.A. (2005). Mechanism and origin of enantiosselectivity in the $Rh_2(OAc)(DPTI)_3$-catalyzed cyclopropenation of alkynes. *J. Am. Chem. Soc.* 127: 6190–6191.

66 Boruta, D.T., Dmitrenko, O., Yap, G.P.A., and Fox, J.M. (2012). $Rh_2(S\text{-}PTTL)_3TPA$–a mixed-ligand dirhodium(II) catalyst for enantioselective reactions of alpha-alkyl-alpha-diazo esters. *Chem. Sci.* 3: 1589–1593.

67 Lindsay, V.N.G., Fiset, D., Gritsch, P.J. et al. (2013). Stereoselective $Rh_2(S\text{-}IBAZ)_4$-catalyzed cyclopropanation of alkenes, alkynes, and allenes: asymmetric synthesis of diacceptor cyclopropylphosphonates and alkylidenecyclopropanes. *J. Am. Chem. Soc.* 135: 1463–1470.

68 Doyle, M.P., Ene, D.G., Peterson, C.S., and Lynch, V. (1999). Macrocyclic cyclopropenes by highly enantioselective intramolecular addition of metal carbenes to alkynes. *Angew. Chem. Int. Ed.* 38: 700–702.

69 Uehara, M., Suematsu, H., Yasutomi, Y., and Katsuki, T. (2011). Enantioenriched synthesis of cyclopropenes with a quaternary stereocenter, versatile building blocks. *J. Am. Chem. Soc.* 133: 170–171.

70 Cui, X., Xu, X., Lu, H. et al. (2011). Enantioselective cyclopropenation of alkynes with acceptor/acceptor-substituted diazo reagents via Co(II)-based metalloradical catalysis. *J. Am. Chem. Soc.* 133: 3304–3307.

71 Zhu, S., Xu, X., Perman, J.A., and Zhang, X.P. (2010). A general and efficient cobalt(II)-based catalytic system for highly stereoselective cyclopropanation of alkenes with alpha-cyanodiazoacetates. *J. Am. Chem. Soc.* 132: 12796–12799.

72 Briones, J.F. and Davies, H.M.L. (2012). Gold(I)-catalyzed asymmetric cyclopropenation of internal alkynes. *J. Am. Chem. Soc.* 134: 11916–11919.

73 Chen, K., Huang, X., Kan, S.B.J. et al. (2018). Enzymatic construction of highly strained carbocycles. *Science* 360: 71.

74 Chen, K. and Arnold, F.H. (2020). Engineering cytochrome P450s for enantioselective cyclopropenation of internal alkynes. *J. Am. Chem. Soc.* 142: 6891–6895.

75 Zhang, F. and Fox, J.M. (2006). Synthesis of cyclopropene alpha-amino acids via enantioselective desymmetrization. *Org. Lett.* 8: 2965–2968.

76 Simaan, S., Masarwa, A., Bertus, P., and Marek, I. (2006). Enantiomerically pure cyclopropenylcarbinols as a source of chiral alkylidenecyclopropane derivatives. *Angew. Chem. Int. Ed.* 45: 3963–3965.

77 Nikitina, L.E., Plemenkov, V.V., Strunskaya, E.I. et al. (1998). Electrochemical synthesis of monobromocyclopropanes of the terpene series as a step on the way to chiral cyclopropenes. *Russ. J. Gen. Chem.* 68: 1748–1751.

78 Liao, L.-a., Yan, N., and Fox, J.M. (2004). Dianion approach to chiral cyclopropene carboxylic acids. *Org. Lett.* 6: 4937–4939.

79 Yang, Z., Xie, X., and Fox, J.M. (2006). Diastereoselective synthesis of methylenecyclopropanes from chiral cyclopropene derivatives. *Angew. Chem. Int. Ed.* 45: 3960–3962.

2

Chiral Heterocycles for Asymmetric Synthesis

Radovan Šebesta and Tibor Peňaška

Comenius University in Bratislava, Faculty of Natural Sciences, Department of Organic Chemistry, Mlynská dolina, Ilkovičova 6, Bratislava 842 15, Slovakia

2.1 Introduction

Heterocycles are prominent structural elements across naturally occurring and manufactured chiral compounds. Many synthetic methods have been developed for the formation and derivatization of heterocycles. From this vast pool of synthetic knowledge, a large portion deals with stereoselective transformations, which lead to chiral compounds. There are two approaches to chiral heterocyclic compounds. The first approach is building on or transforming chiral heterocyclic compounds, using them as building blocks. The other approach is introducing chirality into the molecule via asymmetric synthesis or catalysis. This chapter will deal mainly with the first approach. However, even though synthetic chemistry concerning heterocyclic compounds is vast, (stereo)selective manipulations of heterocyclic compounds are still considered a challenge in fine chemical synthesis [1].

2.2 Small-Ring Heterocycles

The smallest possible heterocyclic compounds comprise a three-membered ring, typically with oxygen or nitrogen as the heteroatom. There are several possible structures with other heteroatoms, such as sulfur, phosphorus, or selenium. However, these derivatives rarely serve as useful building blocks in asymmetric synthesis and will not be considered. Reactive compounds, such as dioxiranes or reactive intermediates, such as halonium, tiiranium, or seleniranium ions, are not separately discussed here neither.

Chiral epoxides are among the most crucial oxygen building blocks. Various asymmetric catalytic epoxidation methods have been developed for their preparation. Sharpless and Jacobsen's epoxidations are the most developed methodologies for obtaining chiral epoxides. Sharpless' method utilizes cheap and readily-available catalysts based on esters of tartaric acid [2]. One of the earliest demonstrations of this method was a total synthesis of disparlure (**3**), a pheromone of female gypsy moths

Chiral Building Blocks in Asymmetric Synthesis: Synthesis and Applications, First Edition.
Edited by Elżbieta Wojaczyńska and Jacek Wojaczyński.

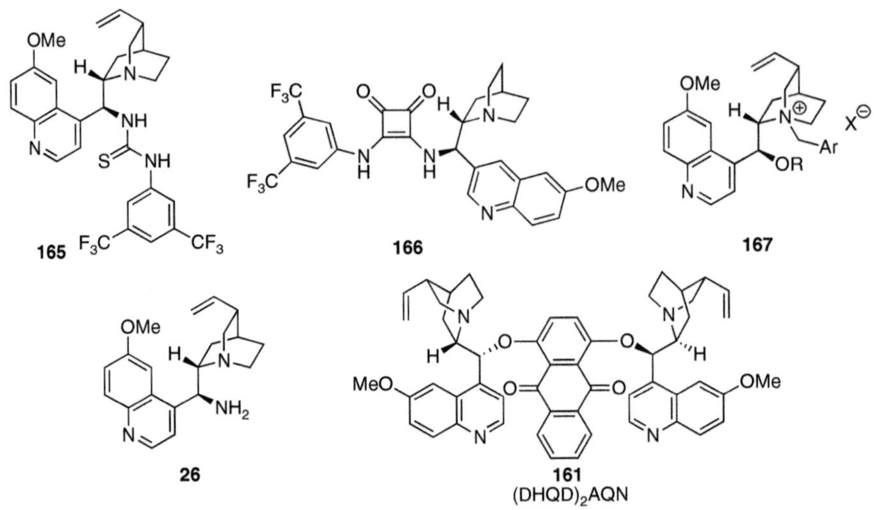

Figure 2.1 Examples of chiral catalysts derived from *Cinchona* alkaloids.

Lymantria dispar [3]. This compound features an oxirane ring, which has been introduced by enantioselective epoxidation of allyl alcohol **1**. Typical conditions for this transformation comprise titanium isopropoxide and diethyl tartrate (DET) as the catalyst and *tert*-butyl hydroperoxide (TBHP) as an oxidant (Scheme 2.1).

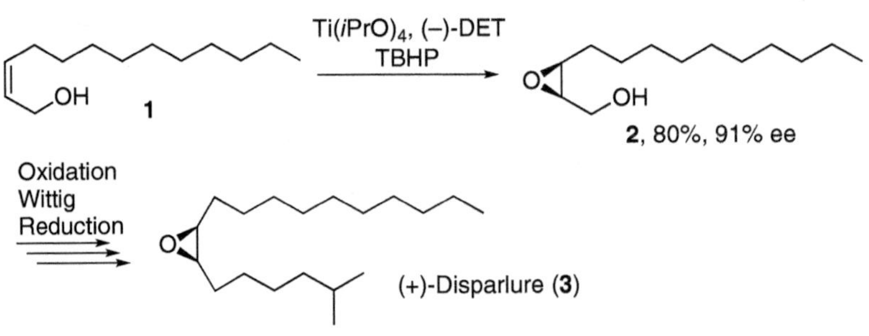

Scheme 2.1 Sharpless' epoxidation in the total synthesis of (+)-disparlure.

Sharpless asymmetric epoxidation is a versatile method for the epoxidation of allylic alcohols. Allyl alcohol **4** is enantioselectively epoxidized to afford glycidol (**5**), an important chiral building block. Because tartrate ligands for Sharpless epoxidation are readily available in both enantiomeric forms, both enantiomers of glycidol (**5**) can be synthesized effectively [4]. Production of enantiomerically enriched (*S*)-glycidol (**5**) has also been commercialized by ARCO (Scheme 2.2). Glycidol (**5**) is used as a building block in the industrial synthesis of glycerol, glycidyl ethers, esters, and amines. For instance, (*R*)-glycidol has been employed as a starting material in synthesizing polycyclic diterpene (+)-intricarene found in *Pseudopterogorgia* corals [5].

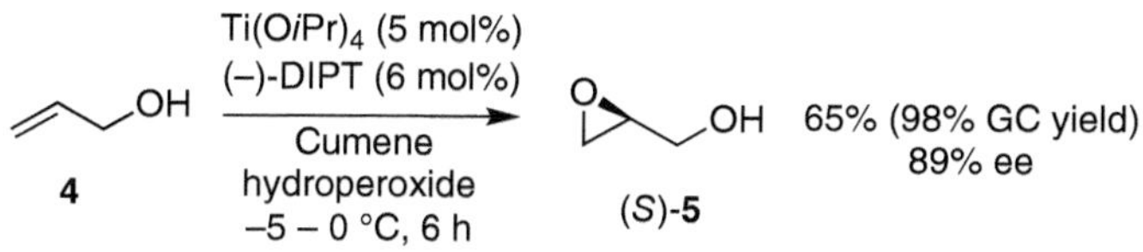

Scheme 2.2 Epoxidation of allyl alcohol yielding an industrial building block glycidol. DIPT, Diisopropyl tartrate; GC, Gas chromatography.

Jacobsen's method for epoxidation of alkenes relies on manganese salen complexes as catalysts. Typically, sodium hypochlorite is an oxidant, but other common oxidation reagents like *meta*-chloroperoxybenzoic acid (*m*-CPBA) can also be employed. The recent total synthesis of bruceol (**9**) is an excellent example of the versatility of the Jacobsen epoxidation [6]. It also demonstrates the use of chiral epoxides in the synthesis, in which they can be opened by nucleophiles and serve as precursors for alcohols (Scheme 2.3).

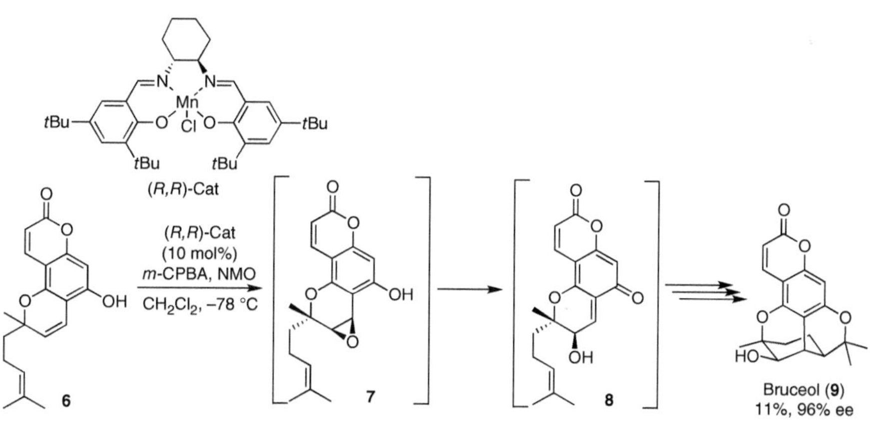

Scheme 2.3 Synthesis of epoxide building block 7 in the total synthesis of bruceol. NMO, 4-methylmorpholine *N*-oxide.

Other methods utilizing different metals like Nb, V, or Cr have also been developed. A biomimetic approach inspired by the Fe-heme oxidation shuttle led to various Fe-catalyzed epoxidations [7]. There are also metal-free methodologies for catalytic epoxidations. Shi described chiral ketones derived from carbohydrates as efficient epoxidation catalysts. Recently, chiral organocatalysis opened new possibilities to chiral epoxides [8].

Among several developments, peptide-catalyzed epoxidation and chiral amine-catalyzed electrophilic epoxidation of unsaturated carbonyl compounds should be mentioned. Julia-Colona epoxidation utilizing polyleucine as catalyst served as an inspiration to develop other peptide-based catalysts. Miller and coworkers developed aspartate-based peptides **12** as catalysts for epoxidations of alkenes **10** (Scheme 2.4) [9]. This concept also proved helpful in the epoxidation of complex peptide-based natural products [10] or polyenes [11].

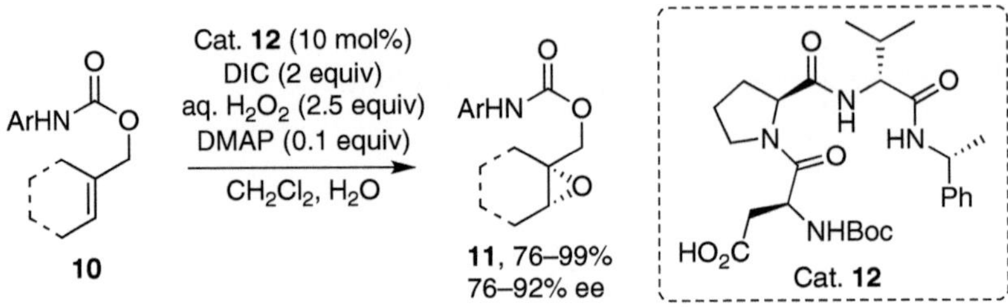

Scheme 2.4 Organocatalytic epoxidation of alkenes. DIC, diisopropylcarbodiimide; DMAP, 4-(Dimethylamino)pyridine. Source: Hanson et al. [4] / American Chemical Society.

Chiral secondary amines can form enamines or iminium ions with unsaturated aldehydes and ketones. This feature opens the possibility for epoxidation of electron-deficient C=C double bond. Jørgensen's discovery that pyrrolidine-based organocatalyst **15** is efficient for enantioselective epoxidation of enals inspired the development of various related epoxidation methods (Scheme 2.5a) [12]. The corresponding epoxy aldehydes **16** and ketones **17** are valuable building blocks applicable in synthesizing various natural products **18–21** (Scheme 2.5b).

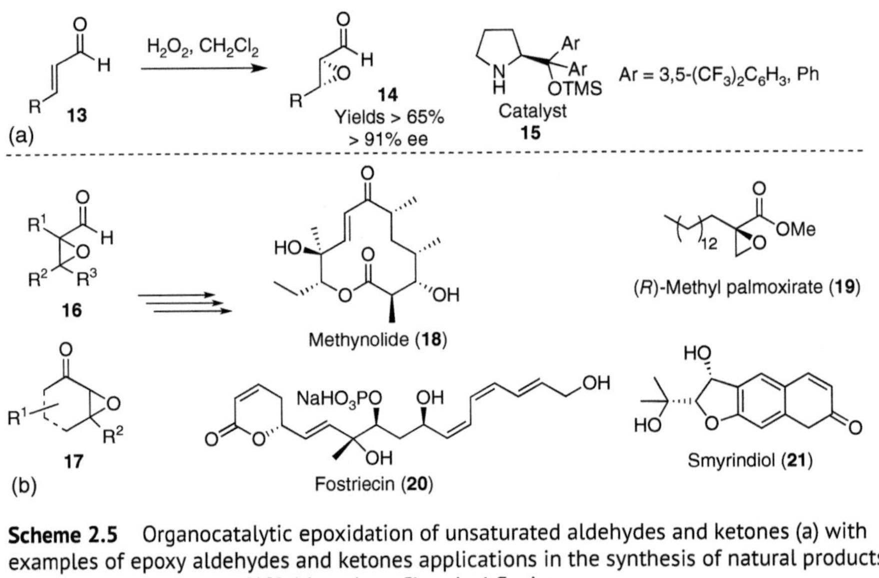

Scheme 2.5 Organocatalytic epoxidation of unsaturated aldehydes and ketones (a) with examples of epoxy aldehydes and ketones applications in the synthesis of natural products (b). Source: Marigo et al. [12] / American Chemical Society.

For activation of cyclic enones, primary amine catalysts **26** are more effective. List and coworkers developed an improved method for epoxidation of cyclic enones **22** [13]. This method has been recently employed in the total synthesis of epicoccin G (**25**) and rostratin [14]. For example, epoxidation of enone **22** afforded an essential chiral building block **23** at the very beginning of the synthetic sequence (Scheme 2.6).

Aziridines are another important type of three-membered heterocyclic compound. They are less widely occurring than epoxides, but they are still among the

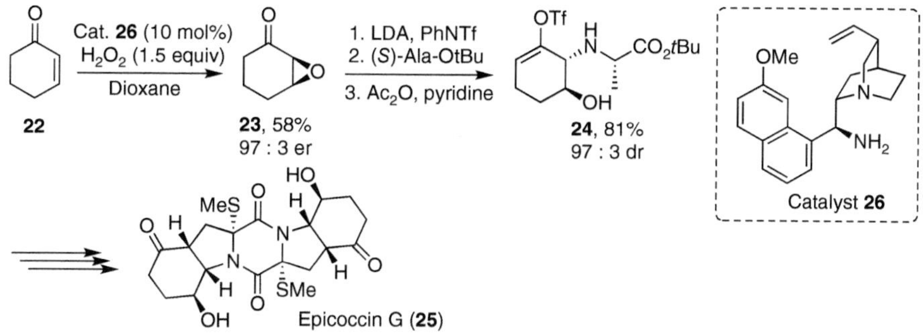

Scheme 2.6 Epoxidation of an unsaturated ketone with a primary amine catalyst.

prominent heterocyclic skeletons. Stereoselective formation of chiral aziridines can be accomplished by adding nitrene precursor to an alkene or carbene to an imine. An interesting strategy to use racemic aziridines is to employ them in catalytic kinetic resolutions [15].

Recently, Baran and coworkers used a chiral aziridine building block **28** in the total synthesis of bacterial alkaloids teleocidins B1–B4 [16]. Building block **28** can be synthesized from serine in five steps [17]. Baran's team used it in the nucleophilic opening with an indole derivative **27** utilizing indole's C-3 nucleophilicity (Scheme 2.7).

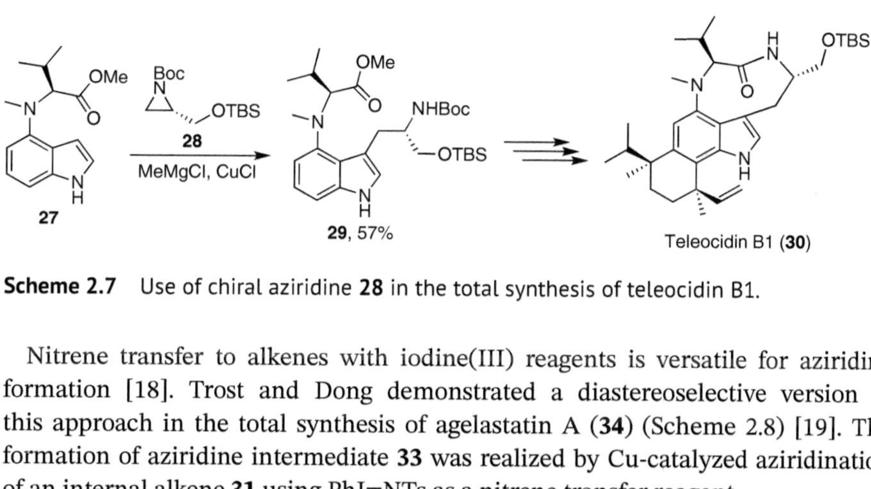

Scheme 2.7 Use of chiral aziridine **28** in the total synthesis of teleocidin B1.

Nitrene transfer to alkenes with iodine(III) reagents is versatile for aziridine formation [18]. Trost and Dong demonstrated a diastereoselective version of this approach in the total synthesis of agelastatin A (**34**) (Scheme 2.8) [19]. The formation of aziridine intermediate **33** was realized by Cu-catalyzed aziridination of an internal alkene **31** using PhI=NTs as a nitrene transfer reagent.

A common strategy to generate aziridines is a nucleophilic displacement of an adjacent leaving group. This approach can be exemplified by the total synthesis of mitomycin K (**37**), where the aziridine moiety was built via azide formation. Then, in ketone **35**, the azide group was reduced by Staudinger reaction. Subsequently, aziridine ring was formed by nucleophilic substitution of the mesyl group (Scheme 2.9) [20].

Aziridination or epoxidation of imines or carbonyls with chiral sulfur ylides has become a valuable alternative to more traditional methods employing nitrene or

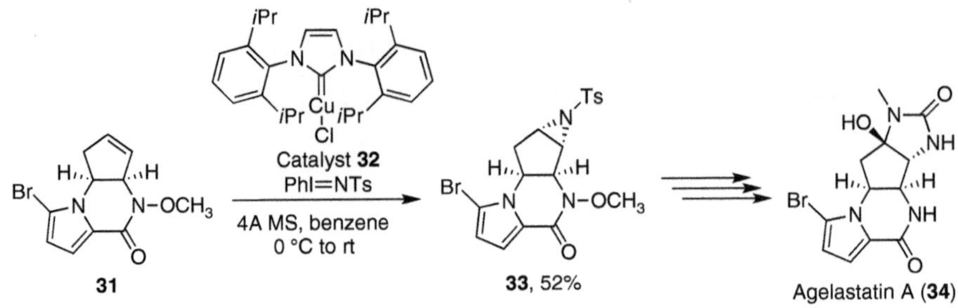

Scheme 2.8 Cu-catalyzed aziridination in the total synthesis of agelastatin A. MS, Molecular Sieve. Source: Trost and Dong [19] / American Chemical Society.

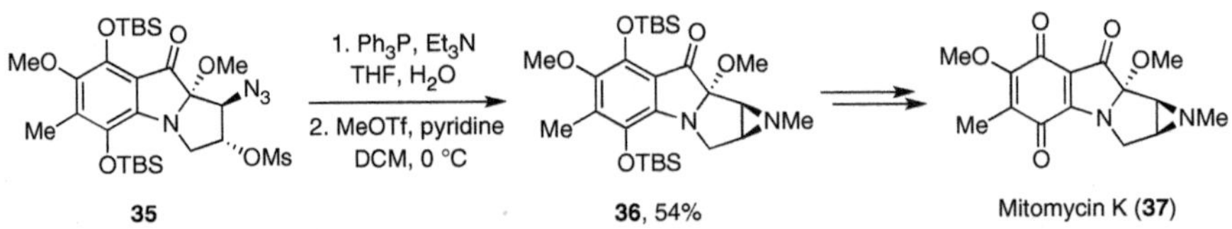

Scheme 2.9 Aziridine formation via nucleophilic displacement in the synthesis of mitomycin K. DCM, Dichloromethane. Source: Peris et al. [9] / John Wiley & Sons.

oxo-additions [21]. Aggarwal demonstrated the viability of this method in the total synthesis of cyclopiazonic acid derivatives **41** (Scheme 2.10) [22]. The reaction of imine **38** with chiral sulfur ylide **39** generated important aziridine intermediate **40**.

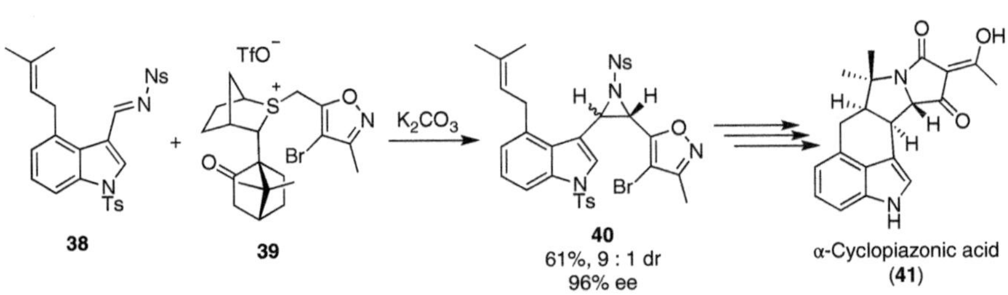

Scheme 2.10 Aziridination of imines with sulfur ylides in the total synthesis of α-cyclopiazonic acid. Source: Zhurakovskyi et al. [22] / John Wiley & Sons.

Among four-membered heterocycles, azetidines and related β-lactams are important structural motifs in organic synthesis. Recently, significant progress has been achieved in the construction and synthetic transformations of azetidines [23]. Two recent examples of new azetidine chemistry are highlighted. Aggarwal and coworkers developed a strategy for building azetidine core by 1,2-metalate rearrangement of highly strained azabicyclo[1.1.0]butyl lithium [24]. This attractive building block was accessed from 2,3-dibromopropan-1-amine (**42**) and transformed into a crystalline aryl sulfoxide derivative **45**, serving as its convenient synthetic equivalent. The reaction of azabicyclo[1.1.0]butyllithium with boranes **46** afforded azetidyl boranes **48**. The reaction was enantiospecific with a complete

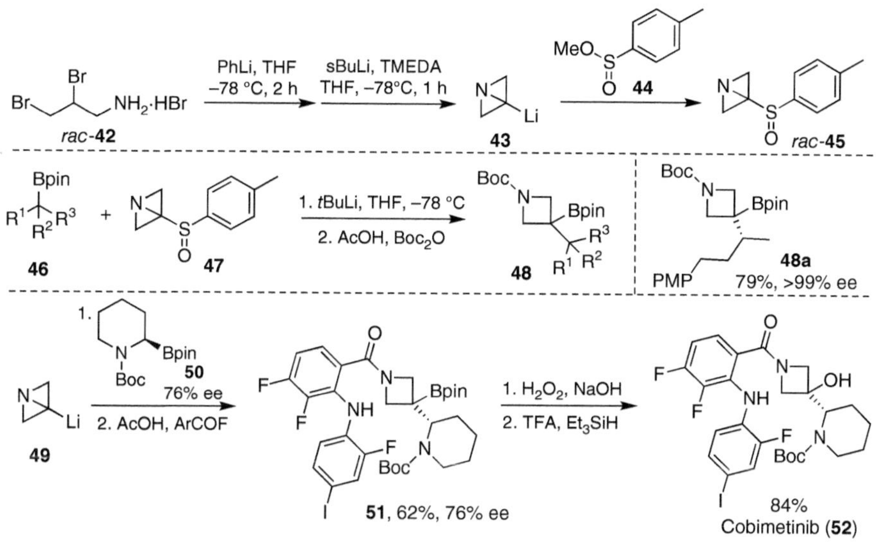

Scheme 2.11 Synthesis of azetidines using azabicyclo[1.1.0]butyllithium. THF, tetrahydrofuran; TMEDA, tetramethylethylenediamine; TFA, Trifluoroacetic acid.

retention of stereogenic information. If enantioenriched boronates were employed, corresponding azetidines were obtained with high enantiomeric purity, e.g. **48a**. Aggarwal's team demonstrated the method's usefulness in the synthesis of kinase inhibitor cobimetinib (**52**) (Scheme 2.11).

Due to their ring strain, azetidines are prone to nucleophilic opening. This feature was utilized by Sun and coworkers in a enantioselective desymmetrization of azetidines **53** using sulfur nucleophiles **54** [25]. This transformation was catalyzed by chiral phosphoric acids **55**, and even though it was limited to 2-mercaprobenzothiazole derivatives **56**, it worked well. The method offered a range of chiral amines in high enantiomeric purities, typically over 90% ee (Scheme 2.12). Sun's group later extended this method to desymmetrizing ring-opening of azetidinium salts with related SPINOL-based organocatalysts [26].

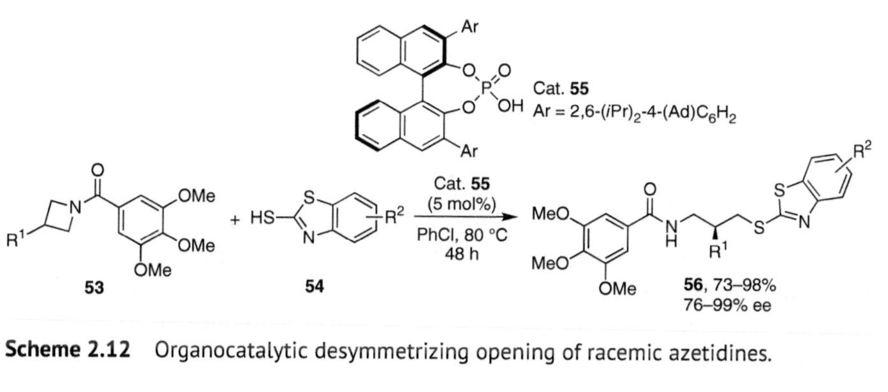

Scheme 2.12 Organocatalytic desymmetrizing opening of racemic azetidines.

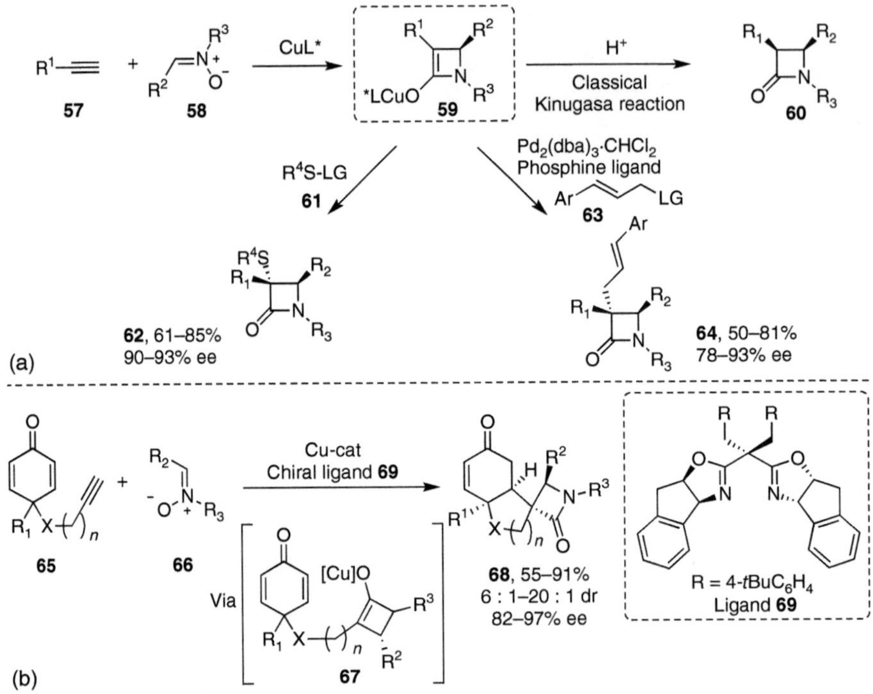

Scheme 2.13 Cu-catalyzed Kinugasa reaction for the synthesis of azetidinones using sulfur electrophiles or Pd-allyl complexes (a), and intramolecular Michael addition via Cu-enolate intermediate (b).

Azetidinones or β-lactams are an essential structural motif, mainly due to their antibacterial properties. Azetidinones are core structures in penicillin, cephalosporin, carbapenem, and carbecephem classes of antibiotics [27, 28]. Primary method for establishing β-lactam core is through Staudinger ketene cycloaddition [29], where an imine or aldehyde reacts with ketene affording the β-lactam or β-lactone.

A modern approach to azetidinones is based on Kinugasa reactions, a cycloaddition of alkyne **57** and nitrone **58**. This transformation brings exciting possibilities for further elaboration of the β-lactam moiety via interception of chiral four-membered copper(I)enolate intermediates **59** with diverse electrophiles. Xu's lab recently demonstrated two extensions using sulfur electrophiles **61** or Pd-allyl complexes **63** (Scheme 2.13a) [30, 31]. Enders and coworkers utilized intramolecular Michael addition of the Cu-enolate intermediate **67** in constructing structurally complex spirocyclic β-lactams **68** (Scheme 2.13b) [32].

Metal-catalyzed C–H activation was also utilized as a building block to construct chiral β-lactams. Lee and coworkers developed a relay Rh–Pd-catalyzed formation of β-lactams **72** comprising a quaternary stereogenic center (Scheme 2.14) [33].

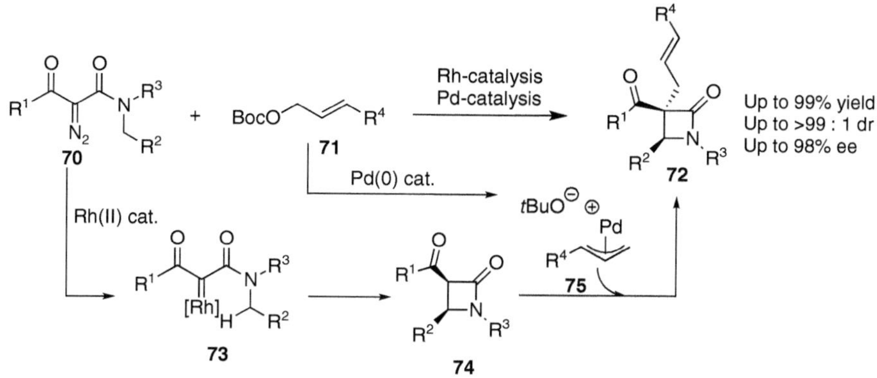

Scheme 2.14 Rh–Pd-catalyzed formation of β-lactams via C-H activation and allylic alkylation. Source: Huang et al. [33] / American Chemical Society.

2.3 Medium-Ring Aliphatic Heterocycles

Five-membered and six-membered chiral building blocks are fundamental cores in synthesizing various bioactive molecules. These types of heterocyclic chiral precursors are primarily prepared via asymmetric catalysis. Therefore, this part is an overview of methods of preparation and application of these molecules in synthesizing specific bioactive compounds.

Clarke and Maddocks published an enantioselective total synthesis of (*R*)-bgugaine (**80**) and (*R*)-irnidine (**81**) in six reaction steps. The key step was the formation of the chiral pyrrolidine derivative **79** from thioacrylic acid S-*p*-tolyl ester **77** and amine **76** catalyzed by a second generation Hoveyda–Grubbs catalyst (first "*clip*" step) and then by the application of a chiral Brønsted acid catalyst (*R*)-TiPSY (**78**) (second "*cycle*" step). Chiral building block **79** was prepared in a good yield of 71% and the enantiomeric ratio of 94 : 6 and used in further reaction steps to afford natural products [34]. Clarke and coworkers also used this synthetic "*clip-cycle*" concept to synthesize other 2,2- and 3,3-disubstituted pyrrolidines and spiropyrrolidines (Scheme 2.15) [35].

Asymmetric dipolar cycloaddition of azomethine ylides is an efficient way to construct pyrrolidine derivatives [36, 37]. The total synthesis of two diastereoisomers of spirotryprostatin A **87** and **88** was described by Gong and coworkers in 2011. A chiral pyrrolidine-based building block **86** of these bioactive molecules was prepared via asymmetric 1,3-dipolar cycloaddition of acrylate **84** with azomethine ylides generated from aldehyde **82** and diester **83** catalyzed by a chiral Brønsted acid **85** (Scheme 2.16) [38].

An asymmetric gold-catalyzed cycloisomerization of a 1,4-diynamide **89** was applied to construct a chiral pyrrolidine moiety in the total synthesis of enantiomerically pure (+)-mesembrine (**91**). The key precursor **90** was obtained in 76% yield and 70% ee and used in the subsequent reaction step (Scheme 2.17) [39].

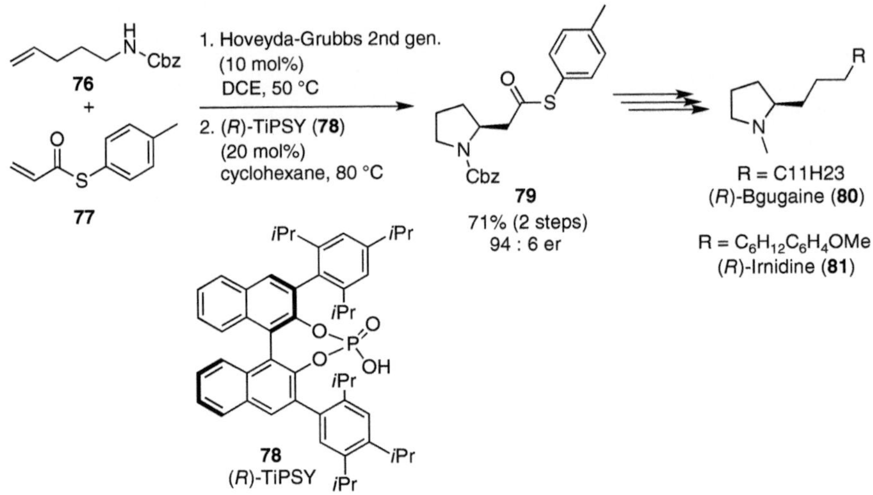

Scheme 2.15 Formation of chiral pyrrolidine building block **79**. DCE, 1,2-Dichloroethane.

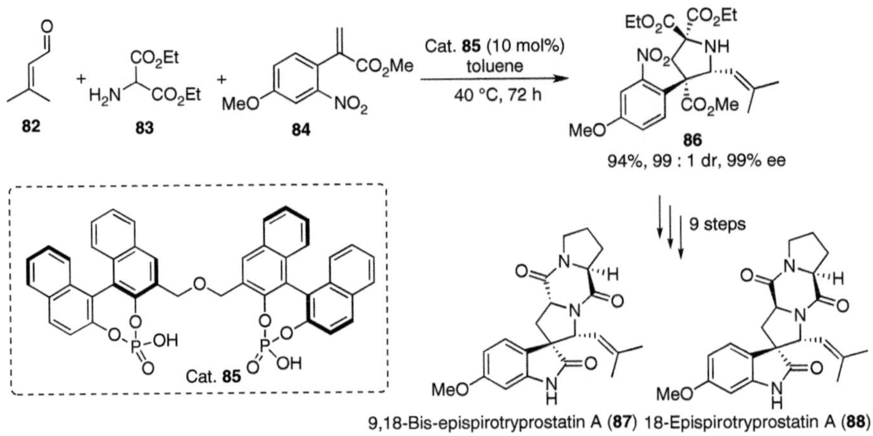

Scheme 2.16 Preparation of chiral pyrrolidine **86** by dipolar cycloaddition. Source: Cheng et al. [38] / American Chemical Society.

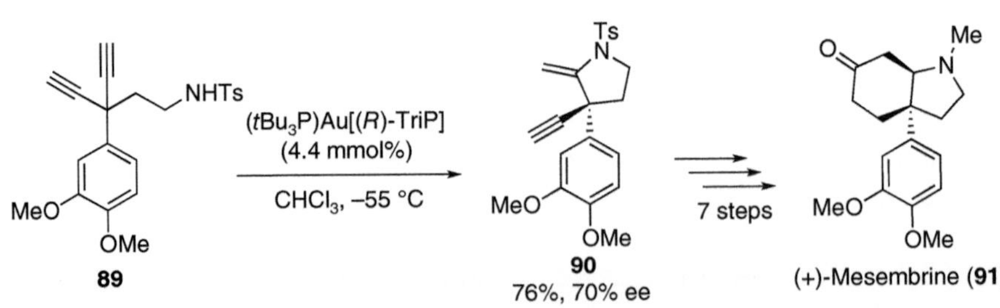

Scheme 2.17 Pyrrolidine building block **90** obtained by Au-catalyzed cycloizomerization. Source: Spittler et al. [39] / American Chemical Society.

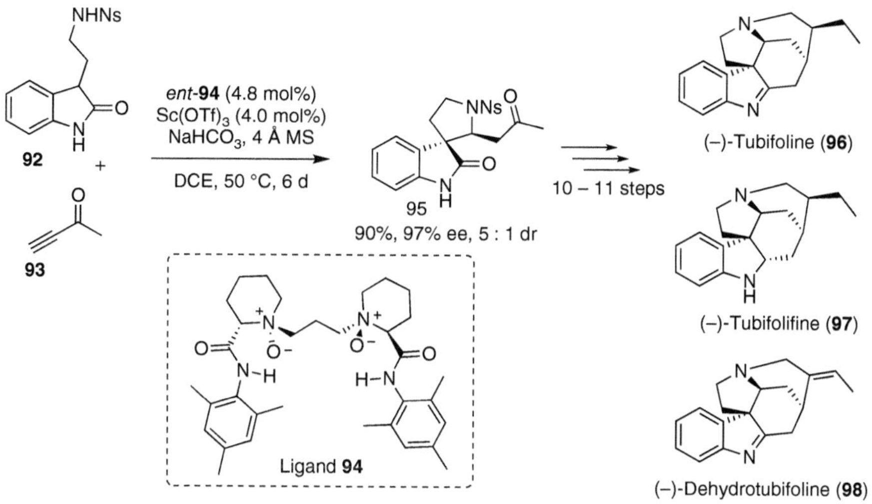

Scheme 2.18 Synthesis of spirooxindol pyrrolidine **95**. DCE, 1,2-Dichloroethane; MS, Molecular Sieve. Source: He et al. [40] / John Wiley & Sons.

An enantioselective preparation of spiropyrrolidine oxindoles developed by Xie and coworkers utilized a tandem Michael addition of tryptamine-derived oxindoles **92** to alkynone **93** promoted by a chiral *N,N′*-dioxide $Sc(OTf)_3$ catalyst. This method was a critical step in a novel strategy for synthesizing *Strychnos* alkaloids. In general, the reaction proceeded in good yields and high selectivities with various substrates. The authors demonstrated the method in the asymmetric synthesis of (−)-tibufoline, (−)-tibufolidine, and (−)-dehydrotubifoline (**96–98**) (Scheme 2.18) [40].

Chiral 3,4-disubstituted pyrrolidines **102** were prepared via organocatalytic Michael addition of enolizable aldehydes **100** to nitroalkene **99** catalyzed by commercially available proline-based Hayashi–Jørgensen catalyst **15**. These chiral *syn*-diastereomeric heterocyclic building blocks **102** were transformed into pyrrolidines **103**, exhibiting a good antibacterial activity (Scheme 2.19) [41].

In 2016, the first interrupted asymmetric hydroaminomethylation reaction was developed by Zhang and coworkers. The reaction of disubstituted alkenes **104** catalyzed by the rhodium/Yanphos system provided chiral pyrrolidines with high selectivity and excellent yields. The product **105** was used for further transformations in the synthesis of pharmaceuticals Vernakalant (**106**) and Enablex (**107**) (Scheme 2.20) [42].

An iridium/phosphoramidite-catalyzed intermolecular asymmetric dearomatization of indoles with allylic alcohol was developed in 2014. Chiral indoline products of the allylic alkylation reaction are synthesized in 71–97% yield with up to 98% ee using the Lewis acid $Fe(OTf)_2$. The authors have shown this method using this step in total synthesis of (−)-debromoflustramine B (**112**) (Scheme 2.21) [43].

A chiral guanidine **115** was used as an organocatalyst in a cascade double Michael addition between a terminal alkynone **113** and 3-substituted oxindole **114**. The

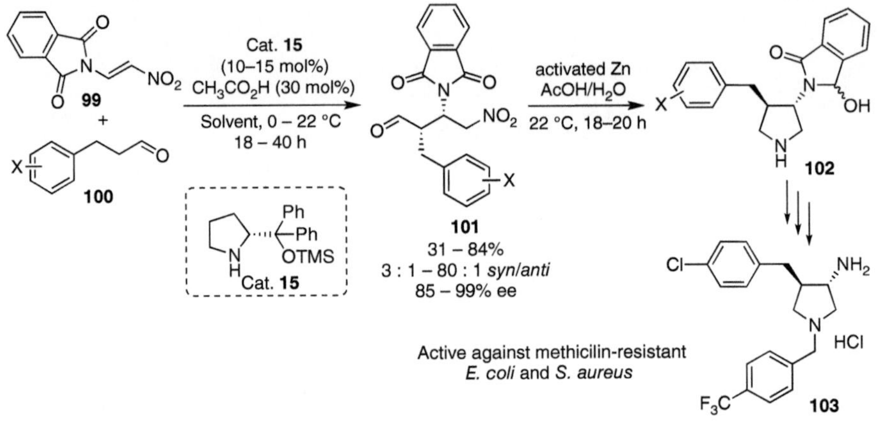

Scheme 2.19 Organocatalytic synthesis of chiral pyrrolidines with antibacterial properties. Source: Rodriguez et al. [41] / John Wiley & Sons.

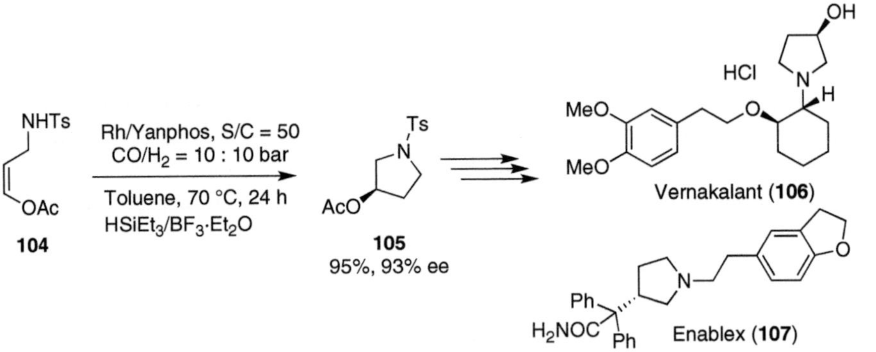

Scheme 2.20 Rh-catalyzed synthesis of pyrrolidine building block **105**. Source: Chen et al. [42] / American Chemical Society.

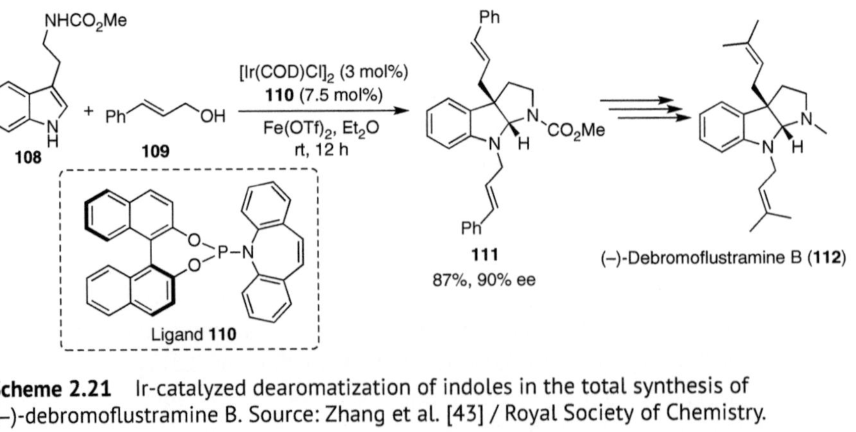

Scheme 2.21 Ir-catalyzed dearomatization of indoles in the total synthesis of (–)-debromoflustramine B. Source: Zhang et al. [43] / Royal Society of Chemistry.

authors prepared a series of spirooxindole derivatives containing dihydrofuran or pyrrolidine with high-level enantiomeric purities up to 95% ee. One of the products of this reaction, spirooxindol **116**, was also applied as a chiral building block in the total synthesis of (−)-salacin (**117**) for the first time (Scheme 2.22) [44].

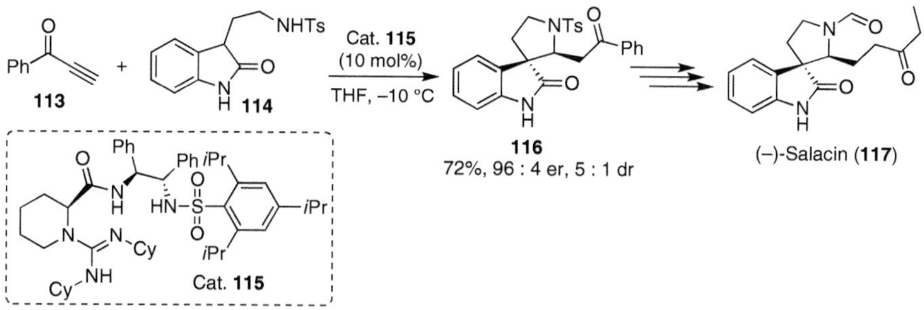

Scheme 2.22 Organocatalytic synthesis of spirooxindol intermediate **116**. Source: Kang et al. [44] / John Wiley & Sons.

A nine-step total synthesis of the anti-tumor alkaloid strychnofoline (**123**) was developed by Xu and coworkers in 2018. The main chiral heterocyclic intermediate **122** was prepared from commercially available starting material – 6-methoxytryptamine (**118**) by adding diketene (**119**), then acrolein derivative **120** with Hayashi–Jørgensen catalyst **15** and acyl chloride. This sequence of reactions (one-pot acylation/asymmetric Michael addition/Pictet–Spengler reaction) led to the quinolizidine precursor **121**, transformed in the next step by oxidative rearrangement to spirooxindole **122**. Reactions proceeded in good yields with excellent enantioselectivities (Scheme 2.23) [45].

Asymmetric deprotonation of the *N*-Boc-pyrrolidine (**124**) followed by diastereoselective addition to carbonyl and then Pictet–Spengler cyclization was used to synthesize two phenantroindolizidine alkaloids – antofine (**129**) and 14-hydroxyantofine (**128**) (Scheme 2.24). Both diastereoisomers of **127** afforded the same cyclization product **128**. Stereoselectivity can be controlled by the choice of an appropriate ligands – sparteine enantiomers, which are commercially available [46].

Chiral tetrahydrofuran derivative **131** was prepared in an efficient four-step synthesis from commercially available pantolactone (**130**) (Scheme 2.25). Chirality in the product can be switched by using both enantiomers of this molecule. The authors used this chiral block to prepare new amprenavir and empagliflozin analogs **132–133** [47].

Glutamic acid (**134**) is an available and cheap amino acid suitable as a natural chiral precursor for further synthetic transformations. An eight-step reaction led from **134** to enantiopure pyrrolidine-based lactam **136** with three contiguous chirality centers (Scheme 2.26). This core was used as a critical chiral building block to synthesize several bioactive molecules. The authors published two papers focused on preparing swainsonine, 1,4-dideoxy-1,4-imino-L-ribitol (LBR) [48], and cerebroside sphingolipid (molecules **137–139**) [49].

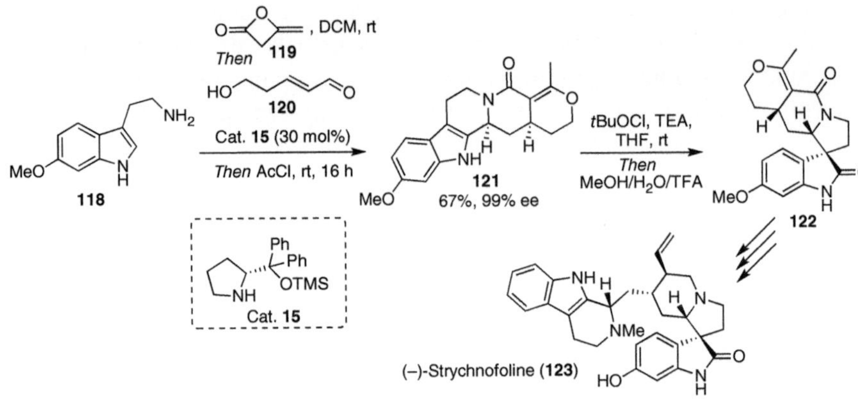

Scheme 2.23 Spirooxindol **122** in the total synthesis of (–)-strychnofoline. Source: Yu et al. [45] / Royal Society of Chemistry.

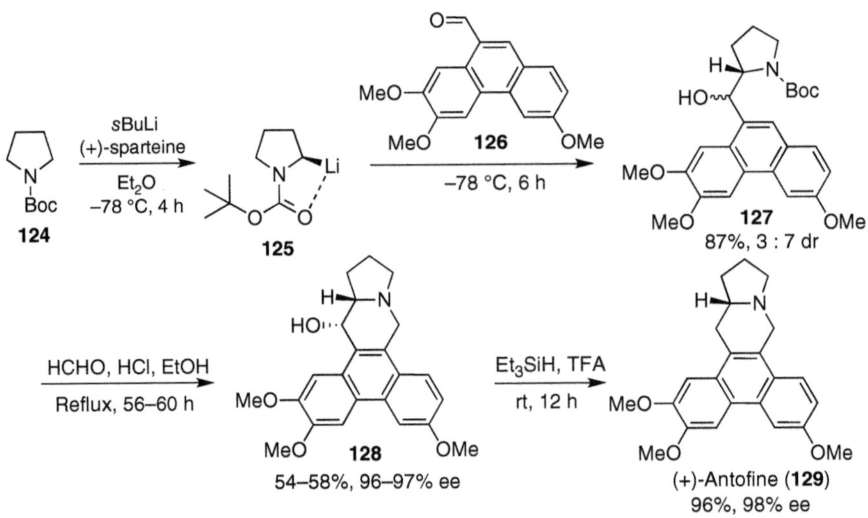

Scheme 2.24 Chiral pyrrolidine building block **125** obtained via enantioselective deprotonation. TFA, Trifluoroacetic acid.

In 2017, Bode's research group described a new pathway in the synthesis of (*S*)-*N*-Boc-5-oxaproline (**143**) as a key chiral heterocyclic element for the synthesis of proteins with the alfa-ketoacid-hydroxylamine (KAHA) ligation **144** (Scheme 2.27). This approach is based on an enantioselective conjugate addition of a hydroxylamine **140** to butenoate **141** catalyzed by Hayashi–Jørgensen catalyst **15**. Related products, 5-hydroxyisoxazolidine or β-amino aldehydes, are consequently transformed in several reaction steps to oxaproline precursor **143** [50].

A catalytic enantioselective 5-*endo* bromocycloetherification is an effective method for preparing chiral bicyclic tetrahydrofuran-fused piperidine building block **146** used in the total synthesis of deoxoapodine (**147**) (Scheme 2.28). The

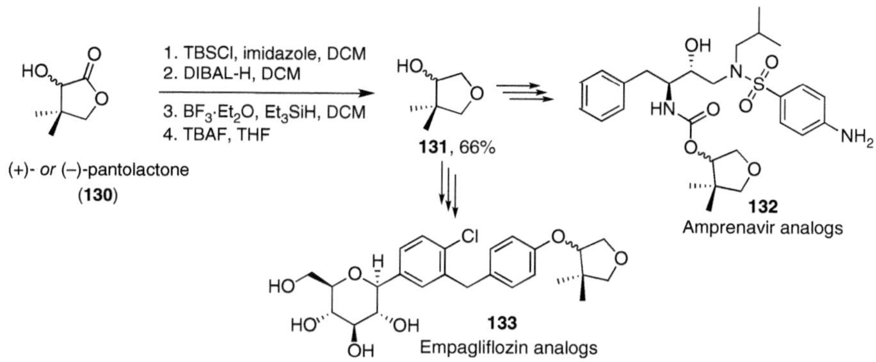

Scheme 2.25 Chiral lactone **130** in the total syntheses of amprenavir and empagliflozin analogs. DCM, Dichloromethane.

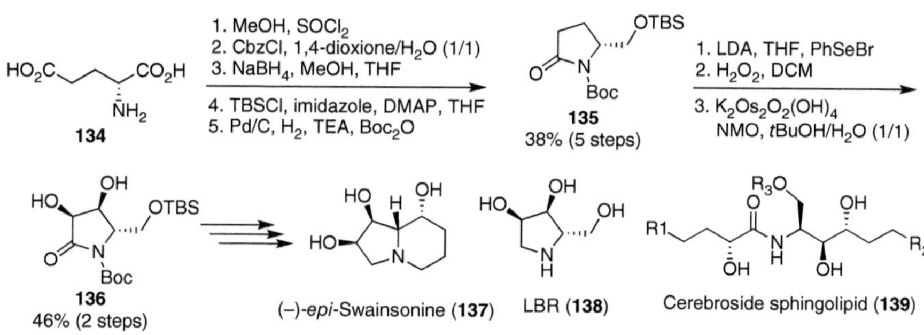

Scheme 2.26 Chiral pyrrolidine **135** in the total synthesis of swainsonine. TEA, Triethylamine; DCM, Dichloromethane; DMAP, 4-(Dimethylamino)pyridine.

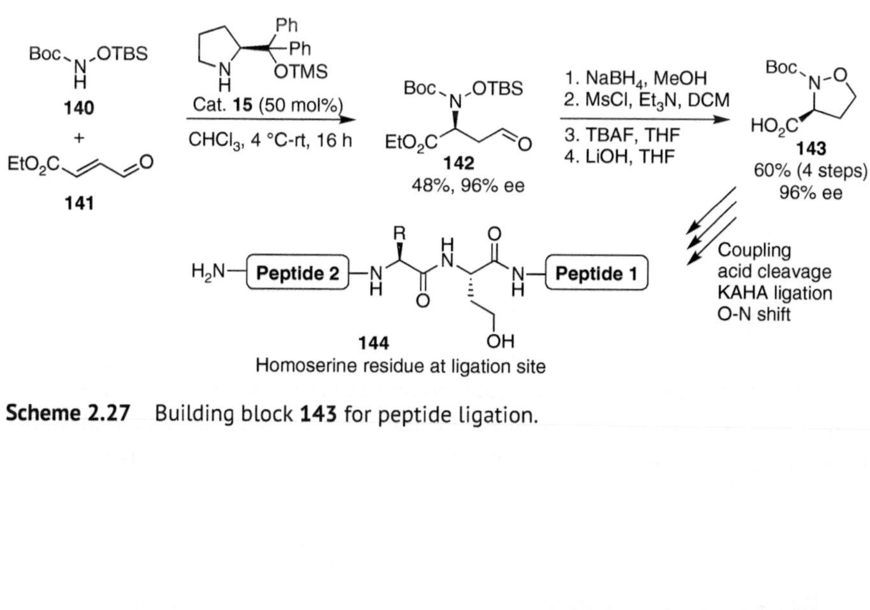

Scheme 2.27 Building block **143** for peptide ligation.

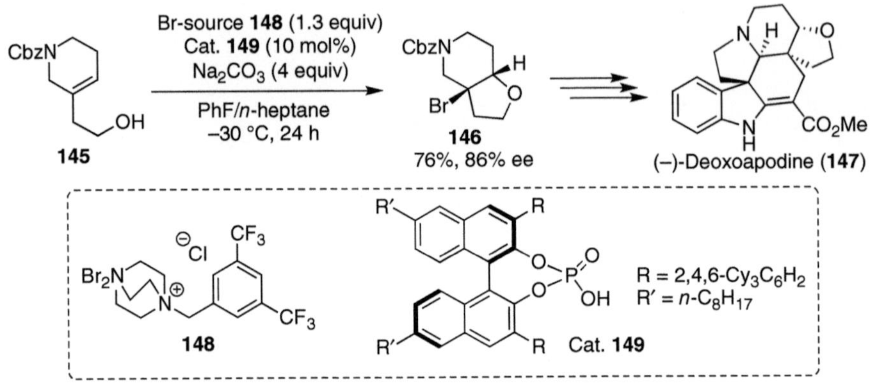

Scheme 2.28 Intermediate **146** in the total synthesis of (–)-deoxoapodine.

final product was obtained in a 10-step synthesis with an overall yield of 8.6% [51]. (–)-Deoxoapodine was also prepared by Peng [52] and coworkers through a domino deprotection-Michael addition-nucleophilic substitution protocol and by Boger and coworker [53] in a divergent total synthesis with a powerful intramolecular [4+2]/[3+2] cycloaddition cascade of a 1,3,4-oxadiazole that provided the chiral functionalized pentacyclic ring system.

Jiao and Wang used a catalytic parallel kinetic resolution to synthesize chiral tetracyclic building block **151** with two quaternary stereocenters. The reaction provides two products, indole derivatives **151** and **152**, due to variability of reactivity caused by ambident nucleophilicity (*C3* vs. *N*) in the starting racemic indole **150** (Scheme 2.29). Intermediate **151** is the key structure in further preparation for 5-*exo*-trig radical cyclization, forming the bridged nitrogen-containing ring [54].

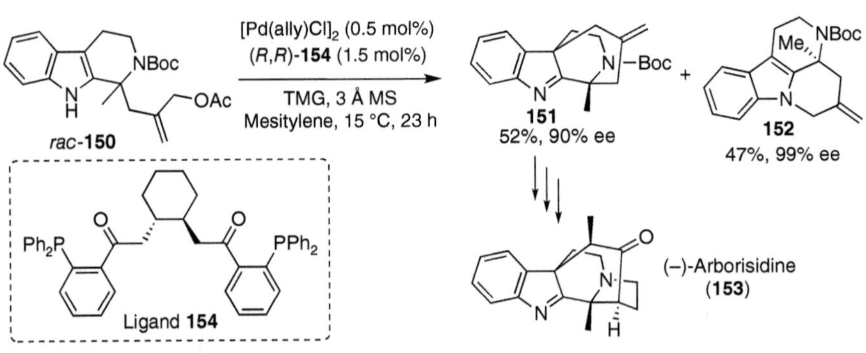

Scheme 2.29 Pd-catalyzed allylic substitution for the synthesis of derivative **151**. TMG, 1,1,3,3-tetramethylguanidine; MS, Molecular Sieve.

A nickel-catalyzed regiospecific and enantioselective reductive cyclization of *N*-alkynones is a practical method for synthesizing pyrrolidines and piperidines with a tertiary allylic alcohol silyl ether moiety (Scheme 2.30). In this reaction, the

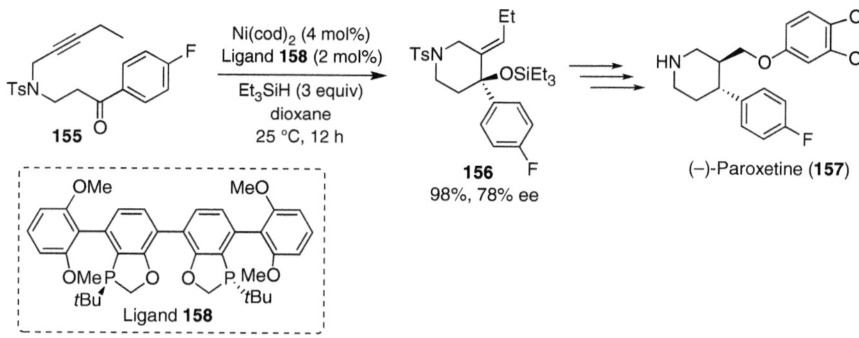

Scheme 2.30 Ni-catalyzed synthesis of piperidine building block **156**.

authors designed and synthesized a new P-chiral bisphosphorus ligand **158**. This procedure synthesized various derivatives in high yields and enantiomeric purities. The authors used one of the products, piperidine **156**, as a chiral building block to synthesize antidepressant medicine paroxetine (**157**) [55].

2.4 Other Heterocyclic Building Blocks

This section discusses quinuclidine, a prominent structure featured in many natural products and manufactured bioactive compounds and chiral catalysts. Substituted quinuclidine core is present in *Cinchona* alkaloids, which have many uses in medicine and chemical synthesis [56]. In metal-catalyzed transformations, enantioselective oxidations of alkenes are among the most prominent applications for *Cinchona* scaffolds. In 1980, Hentges and Sharpless described dihydroquinine and dihydroquinidine acetates as ligands in the osmium tetroxide dihydroxylations of alkenes [57]. However, the real breakthrough came with the recognition of superior effectivity of bisalkaloid ligands, such as $(DHQ)_2PHAL$ and $(DHQD)_2PHAL$, in these transformations (Scheme 2.31a). With these catalysts, a large variety of alkenes can be dihydroxylated with a high degree of enantioselectivity (Scheme 2.31b) [58].

Cinchona alkaloids became a crucial structural motif for various chiral organocatalysts. Several positions within their structure can be targeted and used for building diverse catalysts. Several catalyst types were mentioned as recent examples.

Attachment of *Cinchona* alkaloid to a thiourea moiety was developed independently by four research groups led by Chen, Soós, Connon, and Dixon in 2005 [59–62]. They utilized the transformation of a secondary hydroxyl group into an amino group. These catalysts, exemplified by derivative **165**, were initially utilized for Michael additions to nitroalkenes or enone. Later, *Cinchona* alkaloid-based thioureas became one of the privileged catalysts for many transformations [63, 64]. Similarly, related squaramide organocatalysts **166** were prepared by Du and coworker [65] and Jørgensen and coworker [66]. Quinuclidine nitrogen alkylation affords various quaternary ammonium salts **167**, employed as chiral phase-transfer catalysts. These derivatives were used in the 1980s phase-transfer alkylations

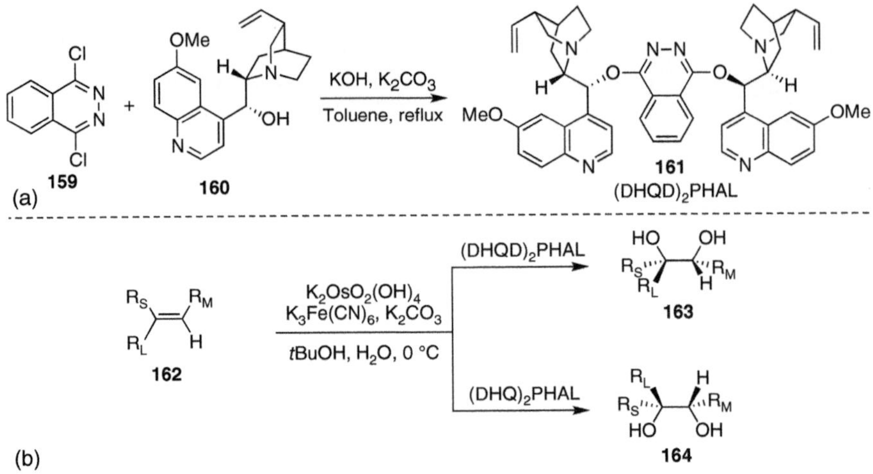

Scheme 2.31 General preparation of bisalkaloid ligands (a) and its application in Os-mediated asymmetric dihydroxylation of alkenes (b).

of indanones. The work culminated in the industrial application by Merck in the synthesis of diuretic medicine indacrinone [67]. Many more applications of *Cinchona* alkaloid-based quaternary ammonium salts were later developed [68]. The hydroxyl group in *Cinchona* alkaloids can be easily transformed into an amino group. Pioneering studies by Chen and Melchiorre revealed that primary amine **26** and its congeners are highly efficient and readily available organocatalysts [69, 70], applied in a diverse array of enamine-catalyzed transformations [71, 72]. Bisalkaloid derivatives **161**, mentioned as catalysts for Os-mediated dihydroxylation, were also useful as chiral organocatalysts. These derivatives, or even parent *Cinchona* alkaloids, are successfully used as chiral bases in Mannich reactions or allylic alkylations [73–75].

2.5 Conclusions

Heterocyclic compounds are widely occurring as natural products, medicines, and other compounds of biological relevance. This chapter highlighted some critically important structures, such as epoxides and aziridines. In addition, there was an emphasis on the growing importance of four-membered compounds, azetidines, and β-lactams. There was also extensive attention to the pyrrolidine motif, which occurs in many diverse structures. From heterocycles with larger ring sizes, quinuclidine, a part of *Cinchona* alkaloids that became one of the prominent scaffolds for building chiral catalysts, was discussed. However, this chapter could not exhaustively cover such a vast topic as chiral heterocyclic building blocks. Instead, the authors would like to highlight crucial structural building blocks and some recent and inspiring examples of their use.

References

1 Campos, K.R., Coleman, P.J., Alvarez, J.C. et al. (2019). *Science* 363: eaat0805.
2 Katsuki, T. and Sharpless, K.B. (1980). *J. Am. Chem. Soc.* 102: 5974–5976.
3 Rossiter, B.E., Katsuki, T., and Sharpless, K.B. (1981). *J. Am. Chem. Soc.* 103: 464–465.
4 Hanson, R.M., Ko, S.Y., and Sharpless, K.B. (1990). Catalytic asymmetric epoxidation. US Patent 4,900,847, 13 February 1990.
5 Tang, B., Bray, C.D., and Pattenden, G. (2006). *Tetrahedron Lett.* 47: 6401–6404.
6 Day, A.J., Lee, J.H.Z., Phan, Q.D. et al. (2019). *Angew. Chem. Int. Ed.* 58: 1427–1431.
7 Gelalcha, F.G. (2014). *Adv. Synth. Catal.* 356: 261–299.
8 Davis, R.L., Stiller, J., Naicker, T. et al. (2014). *Angew. Chem. Int. Ed.* 53: 7406–7426.
9 Peris, G., Jakobsche, C.E., and Miller, S.J. (2007). *J. Am. Chem. Soc.* 129: 8710–8711.
10 Hilton, M.J., Brackett, C.M., Mercado, B.Q. et al. (2020). *ACS Cent. Sci.* 6: 426–435.
11 Lichtor, P.A. and Miller, S.J. (2012). *Nat. Chem.* 4: 990–995.
12 Marigo, M., Franzen, J., Poulsen, T.B. et al. (2005). *J. Am. Chem. Soc.* 127: 6964–6965.
13 Wang, X., Reisinger, C.M., and List, B. (2008). *J. Am. Chem. Soc.* 130: 6070–6071.
14 Thesmar, P. and Baudoin, O. (2019). *J. Am. Chem. Soc.* 141: 15779–15783.
15 Chai, Z. (2020). *Synthesis* 52: 1738–1750.
16 Nakamura, H., Yasui, K., Kanda, Y., and Baran, P.S. (2019). *J. Am. Chem. Soc.* 141: 1494–1497.
17 Colarusso, S., Conte, I., Di Filippo, M. et al. (2011). *Synlett* 2011: 1527–1532.
18 Darses, B., Rodrigues, R., Neuville, L. et al. (2017). *Chem. Commun.* 53: 493–508.
19 Trost, B.M. and Dong, G. (2006). *J. Am. Chem. Soc.* 128: 6054–6055.
20 Gu, Q.-S. and Yang, D. (2017). *Angew. Chem. Int. Ed.* 56: 5886–5889.
21 Aggarwal, V.K. and Winn, C.L. (2004). *Acc. Chem. Res.* 37: 611–620.
22 Zhurakovskyi, O., Türkmen, Y.E., Löffler, L.E. et al. (2018). *Angew. Chem. Int. Ed.* 57: 1346–1350.
23 Mughal, H. and Szostak, M. (2021). *Org. Biomol. Chem.* 19: 3274–3286.
24 Fawcett, A., Murtaza, A., Gregson, C.H.U., and Aggarwal, V.K. (2019). *J. Am. Chem. Soc.* 141: 4573–4578.
25 Wang, Z., Sheong, F.K., Sung, H.H.Y. et al. (2015). *J. Am. Chem. Soc.* 137: 5895–5898.
26 Qian, D., Chen, M., Bissember, A.C., and Sun, J. (2018). *Angew. Chem. Int. Ed.* 57: 3763–3766.
27 Georg, G.I. (1993). *The Organic Chemistry of* β*-Lactams*. Wiley-VCH.
28 Page, M.I. (2012). *The Chemistry of β-Lactams*. Dordrecht: Springer.
29 Konaklieva, M.I. and Plotkin, B.J. (2011). *Amino Acids, Peptides and Proteins in Organic Chemistry*, 293–319. Wiley-VCH.
30 Qi, J., Wei, F., Huang, S. et al. (2021). *Angew. Chem. Int. Ed.* 60: 4561–4565.

31 Qi, J., Wei, F., Tung, C.-H., and Xu, Z. (2021). *Angew. Chem. Int. Ed.* 60: 13814–13818.
32 Shu, T., Zhao, L., Li, S. et al. (2018). *Angew. Chem. Int. Ed.* 57: 10985–10988.
33 Huang, L.-Z., Xuan, Z., Jeon, H.J. et al. (2018). *ACS Catal.* 8: 7340–7345.
34 Maddocks, C.J. and Clarke, P.A. (2021). *Tetrahedron* 78: 131789.
35 Maddocks, C.J., Ermanis, K., and Clarke, P.A. (2020). *Org. Lett.* 22: 8116–8121.
36 Adrio, J. and Carretero, J.C. (2014). *Chem. Commun.* 50: 12434–12446.
37 Pandey, G., Banerjee, P., and Gadre, S.R. (2006). *Chem. Rev.* 106: 4484–4517.
38 Cheng, M.-N., Wang, H., and Gong, L.-Z. (2011). *Org. Lett.* 13: 2418–2421.
39 Spittler, M., Lutsenko, K., and Czekelius, C. (2016). *J. Org. Chem.* 81: 6100–6105.
40 He, W., Hu, J., Wang, P. et al. (2018). *Angew. Chem. Int. Ed.* 57: 3806–3809.
41 Rodriguez, L., Fišera, R., Gaálová, B. et al. (2020). *Eur. J. Org. Chem.* 2565–2575.
42 Chen, C., Jin, S., Zhang, Z. et al. (2016). *J. Am. Chem. Soc.* 138: 9017–9020.
43 Zhang, X., Han, L., and You, S.-L. (2014). *Chem. Sci.* 5: 1059–1063.
44 Kang, T., Zhao, P., Yang, J. et al. (2018). *Chem. Eur. J.* 24: 3703–3706.
45 Yu, Q., Guo, P., Jian, J. et al. (2018). *Chem. Commun.* 54: 1125–1128.
46 Deng, M., Su, B., Zhang, H. et al. (2014). *RSC Adv.* 4: 14979–14984.
47 Athawale, P.R., Kumari, N., Dandawate, M.R. et al. (2019). *Eur. J. Org. Chem.* 2019: 4805–4810.
48 Wang, X.-L., Huang, W.-F., Lei, X.-S. et al. (2011). *Tetrahedron* 67: 4919–4923.
49 Huang, W.-F., Li, Q.-R., Chao, L.-M. et al. (2010). *Tetrahedron Lett.* 51: 4317–4319.
50 Murar, C.E., Harmand, T.J., and Bode, J.W. (2017). *Bioorg. Med. Chem.* 25: 4996–5001.
51 Yoshida, K., Okada, K., Ueda, H., and Tokuyama, H. (2020). *Angew. Chem. Int. Ed.* 59: 23089–23093.
52 Zhou, Y.-G., Wong, H.N.C., and Peng, X.-S. (2020). *J. Org. Chem.* 85: 967–976.
53 Lee, K. and Boger, D.L. (2014). *J. Am. Chem. Soc.* 136: 3312–3317.
54 Wang, F.-Y. and Jiao, L. (2021). *Angew. Chem. Int. Ed.* 60: 12732–12736.
55 Liu, G., Fu, W., Mu, X. et al. (2018). *Commun. Chem.* 1: 90.
56 Song, C.E. (2009). *Cinchona Alkaloids in Synthesis and Catalysis*. Weinheim: Wiley-VCH.
57 Hentges, S.G. and Sharpless, K.B. (1980). *J. Am. Chem. Soc.* 102: 4263–4265.
58 Ager, D.J. (2009). *Cinchona* alkaloids as chiral ligands in asymmetric oxidations. In: *Cinchona Alkaloids in Synthesis and Catalysis: Ligands, Immobilization and Organocatalysis*, 29–71. Wiley.
59 Li, B.-J., Jiang, L., Liu, M. et al. (2005). *Synlett* 2005: 603–606.
60 Vakulya, B., Varga, S., Csámpai, A., and Soós, T. (2005). *Org. Lett.* 7: 1967–1969.
61 McCooey, S.H. and Connon, S.J. (2005). *Angew. Chem. Int. Ed.* 44: 6367–6370.
62 Ye, J., Dixon, D.J., and Hynes, P.S. (2005). *Chem. Commun.* 4481–4483.
63 Connon, S.J. (2008). *Chem. Commun.* 2499–2510.
64 Siau, W.-Y. and Wang, J. (2011). *Catal. Sci. Technol.* 1: 1298–1310.
65 Yang, W. and Du, D.-M. (2010). *Org. Lett.* 12: 5450–5453.
66 Jiang, H., Paixão, M.W., Monge, D., and Jørgensen, K.A. (2010). *J. Am. Chem. Soc.* 132: 2775–2783.

67 Dolling, U.H., Davis, P., and Grabowski, E.J.J. (1984). *J. Am. Chem. Soc.* 106: 446–447.
68 Ooi, T. and Maruoka, K. (2007). *Angew. Chem. Int. Ed.* 46: 4222–4266.
69 Xie, J.-W., Chen, W., Li, R. et al. (2007). *Angew. Chem. Int. Ed.* 46: 389–392.
70 Bartoli, G., Bosco, M., Carlone, A. et al. (2007). *Org. Lett.* 9: 1403–1405.
71 Duan, J. and Li, P. (2014). *Catal. Sci. Technol.* 4: 311–320.
72 Jiang, L. and Chen, Y.-C. (2011). *Catal. Sci. Technol.* 1: 354–365.
73 Sonsona, I.G., Alegre-Requena, J.V., Marqués-López, E. et al. (2020). *Chem. Eur. J.* 26: 5469–5478.
74 Lu, S., Ng, S.V.H., Lovato, K. et al. (2019). *Nat. Commun.* 10: 3061.
75 Kowalczyk, D. and Albrecht, Ł. (2018). *Adv. Synth. Catal.* 360: 406–410.

3

Saturated Heterocycles as Chiral Auxiliaries

Agnieszka Październiok-Holewa[1,2] *and Sebastian Stecko*[3]

[1] *Silesian University of Technology, Department of Organic Chemistry, Bioorganic Chemistry and Biotechnology, B. Krzywoustego 4, 44-100 Gliwice, Poland*
[2] *Biotechnology Center of Silesian University of Technology, B. Krzywoustego 8, 44-100 Gliwice, Poland*
[3] *Institute of Organic Chemistry, Polish Academy of Sciences, Kasprzaka 44/52, 01-224 Warsaw, Poland*

3.1 Introduction

Chiral saturated heterocycles, such as pyrrolidines or piperidines, are frequently used in asymmetric synthesis. They are employed as ligands in metal-catalyzed reactions, catalysts in enantioselective organocatalytic transformations, and chiral auxiliaries in diastereoselective transformations. Compared to six-membered saturated heterocycles, five-membered ones are more frequently employed in asymmetric synthesis. Thanks to the higher rigidity of the five-membered ring, these systems can more efficiently transfer chiral information to the bound achiral substrates. Furthermore, a significant factor is the good availability of heterocyclic scaffolds. Various heterocyclic chiral auxiliaries are frequently used in asymmetric alkylation of carboxylic acid derivatives, pericyclic reactions, Michael addition, and aldol reactions to provide complex chiral intermediates in enantiomerically pure form that are further employed in the synthesis of natural targets of biological and pharmaceutical importance. Among the various auxiliaries, Evans' oxazolidinones and their sulfur analogs, Enders' SAMP/RAMP hydrazones, Oppolzer's camphorsultam, and phosphonamides are the most frequently used asymmetry inducers that have been utilized in the total synthesis of natural products and bioactive molecules. The chapter will provide a brief overview of the applications of heterocyclic auxiliaries in diastereoselective synthesis, and illustrative examples were chosen to spotlight the underlying principles of their use in stereocontrolled synthesis.

3.2 Proline-Derived Chiral Auxiliaries

The first applications of pyrrolidine-based auxiliaries date to Yamada's work in the late 1960s, which employed proline esters as chiral auxiliaries in asymmetric

Chiral Building Blocks in Asymmetric Synthesis: Synthesis and Applications, First Edition.
Edited by Elżbieta Wojaczyńska and Jacek Wojaczyński.

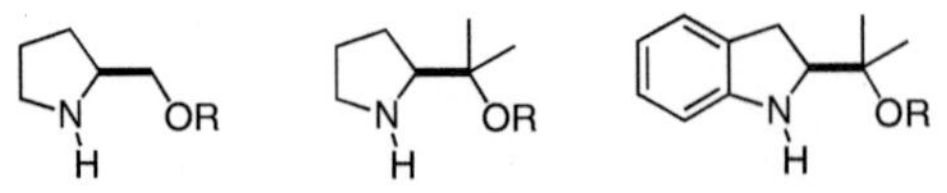

Figure 3.1 Proline-based chiral auxiliaries. Source: Based on Refs. [1–3].

alkylation reactions via enamine derivatives. Since then, several examples that use Yamada's auxiliaries have been reported. In 1980, Evans and Takacs [1] and Sonnet and Heath [2] independently reported that dianions of amides of (*S*)-prolinol **1a** furnished diastereomeric mixtures of alkylated products in ratios of up to 97 : 3. It was also demonstrated that diastereoselectivity could be reversed when amides of methyl, MOM, and TBS ethers (**1b–d**) of (*S*)-prolinol were applied (Figure 3.1) [1, 2]. However, the diastereoselectivities obtained with these ether auxiliaries were lower. Bulkier (*S*)-prolinol analogs **2** (Figure 3.1, R: H, Me), elaborated by Lin et al. [4], have slightly improved the diastereomeric ratios in alkylation reactions of their amide enolates.

Andersson and Hedenström [5] have investigated the alkylation of the enolates of propanoylamides **4** with benzyl bromide and *n*-butyl iodide (Scheme 3.1). Auxiliaries **2a** and **2b** induced opposite selectivity, (*R*)- and (*S*)-configuration, respectively, at the newly created stereogenic center. The stereoselectivities and conversion yields of these alkylations were moderate to excellent, but it was demonstrated that the diastereomeric ratios could be improved using various additives and conditions. With an excess LiCl used as an additive, a significantly higher diastereomeric ratio was obtained. When Cp_2ZrCl_2 was used as an enolate coordinating agent, benzylation of **4b** gave an excellent diastereomeric ratio of 99 : 1. Good to excellent diastereoselectivities were also observed for alkylation of *N*-acylated derivatives of indolinols **3** (Figure 3.1) [3].

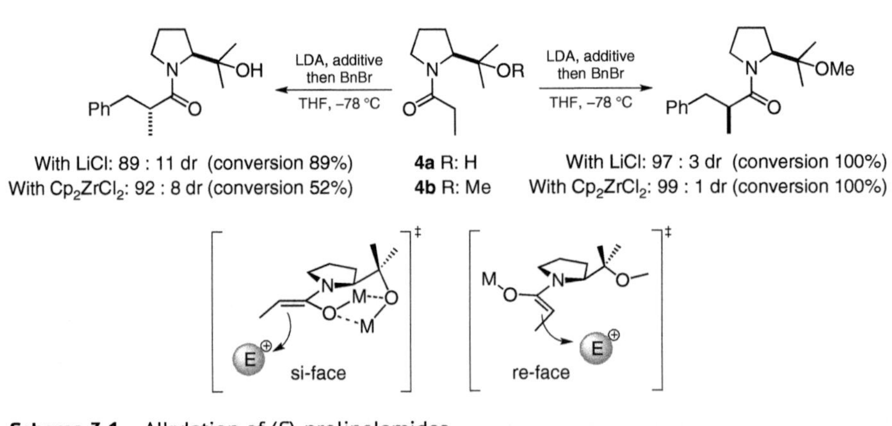

Scheme 3.1 Alkylation of (*S*)-prolinolamides.

Raubo et al. from Merck investigated an auxiliary-mediated acyl-Claisen rearrangement as a key step in synthesizing enantiomerically pure crucial intermediates of NK1-receptor antagonist analogs of aprepitant (Emend®) (Scheme 3.2) [6]. An *O*-methoxy prolinol-derived allylamine **5** was acylated with a chiral acid chloride, and the resulting intermediate was transformed to amide **6** via a Lewis acid (LA)-catalyzed acyl-Claisen rearrangement. Using the chiral prolinol auxiliary and the chiral acid chloride was essential to achieve good diastereoselectivity for the rearrangement into amide **6**, which was a key intermediate for the stereoselective construction of the tetrahydropyran ring in **7**.

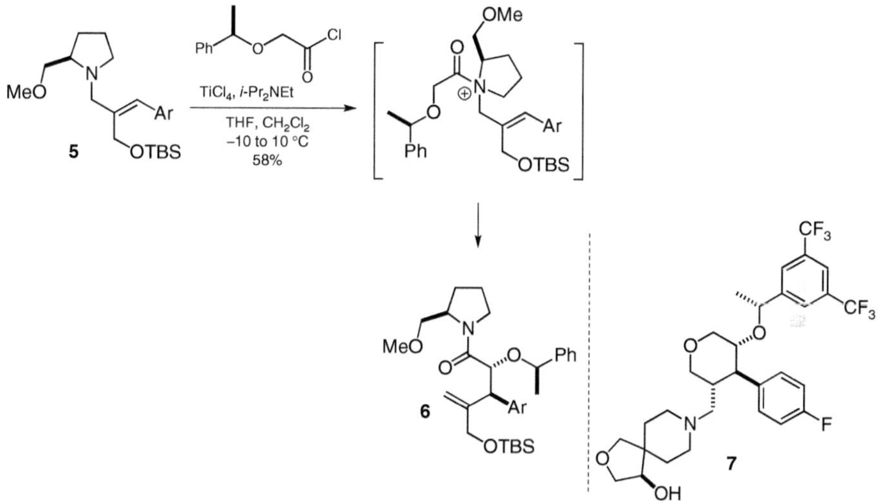

Scheme 3.2 Stereoselective acyl-Claisen rearrangement as a key step in synthesizing NK1-receptor antagonist **7**. Source: Raubo et al. [6]/Georg Thieme Verlag KG.

Another example of auxiliary-controlled Claisen-type rearrangement has been reported by Maulide and coworkers [7], who investigated the properties of ketiminium salts (Scheme 3.3). The treatment of chiral tertiary amide **8** with Tf_2O/collidine provided a ketiminium intermediate that rearranged to substituted butyrolactone **9**. The reaction proceeded most likely via a Claisen-type

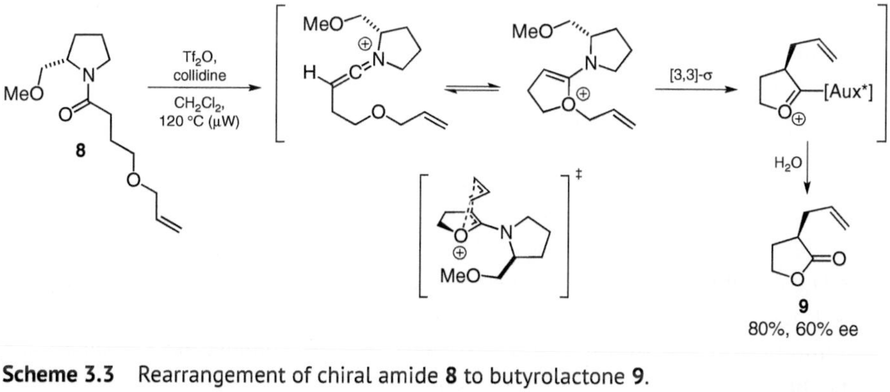

Scheme 3.3 Rearrangement of chiral amide **8** to butyrolactone **9**.

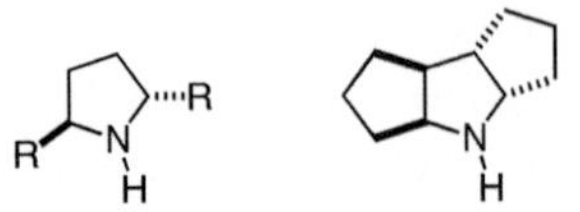

10a, R: Me
10b, R: Ph
10c, R: CH_2OMe

11

Figure 3.2 Examples of 2,5-disubstituted pyrrolidine chiral auxiliaries.

rearrangement and provided lactone **9** with excellent yields but only moderate stereoselectivity.

3.3 2,5-*trans*-Disubstituted Pyrrolidines

trans-2,5-Dimethylpyrrolidine **10a** was the first chiral amine possessing a C_2-symmetry used as a chiral auxiliary in asymmetric synthesis. Since then, many related systems have been developed, including compounds **10**, **11** (Figure 3.2). These amines, which can be regarded as C_2-symmetric analogs of **1a**, often give enhanced stereoselectivity compared to other prolinol derivatives. However, their preparation is more onerous. Typically, they are synthesized by reducing 1,4-diketones, followed by hydroxyl group activation and cyclization with a chiral amine (e.g. (*S*)-1-phenylethylamine) [8, 9]. Alternatively, asymmetric methods for ketone reduction are also available [10, 11].

Typical diastereoselective reactions aided by chiral auxiliaries **10** include enolate alkylation reaction [12, 13], aldol reaction [14, 15], and conjugate addition [16].

For example, pyrrolidine **10b** has been used as a chiral auxiliary for regio- and stereoselective Michael addition to conjugated amides, such as **12** (Scheme 3.4) [17]. Grignard reagents are added to the δ-position of the conjugated double bond system of amide **12**. The resulting enolate intermediate was then trapped with an appropriate electrophile to give the 6-substituted or 3,6-disubstituted amides **13** with excellent diastereoselectivity. The regioselectivity of alkylation can be changed if auxiliaries replace the type **10b** pyrrolidine scaffold with coordinating side chains, thus substrate **14** was alkylated via Michael addition of Grignard reagents to furnish an intermediate **15**, which upon treatment with alkyl halide afforded the 2,3-substituted amide **16** with excellent diastereoselectivity.

Thanks to their C_2-symmetry, 2,5-disubstituted pyrrolidines provide a good facial differentiation in Diels–Alder cycloadditions [18] and 1,3-dipolar cycloadditions [19–21]. For example, Rawal and coworkers have accomplished the total synthesis of (−)-α-elemene (**19**), a naturally occurring terpene, using chiral 1-amino-3-siloxy-1,3-butadiene **17** bearing attached ***ent*-10b** auxiliary as a diene component in the Diels–Alder reaction (Scheme 3.5) [22]. The chiral diene **17** was reacted with methacrolein at 20 °C in toluene to give adduct **18** as the major product (de 80%). The Diels–Alder adduct **18** was then transformed into **19** in six steps. Moreover, Rawal and coworkers [23, 24] have elegantly demonstrated the synthetic utility of aminosiloxydiene in the total synthesis of the *Aspidosperma* alkaloid and tabersonine.

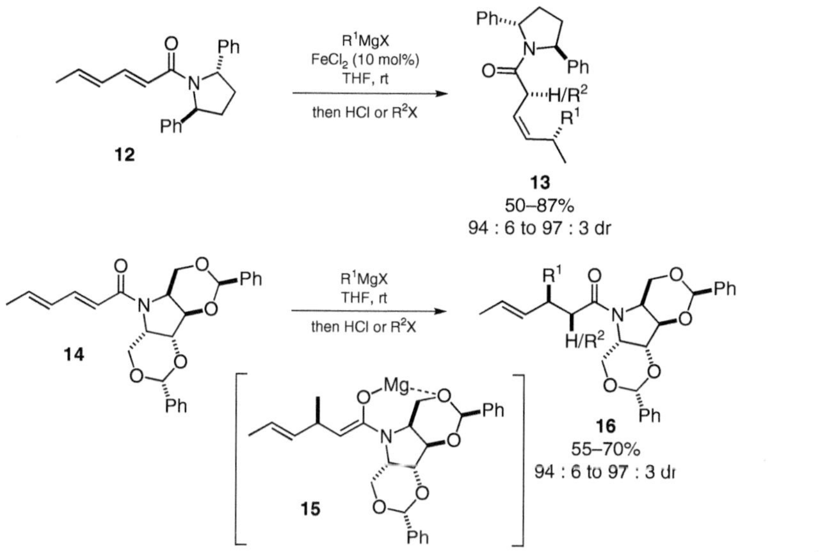

Scheme 3.4 Asymmetric Michael addition to conjugated amides. Source: Okada et al. [17]/John Wiley & Sons.

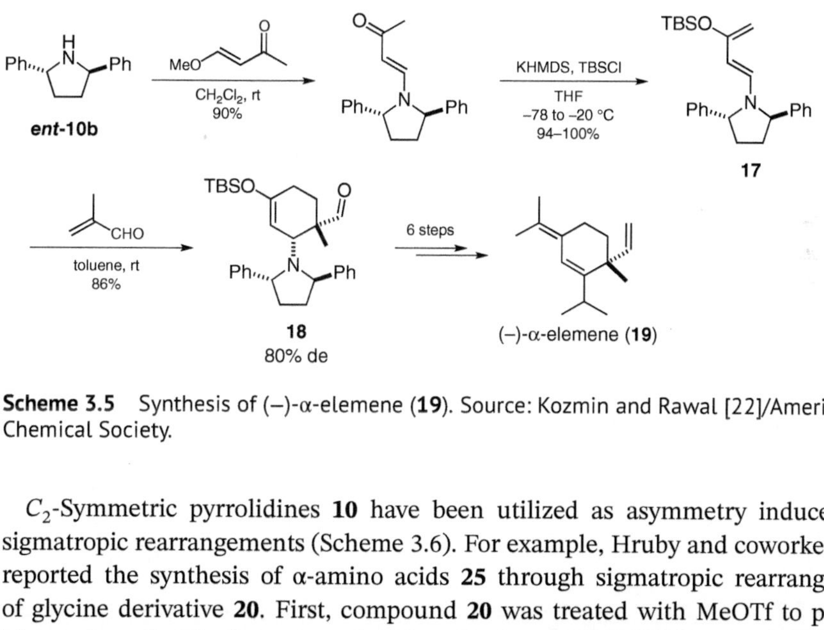

Scheme 3.5 Synthesis of (−)-α-elemene (**19**). Source: Kozmin and Rawal [22]/American Chemical Society.

C_2-Symmetric pyrrolidines **10** have been utilized as asymmetry inducers for sigmatropic rearrangements (Scheme 3.6). For example, Hruby and coworkers [25] reported the synthesis of α-amino acids **25** through sigmatropic rearrangement of glycine derivative **20**. First, compound **20** was treated with MeOTf to provide Meerwein salts **21**, then reacted with various substituted lithium allyloxides to afford **22**. Upon warming to room temperature, intermediates **22** rearranged to amides **23** in good yields and with moderate to good selectivity. Treatment of **23** with iodine and water resulted in removing the auxiliary with iodolactonization to iodide **24**. The subsequent reductive elimination provided the final unsaturated α-amino acids **25** (Scheme 3.6).

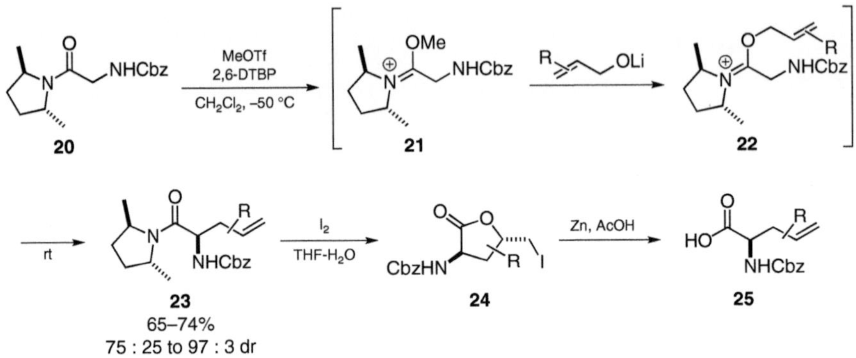

Scheme 3.6 Diastereoselective Eschenmoser–Claisen rearrangement of glycine derivatives. 2,6-DTBP = 2,6-di-*t*-butylpyridine.

The efficiency of the rearrangement was further improved by its thio-variant shown in Scheme 3.7 [26]. Thioamide **26** was treated with an alkyl halide in the presence of iron salt to provide *S*-alkylated intermediate **27**, which after warming up, rearranged to **28** in good yield and with excellent diastereoselectivity. The reaction scope is remarkably broad, and several substituted allyl bromides were successfully used, as depicted in Scheme 3.7. Both rearrangements presented in Schemes 3.6 and 3.7 are complementary because they display the opposite stereochemical outcomes, attributed to slightly different transition states due to longer C—S bonds and larger C—S—C bond angles compared to the oxygen analogs.

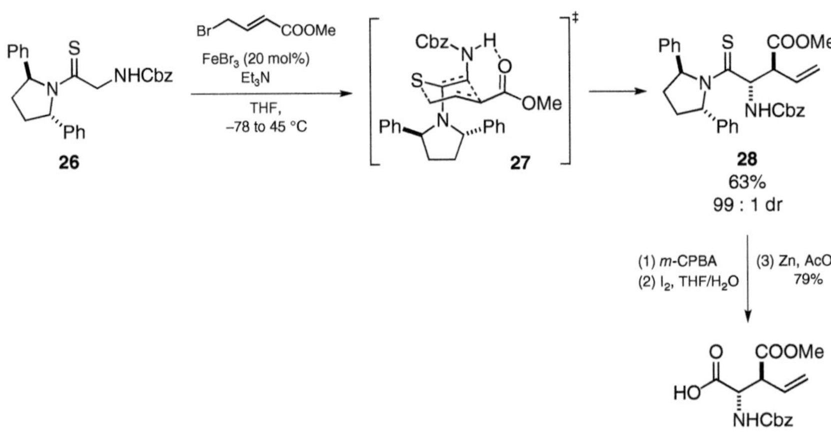

Scheme 3.7 Diastereoselective thio-Claisen rearrangement of glycine derivatives. Source: Liu et al. [26]/American Chemical Society.

Nubbemeyer and coworkers have investigated auxiliary-controlled asymmetric acyl-Claisen rearrangements of chiral allylamines, e.g. **29** (Scheme 3.8) [27]. Its *N*-acylation with acyl fluoride, followed by Lewis acid-mediated rearrangement, provided the corresponding amides like **30** in good yield and excellent diastereoselectivity for the auxiliary **10c**. 2,3-*anti*/2,3-*syn* Diastereoselectivity was almost exclusively favoring the 2,3-*anti*-isomers. Furthermore, it was observed that the relative stereochemistry of the *anti*-configured acyl chain depended on the auxiliary. C_2-Symmetric auxiliary **10c** led to (2*S*,3*S*)-stereoisomer **30**, whereas the other proline and 3-hydroxyproline-derived auxiliaries gave the corresponding (2*R*,3*R*)-diastereoisomer.

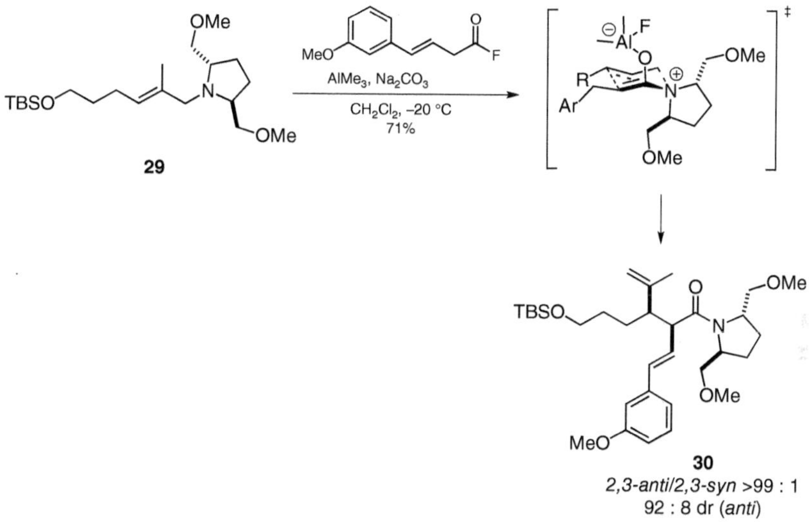

Scheme 3.8 Asymmetric acyl-Claisen rearrangement of allylamines.

C_2-Symmetric pyrrolidines are also useful in various radical reactions. For example, the chiral auxiliary **10a** was used for Giese addition of radicals to acrylates to provide the corresponding product **31** in good diastereomeric excess (Scheme 3.9) [14, 28]. Moderate to good diastereoselectivity was noticed for radical addition to chiral Michael-type acceptor **32** [29].

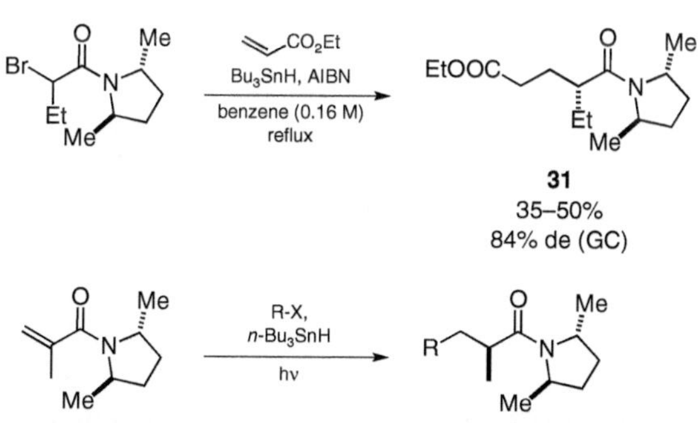

Scheme 3.9 Asymmetric radical reactions. Source: Based on Porter et al. [14] and Stack et al. [28].

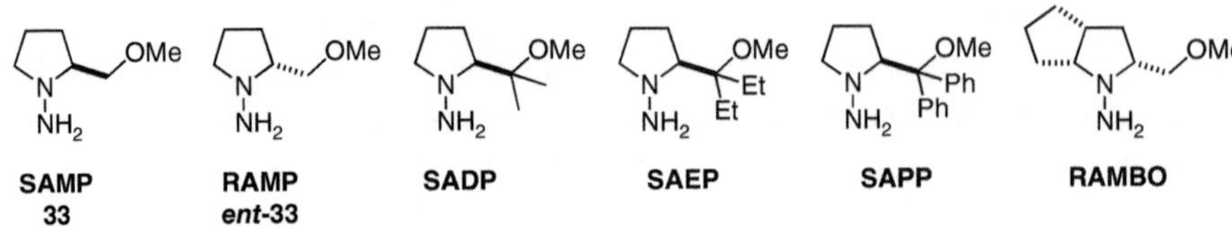

Figure 3.3 Selected analogs of SAMP/RAMP hydrazines. Source: Based on Refs. [34–43].

3.4 Heterocyclic Hydrazones

(*S*)-1-Amino-2-methoxymethylpyrrolidine (SAMP, **33**) and its *R*-enantiomer (RAMP, ***ent*-33**) were introduced as chiral auxiliaries to form hydrazones from aldehydes and ketones by Enders [30–34].

SAMP is available from (*S*)-proline in a four-step synthesis [34–36], and RAMP is prepared from (*R*)-glutamic acid in a six-step procedure [32, 37]. The sterically hindered analogs, for instance, SADP [38–41], SAEP [42, 43], SAPP [43, 44], and RAMBO [45–48] (Figure 3.3), can be prepared in the same manner starting from the corresponding amino acid derivatives.

The chiral hydrazones are prepared by mixing hydrazine **33** or its analogs with aldehydes or ketones under conditions enabling water removal. In the case of aldehydes, the reaction is rapid and proceeds at rt or around 0 °C, often without any solvent. For ketones, reflux is required.

The Enders' hydrazone methodology emerged as a powerful tool for many asymmetric transformations, and there are several detailed review articles [49–57]. Only a brief overview will be provided.

The broad group of transformations encompasses electrophilic α-functionalization of carbonyl derivatives. The typical electrophilic reactions on the α-carbon of the hydrazine **34** require prior deprotonation of the hydrazone C–H acid to provide aza-enolate species **35**. Once the aza-enolate is formed, it can react with many electrophiles.

The aza-enolate **35** may exist in different conformations and configurations, as exemplified for SAMP as a chirality inducer (Scheme 3.10, **35a–d**). Aza-enolate **35a** is the most favored of all intermediates because of the low steric repulsion between the chiral auxiliary and olefin group. The addition of alkyl halide, or any other electrophilic species, to aza-enolate proceeds from the *si*-face (the *re*-face approach is sterically less favored) and gives product **36** as the major isomer.

The known synthetically useful C—C bond-forming reactions (Scheme 3.11) of α-metalated hydrazones include α-alkylation [58–61], Michael-type additions [62], aldol-type reactions [63, 64], and acylations. Other suitable electrophiles are oxiranes, aziridines, and other heteroatomic electrophiles. Moreover, α-phosphinylation [65, 66], α-silylation [67], α-sulfenylation [68], α-selenylation [69], α-amination [70], and α-hydroxylation [71] have been described, too.

After workup, the hydrazone moiety of **36** can be cleaved to restore the initial carbonyl function and provide the corresponding α-substituted aldehyde or ketone

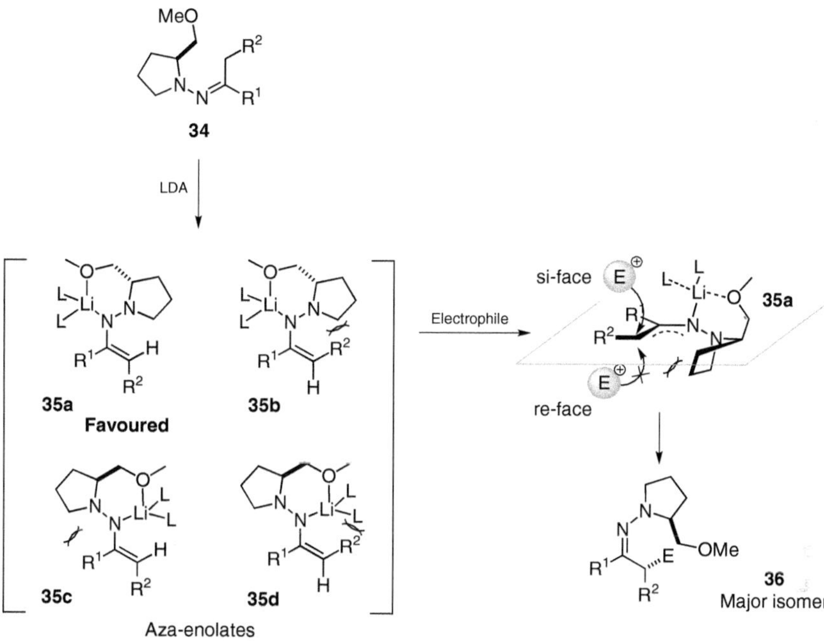

Scheme 3.10 Conformations of aza-enolates and stereochemical model of the reaction of Enders' hydrazones with electrophilic reagents.

(Scheme 3.12). Modification of reaction conditions generates other functionalities than carbonyl ones, including amines [73], nitriles [74–76], and dithianes or dithiolanes [72] (Scheme 3.12).

The SAMP/RAMP method is also an attractive tool for the stereoselective synthesis of amines via nucleophilic addition of organometallic species to C=N double bonds of chiral hydrazones **37** (Scheme 3.13). These additions are most frequently performed with organolithium, Grignard, or organocerium reagents and introduced alkyl [77–81], aryl [77–84], and heteroaryl substituents [85]. The nucleophilic addition of various organometallic reagents proceeded with high diastereoselectivity and good yields to give the corresponding hydrazine derivatives **38**, **39**. Moreover, excellent diastereoselectivity is observed in 1,2-addition of other nucleophilic reagents, such as cyanide (Strecker reaction, **40**) [86, 87]. Moderate diastereoselectivity is observed in the addition of phosphites (**41**) [88] and for SAMP-aided hydride addition [89].

Nucleophilic addition to α-amino, α-thio, and α-alkoxy aldehyde derivatives **42** enables the stereoselective synthesis of hydrazines **43** that are excellent precursors of chiral 1,2-diamines, 1,2-amino alcohols and related multifunctional compounds (Scheme 3.14) [90–95].

Both SAMP-aided α-alkylation of aldehydes and SAMP-mediated addition to the C=N bond have found numerous applications in the total synthesis of natural products and pharmaceuticals [10, 96–107]. Moreover, both transformations are used as elementary key steps in total synthesis in several cases, highlighting the high

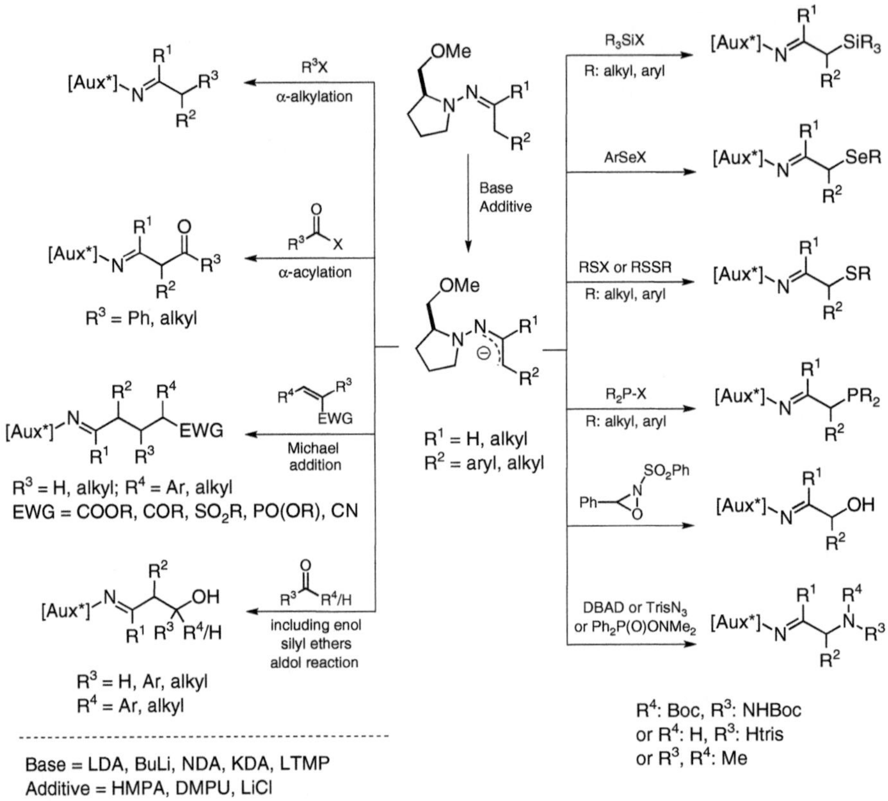

Scheme 3.11 α-Functionalization of Enders' hydrazones with various electrophilic reagents.

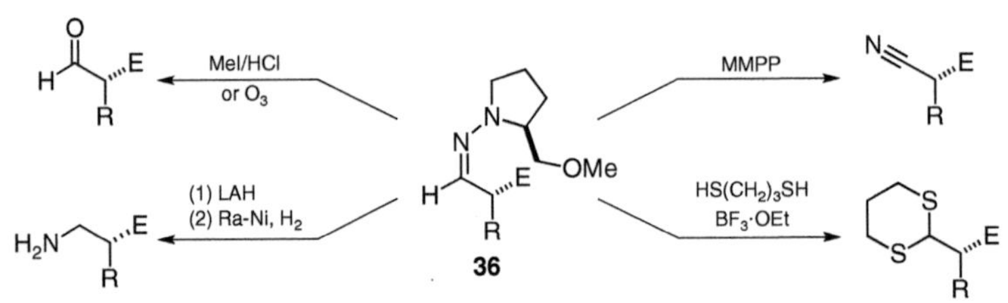

Scheme 3.12 Cleavage of hydrazine moiety. Source: Diez et al. [72]/Elsevier.

versatility and robustness of these methods. For example, Enders and Kirchoff [101] have employed such an approach in the synthesis of 2-*epi*-deoxoprosopinine (**50**) (Scheme 3.15). SAMP-aided nucleophilic addition was used as a key step in synthesizing hydrazine **45** from compound **44**. After removing the auxiliary, hydrazine **45** was converted into iodide **46**. This iodide was then employed as a reagent for SAMP-aided alkylation of the dihydroxyacetone derivative **47** to provide product **48**. The removal of the auxiliary group, followed by an intramolecular reductive amination and deprotection, provided the target piperidine derivative **49**.

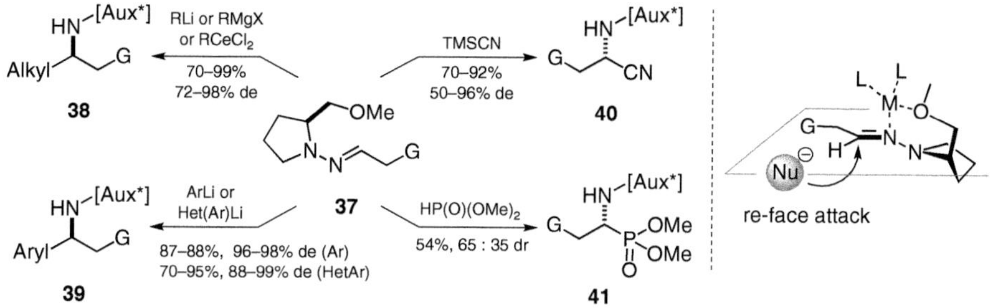

Scheme 3.13 Nucleophilic addition of organometallic species to C=N double bonds of chiral hydrazones.

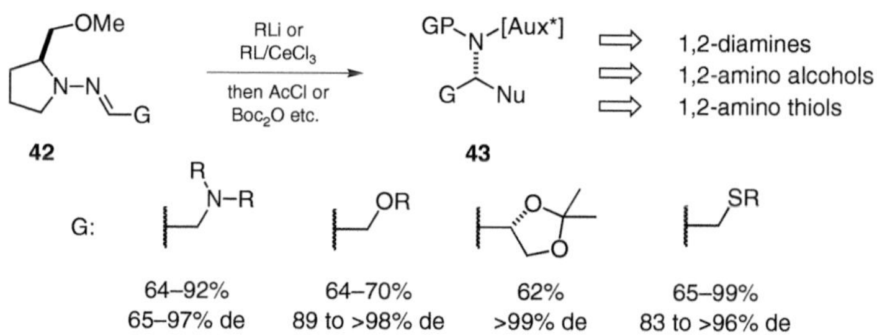

Scheme 3.14 Alkylation of α-heteroatom substituted hydrazones. Source: Based on Refs. [90–95].

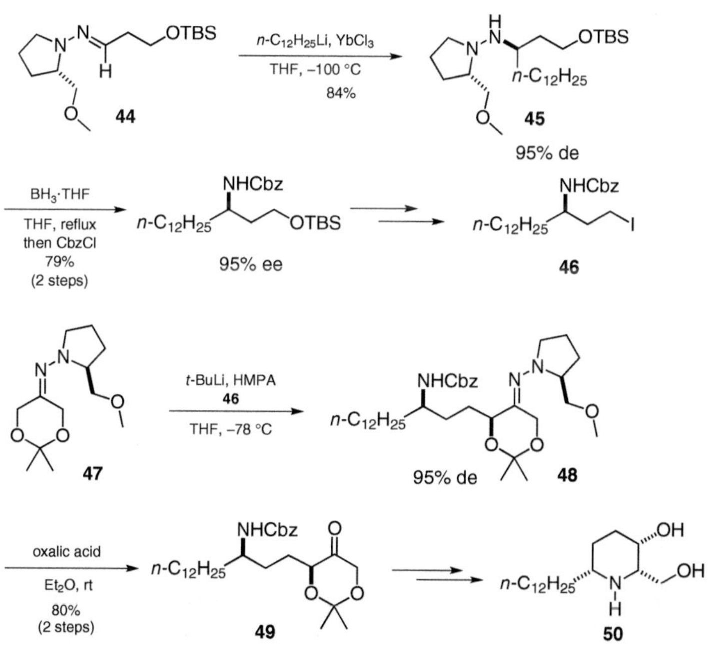

Scheme 3.15 Total synthesis of 2-*epi*-deoxoprosopinine (**50**).

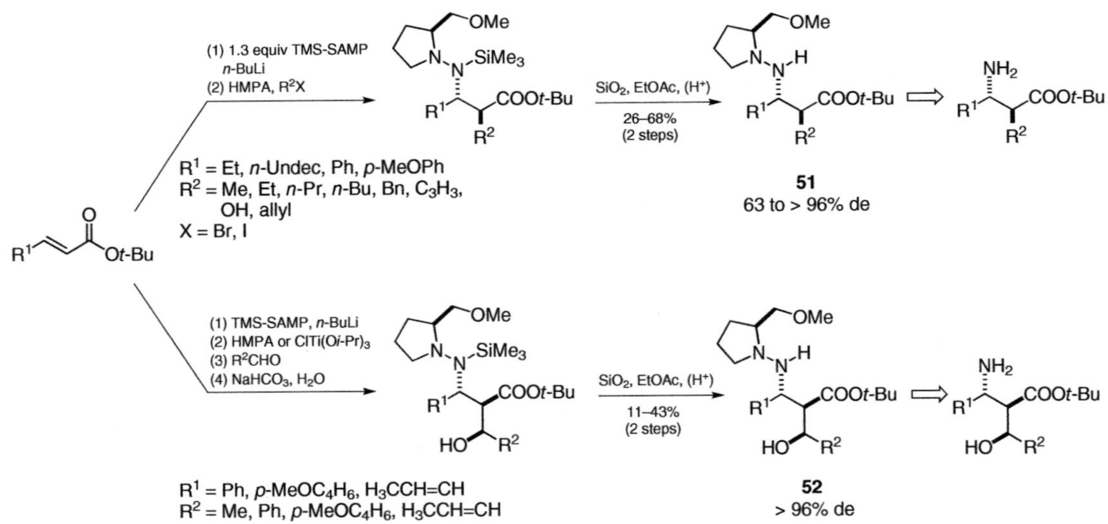

Scheme 3.16 Tandem aza-Michael addition/α-ester enolate alkylation or aldol reaction.

Figure 3.4 Selected analogs of Enders' proline-based hydrazines. Source: Wengryniuk et al. [109]/ American Chemical Society.

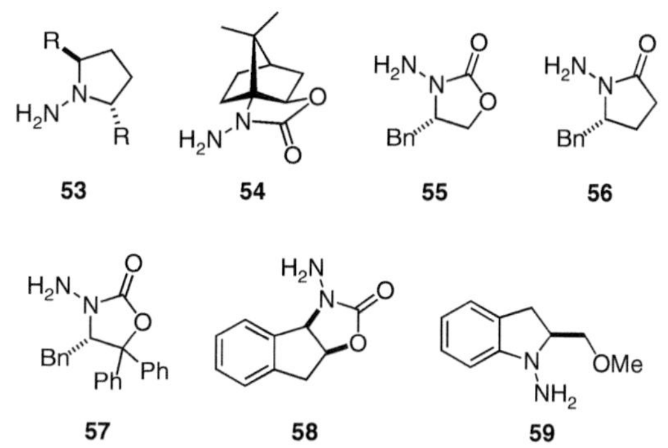

A useful extension of Enders' hydrazine method is the tandem hetero-Michael addition/α-ester enolate alkylation or aldol reaction [108]. As shown in Scheme 3.16, after the initial 1,4-addition of the chiral *N*-nucleophile, α-alkylation/α-aldol reaction using the appropriate halides R^2X or aldehydes R^2CHO leads to the α-substituted β-hydrazine esters **51** and **52** with good to excellent stereoselectivity.

Since its first report in 1976, Enders' hydrazine methodology was refined by other groups, resulting in various heterocyclic hydrazine derivatives (**53**–**58**) as chiral auxiliaries. Some of them are shown in Figure 3.4 [109].

Hydrazones derived from hydrazine **53** undergo highly diastereoselective Staudinger [2+2] cycloadditions with *in situ* generated ketenes (Scheme 3.17). Oxidative cleavage of the N—N bond affords enantiomerically pure β-lactams **60**, **61**, and **62** [42, 110].

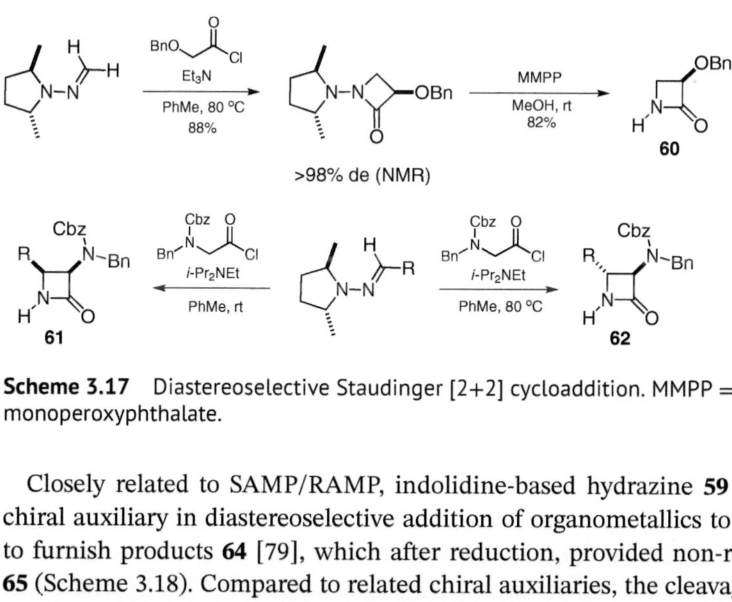

Scheme 3.17 Diastereoselective Staudinger [2+2] cycloaddition. MMPP = magnesium monoperoxyphthalate.

Closely related to SAMP/RAMP, indolidine-based hydrazine **59** was used as a chiral auxiliary in diastereoselective addition of organometallics to hydrazones **63** to furnish products **64** [79], which after reduction, provided non-racemic amines **65** (Scheme 3.18). Compared to related chiral auxiliaries, the cleavage of the N—N bond derived from the indoline-based auxiliary is easier.

Oxazolidinone-based chiral hydrazines **55** have found application in asymmetric nucleophilic and radical addition to imines (Scheme 3.19) [111–116]. The

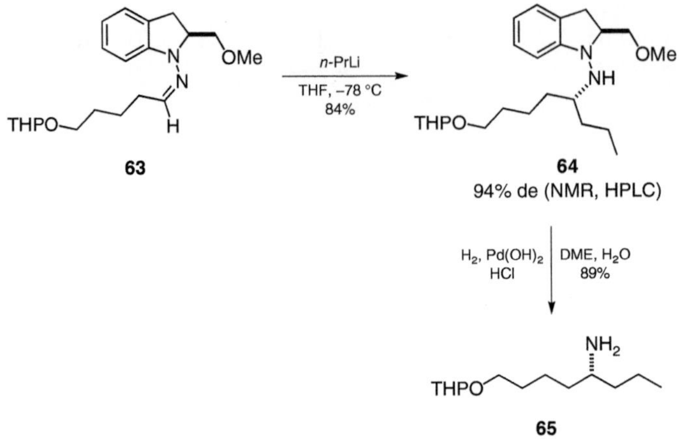

Scheme 3.18 Nucleophilic addition to indoline-derived chiral hydrazone.

additional nitrogen atom in the hydrazine moiety, and the presence of a carbonyl group, enhance the reactivity of imine functionality toward nucleophiles and radical species. Moreover, Lewis acid (LA) chelation additionally increases the electrophilicity of the C=N bond, while the presence of a two-point binding site restricts N—N bond rotation (Figure 3.5). The stereogenic center present in the heterocyclic ring is responsible for the facial differentiation of the imine group. Type **55** auxiliaries are prepared from the corresponding oxazolidinones via electrophilic amination with, e.g. NH_2Cl [112, 117], and the corresponding hydrazones are prepared by condensation of **55** with aldehydes or ketones [113, 114].

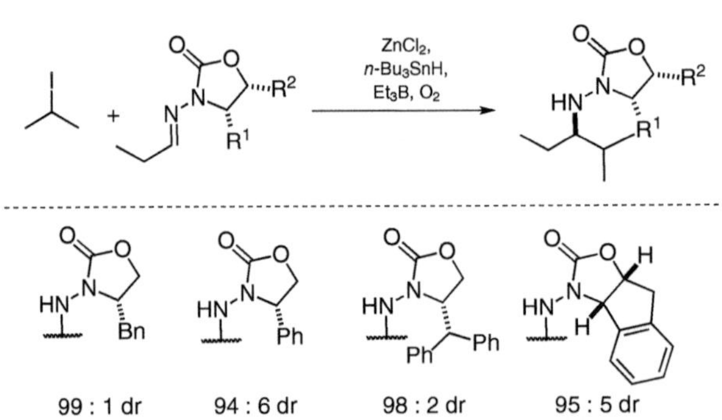

Scheme 3.19 Diastereoselective radical addition to *N*-acyl hydrazones. Source: Based on Refs. [111–116].

Lewis acid-promoted addition of secondary or tertiary radicals to aldhydrazones proceeds with excellent diastereoselectivity to provide the corresponding secondary and tertiary amines after removing the chiral auxiliary [113–120]. Typical Lewis acids are $ZnCl_2$ and $InCl_3$. The corresponding radical species are generated by the combination of $Bu_3SnH/Et_3B/O_2$ (Scheme 3.25) or upon irradiation in $Mn_2(CO)_{10}$ [134, 135]. The Mn-mediated process is suitable for synthesizing heterocyclic

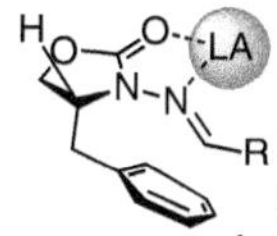

Figure 3.5 Structure of *N*-acyl hydrazones. LA = Lewis acid.

products via radical–polar crossover process involving radical addition to hydrazine and intramolecular cyclization [79, 135, 136]. This approach was employed in the preparation of alkaloids (e.g. quinine [137], coniine [79]), γ-amino acids [138], and α,α-disubstituted α-amino acids [111].

3.5 Oxazolidine-Based Chiral Auxiliaries

Another class of amino acid-derived heterocyclic chiral auxiliaries includes oxazolidinones and their thio-analogs, which have become an integral tool for preparing enantiopure fine chemicals, pharmaceuticals, and intermediates in the synthesis of natural products. The popularity of this class of chiral auxiliaries relies on the ease of preparation, their reliable efficiency, and predictable stereoselectivity, as well as facile installation and removal. Evans et al. first introduced these chiral auxiliaries in 1981 [139]. In addition to the original oxazolidinones introduced by Evans, several other modified derivatives have also been developed over the years and exploited in asymmetric synthesis (Figure 3.6), including monosubstituted ones, e.g. **66** [139], commonly called Evans' auxiliaries, and the highly substituted analogs (e.g. **67**) [140] or pseudoephedrine-based compound **68** [141–143]. This group can also be extended to camphor-derived (e.g. **69**) [144] and carbohydrate-derived [145] oxazolidines (**70**).

Monosubstituted oxazolidinones **66** are easily prepared from chiral 2-amino alcohols by treatment with phosgene, diethyl carbonate, or other carbonyl sources [146]. In the case of highly substituted oxazolidinones, e.g. **67**, their preparation starts from the amino acid, which is first converted into the corresponding ester. A subsequent nucleophilic addition to the ester group installs the substituents at C5. The resulting amino alcohol is then treated with phosgene or diethyl carbonate [147].

The Evans chiral oxazolidinones have been extensively employed in various highly diastereoselective transformations of carboxylic acid derivatives, similar to previously discussed Enders' auxiliaries. The broad portfolio of the possible transformations includes α-alkylations [148–150], cycloadditions [151], Michael additions [152–155], α-aminations [156–159], α-azidations [160–165], α-brominations [166–168], α-hydroxylations [169–174], α-thionylation [175–177], and addition to the C=N bonds [177–179]. However, they are commonly applied in the asymmetric aldol reaction of amides **71** (Scheme 3.20). Since the application of oxazolidinones as chiral auxiliaries has already been reviewed [51, 63, 150, 182–186], the focus will be on the general remarks and some recent developments.

The aldol reaction may lead to four diastereoisomeric products, and the proper choice of reagents and reaction conditions allows access to all of them

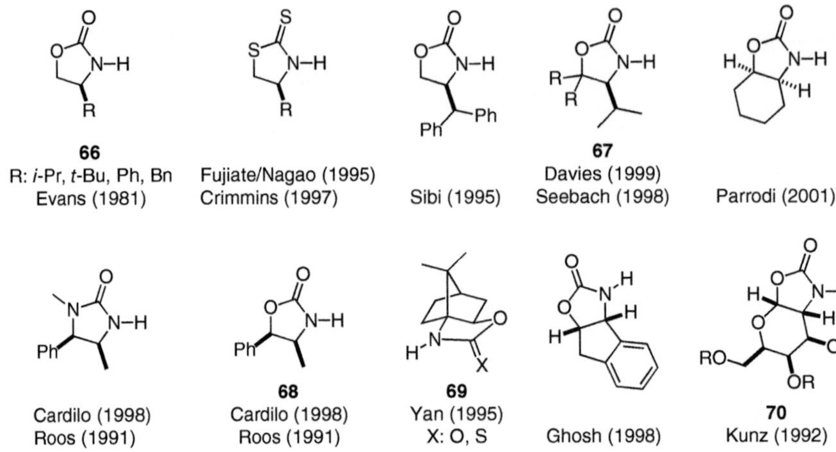

Figure 3.6 Selected Evans-type chiral auxiliaries.

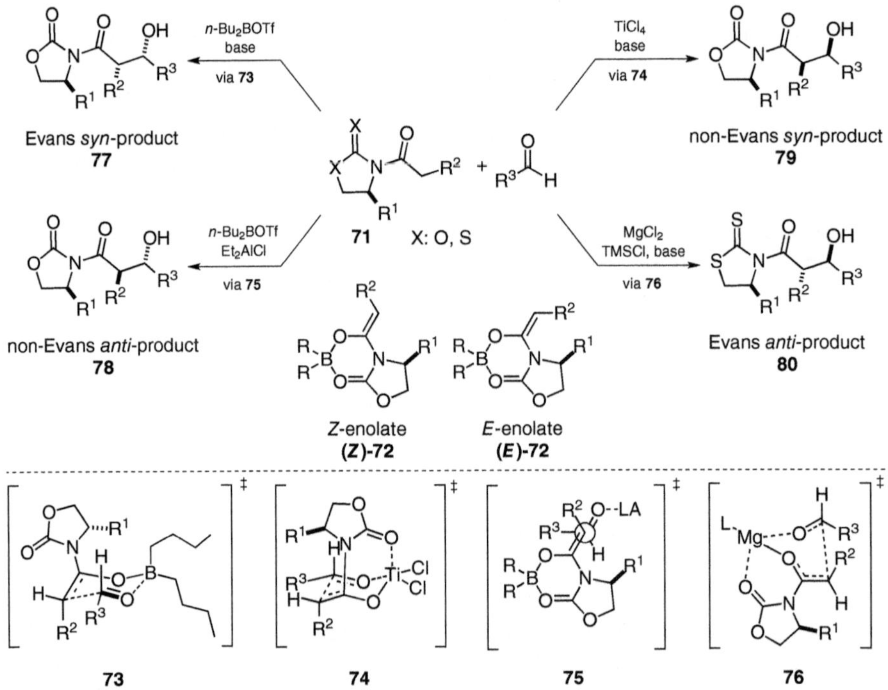

Scheme 3.20 Evans' oxazolidinone as asymmetry inducer in a diastereoselective aldol reaction. Source: Based on Evans et al. [180, 181] and Crimmins and Powell [181].

(Scheme 3.20). However, to ensure the desired *syn*- or *anti*-selectivity, it is necessary to generate the enolate intermediates **72** with the appropriate configuration, as the diastereoselectivity of this addition is directly related to the geometry of these enolates in six-membered transition states. The reaction of an aldehyde with enolate (*Z*)-**72** provides predominantly *syn*-aldol products **77** (also called *syn*-Evans products), whereas the (*E*)-enolate ((*E*)-**72**) leads to *anti* products **78** (Scheme 3.20) [63, 182, 183]. Such preference for both enolates results from minimizing the 1,3-diaxial interactions between the chiral auxiliary and the alkyl gro up R^2 in the corresponding Zimmerman–Traxler-type cyclic transition states (**73**). The synthesis of *syn*-Evans aldols **77** is one of the most popular applications of *N*-acyl oxazolidinones. *Anti*-aldols **78** are formed in bulky boron Lewis acids or *n*-Bu_2BOTf and the addition of a large Lewis acid (e.g. Et_2AlCl), which is responsible for extra electrophile coordination (**75**) [188].

Two other aldol products, **79** and **80**, are usually formed when the aldol reaction is conducted under chelate-controlled conditions. For example, the use of Ti, Zn, Sn, or Li enolates 74 results in the formation of "non-Evans" *syn*-aldols **79**. On the other hand, *anti*-products **80** are formed, in the case of $MgCl_2$-catalyzed reaction, in TMSCl, which allows for the turnover of the metal complex 76 by silylation of the *anti*-aldol products (Scheme 3.20) [180, 189]. The reaction can also be conducted with related *N*-acyl thiazolidinethiones instead of oxazolidinones [189]. The advantage of these sulfur-containing oxazolidinones is the milder cleavage conditions.

The key advantage of Evans-type auxiliaries is that they are easily removed to provide the corresponding free carboxylic acid and directly converted into other reactive functional groups. The requisite carboxylic acid is prepared by hydrolysis with LiOOH under basic conditions. By reducing agents, such as $NaBH_4$, $LiBH_4$, and $LiAlH_4$ [181, 190], the oxazolidinone ring undergoes reductive cleavage to furnish the corresponding alcohols. On the other hand, treatment of *N*-acyl oxazolidinones with alkoxides leads to ester derivatives, whereas in the presence of MeNHOMe, aldols are transformed to Weinreb amides [191, 192].

The adducts of *N*-acyl oxazolidinethiones and thiazolidinethiones can be readily transformed into other functional groups, as in the case of oxazolidinones, but often under milder conditions. Thus, they can be readily transformed into alcohols, esters, and amides, including Weinreb amides [193, 194]. In contrast to oxazolidinones, they can be directly converted to aldehydes upon treatment with DIBAL-H [154]. Furthermore, nucleophilic addition of acetonitrile anion or ethyl acetate enolate to *N*-acyl thiazolidinethiones leads to the corresponding β-ketonitriles and β-ketoesters, respectively [195–199]. Similarly, adding dimethyl methylphosphonate lithium salt yields β-ketophosphonates [200].

Since their development, the Evans oxazolidinones have become the most established and extensively used chiral auxiliaries in the stereoselective formation of carbon-carbon and carbon-heteroatom bonds, with hundreds of examples of their use in the total synthesis of natural products in several excellent reviews [51, 63, 182–186]. To illustrate their importance in the stereoselective synthesis of natural targets, only selected examples of total synthesis of natural products will be discussed, accomplished using Evans-type chiral auxiliaries or related heterocycles in one or more key steps.

Baulamycins A and B are natural products isolated from a marine source. These compounds exhibit *in vitro* activity against non-ribosomal peptide synthetase-independent siderophores, SbnE, and AsbA [201]. In 2017, Goswami and

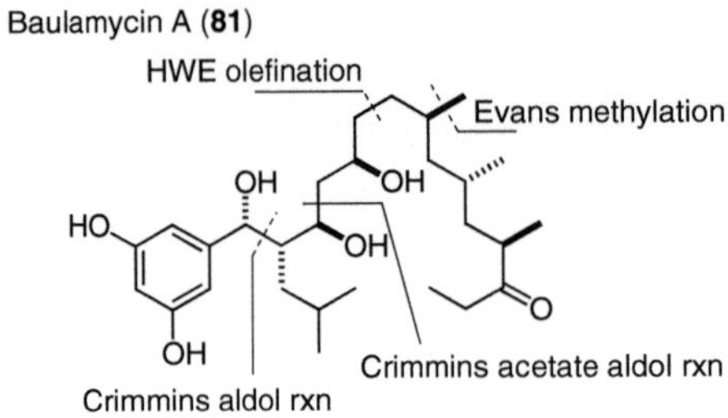

Figure 3.7 Structure of baulamycin A (**81**).

coworkers [202] reported a route for the stereoselective synthesis of baulamycin A (**81**) and its congeners. The key transformations are shown in Figure 3.7 and involve Evans' auxiliary-controlled aldol reaction, HWE olefination, and Evans methylation to introduce the required stereochemistry.

Schemes 3.21 and 3.22 present the synthetic routes of crucial fragments **88** and **95**, respectively. The aldol reaction of aldehyde **83** with the *N*-acyl chiral oxazolidinone **82** in $TiCl_4$, DIPEA, and NMP afforded *syn*-aldol product **84** as a single isomer (Scheme 3.21). After protecting the free hydroxyl group in **84**, the chiral auxiliary was removed reductively, and the resulting alcohol was oxidized to aldehyde **85**. This aldehyde was subjected to another asymmetric aldol reaction with *N*-acetyl thiazolidinethione-derived titanium enolate to provide compound **86** as the major isomer (dr 5 : 1). Its esterification, followed by silylation, gave TES ether **87**, readily converted to phosphonate **88** using $(MeO)_2P(O)Me/n$-BuLi.

In parallel, compound **95** was synthesized from phosphonate oxazolidinone **90** (Scheme 3.22). Phosphonate **90** was subjected to Horner–Wadsworth–Emmons olefination to aldehyde **89**, then the resulting olefin was hydrogenated to provide compound **91**. Treatment of **91** with NaHMDS and MeI furnished compound **92** as a single isomer in good yield. The removal of the auxiliary group afforded alcohol **93**. Subsequent benzylation of the free hydroxyl group of **93**, followed by desilylation reaction, yielded alcohol **94**, which was oxidized to the corresponding aldehyde **95** (Scheme 3.22). Finally, phosphonate **88** was reacted with aldehyde **95** to provide olefin, which in the subsequent seven steps was transformed into the target baulamycin A (**81**).

3.6 Camphor-Based Chiral Heterocyclic Auxiliaries

Camphor and other terpenes can serve as excellent chiral building blocks for natural products synthesis. It is also an exceptional chiral auxiliary for asymmetric transformations. Any additional attachment to the camphor core provides close interaction with the dense chirality of camphor and sites of reactivity in the attachment. As a result, it effectively shields one of the pro-chiral sides of the reactive attachments.

The commonly used heterocyclic camphor-derived auxiliaries are camphorsultam **96** [203–206], known as Oppolzer's sultam, aza-camphor analogs **97–101** [204, 205], and various camphor-based oxazolidinones **102–106** (Figure 3.8) [204, 205].

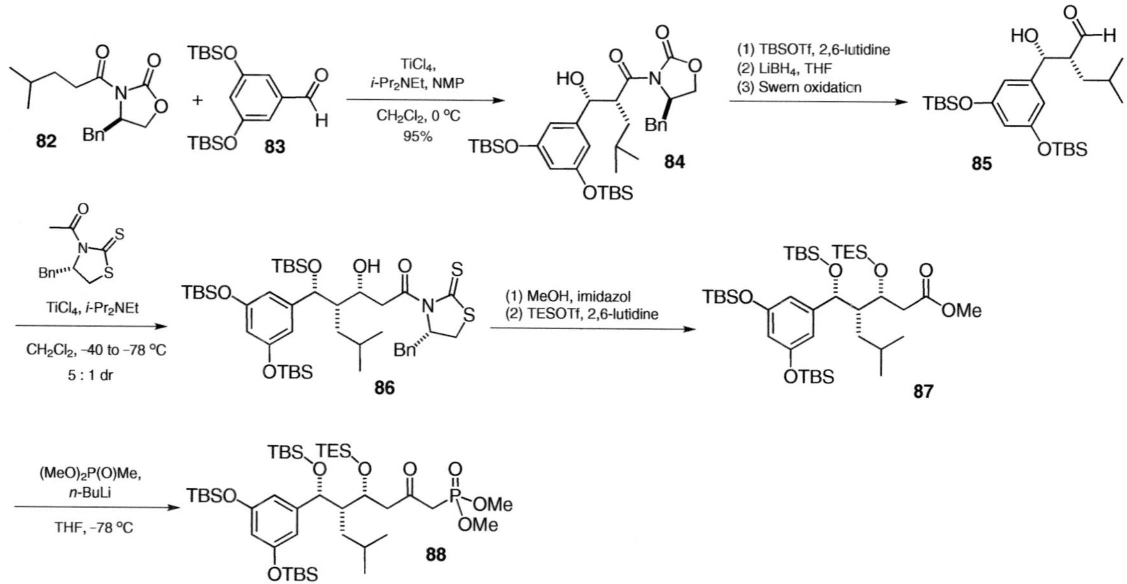

Scheme 3.21 Total synthesis of baulamycin A: synthesis of fragment **88**.

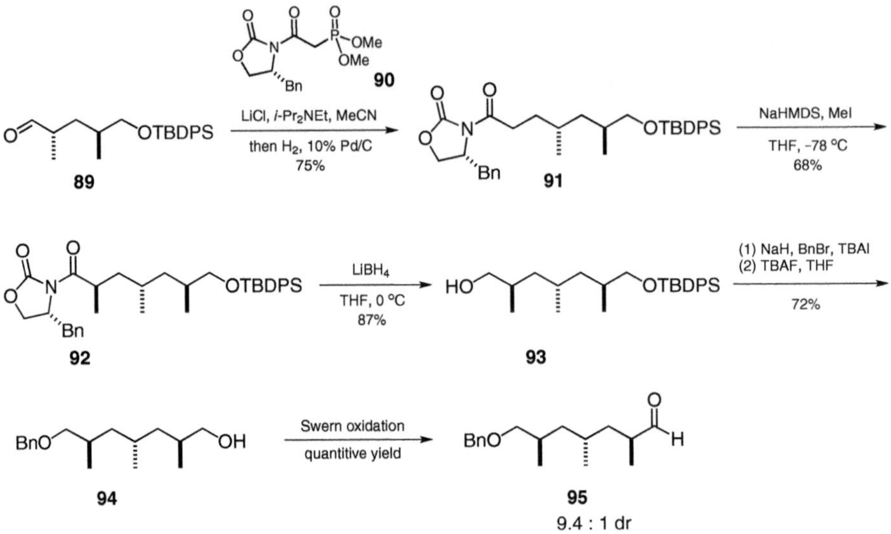

Scheme 3.22 Total synthesis of baulamycin A: synthesis of fragment **95**.

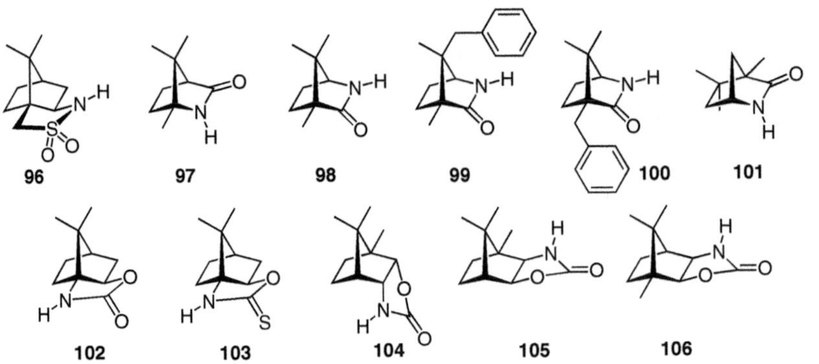

Figure 3.8 Camphor-derived chiral auxiliaries. Source: Based on Refs. [205–208].

In the pioneering work of Oppolzer et al. [209], the sulfonamide derivative **107**, bearing the enone moiety, was utilized as a chiral auxiliary for dienophiles in Diels–Alder cycloaddition (Scheme 3.23). The resultant adducts are formed in excellent yields and good to excellent level of stereochemical control thanks to the highly organized transition state **108**, achieved due to the Lewis acid's bis-coordination, which chelates sulfone carbonyl oxygen. The *s-cis*-conformation of enone is prevalent due to the steric interactions of the β-vinyl hydrogen and the *exo*-methylene hydrogen on C-3 in the *s-trans* conformation. The auxiliary **96** has also been effective as an asymmetry inducer in other pericyclic reactions, including other [4+2] cycloaddition reactions [206, 210–212] and [3+2] cycloaddition reactions [206, 213–216].

Unsaturated amides bearing the sultam moiety are also suitable substrates for diastereoselective 1,4-addition of nucleophilic species, for instance, organometallics (e.g. Grignard reagents) [217]. The resulting enolate **109** can be converted into

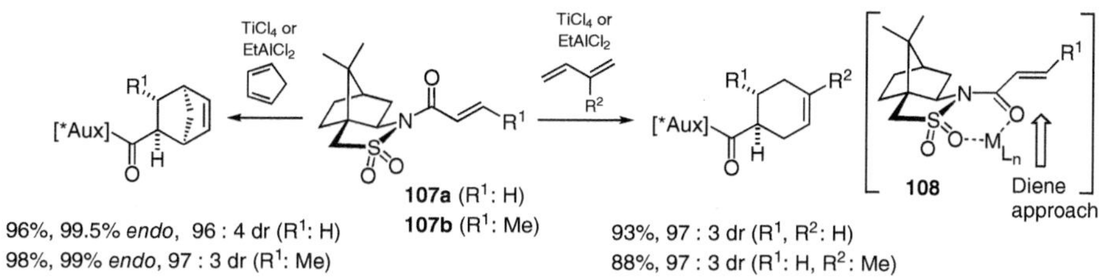

Scheme 3.23 Diels–Alder reaction of dienophile **107**.

3-substituted products **110** by hydrolysis or trapped with a suitable electrophile to provide 2,3-disubstituted carboxylic acid derivatives **111** (Scheme 3.24). Moreover, hydride nucleophiles are suitable reagents for asymmetric conjugate 1,4-hydride addition [218]. The double bond of **107** can also be hydrogenated in a heterogeneous catalyst in a batch [219] or flow [220] mode in excellent yield and with complete diastereoselectivity. Oppolzer's auxiliary is also an effective inducer for Os-catalyzed dihydroxylation [221], cyclopropanation [222], epoxidation [223], and aza-Darzens [224] reaction of type **107** substrates.

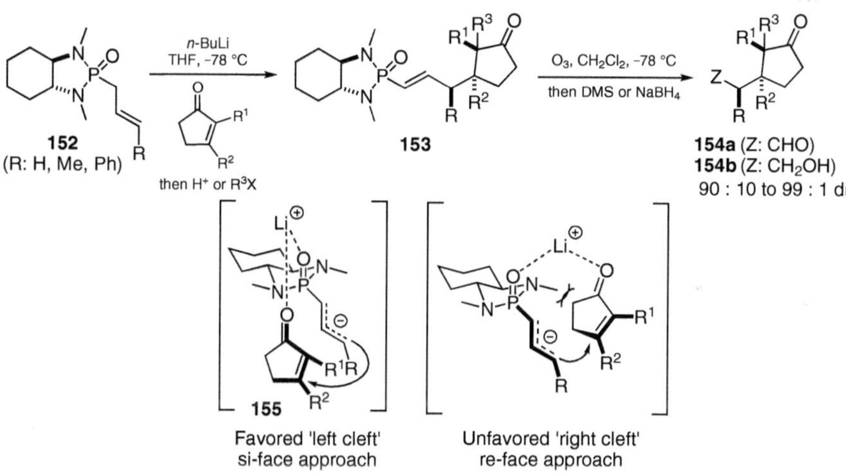

Scheme 3.24 1,4-Addition/enolate trapping of unsaturated amides bearing Oppolzer's auxiliary.

Oppolzer's sultam has been a useful chiral auxiliary for stereoselective aldol reaction of **112** (Scheme 3.25) [121–132, 225]. Depending on the reaction conditions, different aldol stereoisomers can be obtained, as shown in Scheme 3.25. *syn*-Aldols **113** and **114** can be prepared either by using tin or boron enolates of **112**, respectively, with good to excellent diastereoselectivity [225]. Isomer **114** is formed by a dipole-controlled transition state, whereas the tin enolate of **112** reacts through a chelation-controlled transition state to provide *syn*-product **114** [225]. On the other hand, *anti*-aldols **115** can be prepared in good yields and excellent selectivity under Mukaiyama-type conditions by reacting the preformed (*E*)-silyl ketene acetals of **112** with aldehydes in $TiCl_4$ through an "open" transition state [126, 127].

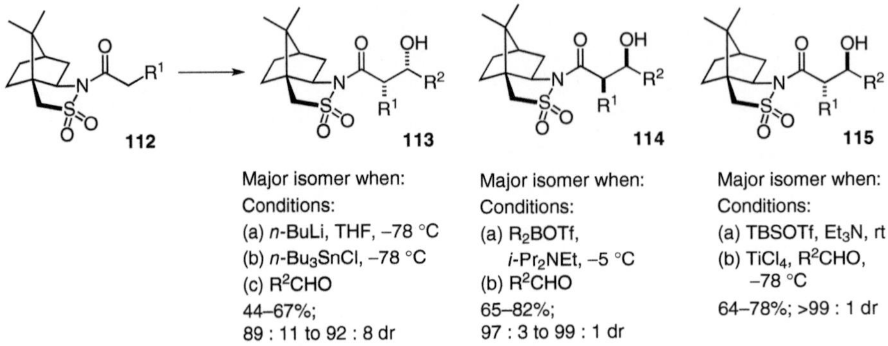

Scheme 3.25 Aldol reaction of N-acylated Oppolzer's sultam with aldehydes. Source: Based on Refs. [121–133].

Sultam **96** delivers good to excellent stereocontrol of electrophilic additions to enolates generated from *N*-acyl derivatives of **112** (Scheme 3.26), including α-alkylation [133, 226–230], α-halogenation [231, 232], α-amination [233, 234], α-azidation [235], α-hydroxylation [236], and α-nitration [237, 238].

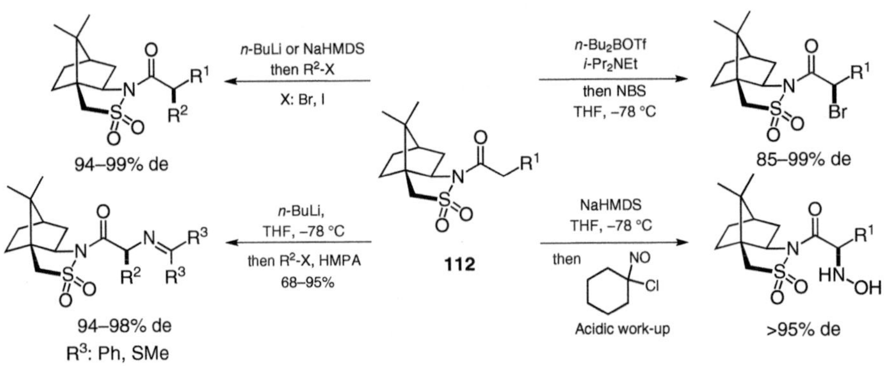

Scheme 3.26 Examples of electrophilic α-addition to N-acylated Oppolzer's sultam.

α-Substituted *N*-α-diazo-acyl (2*R*)-camphorsultam **116** was employed in Brookhart and Templeton's aziridination with *N*-alkoxycarbonyl imines to afford various trisubstituted aziridines, e.g. **117**, in a highly stereo-defined manner (Scheme 3.27) [239]. The chiral auxiliary of the so-obtained aziridine could be easily removed by treatment with TBAOH and H_2O_2 to provide product **118**. Furthermore, the treatment of *N*-Boc aziridines with acid results in rearrangement with the configuration retention at the β-carbon, yielding the trisubstituted *trans*-oxazolidin-2-ones, such as compound **119** as a single isomer. Type **116** chiral diazo compounds were also efficient reagents in an acid-catalyzed ring expansion of cyclohexenones to provide functionalized seven-membered ring products in an almost enantiomerically pure form [187]. Compound **116** was also used as a cyclopropanation reagent [240].

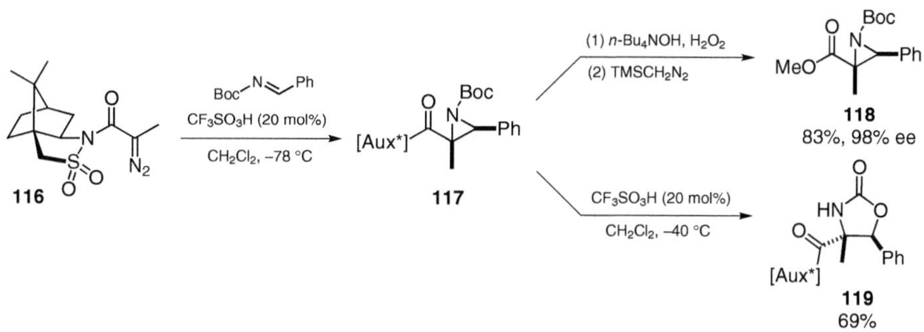

Scheme 3.27 Aziridination with *N*-alkoxycarbonyl imines with diazo reagent **116**. Source: Hashimoto et al. [187]/American Chemical Society.

Two other terpene-derived heterocyclic auxiliaries are 2-aza camphor lactam **97** and 3-aza camphor lactam **98** (Figure 3.8). Both auxiliaries can be prepared from camphoric acid [241]. Besides **97** and **98,** several of their analogs are known, including 8-phenyl-3-aza camphor lactam **99** [242] and 10-phenyl-3-aza lactam **100** (Figure 3.8) [242].

In contrast to Oppolzer's auxiliary **96**, 2-aza lactam **97** is a less effective asymmetry inducer for Diels–Alder cycloaddition [207, 243], giving poor selectivity. Moderate to good stereoselectivity is also observed for Diels–Alder cycloaddition with dienophiles bearing 3-aza lactam auxiliary **98.** The selectivities for cycloaddition to 3-aza-lactam-derived dienophiles complement the selectivities observed for dienophiles derived from the Evans oxazolidinone class of auxiliaries, providing a higher level of diastereoselectivity for simple internally substituted dienes relative to simple terminally substituted dienes [207, 243, 244]. Facial selectivity of cycloaddition can be improved by using a more crowded auxiliary, such as 8-phenyl-3-aza lactam **99** [242], since an additional phenyl ring shields one side of the chelated dienophile more efficiently.

On the other hand, 2-aza lactam **97** is a superior chiral auxiliary for diastereoselective α-alkylation of carboxylic acids (Scheme 3.28). Upon treatment with base (e.g. LDA), *N*-acyl derivative **120** undergoes enolization to exclusively provide (*Z*)-enolate **123.** Intermediate **123** reacts from the *si*-side with various alkyl, allyl, and benzyl halides to provide monoalkylated products **121** in moderate to good yields and with excellent diastereoselectivity [245, 247, 248]. The second deprotonation/alkylation sequence allowed products **122**, bearing quaternary carbon stereocenter, in good yields and high stereoselectivity. The second deprotonation proceeds upon treatment with NDA as a base and leads to non-chelated enolate **124.** Again, enolate **124** reacts with the electrophile from the *re*-side to provide products **122.** Lower de is observed for α-heteroatom-substituted type **120** substrates (e.g. R^1: OR) since the resulting enolates can undergo competing elimination of lactam anion to a ketene intermediate, which disturbs the geometry of the kinetic *Z*-enolate [247]. Acyl chlorides are suitable electrophilic agents to afford acylated products [247]. This asymmetric alkylation method has been employed in the total synthesis of complex bioactive compounds, for instance, antitumor macrocycle (+)-tetronolide [245], and apoptosis-inducing natural product, (−)-rasfonin [248].

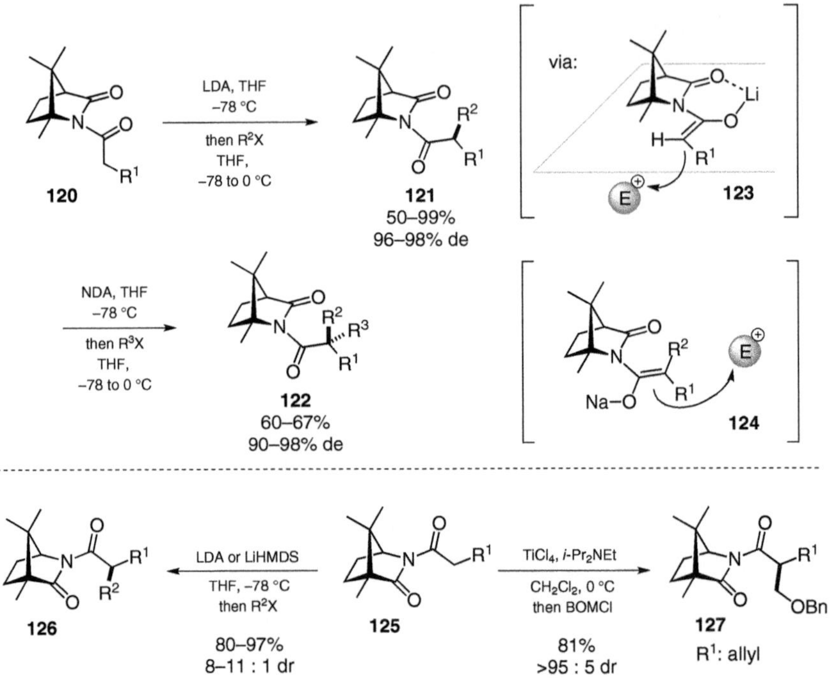

Scheme 3.28 Alkylation of 2-aza (**120**) and 3-aza (**125**) camphor imides. Source: Based on Boeckman et al. [245] and Boeckman and Connell [246].

Similarly, α-alkylation of enolates generated from *N*-acyl 3-aza lactam **125** delivers monosubstituted products **126** in high yield and with very good diastereoselectivity (Scheme 3.28) [247, 248]. In the case of sensitive electrophiles, such as BOMCl, alkylation can be performed efficiently under mild enolization conditions using $TiCl_4$ in *i*-Pr_2NEt (**127**). The *si*-face approach of the electrophile to (*Z*)-enolate is preferred via the analogs of the type **123** transition state.

2-Aza and 3-aza lactams are less efficient chiral auxiliaries for asymmetric aldol reactions. To improve the selectivity, Boeckman and coworkers have developed lactam **101** (Figure 3.8) [242, 246]. The enolization of its *N*-acyl derivative **128** with Et_2BOTf in *i*-Pr_2NEt, followed by the addition of aldehydes, afforded *syn*-aldol **129** products with excellent selectivity (>92 : 8 dr; Scheme 3.29). The high diastereoselectivity was rationalized by transition state **130**, which arose from the favorable dipole alignment and minimized unfavorable steric interactions between the alkyl groups with the boron atom. Lactam **101**-assisted asymmetric alkylation was employed to synthesize some natural bengamides, which show potentially useful antiproliferative activity [208].

Another group of camphor-based auxiliaries is Evans-type oxazolidiniones and oxazolidinethiones (Figure 3.8). Yan auxiliaries **102** and **103** are prepared from ketopinic acid via Curtius rearrangement, followed by cyclization [249].

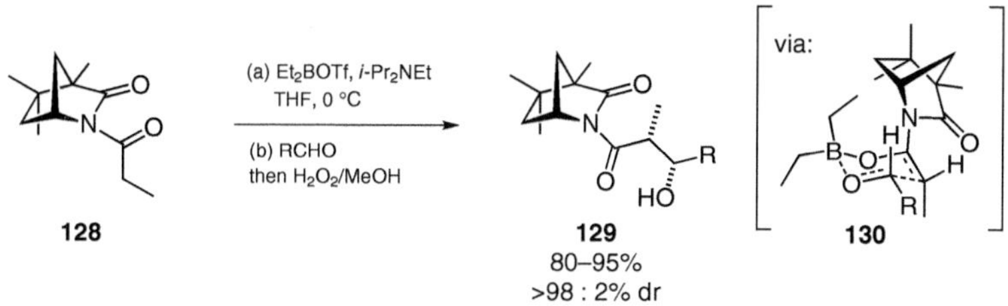

Scheme 3.29 Aldol reaction of *N*-acyl camphor imide **128**.

Alkylation of (*Z*)-enolates **132** derived from **131** with the range of reactive alkyl halides proceeded efficiently and with a high level of diastereoselectivity (**133**, up to 94% de; Scheme 3.30) [249]. In the case of less reactive halides, e.g. EtBr, the addition of HMPA improved reaction yield but eroded diastereoselectivity due to disruption of the key lithium enolate chelate by HMPA. Alkylations of *N*-acylated *endo*-oxazolidine **134** and *N*-acylated *exo*-oxazolidine **135** are also highly selective (>98% de; Scheme 3.30) [250]. Auxiliaries **134** and **135** are prepared from (−)-borneol [250], and (1*R*)-(−)-camphorquinone, respectively [144, 250, 251].

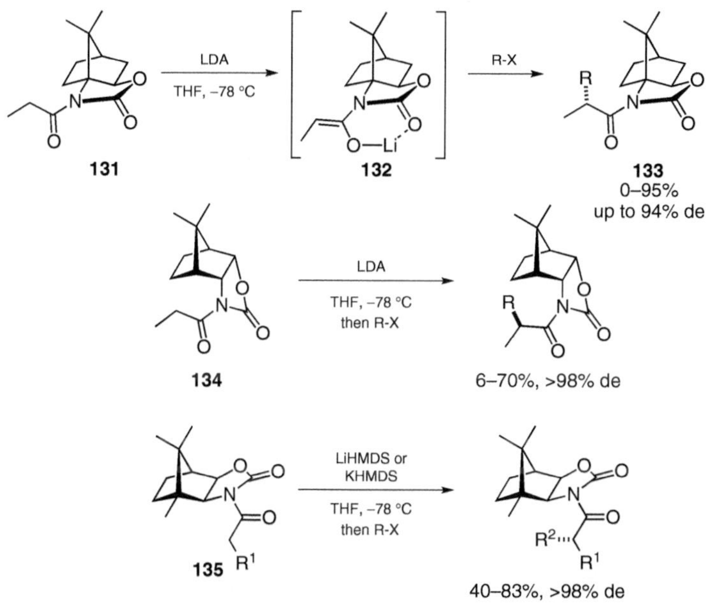

Scheme 3.30 Alkylation reactions of **131**, **134**, and **135** with alkyl halides. Source: Based on Banks et al. [250] and Kouklovsky et al. [251].

As shown in Scheme 3.31, both lithium and boron enolates **136a** provide *syn*-products **137** in aldol reaction with aldehydes [252, 253]. Aldol reaction of aldehydes with titanium enolates also leads to *syn*-products **137**, but with slightly lower diastereoselectivity. The facial selectivity is changed when auxiliary **102** is replaced by thio-congener **103** (Figure 3.8). In this case, the reaction of an aldehyde with titanium enolate (**136b**) yields *syn*-aldol **138** as a major isomer. The reaction of *N*-acyl oxazolidinone with aldehydes can provide *anti*-product **139** when preformed boron enolate is treated with substoichiometric amounts of Lewis acids (e.g. $TiCl_4$ or $SnCl_4$) [254]. *N*-Acetylated **102** and **103** were also successfully employed in acetate aldol reactions with moderate to excellent diastereoselectivity [255–257].

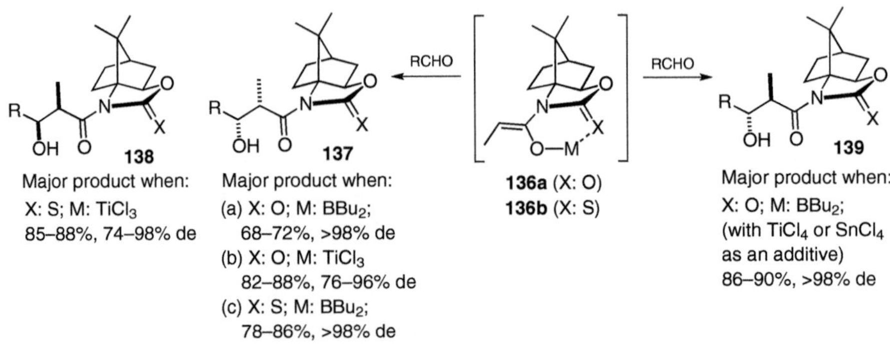

Scheme 3.31 Aldol reaction of *N*-acylated camphor-based oxazolidines.

Once the camphor-based auxiliaries are no longer required for further reactions, they are removed and recovered. Hydrolysis to carboxylic acid may be accomplished under very mild conditions with $LiOH/H_2O_2$. Another approach is the reduction to alcohol using $LiAlH_4$ and DIBAL-H.

3.7 Phosphonamide-Based Chiral Auxiliaries

Phosphonamide [258] derivatives **114–146** (Figure 3.9) bearing the *N*,*N*′-dialkyl-1,2-ethanediamine backbone were first reported as olefination reagents by Corey and Cane [259] and later by Savignac [260, 261], and Hanessian [262] and their coworkers.

Hanessian and coworkers have extended this process toward asymmetric olefination reactions, using a chiral, non-racemic diamine to generate the corresponding

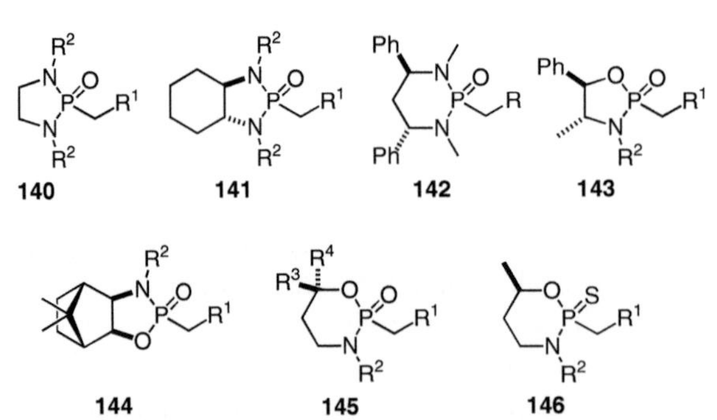

Figure 3.9 Examples of phosphonamide chiral auxiliaries. Source: Corey and Cane [259]/Beilstein Institute for the Advancement of Chemical Sciences.

olefination reagents [263, 264]. The reaction of **145** with ketones leads to the corresponding β-hydroxy phosphonamide intermediates **146**, which undergo elimination of the intermediate oxaphosphetanes to give chiral olefins **147** with moderate to high enantioselectivities (Scheme 3.32). Electrophile attack is favored from the "left cleft" of the anion **148** due to steric and stereoelectronic effects.

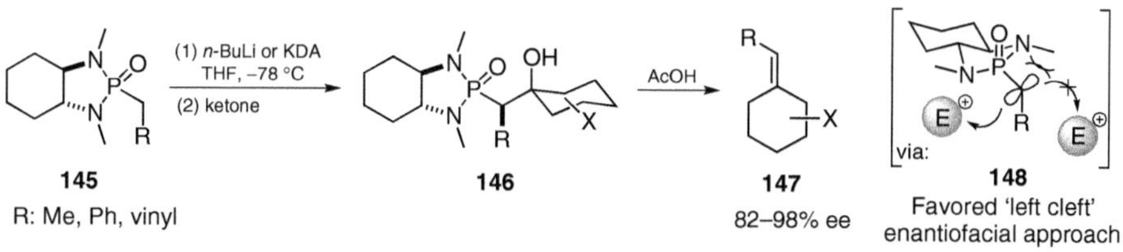

Scheme 3.32 Asymmetric olefination of chiral phosphonamides.

Olefinations based on phosphonamides were employed in the construction of di- and trisubstituted double bonds in the total synthesis of polyoximic acid [265–267], jerangolid A [268], and ambruticin S [269].

Treatment of compound **149**-derived α-phosphoryl carbanion **148** with alkyl halides enabled the diastereoselective synthesis of various α-substituted alkylphosphonic acid derivatives **150** (Scheme 3.33). In the case of the discussed olefination, the attack on the alkyl halide occurs from the "left cleft" side of the anion **148**. Thus, α-substituted products **150** can be afforded with good to excellent diastereoselectivity. The subsequent hydrolysis leads to the corresponding enantioenriched α-substituted phosphonic acids **151** [263, 269, 270]. Other asymmetric alkylation methodologies using chiral phosphonamides have been reported by the groups of Denmark [271, 272] and Steglich [273].

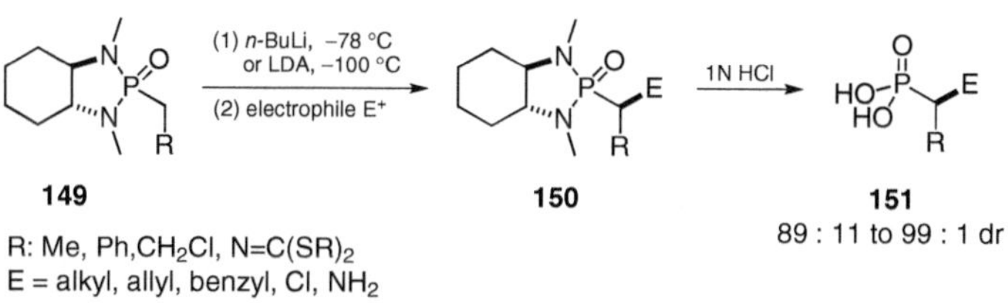

Scheme 3.33 Alkylation of chiral phosphonamides.

The phosphonamide-based auxiliaries were also employed for the stereoselective formation of α-amino-α-alkyl phosphonic acids [274], through amination and azidation of phosphonamide anions, respectively, followed by an acidic workup [275].

Phosphonamides are suitable reagents in asymmetric 1,4-addition and have proved a powerful tool in natural product synthesis (Scheme 3.34) [276–278, 280]. In the conjugate addition of allyl, crotyl, and cinnamyl-derived anions of **152** to Michael acceptors such as enones, lactones, lactams, and α,β-unsaturated esters, up to three contiguous, vicinal or quaternary carbon stereogenic centers can be generated in a single step with a high level of stereocontrol to give adducts **153**.

The stereoselectivity of the reaction can be explained by a lithium-coordinated intermediate **155**, in which chelated Michael acceptors are best accommodated within the "left-cleft" of the reagent **152**. The resulting vinyl phosphonamide product **153** can be cleaved by ozonolysis to the corresponding aldehydes **154a** and the latter reduced to alcohols **154b**, as shown in Scheme 3.34.

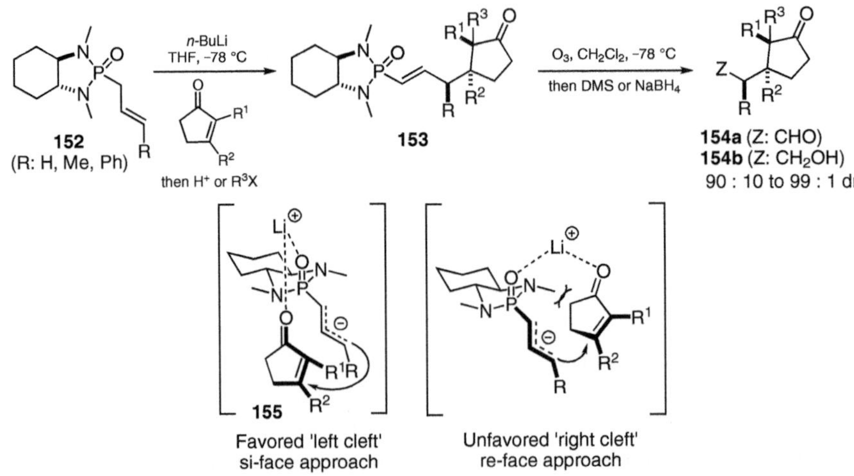

Scheme 3.34 Asymmetric conjugated 1,4-addition of phosphonamides to Michael acceptors. Source: Based on Refs. [276–279].

Denmark [271, 272, 279] and Hua [281, 282] have demonstrated that the selectivity of asymmetric 1,4-addition can be controlled by the P-configuration of P-chiral phosphonoamide scaffolds (Scheme 3.35). Thus, the addition of the Li-anion of **(S_P)-156** ((S_P)-enantiomer) to cyclic enones proceeded with a high level of stereocontrol, providing adducts **157** with up to 98% de. Under the same conditions, P-epimer of **156** gave only poor diastereofacial selection, providing the corresponding 1,4-addition adducts in 28–64% de [282]. Similarly, oxazaphosphorinane **(R_P)-159** delivered adducts **160** with up to 89% de, and poor de values were obtained when diastereomeric oxazaphosphorinane **(S_P)-159** was applied (only 10–15% de). Degradation of the adducts by ozonolysis yielded oxocycloalkane derivatives, which are valuable building blocks, e.g. **158** and **161**.

C_2-Symmetric chiral *N*-phosphonyl group can efficiently control the diastereoselectivity of nucleophilic addition to chiral imines **162** to provide chiral amine derivatives [283]. The scope of nucleophilic agents is shown in Scheme 3.36. The addition of Grignard reagents [284, 285], lithium acetylides [286], cyanide [287], carbamoyl anion [288], lithium phosphites [289], lithiated 1,3-dithianes [290], and ketone [291], ester [292] and Weinreb's amide [293] enolates to aryl and aliphatic imines proceeds in excellent yield and with outstanding diastereoselectivity. Chiral phosphonyl auxiliaries are also suitable asymmetry inducers for asymmetric Mannich reaction [294, 295], aza-Henry reaction [296], aza-Morita–Baylis–Hillman reaction [297, 298], and aza-Darzens reaction [299]. Yang et al. utilized the

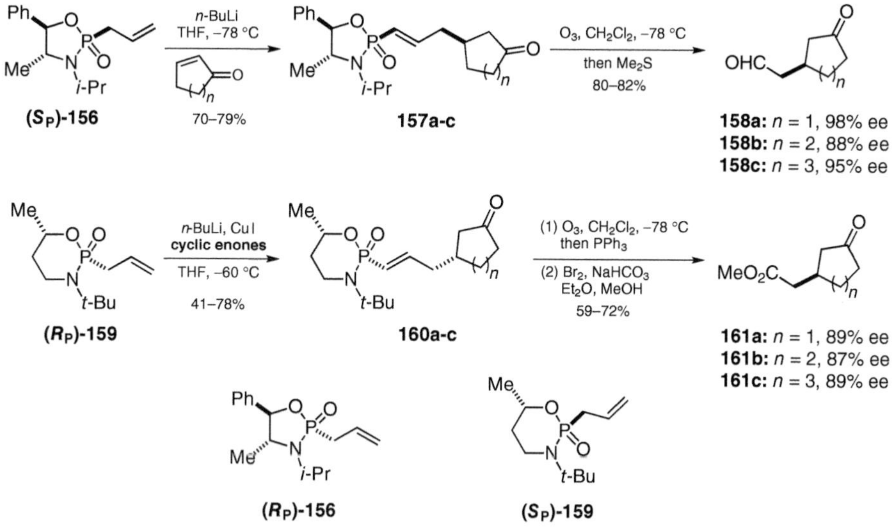

Scheme 3.35 Asymmetric 1,4-addition of P-chiral phosphonoamide to Michael acceptors.

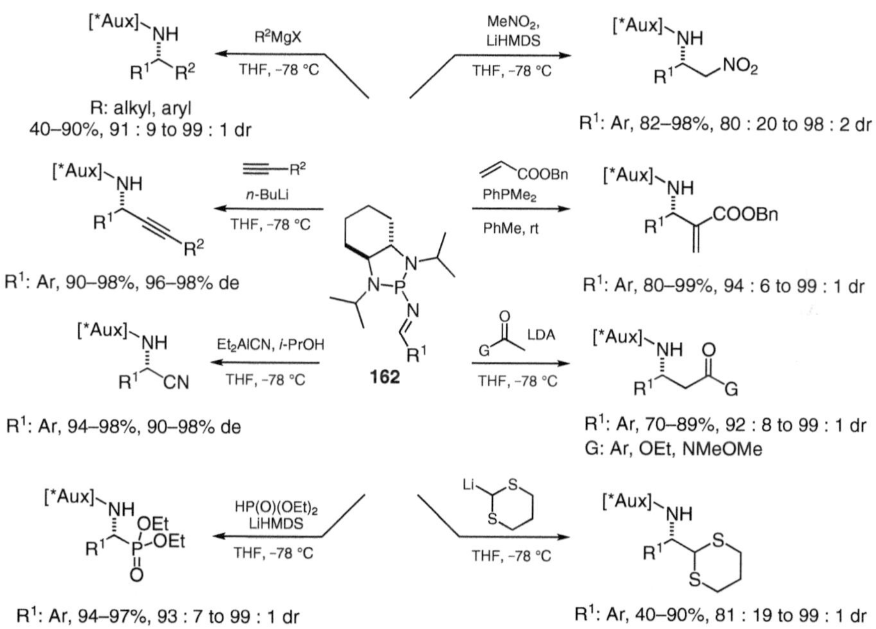

Scheme 3.36 Nucleophilic addition to chiral *N*-phosphonyl imines.

diastereoselective addition of Grignard reagent to type **162** imine as a key step in synthesizing the anti-dementia drug rivastigmine [285].

Recently, the asymmetric borylation closely related to **162** *N*-phosphinyl imines **163** [300] has been studied and utilized to synthesize the anti-cancer drug velcade **164** (Scheme 3.37).

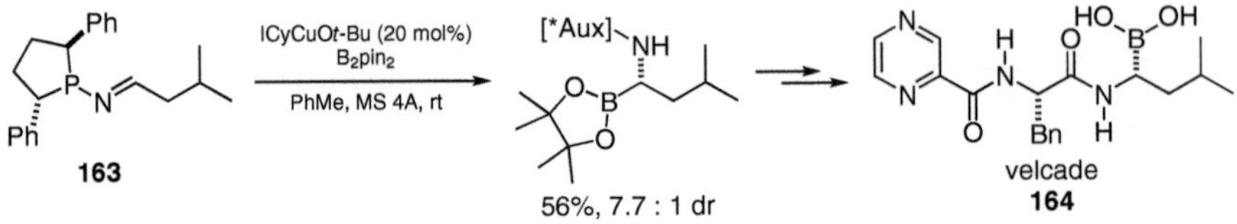

Scheme 3.37 Synthesis of the anti-cancer drug velcade (**164**).

The same chiral *N*-phosphinyl auxiliary appeared as an efficient asymmetry inducer for the asymmetric reduction of chiral *N*-phosphinyl α-imino esters [301] with a range of reducing agents, including Hantzsch esters, silanes, DIBAL, $NaBH_4$, L-selectride (yields up to 98%, up to 99 : 1 dr).

3.8 Schöllkopf's Chiral Auxiliaries and Related Heterocycles for the Asymmetric Synthesis of α-Amino Acids

In the late 1970s, Schöllkopf [302–308], developed a chiral auxiliary-based method to prepare optically pure α-amino acids via diastereoselective alkylation of a masked glycine **166**. Compound **166** is prepared by cyclizing a chiral α-amino acid with glycine to provide bis-lactim **165**. In the next step, it is converted to bis-lactim ether **166** by treatment with Meerwein's salt (Scheme 3.38). Although various amino acid-derived auxiliaries of type **166**, e.g. L-alanine and L-*t*-leucine, have been reported by Schöllkopf, the L-valine derived bis-lactim **166** was the most popular in asymmetric synthesis [302–308].

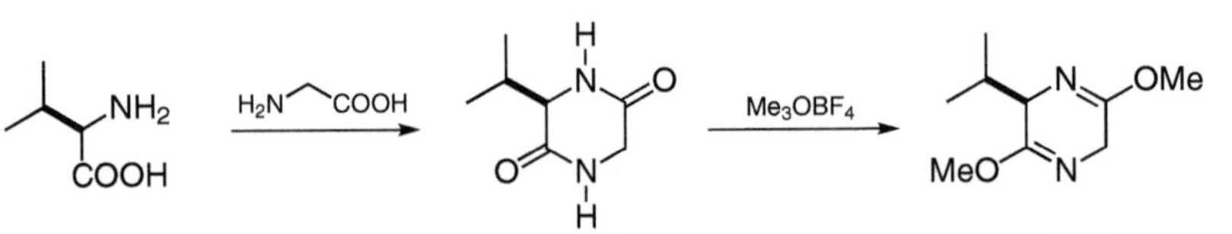

Scheme 3.38 Synthesis of Schöllkopfs chiral auxiliary.

The deprotonation of **166** provides the corresponding enolate **168** at the glycine residue that can be trapped with electrophilic species to give products **167** with high diastereoselectivity, typically >95%. Since enolate **168** is essentially planar, the bulky *i*-Pr group directs the electrophile to attack from the opposite side, affording the *trans* products **167** (Scheme 3.39).

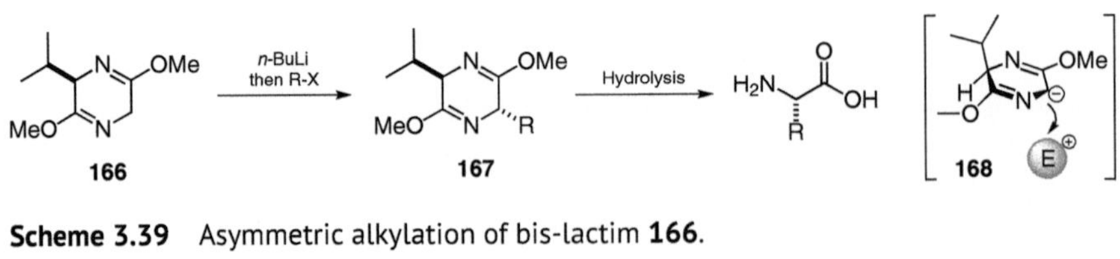

Scheme 3.39 Asymmetric alkylation of bis-lactim **166**.

Chiral heterocyclic auxiliaries for the preparation of (*R*)-amino acids

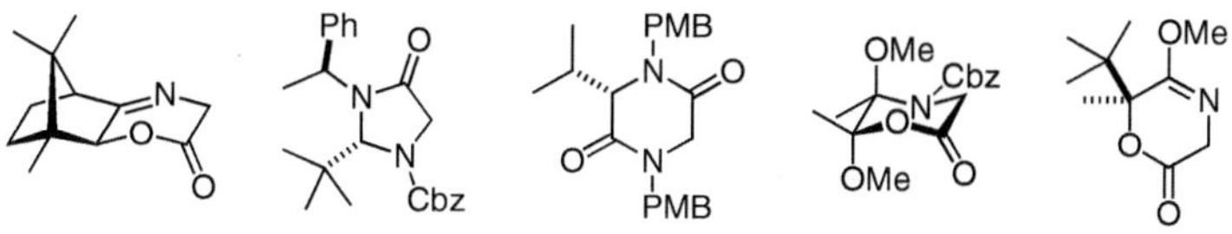

169 170 171 172 173

Chiral heterocyclic auxiliaries for the preparation of (*S*)-amino acids

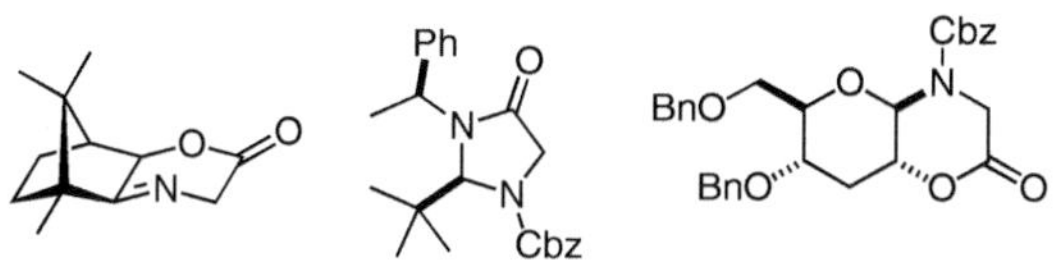

174 175 176

Figure 3.10 Heterocyclic analogs of Schöllkopf's auxiliary. Source: Based on camphor-derived compounds **169** and **174** [323, 324]; imidazolidin-4-ones **170**, **175** [325, 326]; other piperidine **171** [327, 328] and morpholine derivatives **172** [329], **173** [316], **176** [330].

In Schöllkopf's method, a wide range of electrophilic reagents, including alkyl halides [302–307], alkyl sulfonates [309, 310], acyl chlorides [311], aldehydes [312, 313], imines [314], ketones [303], epoxides [39, 315–317], and Micheal acceptors [318–321] can be employed. Subsequent hydrolysis, typically under mildly acidic conditions, affords α-substituted α-amino acids with excellent enantiopurity (ee mostly >95%).

This approach is not limited to Schöllkopf's auxiliary. Other heterocyclic scaffolds, for instance, camphor-derived compounds **169** and **174** [322, 323], imidazolidin-4-ones **170, 175** [324, 325] and other piperidines (e.g. **171**) [326, 327] and morpholine derivatives (**172** [328], **173** [315], **176** [329]), are efficient chiral auxiliaries for asymmetric alkylation (and related reactions with electrophiles) of the masked glycine moiety (Figure 3.10).

Schöllkopf's method and other analogous reactions involving related heterocycles have found widespread use in medicinal chemistry for the preparation of amino acids, peptides, and peptidomimetics [330–333]. It was also applicable in synthesizing α,α-disubstituted α-amino acids through sequential introduction of substituents at the same carbon atom [329, 334–339].

3.9 Conclusions

Although modern methods for asymmetric catalysis can be groundbreaking, the use of chiral heterocyclic auxiliaries is still an essential tool for the stereoselective generation of new stereogenic centers, even if such a method can seem old-fashioned and somewhat inefficient. For many transformations, chiral auxiliaries often represent the only selective method available. Low cost and ready availability, ease of introduction and cleavage of the auxiliaries, high levels of predictability, reliability, time-efficiency, and stereoinduction often render auxiliaries a valuable tool in asymmetric synthesis. Furthermore, in contrast to other methods, facile purification of the auxiliary-substituted products (diastereomers) allows for very high enantiomeric purities of the final products after removal of the auxiliary.

The area of asymmetric synthesis employing chiral heterocyclic auxiliaries, such as Evans', Enders', and Oppolzer's, is well understood and explored. However, there are still some interesting aspects for further investigations and developments. Incoming studies will most likely focus on their application in new and challenging transformations and developments of increasingly efficient synthetic protocols. It would be fascinating to see the methods that would allow the catalytic use of the mentioned chiral auxiliaries.

References

1 Evans, D.A. and Takacs, J.M. (1980). *Tetrahedron Lett.* 21 (44): 4233–4236.
2 Sonnet, P.E. and Heath, R.R. (1980). *J. Org. Chem.* 45 (15): 3137–3139.
3 Andersson, F. and Hedenström, E. (2006). *Tetrahedron: Asymmetry* 17 (13): 1952–1957.
4 Lin, G.Q., Hjalmarsson, M., Högberg, H.E. et al. (1984). *Acta Chem. Scand.* B38: 795–801.
5 Andersson, F. and Hedenström, E. (2004). *Tetrahedron: Asymmetry* 15 (16): 2539–2545.
6 Raubo, P., Giuliano, C., Hill, A.W. et al. (2006). *Synlett* (4): 600–604.
7 Madelaine, C., Valerio, V., and Maulide, N. (2010). *Angew. Chem. Int. Ed.* 49 (9): 1583–1586.
8 Yamamoto, Y., Hoshino, J., Fujimoto, Y. et al. (1993). *Synthesis* (3): 298–302.
9 Zwaagstra, M.E., Meetsma, A., and Feringa, B.L. (1993). *Tetrahedron: Asymmetry* 4 (10): 2163–2172.
10 Chong, J.M., Clarke, I.S., Koch, I. et al. (1995). *Tetrahedron: Asymmetry* 6 (2): 409–418.
11 Sato, M., Gunji, Y., Ikeno, T., and Yamada, T. (2004). *Synthesis* (9): 1434–1438.
12 Zaponakis, G.S. and Katerinopoulos, H.E. (2001). *Tetrahedron Lett.* 42 (36): 6393–6396.
13 Kawanami, Y., Ito, Y., Kitagawa, T. et al. (1984). *Tetrahedron Lett.* 25 (8): 857–860.
14 Porter, N.A., Swann, E., Nally, J., and McPhail, A.T. (1990). *J. Am. Chem. Soc.* 112 (18): 6740–6741.
15 Katsuki, T. and Yamaguchi, M. (1985). *Tetrahedron Lett.* 26 (47): 5807–5810.
16 Kim, B.H., Lee, H.B., Hwang, J.K., and Kim, Y.G. (2005). *Tetrahedron: Asymmetry* 16 (6): 1215–1220.
17 Okada, S., Arayama, K., Murayama, R. et al. (2008). *Angew. Chem. Int. Ed.* 47 (36): 6860–6864.
18 Yasuhiro, K., Tsutomu, K., and Masaru, Y. (1987). *Bull. Chem. Soc. Jpn.* 60 (11): 4190–4192.

19 Nyerges, M., Bendell, D., Arany, A. et al. (2005). *Tetrahedron* 61 (15): 3745–3753.
20 Nyerges, M., Bendell, D., Arany, A. et al. (2003). *Synlett* (7): 0947–0950.
21 Ros, A., Alvarez, E., Dietrich, H. et al. (2005). *Synlett* 2005 (19): 2899–2904.
22 Kozmin, S.A. and Rawal, V.H. (1999). *J. Am. Chem. Soc.* 121: 9562–9573.
23 Kozmin, S.A. and Rawal, V.H. (1998). *J. Am. Chem. Soc.* 120: 13523–13524.
24 Kozmin, S.A., Janey, J.M., and Rawal, V.H. (1999). *J. Org. Chem.* 64: 3039–3052.
25 Qu, H., Gu, X., Liu, Z. et al. (2007). *Org. Lett.* 9 (20): 3997–4000.
26 Liu, Z., Qu, H., Gu, X. et al. (2008). *Org. Lett.* 10 (18): 4105–4108.
27 Friedemann, N.M., Härter, A., Brandes, S. et al. (2012). *Eur. J. Org. Chem.* 2012 (12): 2346–2358.
28 Stack, J.G., Curran, D.P., Geib, S.V. et al. (1992). *J. Am. Chem. Soc.* 114 (18): 7007–7018.
29 Glese, B., Hoffmann, U., Roth, M. et al. (1993). *Tetrahedron Lett.* 34 (15): 2445–2448.
30 Enders, D. and Eichenauer, H. (1976). *Angew. Chem. Int. Ed. Engl.* 15 (9): 549–551.
31 Enders, D. and Eichenauer, H. (1977). *Tetrahedron Lett.* 18 (2): 191–194.
32 Enders, D. (1984). Alkylation of chiral hydrazones. In: *Asymmetric Synthesis*, vol. 3 (ed. M. James), 275–339. Orlando: Academic Press.
33 Fey, P. (1995). Stereoselective synthesis. In: *Houben-Weyl, Methods of Organic Chemistry*, vol. E21a (ed. G. Helmchen, R.W. Hoffman, J. Mulzer and E. Schaumann), 994–1015. Stuttgart: Thieme.
34 Enders, D., Kipphardt, H., and Fey, P. (1987). *Org. Synth.* 65: 183–202.
35 Enders, D. and Eichenauer, H. (1979). *Chem. Ber.* 112 (8): 2933–2960.
36 Enders, D., Eichenauer, H., Baus, U. et al. (1984). *Tetrahedron* 40 (8): 1345–1359.
37 Enders, D., Eichenauer, H., and Pieter, R. (1979). *Chem. Ber.* 112 (11): 3703–3714.
38 Tao, J. (2012). *Lett. Org. Chem.* 9 (5): 325–328.
39 Achatz, O., Grandl, A., and Wanner, K.T. (1999). *Eur. J. Org. Chem.* (8): 1967–1978.
40 Schneider, U., Pannecoucke, X., and Quirion, J.-C. (2002). *Synlett* (10): 1669–1672.
41 Enders, D., Teschner, P., and Raabe, G. (2000). *Synlett* 52 (2): 733–749.
42 Martín-Zamora, E., Ferrete, A., Llera, J.M. et al. (2004). *Chem. Eur. J.* 10 (23): 6111–6129.
43 Enders, D., Kipphardt, H., Gerdes, P. et al. (1988). *Bull. Soc. Chim. Belg.* 97 (8, 9): 691–704.
44 Pareja, C., Martín-Zamora, E., Fernández, R., and Lassaletta, J.M. (1999). *J. Org. Chem.* 64 (24): 8846–8854.
45 Martens, J. and Lübben, S. (1990). *Liebigs Ann. Chem.* 1990 (9): 949–952.
46 Enders, D., Müller, S.F., Raabe, G., and Runsink, J. (2000). *Eur. J. Org. Chem.* (6): 879–892.

47 Enders, D., Müller, S.F., and Raabe, G. (1999). *Angew. Chem. Int. Ed.* 38 (1, 2): 195–197.

48 Enders, D. and Wallert, S. (2002). *Synlett* (2): 304–306.

49 Job, A., Janeck, C.F., Bettray, W. et al. (2002). *Tetrahedron* 58 (12): 2253–2329.

50 Heravi, M.M., Zadsirjan, V., and Daraie, M. (2017). *Curr. Org. Chem.* 14 (1): 61–111.

51 Diaz-Muñoz, G., Miranda, I.L., Sartori, S.K. et al. (2019). *Chirality* 31: 776–812.

52 Yvonne, G. and Frank, G. (2006). *Synthesis* (12): 1899–1930.

53 Lazny, R. and Nodzewska, A. (2010). *Chem. Rev.* 110: 1386–1434.

54 Enders, D. and Thiebes, C. (2001). *Pure Appl. Chem.* 73 (3): 573–578.

55 Enders, D., Wortmann, L., and Peters, R. (2000). *Acc. Chem. Res.* 33 (3): 157–169.

56 Hodgson, D.M. and Charlton, A. (2014). *Tetrahedron* 70 (13): 2207–2236.

57 Kohler, M.C., Wengryniuk, S.E., and Coltart, D.M. Asymmetric α-alkylation of aldehydes, ketones, and carboxylic acids. In: *Stereoselective Synthesis of Drugs and Natural Products* (ed. V. Andrushko and N. Andrushko), 1–31. Hoboken, NJ: Wiley.

58 Enders, D., Del Signore, G., and Raabe, G. (2013). *Turk. J. Chem.* 37: 492–518.

59 Kempf, K., Kempf, O., Orozco, M. et al. (2017). *J. Org. Chem.* 82 (15): 7791–7795.

60 Pancholi, A.K., Geden, J.V., Clarkson, G.J., and Shipman, M. (2016). *J. Org. Chem.* 81: 7984–7992.

61 Geden, J.V., Beasley, B.O., Clarkson, G.J., and Shipman, M. (2013). *J. Org. Chem.* 78: 12243–12250.

62 David, D., Lebrun, S., Couture, A. et al. (2010). *Tetrahedron: Asymmetry* 21 (2): 195–201.

63 Mahrwald, R. (2009). *Aldol Reactions*. London: Springer.

64 Enders, D. (1985). *Chem. Scr.* 25: 139–147.

65 Enders, D., Berg, T., Raabe, G., and Runsink, J. (1997). *Liebigs Ann.* 1997 (2): 345–363.

66 Enders, D., Berg, T., Raabe, G., and Runsink, J. (1996). *Helv. Chim. Acta* 79 (1): 118–122.

67 Enders, D. and Lohray, B.B. (1987). *Angew. Chem. Int. Ed. Engl.* 26 (4): 351–352.

68 Enders, D., Schäfer, T., and Mies, W. (1998). *Tetrahedron* 54 (35): 10239–10252.

69 Ternon, M., Pannecoucke, X., Outurquin, F., and Paulmier, C. (2002). *Tetrahedron* 58 (16): 3275–3279.

70 Enders, D., Joseph, R., and Poiesz, C. (1998). *Tetrahedron* 54 (34): 10069–10078.

71 Enders, D. and Bhushan, V. (1988). *Tetrahedron Lett.* 29 (20): 2437–2440.

72 Díez, E., López, A.M., Pareja, C. et al. (1998). *Tetrahedron Lett.* 39 (43): 7955–7958.

73 Enders, D. and Schubert, H. (1984). *Angew. Chem. Int. Ed. Engl.* 23 (5): 365–366.

74 Enders, D., Janeck, C.F., and Runsink, J. (2000). *Synlett* (5): 641–643.

75 Enders, D. and Plant, A. (1994). *Synlett* (12): 1054–1056.

76 Enders, D., Janeck, C.F., and Raabe, G. (2000). *Eur. J. Org. Chem.* (19): 337–3345.

77 Enders, D. and Funabiki, K. (2001). *Org. Lett.* 3 (10): 1575–1577.

78 Enders, D., Nübling, C., and Schubert, H. (1997). *Liebigs Ann.* 1997 (6): 1089–1100.

79 Kim, Y.H. and Choi, J.Y. (1996). *Tetrahedron Lett.* 37 (31): 5543–5546.

80 Enders, D., Schubert, H., and Nübling, C. (1986). *Angew. Chem. Int. Ed. Engl.* 25 (12): 1109–1110.

81 Enders, D., Moll, A., and Bats, J.W. (2006). *Eur. J. Org. Chem.* 2006 (5): 1271–1284.

82 Enders, D., Braig, V., Boudou, M., and Raabe, G. (2004). *Synthesis* (18): 2980–2990.

83 Boudou, M. and Enders, D. (2005). *J. Org. Chem.* 70 (23): 9486–9494.

84 Enders, D., Braig, V., and Raabe, G. (2001). *Can. J. Chem.* 79 (11): 1528–1535.

85 Enders, D. and Del Signore, G. (2004). *Tetrahedron: Asymmetry* 15 (5): 747–751.

86 Enders, D. and Moser, M. (2003). *Tetrahedron Lett.* 44 (46): 8479–8481.

87 Choi, J.Y. and Kim, Y.H. (1996). *Tetrahedron Lett.* 37 (43): 7795–7796.

88 Rabasso, N. and Fadel, A. (2008). *Synthesis* (15): 2353–2362.

89 Enders, D., Peters, R., Lochtman, R., and Raabe, G. (1999). *Angew. Chem. Int. Ed.* 38 (16): 2421–2423.

90 Enders, D., Moll, A., Schaadt, A. et al. (2003). *Eur. J. Org. Chem.* 2003 (20): 3923–3938.

91 Enders, D., Chelain, E., and Raabe, G. (1997). *ChemInform* 28 (46): https://doi.org/10.1002/chin.199746088.

92 Nicaise, O. and Denmark, S. (1997). *ChemInform* 28 (46): https://doi.org/10.1002/chin.199746042.

93 Enders, D. and Reinhold, U. (1995). *Angew. Chem. Int. Ed. Engl.* 34 (11): 1219–1222.

94 Enders, D. and Reinhold, U. (1996). *Liebigs Ann.* 1996 (1): 11–26.

95 Remuzon, P., Dussy, C., Jacquet, J.-P. et al. (1996). *Tetrahedron: Asymmetry* 7 (4): 1181–1188.

96 Bilodeau, F., Dubé, L., and Deslongchamps, P. (2003). *Tetrahedron* 59 (16): 2781–2791.

97 Dexter, H.R., Allen, E., and Williams, D.M. (2018). *Tetrahedron Lett.* 59: 4323–4325.

98 Enders, D. and Hundertmark, T. (1999). *Eur. J. Org. Chem.* 1999 (4): 751–756.

99 Enders, D. and Niemeier, O. (2005). *Heterocycles* 66: 385–403.

100 Enders, D. and Kaiser, A. (1997). *Heterocycles* 46 (1): 631–635.

101 Enders, D. and Kirchhoff, J.H. (2000). *Synthesis* (14): 2099–2105.

102 Enders, D., Breuer, I., and Nühring, A. (2005). *Eur. J. Org. Chem.* (13): 2677–2683.

103 Enders, D., Vicario, J.L., Job, A. et al. (2002). *Chem. Eur. J.* 8 (18): 4272–4284.

104 Enders, D., David, S., Deckers, K. et al. (2012). *Synthesis* 44 (22): 3483–3488.

105 Enders, D. and Reichenbach, L.F. (2013). *Synthesis* 45 (7): 959–965.

106 Kondaiah, G.C.M., Vivekanandareddy, M., Reddy, A.L. et al. (2011). *Synth. Commun.* 41 (8): 1186–1191.

107 Lebrun, S., Couture, A., Deniau, E., and Grandclaudon, P. (2010). *Chirality* 22: 212–216.

108 Enders, D., Bettray, W., Raabe, G., and Runsink, J. (1994). *Synthesis* (12): 1322–1326.

109 Wengryniuk, S.E., Lim, D., and Coltart, D.M. (2011). *J. Am. Chem. Soc.* 133 (22): 8714–8720.

110 Marqués-López, E., Martín-Zamora, E., Díez, E. et al. (2008). *Eur. J. Org. Chem.* 2008 (17): 2960–02972.

111 Friestad, G.K. and Ji, A. (2008). *Org. Lett.* 10 (11): 2311–2313.

112 Shen, Y. and Friestad, G.K. (2002). *J. Org. Chem.* 67 (17): 6236–6239.

113 Friestad, G.K. and Qin, J. (2000). *J. Am. Chem. Soc.* 122 (34): 8329–8330.

114 Friestad, G.K., Draghici, C., Soukri, M., and Qin, J. (2005). *J. Org. Chem.* 70 (16): 6330–6338.

115 Friestad, G.K. (2005). *Eur. J. Org. Chem.* 2005 (15): 3157–3172.

116 Friestad, G.K. (2014). Control of asymmetry in the radical addition approach to chiral amine synthesis. In: *Stereoselective Formation of Amines* (ed. W. Li and X. Zhang), 1–32. Berlin, Heidelberg: Springer Berlin Heidelberg.

117 Huynh, U., Uddin, M.N., Wengryniuk, S.E. et al. (2016). *Tetrahedron Lett.* 57 (43): 4799–4802.

118 Fernández, M. and Alonso, R. (2003). *Org. Lett.* 5 (14): 2461–2464.

119 Friestad, G.K. and Banerjee, K. (2009). *Org. Lett.* 11 (5): 1095–1098.

120 Cook, G.R., Maity, B.C., and Kargbo, R. (2004). *Org. Lett.* 6 (11): 1741–1743.

121 García-Fortanet, J., Murga, J., Carda, M., and Marco, J.A. (2006). *Org. Lett.* 8 (13): 2695–2698.

122 Sun, Y., Su, X., Zhou, R. et al. (2018). *Tetrahedron* 74 (40): 5955–5964.

123 Garland, K., Pantua, H., Braun, M.-G. et al. (2020). *Bioorg. Med. Chem. Lett.* 30 (20): 127419.

124 García-Fortanet, J., Murga, J., Carda, M. et al. (2007). *Chem. Eur. J.* 13 (18): 5060–5074.

125 Prabhakar, P., Ramesh, D., Rajaram, S. et al. (2011). *Helv. Chim. Acta* 94 (8): 1481–1487.

126 Oppolzer, W., Starkemann, C., Rodriguez, I., and Bernardinelli, G. (1991). *Tetrahedron Lett.* 32 (1): 61–64.

127 Oppolzer, W. and Lienard, P. (1993). *Tetrahedron Lett.* 34 (27): 4321–4324.

128 Yadav, J.S. and Rajender, V. (2010). *Eur. J. Org. Chem.* 2010 (11): 2148–2156.

129 Ma, D., Zou, B., Cai, G. et al. (2006). *Chem. Eur. J.* 12 (29): 7615–7626.

130 Zou, B., Wei, J., Cai, G., and Ma, D. (2003). *Org. Lett.* 5 (19): 3503–3506.

131 Li, M., Han, P., Mao, Z.-Y. et al. (2016). *Tetrahedron Lett.* 57 (50): 5620–5623.

132 Sun, Y., Ding, Y., Li, D. et al. (2017). *Angew. Chem. Int. Ed.* 56 (46): 14627–14631.

133 He, K., Zhu, F., Li, H. et al. (2020). *Org. Prep. Proced. Int.* 52 (6): 510–516.

134 Friestad, G.K. and Qin, J. (2001). *J. Am. Chem. Soc.* 123 (40): 9922–9923.

135 Friestad, G.K., Marié, J.-C., Suh, Y.-S., and Qin, J. (2006). *J. Org. Chem.* 71 (18): 7016–7027.

136 Yamazaki, N. and Kibayashi, C. (1997). *Tetrahedron Lett.* 38 (26): 4623–4626.

137 Korapala, C.S., Qin, J., and Friestad, G.K. (2007). *Org. Lett.* 9 (21): 4243–4246.

138 Friestad, G.K., Marié, J.-C., and Deveau, A.M. (2004). *Org. Lett.* 6 (19): 3249–3252.

139 Evans, D.A., Bartroli, J., and Shih, T.L. (1981). *J. Am. Chem. Soc.* 103 (8): 2127–2129.

140 Bull, S.D., Davies, S.G., Key, M.-S. et al. (2000). *Chem. Commun.* (18): 1721–1722.

141 Jeong, K.E.J., Sung, L.T., Hong, S.K., and Lee, E. (2002). *J. Am. Chem. Soc.* 124 (49): 14655–14662.

142 Nöcker, C., Kaiser, N., Foley, D. et al. (2021). *Tetrahedron* 87: 132118.

143 Kigoshi, H., Suenaga, K., Mutou, T. et al. (1996). *J. Org. Chem.* 61 (16): 5326–5351.

144 Banks, M.R., Blake, A.J., Cadogan, J.I.G. et al. (1996). *Tetrahedron* 52 (11): 4079–4094.

145 Rück, K. and Kunz, H. (1992). *Synlett* (4): 343–344.

146 Gage, J.R. and Evans, D.A. (1990). *Org. Synth.* 68: 77–82.

147 Davies, S.G., Sanganee, H.J., and Szolcsanyi, P. (1999). *Tetrahedron* 55 (11): 3337–3354.

148 Evans, D.A., Ennis, M.D., and Mathre, D.J. (1982). *J. Am. Chem. Soc.* 104 (6): 1737–1739.

149 Tallmadge, E.H. and Collum, D.B. (2015). *J. Am. Chem. Soc.* 137 (40): 13087–13095.

150 Heravi, M.M., Zadsirjan, V., and Farajpour, B. (2016). *RSC Adv.* 6 (36): 30498–30551.

151 Rück-Braun, K. and Kunz, H. (1999). *Chiral Auxiliaries in Cycloadditions.* Wiley.

152 Guduru, S.K.R., Jimmidi, R., Deora, G.S., and Arya, P. (2015). *Org. Lett.* 17 (3): 480–483.

153 Evans, D.A., Bilodeau, M.T., Somers, T.C. et al. (1991). *J. Org. Chem.* 56 (20): 5750–5752.

154 Evans, D.A., Gage, J.R., and Leighton, J.L. (1992). *J. Am. Chem. Soc.* 114 (24): 9434–9453.

155 Esteban, J., Costa, A.M., Gómez, À., and Vilarrasa, J. (2008). *Org. Lett.* 10 (1): 65–68.

156 Wen, S.-J. and Yao, Z.-J. (2004). *Org. Lett.* 6 (16): 2721–2724.

157 Evans, D.A. and Nelson, S.G. (1997). *J. Am. Chem. Soc.* 119 (27): 6452–6453.

158 Evans, D.A., Britton, T.C., Dorow, R.L., and Dellaria, J.F. (1986). *J. Am. Chem. Soc.* 108 (20): 6395–6397.

159 Costanzo, M.J., Almond, H.R., Hecker, L.R. et al. (2005). *J. Med. Chem.* 48 (6): 1984–2008.

160 Beugelmans, R., Bois-Choussy, M., Vergne, C. et al. (1996). *Chem. Commun.* (9): 1029–1030.

161 Vergne, C., Bouillon, J.-P., Chastanet, J. et al. (1998). *Tetrahedron: Asymmetry* 9 (17): 3095–3103.

162 Nieman, J.A., Coleman, J.E., Wallace, D.J. et al. (2003). *J. Nat. Prod.* 66 (2): 183–199.

163 Liu, D.-G., Gao, Y., Wang, X. et al. (2002). *J. Org. Chem.* 67 (5): 1448–1452.

164 Bungard, C.J., Williams, P.D., Ballard, J.E. et al. (2016). *ACS Med. Chem. Lett.* 7 (7): 702–707.

165 Cohen, F., Aggen, J.B., Andrews, L.D. et al. (2019). *ChemMedChem* 14 (16): 1560–1572.

166 Evans, D.A., Britton, T.C., Ellman, J.A., and Dorow, R.L. (1990). *J. Am. Chem. Soc.* 112 (10): 4011–4030.

167 Cheerlavancha, R., Ahmed, A., Leung, Y.C. et al. (2017). *Beilstein J. Org. Chem.* 13: 2316–2325.

168 Lai, P.-S., Dubland, J.A., Sarwar, M.G. et al. (2011). *Tetrahedron* 67 (39): 7586–7592.

169 Gómez-Palomino, A., Romea, P., and Urpí, F. (2018). *Synthesis* 50 (14): 2721–2726.

170 Richardson, M.B. and Williams, S.J. (2013). *Beilstein J. Org. Chem.* 9: 1807–1812.

171 Mohri, T., Ogura, Y., Towada, R., and Kuwahara, S. (2017). *Tetrahedron Lett.* 58 (42): 4011–4013.

172 Zhang, W., Liu, J., Macho, J.M. et al. (2017). *Eur. J. Med. Chem.* 126: 7–14.

173 Fürstner, A., Radkowski, K., Wirtz, C. et al. (2002). *J. Am. Chem. Soc.* 124 (24): 7061–7069.

174 Monma, S., Sunazuka, T., Nagai, K. et al. (2006). *Org. Lett.* 8 (24): 5601–5604.

175 Falck, J.R., Gao, S., Prasad, R.N., and Koduru, S.R. (2008). *Bioorg. Med. Chem. Lett.* 18 (6): 1768–1771.

176 Chachignon, H., Kondrashov, E.V., and Cahard, D. (2018). *Adv. Synth. Catal.* 360 (5): 965–971.

177 Alexander, R.P. and Paterson, I. (1985). *Tetrahedron Lett.* 26 (43): 5339–5340.

178 Davies, S.G., Hunter, I.A., Nicholson, R.L. et al. (2004). *Tetrahedron* 60 (35): 7553–7577.

179 Ishikawa, T., Kawasaki-Takasuka, T., Kubota, T., and Yamazaki, T. (2017). *Beilstein J. Org. Chem.* 13: 2473–2477.

180 Evans, D.A., Tedrow, J.S., Shaw, J.T., and Downey, C.W. (2002). *J. Am. Chem. Soc.* 124 (3): 392–393.

181 Crimmins, M.T. and Powell, M.T. (2003). *J. Am. Chem. Soc.* 125 (25): 7592–7595.

182 Mahrwald, R. (2013). *Modern Methods in Stereoselective Aldol Reactions*. Wiley.

183 Mahrwald, R. and Evans, D.A. (2004). *Modern Aldol Reactions, 2 Volume Set*. Wiley.

184 Heravi, M.M. and Zadsirjan, V. (2013). *Tetrahedron: Asymmetry* 24 (19): 1149–1188.

185 Nazari, A., Heravi, M.M., and Zadsirjan, V. (2021). *J. Organomet. Chem.* 932: 121629.

186 Zadsirjan, V. and Heravi, M.M. (2018). *Curr. Org. Synth.* 15 (1): 3–20.

187 Hashimoto, T., Naganawa, Y., and Maruoka, K. (2010). *Chem. Commun.* 46 (36): 6810–6812.

188 Walker, M.A. and Heathcock, C.H. (1991). *J. Org. Chem.* 56 (20): 5747–5750.

189 Evans, D.A., Downey, C.W., Shaw, J.T., and Tedrow, J.S. (2002). *Org. Lett.* 4 (7): 1127–1130.

190 Crimmins, M.T. and Brown, B.H. (2004). *J. Am. Chem. Soc.* 126 (33): 10264–10266.

191 Evans, D.A., Bender, S.L., and Morris, J. (1988). *J. Am. Chem. Soc.* 110 (8): 2506–2526.

192 Evans, D.A. and Bender, S.L. (1986). *Tetrahedron Lett.* 27 (7): 799–802.

193 Crimmins, M.T. and King, B.W. (1998). *J. Am. Chem. Soc.* 120 (35): 9084–9085.

194 Crimmins, M.T., King, B.W., Tabet, E.A., and Chaudhary, K. (2001). *J. Org. Chem.* 66 (3): 894–902.

195 Steinborn, C., Wildermuth, R.E., David, M. et al. (2020). *Angew. Chem. Int. Ed.* 132: 17435–17438.

196 Nagao, Y., Kumagai, T., Nagase, Y. et al. (1992). *J. Org. Chem.* 57 (15): 4232–4237.

197 Vanderwal, C.D., Vosburg, D.A., Weiler, S., and Sorensen, E.J. (2003). *J. Am. Chem. Soc.* 125 (18): 5393–5407.

198 Fernández-Valparis, J., Romea, P., Urpí, F., and Font-Bardia, M. (2017). *Org. Lett.* 19 (23): 6400–6403.

199 Crimmins, M.T., McDougall, P.J., and Ellis, J.M. (2006). *Org. Lett.* 8 (18): 4079–4082.

200 Delamarche, I. and Mosset, P. (1994). *J. Org. Chem.* 59 (18): 5453–5457.

201 Tripathi, A., Schofield, M.M., Chlipala, G.E. et al. (2014). *J. Am. Chem. Soc.* 136 (4): 1579–1586.

202 Guchhait, S., Chatterjee, S., Ampapathi, R.S., and Goswami, R.K. (2017). *J. Org. Chem.* 82 (5): 2414–2435.

203 Diaz-Muñoz, G., Miranda, I.L., Sartori, S.K. et al. (2019). *Chirality* 31 (10): 776–812.

204 Boeckman, R.K. and Cody, J.A. (2020). Synthesis and utility of hetero and non-heterocyclic chiral auxiliaries derived from terpenes: camphor and pinene. In: *Heterocycles as Chiral Auxiliaries in Asymmetric Synthesis* (ed. M. Braun), 193–251. Cham: Springer International Publishing.

205 Cody, J.A. and Boeckman, R.K. (2012). Chapter 3.3: Terpene derived auxiliaries: camphor and pinene derived auxiliaries. In: *Comprehensive Chirality* (ed. E.M. Carreira and H. Yamamoto), 42–105. Amsterdam: Elsevier.

206 Heravi, M.M. and Zadsirjan, V. (2014). *Tetrahedron: Asymmetry* 25 (15): 1061–1090.

207 Boeckman, R.K., Nelson, S.G., and Gaul, M.D. (1992). *J. Am. Chem. Soc.* 114 (6): 2258–2260.

208 Boeckman, R.K., Clark, T.J., and Shook, B.C. (2002). *Org. Lett.* 4 (12): 2109–2112.

209 Oppolzer, W., Chapuis, C., and Bernardinelli, G. (1984). *Helv. Chim. Acta* 67 (5): 1397–1401.

210 Chapuis, C., Kucharska, A., and Jurczak, J. (2000). *Tetrahedron: Asymmetry* 11 (22): 4581–4591.

211 Achmatowicz, M., Chapuis, C., Rzepecki, P., and Jurczak, J. (1999). *Helv. Chim. Acta* 82 (2): 182–190.

212 Hirama, M., Kato, Y., Seki, C. et al. (2010). *Tetrahedron* 66 (38): 7618–7624.

213 Merino, P., Greco, G., Tejero, T. et al. (2013). *Tetrahedron* 69 (45): 9381–9390.

214 Romański, J., Jóźwik, J., Chapuis, C. et al. (2007). *Tetrahedron: Asymmetry* 18 (7): 865–872.

215 Oppolzer, W., Deerberg, J., and Tamura, O. (1994). *Helv. Chim. Acta* 77 (2): 554–560.

216 Merino, P., Mates, J.A., Revuelta, J. et al. (2002). *Tetrahedron: Asymmetry* 13 (2): 173–190.

217 Oppolzer, W., Poli, G., Kingma, A.J. et al. (1987). *Helv. Chim. Acta* 70 (8): 2201–2214.

218 Oppolzer, W. and Poli, G. (1986). *Tetrahedron Lett.* 27 (39): 4717–4720.

219 Oppolzer, W., Mills, R.J., and Reglier, M. (1986). *Tetrahedron Lett.* 27 (2): 183–186.

220 Sullivan, R.J. and Newman, S.G. (2018). *Chem. Sci.* 9 (8): 2130–2134.

221 Oppolzer, W. and Barras, J.-P. (1987). *Helv. Chim. Acta* 70 (7): 1666–1675.

222 Vallgårda, J., Appelberg, U., Csöregh, I., and Hacksell, U. (1994). *J. Chem. Soc., Perkin Trans.* 1 (4): 461–470.

223 Hajra, S. and Bhowmick, M. (2010). *Tetrahedron: Asymmetry* 21 (18): 2223–2229.

224 Sweeney, J.B., Cantrill, A.A., McLaren, A.B., and Thobhani, S. (2006). *Tetrahedron* 62 (15): 3681–3693.

225 Oppolzer, W., Blagg, J., Rodriguez, I., and Walther, E. (1990). *J. Am. Chem. Soc.* 112 (7): 2767–2772.

226 Brundish, D., Bull, A., Donovan, V. et al. (1999). *J. Med. Chem.* 42 (22): 4584–4603.

227 Oppolzer, W., Moretti, R., and Zhou, C. (1994). *Helv. Chim. Acta* 77 (8): 2363–2380.

228 Oppolzer, W., Darcel, C., Rochet, P. et al. (1997). *Helv. Chim. Acta* 80 (5): 1319–1337.

229 Paintner, F.F., Görler, K., and Voelter, W. (2003). *Synlett* (4): 0522–0526.

230 Lu, W.-C., Cao, X.-F., Hu, M. et al. (2011). *Chem. Biodivers.* 8 (8): 1497–1511.

231 Curran, D.P., Shen, W., Zhang, J., and Heffner, T.A. (1990). *J. Am. Chem. Soc.* 112 (18): 6738–6740.

232 Curran, D.P., Shen, W., Zhang, J. et al. (1994). *Heterocycles* 37: 1773–1788.

233 Harding, J.R., Hughes, R.A., Kelly, N.M. et al. (2000). *J. Chem. Soc., Perkin Trans.* 1 (20): 3406–3416.

234 Marchi, C., Trepat, E., Moreno-Mañas, M. et al. (2002). *Tetrahedron* 58 (28): 5699–5708.

235 Liang, B., Carroll, P.J., and Joullié, M.M. (2000). *Org. Lett.* 2 (26): 4157–4160.

236 Zhang, L., Zhu, L., Yang, J. et al. (2016). *J. Org. Chem.* 81 (9): 3890–3900.

237 Otaka, A., Mitsuyama, E., Kinoshita, T. et al. (2000). *J. Org. Chem.* 65 (16): 4888–4899.
238 Davison, E.C., Fox, M.E., Holmes, A.B. et al. (2002). *J. Chem. Soc., Perkin Trans.* 1 (12): 1494–1514.
239 Hashimoto, T., Nakatsu, H., Watanabe, S., and Maruoka, K. (2010). *Org. Lett.* 12 (8): 1668–1671.
240 Haddad, N. and Galili, N. (1997). *Tetrahedron: Asymmetry* 8 (20): 3367–3370.
241 Kuo, D.L. and Money, T. (1988). *Can. J. Chem.* 66 (8): 1794–1804.
242 Boeckman, R.K. and Reeder, L.M. (2004). *Synlett* 2004 (8): 1399–1403.
243 Boeckman, R.K. and Liu, Y. (1996). *J. Org. Chem.* 61 (23): 7984–7985.
244 Boeckman, R.K. and Wrobleski, S.T. (1996). *J. Org. Chem.* 61 (21): 7238–7239.
245 Boeckman, R.K., Shao, P., Wrobleski, S.T. et al. (2006). *J. Am. Chem. Soc.* 128 (32): 10572–10588.
246 Boeckman, R.K. and Connell, B.T. (1995). *J. Am. Chem. Soc.* 117 (49): 12368–12369.
247 Boeckman, R.K., Boehmler, D.J., and Musselman, R.A. (2001). *Org. Lett.* 3 (23): 3777–3780.
248 Boeckman, R.K., Pero, J.E., and Boehmler, D.J. (2006). *J. Am. Chem. Soc.* 128 (34): 11032–11033.
249 Palomo, C., González, A., García, J.M. et al. (1998). *Angew. Chem. Int. Ed.* 37 (1, 2): 180–182.
250 Banks, M.R., Blake, A.J., Cadogan, J.I.G. et al. (1992). *Tetrahedron* 48 (37): 7979–8006.
251 Kouklovsky, C., Pouilhes, A., and Langlois, Y. (1990). *J. Am. Chem. Soc.* 112 (18): 6672–6679.
252 Yan, T.H., Tan, C.W., Lee, H.C. et al. (1993). *J. Am. Chem. Soc.* 115 (7): 2613–2621.
253 Yan, T.-H., Lee, H.-C., and Tan, C.-W. (1993). *Tetrahedron Lett.* 34 (22): 3559–3562.
254 Wang, Y.-C., Hung, A.-W., Chang, C.-S., and Yan, T.-H. (1996). *J. Org. Chem.* 61 (6): 2038–2043.
255 Yan, T.-H., Hung, A.-W., Lee, H.-C., and Chang, C.-S. (1994). *J. Org. Chem.* 59 (26): 8187–8191.
256 Yan, T.-H., Hung, A.-W., Lee, H.-C. et al. (1995). *J. Org. Chem.* 60 (11): 3301–3306.
257 Wang, Y.-C., Su, D.-W., Lin, C.-M. et al. (1999). *J. Org. Chem.* 64 (17): 6495–6498.
258 Focken, T. and Hanessian, S. (2014). *Beilstein J. Org. Chem.* 10: 1848–1877.
259 Corey, E.J. and Cane, D.E. (1969). *J. Org. Chem.* 34 (10): 3053–3057.
260 Patois, C. and Savignac, P. (1991). *Synlett* 1991 (07): 517–519.
261 Patois, C. and Savignac, P. (1991). *Tetrahedron Lett.* 32 (10): 1317–1320.
262 Hanessian, S., Bennani, Y.L., and Leblanc, Y. (1993). *Heterocycles* 35: 1411–1424.
263 Hanessian, S. and Beaudoin, S. (1992). *Tetrahedron Lett.* 33 (50): 7655–7658.
264 Hanessian, S., Delorme, D., Beaudoin, S., and Leblanc, Y. (1984). *J. Am. Chem. Soc.* 106 (19): 5754–5756.

265 Hanessian, S., Fu, J.-M., Tu, Y., and Isono, K. (1993). *Tetrahedron Lett.* 34 (26): 4153–4156.
266 Hanessian, S. and Beaudoin, S. (1992). *Tetrahedron Lett.* 33 (50): 7659–7662.
267 Hanessian, S. and Fu, J.-m. (2001). *Can. J. Chem.* 79 (11): 1812–1826.
268 Hanessian, S., Focken, T., and Oza, R. (2010). *Org. Lett.* 12 (14): 3172–3175.
269 Hanessian, S., Bennani, Y.L., and Delorme, D. (1990). *Tetrahedron Lett.* 31 (45): 6461–6464.
270 Bennani, Y.L. and Hanessian, S. (1996). *Tetrahedron* 52 (44): 13837–13866.
271 Denmark, S.E. and Chen, C.-T. (1995). *J. Am. Chem. Soc.* 117 (48): 11879–11897.
272 Denmark, S.E. and Kim, J.-H. (2000). *Can. J. Chem.* 78 (6): 673–688.
273 Sting, M. and Steglich, W. (1990). *Synthesis* 1990 (2): 132–134.
274 Ma, J.-A. (2006). *Chem. Soc. Rev.* 35 (7): 630–636.
275 Hanessian, S. and Bennani, Y.L. (1994). *Synthesis* 1994 (12): 1272–1274.
276 Hanessian, S. and Gomtsyan, A. (1994). *Tetrahedron Lett.* 35 (41): 7509–7512.
277 Hanessian, S., Gomtsyan, A., and Malek, N. (2000). *J. Org. Chem.* 65 (18): 5623–5631.
278 Hanessian, S., Griffin, A.M., and Cantin, L.-D. (2000). *Chirality* 12 (5, 6): 342–345.
279 Denmark, S.E. and Kim, J.-H. (1995). *J. Org. Chem.* 60 (23): 7535–7547.
280 Hanessian, S., Gomtsyan, A., Payne, A. et al. (1993). *J. Org. Chem.* 58 (19): 5032–5034.
281 Hua, D.H., King, R.C.Y., McKie, J.A., and Myer, L. (1987). *J. Am. Chem. Soc.* 109 (16): 5026–5029.
282 Hua, D.H., Chen, J.S., Saha, S. et al. (1992). *Synlett* 1992 (10): 817–820.
283 An, G., Seifert, C., and Li, G. (2015). *Org. Biomol. Chem.* 13 (6): 1600–1617.
284 Kattuboina, A., Kaur, P., Nguyen, T., and Li, G. (2008). *Tetrahedron Lett.* 49 (23): 3722–3724.
285 Yang, B., Zhang, C.-Y., Xu, J. et al. (2019). *Chem. Lett.* 48 (9): 1065–1068.
286 Kaur, P., Shakya, G., Sun, H. et al. (2010). *Org. Biomol. Chem.* 8 (5): 1091–1096.
287 Kaur, P., Wever, W., Pindi, S. et al. (2011). *Green Chem.* 13 (5): 1288–1292.
288 Seifert, C.W., Pindi, S., and Li, G. (2015). *J. Org. Chem.* 80 (1): 447–452.
289 Kaur, P., Wever, W., Rajale, T., and Li, G. (2010). *Chem. Biol. Drug Design* 76 (4): 314–319.
290 Kattamuri, P.V., Ai, T., Pindi, S. et al. (2011). *J. Org. Chem.* 76 (8): 2792–2797.
291 Han, J., Ai, T., and Li, G. (2008). *Synthesis* (16): 2519–2526.
292 Chen, Z.-X., Ai, T., Kaur, P., and Li, G. (2009). *Tetrahedron Lett.* 50 (9): 1079–1081.
293 Kaur, P., Nguyen, T., and Li, G. (2009). *Eur. J. Org. Chem.* (6): 912–916.
294 Zhang, H., Yang, Z., Zhao, B.N., and Li, G. (2018). *J. Org. Chem.* 83 (2): 644–655.
295 Ai, T. and Li, G. (2009). *Bioorg. Med. Chem. Lett.* 19 (14): 3967–3969.
296 Kattuboina, A., Kaur, P., Ai, T., and Li, G. (2008). *Chem. Biol. Drug Design* 71 (3): 216–223.
297 Yang, B., Ji, X., Xue, Y. et al. (2016). *Org. Biomol. Chem.* 14 (25): 6024–6035.
298 Liu, Y., Ahmed, S., Qin, X.-Y. et al. (2020). *Chem. Asian J.* 15 (7): 1125–1131.

299 Kattuboina, A. and Li, G. (2008). *Tetrahedron Lett.* 49 (10): 1573–1577.

300 Xie, J.-B., Luo, J., Winn, T.R. et al. (2014). *Beilstein J. Org. Chem.* 10: 746–751.

301 Xiong, Y., Mei, H., Wu, L. et al. (2014). *Beilstein J. Org. Chem.* 10: 653–659.

302 Schöllkopf, U., Groth, U., and Deng, C. (1981). *Angew. Chem. Int. Ed.* 20 (9): 798–799.

303 Schöllkopf, U. (1983). *Tetrahedron* 39 (12): 2085–2091.

304 Schöllkopf, U., Hartwig, W., and Groth, U. (1979). *Angew. Chem. Int. Ed.* 18 (11): 863–864.

305 Schöllkopf, U., Nozulak, J., and Groth, U. (1982). *Synthesis* 1982 (10): 868–870.

306 Schöllkopf, U. and Neubauer, H.-J. (1982). *Synthesis* 1982 (10): 861–864.

307 Chen, J., Corbin, S.P., and Holman, N.J. (2005). *Org. Process Res. Dev.* 9 (2): 185–187.

308 Undheim, K. (2008). *Amino Acids* 34 (3): 357–402.

309 Ma, C., He, X., Liu, X. et al. (1999). *Tetrahedron Lett.* 40 (15): 2917–2918.

310 Ma, C., Liu, X., Li, X. et al. (2001). *J. Org. Chem.* 66 (13): 4525–4542.

311 Jones, E.P., Jones, P., White, A.J.P., and Barrett, A.G.M. (2011). *Beilstein J. Org. Chem.* 7: 1570–1576.

312 Ruiz, M., Ojea, V., and Quintela, J. (1999). *Synlett* 1999 (2): 204–206.

313 Bonnard, I., Rolland, M., Salmon, J.-M. et al. (2007). *J. Med. Chem.* 50 (6): 1266–1279.

314 Cremonesi, G., Dalla Croce, P., Gallanti, M., and La Rosa, C. (2014). *Tetrahedron* 70 (11): 2054–2058.

315 Koch, C.J., Šimonyiová, S., Pabel, J. et al. (2003). *Eur. J. Org. Chem.* (7): 1244–1263.

316 Hammer, K. and Undheim, K. (1998). *Tetrahedron: Asymmetry* 9 (13): 2359–2368.

317 Hammer, K., Rømming, C., and Undheim, K. (1998). *Tetrahedron* 54 (36): 10837–10850.

318 Schöllkopf, U., Pettig, D., Busse, U. et al. (1986). *Synthesis* 1986 (9): 737–740.

319 Fukuari, M., Ichimoto, I., and Kirihata, M. (1996). *Biosci. Biotechnol. Biochem.* 60 (4): 680–682.

320 Schöllkopf, U., Tiller, T., and Bardenhagen, J. (1988). *Tetrahedron* 44 (17): 5293–5305.

321 Jane, D.E., Chalmers, D.J., Howard, J.A.K. et al. (1996). *J. Med. Chem.* 39 (24): 4738–4743.

322 Xu, P.F., Chen, Y.S., Lin, S.I., and Lu, T.J. (2002). *J. Org. Chem.* 67 (7): 2309–2314.

323 Xu, P.F. and Lu, T.J. (2003). *J. Org. Chem.* 68 (2): 658–661.

324 García-Barradas, O. and Juaristi, E. (1997). *Tetrahedron: Asymmetry* 8 (9): 1511–1514.

325 García-Barradas, O. and Juaristi, E. (1995). *Tetrahedron* 51 (12): 3423–3434.

326 Bull, S.D., Davies, S.G., Epstein, S.W., and Ouzman, J.V.A. (1998). *Chem. Commun.* (6): 659–660.

327 Bull, S.D., Davies, S.G., Epstein, S.W. et al. (1998). *J. Chem. Soc., Perkin Trans. 1* (15): 2321–2330.

328 Harding, C.I., Dixon, D.J., and Ley, S.V. (2004). *Tetrahedron* 60 (35): 7679–7692.
329 Keynes, M.N., Earle, M.A., Sudharshan, M., and Hultin, P.G. (1996). *Tetrahedron* 52 (26): 8685–8702.
330 Sparks, S.M., Banker, P., Bickett, D.M. et al. (2009). *Bioorg. Med. Chem. Lett.* 19 (3): 981–985.
331 Jones, P., Altamura, S., Chakravarty, P.K. et al. (2006). *Bioorg. Med. Chem. Lett.* 16 (23): 5948–5952.
332 Hoyt, S.B., London, C., Gorin, D. et al. (2007). *Bioorg. Med. Chem. Lett.* 17 (16): 4630–4634.
333 London, C., Hoyt, S.B., Parsons, W.H. et al. (2008). *Bioorg. Med. Chem. Lett.* 18 (5): 1696–1701.
334 Vassiliou, S. and Magriotis, P.A. (2006). *Tetrahedron: Asymmetry* 17 (11): 1754–1757.
335 Schöllkopf, U., Groth, U., Westphalen, K.O., and Deng, C. (1981). *Synthesis (Germany)* 1981 (12): 969–971.
336 Chinchilla, R., Falvello, L.R., Galindo, N., and Nájera, C. (1997). *Angew. Chem. Int. Ed. Engl.* 36 (9): 995–997.
337 Nájera, C., Abellán, T., and Sansano, J.M. (2000). *Eur. J. Org. Chem.* (15): 2809–2820.
338 Lu, T.-J. and Lin, C.-K. (2008). *J. Org. Chem.* 73 (24): 9527–9534.
339 Xu, P.-F., Li, S., Lu, T.-J. et al. (2006). *J. Org. Chem.* 71 (12): 4364–4373.

4

trans-1,2-Diaminocyclohexane and Its Derivatives in Asymmetric Organocatalysis

Michał Kopyt, Michał P. Głowacki, and Piotr Kwiatkowski

University of Warsaw, Faculty of Chemistry, Biological and Chemical Research Centre, Żwirki i Wigury 101, 02-089 Warsaw, Poland

4.1 Introduction

Enantiomerically pure *trans*-1,2-diaminocyclohexane (**1**, acronym: DACH, Scheme 4.1) belongs to the group of small molecules most frequently used in asymmetric synthesis to design chiral reagents, scaffolds, ligands for catalysis (e.g. salen and Trost ligands), and organocatalysts. The dynamic development of organocatalysis [1, 2] in the last two decades, opened up new possibilities for enantioselective synthesis. This strategy has become the third major branch of asymmetric catalysis, along with catalysis involving chiral metal complexes and biocatalysis. The intensive research in asymmetric organocatalysis has contributed to the development of numerous effective organocatalysts based on readily available *trans*-DACH.

Due to the limited size of this chapter, the most important examples of organocatalysts based on *trans*-DACH will be presented, excluding immobilized catalysts. Moreover, in the second part of this chapter, selected examples of applications of the most effective and interesting derivatives of DACH in asymmetric organocatalysis will be shown.

DACH was obtained for the first time in 1926 by Wieland et al. from cyclohexane-1,2-dicarboxylic acid using Curtius reaction in the final step [3]. Nowadays, mixture of *trans/cis*-DACH is obtained, e.g. as a byproduct during the manufacture of 1,6-hexanediamine, used in the production of Nylon; therefore, this compound is relatively cheap. Enantiomerically pure (1*R*,2*R*)- or (1*S*,2*S*)-*trans*-DACH (**1**) is also readily available and can be easily obtained by resolution of racemate, e.g. using crystallization with L- or D-tartaric acid (Scheme 4.1) [4–6].

Applications of *trans*-DACH in asymmetric synthesis and molecular recognition until 1997 were described in review article by Bennani and Hanessian [7]. Publications presenting the use of enantiomerically pure DACH in asymmetric catalysis at that time were practically limited to reactions in the presence of metal complexes with ligands based on *trans*-DACH, especially with C_2 symmetry. Reports on the use of DACH derivatives as organocatalysts in the twentieth

Chiral Building Blocks in Asymmetric Synthesis: Synthesis and Applications, First Edition.
Edited by Elżbieta Wojaczyńska and Jacek Wojaczyński.

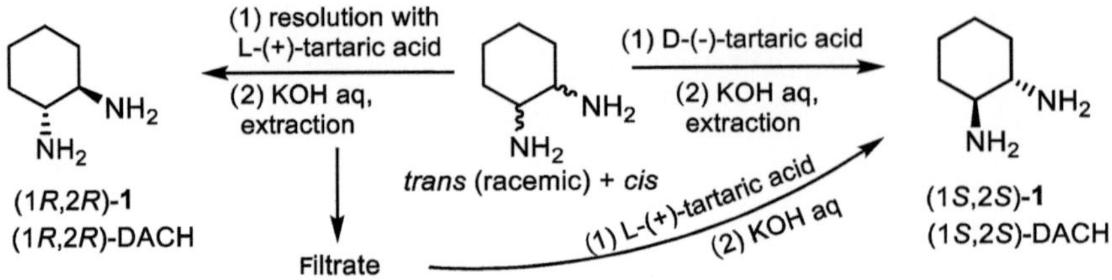

Scheme 4.1 Resolution of the racemic *trans*-1,2-diaminocyclohexane. Source: Based on Refs. [4–6].

century are very rare. In 1988, Shiori and coworkers [8] reported oxidation of 2-methyl-1-tetralone with molecular oxygen in the presence of 50% aqueous sodium hydroxide and 5 mol% of the monobenzyl quaternary ammonium salt derived from (1*R*,2*R*)-*N*,*N*,*N*′,*N*′-tetramethyl-DACH, as a phase transfer catalyst (Figure 4.1a). However, the enantioselectivity of this reaction was very low (4–8% ee). The first example of highly enantioselective reaction in which a diaminocyclohexane derivative was successfully used as an organocatalyst was published in 1998 by Jacobsen and coworker (Figure 4.1b) [9]. Initially, thiourea Schiff bases were tested as ligands with transition metal salts in the reaction of aldimines with hydrogen cyanide (HCN), but finally proved to perform better without metal additives (ee up to 91%). The most efficient organocatalyst for Strecker reaction in terms of enantioselectivity contains in its structure two chiral units: (1*R*,2*R*)-*trans*-DACH and unnatural amino acid L-*tert*-leucine (Figure 4.1b). Moreover, it has no basic nitrogen atoms and its action is based on activation by hydrogen bonding. A very important stage in the development of DACH-based organocatalysts was a successful application of its relatively simple tertiary amine-thiourea derivative in Michael reaction (Figure 4.1c) reported by Takemoto and coworkers in 2003 [10]. This organocatalyst, commonly known as Takemoto's catalyst, and a dozen of its simple modifications turned out to be very good catalysts for numerous asymmetric transformations. All the above-mentioned organocatalysts presented in Figure 4.1 operate by non-covalent interactions such as hydrogen bonds. However, a large group of diaminocyclohexane organocatalysts contain one primary amine. They essentially activate one of the substrates covalently through the formation of enamine or iminium ion. A special case is the direct application of unmodified diaminocyclohexane as an organocatalyst. It has been successfully used in several reactions with the addition

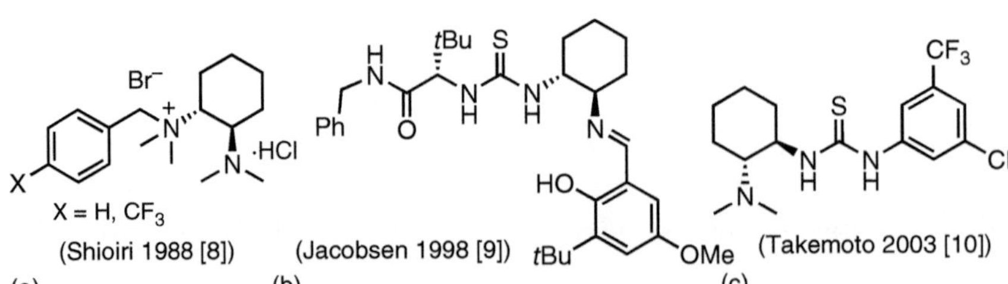

Figure 4.1 Early examples of *trans*-1,2-diaminocyclohexane-based organocatalysts.

of acidic cocatalysts, e.g. in Michael-type reactions [11, 12] and in the aldol reaction [13–15]. However, these examples represent only a very small part compared to applications of other DACH derivatives in organocatalysis.

4.2 *trans*-1,2-Diaminocyclohexane-Based Organocatalysts

Some of DACH-organocatalysts can be obtained directly from enantiomerically pure *trans*-DACH, but most of them require the use of mono-protected precursor. The figure below shows typical examples of protected DACH – e.g. with Boc (**2a**) [16, 17], phthaloyl (**2b**) [18], acetyl (**2c**) [19], and benzoyl (**2d**) [20] protecting groups. Usually, they can be obtained directly from diamine **1** (e.g. **2a** and **2b**) or by ring opening of selected cyclic derivatives, e.g. 4,5-dihydro-1*H*-imidazole **2e** and **2f** [19, 20] or cyclic urea **2g** [21] (Figure 4.2).

Mono-protected DACH derivatives **2a–2d** are very often used for preparation of *N*1,*N*1-dialkyl-cyclohexane-1,2-diamines of type **3** (Figure 4.3) [19, 20, 22, 23]. One of the exceptions is the amine **3c** which can be synthesized directly from DACH by reductive amination with glutaraldehyde [24]. Diamines **3a** [19, 20], **3b** [22, 23], and **3c** [20, 24] are mainly used for further functionalization to obtain non-covalent organocatalysts with basic properties (e.g. Takemoto's catalyst). The remaining diamines **3d–3q** [25–34] usually contain more bulky substituents on one

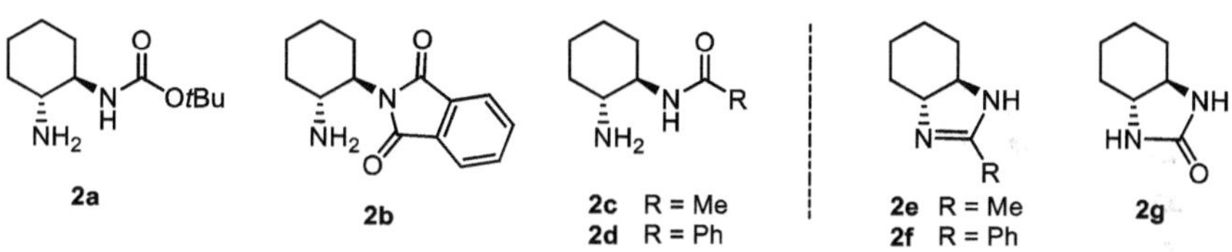

Figure 4.2 Examples of protected DACH derivatives commonly used in the synthesis of *N*1,*N*1-dialkyl-cyclohexane-1,2-diamines. Source: Based on Refs. [16–21].

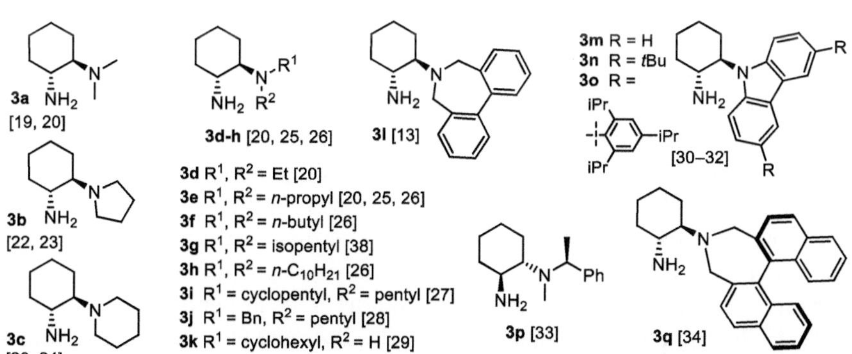

Figure 4.3 Selected *N*1,*N*1-dialkyl-cyclohexane-1,2-diamines used in organocatalysis. Source: Based on Refs. [13, 19, 20, 22–30, 32–34, 38].

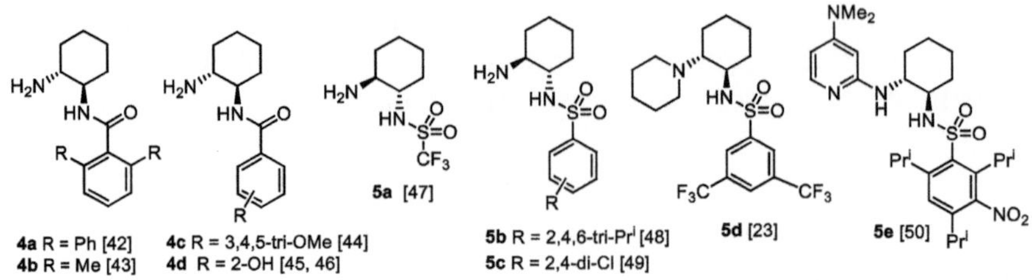

Figure 4.4 Mono-amides (**4**) and sulfonamides (**5**). Source: Based on Refs. [23, 42–50].

of the nitrogen atoms and in the presence of acids as cocatalysts have been used as organocatalysts to activate carbonyl compounds in a covalent manner [35–41]. They include derivatives that have two identical (**3d–3h**) [25, 26] or different substituents (**3i, 3j**) [27, 28] on one of the nitrogen atoms, biphenyl motif in **3l–3o** [19] and diamines containing additional chiral units, e.g. based on α-methylbenzylamine **3p** [33] and 1,1′-binaphthyl **3q** [34]. These diamines in protonated form have been efficiently used as catalysts in various aldol reactions [26–28, 33–41]. In addition, mono-triflate salt of **3k** promotes the Nazarov cyclization enantioselectively; however, it requires a stoichiometric amount of the diamine [29]. Specific amines **3m–3o** that contain the carbazole subunit were used in the presence of benzoic acid as catalysts for iminium ion-mediated radical conjugate addition to β,β-disubstituted cyclic enones, with formation of a quaternary stereocenter [31, 32].

The majority of organocatalysts derived from diaminocyclohexane contain one basic nitrogen atom and at least one hydrogen-bond donor. Among simpler DACH derivatives that belong to this group, mono-amides (**4**) [42–46] and mono-sulfonamides (**5**) [23, 47–50] are found (Figure 4.4). However, their use in organocatalysis is still relatively limited as compared to other derivatives, e.g. thioureas, which are presented in the next paragraph (Figures 4.5 and 4.6).

trans-Diaminocyclohexane mono-thioureas **6** [51–62] and **7** [10, 63–78] containing one basic nitrogen atom belong to the most commonly used and the very diverse group of organocatalysts (Figure 4.5) [79–82]. Compounds **6** and **7** are typically synthesized by coupling of corresponding isothiocyanates with chiral amines [82]. Due to the presence of the basic nitrogen atom, they can be divided into two main groups – primary (**6**) and tertiary (**7a–7f, 7h–7o**) amines. The aforementioned Takemoto's catalyst **7a** [10], which is a member of the second group, is the DACH derivative most widely used in organocatalysis [83–88] which can be explained by simplicity of the structure and versatility. Structures **7b–7o** shown in Figure 4.5 are actually successive modifications of **7a** which perform better in selected chemical transformations, especially in terms of enantioselectivity. Also DACH-thiourea analogs **6a–6f**, which contain a primary amine group, are very important organocatalysts, usually used in catalysis in the presence of acid additives [89]. Typically, these activate one of carbonyl substrates via formation of either iminium or enamine intermediate. Compared to the mentioned types of thiourea derivatives, organocatalysts based on diaminocyclohexane containing a secondary amine group are much less common (e.g. **7g**, **7q**, and **7s**). An important but less

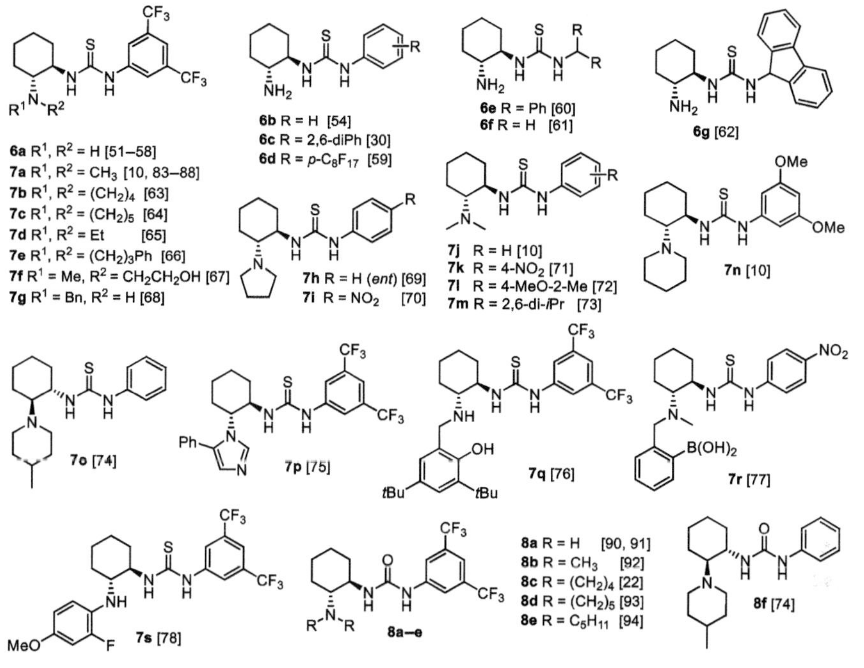

Figure 4.5 Amine-thiourea (**6** and **7**) and urea (**8**) derivatives. Source: Based on Refs. [30, 51–78, 90–94].

frequently used and a smaller group of organocatalysts are also structurally similar urea catalysts **8** (Figure 4.5) [22, 74, 90–94].

The growing interest in catalysis with (thio)ureas in the first years of the twenty first century resulted also in the development of more complex catalytic systems based on *trans*-DACH with primary- (**9**) [51, 95–110], or tertiary (**10**) amino groups [22, 25, 111–126], in particular containing additional chiral units (Figure 4.6), such as amino acids (**9a–9c**, **10a–10e**, especially *tert*-leucine), amino alcohols, monosaccharides, *Cinchona* alkaloids, polycyclic structures (e.g. **9m, 9l** and **10h, 10i**), binaphthyl derivatives, and other chiral amines [51, 95–126]. Among the organocatalysts shown in Figure 4.6, primary amine-thioureas **9a–9o** constitute a special and important group [89]. These types of thioureas were first used independently by Jacobsen (**9a** and **9b**) [51, 95, 96] and Tsogoeva [99–102] in 2006, and proved to be effective catalysts for the nitro-Michael reaction. The primary amine thioureas **9a** and **9b** applied by Jacobsen were known a few years earlier because they were intermediates in syntheses of thiourea Schiff bases – catalysts for Strecker reaction developed in 1998 [9]. Some of catalysts **9** and **10** have resulted in higher enantiomeric excesses over selected simpler catalysts shown in Figure 4.5. Examples of the use of selected thioureas (**6, 7, 9,** and **10**) as catalysts are presented in the second part of this chapter.

The second very important and versatile group of hydrogen-bond donors often used in organocatalysis, in addition to thioureas, are squaramides; however, their importance in the field of asymmetric synthesis was noticed considerably later, at

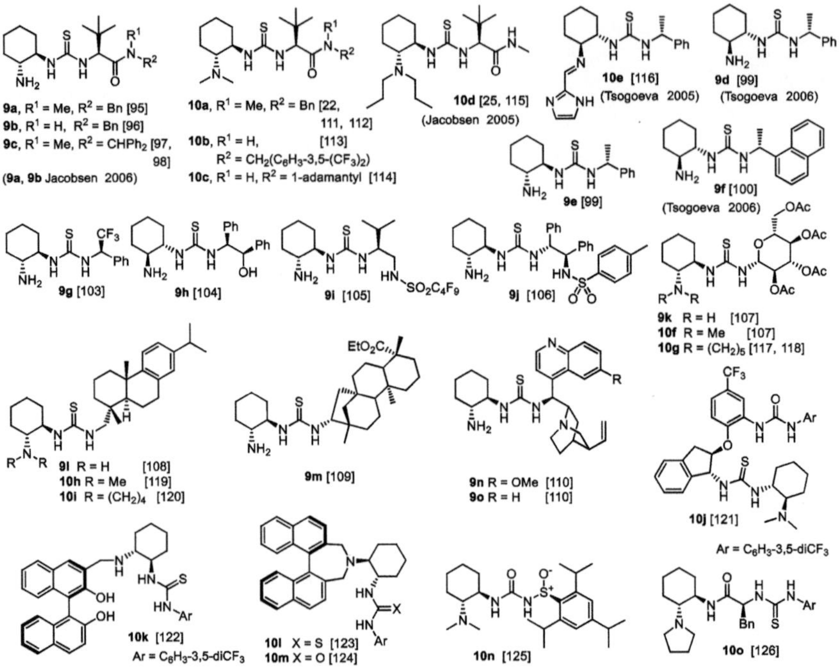

Figure 4.6 Amine-(thio)ureas containing two different chiral units. Source: Based on Refs. [95, 100, 103–126].

the end of the first decade of this century (after 2008) [127]. One of the major differences between these two widely used platforms is a larger distance between two coplanar N—H bonds in squaramides (by 0.6 Å). Selected catalysts containing the squaramide framework and derived from *trans*-DACH are presented in Figure 4.7, including relatively simple structures **11** [128–132] and derivatives **12** [133–139] with an additional chiral unit. A certain limitation in the use of squaramide organocatalysts could be their lower solubility compared to thioureas in less polar

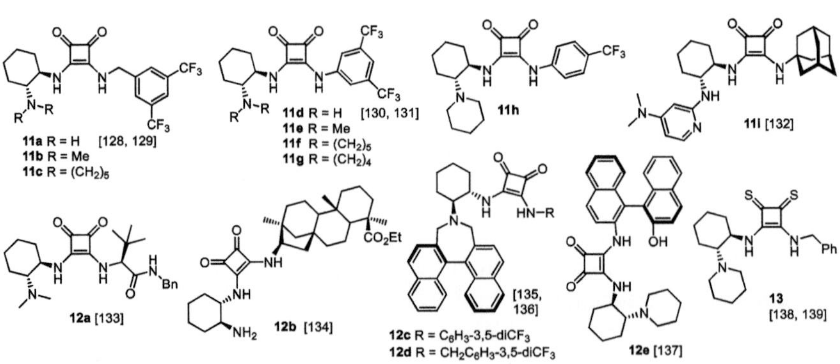

Figure 4.7 Amine-squaramide derivatives. Source: Based on Refs. [128–139].

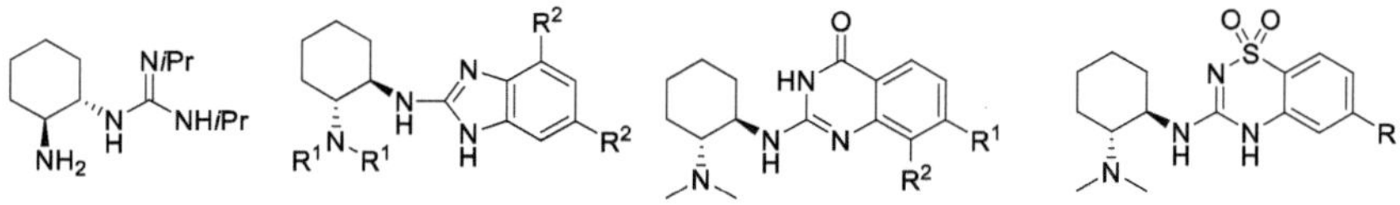

14a [141]
14b R^1 = H, R^2 = H [142]
14c R^1 = Me, R^2 = H [142]
14d R^1 = Me, R^2 = CF_3 [143]
14e R^1 = H, R^2 = H [20, 144]
14f R^1 = CF_3, R^2 = H [20]
14g R^1 = H, R^2 = F [14, 144]
14h R = H [144]
14i R = F [144, 145]

Figure 4.8 Organocatalysts containing a guanidine motif. Source: Based on Refs. [14, 20, 141–145].

solvents, which are often used in organocatalysis. A good example is the large difference in solubility of squaramide **11e** with its thiourea counterpart **7a** (the Takemoto catalyst) in toluene. More recently Rawal introduced more acidic dithiosquaramides of type **13**, which are also more soluble in nonpolar solvents [138, 139].

Another interesting group of DACH-derived organocatalysts with active amine function that contain two hydrogen-bond donors are guanidines [140] (e.g. **14a**) and related structures bearing three nitrogen atoms bonded to the carbon atom (Figure 4.8) [20, 141–145]. This unique structure with hydrogen-bond donors can be found in some heterocyclic derivatives: 2-aminobenzimidazole (**14b–14c**) [142, 143], 2-amino-quinazolin-4-(1*H*)-one (**14e–14g**) [20, 144], and 3-aminobenzothiadiazine-1,1-dioxide (**14h–14j**) [144, 145]. The compounds shown in Figure 4.8 were introduced to the organocatalysis in the years 2009–2012. In this chapter they represent the last group of well-defined unsymmetrical diaminocyclohexane organocatalysts with hydrogen-bond donors in which one of the DACH nitrogen atoms has a basic character.

trans-DACH scaffold has been also used to design various unsymmetrical multiple hydrogen-bonding donor catalysts [146, 147] which do not contain active amine functions and do not exhibit basic properties. The most efficient organocatalysts of this type are (thio)ureas presented in Figure 4.9 [9, 148–170]. A significant number of these structures (**15a–15i**) contain two different chiral subunits, usually including L-*tert*-leucine. Among them are the first known highly enantioselective diaminocyclohexane-based organocatalyst **15a** developed by Jacobsen and coworker in 1998 [9] for Strecker reaction. In the following years, his group optimized the (thio)urea Schiff base organocatalysts (**15b–15e**) [95, 148–153] and also successfully introduced pyrrole-bearing (**15f, 15g**) [154–158] and acetamide **15k** [163] thioureas as well as urea-sulfinamide catalyst **15l** [164] to asymmetric catalysis. Some of the Jacobsen catalysts shown are commercially available.

The second group of catalysts (**16a–16h**) has usually a simpler structure and includes DACH as the only element of chirality [165–170]. All molecules **16** contain at least one additional hydrogen-bond donor in addition to the thiourea moiety. Catalysts presented in Figure 4.9 are usually used for activation of Lewis basic electrophiles such as carbonyl compounds, imines, and nitroolefins in reactions of nucleophilic addition, which usually do not require the presence of a base. Some of the compounds **15** and **16** are also used in asymmetric anion-binding catalysis.

Another group of catalysts synthesized from diaminocyclohexane are quaternary ammonium salts **17** containing additional hydrogen-bond donors [171–174]

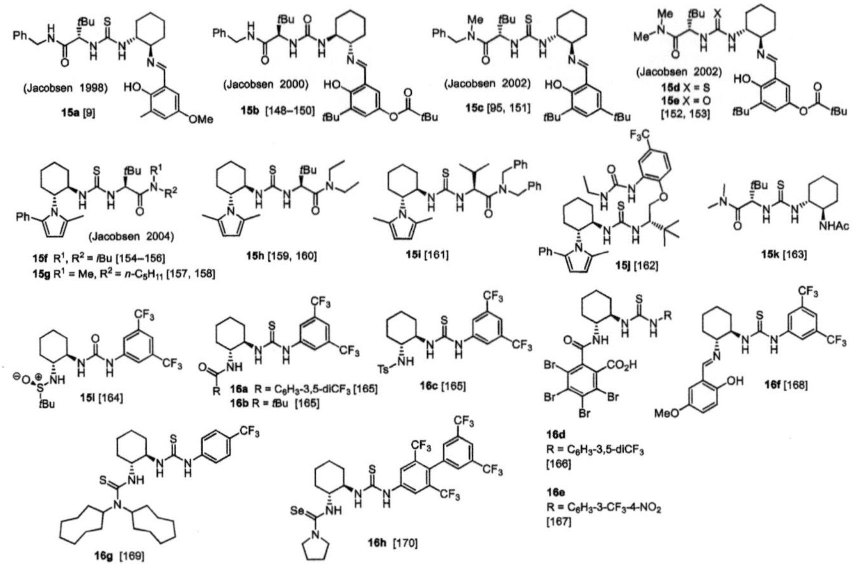

Figure 4.9 Non-basic thiourea and urea organocatalysts activating via hydrogen bonding. Source: Based on Refs. [9, 95, 148–170].

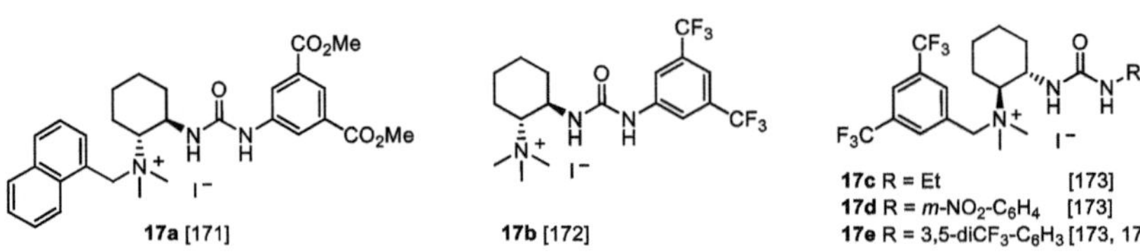

Figure 4.10 Tertiary ammonium salts-ureas **17** for phase transfer catalysis. Source: Based on Refs. [171–174].

(Figure 4.10). Compounds of this type have been used as asymmetric phase transfer catalysts. However, compared to other catalysts of this type (e.g. based on *Cinchona* alkaloids), their importance is still relatively low.

C_2-Symmetric chiral ligands are very common in the preparation of metal complexes for asymmetric catalysis. In contrast, organocatalysts with C_2 symmetry are used sporadically. Among them is *trans*-DACH and its derivatives **18** shown in Figure 4.11 [175–185]. The most commonly used compound of this type is bis-thiourea **18c**, introduced by Nagasawa and coworkers in 2004 [178].

Another particular group of catalysts are compounds bearing two diaminocyclohexane units in one molecule. Examples of such organocatalysts (**19**), usually aminocatalysts, are presented in Figure 4.12 [186–193]. Specific and very interesting catalysts in this group are guanidinium ions **19c** and **19d**, which have been successfully used in enantioselective Claisen rearrangements [188–190].

Organocatalysts containing amino acids (e.g. *tert*-leucine) in their structure have already been presented (e.g. **9a** or **15a**), but they have not formed a direct amide bond with *trans*-diaminocyclohexane. The figure below (Figure 4.13) shows selected

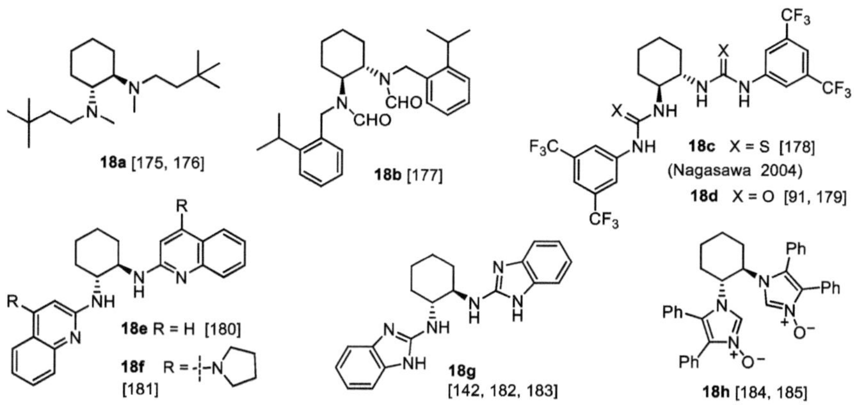

Figure 4.11 C_2-symmetric mono-DACH organocatalysts. Source: Based on Refs. [91, 142, 175–185].

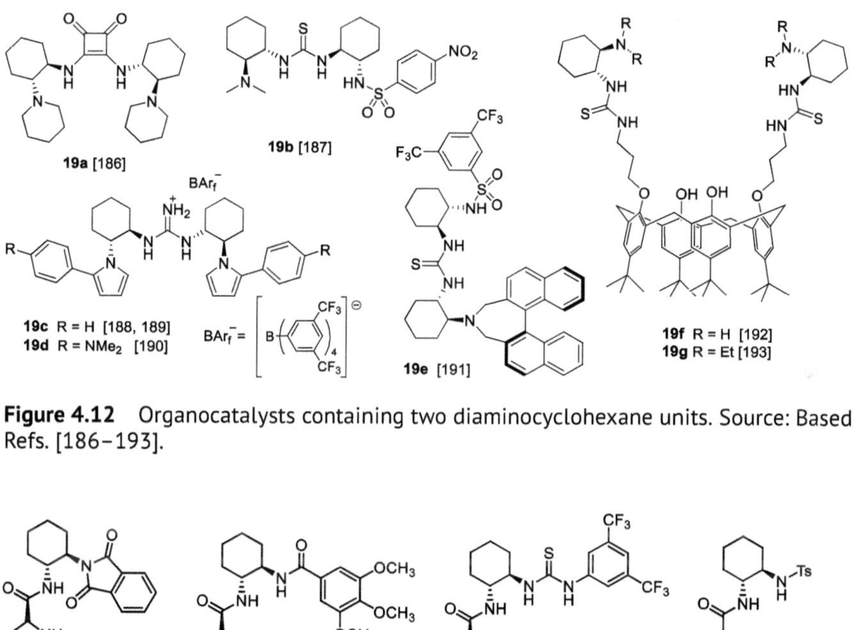

Figure 4.12 Organocatalysts containing two diaminocyclohexane units. Source: Based on Refs. [186–193].

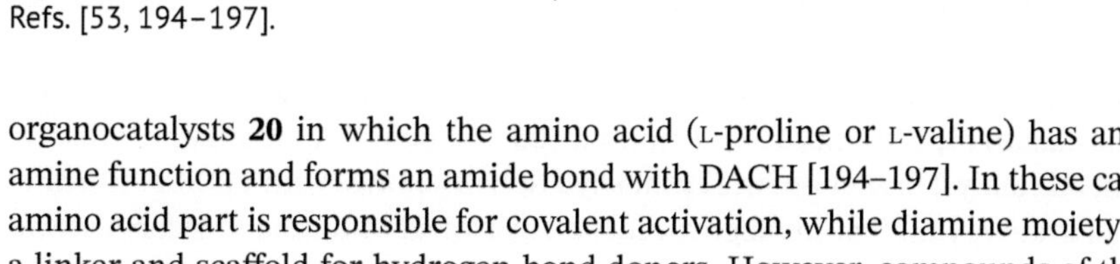

Figure 4.13 Amides of DACH with *N*-unprotected amino acids. Source: Based on Refs. [53, 194–197].

organocatalysts **20** in which the amino acid (L-proline or L-valine) has an active amine function and forms an amide bond with DACH [194–197]. In these catalysts, amino acid part is responsible for covalent activation, while diamine moiety acts as a linker and scaffold for hydrogen-bond donors. However, compounds of this type do not play a significant role in organocatalysis.

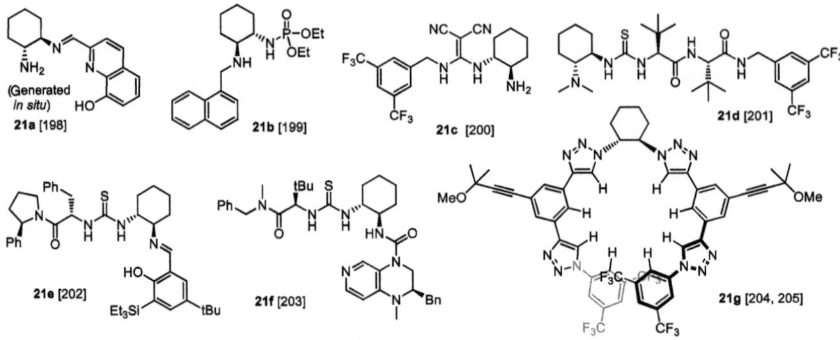

Figure 4.14 Miscellaneous DACH-based organocatalysts (**21**). Source: Based on Refs. [198–205].

Finally, Figure 4.14 presents miscellaneous examples of DACH-organocatalysts that do not meet the criteria of the categories outlined above. Among them are generated *in situ* primary amine-imine **21a**; secondary amine-phosphoramide **21b**; tertiary amine-diaminomethylenemalononitrile **21c**; dipeptide-based bifunctional tertiary amine **21d**; thiourea – Schiff base with three chiral units **21e**; 4-dimethylaminopyridine (DMAP)-thiourea bifunctional catalyst **21f**; and helically chiral oligotriazole **21g** [198–205].

Trans-DACH derivatives are widely used as organocatalysts in various types of enantioselective transformations. There are so many examples of their applications that it is not possible to include all of them in this chapter. We decided to illustrate their usefulness and versatility in selected types of reactions; among them transformations of aldehydes, ketones, imines, and various Michael acceptors, e.g. nitroolefins and α,β-unsaturated carbonyl compounds, are the most numerous group.

4.3 Application of *trans*-1,2-Diaminocyclohexane Derivatives in Asymmetric Organocatalysis

This section presents selected examples of applications of *trans*-DACH-based organocatalysts in asymmetric synthesis. As mentioned in the introduction, the first effective catalyst of this type was reported by Jacobsen and coworker in 1998 [9]. Thiourea **15a** and urea **15b**, containing phenolic hydroxyl as an additional hydrogen-bond donor, proved to be highly enantioselective organocatalysts in Strecker reaction between HCN as the cyanide source and aldimines as well as ketoimines, which leads to a diverse range of substituted amino acid precursors (Scheme 4.2a,b) [9, 148–150, 152]. Thiourea **10d** (Scheme 4.2c) is a highly effective and general catalyst in asymmetric cyanosilylation of ketones with TMSCN as a safer source of cyanide [25]. The analogous reaction of TMSCN addition to aldehydes was catalyzed by imidazole-thiourea catalyst **7p** (Scheme 4.2d) in the presence of benzoic acid as a cocatalyst [75]. The addition of acid was critical to the enantioselectivity

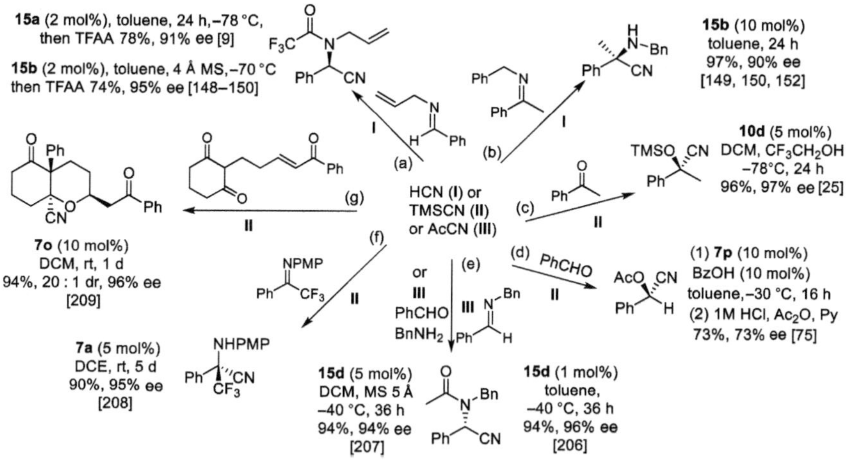

Scheme 4.2 Asymmetric cyanation reactions. TFAA, trifluoroacetic anhydride; MS, molecular sieves; DCE, 1,2-dichloroethane; DCM, dichloromethane. Source: Based on Refs. [9, 25, 75, 148–150, 152, 206–209].

of the process. List and coworkers successfully applied Jacobsen's thiourea catalyst **15d** (Scheme 4.2e) for acylcyanation of imines with commercially available acetyl cyanide instead of dangerous HCN [206]. This catalyst and AcCN were also used in the first example of organocatalytic asymmetric three-component Strecker reaction in which the imine was generated *in situ* [207]. Another example of Strecker reaction reported by Enders et al. was an addition of TMSCN to trifluoromethyl ketimines catalyzed by the Takemoto catalyst **7a** (Scheme 4.2f) [208]. Finally, the organocatalytic enantio- and diastereoselective cycloetherification of 1,3-cyclohexanedione-bearing enones with TMSCN and catalyst **7o** allowed to obtain oxadecalins with two contiguous quaternary stereogenic centers (Scheme 4.2g) [209].

4.3.1 Michael Additions of Malonate Esters to Nitrostyrene

One of the first enantioselective Michael reactions utilizing DACH organocatalysts was the addition of malonates to nitrostyrene (Scheme 4.3) reported by Takemoto and coworkers in 2003 [10]. In this early study, bifunctional thiourea catalyst **7a** was successfully used for the first time [10, 92]. Nowadays, commercially available compound **7a**, known as Takemoto catalyst, is one of the most popular organocatalysts. All of DACH-based catalysts used later in this reaction consist of tertiary amine and double hydrogen-bond donor like squaramide **11b** [210], urea **10m** [124], or heterocyclic systems, e.g. aminobenzoimidazole (**14c, 14d**) [142, 143] or aminoquinazolinone **14f** [20], and allow to obtain generally similar results compared to **7a** (Scheme 4.3).

This type of bifunctional catalysts activates substrates by hydrogen bonding and other noncovalent interactions and shows the best effectiveness in nonpolar solvents. Mechanism of this reaction was elucidated computationally and experimentally by Soós and coworkers [211] who demonstrated deprotonation of

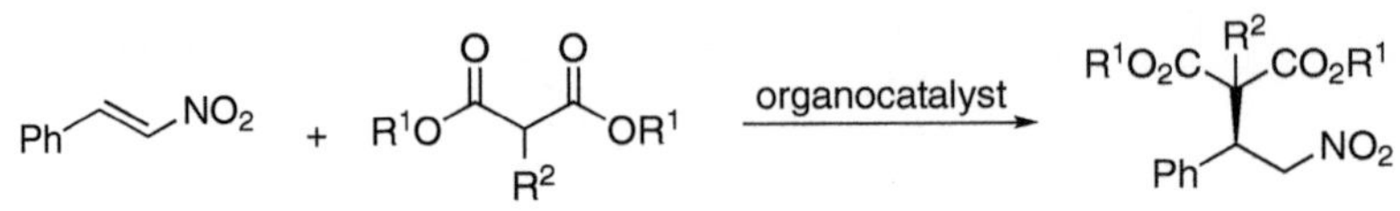

Catalyst (mol%)	R¹	R²	Solvent	Temp.	Time	Yield	ee	Ref.
(1*R*,2*R*)-**7a** (10)	Et	H	Toluene	rt	24 h	86%	93%	[10, 92]
(1*R*,2*R*)-**7a** (10)	Me	Me	Toluene	rt	36 h	82%	93%	[10, 92]
(1*R*,2*R*)-**7a** (10)	Me	OMe	Toluene	rt	28 h	89%	94%	[10, 92]
(1*R*,2*R*)-**14d** (10)	Et	H	Toluene	0 °C	1 d	99%	93%	[143]
(1*R*,2*R*)-**14c**·TFA (10)	Et	H	Toluene	rt	2 d	97%	92%	[142]
(1*R*,2*R*)-**14f** (20)	Et	H	Toluene	rt	1 d	94%	84%	[20]
(1*R*,2*R*)-**11b** (5)	Me	H	DCM	rt	4 d	76%[a)]	86%[a)]	[210]
(1*S*,2*S*)-**10m** (10)	Et	F	DCM	rt	2 d	93%	96%[b)]	[124]

a) *p*-Chloro-β-nitrostyrene was used.
b) Opposite absolute configuration.

Scheme 4.3 Michael additions of malonate to nitrostyrenes. Source: Based on Refs. [10, 92, 124, 142, 143, 210].

nucleophile and binding to protonated catalyst. After that, a slight conformational change allows for interaction between protonated tertiary amine of the catalyst and an electrophile, leading to the preferential formation of one enantiomer. In case of squaramide catalysts an analogous reaction pathway was predicted to be most favorable with additional alternative pathways with slightly higher activation energies [212]. Regardless of the type of hydrogen-bond donor (see **7a**, **10m**, **11b**, and **14c**, **14d**, **14f**), the configuration of the resulting product is determined by the absolute configuration of 1,2-*trans*-diaminocyclohexane unit in the catalyst. Basically all of these noncovalent catalysts, containing tertiary amine and double hydrogen-bond donor, activate the substrates according to a similar mechanism. Organocatalysts with such structural motifs are very common and find application in many other Michael reactions. Products of malonate addition to nitroalkenes are convenient precursors for the synthesis of γ-amino acids, as demonstrated by synthesis of pharmaceutically relevant gabapentinoids, e.g. baclofen [92] and pregabalin [129].

4.3.2 Michael Additions of Other 1,3-Dicarbonyls to Nitrostyrene

Catalysts sharing the same features as above, namely, tertiary amine and double hydrogen-bond donor, can be used with other nucleophiles of similar acidity, especially where similar nucleophile–catalyst complexes could be expected to form (Scheme 4.4a–h) [86, 119, 132, 135, 191, 213–215]. One of the typical examples is addition of acetylacetone to nitrostyrene (Scheme 4.4a) [213], a reaction strictly analogous to malonate addition. In some cases, use of substituted 1,3-dicarbonyls

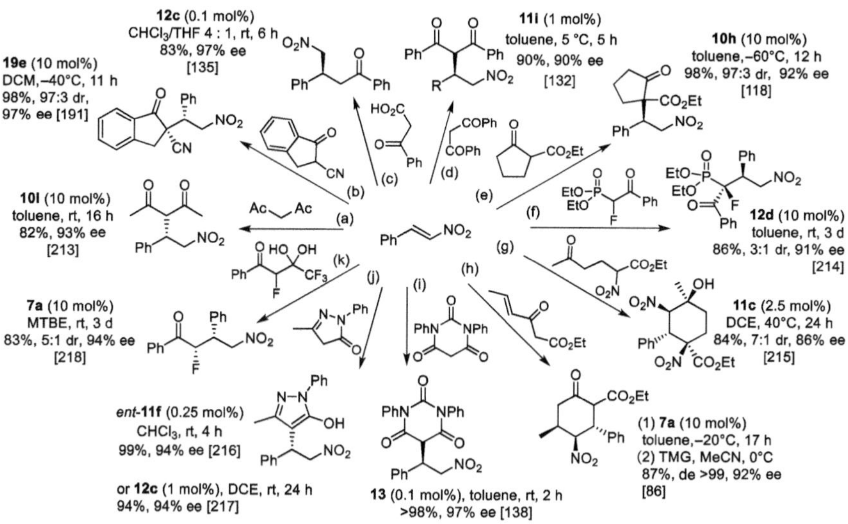

Scheme 4.4 Michael additions of other 1,3-dicarbonyls to nitrostyrene. Source: Based on Refs. [86, 118, 135, 138, 191, 214–218].

allows for introduction of quaternary stereogenic center (Scheme 4.4b,e) [119, 191]. In one case β-ketoacid was used as a nucleophile; a subsequent decarboxylation provided a formal acetophenone adduct with high enantioselectivity (Scheme 4.4c) [135]. As evidenced by the use of catalyst **11i** in reaction of dibenzoylmethane (Scheme 4.4d) [132], the incorporation of 2-amino-DMAP unit in place of the conventional tertiary amine has also proved to be very effective. Selection of appropriate nucleophile, e.g. γ-nitroketone (Scheme 4.4g) [216] or γ,δ-unsaturated-β-ketoesters (Scheme 4.4h) [86], allows for subsequent intramolecular Henry or Michael reaction forming cyclohexane ring with four or three contiguous stereocenters. Takemoto used the double Michael reaction (Scheme 4.4h) for asymmetric synthesis of (−)-epibatidine [86]. The conjugate addition of diphenylbarbituric acid to β-nitrostyrene was carried out with very low loading of unique dithiosquaramide **13** (Scheme 4.4i) [138, 139]. Squaramides **11f** and **12c** allow for very efficient enantioselective addition of phenylpyrazoline (Scheme 4.4j) [216, 217]. In another case, Michael addition followed by elimination of trifluoroacetate provides a product of formal addition of fluoroacetophenone (Scheme 4.4k) [218]. In summary, reactions of easily deprotonated nucleophiles like 1,3-dicarbonyl compounds with β-nitrostyrenes are performed most often using bifunctional catalysts containing a tertiary amine and hydrogen-bond donors.

4.3.3 Michael Additions of Enolizable Ketones and Aldehydes to Nitrostyrene

Enantioselective Michael addition of enolizable carbonyl compounds to nitroalkenes is one of the most general ways to synthesize enantiomerically enriched γ-nitrocarbonyl compounds. First example of acetone addition to β-nitrostyrenes catalyzed by DACH derivatives was published in 2005 by Tsogoeva group (Scheme 4.5) [116]. The reaction was carried out in the presence of catalyst

10e bearing both a thiourea moiety and imine with imidazole group on a chiral scaffold. A year later Tsogoeva and coworkers [99] and Jacobsen and coworkers [51] reported a successful application of primary amine-thioureas **9a** and **9d** in this transformation. Importantly, these are also the first examples of the use of primary amine-thioureas in organocatalysis. The reaction with **9d** is faster and more efficient than with **10e**, and is additionally applicable to acetophenone as a nucleophile. Shortly thereafter Jacobsen published an enantioselective synthesis of a wide range of γ-nitroketones catalyzed by **9a** [51]. Authors proposed the reaction mechanism involving formation of enamine, with nitrostyrene activation by hydrogen bonding to nitro group [51, 99, 219]. This pioneering work has led to the development of other primary amine thiourea organocatalysts (**6e**, **6g**, **9k**, **9j**, **20c**) that effectively promoted the reaction of acetone (or acetophenone) with nitrostyrenes, as shown in Scheme 4.5. Among them is saccharide-thiourea **9k** [107, 220], DACH and 1,2-diphenylethylenediamine-based multiple hydrogen-bonding donor **9j** [106], as well as simpler thioureas **6e** and **6g** [60, 62]. Finally, secondary amine-thiourea catalyst **20c** derived from L-proline was used [221]; however, in this case and for **6g**, the enantioselectivity was lower (75–81% ee).

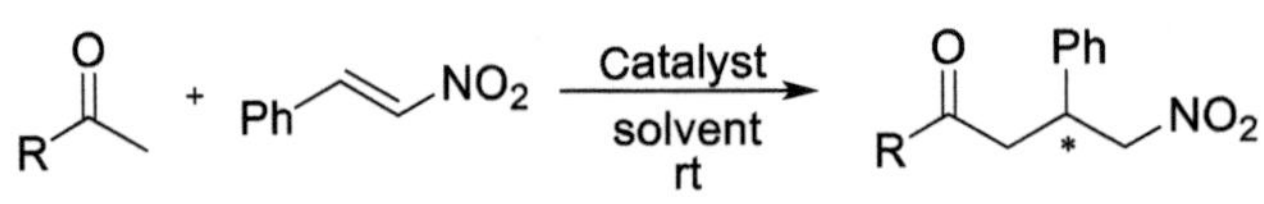

Catalyst (mol%)	R	Solvent	Time	Yield	ee (config.)	Ref.
(*S*,*S*)-**10e** (15)	Me	Toluene	40 h	55%	87% (*R*)	[116]
(*S*,*S*)-**9d**·AcOH (15)	Me	Toluene, 2 equiv H_2O	16 h	85.5%	86% (*R*)	[99]
(*R*,*R*)-**9a**·$PhCO_2H$ (10)	Me	Toluene	24 h	93%	99% (*S*)	[51]
(*S*,*S*)-**9d**·4-NO_2-BzOH (15)	Ph	THF	72 h	95%	99% (*R*)	[219]
(*R*,*R*)-**9k**·AcOH (5)	Me	DCM	60 h	91%	92% (*S*)	[220]
(*R*,*R*)-**9k** (15)	Ph	DCM	96 h	72%	97% (*S*)	[107]
(*R*,*R*)-**6g**·4-Me-BzOH (15)	Me	1,4-Dioxane	24 h	86%	75% (*S*)	[62]
(*R*,*R*)-**6e**·AcOH (10)	Me	Toluene	24 h	72%	93% (*R*)	[60]
(*R*,*R*)-**9j**·AcOH (15)	Ph	DCM	96 h	72%	>99% (*S*)	[106]
(*R*,*R*)-**9j**·AcOH(15)	Me	$CHCl_3$	72 h	92%	96% (*S*)	[106]
(*R*,*R*)-**20c** (20)	Me	Et_2O	48 h	96%	81% (*S*)	[221]

Scheme 4.5 Michael additions of ketones to nitrostyrene. Source: Based on Refs. [51, 60, 62, 99, 106, 107, 116, 219–221].

DACH derivatives have also been tested as catalysts in the asymmetric reactions of nitroolefins with α-branched aldehydes, especially isobutyraldehyde. Readily available thiourea catalyst **6b** is highly effective in this conjugate addition [54]. Organocatalysts based on *trans*-cyclohexane-1,2-diamine with thiourea moiety and isosteviol-derived **9m** or rosin-derived **9l** scaffold have been most efficient for the asymmetric conjugate additions of this aldehyde to β-nitrostyrenes [109, 222]. Primary amine-guanidine **14a** was employed in organocatalytic reaction in aqueous conditions with enantioselectivity up to 80% [223]. A chiral primary amine-salicylamide **4d** was used as an organocatalyst in the reaction performed in deep eutectic solvent (choline chloride and water) [224]. It was also demonstrated that 1,2-cyclohexanediamine derivatives bearing calix[4]arene thiourea moiety (**19f** and **19g**) were effective organocatalysts in this type of reaction, where calixarene skeleton is responsible for stabilization of the complex between β-nitrostyrene and calixarene by van der Waals and C–H···π interactions [225, 226].

4.3.4 Michael Additions of Other Carbon Nucleophiles to Nitrostyrene

Utility of organocatalysis in addition to nitrostyrenes is not limited to 1,3-dicarbonyls and enolizable ketones and aldehydes; other types of nucleophiles can be also used (Scheme 4.6). The first example is an asymmetric alkylation of *ortho*-hydroxybenzophenone imine of glycine alkyl ester with β-nitrostyrene in the presence of Takemoto catalyst (Scheme 4.6a) [227]. The same organocatalyst was used in a cascade double Michael addition (Scheme 4.6b) that resulted in formation of a highly functionalized polycyclic product with five adjacent stereocenters [228]. Takemoto-type thiourea (**7m**) also catalyzed an addition of β-aryl-α-ketophosphonates to nitroalkenes, and the obtained adducts were converted *in situ* to the corresponding α,β-disubstituted γ-nitroamides (Scheme 4.6c) [73].

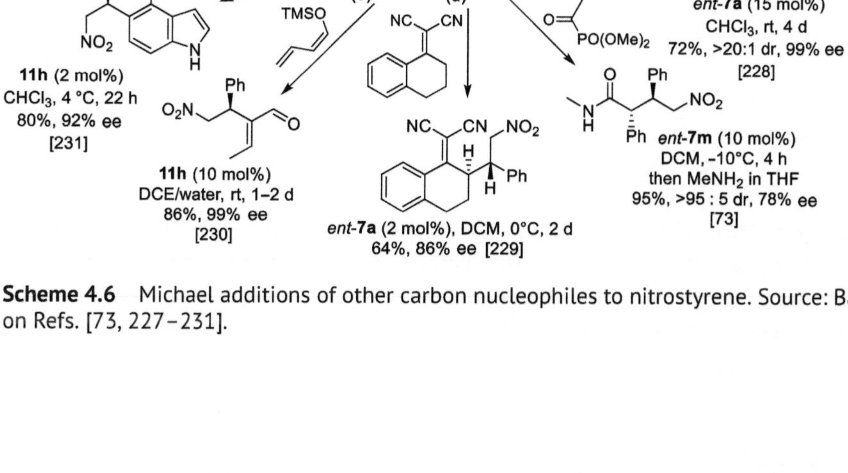

Scheme 4.6 Michael additions of other carbon nucleophiles to nitrostyrene. Source: Based on Refs. [73, 227–231].

A vinylogous Michael addition with selected unsaturated derivatives of dimalononitrile is feasible in the presence of **7a** (Scheme 4.6d) [229]. Analogous squaramide catalyst **11h** can effectively promote Mukaiyama reaction with 1-silyloxybutadiene (Scheme 4.6e) [230] as well as Friedel–Crafts reaction with hydroxyindoles (Scheme 4.6f) [231].

4.3.5 Michael Additions of Heteronucleophiles to Nitrostyrene

Reactions of heteroatom-centered nucleophiles with Michael acceptors also lead to various important groups of chiral products, including heterocycles. Thioacetate can be used as a thiol equivalent, exploiting ability of urea **10n** to complex acetate and similar anions (Scheme 4.7a) [125]. Another sulfur nucleophile successfully used is methyl thioglycolate in the presence of **7a** (Scheme 4.7b) [232]. In case of diphenyl phosphite as a nucleophile reaction can be performed using **11h** or catalyst **18c** without a basic center (Scheme 4.7c) [24, 233]. Finally, oxa- and aza-Michael–Henry cascade reactions of salicylaldimines and 2-aminobenzaldehydes with nitroolefins and thiourea catalysts (**19b, 9d**) resulted in formation of chiral heterocycles – chromanes and 1,2-dihydroquinolines, respectively (Scheme 4.7d,e) [187, 234].

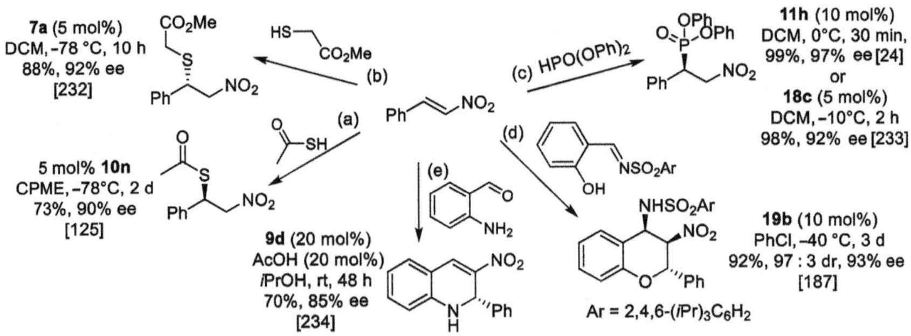

Scheme 4.7 Michael additions of heteroatom-centered nucleophiles to nitrostyrene. CPME, cyclopentyl methyl ether. Source: Based on Refs. [24, 125, 187, 232–234].

4.3.6 Michael Reactions of Enones

Simple ketones are important and extensively studied Michael acceptors for various organocatalytic transformations. DACH derivatives are quite frequently used as organocatalysts for enone activation; in particular covalent catalysis with primary amine (e.g. of types **6** and **9**) has generally found more applications. Noncovalent organocatalysis with tertiary amines is mainly used with aryl enones (e.g. chalcones) as shown in Scheme 4.8a,b [235, 236]. The use of protected hydroxyl malononitriles as nucleophiles and **11f** allows to introduce masked acyl cyanide equivalent in enantioselective addition to the chalcone (Scheme 4.8a) [235]. The reaction of a similar enone with a simple γ-nitro ketone in the presence of thiourea–tertiary amine **7c** promoted an enantioselective cascade process providing a highly functionalized cyclohexane skeleton (Scheme 4.8b) [236]. Another interesting example

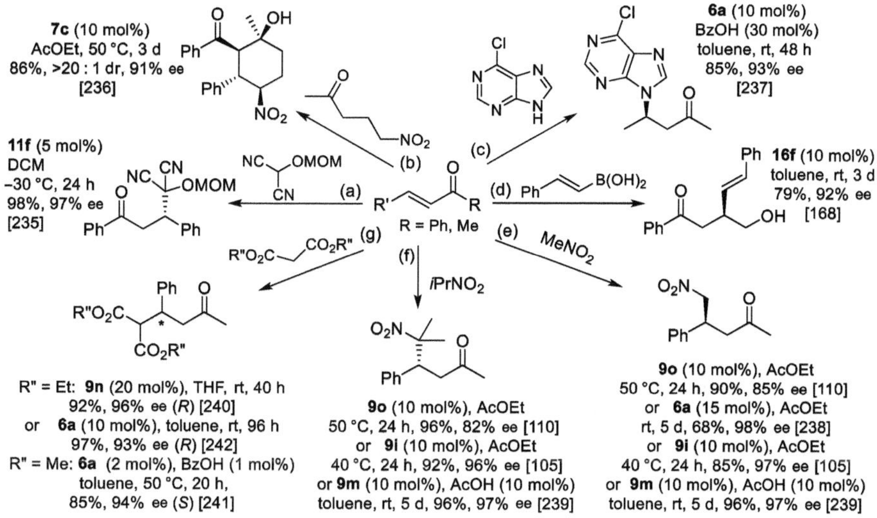

Scheme 4.8 Michael reactions of enones. Source: Based on Refs. [105, 110, 168, 235–242].

is an asymmetric reaction of alkenylboronic acids with specific γ-hydroxy enones, catalyzed by Schiff base thiourea **16f** (Scheme 4.8d) [168]. The remaining examples in Scheme 4.8 show the reactions of enones containing predominantly a methyl (or other alkyl) group on the carbonyl. Their transformations are mainly catalyzed by DACH derivatives containing a primary amine group and often acidic cocatalyst. Nitrogen nucleophiles, as exemplified by 6-chloropurine, react with such enones in the presence of **6a** (Scheme 4.8c) [237]. More typical successful application is an addition of nitromethane (Scheme 4.8e) [105, 110, 238, 239] or branched nitroalkanes, like 2-nitropropane (Scheme 4.8f) [105, 110, 239]. One of the first examples utilized catalyst **9o** containing DACH and *Cinchona* alkaloid in its structure [110], although simpler catalysts, e.g. **6a**, work also efficiently [238]. Malonate esters undergo addition as well (Scheme 4.8g) [240–242]. Similarly, the first DACH catalyst employed was **9n**, [240] and later simpler **6a** was used [241, 242]. Nitromethane addition products were applied in the synthesis of pharmacologically relevant compounds, e.g. (*R*)-baclofen [239].

DACH-derived primary amine-thioureas have also been successfully used in reactions involving cyclic enones, especially 2-cyclohexen-1-one, e.g. in the addition of malonates [240–242]. This reaction is very efficient with a small amount (0.5–2% mol) of the simple catalyst **6a** and is applicable in a multigram-scale synthesis [241].

4.3.7 Michael Reactions of Benzylidene Pyruvates

Benzylidene pyruvates constitute a very interesting and special group of Michael acceptors. They are more active electrophiles than typical enones. With such 1,2-dicarbonyl substrates there is a possibility of forming three-centered hydrogen bond between two carbonyl oxygen atoms and proton from the catalyst, allowing a more favorable interaction and a better-defined transition state. In the case of conjugate additions to benzylidene pyruvates, non-covalent activation with thioureas of type **7** is most often used (Scheme 4.9). Reactions of 1,3-dicarbonyls and related systems (Scheme 4.9a–d) constitute an important group here [23, 243–245]. In the addition of dimedone (Scheme 4.9b) it has been shown that simple tertiary amine-sulfonamide catalyst **5d** was also very effective in this type of reaction [23]. Addition of nitromethane to benzylidene pyruvate was accompanied by the Henry reaction, but the use of *tert*-butyl nitroacetate provided a Michael product (Scheme 4.9c) [244]. A Friedel–Crafts-type reaction with 1-naphthole and squaramide **12c** yielded chromane derivatives (Scheme 4.9e) [246]. The presence of a good carbonyl electrophile in the substrate – namely, α-keto ester group – often resulted in subsequent reactions: aldol (Scheme 4.9d,g) [245, 247] or acetalizations (Scheme 4.9a,b,f) [23, 243, 248]. Moreover, the reaction of β,γ-unsaturated amides in the presence of **7b** lead to dihydropyrans with three adjacent stereocenters (Scheme 4.9h) [249].

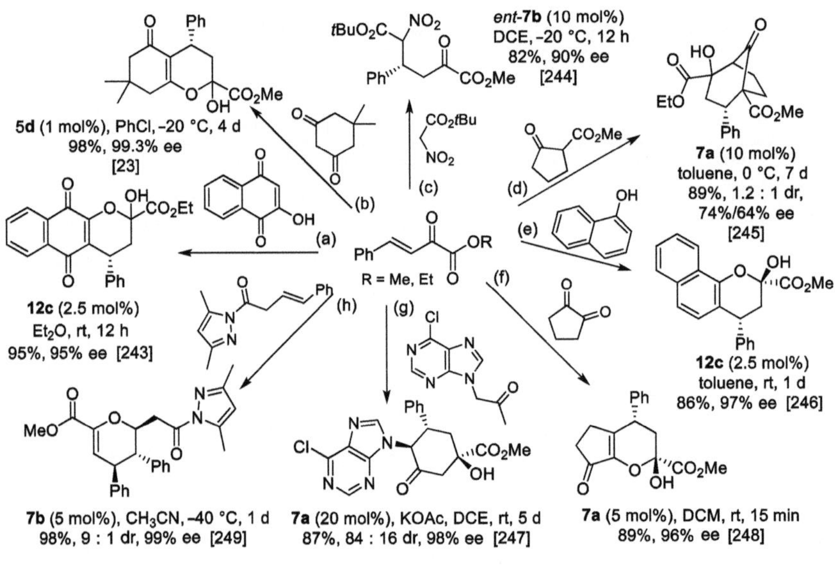

Scheme 4.9 Michael reactions of benzylidene pyruvates. Source: Based on Refs. [23, 243–249].

4.3.8 Reactions of Isatin-Derived Michael Acceptors

Another way of increasing electrophilicity of enone, or of any other conjugated electrophile, is possible by adding an electron-withdrawing group in β-position. α,β-Unsaturated derivatives of isatins are a good example of such compounds (Scheme 4.10). These electrophiles can react in vinylogous Michael reaction (Scheme 4.10a) [250], Mukaiyama–Michael reaction with a unique secondary amine phosphoramide organocatalyst **21b** (Scheme 4.10b) [199], and undergo formal [3+2] cycloaddition in the presence of **7a** (Scheme 4.10c) [251]. The domino Michael/Michael reaction with Takemoto catalyst provides a six-membered cyclic

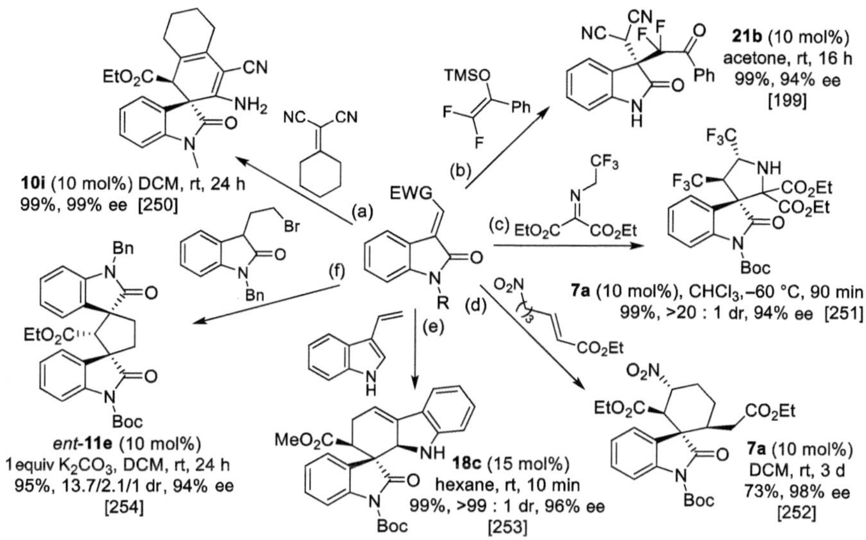

Scheme 4.10 Reactions of isatin-derived Michael acceptors. Source: Based on Refs. [199, 250–254].

product with four consecutive stereogenic centers (Scheme 4.10d) [252]. 3-Vinylindoles undergo very fast Diels–Alder reaction with isatin derivatives in the presence of bis-thiourea **18c** (Scheme 4.10e) [253]. Finally, cascade Michael-alkylation sequence with 2-oxoindole was performed in the presence of **11e** and a stoichiometric amount of K_2CO_3 (Scheme 4.10f) [254].

4.3.9 Michael Reactions of Maleimides

α,β-Unsaturated amides are weaker electrophiles than simple enones, but maleimides can undergo Michael additions relatively easily (Scheme 4.11). An example of this can be the reaction of 1,3-dicarbonyl compounds, e.g. acetylacetone, with C_2-symmetric bis(2-aminobenzimidazole) **18g** (Scheme 4.11a) [255]. α-Angelica lactone undergoes an addition catalyzed by **10c** to form a butenolide

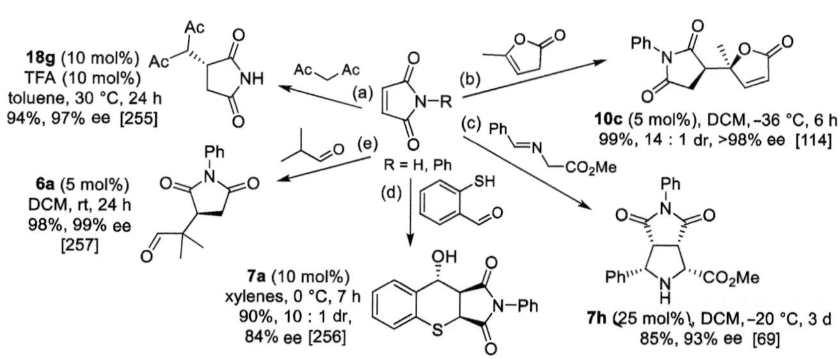

Scheme 4.11 Michael reactions of maleimides. Source: Based on Refs. [69, 114, 255–257].

derivative with a quaternary stereogenic center (Scheme 4.11b) [114]. Glycine aldimino esters react in the presence of **7h** to form pyrrolidines (Scheme 4.11c) [69]. The use of 2-mercaptobenzaldehydes and Takemoto catalyst provides benzothiopyrans (Scheme 4.11d) [256]. Finally, reaction of isobutyraldehyde is effectively catalyzed by primary amine-thiourea **6a** (Scheme 4.11e) [257].

4.3.10 Aldol Reaction and Other Asymmetric Reactions of Aldehydes

The aldol reaction is one of the most important and versatile transformations for the construction of carbon–carbon bonds from readily available carbonyl compounds. This reaction has been extensively studied with a variety of chiral amines, including diaminocyclohexane derivatives as catalysts. In the reaction of aromatic aldehydes with cyclohexanone, it is possible to use directly *trans*-DACH **1** with adipic acid as a cocatalyst (Scheme 4.12) [13]. The major *anti*-product is formed with a high enantioselectivity (93% ee). The use of various *N*1,*N*1-dialkyl-cyclohexane-1,2-diamines: **3e, 3f, 3i, 3j, 3l, 3p, 3q** [26–28, 33, 34, 37, 41] and bifunctional L-prolinamide-DACH catalyst **20a** [194] resulted in a very high *anti* selectivity (≥9 : 1) and also higher

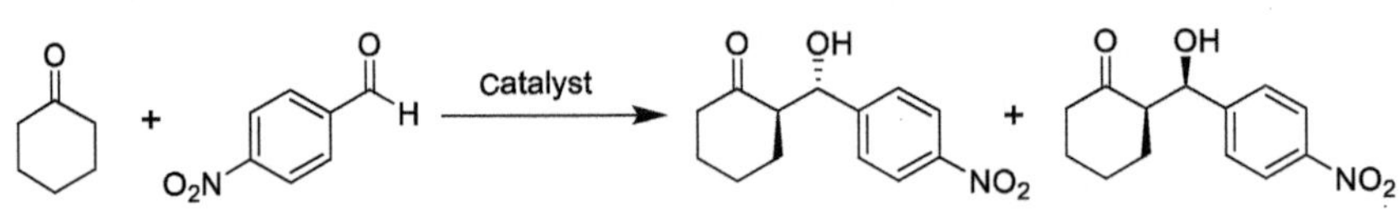

Catalyst (mol%)	Acid (mol%)	Solvent	Temp.	Time	Yield	dr *anti*/*syn*	ee	Ref.
1 (20)	Adipic acid (20)	$MeOH/H_2O$	rt	48 h	75%	80 : 20	93%	[13]
3e (10)	TfOH (10)/ *m*-NO_2BzOH (10)	CH_2Cl_2	rt	12 h	99%	90 : 10	98%[a]	[26]
3e (10)	Succinic acid (10)	Neat	rt	24 h	94%	25 : 75	84% (*syn*)	[40]
3f (5)	TfOH (3.5)	H_2O	rt	2,5 d	95%	95 : 5	95%	[37]
3i (10)	TFA (20)	EtOH	rt	24 h	95%	98 : 2	>99%[b]	[27]
3j (15)	TFA (30)	EtOH	15 °C	48 h	84%	94 : 6	99%	[28]
3l (5)	—	Neat	20 °C	3 h	92%	98 : 2	99%	[41]
3p (10)	TFA (30)	H_2O	rt	24 h	85%	97 : 3[c]	97%[b,c]	[33]
3q (3.5)	TfOH (3.5)	H_2O	rt	10 h	96%	98 : 2	98%[b]	[34]
20a (5)	3-MeBzOH (5)	CH_3CN	−20 °C	30 h	95%	93 : 7	96%[b]	[194]

a) Absolute configuration not confirmed.
b) Opposite absolute configuration.
c) Benzaldehyde was used.

Scheme 4.12 Reaction of 4-nitrobenzaldehyde with cyclohexanone. Source: Based on Refs. [13, 26–28, 33, 34, 37, 40, 41, 194].

enantiomeric excesses (≥95%) (Scheme 4.12). It is also possible to obtain predominantly *syn*-aldol product, e.g. using **3e** with the addition of succinic acid and without solvent [40].

Scheme 4.13a–c shows selected examples of aldol reactions between aromatic aldehyde and other ketones: acetone, [26] hydroxyacetone, [197] tetrahydro-4*H*-pyran-4-one [195] with organocatalysts derived from *trans*-DACH (e.g. **3e**, **20b**, **20d**). C_2-Symmetric bis-thiourea **18c** in combination with a Lewis base (DMAP) was applied by Nagasawa in the enantioselective Morita–Baylis–Hillman reaction (Scheme 4.13d) [178, 258]. Takemoto catalyst **7a** has been used in a synthesis of chiral oxazolidinones, formed in aldol reaction with 2-isocyanatomalonate followed by ring closing to urethane (Scheme 4.13e) [259]. Hydrolysis of this product can afford β-hydroxy-α-amino acids used in total synthesis of mycestericin C [259] and caprazamycin A [260]. Trifunctional squaramide **12e** (2 mol%) promotes the Henry reaction with nitromethane (Scheme 4.13f) [137]. Organocatalyst **12e** tolerates one-pot oxidation of alcohols with MnO_2 to aldehydes and allows for a subsequent Henry reaction [261]. Squaramide **11h** (Scheme 4.13g) catalyzes hydrophosphonylation of aldehydes under mild reaction conditions rendering α-hydroxy phosphonates [262]. A last example is the allylation of aldehydes using allyltrichlorosilane and C_2-symmetric Lewis bases: bis-imidazole *N*-oxide **18h** [185] and bis-formamides **18b** [177], leading to corresponding homoallylic alcohols with moderate enantioselectivity (Scheme 4.13h).

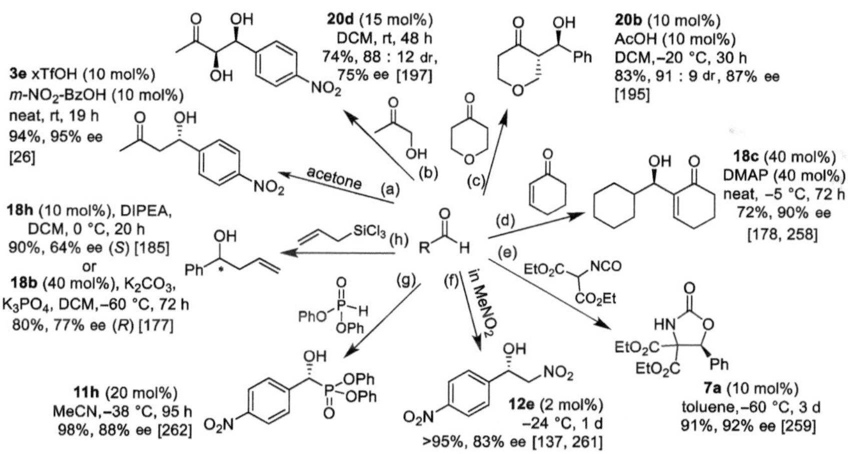

Scheme 4.13 Asymmetric reactions of aldehydes. DIPEA, diisopropylethylamine. Source: Based on Refs. [26, 137, 177, 178, 185, 195, 197, 258, 259, 261, 262].

4.3.11 Reactions of Isatins

An important group of reactions are the additions of various nucleophiles to isatins, a unique group of reactive and rigid cyclic ketones, allowing for well-defined transition states. Enolizable ketones can react with isatin under covalent organocatalytic conditions, e.g. for cyclohexanone depending on the catalyst (**1**/adipic acid or

3p/TFA) and solvent used; the diastereoselectivity of the reaction can be fully controlled (Scheme 4.14a) [263, 264]. Noncovalent organocatalyst **12c** allows for use of β-ketoacids as nucleophiles with a subsequent decarboxylation resulting in formal addition of acetophenone (Scheme 4.14b) [136]. Another interesting example is the vinylogous aldol–cyclization cascade reaction of allyl pyrazoleamides with isatin and 1 mol% of Takemoto catalyst leading to spirocyclic dihydropyranones (Scheme 4.14c) [265]. Finally, urea **8b** catalyzes enantioselective hydroxyalkylation of 4-hydroxyindole with isatins (Scheme 4.14d) [266].

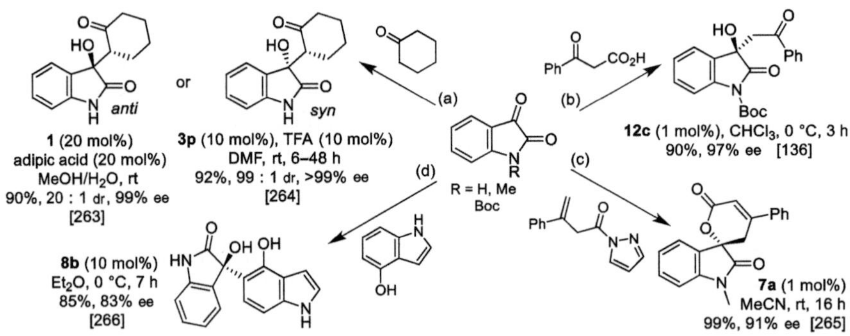

Scheme 4.14 Enantioselective reactions of isatins. Source: Based on Refs. [133, 263–266].

4.3.12 Reactions of Imines

Imines are very important compounds used as building blocks for the enantioselective synthesis of diverse amines and nitrogen heterocycles. At the beginning of this section, key examples of application of imines in the Strecker reaction have been presented (Scheme 4.2a,b,e,f). Scheme 4.15 illustrates other types of reactions involving imines and organocatalysts based on *trans*-DACH. One of the first examples of such enantioselective transformations is the Mukaiyama–Mannich reaction of silyl ketene acetals with *N*-Boc aldimines and thiourea catalyst **15c** reported by Jacobsen and coworkers (Scheme 4.15a) [95]. This is a very efficient method for asymmetric synthesis of optically active β-amino esters. A few years later, thiourea-urea **15j** (1 mol%) with enhanced catalytic efficiency, because of its unique foldamer structure, was used in this reaction (Scheme 4.15a) [162]. Schiff base **15c** was also applied in the aza-Baylis–Hillman reaction (Scheme 4.15b), with high enantioselectivity and moderate yields [151]. A similar catalyst **15d** was used in the enantioselective hydrophosphonylation of a wide range of *N*-benzyl imines (Scheme 4.15c) [153]. Jacobsen also introduced acetamide catalyst **15k** which has proved to be highly effective and selective in the nitro-Mannich reaction of nitroalkanes with *N*-Boc imines (Scheme 4.15d) [163]. This reaction is also efficiently catalyzed by symmetrical chiral Brønsted acids $\mathbf{18f}_2(TfOH)_3$ – polar ionic hydrogen-bond donors (Scheme 4.15d) [181]. Thiourea **10a** with acetic acid additive catalyzes the addition of 2-nitropropionates with generation of quaternary stereocenter (Scheme 4.15e) [267]. Bifunctional urea **10m**, based on the rigid binaphthyl skeleton, is an effective catalyst in the Mannich reaction of α-cyanoketones [123]

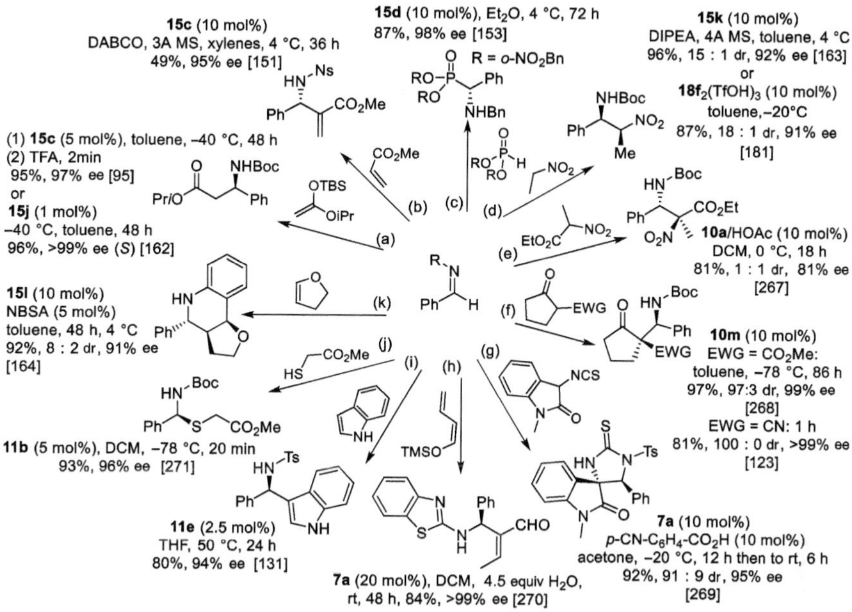

Scheme 4.15 Enantioselective reactions of imines. DABCO, 1,4-diazabicyclo[2.2.2]octane; EWG, electron-withdrawing group. Source: Based on Refs. [95, 123, 131, 151, 153, 162, 164, 181, 267–271].

and β-ketoesters [268] with *N*-Boc-imines (Scheme 4.15f). Takemoto catalyst **7a** was used in the asymmetric cascade Mannich/cyclization reaction of 3-isothiocyanato oxindoles with sulfimides to construct spiro-imidazolidine (Scheme 4.15g) [269], and also in the Mukaiyama-type addition of silyl-dienol ethers to imines to form aza-Baylis–Hillman type products (Scheme 4.15h) [270]. Squaramide **11e** effectively catalyzes the enantioselective Friedel–Crafts reaction of indoles with *N*-tosyl imines (Scheme 4.15i) [131]. A similar catalyst **11b** promotes addition of alkyl thioglycolates to *N*-Boc aldimines (Scheme 4.15j) [271]. Finally, *N*-aryl imines were successfully used in the enantioselective Povarov reaction with electron-rich olefins such as vinyl ethers and lactams catalyzed by bifunctional sulfinamido urea derivative **15l** and *ortho*-nitrobenzenesulfonic acid (NBSA) to afford chiral tetrahydroquinolines (Scheme 4.15k) [164].

4.3.13 Pictet–Spengler-Type Reactions

trans-DACH-derived thioureas have also been successfully used in the Pictet–Spengler-type reactions of tryptamines and aldehydes, to afford chiral tetrahydro-β-carboline, a structural motif found in various natural products. In an enantioselective version of this reaction with imines obtained from aliphatic aldehydes, the cyclization proceeds after generation of *N-acyliminium* ion *in situ* in the presence of thiourea hydrogen-bond donor **15f** (Scheme 4.16a) [154]. This methodology, developed by Jacobsen and coworkers, was used in the enantioselective total synthesis of (+)-yohimbine [156]. Several years later, in an analogous reaction involving

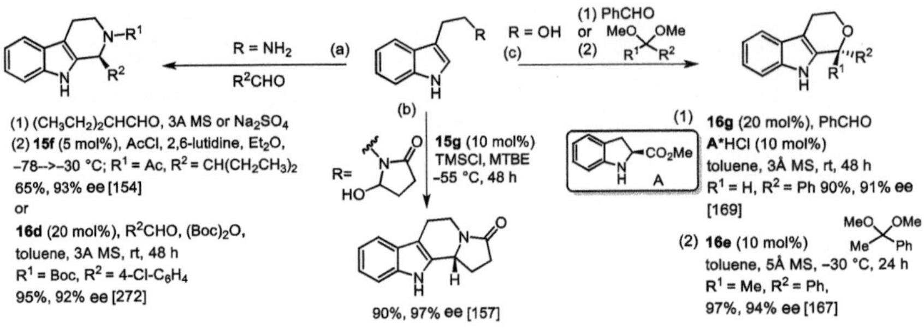

Scheme 4.16 Asymmetric Pictet–Spengler-type reactions. Source: Based on Refs. [154, 157, 167, 169, 272].

aryl imines, Sidel used a thiourea catalyst **16d** containing an additional carboxylic acid group (Scheme 4.16a) [272]. *N*-Acyliminium ion can be also generated from hydroxylactams, via chlorolactam intermediate, and this approach was used in the cyclization with anion-binding catalyst **15g**, allowing for construction of polycyclic indole derivatives (Scheme 4.16b), synthesis of alkaloid (+)-harmicine [157], and the corresponding pyrrole analogs [158]. Bis-thiourea **16g** and amine hydrochloride **A**·HCl accelerate enantioselective oxa-Pictet–Spengler reactions of tryptophol with aldehydes (Scheme 4.16c) [169]. The reaction proceeds through the formation of an oxocarbenium intermediate with anion binding by the catalyst. An analogous reaction with acetophenone dimethyl acetal was effectively catalyzed by carboxylic acid-thiourea **16e** (Scheme 4.16c) [167].

4.3.14 Reactions of Nitrogen Electrophiles

An important method for the synthesis of chiral carbonyl compounds with a nitrogen substituent in the α-position is the reaction using electrophilic azodicarboxylates (Scheme 4.17). A group of these reactions involve the functionalization

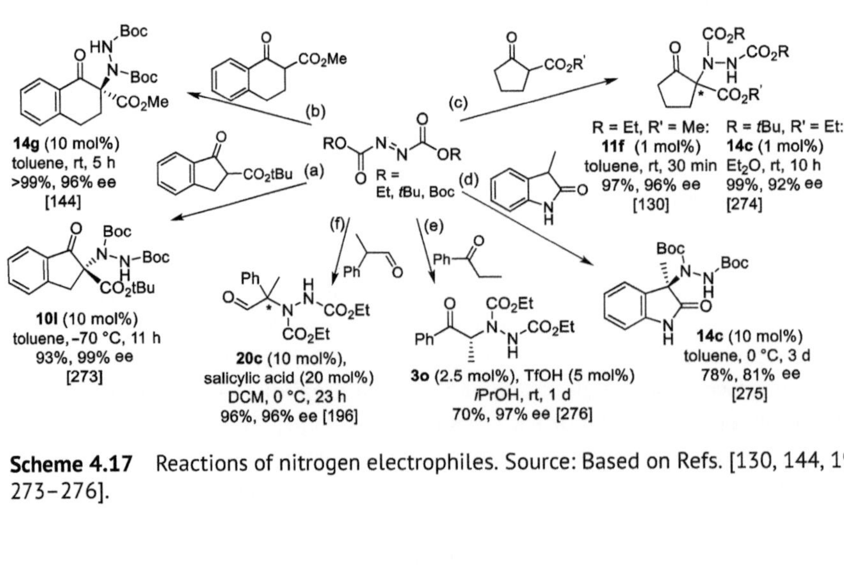

Scheme 4.17 Reactions of nitrogen electrophiles. Source: Based on Refs. [130, 144, 196, 273–276].

of 1,3-dicarbonyl compounds in the presence of tertiary amines – hydrogen-bond donors, e.g. **10l**, **11f**, **14c**, **14g** (Scheme 4.17a–c) [130, 144, 273, 274], although the reaction of oxindole is also possible with a similar catalytic system (Scheme 4.17d) [275]. Use of covalent catalysts (**3o** and **20c**) allows for enantioselective reaction of azodicarboxylates with typical enolizable ketones and aldehydes (Scheme 4.17e,f) [196, 276].

4.3.15 Miscellaneous Reactions

In the last part, we present selected miscellaneous reactions catalyzed by *trans*-DACH-based organocatalysts (Schemes 4.18 and 4.19). Among them is asymmetric dearomative functionalization of quinolines (Scheme 4.18a,b) [67, 204] and isoquinolines [155]. Helical chiral oligotriazole **21g** has proved to be an effective catalyst in the acyl-Mannich reactions of quinolines with silyl ketene acetals (Scheme 4.18a) [204]. Several years earlier, Jacobsen demonstrated that an analogous reaction with isoquinolines could be efficiently promoted by thiourea **15f** and provided 1-substituted dihydroisoquinolines [155]. Another early example is the Petasis-like reaction between quinoline, PhOCOCl, and styrylboronic acid catalyzed by a modified Takemoto thiourea **7f** (Scheme 4.18b) [67]. *trans*-DACH **3o** containing carbazole moiety was used in conjugate additions of benzodioxole-derived radicals

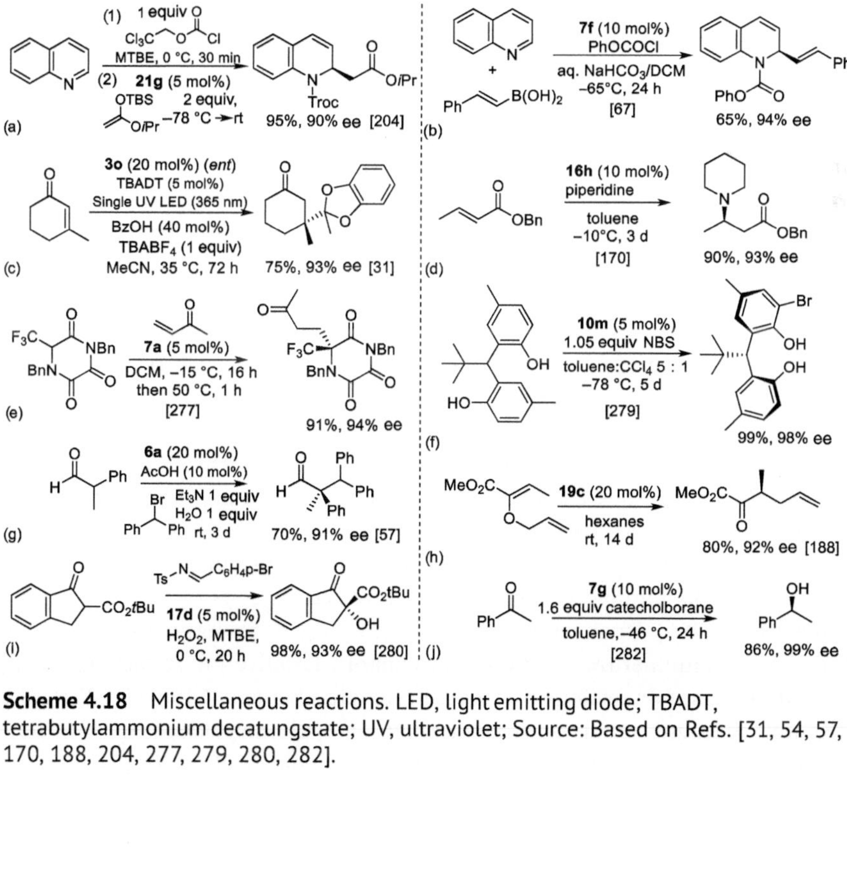

Scheme 4.18 Miscellaneous reactions. LED, light emitting diode; TBADT, tetrabutylammonium decatungstate; UV, ultraviolet; Source: Based on Refs. [31, 54, 57, 67, 170, 188, 204, 277, 279, 280, 282].

to β,β-disubstituted cyclic enones (Scheme 4.18c) [31]. This example shows the successful combination of iminium organocatalysis with photoredox catalysis. Unprecedented selenourea-thiourea **16h** bearing multiple hydrogen-bonding donors found an application in enantioselective conjugate additions of cyclic amines to unactivated α,β-unsaturated esters (Scheme 4.18d) [170]. The conjugate additions shown earlier in this chapter concerned the use of prochiral Michael acceptor. Scheme 4.18e shows an example of stereoconvergent Michael reaction of *rac*-α-CF_3-triketopiperazine and methyl vinyl ketone in the presence of **7a** [277]. A similar type of conjugate addition is the reaction of α-nitrocyclohexanone and acrolein catalyzed by **7c** [278]. An interesting example is the enantioselective desymmetrization of bisphenols by *ortho*-bromination in the presence of tertiary amine-urea of type **10m** (Scheme 4.18f) [279]. A simple primary amine-thiourea **6a** has been used successfully as a catalyst for α-alkylation of α-arylpropionaldehydes with diarylbromomethane via an S_N1 pathway (Scheme 4.18g) [57]. Chiral C_2-symmetric bis-DACH-guanidinium ion **19c** with the noncoordinating counterion was applied as the effective catalyst for the [3,3]-sigmatropic (Claisen) rearrangement of substituted allyl vinyl ethers (Scheme 4.18h) [188]. A chiral quaternary ammonium salt **17e** was used in hydroxylation of β-ketoester by oxaziridine formed *in situ* (Scheme 4.18i) [174, 280]. The same nucleophile was subjected to asymmetric fluorination and chlorination by appropriate mild electrophilic halogenation agents in the presence of DACH-bifunctional ammonium salts [171, 281]. An interesting and unusual example is the secondary amine-thiourea catalyst **7g** that has been successfully used for the enantioselective reduction of ketones with catecholborane (Scheme 4.18j) [282].

Finally, some representative examples of cyclization and cycloaddition reactions catalyzed by DACH derivatives are presented in Scheme 4.19. One possibility is the intramolecular variant of the Michael reaction, which is less common than the intermolecular version discussed extensively earlier in this chapter. The intramolecular Michael addition catalyzed by **7a** was applied, e.g. in enantioselective synthesis of 2,3-disubstituted *trans*-2,3-dihydrobenzofurans (Scheme 4.19a) [283]. Chiral five- and six-membered lactones can be obtained by tertiary aminourea-catalyzed (**8e**) iodolactonization, e.g. of 5-hexenoic acids (Scheme 4.19b) [94]. Jacobsen and coworkers also developed an effective iodolactonization of alkenyl trichloroacetimidates [284]. Chiral pyrazolines were synthesized in an analogous manner with **7b** using β,γ-unsaturated hydrazones as starting materials [285]. Another interesting example is a construction of 4-piperidinone ring in formal aza-Diels–Alder reactions of enones with cyclic imines catalyzed by primary amine-thiourea **9c** (Scheme 4.19c) [97]. Organocatalyst with carboxylic acid group and hydrogen-bond donors **16d** was exploited effectively in Povarov reaction (intramolecular aza-Diels–Alder reaction) providing tetrahydroquinoline with three contiguous stereogenic centers (Scheme 4.19d) [286]. Dual catalytic system consisting of primary amine-thiourea **6c** and achiral Schreiner thiourea promotes enantioselective intramolecular [5+2] cycloadditions based on oxidopyrylium intermediates (Scheme 4.19e) [30, 287]. The last example is enantioselective intramolecular [2+2] photocycloaddition reactions in the presence of C_2-symmetric bisthiourea **18c** (Scheme 4.19f) [288].

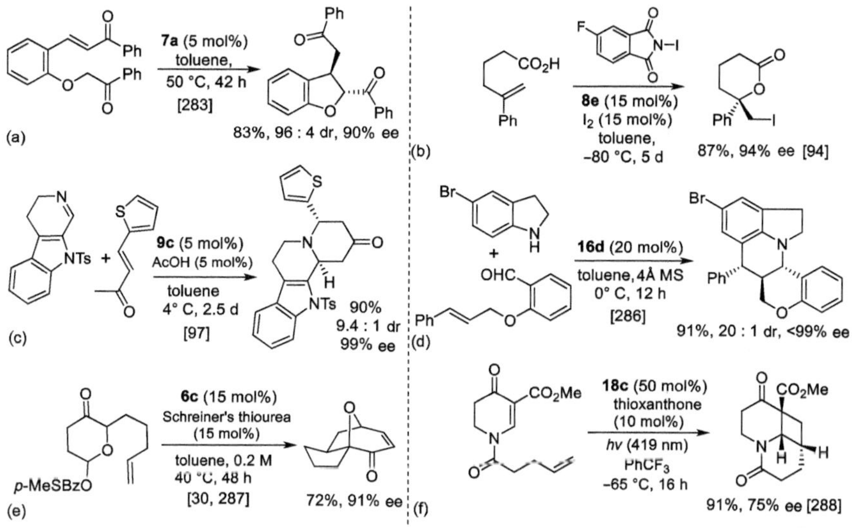

Scheme 4.19 Miscellaneous cyclizations and cycloadditions. Source: Based on Refs. [30, 94, 97, 283, 286–288].

4.4 Conclusions

Trans-DACH is one of the most important chiral building blocks that had a huge impact on development of asymmetric catalysis, including organocatalysis. This easily available chiral scaffold can be found in many effective organocatalysts. Due to the nature of nitrogen atoms in *trans*-DACH, they can be divided into a few general groups: diamines with one primary amine function (e.g. **3**), primary- and tertiary-monoamines containing hydrogen-bond donors (e.g. most of catalysts **4-14**), and exclusively hydrogen-bond donors without amine properties (e.g. **15** and **16**). Synthesis of a large group of these organocatalysts is usually relatively simple. One of the most frequently used compounds of this type is the Takemoto catalyst (**7a**) and its modifications. The DACH derivatives presented in Figures 4.3–4.14 have found application in many types of asymmetric transformations over the past two decades, among which functionalization of Michael acceptor, imines and simple aldehydes, and ketones dominates in the literature.

References

1 MacMillan, D.W.C. (2008). The advent and development of organocatalysis. *Nature* 455 (7211): 304–308. https://doi.org/10.1038/nature07367.

2 Dalko, P.I. (ed.) (2013). *Comprehensive Enantioselective Organocatalysis: Catalysts, Reactions, and Applications*. Weinheim: Wiley-VCH.

3 Wieland, H., Schlichting, O., and von Langsdorff, W. (1926). Untersuchungen über die Gallensäuren. XXV. Mitteilung. Studien zum Abbau von Dicarbonsäuren. *Hoppe-Seyler's Z. Physiol. Chem.* 161 (1–3): 74–79. https://doi.org/10.1515/bchm2.1926.161.1-3.74.

4 Galsbøl, F., Steenbøl, P., Sørensen, B.S. et al. (1972). The preparation, separation, and characterization of the lel3- and ob3-isomers of tris(trans-1,2-cyclohexanediamine)rhodium(III) complexes. *Acta Chem. Scand.* (26): 3605–3611. https://doi.org/10.3891/acta.chem.scand.26-3605.

5 Whitney, T.A. (1980). Direct optical resolution of trans-1,2-diamino cyclohexane from an amine mixture. *J. Org. Chem.* 45 (21): 4214–4216. https://pubs.acs.org/doi/10.1021/jo01309a029.

6 Larrow, J.F. and Jacobsen, E.N. (1998). (*R*,*R*)-*N*,*N*′-Bis(3,5-di-*tert*-butylsalicylidene)-1,2-cyclohexanediamino manganese(III) chloride, a highly enantioselective epoxidation catalyst. *Org. Synth.* (75): 1. https://doi.org/10.15227/orgsyn.075.0001.

7 Bennani, Y.L. and Hanessian, S. (1997). *trans*-1,2-Diaminocyclohexane derivatives as chiral reagents, scaffolds, and ligands for catalysis: applications in asymmetric synthesis and molecular recognition. *Chem. Rev.* 97 (8): 3161–3196. https://pubs.acs.org/doi/10.1021/cr9407577.

8 Masui, M., Ando, A., and Shioiri, T. (1988) New methods and reagents in organic synthesis. 75. Asymmetric synthesis of α-hydroxy ketones using chiral phase transfer catalysts. *Tetrahedron Lett.*, 29 (23), 2835–2838. https://doi.org/10.1016/0040-4039(88)85224-9

9 Sigman, M.S. and Jacobsen, E.N. (1998). Schiff base catalysts for the asymmetric Strecker reaction identified and optimized from parallel synthetic libraries. *J. Am. Chem. Soc.* 120 (19): 4901–4902. https://pubs.acs.org/doi/10.1021/ja980139y.

10 Okino, T., Hoashi, Y., and Takemoto, Y. (2003). Enantioselective Michael reaction of malonates to nitroolefins catalyzed by bifunctional organocatalysts. *J. Am. Chem. Soc.* 125 (42): 12672–12673. https://pubs.acs.org/doi/10.1021/ja036972z.

11 Wang, J., Qi, C., Ge, Z. et al. (2010). Efficient direct asymmetric vinylogous Michael addition reactions of γ-butenolides to chalcones catalyzed by vicinal primary-diamine salts. *Chem. Commun.* 46 (12): 2124–2126. https://doi.org/10.1039/B923925A.

12 Inokoishi, Y., Sasakura, N., Nakano, K. et al. (2010). A new powerful strategy for the organocatalytic asymmetric construction of a quaternary carbon stereogenic center. *Org. Lett.* 12 (7): 1616–1619. https://pubs.acs.org/doi/10.1021/ol100350w.

13 Liu, Y., Wang, J., Sun, Q., and Li, R. (2011) Chiral 1,2-diaminocyclohexane as organocatalyst for enantioselective aldol reaction. *Tetrahedron Lett.*, 52 (28): 3584–3587. https://doi.org/10.1016/j.tetlet.2011.04.116

14 Wu, S., Wang, L., Tang, J. et al. (2013). Direct asymmetric aldol reaction of cyclohexanone with aldehydes catalyzed by chiral *trans*-cyclohexanediamine l-tartrate salt. *Monatsh. Chem.* 144 (3): 381–386. https://doi.org/10.1007/s00706-012-0821-6.

15 Mondal, A. and Bhowmick, K.C. (2019). Asymmetric direct aldol reaction catalyzed by (1*R*, 2*R*)-(+)-1, 2-diammonium cyclohexane-L-tartrate in water. *Curr. Organocatalysis* 6 (2): 165–170. https://doi.org/10.2174/2213337206666181227151140.

16 Lagriffoule, P., Eriksson, M., Jensen, K.K. et al. (1997). Peptide nucleic acids with a conformationally constrained chiral cyclohexyl-derived backbone. *Chem. Eur. J.* 3 (6): 912–919. https://doi.org/10.1002/chem.19970030613.

17 Kulkarni, C., Berrocal, J.A., Lutz, M. et al. (2019). Directing the solid-state organization of racemates via structural mutation and solution-state assembly processes. *J. Am. Chem. Soc.* 141 (15): 6302–6309. https://pubs.acs.org/doi/10.1021/jacs.9b00452.

18 Kaik, M. and Gawroński, J. (2003). Facile monoprotection of *trans*-1,2-diaminocyclohexane. *Tetrahedron: Asymmetry* 14 (11): 1559–1563. https://doi.org/10.1016/S0957-4166(03)00308-2.

19 Mitchell, J.M. and Finney, N.S. (2000). An efficient method for the preparation of *N,N*-disubstituted 1,2-diamines. *Tetrahedron Lett.* 41 (44): 8431–8434. https://doi.org/10.1016/S0040-4039(00)01501-X.

20 Suez, G., Bloch, V., Nisnevich, G., and Gandelman, M. (2012). Design and development of bioinspired guanine-based organic catalyst for asymmetric catalysis. *Eur. J. Org. Chem.* (11): 2118–2122. https://doi.org/10.1002/ejoc.201200118.

21 Fuentes de Arriba, Á.L., Seisdedos, D.G., Simón, L. et al. (2010). Synthesis of monoacylated derivatives of 1,2-cyclohexanediamine. Evaluation of their catalytic activity in the preparation of Wieland–Miescher ketone. *J. Org. Chem.* 75 (23): 8303–8306. https://pubs.acs.org/doi/10.1021/jo101723v.

22 Berkessel, A., Mukherjee, S., Müller, T.N. et al. (2006). Structural optimization of thiourea-based bifunctional organocatalysts for the highly enantioselective dynamic kinetic resolution of azlactones. *Org. Biomol. Chem.* 4 (23): 4319–4330. https://doi.org/10.1039/B607574F.

23 Dajek, M., Kowalczyk, R., and Boratyński, P.J. (2018). *trans*-1,2-Diaminocyclohexane-based sulfonamides as effective hydrogen-bonding organocatalysts for asymmetric Michael–hemiacetalization reaction. *Catal. Sci. Technol.* 8 (17): 4358–4363. https://doi.org/10.1039/C8CY01199K.

24 Zhu, Y., Malerich, J.P., and Rawal, V.H. (2010). Squaramide-catalyzed enantioselective Michael addition of diphenyl phosphite to nitroalkenes. *Angew. Chem. Int. Ed.* 49 (1): 153–156. https://doi.org/10.1002/anie.200904779.

25 Fuerst, D.E. and Jacobsen, E.N. (2005). Thiourea-catalyzed enantioselective cyanosilylation of ketones. *J. Am. Chem. Soc.* 127 (25): 8964–8965. https://pubs.acs.org/doi/10.1021/ja052511x.

26 Luo, S., Xu, H., Li, J. et al. (2007). A simple primary–tertiary diamine–Brønsted acid catalyst for asymmetric direct aldol reactions of linear aliphatic ketones. *J. Am. Chem. Soc.* 129 (11): 3074–3075. https://pubs.acs.org/doi/10.1021/ja069372j.

27 Li, L., Gou, S., and Liu, F. (2013). Highly stereoselective anti-aldol reactions catalyzed by simple chiral diamines and their unique application in configuration switch of aldol products. *Tetrahedron Lett.* 54 (47): 6358–6362. https://doi.org/10.1016/j.tetlet.2013.09.053.

28 Xu, B., Li, L., and Gou, S. (2013). A chiral primary-tertiary-1,2-diamine as an efficient catalyst in asymmetric aldehyde–ketone or ketone–ketone aldol reactions. *Tetrahedron: Asymmetry* 24 (24): 1556–1561. https://doi.org/10.1016/j.tetasy.2013.09.026.

29 Bow, W.F., Basak, A.K., Jolit, A. et al. (2010). Enamine-iminium ion Nazarov cyclization of α-ketoenones. *Org. Lett.* 12 (3): 440–443. https://pubs.acs.org/doi/10.1021/ol9025765.

30 Burns, N.Z., Witten, M.R., and Jacobsen, E.N. (2011). Dual catalysis in enantioselective oxidopyrylium-based [5+2] cycloadditions. *J. Am. Chem. Soc.* 133 (37): 14578–14581. https://pubs.acs.org/doi/10.1021/ja206997e.

31 Murphy, J.J., Bastida, D., Paria, S. et al. (2016). Asymmetric catalytic formation of quaternary carbons by iminium ion trapping of radicals. *Nature* 532 (7598): 218–222. https://doi.org/10.1038/nature17438.

32 Bahamonde, A., Murphy, J.J., Savarese, M. et al. (2017). Studies on the enantioselective iminium ion trapping of radicals triggered by an electron-relay mechanism. *J. Am. Chem. Soc.* 139 (12): 4559–4567. https://pubs.acs.org/doi/10.1021/jacs.7b01446.

33 Raj, M., Parashari, G.S., and Singh, V.K. (2009). Highly enantioselective organocatalytic *syn-* and *anti-*aldol reactions in aqueous medium. *Adv. Synth. Catal.* 351 (9): 1284–1288. https://doi.org/10.1002/adsc.200900122.

34 Peng, F.-Z., Shao, Z.-H., Pu, X.-W., and Zhang, H.-B. (2008). Highly diastereo- and enantioselective direct aldol reactions promoted by water-compatible organocatalysts bearing central and axial chiral elements. *Adv. Synth. Catal.* 350 (14, 15): 2199–2204. https://doi.org/10.1002/adsc.200800335.

35 Luo, S., Xu, H., Zhang, L. et al. (2008). Highly enantioselective direct *syn-* and *anti-*aldol reactions of dihydroxyacetones catalyzed by chiral primary amine catalysts. *Org. Lett.* 10 (4): 653–656. https://pubs.acs.org/doi/10.1021/ol703023t.

36 Luo, S., Xu, H., Chen, L., and Cheng, J.-P. (2008). Asymmetric direct aldol reactions of pyruvic derivatives. *Org. Lett.* 10 (9): 1775–1778. https://pubs.acs.org/doi/10.1021/ol800471b.

37 Lin, J.-H., Zhang, C.-P., and Xiao, J.-C. (2009). Enantioselective aldol reaction of cyclic ketones with aryl aldehydes catalyzed by a cyclohexanediamine derived salt in the presence of water. *Green Chem.* 11 (11): 1750–1753. https://doi.org/10.1039/B916583E.

38 Luo, S., Qiao, Y., Zhang, L. et al. (2009). Asymmetric direct aldol reactions of acetoacetals catalyzed by a simple chiral primary amine. *J. Org. Chem.* 74 (24): 9521–9523. https://pubs.acs.org/doi/10.1021/jo9021259.

39 Luo, S., Zhou, P., Li, J., and Cheng, J.-P. (2010). Asymmetric retro- and transfer-aldol reactions catalyzed by a simple chiral primary amine. *Chem. Eur. J.* 16 (15): 4457–4461. https://doi.org/10.1002/chem.201000181.

40 Gao, J., Bai, S., Gao, Q. et al. (2011). Acid controlled diastereoselectivity in asymmetric aldol reaction of cycloketones with aldehydes using enamine-based organocatalysts. *Chem. Commun.* 47 (23): 6716–6718. https://doi.org/10.1039/C0CC05224H.

41 Lygo, B., Davison, C., Evans, T. et al. (2011). Highly enantioselective aldol reactions using a tropos dibenz[*c,e*]azepine organocatalyst. *Tetrahedron* 67 (52): 10164–10170. https://doi.org/10.1016/j.tet.2011.09.101.

42 Witten, M.R. and Jacobsen, E.N. (2015). A simple primary amine catalyst for enantioselective α-hydroxylations and α-fluorinations of branched aldehydes. *Org. Lett.* 17 (11): 2772–2775. https://pubs.acs.org/doi/10.1021/acs.orglett.5b01193.

43 McLeod, D., Cherubini-Celli, A., Sivasothirajah, N. et al. (2020). Enantioselective 1,3-dipolar [6+4] cycloaddition of pyrylium ions and fulvenes towards cyclooctanoids. *Chem. Eur. J.* 26 (50): 11417–11422. https://doi.org/10.1002/chem.202001369.

44 Tang, L., Luo, Y., Xue, J.-W. et al. (2017). Highly enantioselective Michael-aldol-dehydration reaction for the synthesis of chiral 3,5-diaryl-cyclohexenones catalyzed by primary amine. *Tetrahedron* 73 (8): 1114–1119. https://doi.org/10.1016/j.tet.2017.01.004.

45 Martínez-Guillén, J., Flores-Ferrándiz, J., Gómez, C. et al. (2018). Asymmetric conjugate addition of α,α-disubstituted aldehydes to nitroalkenes organocatalyzed by chiral monosalicylamides from *trans*-cyclohexane-1,2-diamines. *Molecules* 23 (1): 141. https://doi.org/10.3390/molecules23010141.

46 Torregrosa-Chinillach, A., Moragues, A., Pérez-Furundarena, H. et al. (2018). Enantioselective Michael addition of aldehydes to maleimides organocatalyzed by a chiral primary amine-salicylamide. *Molecules* 23 (12): 3299. https://doi.org/10.3390/molecules23123299.

47 Xue, F., Zhang, S., Duan, W., and Wang, W. (2008). A novel bifunctional sulfonamide primary amine-catalyzed enantioselective conjugate addition of ketones to nitroolefins. *Adv. Synth. Catal.* 350 (14, 15): 2194–2198. https://doi.org/10.1002/adsc.200800445.

48 Rasappan, R. and Reiser, O. (2009). Cyclohexane-1,2-diamines: efficient catalysts for the enantioselective conjugate addition of ketones to nitro olefins. *Eur. J. Org. Chem.* (9): 1305–1308. https://doi.org/10.1002/ejoc.200801263.

49 Romero, J.A., Navarrate, A., Servín, F.A. et al. (2014). Oxygen–chlorine interactions in the transition state of asymmetric Michael additions of carbonyl compounds to β-nitrostyrene. *Tetrahedron: Asymmetry* 25 (13, 14): 997–1001. https://doi.org/10.1016/j.tetasy.2014.05.002.

50 Isik, M. and Tanyeli, C. (2013). Cu-catalyzed selective mono-*N*-pyridylation: direct access to 2-aminoDMAP/sulfonamides as bifunctional organocatalysts. *J. Org. Chem.* 78 (4): 1604–1611. https://pubs.acs.org/doi/10.1021/jo302713b.

51 Huang, H. and Jacobsen, E.N. (2006). Highly enantioselective direct conjugate addition of ketones to nitroalkenes promoted by a chiral primary amine–thiourea catalyst. *J. Am. Chem. Soc.* 128 (22): 7170–7171. https://doi.org/10.1021/ja0620890.

52 Procuranti, B. and Connon, S.J. (2007). A reductase-mimicking thiourea organocatalyst incorporating a covalently bound NADH analogue: efficient 1,2-diketone reduction with in situ prosthetic group generation and recycling. *Chem. Commun.* (14): 1421–1423. https://doi.org/10.1039/B618792G.

53 Mei, K., Zhang, S., He, S. et al. (2008). (*S*)-Pyrrolidine sulfonamide catalyzed asymmetric direct aldol reactions of aryl methyl ketones with aryl aldehydes. *Tetrahedron Lett.* 49 (17): 2681–2684. https://doi.org/10.1016/j.tetlet.2008.02.164.

54 Zhang, X., Liu, S., Lao, J. et al. (2009). Asymmetric conjugate addition of carbonyl compounds to nitroalkenes catalyzed by chiral bifunctional thioureas. *Tetrahedron: Asymmetry* 20 (12): 1451–1458. https://doi.org/10.1016/j.tetasy.2009.06.003.

55 Berkessel, A. and Seelig, B. (2009). A simplified synthesis of Takemoto's catalyst. *Synthesis* (12): 2113–2115. https://doi.org/10.1055/s-0029-1216804.

56 Uehara, H. and Barbas, C. (2009). *anti*-Selective asymmetric Michael reactions of aldehydes and nitroolefins catalyzed by a primary amine/thiourea. *Angew. Chem. Int. Ed.* 48 (52): 9848–9852. https://doi.org/10.1002/anie.200905313.

57 Brown, A.R., Kuo, W.-H., and Jacobsen, E.N. (2010). Enantioselective catalytic α-alkylation of aldehydes via an S_N1 pathway. *J. Am. Chem. Soc.* 132 (27): 9286–9288. https://doi.org/10.1021/ja103618r.

58 Yu, F., Jin, Z., Huang, H., Ye, T., Liang, X., and Ye, J. (2010) A highly efficient asymmetric Michael addition of α,α-disubstituted aldehydes to maleimides catalyzed by primary amine thiourea salt. *Org. Biomol. Chem.*, 8 (20): 4767–4774. https://doi.org/10.1039/C0OB00154F

59 Miura, T., Nishida, S., Masuda, A. et al. (2011). Asymmetric Michael additions of aldehydes to maleimides using a recyclable fluorous thiourea organocatalyst. *Tetrahedron Lett.* 52 (32): 4158–4160. https://doi.org/10.1016/j.tetlet.2011.05.145.

60 Günler, Z., Alfonso, I., Jimeno, C., and Pericàs, M. (2016). Concentration effect in the asymmetric Michael addition of acetone to β-nitrostyrenes catalyzed by primary amine thioureas. *Synthesis* 49 (02): 319–325. https://doi.org/10.1055/s-0036-1589408.

61 Gammack Yamagata, A.D., Datta, S., Jackson, K.E. et al. (2015). Enantioselective desymmetrization of prochiral cyclohexanones by organocatalytic intramolecular Michael additions to α,β-unsaturated esters. *Angew. Chem. Int. Ed.* 54 (16): 4899–4903. https://doi.org/10.1002/anie.201411924.

62 Yu, L. and Li, P. (2014). New simple primary amine–thiourea organocatalysts and their application in asymmetric conjugate addition. *Tetrahedron Lett.* 55 (27): 3697–3700. https://doi.org/10.1016/j.tetlet.2014.05.006.

63 Zhao, D., Yang, D., Wang, Y. et al. (2011). Enantioselective Mannich reaction of a highly reactive Horner–Wadsworth–Emmons reagent with imines catalyzed by a bifunctional thiourea. *Chem. Sci.* 2 (10): 1918–1921. https://doi.org/10.1039/C1SC00351H.

64 Nie, S., Hu, Z., Xuan, Y. et al. (2010). Organocatalytic asymmetric conjugate addition of malonates to 3-nitro-2H-chromenes. *Tetrahedron: Asymmetry* 21 (16): 2055–2059. https://doi.org/10.1016/j.tetasy.2010.07.015.

65 Amarasinghe, N.R., Turner, P., and Todd, M.H. (2012). The first catalytic, enantioselective aza-Henry reaction of an unactivated cyclic imine. *Adv. Synth. Catal.* 354 (16): 2954–2958. https://doi.org/10.1002/adsc.201200558.

66 Bui, T., Syed, S., and Barbas, C.F. (2009). Thiourea-catalyzed highly enantio- and diastereoselective additions of oxindoles to nitroolefins: application to the formal synthesis of (+)-physostigmine. *J. Am. Chem. Soc.* 131 (25): 8758–8759. https://doi.org/10.1021/ja903520c.

67 Yamaoka, Y., Miyabe, H., and Takemoto, Y. (2007). Catalytic enantioselective Petasis-type reaction of quinolines catalyzed by a newly designed thiourea catalyst. *J. Am. Chem. Soc.* 129 (21): 6686–6687. https://doi.org/10.1021/ja071470x.

68 Han, B., Liu, Q.-P., Li, R. et al. (2008). Discovery of bifunctional thiourea/secondary-amine organocatalysts for the highly stereoselective nitro-Mannich reaction of α-substituted nitroacetates. *Chem. Eur. J.* 14 (27): 8094–8097. https://doi.org/10.1002/chem.200801170.

69 Bai, J.-F., Wang, L.-L., Peng, L. et al. (2011). Metal-free asymmetric 1,3-dipolar cycloaddition of *N*-arylmaleimides to azomethine ylides catalyzed by chiral tertiary amine thiourea. *Eur. J. Org. Chem.* (23): 4472–4478. https://doi.org/10.1002/ejoc.201100205.

70 Li, L., Klauber, E.G., and Seidel, D. (2008). Catalytic enantioselective aldol additions of α-isothiocyanato imides to aldehydes. *J. Am. Chem. Soc.* 130 (37): 12248–12249. https://doi.org/10.1021/ja804838y.

71 Jiang, Y., Yu, S.-W., Yang, Y. et al. (2018). Synthesis of polycyclic spirooxindoles via an asymmetric catalytic one-pot stepwise aldol/chloroetherification/aromatization procedure. *Org. Biomol. Chem.* 16 (36): 6647–6651. https://doi.org/10.1039/C8OB01713A.

72 Azuma, T., Murata, A., Kobayashi, Y. et al. (2014). A dual arylboronic acid–aminothiourea catalytic system for the asymmetric intramolecular hetero-Michael reaction of α,β-unsaturated carboxylic acids. *Org. Lett.* 16 (16): 4256–4259. https://doi.org/10.1021/ol501954r.

73 Guang, J. and Zhao, J.C.-G. (2013). Organocatalyzed asymmetric Michael reaction of β-aryl-α-ketophosphonates and nitroalkenes. *Tetrahedron Lett.* 54 (42): 5703–5706. https://doi.org/10.1016/j.tetlet.2013.08.015.

74 Kurimoto, Y., Nasu, T., Fujii, Y. et al. (2019). Asymmetric cycloetherification of in situ generated cyanohydrins through the concomitant construction of three chiral carbon centers. *Org. Lett.* 21 (7): 2156–2160. https://doi.org/10.1021/acs.orglett.9b00462.

75 Zhang, Z., Lippert, K.M., Hausmann, H. et al. (2011). Cooperative thiourea–Brønsted acid organocatalysis: enantioselective cyanosilylation of aldehydes with TMSCN. *J. Org. Chem.* 76 (23): 9764–9776. https://doi.org/10.1021/jo201864e.

76 Bai, M., Chen, Y.-Z., Cui, B.-D. et al. (2019). Thiourea-catalyzed asymmetric domino Michael-cyclization reaction of 3-isothiocyanato oxindoles with β,γ-unsaturated α-keto esters for the synthesis of spirocyclic oxindoles. *Tetrahedron* 75 (14): 2155–2161. https://doi.org/10.1016/j.tet.2019.02.036.

77 Hayama, N., Azuma, T., Kobayashi, Y., and Takemoto, Y. (2016). Chiral integrated catalysts composed of bifunctional thiourea and arylboronic acid: asymmetric aza-Michael addition of α,β-unsaturated carboxylic acids. *Chem. Pharm. Bull.* 64 (7): 704–717. https://doi.org/10.1248/cpb.c15-00983.

78 Yoshida, K., Inokuma, T., Takasu, K., and Takemoto, Y. (2010). Catalytic asymmetric synthesis of both enantiomers of 4-substituted 1,4-dihydropyridines with the use of bifunctional thiourea-ammonium salts bearing different counterions. *Molecules* 15 (11): 8305–8326. https://doi.org/10.3390/molecules15118305.

79 Connon, S.J. (2006). Organocatalysis mediated by (thio)urea derivatives. *Chem. Eur. J.* 12 (21): 5418–5427. https://doi.org/10.1002/chem.200501076.

80 Parvin, T., Yadav, R., and Choudhury, L.H. (2020). Recent applications of thiourea-based organocatalysts in asymmetric multicomponent reactions (AMCRs). *Org. Biomol. Chem.* 18 (29): 5513–5532. https://doi.org/10.1039/D0OB00595A.

81 Gandhi, S., Sivadas, V., and Baire, B. (2021). Thiourea–tertiary amine promoted cascade catalysis: a tool for complexity generation. *Eur. J. Org. Chem.* (2): 220–234. https://doi.org/10.1002/ejoc.202001114.

82 Steppeler, F., Iwan, D., Wojaczyńska, E., and Wojaczyński, J. (2020). Chiral thioureas – preparation and significance in asymmetric synthesis and medicinal chemistry. *Molecules* 25 (2): 401. https://www.mdpi.com/1420-3049/25/2/401.

83 Takemoto, Y. and Miyabe, H. (2007). The amino thiourea-catalyzed asymmetric nucleophilic reactions. *CHIMIA* 61 (5): 269–275. https://doi.org/10.2533/chimia.2007.269.

84 Takemoto, Y. (2010). Development of chiral thiourea catalysts and its application to asymmetric catalytic reactions. *Chem. Pharm. Bull.* 58 (5): 593–601. https://doi.org/10.1248/cpb.58.593.

85 Okino, T., Nakamura, S., Furukawa, T., and Takemoto, Y. (2004). Enantioselective aza-Henry reaction catalyzed by a bifunctional organocatalyst. *Org. Lett.* 6 (4): 625–627. https://doi.org/10.1021/ol0364531.

86 Hoashi, Y., Yabuta, T., and Takemoto, Y. (2004). Bifunctional thiourea-catalyzed enantioselective double Michael reaction of γ,δ-unsaturated β-ketoester to nitroalkene: asymmetric synthesis of (−)-epibatidine. *Tetrahedron Lett.* 45 (50): 9185–9188. https://doi.org/10.1016/j.tetlet.2004.10.082.

87 Hoashi, Y., Okino, T., and Takemoto, Y. (2005). Enantioselective Michael addition to α,β-unsaturated imides catalyzed by a bifunctional organocatalyst. *Angew. Chem. Int. Ed.* 44 (26): 4032–4035. https://doi.org/10.1002/anie.200500459.

88 Inokuma, T., Hoashi, Y., and Takemoto, Y. (2006). Thiourea-catalyzed asymmetric Michael addition of activated methylene compounds to α,β-unsaturated imides: dual activation of imide by intra- and intermolecular hydrogen bonding. *J. Am. Chem. Soc.* 128 (29): 9413–9419. https://doi.org/10.1021/ja061364f.

89 Serdyuk, O.V., Heckel, C.M., and Tsogoeva, S.B. (2013). Bifunctional primary amine-thioureas in asymmetric organocatalysis. *Org. Biomol. Chem.* 11 (41): 7051–7071. https://doi.org/10.1039/C3OB41403E.

90 Tan, K.L. and Jacobsen, E.N. (2007). Indium-mediated asymmetric allylation of acylhydrazones using a chiral urea catalyst. *Angew. Chem. Int. Ed.* 46 (8): 1315–1317. https://doi.org/10.1002/anie.200603354.

91 Sohtome, Y., Tanatani, A., Hashimoto, Y., and Nagasawa, K. (2004). Development of novel chiral urea catalysts for the hetero-Michael reaction. *Chem. Pharm. Bull.* 52 (4): 477–480. https://doi.org/10.1248/cpb.52.477.

92 Okino, T., Hoashi, Y., Furukawa, T. et al. (2005). Enantio- and diastereoselective Michael reaction of 1,3-dicarbonyl compounds to nitroolefins catalyzed by a bifunctional thiourea. *J. Am. Chem. Soc.* 127 (1): 119–125. https://doi.org/10.1021/ja044370p.

93 Yoneda, N., Hotta, A., Asano, K., and Matsubara, S. (2014). Asymmetric oxy-Michael addition to γ-hydroxy-α,β-unsaturated carbonyls using formaldehyde as an oxygen-centered nucleophile. *Org. Lett.* 16 (23): 6264–6266. https://doi.org/10.1021/ol503104b.

94 Veitch, G.E. and Jacobsen, E.N. (2010). Tertiary aminourea-catalyzed enantioselective iodolactonization. *Angew. Chem. Int. Ed.* 49 (40): 7332–7335. https://doi.org/10.1002/anie.201003681.

95 Wenzel, A.G. and Jacobsen, E.N. (2002). Asymmetric catalytic Mannich reactions catalyzed by urea derivatives: enantioselective synthesis of *β*-aryl-*β*-amino acids. *J. Am. Chem. Soc.* 124 (44): 12964–12965. https://doi.org/10.1021/ja028353g.

96 Lalonde, M.P., Chen, Y., and Jacobsen, E.N. (2006). A chiral primary amine thiourea catalyst for the highly enantioselective direct conjugate addition of α,α-disubstituted aldehydes to nitroalkenes. *Angew. Chem. Int. Ed.* 45 (38): 6366–6370. https://doi.org/10.1002/anie.200602221.

97 Lalonde, M.P., McGowan, M.A., Rajapaksa, N.S., and Jacobsen, E.N. (2013). Enantioselective formal aza-Diels–Alder reactions of enones with cyclic imines catalyzed by primary aminothioureas. *J. Am. Chem. Soc.* 135 (5): 1891–1894. https://doi.org/10.1021/ja310718f.

98 Rajapaksa, N.S., McGowan, M.A., Rienzo, M., and Jacobsen, E.N. (2013). Enantioselective total synthesis of (+)-reserpine. *Org. Lett.* 15 (3): 706–709. https://doi.org/10.1021/ol400046n.

99 Yalalov, D.A., Tsogoeva, S.B., and Schmatz, S. (2006). Chiral thiourea-based bifunctional organocatalysts in the asymmetric nitro-Michael addition: a joint experimental-theoretical study. *Adv. Synth. Catal.* 348 (7, 8): 826–832. https://doi.org/10.1002/adsc.200505443.

100 Tsogoeva, S.B. and Wei, S. (2006). Highly enantioselective addition of ketones to nitroolefins catalyzed by new thiourea–amine bifunctional organocatalysts. *Chem. Commun.* (13): 1451–1453. https://doi.org/10.1039/B517937H.

101 Wei, S., Yalalov, D.A., Tsogoeva, S.B., and Schmatz, S. (2007). New highly enantioselective thiourea-based bifunctional organocatalysts for nitro-Michael addition reactions. *Catal. Today* 121 (1, 2): 151–157. https://doi.org/10.1016/j.cattod.2006.11.018.

102 Yalalov, D.A., Tsogoeva, S.B., Shubina, T.E. et al. (2008). Evidence for an enol mechanism in a highly enantioselective Mannich-type reaction catalyzed by primary amine-thiourea. *Angew. Chem. Int. Ed.* 47 (35): 6624–6628. https://doi.org/10.1002/anie.200800849.

103 Vazquez-Chavez, J., Luna-Morales, S., Cruz-Aguilar, D.A. et al. (2019). The effect of chiral N-substituents with methyl or trifluoromethyl groups on the catalytic performance of mono- and bifunctional thioureas. *Org. Biomol. Chem.* 17 (47): 10045–10051. https://doi.org/10.1039/C9OB01893J.

104 Guo, Z.-L., Xue, J.-H., Fu, L.-N. et al. (2014). The direct asymmetric alkylation of α-amino aldehydes with 3-indolylmethanols by enamine catalysis. *Org. Lett.* 16 (24): 6472–6475. https://doi.org/10.1021/ol503318c.

105 Kawada, M., Nakashima, K., Hirashima, S. et al. (2017). Asymmetric conjugate addition of nitroalkanes to enones using a sulfonamide–thiourea organocatalyst. *J. Org. Chem.* 82 (13): 6986–6991. https://doi.org/10.1021/acs.joc.7b00835.

106 Sun, Z.-W., Peng, F.-Z., Li, Z.-Q. et al. (2012). Enantioselective conjugate addition of both aromatic ketones and acetone to nitroolefins catalyzed by chiral primary amines bearing multiple hydrogen-bonding donors. *J. Org. Chem.* 77 (8): 4103–4110. https://doi.org/10.1021/jo300011x.

107 Liu, K., Cui, H.-F., Nie, J. et al. (2007). Highly enantioselective Michael addition of aromatic ketones to nitroolefins promoted by chiral bifunctional primary amine-thiourea catalysts based on saccharides. *Org. Lett.* 9 (5): 923–925. https://doi.org/10.1021/ol0701666.

108 Jiang, X., Zhang, Y., Chan, A.S.C., and Wang, R. (2009). Highly enantioselective synthesis of γ-nitro heteroaromatic ketones in a doubly stereocontrolled manner catalyzed by bifunctional thiourea catalysts based on dehydroabietic amine: a doubly stereocontrolled approach to pyrrolidine carboxylic acids. *Org. Lett.* 11 (1): 153–156.

109 Ma, Z.-W., Liu, Y.-X., Zhang, W.-J. et al. (2011). Highly enantioselective Michael additions of isobutyraldehyde to nitroalkenes promoted by amphiphilic bifunctional primary amine-thioureas in organic or aqueous medium. *Eur. J. Org. Chem.* (33): 6747–6754. https://doi.org/10.1002/ejoc.201101086.

110 Li, P., Wang, Y., Liang, X., and Ye, J. (2008). Asymmetric multifunctional organocatalytic Michael addition of nitroalkanes to α,β-unsaturated ketones. *Chem. Commun.* (28): 3302–3304. https://doi.org/10.1039/B804540B.

111 Puglisi, A., Benaglia, M., Annunziata, R., and Rossi, D. (2008). Stereoselective nucleophilic addition to imines catalyzed by chiral bifunctional thiourea organocatalysts. *Tetrahedron: Asymmetry* 19 (19): 2258–2264. https://doi.org/10.1016/j.tetasy.2008.09.030.

112 Ding, R. and Wolf, C. (2018). Organocatalytic asymmetric synthesis of α-oxetanyl and α-azetidinyl tertiary alkyl fluorides and chlorides. *Org. Lett.* 20 (3): 892–895. https://doi.org/10.1021/acs.orglett.8b00039.

113 Manna, M.S. and Mukherjee, S. (2014). Remarkable influence of secondary catalyst site on enantioselective desymmetrization of cyclopentenedione. *Chem. Sci.* 5 (4): 1627–1633. https://doi.org/10.1039/C3SC53102C.

114 Manna, M.S. and Mukherjee, S. (2012). Catalytic asymmetric direct vinylogous Michael addition of deconjugated butenolides to maleimides for the construction of quaternary stereogenic centers. *Chem. Eur. J.* 18 (48): 15277–15282. https://doi.org/10.1002/chem.201203180.

115 Zuend, S.J. and Jacobsen, E.N. (2007). Cooperative catalysis by tertiary amino-thioureas: mechanism and basis for enantioselectivity of ketone cyanosilylation. *J. Am. Chem. Soc.* 129 (51): 15872–15883. https://pubs.acs.org/doi/pdf/10.1021/ja0735352.

116 Tsogoeva, S.B., Yalalov, D.A., Hateley, M.J. et al. (2005). Asymmetric organocatalysis with novel chiral thiourea derivatives: bifunctional catalysts for the Strecker and nitro-Michael reactions. *Eur. J. Org. Chem.* (23): 4995–5000. https://doi.org/10.1002/ejoc.200500420.

117 Yuan, H.-N., Wang, S., Nie, J. et al. (2013). Hydrogen-bond-directed enantioselective decarboxylative Mannich reaction of β-ketoacids with ketimines: application to the synthesis of anti-HIV Drug DPC 083. *Angew. Chem. Int. Ed.* 52 (14): 3869–3873. https://doi.org/10.1002/anie.201210361.

118 Yuan, H.-N., Li, S., Nie, J. et al. (2013). Highly enantioselective decarboxylative Mannich reaction of malonic acid half Oxyesters with cyclic trifluoromethyl ketimines: synthesis of β-amino esters and anti-HIV Drug DPC 083. *Chem. Eur. J.* 19 (47): 15856–15860. https://doi.org/10.1002/chem.201303307.

119 Jiang, X., Zhang, Y., Liu, X., Zhang, G., Lai, L., Wu, L., Zhang, J., and Wang, R. (2009) Enantio- and diastereoselective asymmetric addition of 1,3-dicarbonyl compounds to nitroalkenes in a doubly stereocontrolled manner catalyzed by bifunctional rosin-derived amine thiourea catalysts. *J. Org. Chem.*, 74 (15): 5562–5567. https://doi.org/10.1021/jo9009276

120 Jiang, X., Zhang, G., Fu, D. et al. (2010). Direct organocatalytic asymmetric aldol reaction of α-isothiocyanato imides to α-ketoesters under low ligand loading: a doubly stereocontrolled approach to cyclic thiocarbamates bearing chiral quaternary stereocenters. *Org. Lett.* 12 (7): 1544–1547. https://doi.org/10.1021/ol1002829.

121 Probst, N., Madarász, Á., Valkonen, A. et al. (2012). Cooperative assistance in bifunctional organocatalysis: enantioselective Mannich reactions with aliphatic and aromatic imines. *Angew. Chem. Int. Ed.* 51 (34): 8495–8499. https://doi.org/10.1002/anie.201203852.

122 Han, W.-Y., Wu, Z.-J., Zhang, X.-M., and Yuan, W.-C. (2012). enantioselective organocatalytic three-component Petasis reaction among salicylaldehydes, amines, and organoboronic acids. *Org. Lett.* 14 (4): 976–979. https://doi.org/10.1021/ol203109a.

123 Lee, J.H. and Kim, D.Y. (2009). Enantio- and diastereoselective Mannich-type reactions of α-cyano ketones with *N*-Boc aldimines catalyzed by chiral bifunctional urea. *Adv. Synth. Catal.* 351 (11, 12): 1779–1782. https://doi.org/10.1002/adsc.200900268.

124 Kwon, B.K., Kim, S.M., and Kim, D.Y. (2009). Highly enantioselective conjugate addition of fluoromalonates to nitroalkenes using bifunctional organocatalysts. *J. Fluorine Chem.* 130 (8): 759–761. https://doi.org/10.1016/j.jfluchem.2009.06.002.

125 Kimmel, K.L., Robak, M.T., and Ellman, J.A. (2009). Enantioselective addition of thioacetic acid to nitroalkenes via *N*-sulfinyl urea organocatalysis. *J. Am. Chem. Soc.* 131 (25): 8754–8755. https://doi.org/10.1021/ja903351a.

126 Chen, H., Liu, H., Zhao, S.-H. et al. (2019). Enantioselective arylation of 3-carboxamide oxindoles with quinone monoimines and synthesis of chiral spirooxindole-benzofuranones. *Synlett* 30 (9): 1067–1072. https://doi.org/10.1055/s-0037-1611782.

127 Alemán, J., Parra, A., Jiang, H., and Jørgensen, K.A. (2011). Squaramides: bridging from molecular recognition to bifunctional organocatalysis. *Chem. Eur. J.* 17 (25): 6890–6899. https://doi.org/10.1002/chem.201003694.

128 Mailhol, D., Duque M. del M.S., Raimondi, W. et al. (2012). Enantioselective organocatalytic Michael addition of cyclobutanones to nitroalkenes. *Adv. Synth. Catal.* 354 (18): 3523–3532. https://doi.org/10.1002/adsc.201200658.

129 Baran, R., Veverková, E., Škvorcová, A., and Šebesta, R. (2013) Enantioselective Michael addition of 1,3-dicarbonyl compounds to a nitroalkene catalyzed by chiral squaramides – a key step in the synthesis of pregabalin. *Org. Biomol. Chem.*, 11 (44): 7705–7711. https://doi.org/10.1039/C3OB41709C.

130 Konishi, H., Lam, T.Y., Malerich, J.P., and Rawal, V.H. (2010). Enantioselective α-amination of 1,3-dicarbonyl compounds using squaramide derivatives as hydrogen bonding catalysts. *Org. Lett.* 12 (9): 2028–2031. https://doi.org/10.1021/ol1005104.

131 Qian, Y., Ma, G., Lv, A. et al. (2010). Squaramide-catalyzed enantioselective Friedel–Crafts reaction of indoles with imines. *Chem. Commun.* 46 (17): 3004–3006. https://doi.org/10.1039/B922120D.

132 Işık, M., Unver, M.Y., and Tanyeli, C. (2015). Modularly evolved 2-aminoDMAP/squaramides as highly active bifunctional organocatalysts in Michael addition. *J. Org. Chem.* 80 (2): 828–835. https://doi.org/10.1021/jo5022597.

133 Ray, B. and Mukherjee, S. (2019). Catalytic enantioselective Michael addition of deconjugated butyrolactams to maleimides. *Tetrahedron* 75 (24): 3292–3298. https://doi.org/10.1016/j.tet.2019.04.049.

134 Ma, Z., Liu, X., Liu, J. et al. (2017). Highly enantioselective Michael addition of α,α-disubstituted aldehydes to maleimides catalyzed by new primary amine-squaramide bifunctional organocatalysts. *Tetrahedron Lett.* 58 (48): 4487–4490. https://doi.org/10.1016/j.tetlet.2017.10.026.

135 Moon, H.W. and Kim, D.Y. (2012). Enantioselective decarboxylative Michael addition of β-ketoacids to nitroalkenes catalyzed by binaphthyl-derived organocatalysts. *Tetrahedron Lett.* 53 (48): 6569–6572. https://doi.org/10.1016/j.tetlet.2012.09.100.

136 Suh, C.W., Chang, C.W., Choi, K.W. et al. (2013). Enantioselective decarboxylative aldol addition of β-ketoacids to isatins catalyzed by binaphthyl-modified organocatalyst. *Tetrahedron Lett.* 54 (28): 3651–3654. https://doi.org/10.1016/j.tetlet.2013.04.132.

137 Alegre-Requena, J.V., Marqués-López, E., and Herrera, R.P. (2016) Trifunctional squaramide catalyst for efficient enantioselective Henry reaction activation. *Adv. Synth. Catal.*, 358 (11): 1801–1809. https://doi.org/10.1002/adsc.201600046.

138 Rombola, M., Sumaria, C.S., Montgomery, T.D., and Rawal, V.H. (2017). Development of chiral, bifunctional thiosquaramides: enantioselective Michael additions of barbituric acids to nitroalkenes. *J. Am. Chem. Soc.* 139 (15): 5297–5300. https://doi.org/10.1021/jacs.7b01115.

139 Rombola, M. and Rawal, V.H. (2018). Dicyclopentyl dithiosquarate as an intermediate for the synthesis of thiosquaramides. *Org. Lett.* 20 (3): 514–517. https://doi.org/10.1021/acs.orglett.7b03549.

140 Dong, S., Feng, X., and Liu, X. (2018). Chiral guanidines and their derivatives in asymmetric synthesis. *Chem. Soc. Rev.* 47 (23): 8525–8540. https://doi.org/10.1039/C7CS00792B.

141 Avila, A., Chinchilla, R., and Nájera, C. (2012). Enantioselective Michael addition of α,α-disubstituted aldehydes to maleimides organocatalyzed by chiral primary amine-guanidines. *Tetrahedron: Asymmetry* 23 (24): 1625–1627. https://doi.org/10.1016/j.tetasy.2012.11.002.

142 Almaşi, D., Alonso, D.A., Gómez-Bengoa, E., and Nájera, C. (2009). Chiral 2-aminobenzimidazoles as recoverable organocatalysts for the addition of 1,3-dicarbonyl compounds to nitroalkenes. *J. Org. Chem.* 74 (16): 6163–6168. https://doi.org/10.1021/jo9010552.

143 Lee, M., Zhang, L., Park, Y., and Park, H. (2012). Chiral 2-aminobenzimidazole bifunctional organocatalysts: effect of di-CF3 and TFA on catalytic mechanisms. *Tetrahedron* 68 (5): 1452–1459. https://doi.org/10.1016/j.tet.2011.12.021.

144 Inokuma, T., Furukawa, M., Uno, T. et al. (2011). Bifunctional hydrogen-bond donors that bear a quinazoline or benzothiadiazine skeleton for asymmetric organocatalysis. *Chem. Eur. J.* 17 (37): 10470–10477. https://doi.org/10.1002/chem.201101338.

145 Kobayashi, Y., Taniguchi, Y., Hayama, N. et al. (2013). A powerful hydrogen-bond-donating organocatalyst for the enantioselective intramolecular oxa-Michael reaction of α,β-unsaturated amides and esters. *Angew. Chem. Int. Ed.* 52 (42): 11114–11118. https://doi.org/10.1002/anie.201305492.

146 Taylor, M.S. and Jacobsen, E.N. (2006). Asymmetric catalysis by chiral hydrogen-bond donors. *Angew. Chem. Int. Ed.* 45 (10): 1520–1543. https://doi.org/10.1002/anie.200503132.

147 Doyle, A.G. and Jacobsen, E.N. (2007). Small-molecule H-bond donors in asymmetric catalysis. *Chem. Rev.* 107 (12): 5713–5743. https://doi.org/10.1021/cr068373r.

148 Sigman, M.S., Vachal, P., and Jacobsen, E.N. (2000). A general catalyst for the asymmetric Strecker reaction. *Angew. Chem. Int. Ed.* 39 (7): 1279–1281. https://doi.org/10.1002/(SICI)1521-3773(20000403)39:7<1279::AID-ANIE1279>3.0.CO;2-U.

149 Vachal, P., and Jacobsen, E.N. (2000) Enantioselective catalytic addition of HCN to ketoimines. Catalytic synthesis of quaternary amino acids. *Org. Lett.*, 2 (6): 867–870. https://doi.org/10.1021/ol005636

150 Su, J.T., Vachal, P., and Jacobsen, E.N. (2001). Practical synthesis of a soluble schiff base catalyst for the asymmetric Strecker reaction. *Adv. Synth. Catal.* 343 (2): 197–200. https://doi.org/10.1002/1615-4169(20010226)343:2<197::AID-ADSC197>3.0.CO;2-8.

151 Raheem, I.T. and Jacobsen, E.N. (2005). Highly enantioselective aza-Baylis–Hillman reactions catalyzed by chiral thiourea derivatives. *Adv. Synth. Catal.* 347 (11–13): 1701–1708. https://doi.org/10.1002/adsc.200505230.

152 Vachal, P. and Jacobsen, E.N. (2002). Structure-based analysis and optimization of a highly enantioselective catalyst for the Strecker reaction. *J. Am. Chem. Soc.* 124 (34): 10012–10014. https://doi.org/10.1021/ja027246j.

153 Joly, G.D. and Jacobsen, E.N. (2004). Thiourea-catalyzed enantioselective hydrophosphonylation of imines: practical access to enantiomerically enriched α-amino phosphonic acids. *J. Am. Chem. Soc.* 126 (13): 4102–4103. https://doi.org/10.1021/ja0494398.

154 Taylor, M.S. and Jacobsen, E.N. (2004). Highly enantioselective catalytic acyl-Pictet–Spengler reactions. *J. Am. Chem. Soc.* 126 (34): 10558–10559. https://doi.org/10.1021/ja046259p.

155 Taylor, M.S., Tokunaga, N., and Jacobsen, E.N. (2005). Enantioselective thiourea-catalyzed acyl-Mannich reactions of isoquinolines. *Angew. Chem. Int. Ed.* 44 (41): 6700–6704. https://doi.org/10.1002/anie.200502277.

156 Mergott, D.J., Zuend, S.J., and Jacobsen, E.N. (2008). Catalytic asymmetric total synthesis of (+)-Yohimbine. *Org. Lett.* 10 (5): 745–748. https://doi.org/10.1021/ol702781q.

157 Raheem, I.T., Thiara, P.S., Peterson, E.A., and Jacobsen, E.N. (2007). Enantioselective Pictet–Spengler-type cyclizations of hydroxylactams: H-bond donor catalysis by anion binding. *J. Am. Chem. Soc.* 129 (44): 13404–13405. https://doi.org/10.1021/ja076179w.

158 Raheem, I.T., Thiara, P.S., and Jacobsen, E.N. (2008). Regio- and enantioselective catalytic cyclization of pyrroles onto *N*-acyliminium ions. *Org. Lett.* 10 (8): 1577–1580. https://doi.org/10.1021/ol800256j.

159 Martin, N.J.A., Ozores, L., and List, B. (2007). Organocatalytic asymmetric transfer hydrogenation of nitroolefins. *J. Am. Chem. Soc.* 129 (29): 8976–8977. https://doi.org/10.1021/ja074045c.

160 Martin, N.J.A., Cheng, X., and List, B. (2008). Organocatalytic asymmetric transferhydrogenation of β-nitroacrylates: accessing β_2-amino acids. *J. Am. Chem. Soc.* 130 (42): 13862–13863.

161 Massolo, E., Benaglia, M., Orlandi, M. et al. (2015). Enantioselective organocatalytic reduction of β-trifluoromethyl nitroalkenes: an efficient strategy for the synthesis of chiral β-trifluoromethyl amines. *Chem. Eur. J.* 21 (9): 3589–3595. https://doi.org/10.1002/chem.201405730.

162 Jones, C.R., Dan Pantoş, G., Morrison, A.J., and Smith, M.D. (2009). Plagiarizing proteins: enhancing efficiency in asymmetric hydrogen-bonding catalysis through positive cooperativity. *Angew. Chem. Int. Ed.* 48 (40): 7391–7394. https://doi.org/10.1002/anie.200903063.

163 Yoon, T.P. and Jacobsen, E.N. (2005). Highly enantioselective thiourea-catalyzed nitro-Mannich reactions. *Angew. Chem. Int. Ed.* 44 (3): 466–468. https://doi.org/10.1002/anie.200461814.

164 Xu, H., Zuend, S.J., Woll, M.G. et al. (2010). Asymmetric cooperative catalysis of strong Brønsted acid–promoted reactions using chiral ureas. *Science* 327 (5968): 986–990. https://doi.org/10.1126/science.1182826.

165 Klauber, E.G., De, C.K., Shah, T.K., and Seidel, D. (2010). Merging nucleophilic and hydrogen bonding catalysis: an anion binding approach to the kinetic resolution of propargylic amines. *J. Am. Chem. Soc.* 132 (39): 13624–13626. https://doi.org/10.1021/ja105337h.

166 Min, C., Mittal, N., Sun, D.X., and Seidel, D. (2013). Conjugate-base-stabilized Brønsted acids as asymmetric catalysts: enantioselective Povarov reactions with secondary aromatic amines. *Angew. Chem. Int. Ed.* 52 (52): 14084–14088. https://doi.org/10.1002/anie.201308196.

167 Zhu, Z., Odagi, M., Zhao, C. et al. (2020). Highly acidic conjugate-base-stabilized carboxylic acids catalyze enantioselective oxa-Pictet–Spengler reactions with ketals. *Angew. Chem. Int. Ed.* 59 (5): 2028–2032. https://doi.org/10.1002/anie.201912677.

168 Inokuma, T., Takasu, K., Sakaeda, T., and Takemoto, Y. (2009). Hydroxyl group-directed organocatalytic asymmetric Michael addition of α,β-unsaturated ketones with alkenylboronic acids. *Org. Lett.* 11 (11): 2425–2428. https://doi.org/10.1021/ol9006053.

169 Zhao, C., Chen, S.B., and Seidel, D. (2016). Direct formation of oxocarbenium ions under weakly acidic conditions: catalytic enantioselective oxa-Pictet–Spengler reactions. *J. Am. Chem. Soc.* 138 (29): 9053–9056. https://doi.org/10.1021/jacs.6b05225.

170 Lin, Y., Hirschi, W.J., Kunadia, A. et al. (2020). A selenourea-thiourea Brønsted acid catalyst facilitates asymmetric conjugate additions of amines to α,β-unsaturated esters. *J. Am. Chem. Soc.* 142 (12): 5627–5635. https://doi.org/10.1021/jacs.9b12457.

171 Novacek, J. and Waser, M. (2014). Syntheses and applications of (thio)urea-containing chiral quaternary ammonium salt catalysts: (thio)urea-containing chiral quaternary ammonium salt catalysts. *Eur. J. Org. Chem.* (4): 82–809. https://doi.org/10.1002/ejoc.201301594.

172 Perillo, M., Di Mola, A., Filosa, R. et al. (2014). Cascade reactions of glycine Schiff bases and chiral phase transfer catalysts in the synthesis of α-amino acids 3-substituted phthalides or isoindolinones. *RSC Adv.* 4 (9): 4239–4246. https://doi.org/10.1039/C3RA46268D.

173 Tiffner, M., Novacek, J., Busillo, A. et al. (2015). Design of chiral urea-quaternary ammonium salt hybrid catalysts for asymmetric reactions of glycine Schiff bases. *RSC Adv.* 5 (96): 78941–78949. https://doi.org/10.1039/C5RA14466C.

174 Novacek, J., Izzo, J.A., Vetticatt, M.J., and Waser, M. (2016). bifunctional ammonium salt catalyzed asymmetric α-hydroxylation of β-ketoesters by simultaneous resolution of oxaziridines. *Chem. Eur. J.* 22 (48): 17339–17344. https://doi.org/10.1002/chem.201604153.

175 Gu, X., Tang, Y., Zhang, X. et al. (2016). Organocatalytic Knoevenagel condensation by chiral C_2-symmetric tertiary diamines. *New J. Chem.* 40 (8): 6580–6583. https://doi.org/10.1039/C6NJ00613B.

176 Kizirian, J.-C., Caille, J.-C., and Alexakis, A. (2003). Conceptually new chiral tertiary C_2 symmetric diamines in asymmetric synthesis. *Tetrahedron Lett.* 44 (49): 8893–8895. https://doi.org/10.1016/j.tetlet.2003.09.171.

177 Tanimura, Y. and Ishimaru, K. (2012). Asymmetric allylation reactions of aromatic aldehydes with allyltrichlorosilane catalyzed by chiral bisformamide-type organocatalysts. *Tetrahedron: Asymmetry* 23 (5): 345–349. https://doi.org/10.1016/j.tetasy.2012.02.015.

178 Sohtome, Y., Tanatani, A., Hashimoto, Y., and Nagasawa, K. (2004). Development of bis-thiourea-type organocatalyst for asymmetric Baylis–Hillman reaction. *Tetrahedron Lett.* 45 (29): 5589–5592. https://doi.org/10.1016/j.tetlet.2004.05.137.

179 Collar, A.G., Trujillo, C., Lockett-Walters, B. et al. (2019). Catalytic asymmetric γ-lactam synthesis from enolisable anhydrides and imines. *Chem. Eur. J.* 25 (30): 7275–7279. https://doi.org/10.1002/chem.201901028.

180 Nugent, B.M., Yoder, R.A., and Johnston, J.N. (2004). Chiral proton catalysis: a catalytic enantioselective direct aza-Henry reaction. *J. Am. Chem. Soc.* 126 (11): 3418–3419. https://pubs.acs.org/doi/10.1021/ja031906i.

181 Davis, T.A., Wilt, J.C., and Johnston, J.N. (2010). Bifunctional asymmetric catalysis: amplification of Brønsted basicity can orthogonally increase the reactivity of a chiral Brønsted acid. *J. Am. Chem. Soc.* 132 (9): 2880–2882. https://doi.org/10.1021/ja908814h.

182 Gómez-Torres, E., Alonso, D.A., Gómez-Bengoa, E., and Nájera, C. (2013). Enantioselective synthesis of succinimides by Michael addition of 1,3-dicarbonyl compounds to maleimides catalyzed by a chiral bis(2-aminobenzimidazole) organocatalyst: enantioselective synthesis of succinimides. *Eur. J. Org. Chem.* (8): 1434–1440. https://doi.org/10.1002/ejoc.201201046.

183 Trillo, P., Baeza, A., and Nájera, C. (2014). Bis(2-aminobenzoimidazole)-organocatalyzed asymmetric alkylation of activated methylene compounds with benzylic and allylic alcohols. *Synthesis* 46 (24): 3399–3414. https://dx.doi.org/10.1055/s-0034-1378618.

184 Mucha, P., Mlostoń, G., Jasiński, M. et al. (2008). A new approach to enantiomerically pure bis-imidazoles derived from trans-1,2-diaminocyclohexane. *Tetrahedron: Asymmetry* 19 (13): 1600–1607. https://doi.org/10.1016/j.tetasy.2008.06.015.

185 Kwiatkowski, P., Mucha, P., Mlostoń, G., and Jurczak, J. (2009). Novel chiral C_2-symmetric bisimidazole-*N*-oxides as promising organocatalysts for enantioselective allylation of aromatic aldehydes. *Synlett* (11): 1757–1760. https://dx.doi.org/10.1055/s-0029-1217365.

186 Kucherenko, A.S., Kostenko, A.A., Komogortsev, A.N. et al. (2019). C_2-Symmetric chiral squaramide, recyclable organocatalyst for asymmetric Michael reactions. *J. Org. Chem.* 84 (7): 4304–4311. https://doi.org/10.1021/acs.joc.9b00252.

187 Hou, W., Zheng, B., Chen, J., and Peng, Y. (2012). Asymmetric synthesis of polysubstituted 4-amino- and 3,4-diaminochromanes with a chiral multifunctional organocatalyst. *Org. Lett.* 14 (9): 2378–2381. https://doi.org/10.1021/ol300798t.

188 Uyeda, C. and Jacobsen, E.N. (2008). Enantioselective Claisen rearrangements with a hydrogen-bond donor catalyst. *J. Am. Chem. Soc.* 130 (29): 9228–9229. https://doi.org/10.1021/ja803370x.

189 Uyeda, C., Rötheli, A.R., and Jacobsen, E.N. (2010). Catalytic enantioselective Claisen rearrangements of o-allyl β-ketoesters. *Angew. Chem. Int. Ed.* 49 (50): 9753–9756. https://doi.org/10.1002/anie.201005183.

190 Uyeda, C. and Jacobsen, E.N. (2011). Transition-state charge stabilization through multiple non-covalent interactions in the guanidinium-catalyzed enantioselective Claisen rearrangement. *J. Am. Chem. Soc.* 133 (13): 5062–5075. https://doi.org/10.1021/ja110842s.

191 Lee, H.J., Woo, S.B., and Kim, D.Y. (2012). Enantio- and diastereoselective Michael addition reactions of α-cyanoketones to nitroalkenes catalyzed by binaphthyl-derived organocatalyst. *Tetrahedron Lett.* 53 (26): 3374–3377. https://doi.org/10.1016/j.tetlet.2012.04.095.

192 Durmaz, M. and Sirit, A. (2013). Calixarene-based chiral primary amine thiourea promoted highly enantioselective asymmetric Michael reactions of α,α-disubstituted aldehydes with maleimides. *Tetrahedron: Asymmetry* 24 (23): 1443–1448. https://doi.org/10.1016/j.tetasy.2013.09.010.

193 Genc, H.N. and Sirit, A. (2016). Novel efficient bifunctional calixarene thiourea organocatalysts: synthesis and application in the direct enantioselective aldol reactions. *Tetrahedron: Asymmetry* 27 (4, 5): 201–207. https://doi.org/10.1016/j.tetasy.2016.01.011.

194 Yang, Y., He, Y.-H., Guan, Z., and Huang, W.-D. (2010). A highly efficient large-scale asymmetric direct intermolecular aldol reaction employing L-prolinamide as a recoverable catalyst. *Adv. Synth. Catal.* 352 (14, 15): 2579–2587. https://doi.org/10.1002/adsc.201000355.

195 Guan, Z., Luo, Y., Zhang, B.-Q. et al. (2014). The application of a structurally simple, recyclable, and large-scale L-prolinamide catalyst for asymmetric aldol reactions. *Tetrahedron: Asymmetry* 25 (10, 11): 802–812. https://doi.org/10.1016/j.tetasy.2014.04.009.

196 Fu, J.-Y., Xu, X.-Y., Li, Y.-C. et al. (2010). Effective construction of quaternary stereocenters by highly enantioselective α-amination of branched aldehydes. *Org. Biomol. Chem.* 8 (20): 4524–4526. https://doi.org/10.1039/C0OB00406E.

197 Sarkar, D., Harman, K., Ghosh, S., and Headley, A.D. (2011). Chiral amine organocatalysts for the *syn*-aldol reaction involving substituted benzaldehydes and hydroxyacetone. *Tetrahedron: Asymmetry* 22 (10): 1051–1054. https://doi.org/10.1016/j.tetasy.2011.05.021.

198 Zhu, X., Lin, A., Fang, L. et al. (2011). In situ formed bifunctional primary amine-imine catalyst for asymmetric aldol reactions of α-keto esters. *Chem. Eur. J.* 17 (30): 8281–8284. https://doi.org/10.1002/chem.201100200.

199 Yu, J.-S., Liao, F.-M., Gao, W.-M. et al. (2015). Michael addition catalyzed by chiral secondary amine phosphoramide using fluorinated silyl enol ethers: formation of quaternary carbon stereocenters. *Angew. Chem. Int. Ed.* 54 (25): 7381–7385. https://doi.org/10.1002/anie.201501747.

200 Kanada, Y., Yuasa, H., Nakashima, K. et al. (2013). Asymmetric conjugate addition of aldehydes to vinyl sulfone using a diaminomethylenemalononitrile organocatalyst. *Tetrahedron Lett.* 54 (36): 4896–4899. https://doi.org/10.1016/j.tetlet.2013.06.141.

201 You, Z.-H., Chen, Y.-H., Tang, Y., and Liu, Y.-K. (2019). Application of E1cB elimination in asymmetric organocatalytic cascade reactions to construct polyheterocyclic compounds. *Org. Lett.* 21 (20): 8358–8363. https://doi.org/10.1021/acs.orglett.9b03138.

202 Peterson, E.A. and Jacobsen, E.N. (2009). Enantioselective, thiourea-catalyzed intermolecular addition of indoles to cyclic *N*-acyl iminium ions. *Angew. Chem. Int. Ed.* 48 (34): 6328–6331. https://doi.org/10.1002/anie.200902420.

203 Mei, M.-S., Wang, Y.-H., Hu, Q. et al. (2020). Enantioselective [3+2] annulation of isatin-derived MBH-carbonates and 3-nitroindoles enabled by a bifunctional DMAP-thiourea. *Chem. Commun.* 56 (73): 10718–10721. https://doi.org/10.1039/D0CC04462H.

204 Zurro, M., Asmus, S., Beckendorf, S. et al. (2014). Chiral helical oligotriazoles: new class of anion-binding catalysts for the asymmetric dearomatization of electron-deficient *N*-heteroarenes. *J. Am. Chem. Soc.* 136 (40): 13999–14002. https://doi.org/10.1021/ja507940k.

205 Fischer, T., Bamberger, J., Gómez-Martínez, M. et al. (2019). Helical multi-coordination anion-binding catalysts for the highly enantioselective dearomatization of pyrylium derivatives. *Angew. Chem. Int. Ed.* 58 (10): 3217–3221. https://doi.org/10.1002/anie.201812031.

206 Pan, S.C., Zhou, J., and List, B. (2007). Catalytic asymmetric acylcyanation of imines. *Angew. Chem. Int. Ed.* 46 (4): 612–614. https://doi.org/10.1002/anie.200603630.

207 Pan, S.C. and List, B. (2007). Catalytic asymmetric three-component acyl-Strecker reaction. *Org. Lett.* 9 (6): 1149–1151. https://doi.org/10.1021/ol0702674.

208 Enders, D., Gottfried, K., and Raabe, G. (2010). Organocatalytic enantioselective Strecker synthesis of α-quaternary α-trifluoromethyl amino acids. *Adv. Synth. Catal.* 352 (18): 3147–3152. https://doi.org/10.1002/adsc.201000666.

209 Wada, Y., Murata, R., Fujii, Y. et al. (2020). Enantio- and diastereoselective construction of contiguous tetrasubstituted chiral carbons in organocatalytic oxadecalin synthesis. *Org. Lett.* 22 (12): 4710–4715. https://doi.org/10.1021/acs.orglett.0c01501.

210 Veverková, E., Bilka, S., Baran, R., and Šebesta, R. (2016). Squaramide-catalyzed Michael addition as a key step for the direct synthesis of GABAergic drugs. *Synthesis* 48 (10): 1474–1482. https://doi.org/10.1055/s-0035-1560420.

211 Hamza, A., Schubert, G., Soós, T., and Pápai, I. (2006). Theoretical Studies on the bifunctionality of chiral thiourea-based organocatalysts: competing routes to

C–C bond formation. *J. Am. Chem. Soc.* 128 (40): 13151–13160. https://doi.org/10.1021/ja063201x.

212 Kótai, B., Kardos, G., Hamza, A. et al. (2014). On the mechanism of bifunctional squaramide-catalyzed organocatalytic Michael addition: a protonated catalyst as an oxyanion hole. *Chem. Eur. J.* 20 (19): 5631–5639. https://doi.org/10.1002/chem.201304553.

213 Peng, F.-Z., Shao, Z.-H., Fan, B.-M. et al. (2008). Organocatalytic enantioselective Michael addition of 2,4-pentandione to nitroalkenes promoted by bifunctional thioureas with central and axial chiral elements. *J. Org. Chem.* 73 (13): 5202–5205. https://doi.org/10.1021/jo800774m.

214 Jeong, H.J. and Kim, D.Y. (2015). Organocatalytic asymmetric Michael addition of α-fluoro β-ketophosphonate to nitroalkenes. *Bull. Korean Chem. Soc.* 36 (12): 2936–2939. https://doi.org/10.1002/bkcs.10580.

215 Martínez, J.I., Villar, L., Uria, U. et al. (2014). Bifunctional squaramide catalysts with the same absolute chirality for the diastereodivergent access to densely functionalised cyclohexanes through enantioselective domino reactions. Synthesis and mechanistic studies. *Adv. Synth. Catal.* 356 (17): 3627–3648. https://doi.org/10.1002/adsc.201400502.

216 Li, J.-H. and Du, D.-M. (2013). Squaramide-catalysed enantioselective Michael addition of pyrazolin-5-ones to nitroalkenes. *Org. Biomol. Chem.* 11 (36): 6215–6223. https://doi.org/10.1039/C3OB41045E.

217 Kim, Y.H., Yoon, J.H., Lee, M.Y., and Kim, D.Y. (2017). Enantioselective conjugate addition of pyrazolones to nitroalkenes catalyzed by binaphthyl-modified squaramide organocatalyst. *Bull. Korean Chem. Soc.* 38 (10): 1242–1245. https://doi.org/10.1002/bkcs.11241.

218 Saidalimu, I., Fang, X., Lv, W. et al. (2013). Organocatalytic asymmetric Michael addition/carbon-carbon bond cleavage of trifluoromethyl α-fluorinated *gem* -diols to nitroolefins. *Adv. Synth. Catal.* 355 (5): 857–863. https://doi.org/10.1002/adsc.201200757.

219 Li, B.-L., Wang, Y.-F., Luo, S.-P. et al. (2010). Enantioselective Michael addition of aromatic ketones to nitroolefins catalyzed by bifunctional thioureas and mechanistic insight. *Eur. J. Org. Chem.* (4): 656–662. https://doi.org/10.1002/ejoc.200900932.

220 Gu, Q., Guo, X.-T., and Wu, X.-Y. (2009). Highly enantioselective Michael addition of acetone to nitroolefins catalyzed by chiral bifunctional primary amine-thiourea catalysts with acetic acid. *Tetrahedron* 65 (27): 5265–5270. https://doi.org/10.1016/j.tet.2009.04.087.

221 Wang, Q., Peng, L., Fu, J. et al. (2010). Effective asymmetric Michael addition of acetone to nitroalkenes promoted by chiral proline amide-thiourea bifunctional catalysts. *Arkivoc* 2010 (2): 340–351. https://doi.org/10.3998/ark.5550190.0011.229.

222 Guo, X.-T., Sha, F., and Wu, X.-Y. (2016). Highly enantioselective Michael addition of α,α-disubstituted aldehydes to nitroolefins. *Res. Chem. Intermed.* 42 (7): 6373–6380. https://doi.org/10.1007/s11164-016-2468-0.

223 Avila, A., Chinchilla, R., Fiser, B. et al. (2014). Enantioselective Michael addition of isobutyraldehyde to nitroalkenes organocatalyzed by chiral primary amine-guanidines. *Tetrahedron: Asymmetry* 25 (5): 462–467. https://doi.org/10.1016/j.tetasy.2014.02.006.

224 Torregrosa-Chinillach, A., Sánchez-Laó, A., Santagostino, E., and Chinchilla, R. (2019). Organocatalytic asymmetric conjugate addition of aldehydes to maleimides and nitroalkenes in deep eutectic solvents. *Molecules* 24 (22): 4058. https://doi.org/10.3390/molecules24224058.

225 Durmaz, M. and Sirit, A. (2013). Calixarene-based highly efficient primary amine–thiourea organocatalysts for asymmetric Michael addition of aldehydes to nitrostyrenes. *Supramol. Chem.* 25 (5): 292–301. https://doi.org/10.1080/10610278.2013.773331.

226 De Simone, N.A., Meninno, S., Talotta, C. et al. (2018). Solvent-free enantioselective Michael reactions catalyzed by a calixarene-based primary amine thiourea. *J. Org. Chem.* 83 (17): 10318–10325. https://doi.org/10.1021/acs.joc.8b01454.

227 Guerrero-Corella, A., Esteban, F., Iniesta, M. et al. (2018). 2-Hydroxybenzophenone as a chemical auxiliary for the activation of ketiminoesters for highly enantioselective addition to nitroalkenes under bifunctional catalysis. *Angew. Chem. Int. Ed.* 57 (19): 5350–5354. https://doi.org/10.1002/anie.201800435.

228 Zuo, X., Liu, X.-L., Wang, J.-X. et al. (2019). Organocatalytic reaction of chromone-oxindole synthon: access to chromanone-based spirocyclohexaneoxindoles with five adjacent stereocenters. *J. Org. Chem.* 84 (11): 6679–6688. https://doi.org/10.1021/acs.joc.9b00326.

229 Jiang, L., Zheng, H.-T., Liu, T.-Y. et al. (2007). Asymmetric direct vinylogous carbon–carbon bond formation catalyzed by bifunctional organocatalysts. *Tetrahedron* 63 (24): 5123–5128. https://doi.org/10.1016/j.tet.2007.04.011.

230 Frias, M., Mas-Ballesté, R., Arias, S. et al. (2017). Asymmetric synthesis of Rauhut–Currier type products by a regioselective Mukaiyama reaction under bifunctional catalysis. *J. Am. Chem. Soc.* 139 (2): 672–679. https://doi.org/10.1021/jacs.6b07851.

231 Vila, C., Rostoll-Berenguer, J., Sánchez-García, R. et al. (2018). Enantioselective synthesis of 2-amino-1,1-diarylalkanes bearing a carbocyclic ring substituted indole through asymmetric catalytic reaction of hydroxyindoles with nitroalkenes. *J. Org. Chem.* 83 (12): 6397–6407. https://doi.org/10.1021/acs.joc.8b00612.

232 Belavagi, N.S., Panchamukhi, S.I., Deshapande, N. et al. (2015). Enantioselective synthesis of 6-substituted thiomorpholin-3-ones via asymmetric thia-Michael addition reactions using chiral bifunctional thiourea catalysts followed by intramolecular reductive cyclization. *Tetrahedron: Asymmetry* 26 (15, 16): 829–834. https://doi.org/10.1016/j.tetasy.2015.06.009.

233 Nazish, M., Jakhar, A., Gupta, N. et al. (2018). Enantioselective synthesis of β-nitro phosphonates catalyzed by a secondary amine bisthiourea. *Synlett* 29 (10): 1385–1389. https://doi.org/10.1055/s-0037-1609683.

234 Wang, Y., Zhang, W., Luo, S. et al. (2009). One-pot organocatalytic asymmetric synthesis of 3-nitro-1,2-dihydroquinolines by a dual-activation protocol. *Chem. Asian J.* 4 (12): 1834–1838. https://doi.org/10.1002/asia.200900298.

235 Yang, K.S., Nibbs, A.E., Türkmen, YE., and Rawal, V.H. (2013). Squaramide-catalyzed enantioselective Michael addition of masked acyl cyanides to substituted enones. *J. Am. Chem. Soc.* 135 (43): 16050–16053. https://doi.org/10.1021/ja409012q.

236 Yu, L., Yang, Z., Peng, J., and Li, P. (2016). Substrate-controlled synthesis of functionalized cyclohexanes with four stereocenters by organocatalytic asymmetric domino reactions between γ-nitro ketone and enone: synthesis of functionalized cyclohexanes with four stereocenters. *Eur. J. Org. Chem.* (3): 535–540. https://doi.org/10.1002/ejoc.201501303.

237 Wu, H., Tian, Z., Zhang, L. et al. (2012). Organocatalytic enantioselective aza-Michael addition of purine bases to α,β-unsaturated ketones. *Adv. Synth. Catal.* 354 (16): 2977–2984. https://doi.org/10.1002/adsc.201200488.

238 Mei, K., Jin, M., Zhang, S. et al. (2009). Simple cyclohexanediamine-derived primary amine thiourea catalyzed highly enantioselective conjugate addition of nitroalkanes to enones. *Org. Lett.* 11 (13): 2864–2867. https://doi.org/10.1021/ol9010322.

239 Guo, X.-T., Shen, J., Sha, F., and Wu, X.-Y. (2015). Highly enantioselective Michael addition of nitroalkanes to enones and its application in syntheses of (*R*)-Baclofen and (*R*)-Phenibut. *Synthesis* 47 (14): 2063–2072. https://doi.org/10.1055/s-0034-1380203.

240 Li, P., Wen, S., Yu, F. et al. (2009). Enantioselective organocatalytic Michael addition of malonates to α,β-unsaturated ketones. *Org. Lett.* 11 (3): 753–756. https://doi.org/10.1021/ol802892h.

241 Dudziński, K., Pakulska, A.M., and Kwiatkowski, P. (2012). An efficient organocatalytic method for highly enantioselective Michael addition of malonates to enones catalyzed by readily accessible primary amine-thiourea. *Org. Lett.* 14 (16): 4222–4225. https://doi.org/10.1021/ol3019055.

242 Moritaka, M., Miyamae, N., Nakano, K. et al. (2012). Highly efficient asymmetric Michael addition reaction of malonates to α,β-unsaturated ketones promoted by a chiral thiourea/PPY dual-catalyst system. *Synlett* 23 (17): 2554–2558. https://doi.org/10.1055/s-0032-1317317.

243 Lee, J.H. and Kim, D.Y. (2013). Enantioselective Michael addition of 2-hydroxy-1,4-naphthoquinone to β,γ-unsaturated α-keto esters catalyzed by binaphthyl-modified squaramide. *Bull. Korean Chem. Soc.* 34 (6): 1619–1620. https://doi.org/10.5012/bkcs.2013.34.6.1619.

244 Lu, R., Wei, W., Wang, J. et al. (2012). Organocatalytic conjugate addition of α-nitroacetates to β,γ-unsaturated α-keto esters and subsequent decarboxylation: synthesis of optically active δ-nitro-α-keto esters. *Tetrahedron* 68 (46): 9397–9404. https://doi.org/10.1016/j.tet.2012.09.014.

245 Lefranc, A., Gremaud, L., and Alexakis, A. (2014). Construction of bicyclo[3.2.1]octanes with four stereogenic centers by organocatalytic domino

Michael/aldol reaction. *Org. Lett.* 16 (20): 5242–5245. https://doi.org/10.1021/ol502171h.

246 Kim, D.Y. (2013). Organocatalytic enantioselective Friedel–Crafts reaction of naphthol with β,γ-unsaturated α-keto esters. *Bull. Korean Chem. Soc.* 34 (12): 3539–3540. https://doi.org/10.5012/bkcs.2013.34.12.3539.

247 Huang, K.-X., Xie, M.-S., Zhang, Q.-Y. et al. (2018). Synthesis of chiral six-membered carbocyclic purine nucleosides via organocatalytic enantioselective [3+3] annulation. *Org. Lett.* 20 (17): 5398–5401. https://doi.org/10.1021/acs.orglett.8b02309.

248 Preegel, G., Ilmarinen, K., Järving, I. et al. (2015). Enantioselective organocatalytic Michael addition–cyclization cascade of cyclopentane-1,2-dione with substituted (*E*)-2-oxobut-3-enoates. *Synthesis* 47 (23): 3805–3812. https://doi.org/10.1055/s-0035-1560347.

249 Qin, J., Zhang, Y., Liu, C. et al. (2019). Asymmetric inverse-electron-demand Diels–Alder reaction of β,γ-unsaturated amides through dienolate catalysis. *Org. Lett.* 21 (18): 7337–7341. https://doi.org/10.1021/acs.orglett.9b02629.

250 Shi, X.-M., Dong, W.-P., Zhu, L.-P. et al. (2013). Asymmetric vinylogous Michael addition/cyclization cascade reaction for the construction of diversely structured spiro-oxindole skeletons. *Adv. Synth. Catal.* 355 (16): 3119–3123. https://doi.org/10.1002/adsc.201300329.

251 Zhu, W.-R., Zhang, Z.-W., Huang, W.-H. et al. (2019). Asymmetric synthesis of vicinally bis(trifluoromethyl)-substituted 3,3′-pyrrolidinyl spirooxindoles via organocatalytic 1,3-dipolar cycloaddition reactions. *Synthesis* 51 (09): 1969–1979. https://doi.org/10.1055/s-0037-1612089.

252 Monari, M., Montroni, E., Nitti, A. et al. (2015). Highly stereoselective [4+2] and [3+2] spiroannulations of 2-(2-oxoindolin-3-ylidene)acetic esters catalyzed by bifunctional thioureas. *Chem. Eur. J.* 21 (31): 11038–11049. https://doi.org/10.1002/chem.201500676.

253 Tan, B., Hernández-Torres, G., and Barbas, C.F. (2011). Highly efficient hydrogen-bonding catalysis of the Diels–Alder reaction of 3-vinylindoles and methyleneindolinones provides carbazolespirooxindole skeletons. *J. Am. Chem. Soc.* 133 (32): 12354–12357. https://doi.org/10.1021/ja203812h.

254 Sun, W., Zhu, G., Wu, C. et al. (2012). An organocatalytic cascade strategy for the enantioselective construction of spirocyclopentane bioxindoles containing three contiguous stereocenters and two spiro quaternary centers. *Chem. Eur. J.* 18 (22): 6737–6741. https://doi.org/10.1002/chem.201200478.

255 Gómez-Torres, E., Alonso, D.A., Gómez-Bengoa, E., and Nájera, C. (2011). Conjugate addition of 1,3-dicarbonyl compounds to maleimides using a chiral C_2-symmetric bis(2-aminobenzimidazole) as recyclable organocatalyst. *Org. Lett.* 13 (22): 6106–6109. https://doi.org/10.1021/ol202599h.

256 Zu, L., Xie, H., Li, H. et al. (2007). Chiral amine thiourea-promoted enantioselective domino Michael-aldol reactions between 2-mercaptobenzaldehydes and maleimides. *Adv. Synth. Catal.* 349 (11, 12): 1882–1886. https://doi.org/10.1002/adsc.200700158.

257 Bai, J.-F., Peng, L., Wang, L. et al. (2010). Chiral primary amine thiourea promoted highly enantioselective Michael reactions of isobutylaldehyde with maleimides. *Tetrahedron* 66 (46): 8928–8932. https://doi.org/10.1016/j.tet.2010.09.044.

258 Sohtome, Y., Takemura, N., Takagi, R. et al. (2008). Thiourea-catalyzed Morita–Baylis–Hillman reaction. *Tetrahedron* 64 (40): 9423–9429. https://doi.org/10.1016/j.tet.2008.07.087.

259 Sakamoto, S., Kazumi, N., Kobayashi, Y. et al. (2014). Asymmetric synthesis of trisubstituted oxazolidinones by the thiourea-catalyzed aldol reaction of 2-isocyanatomalonate diester. *Org. Lett.* 16 (18): 4758–4761. https://doi.org/10.1021/ol502198e.

260 Nakamura, H., Tsukano, C., Yoshida, T. et al. (2019). Total synthesis of caprazamycin A: practical and scalable synthesis of *syn*-β-hydroxyamino acids and introduction of a fatty acid side chain to 1,4-diazepanone. *J. Am. Chem. Soc.* 141 (21): 8527–8540. https://doi.org/10.1021/jacs.9b02220.

261 Alegre-Requena, J.V., Marqués-López, E., and Herrera, R.P. (2018). Organocatalyzed enantioselective aldol and Henry reactions starting from benzylic alcohols. *Adv. Synth. Catal.* 360 (1): 124–129. https://doi.org/10.1002/adsc.201701351.

262 Alegre-Requena, J.V., Marqués-López, E., Sanz Miguel, P.J., and Herrera, R.P. (2014). Organocatalytic enantioselective hydrophosphonylation of aldehydes. *Org. Biomol. Chem.* 12 (8): 1258–1264. https://doi.org/10.1039/C3OB42403K.

263 Liu, Y., Gao, P., Wang, J. et al. (2012). Primary 1,2-diamine catalysis (V): efficient asymmetric aldol reactions of isatins with cyclohexanone. *Synlett* 23 (07): 1031–1034. https://doi.org/10.1055/s-0031-1290507.

264 Raj, M., Veerasamy, N., and Singh, V.K. (2010). Highly enantioselective synthesis of 3-cycloalkanone-3-hydroxy-2-oxindoles, potential anticonvulsants. *Tetrahedron Lett.* 51 (16): 2157–2159. https://doi.org/10.1016/j.tetlet.2010.02.082.

265 Li, T.-Z., Jiang, Y., Guan, Y.-Q. et al. (2014). Direct enantioselective vinylogous aldol–cyclization cascade reaction of allyl pyrazoleamides with isatins: asymmetric construction of spirocyclic oxindole–dihydropyranones. *Chem. Commun.* 50 (74): 10790–10792. https://doi.org/10.1039/C4CC04235B.

266 Wu, H., Wang, L., Zhang, J., and Jin, Y. (2019). Urea-derivative catalyzed enantioselective hydroxyalkylation of hydroxyindoles with isatins. *Molecules* 24 (21): 3944. https://doi.org/10.3390/molecules24213944.

267 Puglisi, A., Raimondi, L., Benaglia, M. et al. (2009). Enantioselective catalytic addition of nitroesters to N-carboalkyloxy imines: a route to quaternary stereocenters. *Tetrahedron Lett.* 50 (30): 4340–4342. https://doi.org/10.1016/j.tetlet.2009.05.036.

268 Kang, Y.K. and Kim, D.Y. (2009). Organocatalytic highly enantio- and diastereoselective Mannich reaction of β-ketoesters with *N*-Boc-aldimines. *J. Org. Chem.* 74 (15): 5734–5737. https://doi.org/10.1021/jo900880t.

269 Cai, H., Zhou, Y., Zhang, D. et al. (2014). A Mannich/cyclization cascade process for the asymmetric synthesis of spirocyclic thioimidazolidineoxindoles. *Chem. Commun.* 50 (94): 14771–14774. https://doi.org/10.1039/C4CC06000H.

270 Frías, M., Carrasco, A.C., Fraile, A., and Alemán, J. (2018). A general asymmetric formal synthesis of aza-Baylis–Hillman type products under bifunctional catalysis. *Chem. Eur. J.* 24 (13): 3117–3121. https://doi.org/10.1002/chem.201705218.

271 Feng, B.-X., Wang, B., and Li, X. (2016). Asymmetric additions of thioglycolates and *N*-Boc aldimines catalyzed by a bifunctional tertiary-amine squaramide. *Org. Biomol. Chem.* 14 (39): 9206–9209. https://doi.org/10.1039/C6OB01645F.

272 Mittal, N., Sun, D.X., and Seidel, D. (2014). Conjugate-base-stabilized Brønsted acids: catalytic enantioselective Pictet–Spengler reactions with unmodified tryptamine. *Org. Lett.* 16 (3): 1012–1015. https://doi.org/10.1021/ol403773a.

273 Jung, S.H. and Kim, D.Y. (2008). Catalytic enantioselective electrophilic α-hydrazination of β-ketoesters using bifunctional organocatalysts. *Tetrahedron Lett.* 49 (38): 5527–5530. https://doi.org/10.1016/j.tetlet.2008.07.041.

274 Trillo, P., Gómez-Martínez, M., Alonso, D., and Baeza, A. (2014). 2-Aminobenzimidazole organocatalyzed asymmetric amination of cyclic 1,3-dicarbonyl compounds. *Synlett* 26 (01): 95–100. https://doi.org/10.1055/s-0034-1379494.

275 Benavent, L., Baeza, A., and Freckleton, M. (2018). Chiral 2-aminobenzimidazole as bifunctional catalyst in the asymmetric electrophilic amination of unprotected 3-substituted oxindoles. *Molecules* 23 (6): 1374. https://doi.org/10.3390/molecules23061374.

276 Lim, Y.J. and Kim, D.Y. (2013). Enantioselective direct α-amination of aromatic ketones catalyzed by binaphthyl-modified primary amine. *Bull. Korean Chem. Soc.* 34 (7): 1955–1956. https://doi.org/10.5012/bkcs.2013.34.7.1955.

277 Foster, R.W., Lenz, E.N., Simpkins, N.S., and Stead, D. (2017). Organocatalytic stereoconvergent synthesis of α-CF_3 amides: triketopiperazines and their heterocyclic metamorphosis. *Chem. Eur. J.* 23 (37): 8810–8813. https://doi.org/10.1002/chem.201701548.

278 Ding, X.-H., Cui, W.-C., Li, X. et al. (2013). Organocatalytic enantioselective conjugate addition of 2-nitrocyclohexanone to acrylaldehyde: a concise two-step synthesis of chiral building block 1-azaspiro[4.5]decan-6-one. *Tetrahedron Lett.* 54 (15): 1956–1959. https://doi.org/10.1016/j.tetlet.2013.01.114.

279 Xiong, X., Zheng, T., Wang, X. et al. (2020). Access to chiral bisphenol ligands (BPOL) through desymmetrizing asymmetric ortho-selective halogenation. *Chem* 6 (4): 919–932. https://doi.org/10.1016/j.chempr.2020.01.009.

280 Mairhofer, C., Novacek, J., and Waser, M. (2020). Synergistic ammonium (hypo)iodite/imine catalysis for the asymmetric α-hydroxylation of β-ketoesters. *Org. Lett.* 22 (15): 6138–6142. https://doi.org/10.1021/acs.orglett.0c02198.

281 Novacek, J., Monkowius, U., Himmelsbach, M., and Waser, M. (2016). Asymmetric α-chlorination of β-ketoesters using bifunctional ammonium salt catalysis. *Monatsh. Chem.* 147 (3): 533–538. https://doi.org/10.1007/s00706-015-1604-7.

282 Li, D.R., He, A., and Falck, J.R. (2010). Enantioselective, organocatalytic reduction of ketones using bifunctional thiourea-amine catalysts. *Org. Lett.* 12 (8): 1756–1759. https://doi.org/10.1021/ol100365c.

283 Barrios Antúnez, D.-J., Greenhalgh, M.D., Fallan, C. et al. (2016). Enantioselective synthesis of 2,3-disubstituted trans-2,3-dihydrobenzofurans using a Brønsted base/thiourea bifunctional catalyst. *Org. Biomol. Chem.* 14 (30): 7268–7274. https://doi.org/10.1039/C6OB01326K.

284 Brindle, C.S., Yeung, C.S., and Jacobsen, E.N. (2013). Chiral β-iodoamines by urea-catalysed iodocyclization of trichloroacetimidates. *Chem. Sci.* 4 (5): 2100–2104. https://doi.org/10.1039/C3SC50410G.

285 Tripathi, C.B. and Mukherjee, S. (2014). Catalytic enantioselective iodoaminocyclization of hydrazones. *Org. Lett.* 16 (12): 3368–3371. https://doi.org/10.1021/ol5013982.

286 Min, C., Lin, C.-T., and Seidel, D. (2015). Catalytic enantioselective intramolecular aza-Diels-Alder reactions. *Angew. Chem. Int. Ed.* 54 (22): 6608–6612. https://doi.org/10.1002/anie.201501536.

287 Witten, M.R. and Jacobsen, E.N. (2014). Catalytic asymmetric synthesis of 8-oxabicyclooctanes by intermolecular [5+2] pyrylium cycloadditions. *Angew. Chem. Int. Ed.* 53 (23): 5912–5916. https://doi.org/10.1002/anie.201402834.

288 Mayr, F., Brimioulle, R., and Bach, T. (2016). A chiral thiourea as a template for enantioselective intramolecular [2+2] photocycloaddition reactions. *J. Org. Chem.* 81 (16): 6965–6971. https://doi.org/10.1021/acs.joc.6b01039.

5

Diketopiperazines as Chiral Building Blocks

Juan Domingo Sánchez, Juan Francisco González, and José Carlos Menéndez

Unidad de Química Orgánica y Farmacéutica, Departamento de Química en Ciencias Farmacéuticas, Facultad de Farmacia, Universidad Complutense, Plaza de Ramón y Cajal, s.n., 28040 Madrid, Spain

5.1 Introduction

2,5-Diketopiperazines (2,5-DKPs) are cyclodipeptides arising from the double condensation of two amino acids. They are the simplest cyclic peptides and have attracted much attention from the synthetic community due to their presence in many natural products. Some representative examples (Figure 5.1) include brevianamide F, i.e. cyclo-(L-Trp-L-Pro), which has been shown to have antibacterial and antifungal properties [1], rhodotorulic acid, a hydroxamate siderophore [2], phenylahistin, a mitosis inhibitor [3] and cairomycin B, a bridged system arising from cyclo(L-Lys-L-Asp) and endowed with antibiotic properties. 2,5-DKP cores are often present in natural products as a subunit that can be found embedded in complex frameworks. Two examples are spirotryprostatin B, an inhibitor of the cell cycle at the G2/M interphase [4], and verroculugen, an inhibitor of the Slo-1 calcium-activated potassium channel of nematodes [5].

The 2,5-DKP nucleus is also considered a privileged structure in drug discovery because its presence in molecules enables them to bind to a broad range of pharmacological targets [6]. 2,5-DKPs have additional attractive features that include their ability to mimic certain peptide conformations while showing high metabolic stability, the presence of both H-bond acceptor and H-bond donor sites, and the high degree of three-dimensionality of their derivatives.

Due to the importance of DKPs, their chemistry has been reviewed several times [7–11]. This chapter will focus on using DKPs as chiral building blocks to prepare drug molecules and natural products.

Natural and synthetic α-amino acids, which embody a stereogenic center, are an important source of chirality for synthetic endeavors and are in most cases available in both enantiomeric forms. DKPs are constituted by two amino acids, incorporating one or two chiral residues and introducing structural diversity at four positions (nitrogens N_1 and N_4 and carbons C_3 and C_6). This results in a chiral pool of $\frac{1}{2}(n(n+1))$ compounds, where "n" represents the number of chiral amino acids.

Chiral Building Blocks in Asymmetric Synthesis: Synthesis and Applications, First Edition.
Edited by Elżbieta Wojaczyńska and Jacek Wojaczyński.

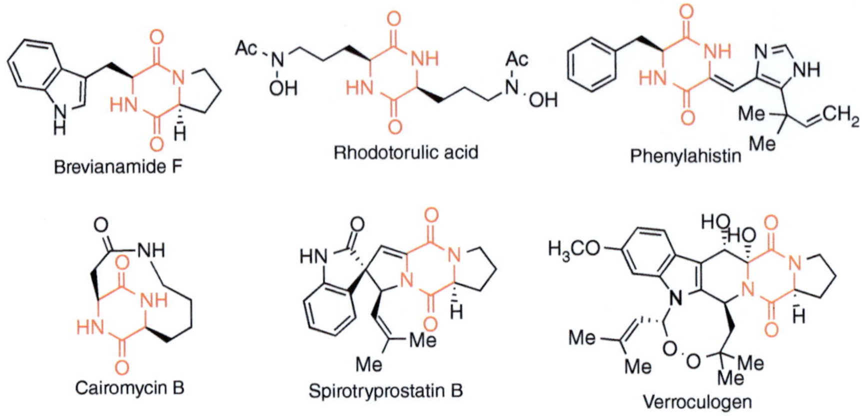

Figure 5.1 Representative natural products containing 2,5-diketopiperazine cores.

Additionally, the existence of *cis*- and *trans*-diastereoisomers doubles the number of possible enantiomerically pure 2,5-DKPs.

5.2 A Brief Overview of the Synthesis of 2,5-Diketopiperazines

The most widely used strategies for generating 2,5-DKPs involve a key step that creates one or two bonds and are summarized in Figure 5.2.

5.2.1 Methods that Create the N_1-C_2 Bond

5.2.1.1 Cyclization of Dipeptides

The most general way to obtain 2,5-DKPs starts with the condensation of an *N*-protected amino acid **1** with a second amino acid in the form of an ester **2** to give a dipeptide **3**, followed by the removal of the amine protecting group to yield **4** and a final cyclization that creates the piperazine ring and furnishes compound **5** (Figure 5.3). The cyclization step needs a cis orientation of the amide bond. It requires heating under reflux with high boiling solvents [12] or microwave irradiation under solvent-free conditions [13] or in water [14]. Acids and bases may

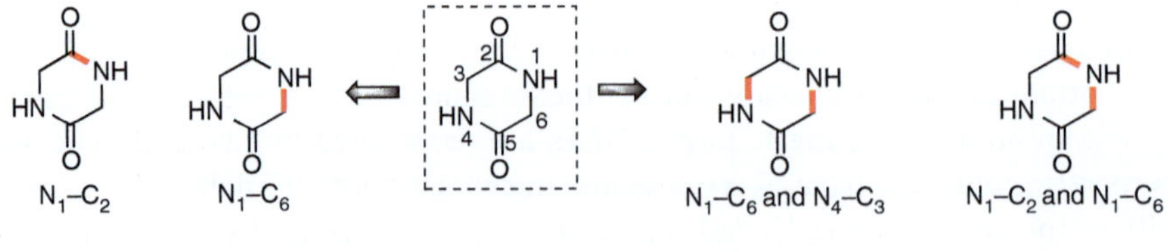

Figure 5.2 Main strategies for the synthesis of 2,5-DKPs.

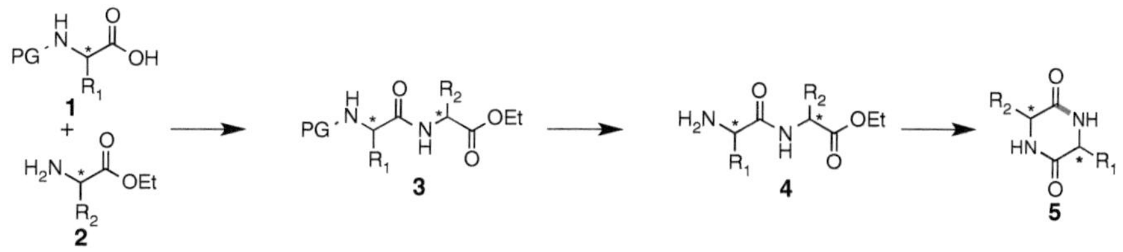

Figure 5.3 Synthesis of 2,5-DKPs from dipeptide esters.

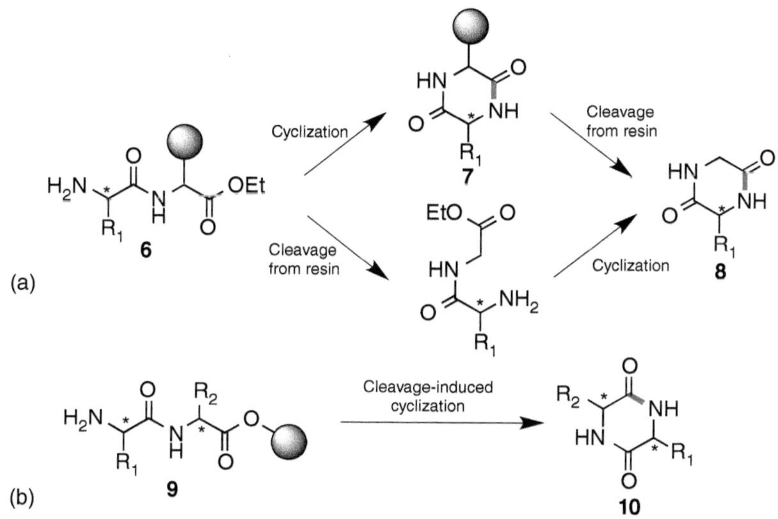

Figure 5.4 Strategies for the synthesis of 2,5-DKPs on solid supports. (a) Strategies involving cleavage from the resin prior to the final step. (b) Strategy based on a final cleavage-induced cyclization.

catalyze the cyclization reaction, but, especially in the latter case, racemization of the C-3 and C-6 stereocenters in the final compound can sometimes be observed.

Solid-phase synthesis techniques have also been applied for the preparation of 2,5-DKPs. Three approaches have been employed: solid-state preparation of a linear dipeptide **6**, followed by cyclization to the C_5-unsubstituted, resin-bound 2,5-DKP **7**, concluding with release from the resin to form the final product **8**. The order of these operations may be reversed (Figure 5.4a). Alternatively, solid-phase synthesis can be applied to prepare dipeptide **9**, followed by its cleavage-induced cyclization to **10** (Figure 5.4b).

5.2.1.2 Cyclization of Ugi Reaction Products

The Ugi reaction is a four-component process starting from isonitriles, aldehydes, carboxylic acids, and amines. Using an α-amino acid as the carboxylic component affords a dipeptide **11** that, following its *N*-deprotection to **12**, can act as a potential DKP precursor. The cyclization step normally requires harsh thermal conditions since it involves the attack of an amine onto an unreactive amide (Figure 5.5).

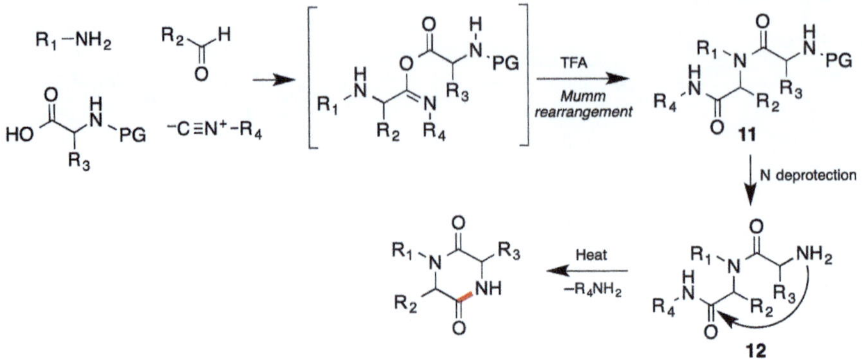

Figure 5.5 Synthesis of 2,5-DKPs via the Ugi reaction: general concept.

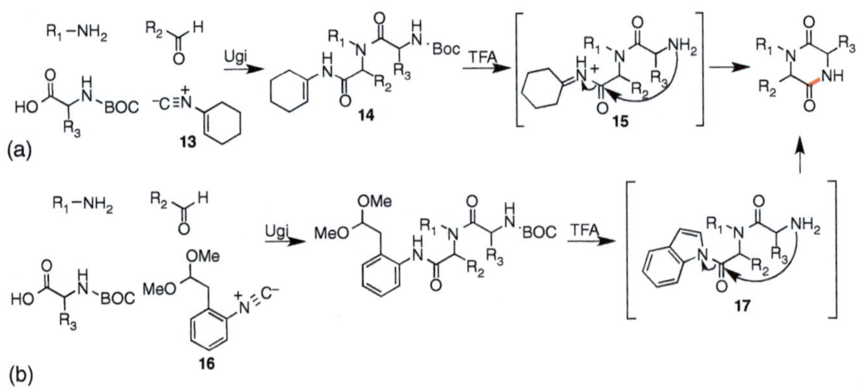

Figure 5.6 Synthesis of 2,5-DKPs via the Ugi reaction: representative practical protocols.

One approach to improve this step involves the use of 1-cyclohexenyl isonitrile **13**, which favors cyclization because of the generation of a highly reactive intermediate acyliminium cation **15** by protonation in the course of the BOC deprotection of dipeptide **14** (Figure 5.6a) [15]. Another isonitrile derivative that can produce a reactive group that favors cyclization to form 2,5-DKPs is 1-isocyano-2-(2,2-dimethoxyethyl)benzene **16**, which allows a one-pot protocol, with retention of stereocenters, via an indolamide intermediate **17** (Figure 5.6b) [16]. More recently, *N*-*tert*-butyl amides, obtained via Ugi chemistry from *tert*-butyl isonitrile, have been shown to yield DKPs by a one-step BOC-deprotection/lactamisation in acetic acid [17].

5.2.1.3 Staudinger/Intramolecular Aza-Wittig Sequences

Another approach to 2,5-DKPs is summarized in Figure 5.7. Acylation of amino acid with chloroacetyl chloride gives intermediate **18**, and is followed by chloride displacement by sodium azide to provide **19**. A subsequent Staudinger reaction by treatment with PPh_3 yields the corresponding iminophosphorane **20**, which undergoes

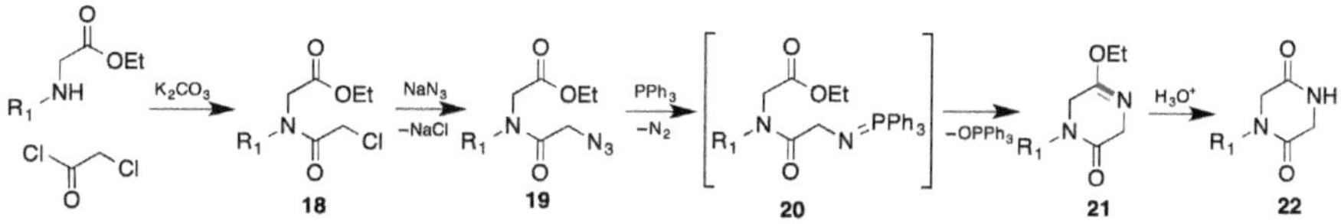

Figure 5.7 Synthesis of 2,5-DKPs by intramolecular aza-Wittig reactions. Source: Hulme et al. [15]/Elsevier.

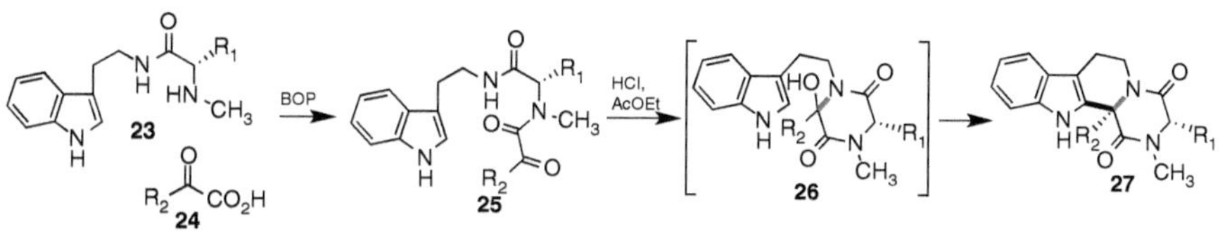

Figure 5.8 Synthesis of 2,5-DKPs coupled to Pictet–Spengler reactions.

an intramolecular aza-Wittig reaction to furnish iminoether **21**. Finally, the hydrolysis of **21** yields the target 2,5-DKPs **22** [18].

5.2.1.4 DKP Formation Coupled to Additional Cyclization Reactions

The use of bivalent electrophiles may allow access to complex heterocyclic frameworks by combining the formation of a 2,5-DKP with an additional cyclization event. For example, the coupling of aminoacyl-tryptamine derivatives **23** with pyruvic acid derivatives **24** in the presence of BOP gave ketoamides **25**, which were cyclized in an acidic medium to give a 2,5-DKP **26**. This non-isolated intermediate afforded a tetrahydro-β-carboline **27** through a Pictet–Spengler condensation, as shown in Figure 5.8 [19].

5.2.2 Methods that Create the N_1-C_6 Bond

5.2.2.1 From α-Haloacyl Amino Acids

2,5-DKPs can be synthesized via intramolecular amide *N*-alkylation reaction. This method starts with an Ugi reaction where α-chloroacetic acid **28** is the carboxyl component and yields intermediate **29**, further cyclized via *N*-alkylation in the presence of a base (Figure 5.9). The main disadvantage of this method is the possibility of epimerization under strongly basic conditions required for the second step [20].

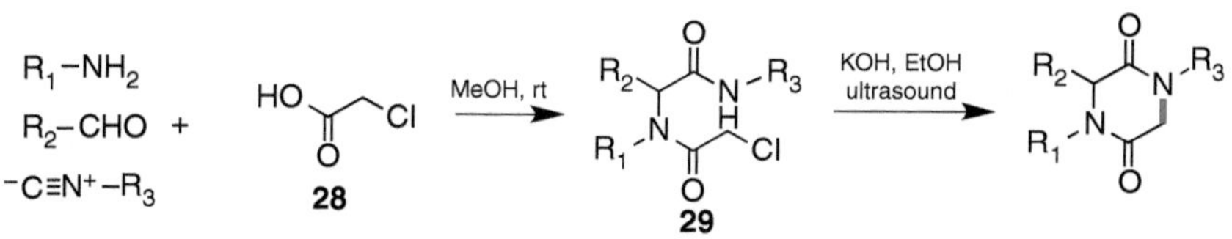

Figure 5.9 Synthesis of 2,5-DKPs by intramolecular amide alkylation.

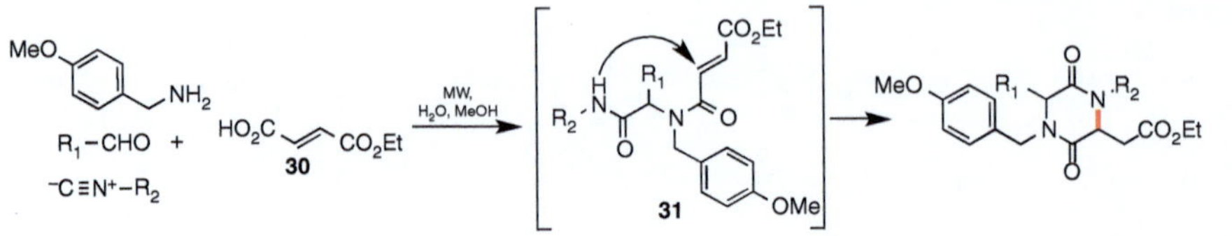

Figure 5.10 Synthesis of 2,5-DKPs by a domino Ugi-aza-Michael sequence. Source: Santra and Andreana [21]/American Chemical Society.

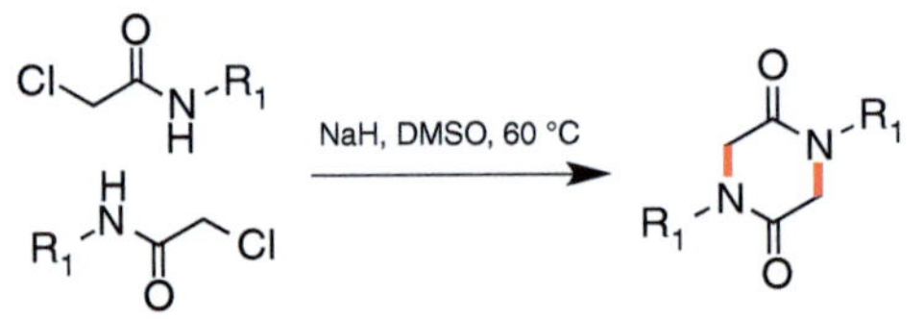

Figure 5.11 Synthesis of 2,5-DKPs by dimerization of α-haloacetamides.

5.2.2.2 By Aza-Michael Additions

Another variation of the Ugi reaction that can be applied to the synthesis of 2,5-DKPs involves using a monoester of maleic acid **30** as the carboxyl component. Under microwave irradiation, the intermediate Ugi product **31** cyclizes via an intramolecular aza-Michael reaction to furnish a DKP in a one-pot domino procedure (Figure 5.10) [21].

5.2.3 Methods that Create Two Bonds

5.2.3.1 Formation of N_1-C_6 and N_4-C_3 Bonds

2,5-DKPs can be obtained from the one-pot dimerization of α-haloacetamides in the presence of a strong base (Figure 5.11). The reaction gives good yields, except for the case of aryl groups with electron-withdrawing substituents and, although it generally affords non-chiral compounds, is compatible with the presence of a stereogenic center at R_1 benzylic side chains [22]. Alternatively, this transformation can be performed using phase transfer catalysis [23].

5.2.3.2 Formation of N_1-C_2 and N_1-C_6 Bonds

The formation of $N_1—C_2$ and $N_1—C_6$ bonds of 2,5-DKPs in a single operation can be performed from *N*-haloacyl derivatives of α-aminoesters and primary amines. One example was described in the research on synthesizing stereoisomeric 2,5-diazabicyclo[2.2.2]octanes as ligands of the σ_1 receptor, which required the

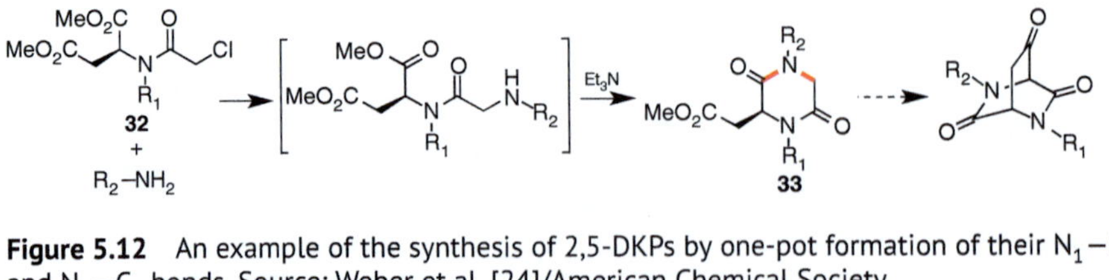

Figure 5.12 An example of the synthesis of 2,5-DKPs by one-pot formation of their N_1-C_2 and N_1-C_6 bonds. Source: Weber et al. [24]/American Chemical Society.

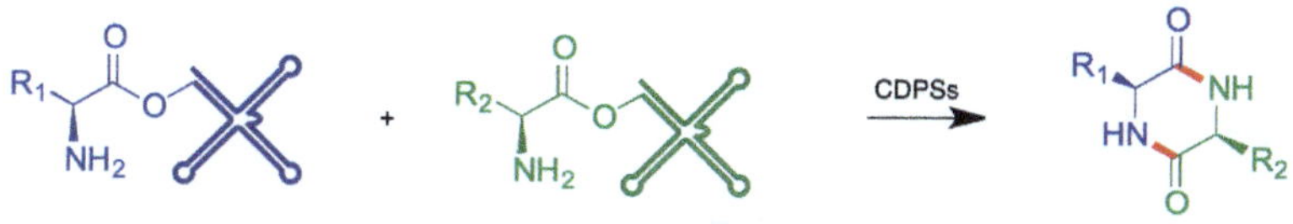

Figure 5.13 Biosynthesis of 2,5-DKPs from aminoacyl-tRNAs catalyzed by cyclodipeptide synthases.

preparation of homochiral DKPs as starting materials. For instance, the aspartic acid derivative **32** was transformed into DKPs **33** by treatment with primary amines in the presence of triethylamine (Figure 5.12) [24].

This strategy was also employed to synthesize phosphodiesterase 5 inhibitors, which eventually led to the development of tadalafil (see Section 5.3.1 below).

5.2.4 Biotechnological Methods

Two different enzyme families can carry out the biosynthesis of 2,5-DKP, namely non-ribosomal peptidase synthetases and the recently discovered cyclodipeptide synthases (CDPSs), which catalyze the formation of cyclodipeptides using as substrates aminoacyl-tRNA (Figure 5.13). The latter enzymes are highly promising biotechnological tools for synthesizing structurally diverse 2,5-DKPs [25, 26]. Recent work discussed reprogramming *Escherichia coli* to produce prenylated indole DKP alkaloids by co-expressing bacterial CDPSs to assemble the DKP ring and fungal prenyltransferases (PTs) to attach the prenyl substituents [27]. Furthermore, 2,5-DKPs can serve as precursors for structures that can serve as substrates of various enzymes, including methyltransferase, prenyltransferase, cyclodipeptide oxidase, and cytochrome P450, leading to the formation of secondary metabolites [28].

5.3 2,5-Diketopiperazines in Drug Synthesis

2,5-DKPs are amongst the most extensively used chiral scaffolds in drug discovery. There are many literature examples where they are employed as a template to build enantiomerically pure bioactive molecules. Below are some representative examples.

5.3.1 Synthesis of Tadalafil

Tadalafil, a potent and selective inhibitor of the phosphodiesterase V enzyme, is an orally active drug marketed as Cialis® by Eli Lilly to treat male erectile dysfunction. It is characterized by a longer half-life compared to related drugs. Moreover, in 2009, it was approved to treat pulmonary hypertension. Its structure is characterized by a tetrahydro-β-carboline heterocyclic ring fused to a DKP and contains two stereocentres, one of which is embedded in the DKP ring.

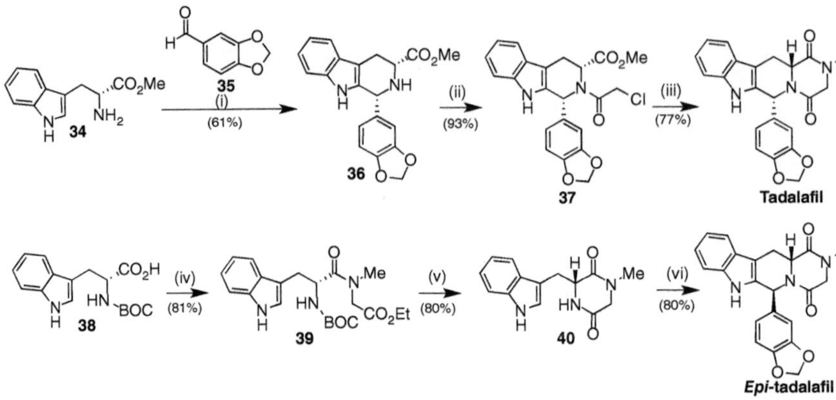

Figure 5.14 Two synthetic approaches to tadalafil. Reagents and conditions: (i) Trifluoroacetic acid, toluene; (ii) $ClCH_2COCl$, $NaHCO_3$, $CHCl_3$; (iii) $MeNH_2$, MeOH; (iv) (a) Ethyl chloroformate, *N*-methylmorpholine (NMM), THF. (b) Ethyl sarcosinate hydrochloride, NMM; (v) (a) HCl, dioxane. (b) $NaHCO_3$, dichloromethane, reflux. (vi) Compound **35**, HCl, dioxane-water, 100 °C.

Many methods have been used to prepare tadalafil, most of which employ tryptophan as a source of chirality and generate the DKP at the last step [29]. In Figure 5.14, one of these synthetic routes has a key intermediate tetrahydro-β-carboline **36**, obtained from D-trytophan methyl ester **34** and piperonal **35** by a diastereoselective Pictet–Spengler cyclization under acidic conditions, with no racemization (pathway A). Treatment of **36** with chloroacetyl chloride leads to the formation of amide **37**, which cyclizes to tadalafil by reaction with methylamine [30, 31]. Similar Pictet–Spengler chemistry has been performed from DKP intermediates. Thus, treatment of *N*-BOC-D-Trp with ethyl sarcosinate, with prior activation with ethyl chloroformate, afforded dipeptide **39**, which was deprotected and cyclized thermally to cyclo(D-Trp-L-Sar) **40**. Treatment of this DKP with aldehyde **35** under acidic catalysis afforded the epimer of tadalafil at the piperidine stereocenter [32]. This example showcases how the reactions starting from DKPs often have different stereochemical outcomes than those starting from the corresponding dipeptides.

5.3.2 Synthesis of Trofinetide (NNZ-2591)

DKPs have attracted interest as synthetic analogs of endogenous neuropeptides, demonstrating neuroprotection against various neuronal diseases and beneficial memory effects [33]. In this context, the allyl DKP trofinetide (NNZ2591) was designed as an analog of cyclo(glycine-proline) to achieve increased half-life, stability, and oral bioavailability. This compound is neuroprotective after ischaemic brain injury and in Parkinson's disease models. It is under clinical development by Neuren Pharmaceuticals for the treatment of the Angelman, Phelan-McDermid, and Pitt Hopkins syndromes, receiving orphan drug designation from the FDA in October 2019.

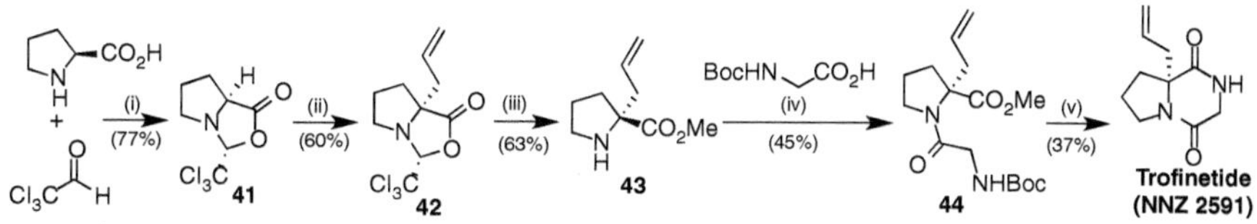

Figure 5.15 Synthesis of NNZ-2591. Reagents and conditions: (i) $CHCl_3$, reflux; (ii) allyl bromide, LDA; (iii) acetyl chloride, CH_3OH, reflux; (iv) BOP-Cl, Et_3N, dichloromethane; (v) trifluoroacetic acid, dichloromethane, then Et_3N.

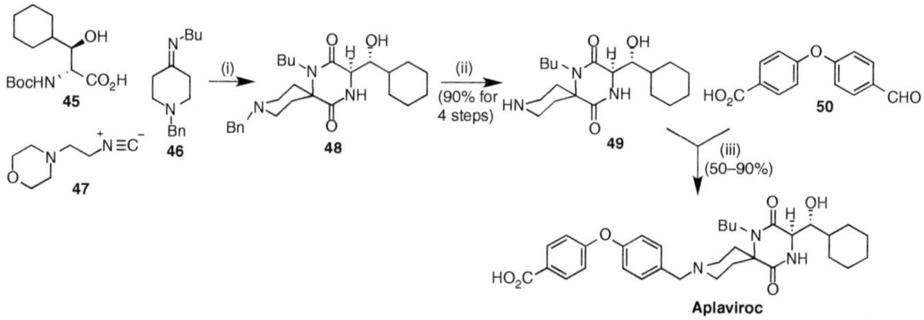

Figure 5.16 Synthesis of aplaviroc. Reagents and conditions: (i) (a) MeOH, 55 °C, (b) concentrated HCl, 55 °C; (ii) (a) H_2, $Pd(OH)_2$/C, (b) 4 M HCl, EtOAc; (iii) (a) $NaBH(AcO)_3$, AcOH, DMF, (b) 4 M HCl/AcOEt.

Figure 5.15 illustrates the synthesis of trofinetide from L-proline. Its reaction with trichloroacetaldehyde led to derivative **41** via cyclization of a hemiaminal-type intermediate. The diastereoselective C-alkylation of **41** with allyl bromide in the presence of lithium diisopropylamide afforded compound **42**, which was transformed into the allylproline methyl ester **43** by treatment with acetyl chloride and methanol. Finally, the coupling reaction with Boc-glycine followed by the removal of the BOC protecting group by trifluoroacetic acid treatment led to trofinetide [34].

5.3.3 Synthesis of Aplaviroc

Aplaviroc (GSK-873140) is a potent non-competitive CCR5 antagonist with antiviral effect, which was developed to treat HIV by GlaxoSmithKline. Its synthesis has the key step in the Ugi reaction between *N*-Boc-amino acid **45**, imine **46**, and isocyanide **47**, followed by removing the BOC protecting group to promote the formation of the spirocyclic DKP **48**. The next step was debenzylation by catalytic hydrogenation to obtain compound **49**. Finally, reductive amination of the phenoxybenzaldehyde derivative **50** with **49** yielded aplaviroc [35]. All four possible enantiomers of aplaviroc were synthesized in optically pure form using this approach. Their pharmacological study showed that the 3*R*,1′*R* isomer exhibited stronger inhibitory activities in binding assays (Figure 5.16).

5.3.4 Synthesis of Retosiban

Retosiban (GSK-221,149-A) is an orally active oxytocin receptor antagonist developed by GlaxoSmithKline to prevent preterm birth [36]. Its synthesis

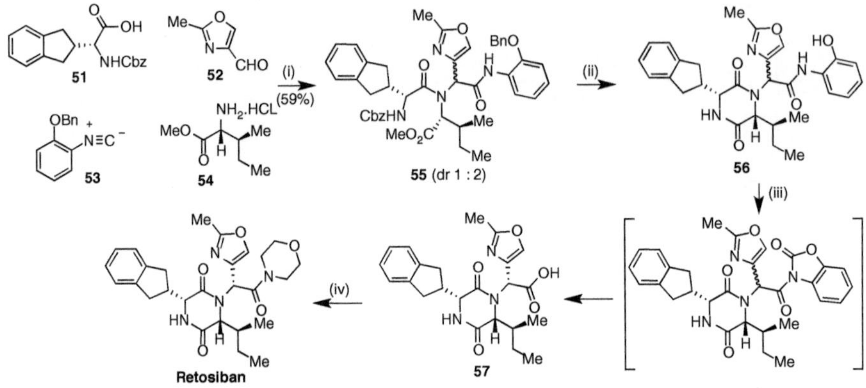

Figure 5.17 Synthesis of retosiban. Reagents and conditions: (i) Triethylamine, MeOH; (ii) H_2, Pd/C, ethanol/acetic acid; (iii) (a) carbonyl diimidazole, dichloromethane, (b) 2 M HCl, acetone; (iv) PyBOP, dichloromethane, then morpholine. Source: Sollis [37]/American Chemical Society.

starts with the Ugi reaction between the protected (*R*)-indanylglycine derivative **51**, 2-methyloxazole-4-carboxaldehyde **52**, 2-benzyloxyphenylisonitrile **53**, and D-allo-isoleucine methyl ester **54** that yields the linear dipeptidic intermediate **55**. This compound was obtained as a mixture of diastereoisomers in which the required one was obtained as a minor product in 1 : 2 ratio [37]. Catalytic hydrogenation led to deprotection of the *O*-benzyl and *Cbz* groups, with concomitant cyclization to afford the DKP system (compound **56**). The selective hydrolysis of the phenolic amide was achieved by reaction with carbonyl diimidazole (CDI), followed by the addition of aqueous hydrochloric acid to give the carboxylic acid **57**, which was transformed into retosiban by activation of the acid with the peptide coupling reagent PyBOP (benzotriazol-1-yloxytripyrrolidinophosphonium hexafluorophosphate) and addition of morpholine. Although intermediates **55** and **56** were obtained as mixtures of diastereoisomers when this mixture was activated with carbonyl diimidazole followed by acidic hydrolysis, epimerization at the position α to the exocyclic amide took place, and the desired diastereoisomer was obtained as the primary product (Figure 5.17) [37].

5.4 Natural Product Synthesis

Chiral DKPs are useful templates for the asymmetric synthesis of amino acids. Furthermore, DKPs have been employed frequently as chiral pool starting materials to synthesize various alkaloids, a valuable source of promising compounds for drug discovery due to their varied bioactivities.

5.4.1 Asymmetric Synthesis of Amino Acids

One of the oldest and best-known applications of DKPs in asymmetric synthesis is the Schöllkopf bis-lactim ether method for synthesizing unnatural amino acids.

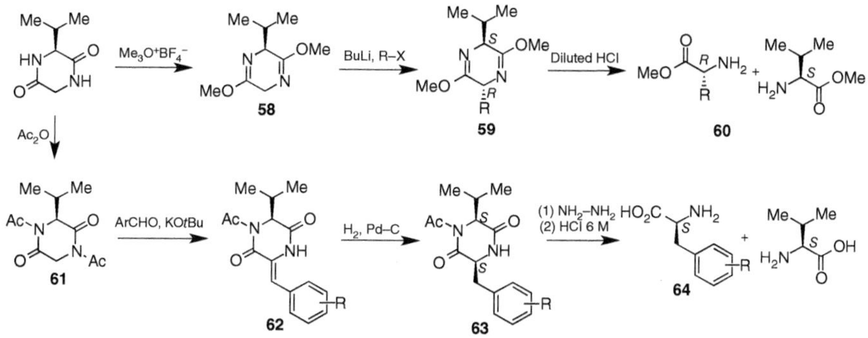

Figure 5.18 Asymmetric synthesis of amino acids from 2,5-DKPs. Source: Mollica et al. [39]/John Wiley & Sons.

Starting from cyclo(L-Val-Gly), the corresponding bis-lactim ether **58** is prepared by *O*-methylation with Meerwein's salt. Regioselective deprotonation of the Gly methylene with BuLi followed by alkylation with a suitable alkyl halide from the opposite side to the Val substituent affords compound **59** in an excellent diastereomeric excess. Final acid hydrolysis under mild conditions affords amino acid **60**, with opposite configuration to the valine chiral auxiliary, together with recovered Val-OMe, which can be recycled to the starting DKP. Cleavage of compounds **59** to dipeptides has also been described [38].

An alternative approach to amino acids involves the *N*-diacetylation of the starting DKP to give compound **61**. Its aldol condensation with aromatic aldehydes in the presence of a base with concomitant deacetylation of the nitrogen atom adjacent to the reaction site yields **62**. Its catalytic hydrogenation takes place opposite the valine isopropyl with full chiral induction. It affords **63**, which can be hydrolyzed to the aromatic amino acid **64**, with the same *S* configuration as the Val auxiliary (Figure 5.18) [39].

5.4.2 Synthesis of Alkaloids Containing a 2,5-Diketopiperazine Core

5.4.2.1 Tryprostatin B

Tryprostatin B, an inhibitor of the mammalian cell cycle, is a prenylindole alkaloid isolated from *Aspergillus fumigatus*. A synthetic approach to this compound from cyclo(L-Trp-L-Pro) **65**, starting with a domino C_3-alkylation-*N*-arylation process induced by treatment with prenyl bromide in an aqueous buffer containing magnesium nitrate, was reported. This reaction afforded a mixture of the target tryprostatin B and the pentacyclic compound **66**, isolated as a mixture of two diastereomers. The prenyl chain in this compound was transferred to the indole C-2 position by treatment with trifluoroacetic acid, which also was added to the double bond to give **67**. For this reason, a final elimination step was required. It was induced by treatment of **67** with triethylamine to give tryprostatin B in a 32% overall yield. The rearrangement could also be carried out directly by treating with ytterbium triflate (Figure 5.19). The very mild conditions suggest that this route could form a biosynthetic pathway for tryprostatins [40].

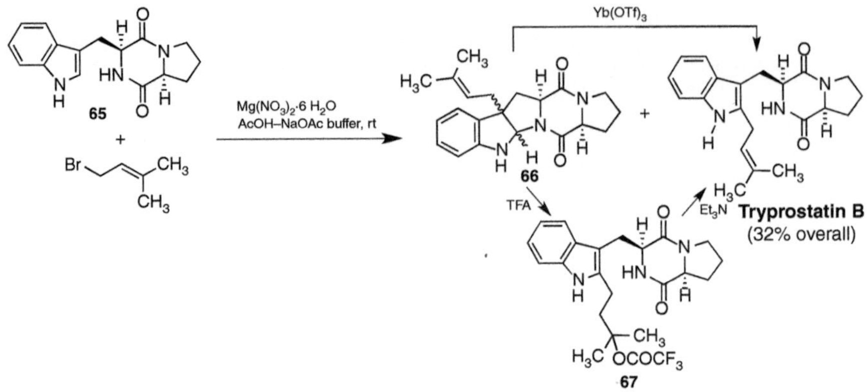

Figure 5.19 Synthesis of tryprostatin B from cyclo(L-Trp-L-Pro).

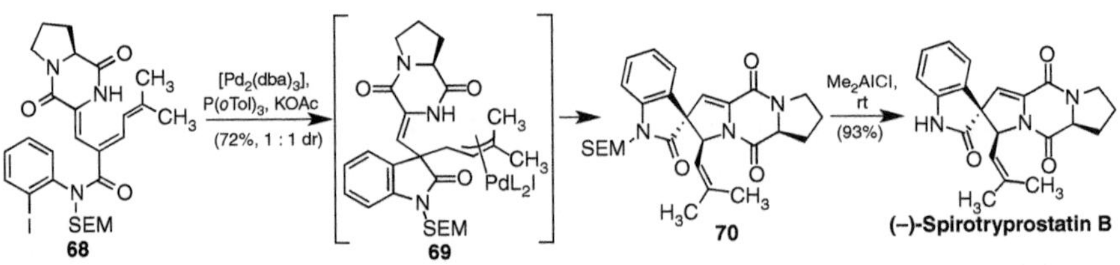

Figure 5.20 Synthesis of spirotryprostatin based on an intramolecular Heck reaction of a functionalized iodoaniline. Source: Overman and Rosen [42]/Elsevier.

5.4.2.2 Spirotryprostatin B

The related spirotryprostatin B is another fungal metabolite from *A. fumigatus* that shows antimitotic properties and has been the subject of much synthetic work [41]. An example of these total syntheses is the method developed by Overman involving a palladium-catalyzed domino process initiated by an intramolecular Heck reaction of the functionalized iodoaniline **68**. The intermediate palladium-allyl species **69** reacted with the amide nitrogen under the same reaction conditions to generate the *N*-SEM derivative of the alkaloid (compound **70**), which was finally deprotected in the presence of Me_2AlCl (Figure 5.20) [42].

5.4.2.3 Stephacidin B

Many DKP natural products contain 2,5-diazabicyclo[2.2.2]octane structural fragments. One interesting example of such a molecular architecture is the antitumor indole alkaloid stephacidin B, isolated from the fungus *Aspergillus ochraceus*. The Baran group has reported a concise route to this heptacyclic framework that involved cyclization of the cyclo-(Pro-Trp) derivative **71** by treatment with LDA and iron acetylacetonate. This radical carbon–carbon bond forming event afforded the bridged intermediate **72**. Some standard functional group manipulations gave compound **73**, which was heated at high temperature to induce a thermal retro-ene reaction that removed the BOC protective group in the form of a molecule of

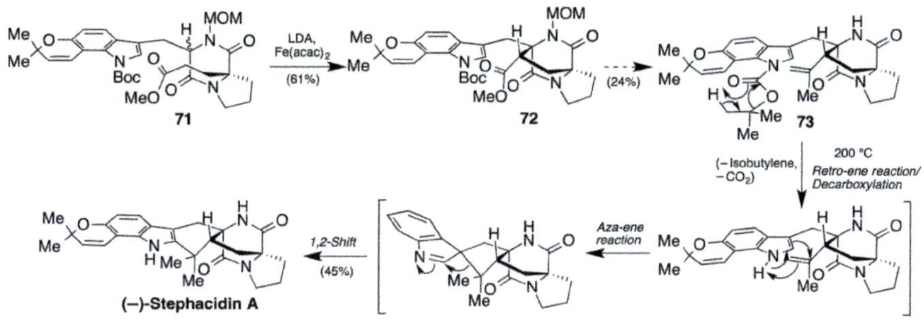

Figure 5.21 Synthesis of stephacidin B from a functionalized diketopiperazine. Source: Mollica et al. [39]/John Wiley & Sons.

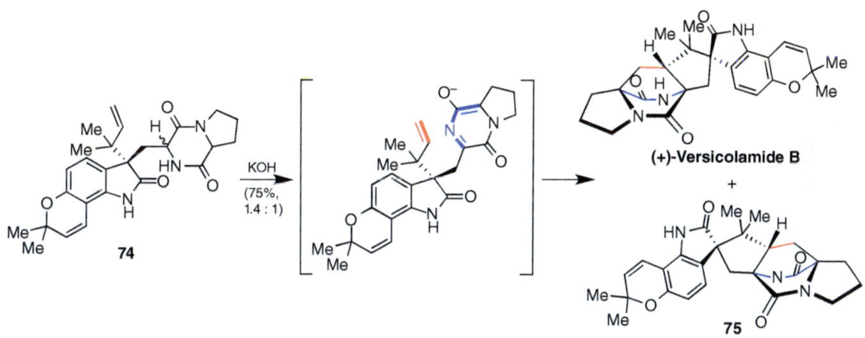

Figure 5.22 Synthesis of versicolamide B via an intramolecular Diels–Alder reaction.

isobutylene and a molecule of carbon dioxide to furnish a deprotected indole. A thermally-induced intramolecular aza ene-type reaction with the participation of the indole N–H generated an unstable spiroindolenine that underwent a thermal 1,2-shift that furnished stephacidin A as a single enantiomer (Figure 5.21) [43].

5.4.2.4 Versicolamide B

Another approach to DKP alkaloids containing 2,5-diazabicyclo[2.2.2]octane moieties is based on intramolecular hetero Diels–Alder reactions. One example of such transformation is summarized in Figure 5.22. It constitutes the key step of the total synthesis reported by Williams of the *Aspergillus* alkaloid (+)-versicolamide B from the advanced DKP precursor **74** via an enolate intermediate [44]. The target alkaloid was isolated, together with its diastereomer **75**, in a 1.4 : 1 ratio.

5.4.2.5 Variecolortide A

3-Methylene-2,5-piperazinediones can be employed as the dienophile partner of cycloadditions. An oxa Diels–Alder reaction based on this concept has been employed by Zipse and Trauner as the key step of a total synthesis of variecolortide A, a fungal metabolite with modest cytotoxic effects. This architecturally complex alkaloid was generated in one step by reaction of the natural products hydroxyviocristin and isoechinulin A at high temperature. This transformation

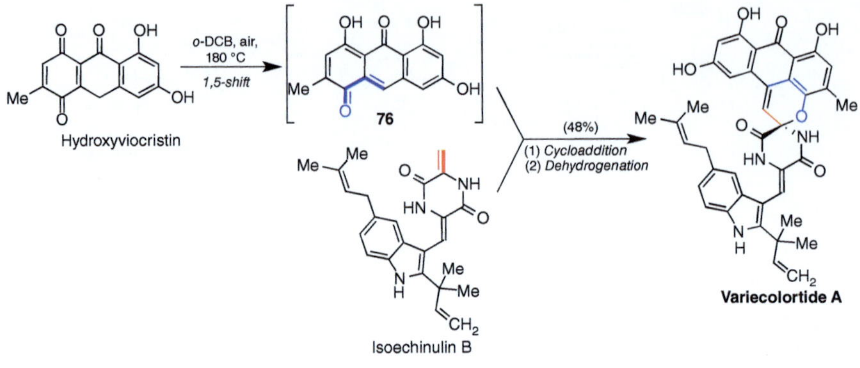

Figure 5.23 Synthesis of variecolortide A via an intermolecular Diels–Alder reaction of isoechinulin B. Source: Baran et al. [43]/John Wiley & Sons.

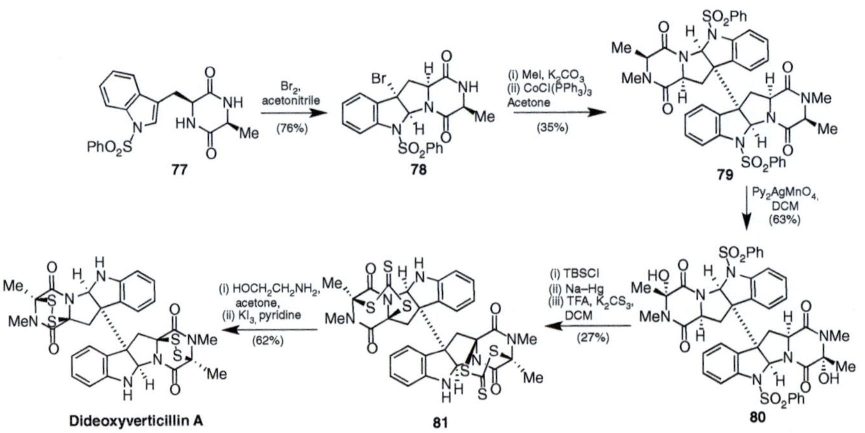

Figure 5.24 Synthesis of dideoxyverticillin A from a protected cyclo(L-Trp-L-Ala).

was interpreted as a domino process comprising an initial 1,5-hydrogen shift in hydroxyviocristin to generate intermediate **76**, containing an α,β-unsaturated ketone moiety that acts as an oxadiene in a hetero Diels–Alder reaction with the exocyclic double bond of isoechinulin B that creates the key spirocyclic N,O acetal moiety. A final air-induced dehydrogenation completed the synthesis of the natural product (Figure 5.23) [45].

5.4.2.6 Dideoxyverticillin

Dideoxyverticillin is a member of the epidithiodiketopiperazine family of alkaloids, isolated from marine *Penicillium* sp., and showed cytotoxic activity [46]. Structurally, this alkaloid is a dimeric molecule derived from a highly functionalized bicyclic dithiodiketopiperazine core. One synthesis of this structure that employed a DKP as a template was reported by Movassaghi and coworkers and is summarized in Figure 5.24. The protected DKP **77** was prepared from *N*-Boc-L-tryptophan, L-alanine methyl ester, and phenylsulfonyl chloride to modulate the activity of the indole ring.

Treatment of **77** with bromine in acetonitrile led to the tetracyclic intermediate **78**, in a regio- and stereoselective fashion. This heterocyclic system was *N*-methylated and submitted to radical reductive dimerization with a tris(triphenylphosphine) cobalt complex in acetonitrile to obtain the dimeric octacyclic system **79**. This key dimerization occurred with high diastereoselectivity among the generated quaternary centers (C3–C3′), giving mainly the *cis*-stereoisomer. The hydroxylation of the α-methyl positions in **79** with bis-(pyridine)-silver(I) permanganate afforded the diol **80** as a single diastereomer. After hydroxyl protection as TBS ethers, the benzenesulfonyl protecting groups were removed with sodium amalgam. Treatment with trithiocarbonate led to form the two bicyclic systems of compounds **81** via a double nucleophilic attack. The addition of ethanolamine led to the cleavage of the bicyclo system with the formation of an unstable tetrathiol intermediate, which was oxidized with potassium triiodide to generate the disulfide bridges characteristic of the final product [47].

5.4.3 Synthesis of Alkaloids Containing a Modified 2,5-Diketopiperazine Core

5.4.3.1 Ardeemin

The ardeemins are a family of fused quinazoline alkaloids that have shown potent activity as inhibitors of glycoprotein P-170, responsible for multidrug resistance to anticancer agents (MDR). Thus, *N*-acetylardeemin and *N*-acetyl-8-demethylardeemin were shown to reverse *in vitro* resistance to vinblastine or taxol 700-fold at relatively noncytotoxic concentrations [48]. The first total synthesis of these alkaloids, reported by Danishefsky, involved the preparation of the reverse-prenylated hexahydropyrrolo[2,3-*b*]indole derivative **82**, which was coupled to alanine methyl ester to yield **83**. BOC deprotection in the presence of trimethylsilyl iodide followed by the addition of ammonia in dimethylaminopyridine allowed the formation of the DKP ring, affording compound **84** as a single enantiomer. Its acylation with *o*-azidobenzoyl chloride gave **85**, transformed into ardeemin by an intramolecular Staudinger-aza Wittig reaction. Its transformation into *N*-acetylardeemin was achieved by treatment of **85** with acetyl chloride in the presence of LDA [49] (Figure 5.25).

5.4.3.2 Phakellin

The phakellin group of natural products belongs to the pyrrole–imidazole family of alkaloids, isolated from marine sponges. Total synthesis of (+)-phakellin reported by Nasagawa started from 4-hydroxy-L-proline, which was transformed into the 2-acylpyrrole intermediate **87** and DKP **88**, which in turn became **89** after some straightforward functional group manipulation. Next, treatment of this compound with trichloroacetonitrile gave intermediate **90**, which underwent an enamide-type Overman rearrangement to afford compound **91**, containing the key quaternary stereocenter attached to a nitrogen function. Finally, some additional manipulation furnished the target alkaloid (Figure 5.26) [50].

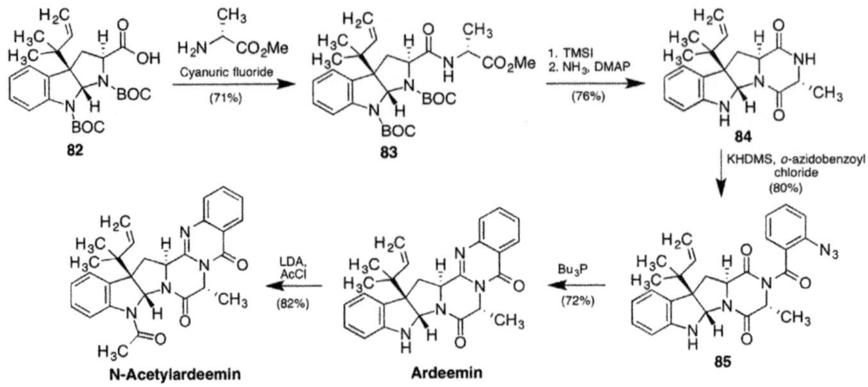

Figure 5.25 Total synthesis of ardeemin and *N*-acetylardeemin.

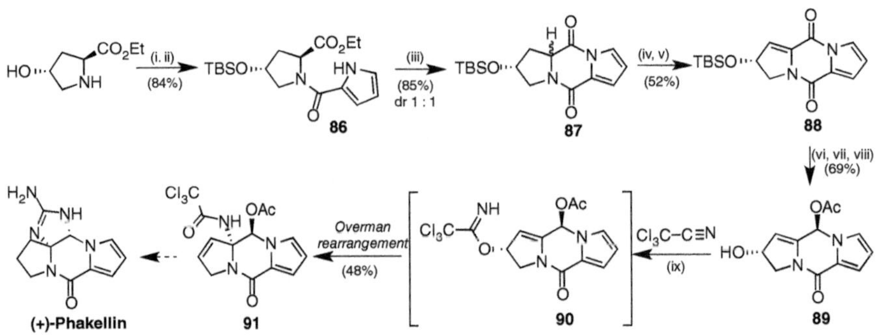

Figure 5.26 Total synthesis of (+)-phakellin based on an Overman rearrangement. Reagents and conditions: (i) TBSCl, imidazole, dichloromethane, rt; (ii) pyrrole-2-carboxylic acid, EDCI, 4-dimethylaminopyridine (DMAP), dichloromethane, rt; (iii) NaH, THF, 0 °C (85%, dr 1 : 1); (iv) IBX·Et_3N N-oxide, DMSO, rt; (v) MsCl, Et_3N, dichloromethane, rt; (vi) $NaBH_4$,$CeCl_3$·$7H_2O$, EtOH/THF 0 °C; (vii) Ac_2O, pyridine, dichloromethane, rt; (viii) HF·Et_3N, THF, rt (75% over two steps, dr 1 : 1); (ix) DBU, dichloromethane, rt.

5.4.3.3 Sarcodonins

Sarcodonin ε and phellodonin are two of the sarcodonin family of natural products. They have unprecedented architectures containing a tricyclic benzodioxazine core and an *N*-oxide at the ring junction. Baran has disclosed a divergent synthesis of these two compounds having as the key step a biomimetic hetero-Diels–Alder reaction between a pyrazine *N*-oxide and an *ortho*-quinone. Thus, cyclo(L-Ile-L-Ile) **92** was transformed into the piperazine-bis-*N*-oxide **93** and then into its derivative **94**, which was the precursor to the non-isolated **95** expected to act as the Diels–Alder dienophile in the key step of the route. On the other hand, *ortho*-quinone **96** was prepared from dichlorobenzoquinone by a 10-step route involving two arylation reactions. When **95** and **96** were reacted together, the products of two hetero Diels–Alder reactions having as the dienophile the double bonds in positions (a) and (b) were obtained (compounds **97** and **98**). Their final debenzylation under

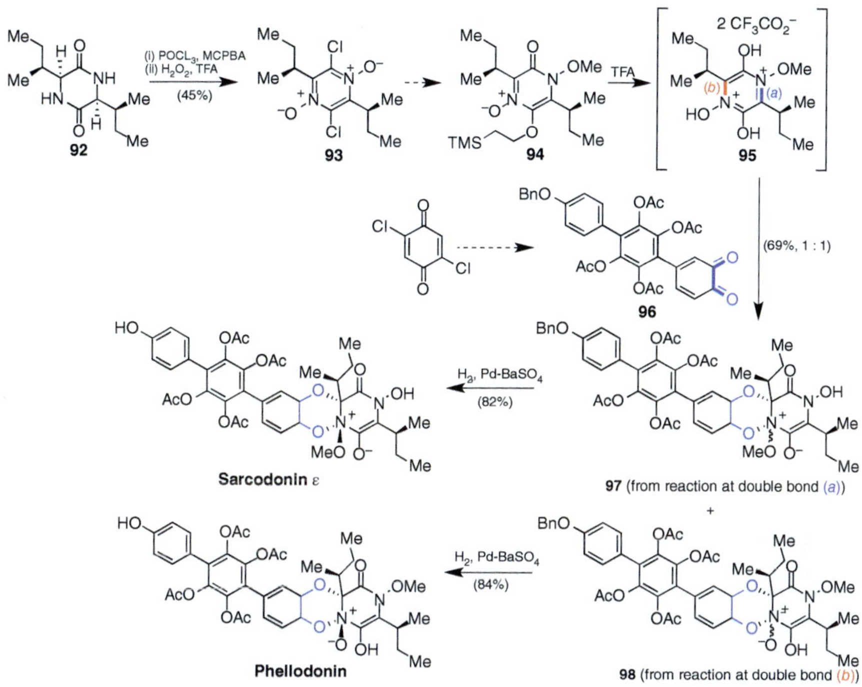

Figure 5.27 Synthesis of sarcodonin ε and phellodonin from a common diketopiperazine precursor.

catalytic hydrogenation conditions afforded the target sarcodonin ε and phellodonin, respectively [51] (Figure 5.27).

5.4.3.4 Ecteinascidin 743

This alkaloid is a metabolite isolated from the Caribbean tunicate *Ecteinascidia turbinata* and was shown to possess highly potent activity (1–100 pM range) against several cancer cell lines such as colon, central nervous system, melanoma, kidney, and breast cancer [52, 53]. Under the INN name trabectedin, it has been approved as an anticancer therapy for several tumors and marketed under the name Yondelis®. The chemical structure of trabectedin is highly complex. In addition to the characteristic pentacyclic template of the tetrahydroisoquinoline family of natural products, a further tetrahydroisoquinoline moiety is attached to the B-ring by a 10-membered lactone. Its low availability from natural sources limited the clinical development of trabectedin. However, this problem was overcome by developing a semisynthetic route from the readily available cyanosafracin B by scientists at PharmaMar [54]. Due to the involvement of a DKP intermediate, we will discuss the total synthesis developed by Fukuyama [55], which is summarized in Figure 5.28. The highly functionalized chiral starting materials **99** and **100**, together with *p*-methoxyphenyl isocyanide and acetaldehyde, were heated in methanol to furnish dipeptide **101** through an Ugi reaction, which provided the carbon atoms necessary to form at a later stage the pentacyclic core of the alkaloid.

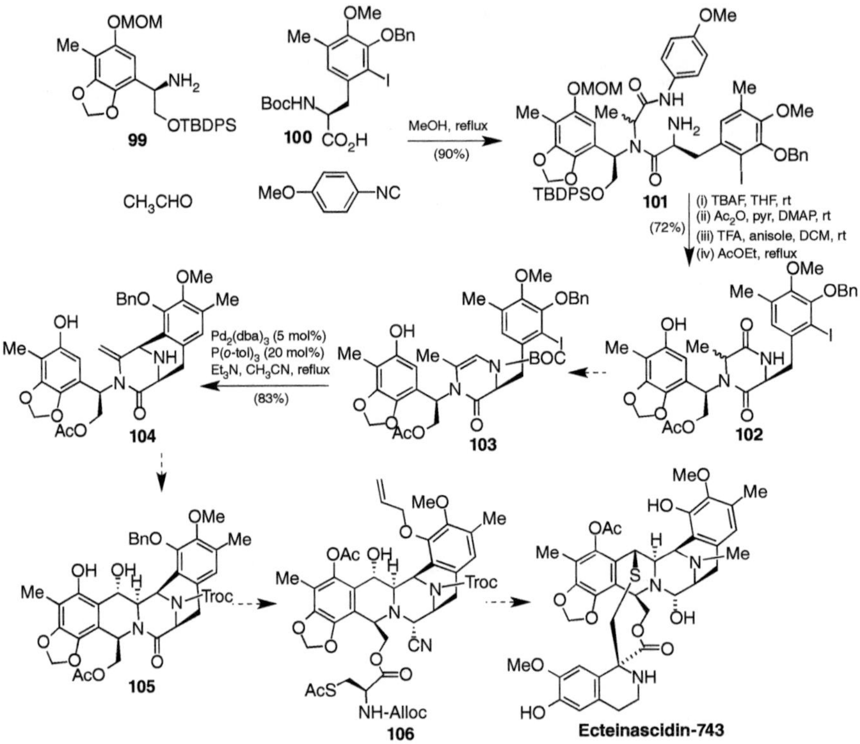

Figure 5.28 Fukaiyama's total synthesis of ecteinascidin 743 from an Ugi-derived DKP.

This dipeptide was cyclized to DKP **102** by *N*-deprotection followed by heating. Subsequent steps were directed at forming cyclic enamide **103** by partial reduction of the carbamate-activated DKP carbonyl followed by elimination. Hence, this transformation set the stage for an intramolecular Heck reaction that afforded the tricyclic compound **104**. The preparation of the pentacyclic structure **105** was achieved using an oxidative cyclization based on the reactivity of the exocyclic double bond at C-3. After appropriate modification of the functional and protective groups, the 10-membered sulfide ring was formed by intramolecular ring closure of intermediate **106**. Finally, a Pictet–Spengler reaction was performed to incorporate the tetrahydroisoquinoline moiety and provided ecteinascidin 743.

5.5 Conclusions

2,5-Piperazinediones (DKPs) are readily available, densely functionalized chiral cyclic dipeptides that are excellent building blocks for the generation of structurally diverse heterocyclic systems, including drug molecules and bioactive natural products.

References

1 Nishanth, K.S., Nath, V.S., Pratap Chandran, R., and Bala, N. (2014). Cyclic dipeptides from rhabditid entomopathogenic nematode-associated *Bacillus cereus* have antimicrobial activities. *World J. Microbiol. Biotechnol.* 30 (2): 439–449.

2 Andersen, D., Renshaw, J.C., and Wiebe, M.G. (2003). Rhodotorulic acid production by *Rhodotorula mucilaginosa*. *Mycol. Res.* 107 (8): 949–956.

3 Hayashi, Y., Yamazaki-Nakamura, Y., and Yakushiji, F. (2013). Medicinal chemistry and chemical biology of diketopiperazine-type antimicrotubule and vascular-disrupting agents. *Chem. Pharm. Bull.* 61 (9): 889–901.

4 Cui, C.B., Kakeya, H., and Osada, H. (1996). Spirotryprostatin B, a novel mammalian cell cycle inhibitor produced by *Aspergillus fumigatus*. *J. Antibiot.* 49 (8): 832–835.

5 Kulke, D., von Samson-Himmelstjerna, G., Miltsch, S.M. et al. (2014). Characterization of the Ca^{2+}-gated and voltage-dependent K^{+}-channel Slo-1 of nematodes and its interaction with emodepside. *PLoS Negl.Trop. Dis.* 8: e3401.

6 Yet, L. (2018). *Privileged Structures in Drug Discovery: Medicinal Chemistry and Synthesis*. Hoboken, NJ, USA: Wiley.

7 Martins, M.B. and Carvalho, I. (2007). Diketopiperazines: biological activity and synthesis. *Tetrahedron* 63 (40): 9923–9932.

8 González, J.F., Ortín, I., de la Cuesta, E., and Menéndez, J.C. (2012). Privileged scaffolds in synthesis: 2,5-piperazinediones as templates for the preparation of structurally diverse heterocycles. *Chem. Soc. Rev.* 41 (21): 6902–6915.

9 Borthwick, A.D. (2012). 2,5-Diketopiperazines: synthesis, reactions, medicinal chemistry, and bioactive natural products. *Chem. Rev.* 112 (7): 3641–3716.

10 Wang, Y., Wang, P., Ma, H., and Zhu, W. (2013). Developments around the bioactive diketopiperazines: a patent review. *Exp. Opin. Therap. Pat.* 23 (11): 1415–1433.

11 Scarel, M. and Marchesan, S. (2021). Diketopiperazine gels: new horizons from the self-assembly of cyclic dipeptides. *Molecules* 26: 3376.

12 Bull, S.D., Davies, S.G., and Moss, W.O. (1998). Practical synthesis of Schöllkpf's bis-lactim ether chiral auxiliary: (3*S*)-3,6-dihydro-2,5-dimethoxy-3-isopropylpyrazine. *Tetrahedron: Asymmetry* 9: 321–327.

13 López-Cobeñas, A., Cledera, P., Sánchez, J.D. et al. (2005). Microwave-assisted synthesis of 2,5-piperazinediones under solvent-free conditions. *Synthesis* 2005, 19: 3412–3422.

14 Pérez-Picaso, L., Escalante, J., Olivo, H.F., and Ríos, M.Y. (2009). Efficient microwave assisted syntheses of 2,5-diketopiperazines in aqueous media. *Molecules* 14 (8): 2836–2849.

15 Hulme, C., Morrisette, M.M., Volz, F.A., and Burns, C.J. (1998). The solution phase synthesis of diketopiperazine libraries via the Ugi reaction: novel application of Armstrong's convertible isonitrile. *Tetrahedron Lett.* 39 (10): 1113–1116.

16 Rhoden, C.R., Rivera, D.G., Kreye, O. et al. (2009). Rapid access to N-substituted diketopiperazines by one pot Ugi 4CR/deprotection + activation/cyclization (UDAC). *J. Comb. Chem.* 11 (6): 1078–1082.

17 Jida, M. and Ballet, S. (2018). An efficient one-pot synthesis of chiral *N*-protected 3-substituted (diketo)piperazines via Ugi-4CR/de-Boc/cyclization process. *ChemistrySelect* 3 (4): 1027–1031.

18 Majumdar, K.C., Ray, K., and Ganai, S. (2010). Intramolecular Aza-Wittig reaction: a new efficient tool for the construction of piperazine 2,5-dione derivatives. *Synlett* 2010 (14): 2122–2124.

19 Siwicka, A., Wojtasieicz, K., Rosiek, B. et al. (2005). Diastereodivergent synthesis of 2,5-diketopiperazine derivatives of β-carboline and isoquinoline from L-amino acids. *Tetrahedron: Asymmetry* 16 (5): 975–993.

20 Marcaccini, S., Pepino, R., and Pozo, M.C. (2001). A facile synthesis of 2,5-diketopiperazines based on isocyanide chemistry. *Tetrahedron Lett.* 42 (14): 2727–2728.

21 Santra, S. and Andreana, P.R. (2007). A one-pot, microwave-influenced synthesis of diverse small molecules by multicomponent reaction cascades. *Org. Lett.* 9 (24): 5035–5038.

22 Cho, S.D., Song, S.Y., Kim, K.H. et al. (2004). One-pot synthesis of symmetrical 1,4-disubstituted piperazine-2,5-diones. *Bull. Korean Chem. Soc.* 25 (3): 415–416.

23 O'Reilly, E., Lestini, E., Balducci, D., and Paradisi, F. (2009). One-step diketopiperazine synthesis using phase transfer catalysis. *Tetrahedron Lett.* 50 (15): 1748–1750.

24 Weber, F., Brune, S., Boergel, F. et al. (2016). Rigidity versus flexibility: is this an issue in σ_1 receptor ligand affinity and activity? *J. Med. Chem.* 59 (11): 5505–5519.

25 Borgman, P., Lopez, R.D., and Lane, A.L. (2019). The expanding spectrum of diketopiperazine natural product biosynthetic pathways containing cyclodipeptide synthases. *Org. Biomol. Chem.* 17 (9): 2305–2314.

26 Canu, N., Moutiez, M., Belin, P., and Gondry, M. (2020). Cyclodipeptide synthases: a promising biotechnological tool for the synthesis of diverse 2,5-diketopiperazines. *Nat. Prod. Rep.* 37 (3): 312–321.

27 Dubois, P., Correia, I., Le Chevalier, F. et al. (2019). Reprogramming *Escherichia coli* for the production of prenylated indole diketopiperazine alkaloids. *Sci. Rep.* 9: 9208.

28 Harken, L. and Shu-Ming, L. (2021). Modifications of diketopiperazines assembled by cyclodipeptide synthases with cytochrome P450 enzymes. *Appl. Microbiol. Biotechnol.* 105 (6): 2277–2285.

29 Gouda, M.A. (2017). Overview of the synthetic routes to tadalafil and its analogues. *Synth. Commun.* 47 (24): 2269–2304.

30 Maw, G.N., Allerton, C.M.N., Gbekor, E., and Million, W.A. (2003). Design, synthesis and biological activity of β-carboline-based type-5 phosphodiesterase inhibitors. *Bioorg. Med. Chem. Lett.* 13 (8): 1425–1428.

31 Daugan, A., Grondin, P., Ruault, C. et al. (2003). The discovery of tadalafil: a novel and highly selective PDE5 inhibitor. 2: 2,3,6,7,12,12a-hexahydropyrazino[1′,2′:1,6]-pyrido[3,4-*b*]indole-1,4-dione analogues. *J. Med. Chem.* 46 (21): 4533–4542.

32 Vedantham, R., Shanmugam, S., Vetukuri, P.V.N.K.V. et al. (2013). First highly stereocontrolled synthesis of tetrahydro *trans*-β-carboline derivatives by exploiting the influence of a cyclic amide. *Arkivoc* 2013 (ii): 22–32.

33 Faden, A.I., Movsesyan, V.A., Knoblach, S.M. et al. (2005). Neuroprotective effects of novel small peptides in vitro and after brain injury. *Neuropharmacology* 49 (3): 410–424.

34 Guan, J., Mathai, S., Harris, P. et al. (2007). Peripheral administration of a novel diketopiperazine, NNZ 2591, prevents brain injury and improves somatosensory-motor function following hypoxia–ischemia in adult rats. *Neuropharmacology* 53 (6): 749–762.

35 Nishizawa, R., Nishiyama, T., Hisaichi, K. et al. (2007). Spirodiketopiperazine-based CCR5 antagonists: lead optimization from biologically active metabolite. *Bioorg. Med. Chem. Lett.* 17 (3): 727–731.

36 Liddle, J., Allen, M.J., Borthwick, A.D. et al. (2008). The discovery of GSK221149A: a potent and selective oxytocin antagonist. *Bioorg. Med. Chem. Lett.* 18 (1): 90–94.

37 Sollis, S.V. (2005). Short and novel stereospecific synthesis of trisubstituted 2,5-diketopiperazines. *J. Org. Chem.* 70 (12): 4735–4740.

38 Hutchby, M., Sedgwick, A.C., and Bull, S.D. (2016). Orthogonally protected Schöllkopf's bis-lactim ethers for the asymmetric synthesis of α-amino acid derivatives and dipeptide esters. *Synthesis* 48 (13): 2036–2049.

39 Mollica, A., Constante, R., Mirzaie, S. et al. (2015). Preparation of constrained unnatural aromatic amino acids *via* unsaturated diketopiperazine intermediate. *J. Heterocyclic Chem.* 53 (6): 2106–2110.

40 Caballero, E., Avendaño, C., and Menéndez, J.C. (2003). Brief total synthesis of the cell cycle inhibitor tryprostatin B and related preparation of its alanine analogue. *J. Org. Chem.* 68 (18): 6944–6951.

41 Ma, Y. and Fan, C. (2016). Progress in the synthesis of spirotryprostatin alkaloids. *Chin. J. Org. Chem.* 36 (19): 2380–2396.

42 Overman, L.E. and Rosen, M.D. (2010). Terminating catalytic asymmetric Heck cyclizations by stereoselective intramolecular capture of η-allylpalladium intermediates: total synthesis of (−)-spirotryprostatin B and three stereoisomers. *Tetrahedron* 66 (33): 6514–6525.

43 Baran, P.S., Guerrero, C.A., Ambhaikar, N.B., and Hafensteiner, B.D. (2005). Short, enantioselective total synthesis of stephacidin A. *Angew. Chem. Int. Ed.* 44 (4): 606–609.

44 Miller, K.A., Tsukamoto, S., and Williams, R.M. (2009). Asymmetric total syntheses of (+)- and (−)-versicolamide B and biosynthetic implications. *Nat. Chem.* 1 (1): 63–68.

45 Kuttruff, C.A., Zipse, H., and Trauner, D. (2011). Concise total syntheses of variecolortides A and B through an unusual hetero-Diels–Alder reaction. *Angew. Chem. Int. Ed.* 50: 1402–1405.

46 Boyer, N., Morrison, K.C., Kim, J. et al. (2013). Synthesis and anticancer activity of epipolythiodiketopiperazine alkaloids. *Chem. Sci.* 4 (4): 1646–1657.

47 Kim, J., Ashenhurst, J.A., and Movassaghi, M. (2009). Total synthesis of (+)-11,11′-dideoxyverticillin A. *Science* 324 (5924): 238–241.

48 Chou, T.C., Depew, K.M., Zheng, Y.H. et al. (1998). Reversal of anticancer multidrug resistance by the ardeemins. *Proc. Natl. Acad. Sci. U.S.A.* 95: 8369–8374.

49 Marsden, S.P., Depew, K.M., and Danishefsky, S.J. (1994). Stereoselective total syntheses of amauromine and 5-N-acetylardeemin. A concise route to the family of "reverse-prenylated" hexahydropyrroloindole alkaloids. *J. Am. Chem. Soc.* 116 (24): 11143–11144.

50 Imaoka, T., Iwamoto, O., Noguchi, K., and Nagasawa, K. (2009). Total synthesis of (+)-dibromophakellin. *Angew. Chem. Int. Ed.* 48 (21): 3799–3801.

51 Usui, I., Lin, D.W., Masuda, T., and Baran, P.S. (2013). Convergent synthesis and structural confirmation of phellodonin and sarcodonin ε. *Org. Lett.* 15 (9): 2080–2083.

52 Rinehart, K.L. (2000). Antitumor compounds from tunicates. *Med. Chem. Rev.* 20 (1): 1–27.

53 Le, V.H., Inai, M., Williams, R.M., and Kan, T. (2015). Ecteinascidins. A review of the chemistry, biology and clinical utility of potent tetrahydroisoquinoline antitumor antibiotics. *Nat. Prod. Rep.* 32 (2): 328–347.

54 Cuevas, C., Pérez, M., Martín, M.J. et al. (2000). Synthesis of ecteinascidin ET-743 and phthalascidin Pt-650 from cyanosafracin B. *Org. Lett.* 2 (16): 2545–2548.

55 Endo, A., Yanagisawa, A., Abe, M. et al. (2002). Total synthesis of ecteinascidin 743. *J. Am. Chem. Soc.* 124 (23): 6552–6554.

6

Amino Acids as Chiral Building Blocks

Elisabete P. Carreiro[1,2] *and Anthony J. Burke*[1,3,4,5]

[1] *University of Évora, Institute for Research and Advanced Studies, LAQV-REQUIMTE, Rua Romão Ramalho, 59, 7000-671 Évora, Portugal*
[2] *Chiratecnics Lda, PO Box 59, Rossio, 7006 802 Évora, Portugal*
[3] *University of Évora, School of Science and Technology, Chemistry and Biochemistry Department, Rua Romão Ramalho 59, 7000-671 Évora, Portugal*
[4] *University of Coimbra, Faculty of Pharmacy, Pólo das Ciências da Saúde, Azinhaga de Santa Comba, 3000-548 Coimbra, Portugal*
[5] *Coimbra Chemistry Centre, Institute for Molecular Sciences, Chemistry Department, Faculty of Science and Technology, University of Coimbra, 3004-535 Coimbra, Portugal*

6.1 Introduction

Amino acids have had an extensive history in organic synthesis and most particularly for obtaining valuable economic targets, such as pharmaceuticals, food additives, agrochemicals, and fragrances, where chirality is a frequently important feature. They are natural (many of them including the 20 canonical examples), renewable, and generally non-toxic. The beauty of using amino acids is that they can be used as part of the structure of the target compound (chiral pool or reagent approach) or be part of the machinery that produces the target, like the catalyst (organocatalyst, a very timely topic considering the award of the Nobel prize in 2021 for the development of asymmetric organocatalysis, *vide infra*), the ligand (for metal coordination), or a chiral auxiliary. The literature is replete with numerous examples which illustrate the application of these building blocks at obtaining such targets. In this chapter we will look at some of the main strategies that have been used by chemists to access important targets, such as biologically active compounds, by using chiral amino acid tools, whether they be building blocks or reagents, or the tools used to introduce the appropriate bonds with the correct stereochemistry, by being a chiral auxiliary or an organocatalyst.

6.2 Chiral Substrates/Reagents

6.2.1 Chiral Amino Acids

The use of α-amino acids as a chiral reagent/pool/building blocks is a very common strategy for the asymmetric synthesis of a wide variety of natural products, as well as

Chiral Building Blocks in Asymmetric Synthesis: Synthesis and Applications, First Edition.
Edited by Elżbieta Wojaczyńska and Jacek Wojaczyński.

pharmaceuticals, agrochemicals, and fragrances. This strategy is advantageous since amino acids are naturally occurring, inexpensive starting materials, and have interesting functionalities that facilitate a wide variety of chemical transformations [1–8]. α-Amino acids can also be used as chiral building blocks for the asymmetric synthesis of a wide variety of important heterocycles with pharmacological interest [9].

Proline is a non-polar cyclic α-amino acid consisting of a pyrrolidine ring, which is part of the structure of many natural products and drugs, making it an important starting material [1–9]. In Section 6.5.1 the application of proline and derivatives as organocatalysts is discussed.

Harmicine is a natural product which is part of the tetrahydro-β-carboline subgroup of indole alkaloids and found in the leaf extract of the Malaysian plant *Kopsia griffithii*. It possesses antileishmanial activity and is an antinociceptive. In 2014 Lood and Koskinen described the synthesis of both (*R*)- and (*S*)-harmicine in nine steps from L- and D-proline (19% overall yield in a total of nine steps and >99% ee) (Scheme 6.1) [10]. One of the principal problems encountered during the synthesis was partial racemization of the α-amino carbonyl stereocenter, which occurred during the protection of L-proline with 9-phenyl-9-fluorenyl. This was controlled by using *N*-methylmorpholine instead of using triethylamine (TEA) or diisopropylamine (DIPEA), affording the protected L-proline in 82% yield and >99% ee. There was also a racemization issue during the amidation reaction of the protected *N*-proline intermediate that formed a Weinreb amide. In this case the partial racemization was minimized by using the protected proline methyl ester. The other key steps involved a lateral lithiation, α-alkylation of an enolate, cyclizations, and a reduction.

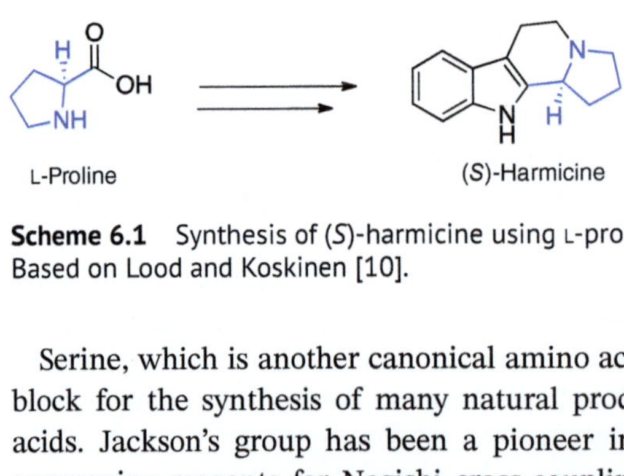

Scheme 6.1 Synthesis of (*S*)-harmicine using L-proline as a chiral pool substrate. Source: Based on Lood and Koskinen [10].

Serine, which is another canonical amino acid, has been used as chiral building block for the synthesis of many natural products, drugs, and unnatural amino acids. Jackson's group has been a pioneer in the application of serine-derived organozinc reagents for Negishi cross-coupling reactions with aryl halides [11]. After optimization of the reaction conditions they prepared 20 alanine derivatives in moderate-to-very-good yields [12]. Later, they adopted this methodology for the synthesis of a novel 4-phosphothiophen-2-yl alanine, an analogue of phosphotyrosine, but which is stable under physiological conditions, and is a substrate of protein tyrosine phosphatases, important for regulating the phosphoproteome in cells (Scheme 6.2) [11]. In 2016, the same group reported the synthesis of omega-oxo amino acid analogues of 8-oxo-2-aminodecanoic acid, an important intermediate of the cyclic tetrapeptide apicidin (a potent inhibitor of histone deacetylase) from

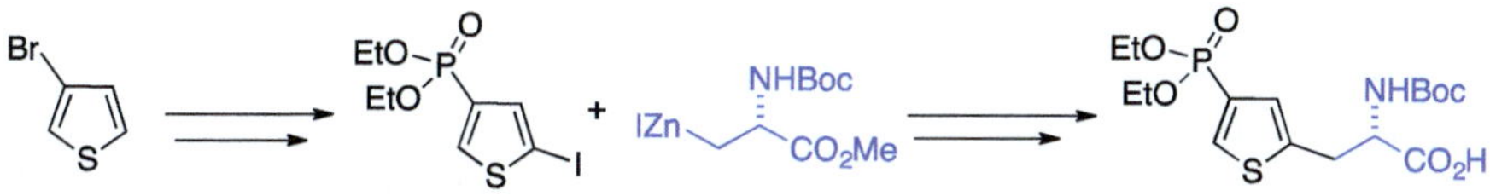

Scheme 6.2 Synthesis of 4-phosphothiophen-2-yl alanine using a serine-derived organozinc reagent as described by Jackson and coworkers. Source: Based on Lilley et al. [11].

the serine-derived organozinc reagent (aspartic and glutamic acid were also used). The synthesis involved a copper-catalyzed allylation, cross-metathesis, and a hydrogenation reaction [13]. In 2019, the same group described the synthesis of enantiomerically pure non-canonical amino acids containing fluorine, azido, and hydroxyl groups (13 examples) [14]. The key step leading to these targets involved a Pd-catalyzed Negishi coupling between a serine-derived organozinc reagent and cycloalkenyl triflates or vinyl bromides, or a Cu(I)-catalyzed allylation of a serine-derived organozinc reagent, followed by radical hydrofluorination (to introduce the fluoride), hydroazidation (to introduce the azide), and hydration (to introduce the hydroxide). These products were obtained with good-to-excellent yields [14].

The antiepileptic drug, (*R*)-lacosamide (Vimpat), was developed by UCB Pharma for the treatment of partial seizures in patients with epilepsy and also as a complementary treatment in patients with brain tumors [15]. D-Serine was considered as a building block for this target, but there were problems of partial racemization of the product and low yields. Later, Yang et al. improved the methodology of UCB Pharma, overcoming the partial racemization and increasing the yield, to give (*R*)-lacosamide with 64% overall yield (in four steps) and 100% ee [16]. The (*R*)-lacosamide was prepared via methylation of Boc-D-serine, followed by its amidation, Boc deprotection, and acetylation. The researchers increased the chemical and optical purity of the compound by using 85% H_3PO_4 instead of 36% HCl in the deprotection of the Boc, the formed phosphate salt was easily precipitated and purified. In 2015, Aratikatla and Bhattacharya described another short and direct route for the synthesis of (*R*)-lacosamide, in only five steps (with an overall yield of 54% and ee >99.9%) starting from *N*-Boc-*N*,*O*-isopropylidene-L-serinol (Scheme 6.3). Furthermore, the *N*-Boc-*N*,*O*-isopropylidene-L-serinol was prepared from L-serine in only four steps [17].

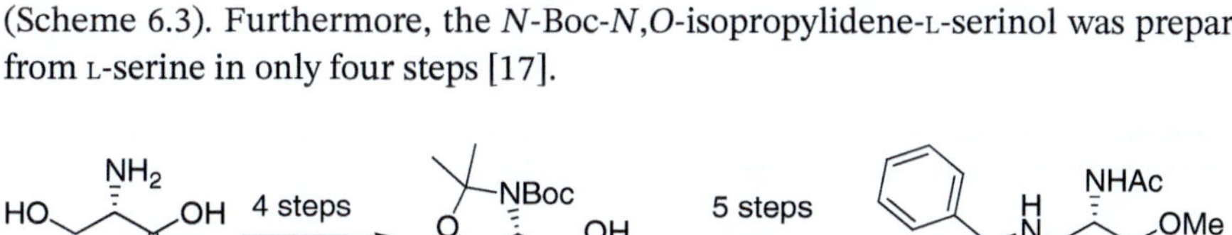

L-Serine

N-Boc-N,O-isopropylidene-L-serinol

(*R*)-Lacosamide
Overall yield 54%
ee >99.9%

Scheme 6.3 Synthetic pathway for the asymmetric synthesis of the (*R*)-lacosamide using L-serine as described by Aratikatla and Bhattacharya. Source: Based on Aratikatla and Bhattacharya [17].

In 2018, D'hooghe and coworkers synthesized a new family of nocardicin-like enantioenriched 2-{3-[2-(2-aminothiazol-4-yl)-2-(me-thoxyimino)acetamido]-2-oxoazetidin-1-yl}acetic acids using L- and D-serine as starting materials [18]. They were obtained in satisfactory-to-excellent yields and enantioselectivities >92% ee. Briefly, these compounds were prepared through the α-functionalization of 3-(di-Boc-amino)-1-(methoxycarbonylmethyl)azetidin-2-ones, the key intermediate, which were prepared in six steps, including the *N*-Boc protection of serine, Miller's hydroxamate synthesis, Mitsunobu reaction (cyclization to obtain the β-lactam ring), *N*-Boc protection, hydrogenolysis (using Raney-Ni as catalyst), and alkylation. This new family of monocyclic-3-amino-β-lactams showed an interesting inhibitory activity against penicillin-binding proteins of various (resistant) bacteria.

Threonine is a polar and essential α-amino acid with two stereogenic centers, existing in four possible stereoisomeric forms with the following configurations: (2*S*,3*R*) (L-threonine), (2*R*,3*S*) (D-threonine), (2*S*,3*S*) (L-allothreonine), and (2*R*,3*R*)-2-amino-3-hydroxybutanoic acid (D-allothreonine). This amino acid building block has been used to synthesize a variety of different biologically active compounds, like FR901464, vulnibactin, and pactamycin [19–22]. Lee et al. reported the asymmetric synthesis of a new thiazoline core possessing two stereogenic centers, using L-threonine (Scheme 6.4) [23]. This thiazoline derivative is an analogue of pulicatins A, C–F, anthiatins A–C, and endurobol (GW501516), which are recognized potent peroxisome proliferator-activated delta receptor agonists. The synthetic pathway, which involved a number of functional group modifications, including a crucial cyclization with diethylaminosulfur trifluoride (DAST) to give the thiazoline unit, involved nine steps starting from Boc-protected L-threonine to give the synthetic target in 3.6% overall yield.

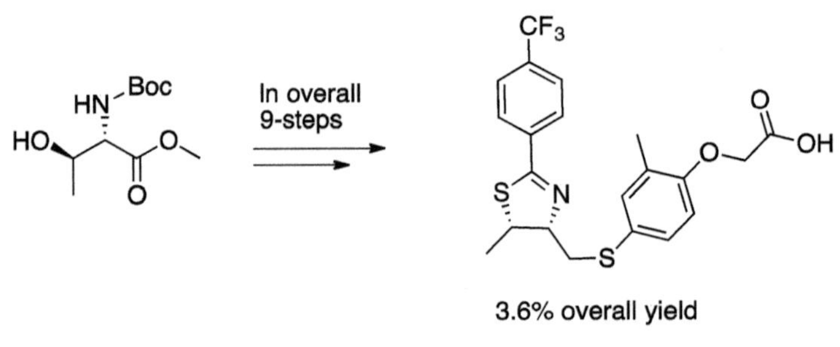

Scheme 6.4 Synthesis of a novel thiazoline derivative starting from L-threonine as reported by Chin's group. Source: Based on Lee et al. [23].

Tryptophan is an α-amino acid possessing an indole ring in the side chain; this ring is an important pharmacophore present in several pharmacologically active compounds. It has been used to synthesize such important drugs as tadalafil (Lilly ICOS) [24], (−)-brevicompanine B [25], verruculogen, and fumitremorgin [26]. Recently, Zhang and coworkers reported an asymmetric total synthesis of sarpagine and koumine alkaloids using L-tryptophan. Although koumine alkaloids have interesting biological properties such as antitumor, anti-inflammatory, and analgesic

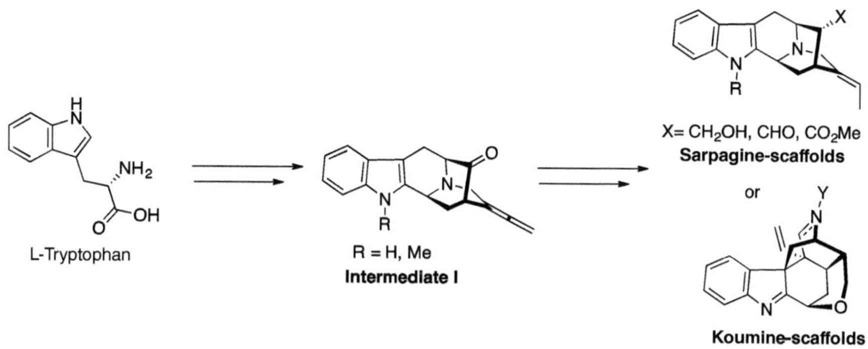

Scheme 6.5 Asymmetric synthesis of sarpagine and koumine alkaloids using L-tryptophan as starting material as reported by Zhang and coworkers. Source: Based on Yang et al. [27].

activities, the properties of the sarpagine alkaloids remain unknown due to their low natural abundance. These two complex alkaloids were successfully synthesized from L-tryptophan (Scheme 6.5) [27]. The key intermediate **I** with ketone and allene groups was a precursor to both families, and was prepared in a very good yield and enantioselectivity (>99% ee) in six steps. Intermediate **I** with a caged core framework was found to be a crucial branching point for the synthesis of a wide variety of sarpagine and koumine alkaloids (Scheme 6.5). The principal synthetic strategy involved a tandem sequential oxidative cyclopropanol ring-opening cyclization and a cooperative organo/metal-assisted ketone α-allenylation that gave key bicyclo[3.3.1]nonane and bicyclo[2.2.2]octane skeletons.

The complexity and the existence of several stereogenic centers in some natural and pharmaceutical products force synthetic chemists to use more than one amino acid as building blocks. In 2020 Long et al. described a synthetic route for the preparation of tubulysin U and N^{14}-desacetoxytubulysin *H*, using L-valinol and L-phenylalaninol as starting materials (Scheme 6.6). The target compounds were successfully prepared in total yields of 7.7% (27 steps, 1.51 g scale) and 6.5% (28 steps, 1.83 g scale). These strategies present several advantages, such as a scalable synthesis using cheap reagents (amino alcohols obtained from natural α-amino acids!) and high stereoselectivities [28].

6.3 Chiral Auxiliaries

Asymmetric synthesis of biologically active compounds can be achieved by the chiral auxiliary approach. This approach involves the attachment of an appropriate chiral appendage known as a *chiral auxiliary* to the substrate, diastereoselective functionalization, and the removal of the auxiliary from the enantioenriched product [29–32]. Various chiral auxiliaries have been developed and most of them are prepared from natural precursors, like natural α-amino acids [33, 34]. α-Amino acids and their derivatives (such as amino alcohols) have had an important role as chiral auxiliaries in asymmetric synthesis of many important target compounds.

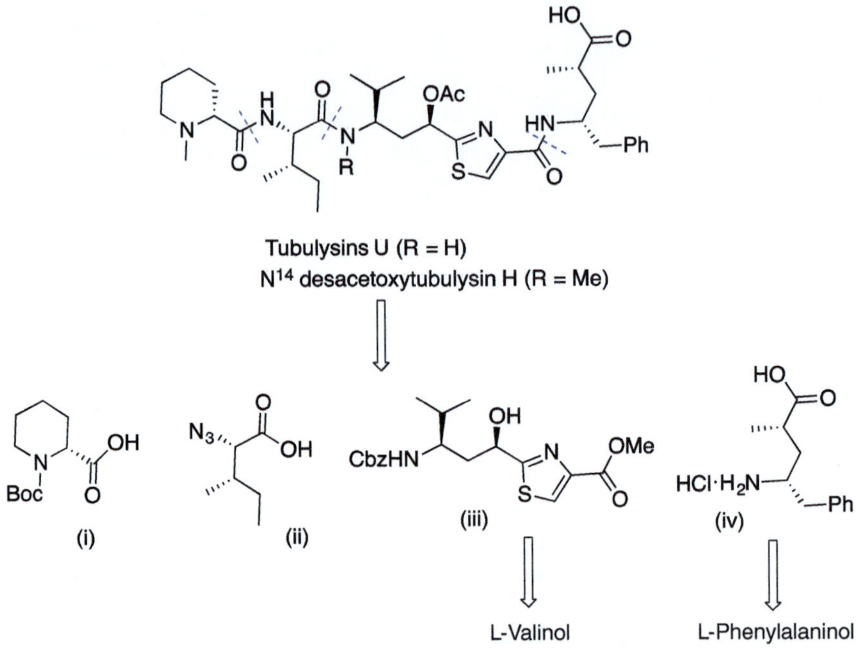

Scheme 6.6 Retrosynthetic strategy to tubulysins starting from L-valinol and L-phenylalaninol as described by Wu and coworkers. Source: Based on Long et al. [28].

6.3.1 Chiral Pyrrolidine Auxiliaries

6.3.1.1 Yamada Auxiliary

The first chiral auxiliary based on (*S*)-proline derivatives was developed in 1969 by Yamada and was used in various stereoselective alkylations that proceeded through the formation of chiral enamine intermediates; subsequently, the chiral auxiliary is removed by hydrolysis [35]. The pioneering reaction described by Yamada was the asymmetric α-alkylation of cyclohexanone using a proline ester as chiral auxiliary, forming the chiral cyclohexanone enamine as intermediate, which then reacts with ketoesters via Michael reaction, furnishing the desired product in moderate yield and diastereoselectivity after hydrolysis. Yamada chiral auxiliaries were used for the synthesis of the bioactive compounds, (+)-mesembrine [36], and (*S*)-α-damascone [37].

6.3.1.2 Enders Auxiliaries

In 1976 Enders and Corey, independently, introduced (*S*)-1-amino-2-methoxymethylpyrrolidine (SAMP) and (*R*)-1-amino-2-methoxymethylpyrrolidine (RAMP) chiral auxiliaries (obtained from (*S*)-proline and (*R*)-glutamic acid) [38, 39]. These two chiral auxiliaries were much more versatile than Yamada's chiral auxiliaries, as they gave better yields and selectivities in several reactions, such as α-alkylations, aldol condensations, Michael additions, rearrangements, nucleophilic additions to C=N bonds, cycloadditions, and Birch reductions. Chiral hydrazine-type auxiliaries SAMP and RAMP react easily with several carbonyl derivatives forming the corresponding hydrazone intermediates, which make them ideal for several reactions, and to be a powerful tool in the synthesis of numerous naturally occurring, and bioactive compounds, e.g. (+)-synargentolide A, (−)-callystatin A, (+)-aspicilin, and zaragozic acid A [40, 41]. In 2018, Williams and coworkers reported the asymmetric synthesis of (2*R*)-pterosin B, which is a natural product found in bracken fern (*Pteridium aquilinum*) and possesses several pharmacological properties. In fact, the development of a stereoselective synthetic route to pterosin B is crucial, since its concentrations in bracken are very low and most syntheses described in the literature are non-stereoselective. The stereoselective synthetic route was based on the Enders' chiral hydrazone methodology using SAMP as chiral auxiliary. The synthetic pathway for the preparation of (2*R*)-pterosin B had as first step a Suzuki–Miyaura cross-coupling between the indanone and benzyl-protected trifluoroborate salt, which gave a functionalized indanone in 54% yield (Scheme 6.7). This was followed by the synthesis of the key hydrazone, in 78% yield, from functionalized indanone and SAMP. The hydrazone intermediate was methylated with MeI, and the methyl group with the correct *R*-configuration was introduced to the bottom face of the aza-enolate. Moreover, the reaction yield was 61% and it took place with a high diasteroselectivity (85 : 15 dr). In the final two steps both the chiral auxiliary and the benzyl group were removed (the chiral auxiliary with ozone) to give the (2*R*)-pterosin B as a single enantiomer in 68% yield [42].

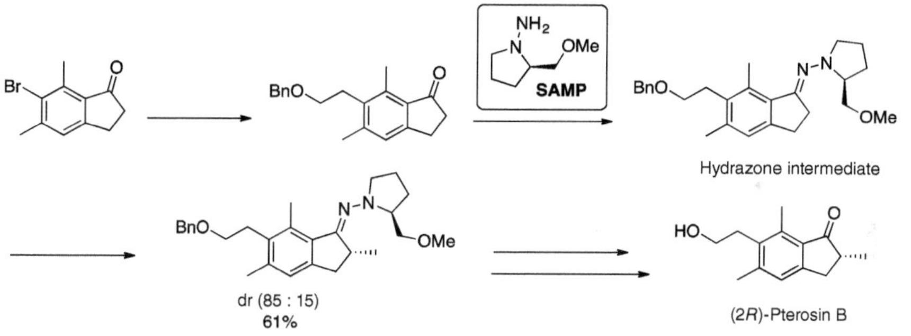

Scheme 6.7 Stereoselective syntheses of (2*R*)-pterosin B using the chiral auxiliary SAMP.

6.3.2 Schöllkopf's Bis-lactim Ethers (Schöllkopf Chiral Auxiliaries): Glycine Enolate Equivalents

In 1979 Schöllkopf introduced a chiral-auxiliary-based method for the preparation of optically pure α-amino acids via diastereoselective alkylation of glycine [43, 44]. The methodology consisted in the condensation of a chiral α-amino acid (L-valine, L-alanine, L-*t*-leucine, and L-*O*,*O*-dimethyl-α-methyldopa) with glycine to yield a bis-lactam, which was reacted with Meerwein's salt (trimethyloxonium tetrafluoroborate) to furnish the bis-lactim ether or (*R*)-2,5-dihydro-3,6-dimethoxy-2-isopropylpyrazine. Undoubtedly the Schöllkopf chiral bis-lactim ether is one of the most popular glycine enolate equivalents for the asymmetric synthesis of α-amino acid derivatives [45, 46]. Barrett's group reported the enantioselective synthesis of a wide diversity of α-arylglycines via an α-arylation reaction of Schollkopf's bis-lactim ether with arynes as electrophiles, obtaining the aryl bis-lactim ethers with moderate-to-very-good yields and diastereoselectivities, which after hydrolysis furnished the desired products in high yields [47]. Later, the same research group described the enantioselective synthesis of aryl quaternary amino acid derivatives by means of a single-pot reaction of three components between Schöllkopf bis-lactim ether, allyl, aryl, or alkyl halide electrophiles, giving the quaternary aryl bis-lactim ethers with good-to-excellent yields and high diastereoselectivities, which afforded the desired amino acids upon hydrolysis. The quaternary aryl bis-lactim ethers with bulkier substituents provided derivatives of the valine dipeptides rather than the desired products [48]. Other researchers reported the synthesis of α-aminoadipic acid and its esters based on imidazo[2,1-*b*][1,3]thiazole and stereoselective radiosynthesis of L- and D-3-[^{18}F]fluoro-α-methyltyrosine using Schöllkopf's chiral auxiliary followed by acidic hydrolysis to generate the desired labeled amino acids [49, 50]. An interesting example of the use of Schöllkopf's chiral auxiliary for the preparation of novel analogues of the immunosuppressant drug FTY720 (fingolimod, Gilenya) used for treatment of multiple sclerosis, was reported by Morris and coworkers [51]. The synthesis of these novel chiral amino alcohols was carried out by alkylation of bis-lactim ether intermediates, followed by the hydrolysis and reduction giving the chiral amino alcohols with moderate yields (Scheme 6.8). Furthermore, these novel compounds showed promising activity against acute myeloid leukemia cell lines.

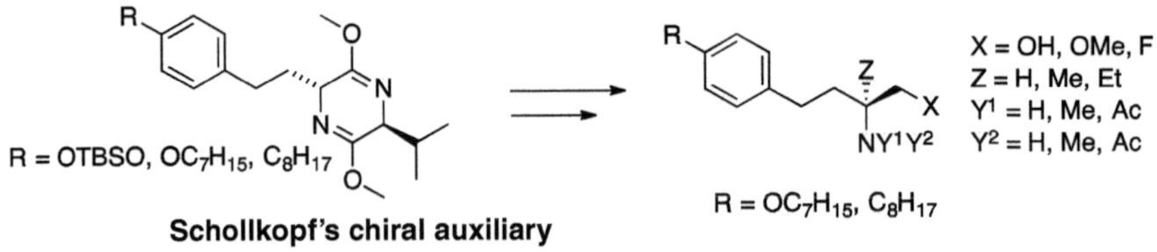

Scheme 6.8 Stereoselective synthesis of novel FTY720 analogues using Schöllkopf's chiral auxiliary.

6.3.3 Oxazolidinones

6.3.3.1 Evans' Chiral Auxiliary

In the early 1980s, Evans introduced the oxazolidinone (4-substituted oxazolidin-2-one) chiral auxiliary, which became one of the most versatile and attractive chiral auxiliaries in asymmetric synthesis to date [52]. It offers good diastereocontrol in α-alkylation, *anti*- and *syn*-aldol, Michael additions, 1,4-additions, and intra-molecular Diels–Alder cycloaddition reactions [53]. In a sense these molecular appendages were the fore-runners to the chiral ligands and organocatalysts that have bedecked the literature in subsequent years culminating in the award of the Nobel Prize in 2021. In addition to inducing good diastereocontrol, these chiral auxiliaries have other advantages that include: (i) easy preparation from cheap natural α-amino acids; (ii) the N-acylation works well with a series of activated carboxylic acid substrates; and (iii) after the transformation the chiral auxiliary can be cleaved and recycled. Due to the exceptional characteristics of these oxazolidinones, several small modifications in their structure were made to overcome some of their limitations, as was the case with the SuperQuat oxazolidinones discussed in Section 6.3.3.2. The Evans' oxazolidinones have played a prominent role in various asymmetric alkylation and aldol reactions leading to the total synthesis of natural products and other complex compounds of pharmacologic interest [54–56]. In 2019, the asymmetric syntheses of the natural products aldgamycin-M and lycopoclavamine-A were described, using chiral oxazolidinone auxiliaries [57, 58]. Another successful example of the application of Evans' chiral auxiliary was in the asymmetric synthesis of (+)-ambrisentan, which is used for the treatment of pulmonary arterial hypertension. This chiral auxiliary-controlled synthesis occurred in five steps from the chiral imide precursor chloroacetyl-(*R*)-4-benzyloxazolidin-2-one, which was reacted with benzophenone under the Crimmins-modified Evans' aldol reaction conditions, giving the preferred diastereoisomer in 85% yield. This diastereoisomer was transformed to the respective epoxide using the Azerad protocol (with consequent removal of the chiral auxiliary), after photochemical epoxide ring opening and finally alkylation to give (+)-ambrisentan in high yield and stereoselectivity (Scheme 6.9) [59].

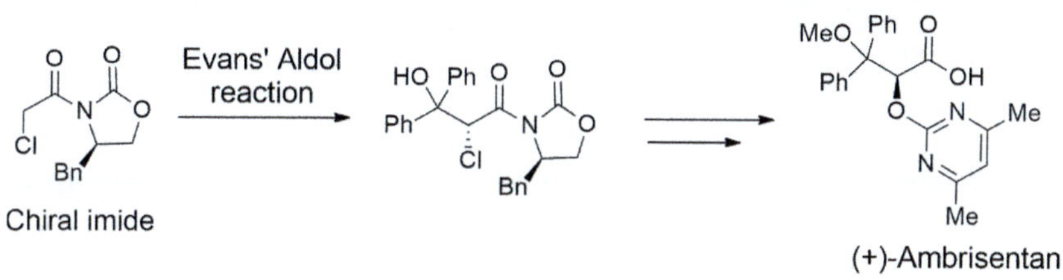

Scheme 6.9 Synthesis of (+)-ambrisentan using the diastereoselective Evans' aldol reaction. Source: Based on Madhu et al. [59].

In 2021, Kakeya and coworkers reported the first total synthesis of presaccharotriolide Z, which was impossible to isolate due to its instability. This macrolide showed promising antiproliferative inhibition against human fibrosarcoma HT1080 cells. Retrosynthetic analysis (Scheme 6.10) indicated that this target could be derived from the chiral vinyl iodide and a chiral 5-carbon backbone aldehyde via a Nozaki–Hiyama–Kishi reaction forming a linear precursor, which could then undergo an intramolecular Shiina macrolactonization creating the desired

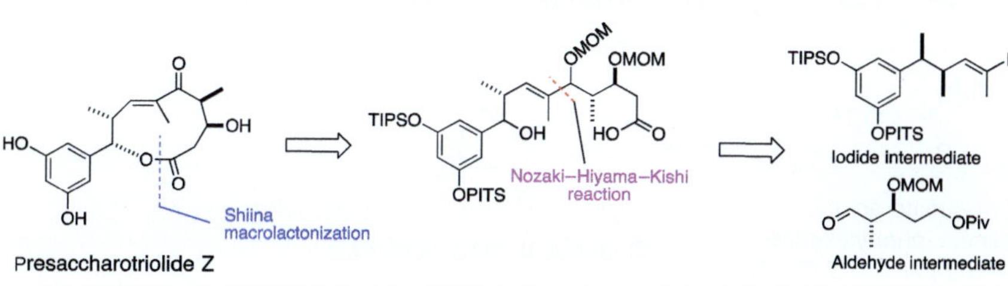

Scheme 6.10 Retrosynthesis of presaccharotriolide Z.

macrolide [60]. Gratifyingly both the chiral aldehyde and the chiral vinyl iodide were obtained through stereoselective routes involving the Evans' chiral auxiliary (Scheme 6.11).

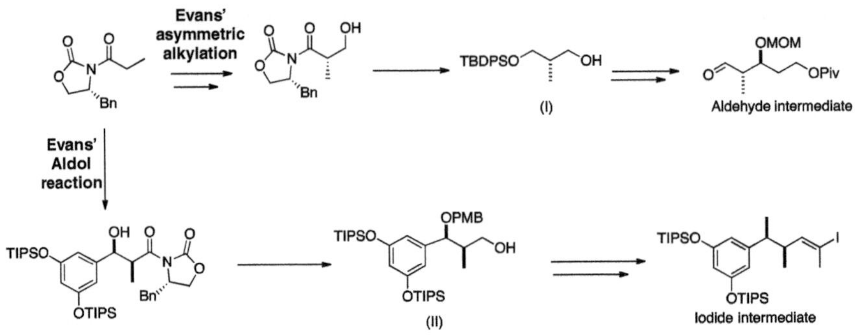

Scheme 6.11 Synthesis of the chiral aldehyde and vinyl iodide intermediates using the Evans' asymmetric alkylation and aldol reactions as described by Kakeya and coworkers. Source: Based on Kuranaga et al. [60].

The chiral aldehyde intermediate (Scheme 6.11) was prepared by Evans' asymmetric alkylation between propionyl oxazolidinone and $BnOCH_2Cl$ using lithium diisopropylamide (LDA). The synthesis of this intermediate also involved Brown's asymmetric allylation. In the case of the chiral vinyl iodide intermediate, this was prepared using an asymmetric Evans' aldol reaction (Scheme 6.11) furnishing the *syn*-adduct as a single diastereomer.

6.3.3.2 SuperQuat Chiral Auxiliaries

SuperQuat chiral auxiliaries were developed by Davies' group in the 1990s to overcome some limitations of Evans' chiral auxiliaries [61, 62]. SuperQuats are basically 4-substituted oxazolidin-2-ones possessing a geminal dimethyl substitution at C(5) (Scheme 6.12). These were designed for the following reasons: (i) to improve the diastereofacial selectivity in a variety of transformations and (ii) to prevent nucleophilic attack at the endocyclic carbonyl group, improving the cleavage properties, thus enabling the recovery and recyclability of the auxiliary. SuperQuats are prepared from the α-amino acids like L-valine, L-phenylglycine, L-phenylalanine, and L-tyrosine using a well-established protocol [63–66]. SuperQuat auxiliary methodologies usually involve N-acylation (preparation of an *N*-Acyl SuperQuat derivative),

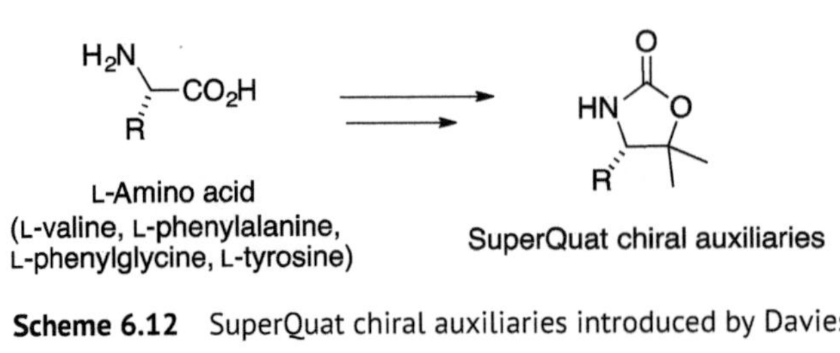

Scheme 6.12 SuperQuat chiral auxiliaries introduced by Davies' group.

subsequent diastereoselective functionalization and cleavage to obtain enantiopure products. SuperQuat chiral auxiliaries have proven to be effective in several types of diasteroselective transformations, and can ensure the creation of one or two new stereogenic centers in the product. These chiral auxiliaries were used for the synthesis of natural products, as well as APIs like: (−)-epothilone A, sintokamide A, B, and E, and nakinadine B and C [67].

In 2019 Wong et al. reported the successful asymmetric synthesis of some chiral bicyclo[1.1.1]pentanes (BCP), through diastereoselective asymmetric alkylation of SuperQuat oxazolidinone (Scheme 6.13) [68]. They prepared a wide range of BCPs with very good diastereoselectivities and yields. The final chiral BCP building blocks were obtained in high enantioselectivity and yields through cleavage of the chiral auxiliary by hydrolysis, reduction, and transesterification. Furthermore, the chiral auxiliary was efficiently recovered. The authors also applied this methodology for the preparation of the BCP analogue of tarenflurbil ((*R*)-enantiomer of

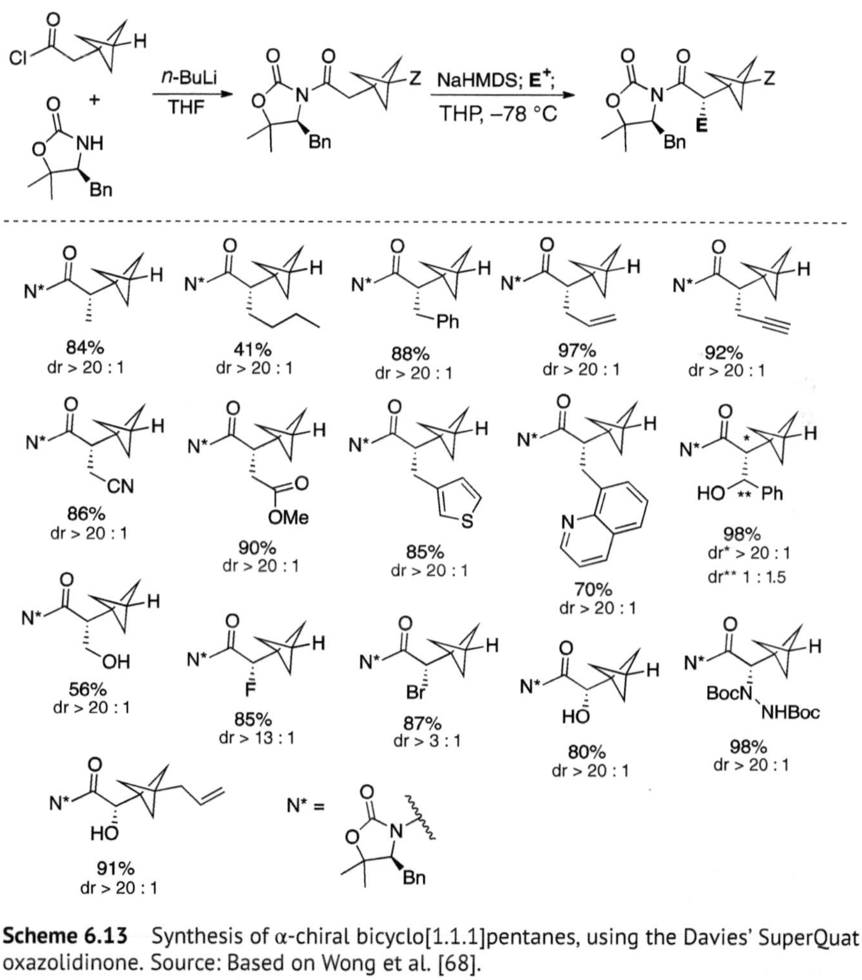

Scheme 6.13 Synthesis of α-chiral bicyclo[1.1.1]pentanes, using the Davies' SuperQuat oxazolidinone. Source: Based on Wong et al. [68].

the nonsteroidal anti-inflammatory drug flurbiprofen), which was successfully prepared in very good yield and stereoselectivity.

6.4 Oxazaborolidines

The initial pioneering studies on the asymmetric catalytic reduction of ketones with oxazaborolidine compounds derived from simple amino acids like valine were conducted by Itsuno et al. in the early 1980s [69]. This was followed by a significant advance of this methodology when oxazaborolidine catalysts derived from proline were effectively deployed in the mid-1980s by Corey et al., leading to the development of the eponymous Corey–Bakshi–Shibata (CBS) reduction of ketones (Scheme 6.14) [70], which in fact has many industrial applications [71].

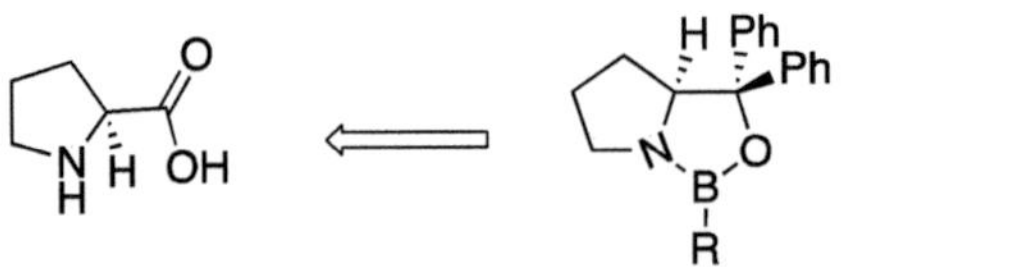

Scheme 6.14 The proline-derived Corey–Bakshi–Shibata (CBS) catalyst. Source: Based on Corey et al. [70].

Due to space limitations, we will not discuss the use of chiral β-amino acid substrates, but for excellent reviews on this topic the reader is referred to the papers from Davies' group [72–74]. We will also not cover the use of chiral ligands derived from α-amino acids, like the Evans' bis-oxazoline (BOX) ligands [75], the Nishiyama PyBox ligands [76], and Burke's ArylideneBOX and IsButBOX [77–79] ligands. The reader is encouraged to consult the numerous excellent reviews on these types of ligands, like the one by Guiry in 2014 [80].

6.5 Organocatalysts

Over the past two decades the application of small-molecule organic catalysts (known as *organocatalysts* or metal-free catalysts) to solve synthetic problems has become important in both industrial and academic settings [81–84]. A wide and expanding range of catalyst classes and diverse activation modes in combination with increased mechanistic understanding and practical applicability has catapulted their use across many areas of synthesis. Small-molecule organic catalysts offer a range of practical advantages over macromolecular (enzyme) or precious metal catalysts, that includes air and water stability, low cost, availability from renewable resources, and relatively low toxicity. Organocatalysis is recognized as an inherently green technology.

Organocatalysis made the most important impact it could possibly make in October 2021, with the award of the Nobel prize in chemistry to Ben List and David

Macmillan for their seminal contributions to the field of asymmetric organocatalysis (see Section 6.5.1 for further details) [85, 86]. Amino acids, like proline, have had a key role in the development of stereoselective organocatalysis. Within the list of small molecules that function as organocatalysts, the amino acids and peptides have a very important role to play. In this section we will discuss these molecules (our discussion is not exhaustive and does not encompass all amino-based organocatalysts, but a selection that we find quite useful).

6.5.1 Amino Acid and Peptide-Based Organocatalysts

In the 1970s two teams of industrial chemists (from Hoffmann–La Roche and Schering AG) made a significant advance in the area of organic synthesis when they demonstrated the feasibility of carrying out the asymmetric version of the classical Robinson annulation reaction using proline as the organocatalyst; this reaction is commonly known in classrooms and in textbooks as the Hajos, Parrish (Hoffmann–La Roche), Wiechert, Eder, and Sauer (Schering AG) reaction (Figure 6.1) [88]. Unfortunately (as the way things go in science), the reaction was ignored for almost 25 years, until List, Barbas III, and Lerner showed that proline could also successfully catalyze intermolecular aldol reactions with enantioselectivities of up to 96% ee [85, 86, 89–91], and McMillian showed that an imidazolidinone organocatalyst based on phenylalanine – which is very cheaply made – could efficiently catalyze an asymmetric Diels–Alder reaction (among many others) with high enantioselectivity (Figure 6.1) [87], in both cases aminocatalysis is at the heart of the reaction mechanism. This was the seminal work that contributed to the Nobel Prize committee's decision to award the prize to these scientists in 2021.

From a historical perspective other key players involved in the development of amino-acid-based organocatalysts have been Jorgensen [92–94] (diarylprolinols were used initially but were then later superseded by their superior trimethylsilane (TMS) ether analogues, as their reactions experienced poor catalytic turnover due to their transformation into a "parasitic" non-reactive oxazolidine species [94], the later version showing superior face differentiation and enantiocontrol) and independently by Hayashi et al. [95]. As we see below this catalyst has shown incredible application in medicinal chemistry. Other catalysts include Jacobsen's isoleucine containing amido-thiourea organocatalyst used in asymmetric Strecker reactions and others [96–98], Wennemers diproline tripeptides [99], and Birman's benzotetramisole compact organocatalysts that are used in enantioselective acyl transfer reactions, namely the kinetic resolution of secondary benzylic alcohols [100, 101].

In this section we will look at the main advances that have been made in the area of asymmetric synthesis with amino acid and peptide-derived organocatalysts. In addition to this work, there have been some good general reviews that cover amino-acid-based organocatalysts published in the last 5 years or so, which can serve as a complementary source of information [102, 103].

This section is organized into the different types of amino acid organocatalyst (generally proline or derivatives) and then subdivided into the types of reaction, which will include examples of usage of enabling technologies (which includes

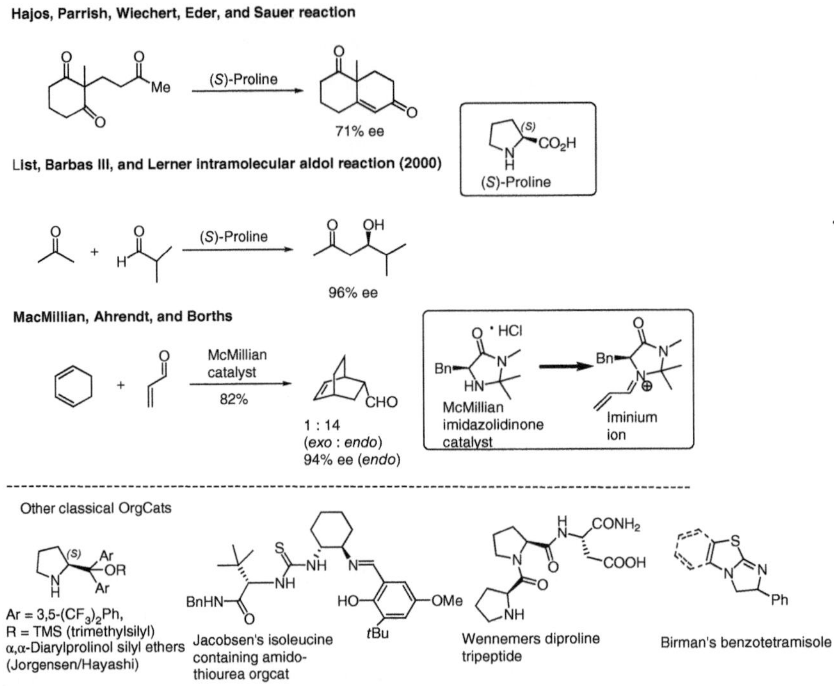

Figure 6.1 The original Hajos, Parrish, Wiechert, Eder, and Sauer reaction, its innovation to the intermolecular proline-catalyzed aldol reaction by List, Barbas III, and Lerner, and the first organocatalyzed Diels–Alder reaction described by MacMillian and coworkers. At the bottom are shown some other important organocatalyst families. Source: Based on Ahrendt et al. [87].

some flow-chemistry examples). We choose generally those examples that lead to socio-economically important target molecules, like pharmaceuticals and pesticides.

6.5.2 Organocatalysts Based on Proline

Proline has been used in a multitude of total and target syntheses.

6.5.2.1 Proline

Some very interesting recent asymmetric catalytic transformations achieved using simple proline are discussed here. Upon doing a search with the key words "proline," and "organocatalysis," a staggering 962 papers were obtained with 41 134 citations for the period 2002–2021. Some recent examples on the use of this catalyst are discussed here.

In 2018, Aggarwal and coworkers reported on the reoptimization of a proline-catalyzed double aldol domino process to a key enal intermediate and its application in the total synthesis of Δ^{12}-Prostaglandin J_3 which is a prostanoid manifesting high activity against cancer stem cells highly resistant toward chemotherapy (Scheme 6.15) [104]. In this approach the key reaction was the

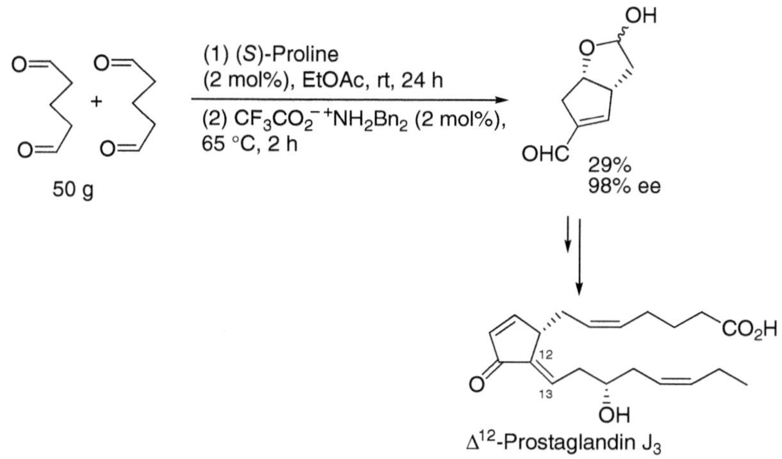

Scheme 6.15 The key proline-catalyzed aldol reaction in the total synthesis of Δ^{12}-Prostaglandin J_3 by Aggarwal and coworkers. Source: Based on Pelšs et al. [104].

L-proline (2 mol%)-catalyzed aldol condensation of succinaldehyde at room temperature in Me-THF for 24 hours, followed by the addition of dibenzylammonium trifluoroacetate (2 mol%). L-Proline catalyzes the first step, and the second steps. The catalysts had to be added sequentially as the dibenzylammonium trifluoroacetate inhibits the first aldol reaction. The other problem encountered by these workers was that the reaction of proline with succinaldehyde leads to oligomerization over time. An initial yield of 14% for the lactol-enal was obtained, but it could be stretched to 33% (NMR). Using these optimized conditions, the reaction was run at 50 g scale, giving 12.8 g of the enal (29% isolated) as a 4 : 1 mixture of epimers and with an enantiomeric excess (ee) of 98%.

Sticking with the aldol condensation in the context of heterogenous catalysis, in 2018 Kitaoka and coworkers reported the cooperative catalysis of a cellulose nanofiber and (*S*)-proline for a number of simple aldol reactions [105]. The cellulose was of a tetramethylpiperidine 1-oxyl (TEMPO)-oxidized type (TOCN). The work was an attempt to improve the main limitation of proline that includes high catalyst loadings. The authors could achieve yields of up to 86% for the reaction of *p*-nitroacetophenone (4-NA) with acetone using (*S*)-proline (10 mol%) in the presence of TOCN (ratio of TOCN to 4-NA; 1 : 1) in MeOH at 40 °C for four hours. Unfortunately, the enantioselectivity was low – 10% ee in favor of the *R*-enantiomer. This was not discussed by the authors, but there may have been opposing enantiofacial discriminating forces at work from the two chiral catalysts involved, namely proline and TOCN.

The use of deep eutectic solvents (DESs) as suitable and alternative solvents has become more common for catalysis over the last five to six years. Many nice examples of their application in organocatalysis are given in the recent review by del Amo and Fanjul-Mosterín [106]. One nice example is that from the lab of Ramon who carried out a number of crossed-aldol reactions with (*S*)-proline (30% loading) with a variety of aldehydes and ketones in the novel DES formed by mixing D-glucose with *rac*-malic acid in a 1 : 1 ratio at room temperature over 48 hours [107]. When the reaction was carried out between cyclohexanone and 4-nitrobenzaldehyde, the aldol product was obtained with a conversion of 95%, a dr of 9 : 1 (*anti*:*syn*), and with an ee for the major diastereomer of 94%. The problem was the high catalyst loading of 30 mol%, which is a general disadvantage of organocatalysis.

Proline can also be used in other reactions besides the aldol condensation – like the asymmetric conjugate addition to nitroalkenes and the asymmetric Mannich reaction. In the case of the conjugate addition to nitroalkenes, some nice examples of the use of proline derivatives can be found in Alonso's recent 2017 review [108].

Hayashi et al. reported on an elegant asymmetric Mannich reaction of alkynyl imine to afford useful chiral propargyl amine derivatives with excellent *syn*-selectivity and almost total stereocontrol (Scheme 6.16) [109]. The catalyst used was a prolinate salt that in fact showed better stereoselectivity than the parent proline. Potassium L-prolinate was found to be the best catalyst at a loading of 10 mol%, and the reaction was performed in *N*-methylpyrrolidine (NMP, highly polar) at room temperature followed by a further spell at −20 °C, before carrying out the reduction of the aldehyde component with $NaBH_4$. Various counter-cations can be used like Li, Na, K, Rb, Cs, Mg, Ca, Sr, Ba, and tetrabutylammonium, but with Mg the enantioselectivity drops off to 90% ee. The reaction showed broad scope with both the alkynyl aldehyde and the alkyl aldehyde (Scheme 6.16). In all cases the ee was >98%, the de was between 4 : 1 and 20 : 1, and the yields in the range 66–90% for a reaction time of between 5 and 12 hours. With regard to the mechanism, the authors proposed that water produced in the condensation reaction forms a H-bonded bridge between the carboxylate oxygen and the *in situ*–formed alkynyl imine (see inset, Scheme 6.16). The means of determining the configuration of the major enantiomer was not discussed, but we suspect that it was through comparison with literature high performance liquid chromatography (HPLC) data.

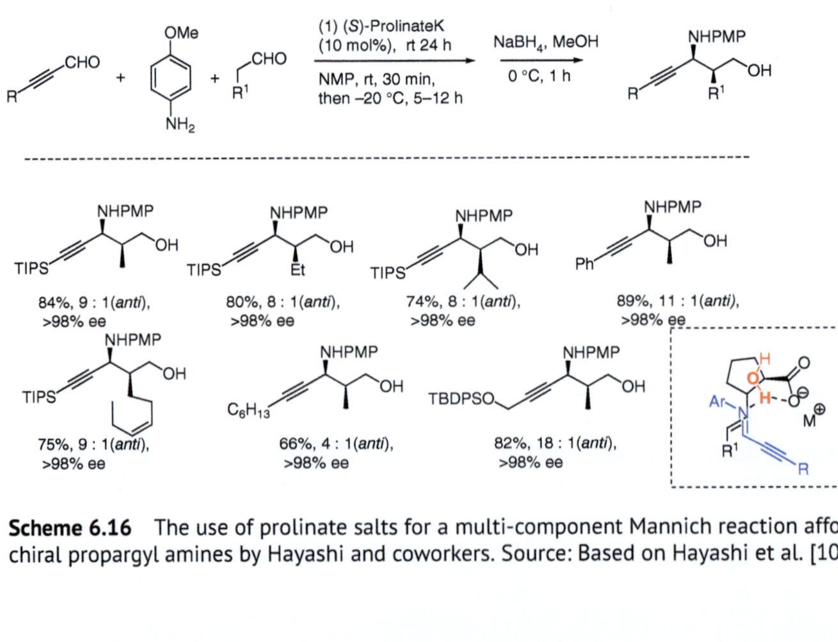

Scheme 6.16 The use of prolinate salts for a multi-component Mannich reaction affording chiral propargyl amines by Hayashi and coworkers. Source: Based on Hayashi et al. [109].

6.5.2.2 Proline Amides and Peptides

The use of both organocatalysis and biocatalysis is a very powerful and sustainable means of accessing key chemicals of important economic and social value; however, combining the two in a sequential or a tandem fashion is an even more powerful approach [110].

Working along these lines, Gröger and Berkessel reported on a tandem-type one-pot process that combined a organocatalyzed aldol condensation followed by an alcohol dehydrogenase (ADH) reduction to a chiral 1,3-diol with high conversions, diasteroselectivities, and enantioselectivities (Scheme 6.17) [111]. It should be noted that in a tandem reaction the two catalytic processes should be compliant with the same reaction conditions. The reaction was carried out at a 1.4 g scale using 3-chlorobenzaldehyde and acetone, a prolinamide organocatalyst (0.5 mol%) and the (*S*)-ADH derived from an *Rhodococcus* sp., with NAD^+ as cofactor, and 9 equiv of isopropanol for cofactor regeneration. The reaction was run at room temperature over 24 hours to give the (1*R*,3*S*)-diol in 50% yield and the (*R*)-aldol intermediate in 40% yield, respectively (Scheme 6.17). The authors did not comment on the stereoselectivity, but all the products were presumed to be obtained with high optical purity.

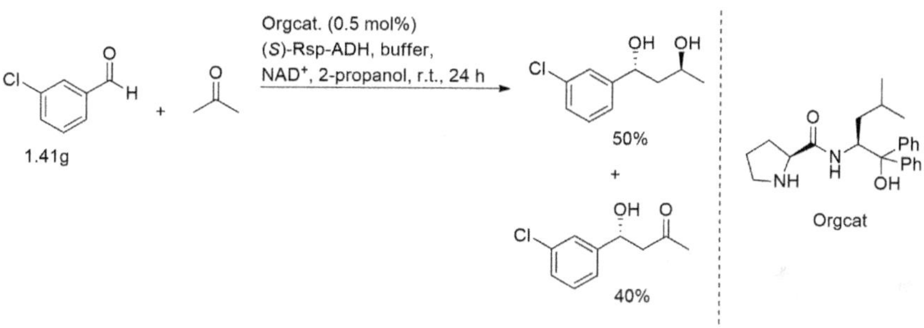

Scheme 6.17 Tandem-type one-pot process of an aldol proline-amide-catalyzed step followed by an ADH reduction. Source: Based on Rulli et al. [111].

With an eye on the development of sustainable processes hinging on enabling methods, like mechanochemistry, the group of Juaristi used a ball-milling-assisted procedure for the organocatalyzed aldol reaction between 5-bromoisatin and acetone with a proline-based thiodipeptide (Scheme 6.18) [112]. The reaction was run at −20 °C over a four hour period using 10 mol% of the catalyst, and a ball-milling speed of 2760 rpm to furnish the aldol product in 61% yield and 86% ee.

The organocatalyzed asymmetric Michael addition of aldehydes and other compounds to nitroolefins is a useful method for accessing nitroolefins that can be processed downstream into bioactive pyrrolidines and γ-butyrolactams. Over the last 10 years the group of Wennemers has developed novel tripeptide proline-based catalysts for catalyzing this reaction [113, 114]. One longstanding problem in organic chemistry has been the *anti*-selective conjugate addition reaction between aldehydes and nitroolefins. In 2020 Wennemers and coworkers reported on an organocatalyzed *anti*-selective Michael addition to give γ-nitroaldehydes (Scheme 6.19) [115]. In a

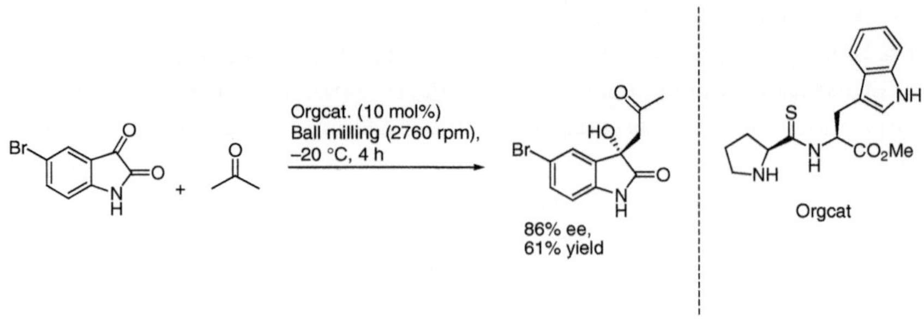

Scheme 6.18 The use of ball milling in the proline-based thiodipeptide catalysis of an aldol reaction with a bromoisatin substrate. Source: Based on Avila-Ortiz et al. [112].

nutshell this was achieved by the installation of substituents at Cδ of the reactive pyrroline unit of the peptide, which enabled the formation of the key *s-cis*-enamine intermediate (inset, Scheme 6.19). After some optimization work, the tripeptide H-D-dmP-Oic-Gln-OH (which contains an all *cis*-octahydroindole-2-carboxylic acid [Oic residue]) was identified as the best catalyst. Each of the catalysts was endowed with a *gem*-dimethyl unit at the Cδ locus of the terminal reactive proline pyrrolidine unit. The *gem*-dimethyl unit favors the more stable *s-cis* conformation leading selectively to the *anti*-product (see inset, Scheme 6.19). The *s-trans* conformation will lead to a higher activation energy transition state in the reaction due to steric hindrance with the *gem*-dimethyl unit. The authors also showed by way of density-functional

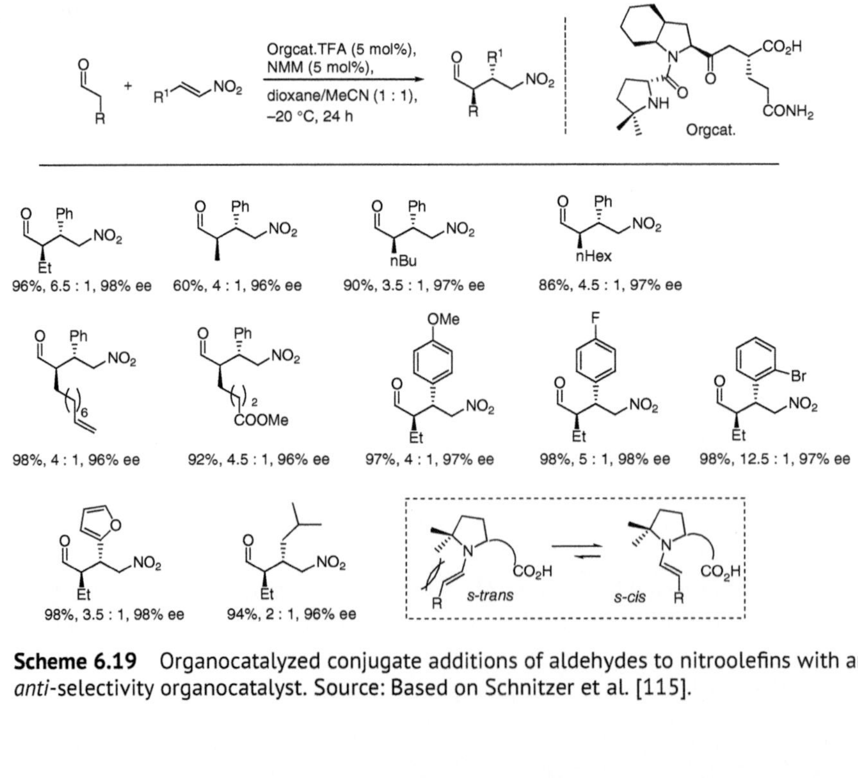

Scheme 6.19 Organocatalyzed conjugate additions of aldehydes to nitroolefins with an *anti*-selectivity organocatalyst. Source: Based on Schnitzer et al. [115].

theory (DFT) calculations that the favored *anti*-product was indeed thermodynamically less favored. It should be noted that the introduction of *gem*-dimethyl units into rings to control reactivity is nothing new, for example it was part of the design concept employed by the Davies' group for their "SuperQuat" chiral auxiliaries (also see above in Section 6.3.3.2) [116] and is also a very useful motief in medicinal chemistry. The reaction is best run with 5 mol% of the catalyst as the triflouroacetate (TFA) salt, with 5 mol% *N*-methylmorpholine (NMM) in dioxane/MeCN at −20 °C for one day. The reaction showed broad scope with both the aldehyde and nitroolefin components with yields in the range 60–98%, drs of 2 : 1 to >20 : 1, and ees of up to 98%, and could tolerate various functionalities.

6.5.2.3 Diarylprolinol Silyl Ether and Analogous Systems

As mentioned above, the diarylprolinol silyl ether system is a formidable catalyst, superior to proline in many respects, that has been effectively used in various organocatalyzed reactions – in fact one can say that it is a truly privileged organocatalyst. In this section we will consider some of the recent applications of this catalytic system.

In 2017, Lutete and Ikemoto reported the synthesis and use of novel diarylprolinol-derived amino perfluoroalkanesulfonamide catalysts in a key aldol reaction that leads to a bis-THF alcohol, which is commonly used in the design of HIV protease inhibitors such as Darunavir (Scheme 6.20) [117]. The best catalysts are those shown in Scheme 6.20, which gave yields between 83% and 87%, diastereoselectivities of >99 : 1 (*anti*), and a remarkable enantioselectivity of >99% ee. The best solvent was NMP. In fact, this catalytic system was found to be more efficient than the standard diphenylprolinol which afforded the same target

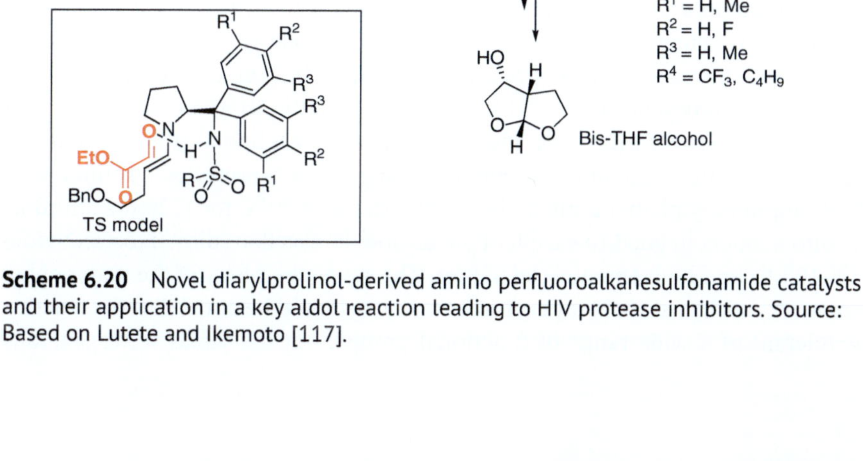

Scheme 6.20 Novel diarylprolinol-derived amino perfluoroalkanesulfonamide catalysts and their application in a key aldol reaction leading to HIV protease inhibitors. Source: Based on Lutete and Ikemoto [117].

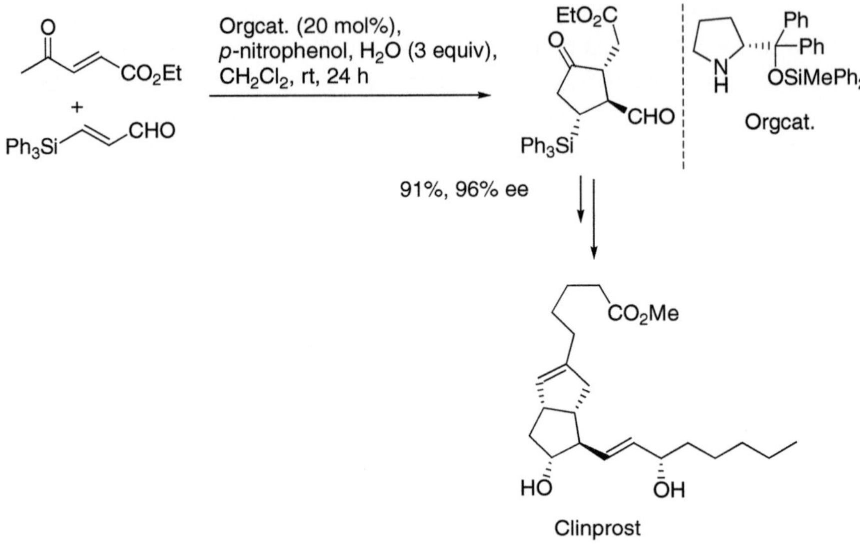

Scheme 6.21 The total synthesis of clinprost via a key diarylprolinol silyl ether catalyzed asymmetric Michael/Michael reaction. Source: Based on Umekubo and Hayashi [118].

aldol product with a dr of 94 : 6 and an ee of 95%. A transition state model to explain the enantiofacial selectivity is shown in the inset.

In 2020 Umekubo and Hayashi reported a route to the potent neuroprotective compound clinprost and an asymmetric Michael/Michael reaction catalyzed by the diphenylprolinol silyl ether constituted the key reaction step (Scheme 6.21) [118]. The catalyst was used at a loading of 20 mol% and the reaction conducted in a water/dichloromethane mixture at room temperature for 24 hours to afford the cyclopentanone derivative in 91% yield and 96% ee. This product could then be transformed to the clinprost target in a multi-step sequence – an amazing 7-step synthesis – with an excellent total yield of 17%.

These authors also reported on an asymmetric one-pot Mukaiyama Michael/Michael reaction catalyzed by the same diphenylprolinol silyl ether which could afford bicyclo[2.2.2]octanone as a single isomer in almost optically pure form via a two-step, one-pot Mukaiyama Michael/intramolecular Michael reaction [119].

Spirocompounds are a very important class of molecules that have had immense application in chemistry over the last decades, notably in medicinal chemistry. In 2017 Mondal et al. reported a vinylogous Michael-aldol cascade for the synthesis of spirocyclic adducts (Scheme 6.22) [120]. In their approach they combined a variety of propenals with α-arylidene pyrazolinone (the reaction was vinylogous on this reaction component) using a TMS-protected prolinol with 3,5-trifluoromethylphenyl units in dichloromethane at 30 °C for 12 hours affording the spiro adducts in good-to-excellent yields, low-to-excellent diastereoselectivities (1 : 1–10 : 1), and high enantioselectivities. The stereochemistry of the major enantiomer was confirmed via X-ray crystallography. It should be noted that the reaction was tolerant of a wide range of functional groups, but the yields were generally

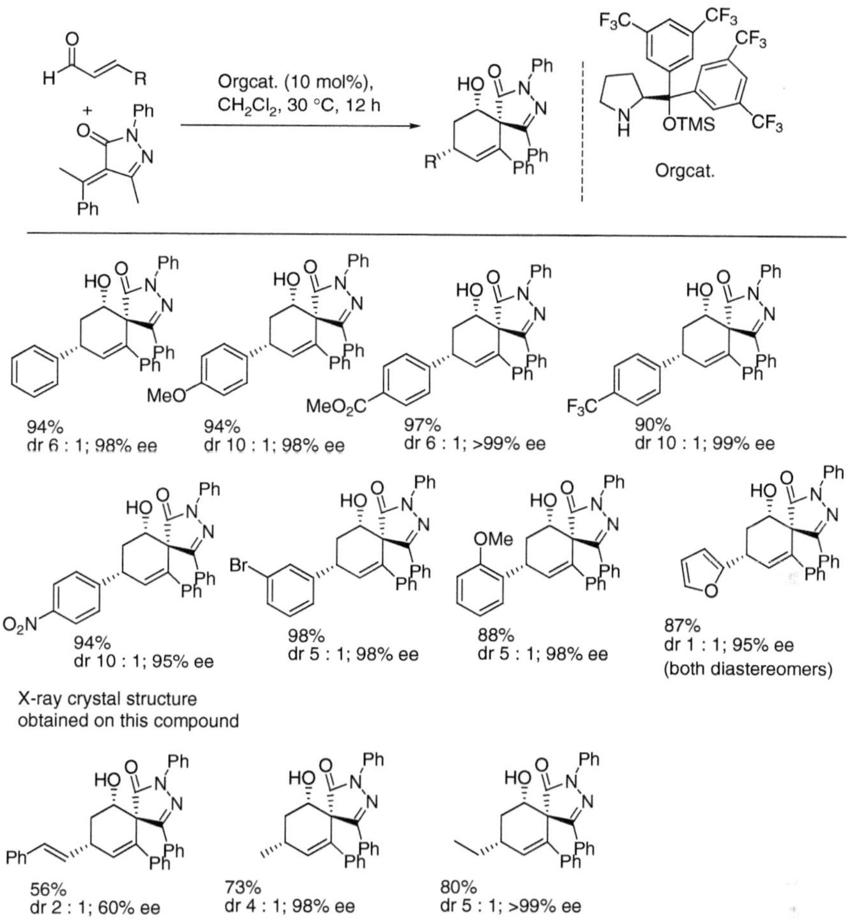

Scheme 6.22 Vinylogous Michael Aldol cascade for the synthesis of spirocyclic adducts catalyzed by a trifluoromethyl-substituted diarylprolinol silyl ether. The structures shown above are the major enantiomers of the main diastereomers. Source: Based on Mondal et al. [120].

lower with alkyl and vinyl substituents in the propenal component (53–80%). The scope on the α-arylidene pyrazolinone was also checked and similar results were again achieved.

The use of radicals in synthesis has made a significant comeback in the last 10 years or so. Organocatalysts can be combined easily with photocatalysts. Melchiorre and coworkers have developed an enantioselective radical conjugate in addition to enals using a *gem*-difluorinated pyrrolidine analogue of the catalyst used in Scheme 6.23 along with an organic photocatalyst that absorbs blue light to generate radicals from stable precursors (Scheme 6.23a) [121]. This was basically a general iminium-ion-based catalytic method that involved the conjugate addition of carbon-centered radicals to aliphatic and aromatic enals. The reactions were generally performed in acetonitrile at −10 °C using a blue LED (λ_{max} = 460 nm)

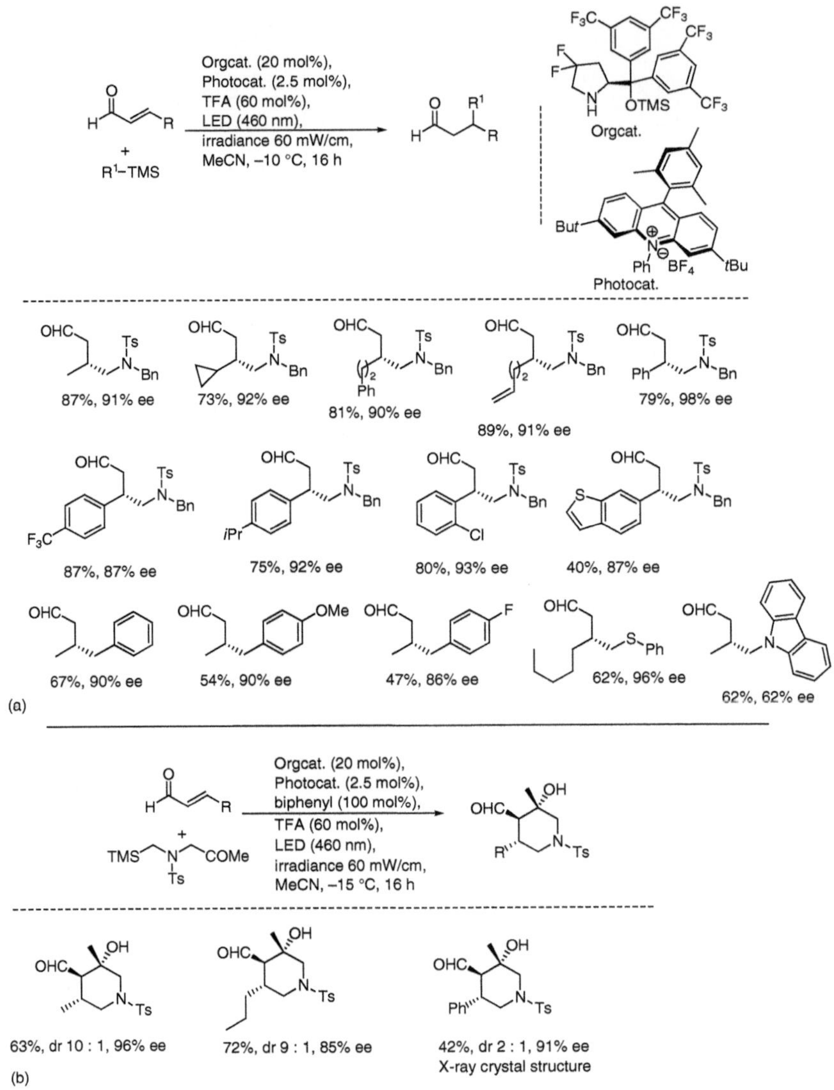

Scheme 6.23 Enantioselective radical conjugate addition to enals using a organocatalyst/photocatalyst combination. (a) Using a simple α-silyl amine radical precursor, (b) using an α-silyl-*N*-tosylamine methyl ester radical precursor. Source: Based on Le Saux et al. [121].

with an irradiance at 60 mW cm^{-2}. They showed good functional group tolerance on both the enal and the radical precursors, with yields in the range 35–87% and ees from 90% to 98%. The authors provided a plausible mechanism for this reaction (not discussed here). By judicious choice of the radical-forming component the authors successfully synthesized a series of stereochemically dense piperidines via a cascade process terminating in an aldol reaction. The stereochemistry of the major diastereomer of the product was unambiguously assigned by single-crystal X-ray

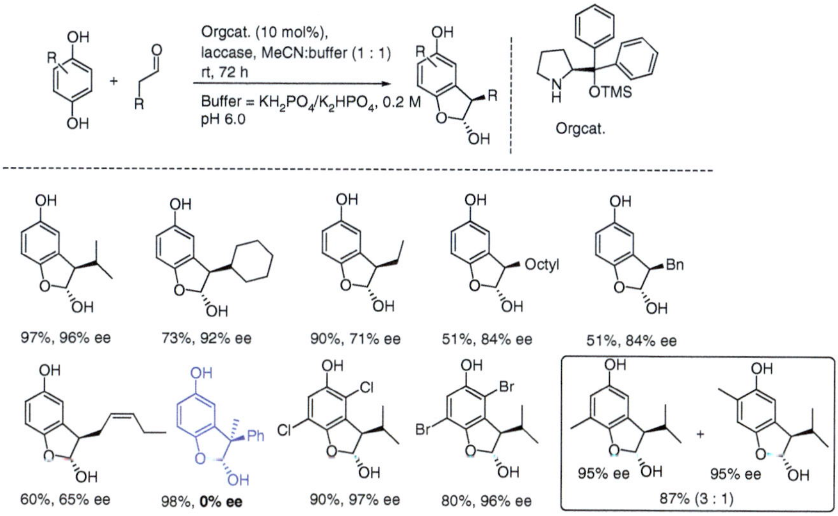

Scheme 6.24 One-pot sequential bio-catalyzed oxidation/organocatalyzed α-arylation using a simple TMS-substituted diphenylprolinol catalyst to give 2,3-dihydrobenzofuran-2,5-diols. Source: Based on Suljić et al. [122].

crystallography (Scheme 6.23b). In addition, a biphenyl additive was required as an electron mediator.

Worgull and coworkers described a one-pot sequential bio-catalyzed oxidation/organocatalyzed α-arylation using a simple TMS-substituted diphenylprolinol catalyst to give 2,3-dihydrobenzofuran-2,5-diols in good-to-excellent yields and good-to-excellent enantioselectivities (Scheme 6.24) [122]. The initial optimization studies showed that the best conditions consisted of 10 mol% of the catalyst and the laccase derived from *Agaricus bisporus* at a concentration of 45 U (from commercial quantities of 6.8 U/mg) at room temperature in a water:buffer (1 : 1) medium (see Scheme 6.24 for buffer conditions). The scope on the aldehyde and the phenol components was then studied to afford a variety of 2,3-dihydrobenzofuran-2,5-diols in yields ranging from 51% to 97% and ees in the range 65–96%. Furthermore, when the only chiral aldehyde (*rac*-2-phenylpropanal) was used, the product with the quaternary center was obtained as a racemic mixture, indicating that there was no kinetic resolution at play during the Michael addition (of the enamine intermediate) to the quinone intermediate. This was not surprising as this aldehyde is in fact a challenging substrate in asymmetric organocatalyzed reactions using secondary amine organocatalysts. In the case of the reaction with the 2-methylcatechol substrate, a mixture of inseparable regioisomers was obtained, although their enantiopurities were determined.

Jørgensen's group reported a high-yielding and highly stereoselective [4+2]-cycloaddition reaction of activated 3-olefinic oxindoles with a tetraenamine derivative using the TMS-protected prolinol with 3,5-trifluoromethylphenyl units as was used in Scheme 6.22 (Scheme 6.25) [123]. A tetraenamine intermediate was formed between the prolinol catalyst and the 2-(cyclohepta-1,3,5-trien-1-yl)acetaldehyde substrate which reacted with the 3-olefinic oxindole to give an addition product, from which the catalyst was liberated and which then underwent the [4+2]-cycloaddition reaction. The best conditions found included using 20 mol% catalyst, with 20 mol% of *o*-fluorobenzoic acid (*o*-FBA) and Na_2SO_4 (5 equiv) in dry $CHCl_3$ at room temperature for 24 hours. Interestingly, sodium sulfate was added to achieve high diastereoselectivities. The first experiment was conducted in an NMR tube in $CDCl_3$ and afforded the spirocyclic oxindole adduct in 93% yield. Unfortunately, the reaction was limited to just one cycloheptatrienyl aldehyde substrate.

The reaction showed broad functional group tolerance on the oxindole reagent, and the stereochemical configuration was assigned by way of obtaining an X-ray crystal structure on one of the brominated products. In most cases the oxindole was Boc protected, but the reaction also proceeded with non-Boc protected oxindoles. In one particular case a benzofuranone substrate was used, but the yield of the spirobenzofuranone product was lower than most of the examples with oxindoles (59%) and the enantiopurity of the major diastereomer was only 68% ee.

Ichikawa and his group used the enantiopode of the prolinol catalyst used by Jørgensen and coworkers [123] in Scheme 6.25 to carry out a key highly diasteroselective asymmetric nucleophilic epoxidation to afford a key γ,δ-diol intermediate in the synthesis of the 10-keto-acaudiol A, a macrolide natural product with reputed antiosteoporosis activity [124]. The reaction gave a product yield of 62%, a dr of >20 : 1, and an ee of 95% for the major enantiomer (Scheme 6.26).

To finalize this section we include an example of an organocatalyzed reductive amination to give the biologically active chiral amine known as calcimimetic (*R*)-(+)-NPS R-568 used for the efficient control of uremic hyperparathyroidism (Scheme 6.27) [125]. The organocatalyst used was a carbonylimidazole diphenylprolinol, with a remarkable low loading of 1 mol%, and the medicinally relevant target could be obtained in 67% yield with an ee of 89%, within a time frame of eight hours for the ketimine reduction at 0 °C. The reaction conditions were previously optimized and the reaction was shown to be broad in scope.

Many nice examples of the use of the diphenylprolinol silyl ether organocatalyst are reviewed by Hayashi, particularly their application in the synthesis of a number of biologically active molecules like baclofen, telcagepant, oseltamivir, and horsfoline [126].

6.5.2.4 Other Proline Analogues

In 2020, Maruoka and coworkers reported the use of amino-triflate-substituted hydroxyproline derivatives as organocatalysts in a variety of interesting aldol reactions (Scheme 6.28) [127]. A handful of organocatalysts were synthesized and it was the *tert*-butyldimethylsilyl (TBDMS)-containing catalyst (10 mol%) that gave good results (83% yield, 83 (*anti*) : 17 dr, and 94% ee (*anti*)), upon optimizing the conditions in the reaction of cyclohexanone with 4-nitrobenzaldehyde in cyclopropylmethyl ether (a green solvent) over a 48 hour period at room temperature. The reaction showed a reasonably broad scope, being able to accommodate a variety of different cycloalkanones (including those containing heteroatoms) and benzaldehydes. Furthermore, it was found that in the reactions with *p*-tolylaldehyde, 4- and 2-methoxybenzaldehyde were very sluggish, obviously due to the electron-rich

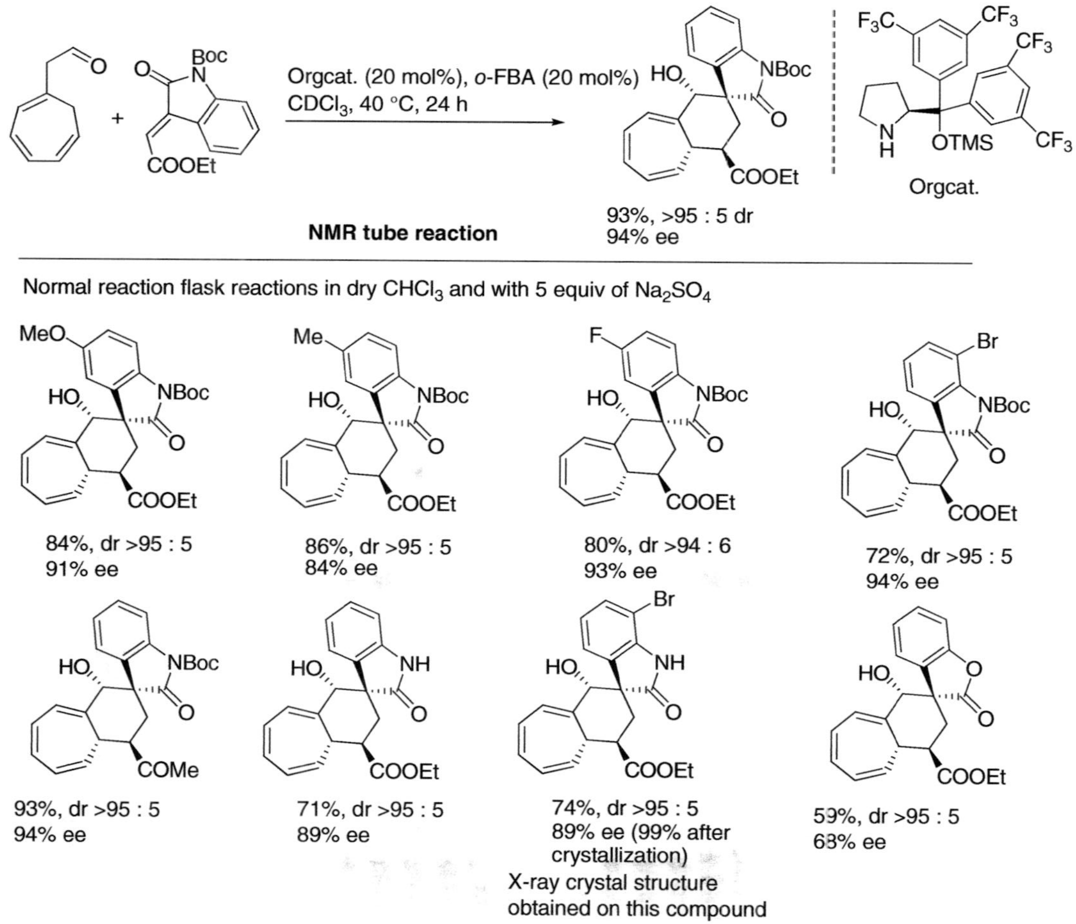

Scheme 6.25 Highly stereoselective [4+2]-cycloaddition reaction of activated 3-olefinic oxindoles with a tetraenamine derivative. Source: Based on Stiller et al. [123].

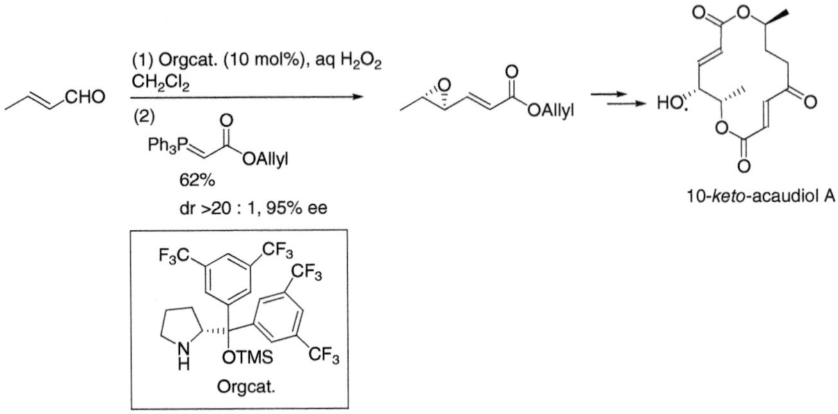

Scheme 6.26 The key highly stereoselective organocatalyzed asymmetric epoxidation step in the synthesis of the macrodiolide 10-keto-acaudiol. Source: Based on Hirabayashi et al. [124].

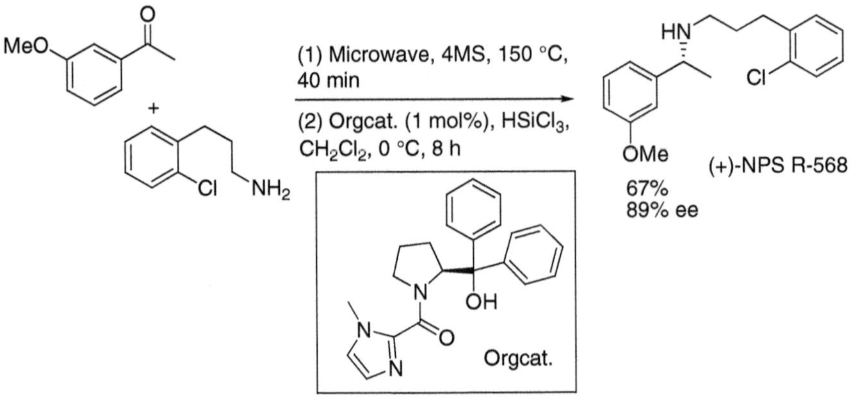

Scheme 6.27 The synthesis of calcimimetic (*R*)-(+)-NPS R-568 via a one-pot imination organocatalyzed asymmetric reductive amination. Source: Based on Gautier et al. [125].

nature of the attached substituents. The ees for the cyclopentanone reaction were inferior to those with the cyclohexanones, which was probably a structure related effect.

Being able to carry out reactions in water is highly desirable, particularly from a green chemistry perspective.

In 2019 Vila and coworkers reported the use of a visible-light organophotoredox strategy for an enantioselective Mannich reaction of dihydroquinoxalinones with ketones (Scheme 6.29) [128]. Initial studies showed that the Mannich reaction between 3,4-dihydroquinoxalin-2-one and acetone in the presence of (*S*)-proline (20 mol%) and the photocatalyst tris[2-phenylpyridine]iridium ($Ir(ppy)_3$) (1 mol%) under irradiation of 5 W white LED in MeCN at room temperature for 74 hours gave the chiral Mannich product with a yield of 34% and an ee of 99%. Key modifications were made by separating the catalytic steps, first the photocatalytic step was run for

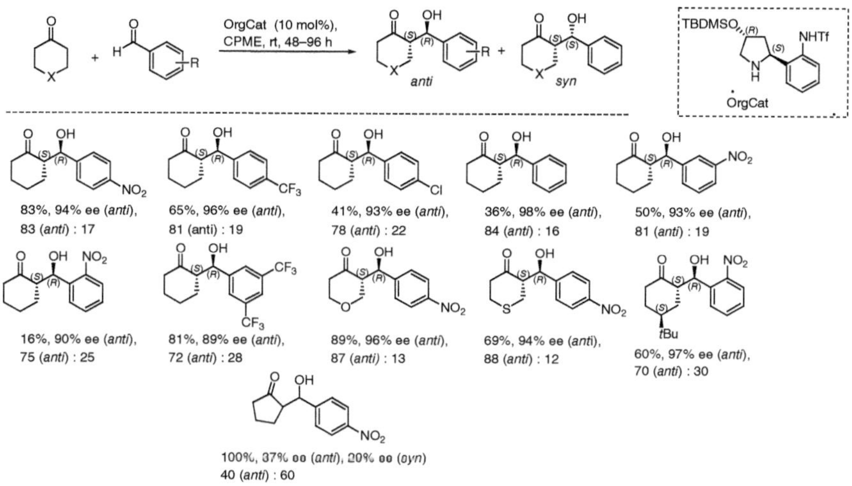

Scheme 6.28 Direct asymmetric aldol reaction between cycloalkanones and substituted benzaldehydes catalyzed by a proline-derived organocatalyst. Source: Based on Lu et al. [127].

48 hours to give the cyclic imine quinoxaline-2-one oxidation product, which was followed by the addition of acetone and proline to form the final product in 77% yield and 99% ee after 80 hours in the dark. Full optimization was achieved by using Eosin Y (2 mol%) in DMF with blue LED irradiation in the first step (7 hours), followed by the addition of proline and acetone to give the final product after 21 hours, in 74% yield and 99% ee. The authors then evaluated the scope of the reaction, varying both the dihydroquinoxalin-2-one structure and the ketone structure (Scheme 6.29). The reaction showed good substrate scope, accommodating various ketone substrates, *N*-substituted dihydroquinoxalin-2-ones, and both electron-donating and electron-withdrawing substituents in the dihydroquinoxalin-2-one's aromatic unit. The yields were good to excellent and the enantiofacial selectivity always high. In the case of those products containing two stereocenters, the diasteroselectivity was poor (1.4 to 2 : 1). Remarkably, the reaction could be run under sunlight with the prototype substrate at a 5 mmol scale (740 mg) to give the product in 67% yield and 99% ee. With regard to the mechanism, the authors carried out a variety of experiments to show that under irradiation with visible light, in the presence of Eosin Y and oxygen, the dihydroquinoxalin-2-one is oxidized to quinoxaline-2-one. A putative mechanism is shown in Scheme 6.30, which invokes the usual proline-type asymmetric induction via H-bonding of its carboxyl hydrogen to the intermediate imine nitrogen; the product is presumed to have the *R*-configuration based on *Re* face attack of the enamine on the cyclic imine.

6.5.3 Imidazolidinones

It was McMillian that made a significant impact in the development of this type of organocatalyst as we have seen above in the introduction. Some recent highlights

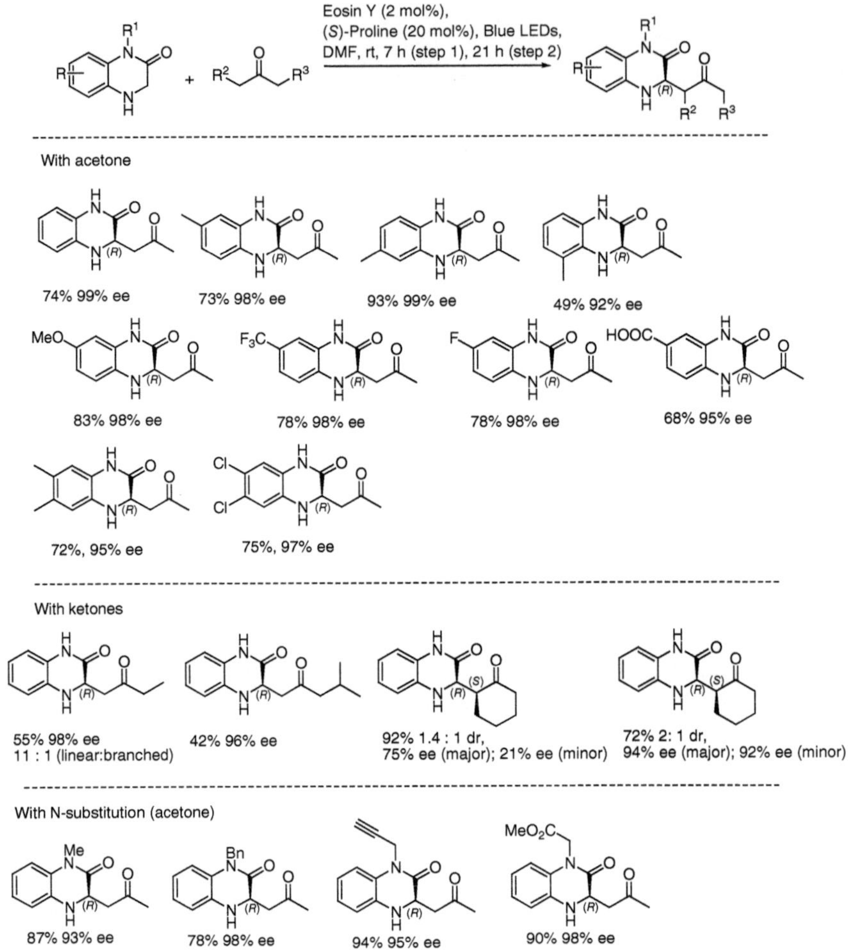

Scheme 6.29 Proline and visible-light organophotoredox strategy for an enantioselective Mannich reaction of dihydroquinoxalinones with ketones by Vila and coworkers. Source: Based on Rostoll-Berenguer et al. [128].

are discussed here. In 2021, Tanimori's group reported the enantiodivergent synthesis using the McMillian catalyst of both antipodes of the isoflavandiol compound known as equol (3-(4-hydroxyphenyl)-7-chromanol) in which both enantiomers – particularly the natural (3*S*)-isomer – have very strong biological activity profiles [129]. The key step in the synthesis of (3*S*)-equol was an aldehyde α-arylation (a very challenging undertaking) using a (2*R*,5*R*)-configured McMillian catalyst in salt form at 20 mol% loading and an Olofsson diaryliodonium salt in the presence of 20 mol% CuBr (Scheme 6.31). The authors did not discuss the yield or the enantioselectivity as the product was reduced and brominated before undergoing the cyclization. The enantiopode was accessed using the corresponding (2*S*,5*S*)-configured catalyst.

In 2015 Šebesta and coworkers reported a highly stereoselective 1,3-dipolar cycloaddition of a nitrone with an α,β-unsaturated aldehyde to afford isoxazolins

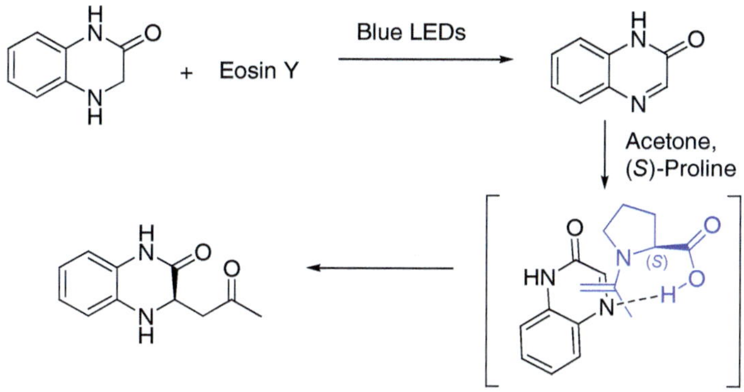

Scheme 6.30 Postulated mechanism for the enantioselective Mannich reaction of dihydroquinoxalinones with ketones. Source: Based on Rostoll-Berenguer et al. [128].

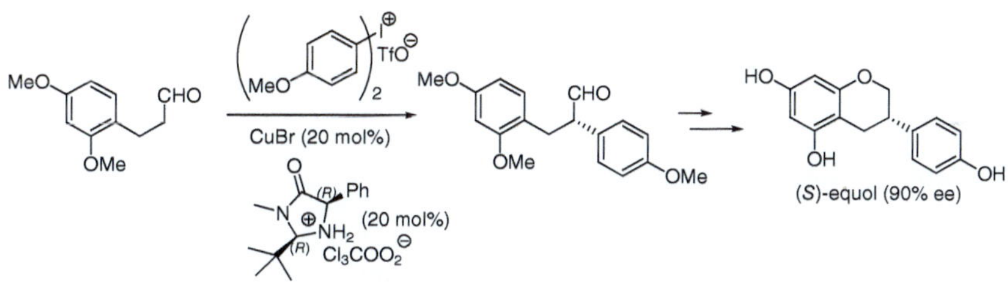

Scheme 6.31 The key α-arylation using an Olofsson diaryliodonium salt and a McMillian catalyst in the synthesis of (3*S*)-equol. Source: Based on Uemura et al. [129].

using the McMillian catalyst using enabling technologies that included ionic liquids and a flow reactor [130]. The reaction conducted in a flow microreactor was particularly interesting as it was possible to form almost exclusively the *endo* isomer with an ee of 94% (Scheme 6.32). Unfortunately, the yield was only moderate (28%); however, this was an improvement on a similar reaction conducted at 60 °C (22%) and in CH_2Cl_2 at room temperature (14%). While the diasteroselectivity was the same for the reaction carried out at 60 °C, the diastereoselectivity dropped to 66 : 34 for the reaction at room temperature. The enantioselectivity for the reaction at 60 °C was 88% ee, while the room-temperature reaction was only 62% ee.

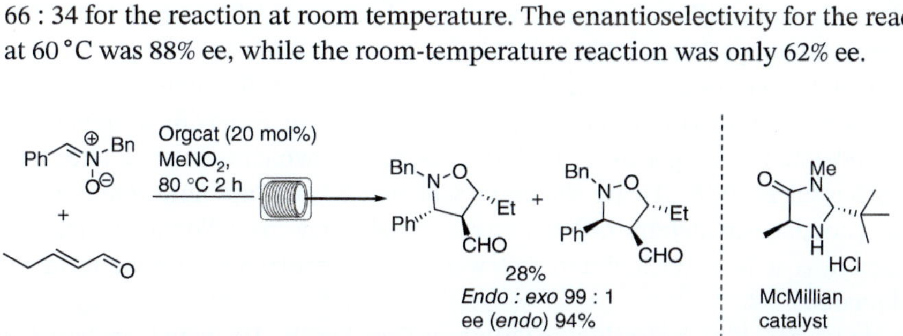

Scheme 6.32 A flow-chemistry-assisted highly stereoselective 1,3-dipolar cycloaddition of a nitrone with an α,β-unsaturated aldehyde to afford isoxazolins using the McMillian catalyst. Source: Based on Mojzesová et al. [130].

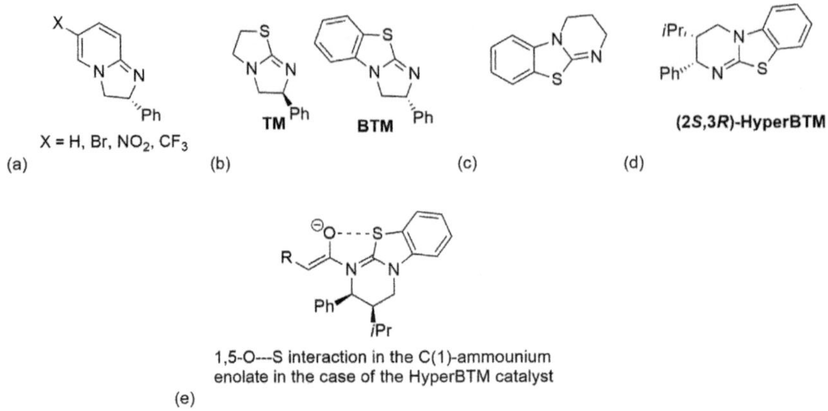

Figure 6.2 (a) Birmann's first-generation 2,3-dihydroimidazo[1,2-*a*]pyrimidine catalysts, (b) second-generation tetramisole catalysts, (c) Okamoto's catalyst, (d) Smith's HyperBTM isothiourea catalyst, and (e) C(1)-ammonium enolate stabilization via a 1,5-O···S interaction. Source: Based on Lood et al. [10].

6.5.4 Amidine-Based Organocatalysts

To finalize this section on the use of amino-acid-derived organocatalysts, we discuss briefly the 2,3-dihydroimidazo[1,2-*a*]pyrimidines that were originally pioneered by Birmann as an acyl transfer catalyst (Figure 6.2) [131]. These catalysts are derived from (*R*)-phenylglycinol.

The catalysts were successfully applied in the kinetic resolution of a plethora of chiral secondary alcohols, at 0 °C during a period that ranged from 1 to 30 hours, affording the chiral alcohol with high enantioselectivity. The best performing catalyst was that containing the CF_3 unit (Figure 6.2a). These catalysts became known as amidine-based catalysts (ABCs). A follow-up study, also reported in 2006, showed that secondary benzylic alcohols could be resolved with very high selectivity using benzotetramisole (BMT) [100]. Again, in 2006, Okamoto developed a similar type of catalyst, namely 3,4-dihydro-2*H*-9-thia-1,4a-diazafluorene (Figure 6.2c) which was used to tremendous effect as an acyl transfer catalyst and which was even faster than 4-dimethylaminopyridine (DMAP) [132].

In 2020, Birmann's group investigated the use of their ABC catalysts in cascade cyclizations of α,β-unsaturated thioesters leading to indane-fused dihydropyranones with moderate-to-very-good ees [133], the best results, which were conducted with 10 mol% catalyst (BTM II), are shown in Scheme 6.33. The absolute configuration of the product was determined by single-crystal X-ray crystallography. The key reaction was a Michael addition followed by a lactonization step to form the final product.

Lastly, we consider the isothiourea catalysts from Smith's lab, which are based on the Birman/Okamoto systems (Figure 6.2) [134]. Besides being used successfully in the kinetic resolution of axially chiral biaryl diols, (*R*)-BTM was found to be particularly efficient giving ees of up to 90%, *s* values of up to 40 [s = selectivity value and

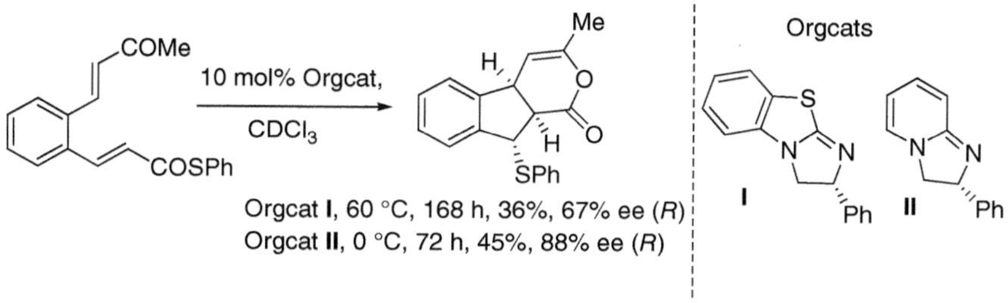

Scheme 6.33 Cascade cyclizations of α,β-unsaturated thioesters leading to indane-fused dihydropyranones using Birman catalysts. Source: Based on Ahlemeyer et al. [133].

is equal to kR/kS for the reaction]. Also a variety of different anhydrides were used, including mixed anhydride reagents [135], that were used with tremendous effect for a variety of other reactions that involve the intermediacy of C(1)-ammonium enolates [134], including lactonization, lactamization, cycloaddition reactions, and rearrangements involving ammonium ylides. The key to the success of these reactions is the preferential formation of the (*Z*)-enolate, by way of a 1,5-O· · ·S interaction between the enolate oxygen and the S atom of the catalyst (Figure 6.2e).

6.6 Conclusions

In this chapter we have seen the power and amazing versatility of amino acids (and some derivatives) as building blocks for the construction of important chiral targets, whether this is achieved with chiral amino acid reagents, chiral amino-acid-based auxiliaries, or amino-acid-derived organocatalysts. We have seen how they have been effectively harnessed for the synthesis of many important biologically active targets, such as pharmaceuticals and other important compounds. Over the last 50 years, there has been a gradual evolution in the manner of application of these building blocks, from their use as chiral substrates in the chiral pool approach, through their use as chiral reagents and auxiliaries, to their more recent exploitation as organocatalysts over the last 20 years. There is still much more ground to be covered by this interesting building block, and we expect amino acids to continue to have an important role as key tools in the asymmetric synthesis of specific important targets over the next decades.

References

1 Muzler, J.C. (2012). Pool synthesis: from α-amino acids and derivatives. In: *Comprehensive Chirality*, vol. 2 (ed. E.M. Carreira and H. Yamamoto), 122–162. Elsevier.
2 Paek, S.-M., Jeong, M., Jo, J. et al. (2016). *Molecules* 21: 951–963.
3 Singh, P., Lavanya, S., Manda, K. et al. (2017). *Tetrahedron* 73: 1911–2008.

4 Yin, Z.; Hu, W.; Zhang, W.; H. Konno, Moriwaki, H; Izawa, K.; Han, J.; Soloshonok, V.A. (2020) *Amino Acids* 52, 1227–1261. (https://doi.org/10.1007/s00726-020-02887-4).
5 Piotrowska, D.G., Głowacka, I.E., Wróblewski, A.E., and Lubowiecka, L. (2019). *Beilstein J. Org. Chem.* 15: 236–255.
6 Liu, J., Han, J., Izawa, K. et al. (2020). *Eur. J. Med. Chem.* 208: 112736 (1–74). (https://doi.org/10.1016/j.ejmech.2020.112736.
7 Han, J., Konno, H., Sato, T. et al. (2021). *Eur. J. Med. Chem.* 220: 113448 (1–49).
8 Osberger, T.J., Rogness, D.C., Kohrt, J.T. et al. (2016). *Nature* 537: 214–219.
9 Singh, P., Samanta, K., Dasa, S.K., and Panda, G. (2014). *Org. Biomol. Chem.* 12: 6297–6339.
10 Lood, C.S. and Koskinen, A.M.P. (2014). *Eur. J. Org. Chem.* 2357–2364.
11 Lilley, M., Mambwe, B., Jackson, R.F.W., and Muimo, R. (2014). *Chem. Commun.* 50: 9343–9345.
12 Ross, A.J., Lang, H.L., and Jackson, R.F.W. (2010). *J. Org. Chem.* 75: 245–248.
13 Salih, N., Adams, H., and Jackson, R.F.W. (2016). *J. Org. Chem.* 81 (18): 8386–8393.
14 Reeve, P.A.P., Grabowska, U., Oden, L.S. et al. (2019). *ACS Omega* 4: 10854–10865.
15 Riedner, J. (2006). Improved synthesis scheme for lacosamide. EP1642889A1, filed 02 October 2004.
16 Yang, A., Hu, F., Li, Z. et al. (2019). *Org. Process Res. Dev.* 23: 818–824.
17 Aratikatla, E.K. and Bhattacharya, A.K. (2015). *Tetrahedron Lett.* 56: 5802–5803.
18 Decuyper, L., Deketelaere, S., Vanparys, L. et al. (2018). *Chem. Eur. J.* 24: 15254–15266.
19 Horigome, M., Motoyoshi, H., Watanabe, H., and Kitahara, T. (2001). *Tetrahedron Lett.* 42: 8207–8210.
20 Bergeron, R.J., Bharti, N., and Singh, S. (2007). *Synthesis* 1033–1037.
21 Hanessian, S., Vakiti, R.R., Dorich, S. et al. (2011). *Angew. Chem. Int. Ed.* 50: 3497–3500.
22 Sengoku, T., Wierzejska, J., Takahashi, M., and Hidemi, M.T. (2010). *Synlett* 2944–2946.
23 Lee, S.-J., Samala, M., Woo, S.Y. et al. (2018). *ACS Omega* 3: 1970–1976. https://doi.org/10.1021/acsomega.7b01689.
24 Orme, M.W., Martinelli, M.J., Doecke, C.W., et al. (2004). Modified Pictet–Spengler reaction and products prepared therefrom. WO2004/011463 A1, filed 14 July 2003.
25 Ruchti, J. and Carreira, E.M. (2014). *J. Am. Chem. Soc.* 136: 16756–16759.
26 Feng, Y., Holte, D., Zoller, J. et al. (2015). *J. Am. Chem. Soc.* 137: 10160–10163.
27 Yang, Z., Tan, Q., Jiang, Y. et al. (2021). *Angew. Chem. Int. Ed.* 60: 13105–13111.
28 Long, B., Tao, C., Li, Y. et al. (2020). *Org. Biomol. Chem.* 18: 5349–5353.
29 Gnas, Y. and Glorius, F. (2006). *Synthesis* 12: 1899–1930.
30 Gawley, R.E. and Aubé, J. (ed.) (2012). *Principles of Asymmetric Synthesis*, 2e. *Elsevier*.

31 Lin, G.-Q., Li, Y.-M., and Chan, A.S.C. (ed.) (2001). *Principles and Applications of Asymmetric Synthesis*. Wiley.

32 Diaz-Muñoz, G., Miranda, I.L., Sartori, S.K. et al. (2019). *Chirality* 31: 776–812.

33 Maison, W. (2012). Amino acid derived auxiliaries: amino acids as chiral auxiliaries, Chapter 3.1. In: *Comprehensive Chirality* (ed. E.M. Carreira and H. Yamamoto), 1–18. Elsevier.

34 Boeckman, R.K. and Cody, J.A. (ed.) (2020). *Heterocycles as Chiral Auxiliaries in Asymmetric Synthesis*, vol. 55. Cham: Springer.

35 Yamada, S.-I., Hiroi, K., and Achiwa, K. (1969). *Tetrahedron Lett.* 10: 4233–4236.

36 Yamada, S.-I. and Otani, G. (1971). *Tetrahedron Lett.* 12: 1133–1136.

37 Yamada, S.-I., Shibasaki, M., and Terashima, S. (1973). *Tetrahedron Lett.* 5: 381–384.

38 Enders, D. and Eichenauer, H. (1976). *Angew. Chem. Int. Engl. Ed.* 15 (9): 549–551.

39 Corey, E.J. and Enders, D. (1976). *Tetrahedron Lett.* 17 (1): 3–6.

40 Maison, W. (2019). Pyrrolidines as chiral auxiliaries. In: *Heterocycles as Chiral Auxiliaries in Asymmetric Synthesis*, Topics in Heterocyclic Chemistry, vol. 55 (ed. M. Braun). Cham: Springer, 157–192.

41 Heravi, M.M., Zadsirjan, V., and Daraie, M. (2017). *Curr. Org. Synth.* 14: 61–111.

42 Dexter, H.R., Allen, E., and Williams, D.M. (2018). *Tetrahedron Lett.* 59 (49): 4323–4325.

43 Schöllkopf, U., Hartwig, W., and Groth, U. (1979). *Angew. Chem. Int. Ed.* 18: 863–864.

44 Schöllkopf, U., Groth, U., and Deng, C. (1981). *Angew. Chem. Int. Ed. Engl.* 20 (9): 798–799.

45 Schöllkopf, U. (1983). *Tetrahedron* 39: 2085.

46 Sano, S. and Nakao, M. (2015). *Heterocycles* 91: 1349–1375.

47 Jones, E.P., Jones, P., and Barrett, A.G.M. (2011). *Org. Lett.* 13 (5): 1012–1015.

48 Jones, E.P., Jones, P., White, A.J.P., and Barrett, A.G.M. (2011). *Beilstein J. Org. Chem.* 7: 1570–1576.

49 Šačkus, A., Bričkutė, D., Paliulis, O., and Sløk, F.A. (2015). *J. Heterocycl. Chem.* 52 (4): 1032–1036.

50 Meleán, J.C., Humpert, S., Ermert, J., and Coenen, H.H. (2015). *J. Fluorine Chem.* 178: 202–207.

51 Toop, H.D., Dun, M.D., Ross, B.K. et al. (2016). *Org. Biomol. Chem.* 14: 4605–4616.

52 Evans, D.A., Bartroli, J., and Shih, T.L. (1981). *J. Am. Chem. Soc.* 103: 2127–2129.

53 Mourad, A.K. and Czekelius, C. (2020). *Top. Heterocycl. Chem.* 55: 113–156.

54 Heravi, M.M. and Zadsirjan, V. (2013). *Tetrahedron: Asymmetry* 24: 1149–1188.

55 Heravi, M.M., Zadsirjan, V., and Farajpour, B. (2016). *RSC Adv.* 6: 30498–30551.

56 Nazari, A., Heravi, M.M., and Zadsirjan, V. (2021). *J. Organomet. Chem.* 932: 121629.

57 Muralikrishna, K., Satyanarayana, V., Kumar, G.C., and Yadav, J.S. (2019). *ChemistrySelect* 4: 3002–3005.

58 Kaneko, H., Takahashi, S., Kogure, N. et al. (2019). *J. Org. Chem* 84 (9): 5645–5654.
59 Madhu, M., Doda, S.R., Begari, P.K. et al. (2021). *J. Heterocycl. Chem.* 58: 942–946.
60 Kuranaga, T., Tamura, M., Ikeda, H. et al. (2021). *Org. Lett.* 23: 7106–7111.
61 Davies, S.G. and Sanganee, H.J. (1995). *Tetrahedron: Asymmetry* 6: 671–674.
62 Bull, S.D., Davies, S.G., Jones, S. et al. (1998). *Synlett* 519–521.
63 Brenner, M., La Vecchia, L., Leutert, T., and Seebach, D. (2003). *Org. Synth.* 80: 57–65.
64 Davies, S.G., Mortimer, D.A.B., Mulvaney, A.W. et al. (2008). *Org. Biomol. Chem.* 6: 1625–1634.
65 Bew, S.P., Bull, S.D., and Davies, S.G. (2000). *Tetrahedron Lett.* 41: 7577–7581.
66 Bew, S.P., Bull, S.D., Davies, S.G. et al. (2002). *Tetrahedron* 58: 9387–9401.
67 Davies, S., Fletcher, A.M., Roberts, P.M., and Thomson, J.E. (2019). *Org. Biomol. Chem.* 17: 1322–1335; and references therein.
68 Wong, M.L.J., Mousseau, J.J., Mansfield, S.J., and Anderson, E.A. (2019). *Org. Lett.* 21: 2408–2411.
69 Itsuno, S., Nakano, M., Miyazaki, K. et al. (1983). *J. Chem. Soc., Perkin Trans.* 1: 1673–1676.
70 Corey, E.J., Bakshi, R.K., and Shibata, S. (1987). *J. Am. Chem. Soc.* 109: 5551–5553.
71 Burke, A.J., Marques, C.S., Hermann, G.J., and Turner, N. (2018). *Active Pharmaceutical Ingredients in Synthesis, Catalytic Processes in Research and Development*. Weinheim, Germany: Wiley-VCH.
72 Davies, S.G., Smith, A.D., and Price, P.D. (2005). *Tetrahedron: Asymmetry* 16 (17): 2833–2891.
73 Davies, S.G., Fletcher, A.M., Roberts, P.M., and Thomson, J.E. (2012). *Tetrahedron: Asymmetry* 23 (15, 16): 1111–1153.
74 Davies, S.G., Fletcher, A.M., Roberts, P.M., and Thomson, J.E. (2017). *Tetrahedron: Asymmetry* 28 (12): 1842–1868.
75 Evans, D.A., Woerpel, K.A., Hinmann, M.M., and Faul, M.M. (1991). *J. Am. Chem. Soc.* 113: 726–728.
76 Nishiyama, H., Itoh, Y., Matsumoto, H. et al. (1994). *J. Am. Chem. Soc.* 116: 2223–2224.
77 Chercheja, S., Carreiro, E.P., Burke, A.J. et al. (2005). *J. Mol. Catal. A: Chem.* 236: 38–45.
78 Burke, A.J., Carreiro, E.P., Chercheja, S. et al. (2006). *Inorg. Chem. Commun.* 9: 823–826.
79 Burke, A.J., Carreiro, E.P., Chercheja, S. et al. (2007). *J. Organomet. Chem.* 692: 4863–4874.
80 O'Reilly, S. and Guiry, P.J. (2014). *Synthesis* 46: 722–739.
81 Smith, M.B. (2020). *March's Advanced Organic Chemistry*, 8e. Hoboken, NJ: Wiley.
82 Rothenberg, G. (2017). *Catalysis – Concepts and Green Applications*, 2e. Weinheim, Germany: Wiley-VCH.

83 Burke, A.J. and Hermann, G.J. (2021). *Amino-Cinchona derivatives in organocatalysis.* In: *Organocatalysis: Stereoselective Reactions and Applications in Organic Synthesis*, Physical Sciences Reviews (ed. M. Benaglia), 85–175. Walter de Gruyter GmbH.

84 Xiang, S.-H. and Tan, B. (2020). *Nat. Commun.* 11: 3786 (1–5).

85 Durrani, J. (2021). How organocatalysis won the Nobel prize. *Chemistryworld* (15 Oct 2021), pp. 22–27.

86 Rouhi, A.M. (2004). A renaissance in organocatalysis. *Chemical & Engineering News* (September 6), pp. 41–45.

87 Ahrendt, K.A., Borths, C.J., and McMillian, D.W.C. (2000). *J. Am. Chem. Soc.* 122: 4243–4244.

88 Eder, U. (1971). *Angew. Chem. Int. Engl. Ed.* 10: 496–497. For a comprehensive review of its mechanism see: List, B., Hoang, L., and Martin, H.J. (2004). *Proc. Nat. Acad. Sci.* 101: 5839–5842.

89 List, B. (2002). *Tetrahedron* 58: 5573–5590.

90 For applications in flow see:Rodríguez-Escrich, C. and Pericàs (2019). *Chem. Rec.* 19: 1872–1890.

91 For applications in flow see:Atodiresei, I., Vila, C., and Rueping, M. (2015). *ACS Catal.* 5: 1972–1985.

92 Franzén, J., Marigo, M., Fielenbach, D. et al. (2005). *J. Am. Chem. Soc.* 127: 18296–18304.

93 For further references see:Jiang, H., Falcicchio, A., Jensen, K.L. et al. (2009). *J. Am. Chem. Soc.* 131: 7153–7157.

94 For a good review:Jensen, K.L., Dickmeiss, G., Jiang, H. et al. (2012). *Acc. Chem. Res.* 45: 248–264; (this paper has already received 543 citations (7-5-2021)).

95 Hayashi, Y., Gotoh, H., Hayashi, T., and Shoji, M. (2005). *Angew. Chem. Int. Ed.* 44: 4212–4215.

96 Sigman, M.S. and Jacobsen, E.N. (1998). *J. Am. Chem. Soc.* 120: 4901–4902.

97 Reisman, S.E., Doyle, A.G., and Jacobsen, E.N. (2008). *J. Am. Chem. Soc.* 130: 4901–4902.

98 Fuerst, D.E. and Jacobsen, E.N. (2005). *J. Am. Chem. Soc.* 127: 8964–8965.

99 Krattiger, P., Kovasy, R., Revell, J.D. et al. (2005). *Org. Lett.* 7: 1101–1103.

100 Birman, V.B. and Li, X. (2006). *Org. Lett.* 8: 1351–1354.

101 For a recent review see:McLaughlin, C. and Smith, A.D. (2021). *Chem. Eur. J.* 27: 1533–1555.

102 Krištofílová, D., Modrocká, V., Mečiarová, M., and Šebesta, R. (2020). *ChemSusChem* 13: 2828–2858.

103 Atodiresei, I., Vila, C., and Rueping, M. (2015). *ACS Catal.* 5: 1972–1985.

104 Pelšs, A., Gandhamsetty, N., Smith, J.R. et al. (2018). *Chem. Eur. J.* 24: 9542–9545. For a complete history of the project see: Bennett, S.H., Coulthard, G., and Aggarwal, V.K. (2020) *Chem. Rec.* 20: 936–947.

105 Kanomata, K., Tatebayashi, N., Habaki, X., and Kitaoka, T. (2018). *Sci. Rep.* 8: 4098 (1–6).

106 Fanjul-Mosteirín, N. and del Amo, V. (2021). *Tetrahedron* 84: 131967 (1–13).

107 Martínez, R., Berbegal, L., Guillena, G., and Ramón, D. (2016). *J. Green Chem.* 18: 1724–1730.
108 Alonso, D.A., Baeza, A., Chinchilla, R. et al. (2017). *Molecules* 22: 895 (1–51).
109 Hayashi, Y., Yamazaki, T., Kawauchi, G., and Sato, I. (2018). *Org. Lett.* 20: 2391–2394.
110 Rudroff, F., Mihovilovic, M.D., Gröger, H. et al. (2018). *Nat. Catal.* 1: 12–22.
111 Rulli, G., Duangdee, N., Hummel, W. et al. (2017). *Eur. J. Org. Chem.* 812–817.
112 Avila-Ortiz, C.G., Pérez-Venegas, M., Vargas-Caporali, J., and Juaristi, E. (2019). *Tetrahedron Lett.* 60: 1749–1757.
113 Duschmalá, J. and Wennemers, H. (2012). *Chem. Eur. J.* 18: 1111–1120.
114 Grünenfelder, C.E., Kisunzu, J.K., and Wennemers, H. (2016). *Angew. Chem. Int. Ed.* 55: 8571–8574.
115 Schnitzer, T., Budinská, A., and Wennemers, H. (2020). *Nat. Catal.* 3: 143–147.
116 Davies, S.G., Sanganee, H.J., and Szolcsanyl, P. (1999). *Tetrahedron* 55: 3337–3354.
117 Lutete, L.M. and Ikemoto, T. (2017). *Chem. Lett.* 46: 577–579.
118 Umekubo, N. and Hayashi, Y. (2020). *Org. Lett.* 22: 9365–9370.
119 Umekubo, N. and Hayashi, Y. (2020). *Eur. J. Org. Chem.* 5596–5600.
120 Mondal, S., Mukherjee, S., Yetra, S.R. et al. (2017). *Org. Lett.* 19: 4367–4370.
121 Le Saux, E., Ma, D., Bonilla, P. et al. (2021). *Angew. Chem. Int. Ed.* 60: 5357–5362.
122 Suljić, S., Pietruszka, J., and Worgull, D. (2015). *Adv. Synth. Catal.* 357: 1822–1830.
123 Stiller, J., Poulsen, P.H., Cruz, D.C. et al. (2014). *Chem. Sci.* 5: 2052–2056.
124 Hirabayashi, J., Yakushiji, F., Katsuyama, A., and Ichikawa, S. (2020). *Org. Lett.* 22: 5545–5549.
125 Gautier, F.M., Jones, S., Li, X., and Martin, S.J. (2011). *Org. Biomol. Chem.* 9: 7860–7868.
126 Hayashi, Y. (2021). Domino and one-pot syntheses of biologically active compounds using diphenylprolinol silyl ether. Chapter 1. In: *Organocatalysis: Stereoseletive Reactions and Applications in Organic Synthesis* (ed. M. Benaglia), 1–27. Berlin/Boston: Walter de Gruyter GmbH.
127 Lu, H., Lv, J., Zhou, C. et al. (2020). *Asian J. Org. Chem.* 9: 206–209.
128 Rostoll-Berenguer, J., Blay, G., Muñoz, M.C. et al. (2019). *Org. Lett.* 21: 6011–6015.
129 Uemura, T., Saito, Y., Sonoda, M., and Tanimori, S. (2021). *Synlett* 32: 693–696.
130 Mojzesová, M., Mečiarová, M., Almássy, A. et al. (2015). *Chem. Pap.* 69: 737–746.
131 Birman, V.B., Uffman, E.W., Jiang, H. et al. (2004). *J. Am. Chem. Soc.* 126: 12226–12227.
132 Kobayashi, M. and Okamoto, S. (2006). *Tetrahedron Lett.* 47: 4347–4350.
133 Ahlemeyer, N.A., Straub, M.R., Leace, D.M. et al. (2020). *J. Org. Chem.* 86: 1191–1197.
134 McLaughlin, C. and Smith, A.D. (2021). *Chem. Eur. J.* 27: 1533–1555.
135 Qu, S., Greenhalgh, M.D., and Smith, A.D. (2019). *Chem. Eur. J.* 25: 2816–2823.

7

Carbohydrate-Based Catalysts and Auxiliaries in Organic Synthesis

Sebastian Baś[1], Szymon Buda[1], and Jacek Młynarski[2]

[1]Jagiellonian University, Faculty of Chemistry, Gronostajowa 2, 30-387 Kraków, Poland
[2]Institute of Organic Chemistry, Polish Academy of Sciences, Kasprzaka 44/52, 01-224 Warsaw, Poland

7.1 Chiral Auxiliaries

The chiral auxiliary concept – involving the temporary incorporation of an optically pure unit into a molecule in order to control the stereochemistry of desired reaction – is a commonly utilized approach in asymmetric synthesis. Since the first introduction by Corey and Ensley in 1975 [1] it has become one of the most fundamental techniques for the preparation of enantiomerically pure compounds. Nowadays, a plethora of examples of chiral auxiliaries based on amino acids, terpenes, and alkaloids are reported as a useful tool for the construction of complex structures [2]. Sugars are distinguishable by their high availability in optically pure form, polyfunctional nature, and low costs, yet were somehow neglected in this strategy for a long time. Apart from the obvious advantages mentioned above carbohydrates also exhibit the anomeric and *exo*-anomeric effects which often have an essential impact on stereochemistry control by influencing the spatial orientation of the substituent linked to the anomeric position. Although carbohydrates exist mainly in only one enantiomeric form of D-sugars (L-sugars are quite rare and expensive) they are often used in the synthesis of both enantiomeric forms of desired products since they can act as pseudo-enantiomers. The most prominent example of such pseudo-enantiomeric effect for carbohydrate-derived auxiliaries is exhibited by derivatives of D-galactose and D-arabinose (Figure 7.1) which predominantly result in products with opposite configurations.

The additional advantage of sugar auxiliaries implementation is the convenient separation of obtained diastereoisomers by simple column chromatography or recrystallization. Compounds thus obtained in pure diastereomeric form are prone to form high-quality crystals appropriate for further X-ray analysis that allows establishing the absolute configuration of products. Subsequent detachment of sugar-auxiliary from the optically pure product is a generally straightforward and efficient process. Moreover, in many cases, the recycling of sugar-auxiliary is also possible. The first example of carbohydrates introduction as chiral auxiliaries

Chiral Building Blocks in Asymmetric Synthesis: Synthesis and Applications, First Edition.
Edited by Elżbieta Wojaczyńska and Jacek Wojaczyński.

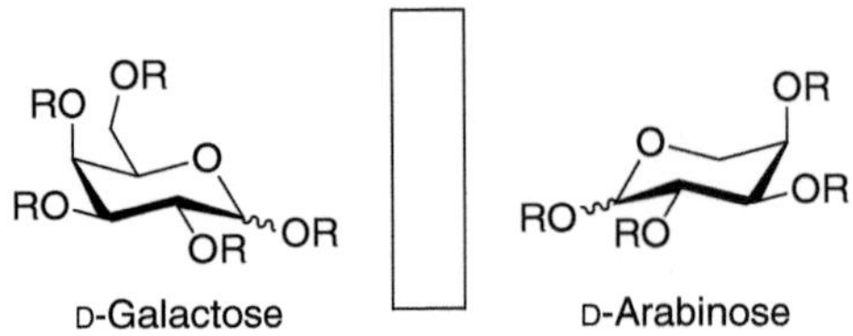

Figure 7.1 The pseudo-enantiomeric relation of D-galactose and D-arabinose.

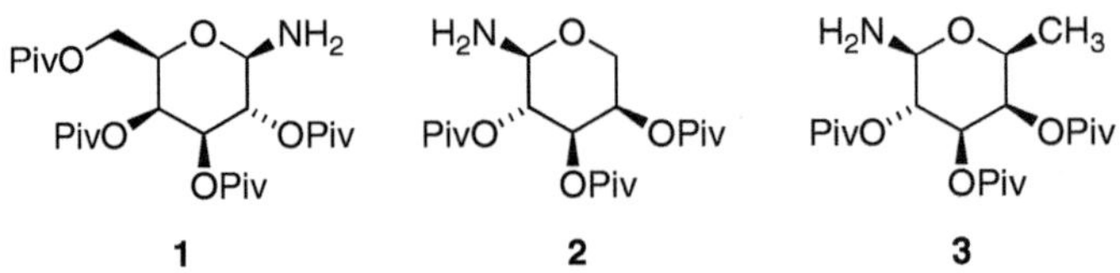

Figure 7.2 The examples of Kunz auxiliaries based on D-galactose (**1**), D-arabinose (**2**) and L-fucose (**3**).

in stereoselective [3+2] cycloaddition was reported by Vasella in 1977 [3]. With the proven potential in the auxiliary-approach, sugars gradually increase their recognition as stereodifferentiating auxiliaries in numerous asymmetric reactions.

7.1.1 Chiral Imines

The most recognizable type of carbohydrate derived auxiliary is per-*O*-pivaloylated D-galactosyl amine (**1**) often called the Kunz auxiliary (Figure 7.2), that can be easily prepared from *O*-pivaloylated or acetylated galactose, by the formation of galactosyl azide in reaction with trimethylsilyl azide in the presence of tin tetrachloride followed by azide reduction. In the same manner, derivatives of D-arabinose (**2**)and L-fucose (**3**) can be also achieved.

The utility of Kunz auxiliaries in the diastereoselective reaction of imines has been demonstrated for the variety of asymmetric reactions (Scheme 7.1) [4]. In 1987 Kunz reported the stereoselective Strecker reaction of imine **4** with trimethylsilyl cyanide (TMSCN). In the developed protocol per-*O*-pivaloyl-β-D-galactopyranose (**1**) underwent condensation with aliphatic and aromatic aldehydes to afford imine **4** that in the presence of a Lewis acid in a polar solvent was attacked by cyanide [5]. Among all tested parameters, combinations of $SnCl_4$ with THF and $ZnCl_2$ with 2-propanol provided the highest diastereoselectivity (dr 7 : 1–13 : 1) with a predominant formation of product (*R*)-**5**, an optically pure precursor of α-amino acids. The observed selectivity was explained by the combination of steric and stereoelectronic effects [6]. The proposed transition state assumed coordination of a Lewis acid to the 2-*O*-pivaloyl group and the imine nitrogen (**TS1**, Scheme 7.1); the subsequent approach of cyanide is favored from the less hindered *Si*-face of the imine. Interestingly, the change of the solvent polarity has a vital impact on the reaction selectivity since for $ZnCl_2$ in chloroform the (*S*)-**5** predominates with a slightly lower diastereomer ratio (dr 3 : 1–9 : 1) [7]. This phenomenon was explained by the difference in the transition complex structure and the nature of the nucleophile. While in polar solvents free cyanide is a predominant form, in

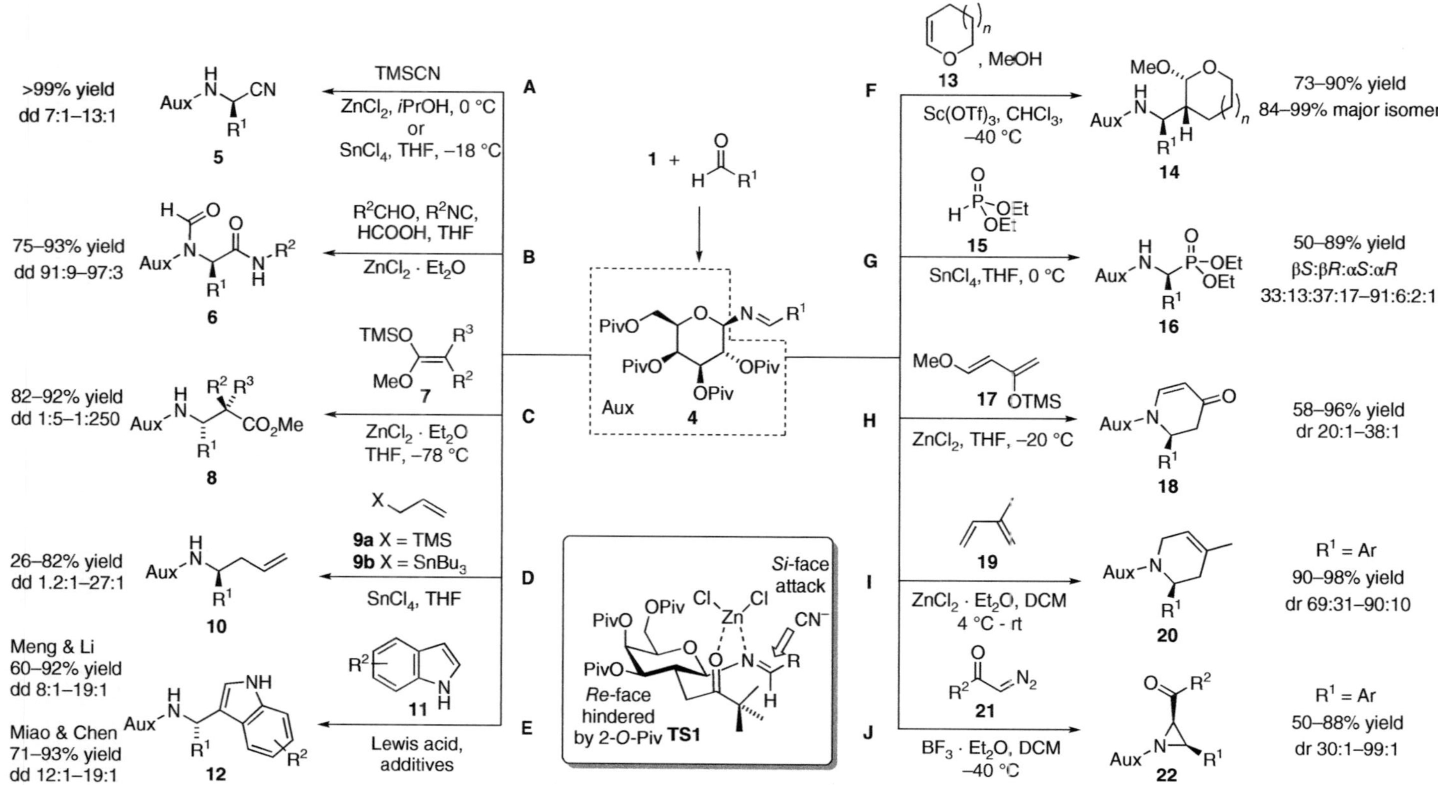

Scheme 7.1 The variety of asymmetric reactions involving Kunz auxiliary as a stereodifferentiating agent.

chloroform the additional interaction between the trimethylallylsilane (TMS) part of the reactant and the chloride ligand from the $ZnCl_2$-**4** complex is essential to release the active form of the nucleophile that attacks from the *Re*-side. The access to (*S*)-amines with improved selectivity (dr 7 : 1– 10 : 1) can be achieved by the introduction of arabinosylamine (**2**) as a chiral auxiliary. Separation of the side diastereoisomer may be achieved by crystallization or column chromatography to afford optically pure (*R*)-**5** for galactosylamine (**1**) and (*S*)-**5** for arabinosylamine (**2**) [8]. Subsequent hydrolysis in acidic conditions leads to the formation of D and L-amino acids. In a similar manner the four-component, one-pot Ugi reaction of **4**, aromatic or aliphatic aldehyde, formic acid and isocyanide in the presence of $ZnCl_2$ leads to formation of (*R*)-**6** with excellent yields and diastereoselectivity (Scheme 7.1, Path **B**) [9]. The enhancement of steric interactions caused by a larger nucleophile improves the selectivity of the Ugi reaction (dr 91 : 9–97 : 3). The previously described stereoselectivity inversion in non-polar solvents was not observed in this case [7], though the opposite configuration can be achieved by employment of the pseudo-enantiomeric auxiliary arabinosylamine (**2**). In comparison to **1**, **2** gives enhanced imine reactivity, resulting in nearly quantitative yields (>95%) and excellent diastereoselectivity (dr 96 : 4–98 : 2) [10]. Kunz and coworkers also performed the Ugi reaction with **3** immobilized on a solid phase through hydroxyl at the C-6 position [11]. Products linked to polymer support were obtained with high selectivity. Further release from the solid phase and auxiliary detachment provide optically pure Ugi products and recycled galactosamine. Imine **3** was explored in the Mannich reaction with silyl ketene acetals (**7**) catalyzed by $ZnCl_2$ leading to the formation of *N*-galactosyl-β-amino acid esters **8** (Scheme 7.1, Path **C**) [12]. The observed diastereoselectivity, favoring the *S*-configuration at the newly formed stereogenic center, varied from 5 : 1 to 250 : 1 for methyl isobutyrate derivative (**7**, $R^2 = R^3 = Me$).

Moreover, the implementation of **4** into the Mannich reaction of lithium enolates (**23**) and bis-silyl ketene acetals (**24**) opens the routes for *threo* (**25a**) and *erythro* (**25b**) diastereomers, respectively with excellent diastereoselectivity: only two out of four possible isomers were obtained with significant predomination of one product (Scheme 7.2) [13].

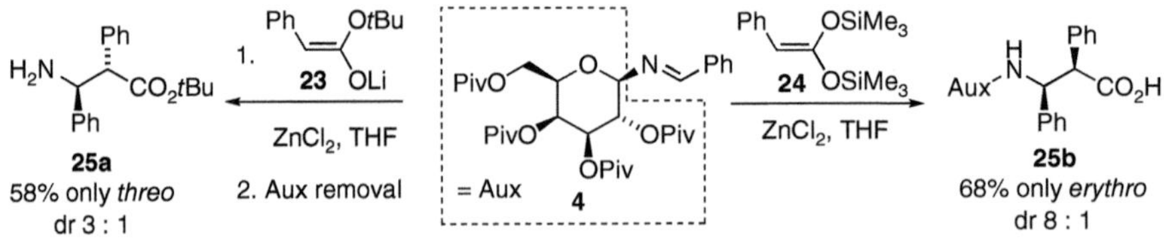

Scheme 7.2 The Mannich reaction selectivity difference in relation to the type of nucleophile. Source: Based on Kunz et al. [13].

The vinylogous Mannich reaction between **4** and silyl dienolate activated by $FeCl_3$ in THF has been reported by Miao and Chen. Chiral δ-lactams as final products were obtained with high yields and enantioselectivities [14]. Optically pure homoallylamines (**10**) can be also accessed by Schiff base **4** interactions with allyl-trimethylsilane (**9a**, X = TMS) or allyl-tributylstannane (**9b**, $X = SnBu_3$,

Scheme 7.1, Path **D**) [15]. In general, allylamines **10** were obtained with only moderate yields and selectivity (dr 1.2 : 1–27 : 1) in the reaction of aromatic aldehydes imines and **9a**. Imines derived from electron-rich aldehydes or aliphatic aldehydes undergo this reaction only with **9b** with 32–49% yield, whereas reactions with **9a** caused the only anomierization of **4**. To obtain enantiomeric allylamines *O*-pivaloyl protected L-fucosylimine (**4**) has been introduced to the reaction with **9a** to afford (*R*)-configured products with high diastereomeric discrimination (dr 3.3 : 1–49 : 1). Auxiliary **2** was unreactive in this case [16]. The additional advantage of this protocol is that *N*-fucosyl-homoallyl-amines are prone to crystalize in pure (*R*)-diastereomer form.

The asymmetric aza-Friedel–Crafts reaction between imine **4** and indole derivatives **12** has been independently reported by Meng and coworkers [17] and Miao with Chen [18, 19]. In the first case, $InCl_3$ in DCM with the addition of molecular sieves provides the highest diastereoselectivity of **11**, whereas the second methodology involves $SnCl_4$ in DCM and 0.2 equiv of *t*-butylammonium iodide (TBAI) to prevent the formation of the double addition side product. In both cases yield of **11** was high and the diastereoselectivity does not decrease below dr 8 : 1 (Scheme 7.1, Path **E**). Another type of nucleophiles undergoing the selective three-component Povaroy-type reaction with chiral imine **4** are enol-ethers **13** and alcohols (Scheme 7.1, Path **F**) [20]. Scandium triflate in chloroform was a catalyst of choice that provided the best results at −40 °C for the selective formation of **14**. In an approach similar to the Strecker reaction diethyl phosphite (**15**) reacts with imine **4** in the presence of $SnCl_4$ in THF to afford (*S*)-**16** with satisfying selectivity. The side anomerizaton of a chiral auxiliary has been noticed in this reaction (Scheme 7.1, Path **G**). Exchange of **1** with **2** or **3** provides the enantiomeric products with (*R*)-configuration of final amino phosphonic ester derivatives [21]. The tandem Mannich–Michael reaction of **4** and Danishefsky's diene (**17**) reported by Kunz and Pfrengle in 1989 opens the synthetic route to optically pure piperidine scaffolds **18** [22]. At the first stage, the Mannich reaction between **4** and silyl-enol ether part of **17** takes place in the presence of $ZnCl_2$ in THF at −20 °C leading to the formation of a Mannich base with remarkable stereoselectivity. Thus formed, the amine promptly reacts with the remaining double bond, resulting in cyclic Michael product **18** (Scheme 7.1 Path **H**). Depending on the type of aldehyde used in the formation of **4** and reaction work-up, it is possible to prevent the second stage and obtain a Mannich base with high diastereoselectivity (dr 20 : 1–50 : 1). Dihydropiperidones **18** obtained via this route way were utilized in subsequent asymmetric transformation leading to the synthesis of various natural products [23]. Mentioned tandem Mannich–Michael reaction at the beginning was mistakenly taken as [4+2] cycloaddition, however, the proper hetero-Diels–Alder reaction between **4** and isoprene **19** is also known (Scheme 7.1, Path **I**) [24]. This process required activation by zinc chloride in DCM at a relatively high temperature of 4–20 °C, which also causes the drop of diastereoselectivity. Although products **20** were obtained with moderate diastereomeric ratios, the major product can be easily isolated by flash chromatography. *N*-Galactosyl aziridines **22** can be prepared from **4** by the addition of diazo carbonyl compounds **21** promoted by BF_3

with good yields and high selectivity (Scheme 7.1, Path **J**) [25]. The subsequent aziridine ring-opening leads to the formation of enantiomerically pure amino acids accompanied by the release of the sugar moiety.

7.1.2 Chiral Oxazolidinones

Sugar-derived auxiliaries have been successfully exploited in form of oxazolidinones to provide stereoselective 1,4-addition, [4+2] or [3+2] cycloadditions, epoxidation as well as α-addition to enolates and aldol reaction. The pioneer application of sugar derived oxazolidinone **26** was the 1,4-addition of diethyl aluminum chloride to cinnamoyl moiety reported by Kunz and Pees in 1989 (Table 7.1, entry 1) [26]. The reaction proceeded with a high selectivity in favor of the (*S*)-configured isomer which was confirmed by comparison of optical rotation of final product after the auxiliary (**Aux^1**) detachment.

Table 7.1 The 1,4-addition to oxazolidinone-conjugated sugar-auxiliaries.

	Compound	Auxiliary	R^1	R^2	Conditions	Yield	dr
1	**26**	**Aux^1**	Ph	Et	DCM, −80 °C	>90%	87 : 13
2	**27a**	**Aux^2**	Ph	Alkyl	Toluene, −40 °C	74–87%	96 : 4–99 : 1
3	**27a**	**Aux^2**	Ph	Me	Toluene, −40 °C, 5 h irradiation	82%	>98 : 2
4	**27b**	**Aux^2**	Me	Alkyl	Toluene, −40 °C	38–70%	90 : 10–98 : 2
5	**28a**	**Aux^3**	Ph	Alkyl	Toluene, −55 °C	54–76%	29 : 71–96 : 4
6	**28b**	**Aux^3**	Me	Alkyl	Toluene, −55 °C	41–75%	35 : 65–73 : 27

Source: Based on Refs. [26–29].

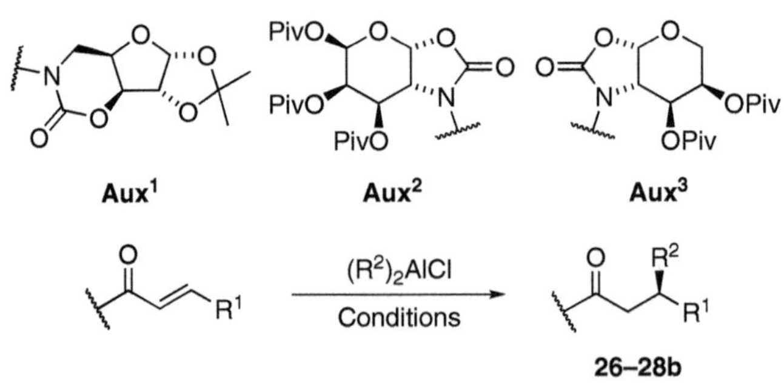

Remarkable results were also obtained for oxazolidinones derived from compounds **27-28** (Table 7.1, entries 2–6) [27–29]. In particular, galactosamine derivative **27** provided β-branched amides in nearly diastereomerically pure form (dr 90 : 10 to 99 : 1), however, the reaction time highly depends on the type of type of alkylaluminum compound (Table 7.1, entries 2–4). The absolute configuration of the newly formed stereocenter was established to be (*S*)-adduct after auxiliary removal. To access the opposite configuration compounds **28a,b** have to be introduced to afford

(*R*)-β-adducts with moderate to excellent diastereoselectivity (Table 7.1, entries 5, 6) [27]. Further investigation of this process revealed that dialkylaluminum chlorides required different reaction conditions (Table 7.1, entry 2) than dimethylaluminum chloride (Table 7.1, entry 3) activated by UV-light irradiation or radical initiation at the first stage. Moreover, two equivalents of the diorganylaluminum chloride reactant are necessary to accomplish the reaction with satisfying yields. On this basis, a transition state model has been proposed that explains the observed selectivity and reaction requirements: The transfer of higher alkyl chains (ethyl and longer) is assisted by σ-bond of β-carbon-hydrogen bond and empty *p* orbital of aluminum overlap. Such an interaction cannot take place in the case of dimethylaluminum chloride, thus this reactant required an additional activation. The subsequent intermediate of 1,4-addition enolate **30** can be trapped by electrophilic reactants by a selective *exo*-attack. Nevertheless, the mismatch effect between previously formed stereogenic center and chiral auxiliary (**Aux^2**) has been observed. To overcome this effect, a glucosamine derived auxiliary (**Aux^4**) was developed to merge both stereodifferentiating effects and it achieved the highest selectivity for dual-functionalization of the double bond (Scheme 7.3) [28].

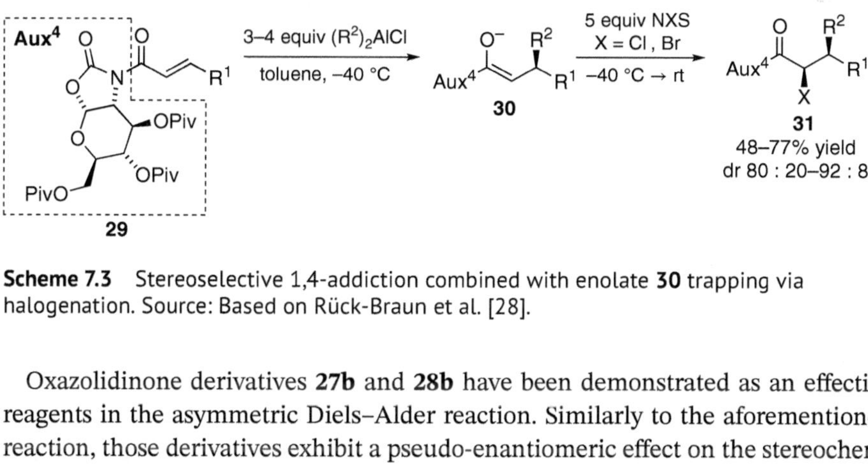

Scheme 7.3 Stereoselective 1,4-addiction combined with enolate **30** trapping via halogenation. Source: Based on Rück-Braun et al. [28].

Oxazolidinone derivatives **27b** and **28b** have been demonstrated as an effective reagents in the asymmetric Diels–Alder reaction. Similarly to the aforementioned reaction, those derivatives exhibit a pseudo-enantiomeric effect on the stereochemistry control leading to the formation of both enantiomeric series of products. The Diels–Alder reaction between **27b** or **28b** and cyclopentadiene **32** proceeds in DCM in the presence of Et_2AlCl to provide cycloadducts **33** with high *endo*- and diastereoselectivity (Scheme 7.4) [4].

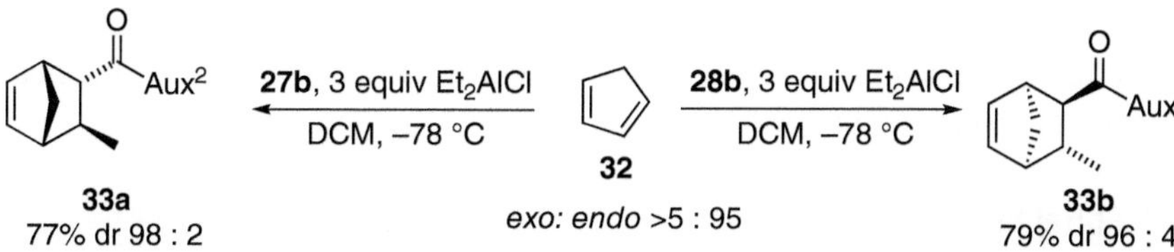

Scheme 7.4 Stereoselective Diels-Alder reaction of **27b** and **28b.** Source: Based on Kunz and Stoye [4].

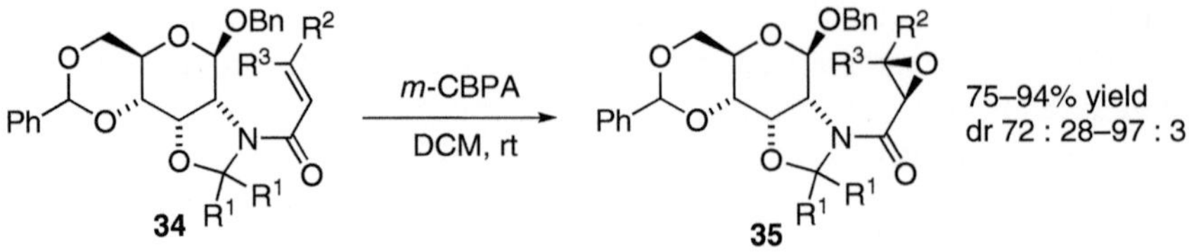

Scheme 7.5 Stereoselective epoxidation of oxazolidine-based alkene **34.** Source: Based on Vega-Pérez et al. [30].

The oxazolidine derivative **34** was successfully used in the oxidation of conjugated double bonds (Scheme 7.5) [30]. In comparison to other tested chiral derivatives compound **34** exhibits the highest selectivity and solubility in DCM. Synthesized oxiranes **35** were characterized by high yields and good stereoselectivity.

The xylose-based auxiliary **36** prepared by Köll and Lützen was utilized in α-functionalizations of amides through the chiral lithium-enolate formation [31–33]. Products of alkylations **37** were obtained with average yields of 30–67% and diastereomer ratio 5 : 1 to >99 : 1 (Scheme 7.6). Interestingly, the type of the R^1 substituent affects the selectivity: the change between aliphatic and aromatic substituents results in the inversion of the configuration on α position [31]. The authors explain this phenomenon by the preferred formation of (*Z*)-enolate of aliphatic esters that exposed the *Si*-face for electrophile, whereas the (*E*)-enolate is predominating for aromatic esters stabilized by the interaction between auxiliary and aromatic substituent.

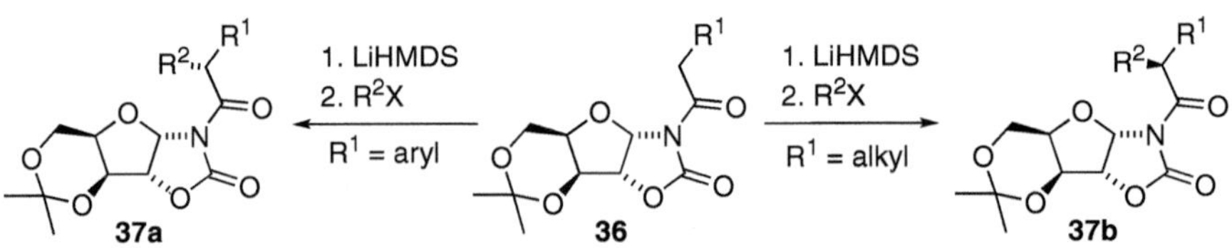

Scheme 7.6 The reversed selectivity for **36** that depends on R^1 substituent type.

This methodology was extended for acylation and halogenation reactions, which also exhibit the same effect of selectivity dependence from the type of substituent. [32] In the case of aldol reaction compound **36** results in the predominant formation of *syn*-products, however, with inversed configurations on both centers when the R^1 group was changed from aliphatic to aromatic [33]. It is noteworthy that aliphatic aldehydes provide higher diastereoselectivity (1 : 5–1 : 15) than aromatic counterparts, but both types suffer from low reactivity. This observation was rationalized by different transition states – chair-like conformation for aliphatic aldehydes and boat or twisted-boat-like for aromatic.

7.1.3 Chiral Esters

Sugar scaffolds attached to prochiral acids by the ester group are popular and versatile types of carbohydrate-derived auxiliaries. The polyhydroxylated structure of carbohydrates generates a huge number of possibilities to bind a carboxylic acid to an auxiliary (Figure 7.3). A plethora of examples of sugars bearing esters is

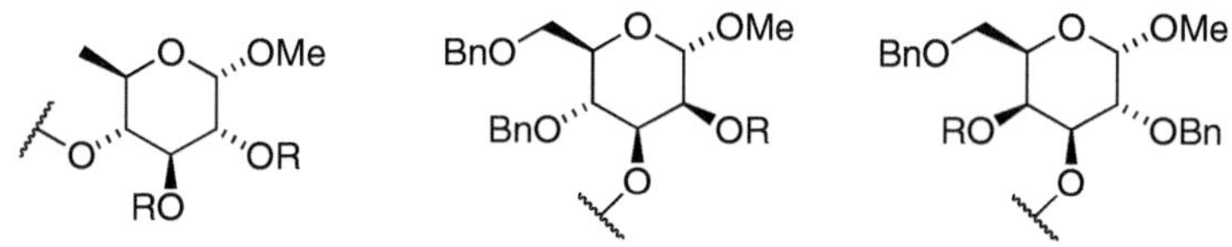

Figure 7.3 Examples of sugar-derived auxiliaries used in ester forms.

already reported in the literature; among them the most common are derivatives of α,β-unsaturated compounds, aliphatic esters and their silyl-enol ethers. Crotonic esters with a spectrum of carbohydrates derivatives (**Aux**[5]–**Aux**[7]) attached at C3 or C4 positions have been investigated in 1,4-additions of organocopper reagents created *in situ* from Grignard reactants and alkyl lithium reagents [34].

D-Glucopyranoside **Aux**[5] exhibits the best efficiency of all sugar derivatives verified to date (Table 7.2). The Michael addition products were obtained with high yield and selectivity [35, 36]. Moreover, the configuration at the β position of the crotonic ester can be controlled by the type of organometallic reagent used in reaction. The (*R*)-configured products were obtained with copper reactants (Table 7.2, entries 1–3), whereas organolithium reagents provided the opposite (*S*)-configuration (Table 7.2, entry 4). The reason for this complementary behavior is related to the coordination ability of metal counter ion to the carboxylic group which changes the conformation of crotonic part from s-*cis*, *syn* when Li is used, to s-*trans*, *syn* when complexes with more sterically demanding ions (magnesium or copper) are formed (Scheme 7.7).

This methodology was extended by a subsequent trapping of the enolate resulted from metalorganic compound 1,4-addition with methyl iodide. This protocol selectively provides 2,3-*syn*-disubstituted carboxylic esters when organocopper reactants were used. On the other hand, organolithium reagents result in formation of 2,3-*anti*-products with high yield and excellent stereochemical purity [36].

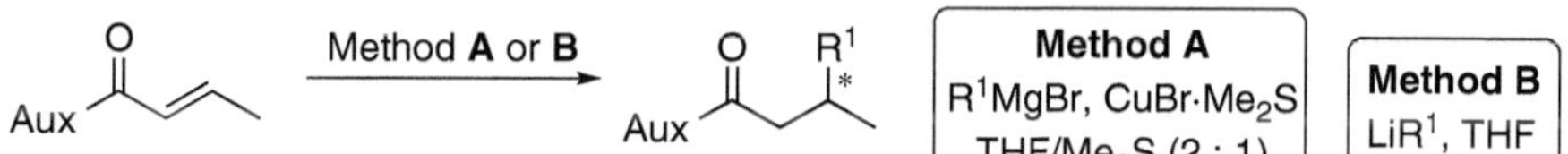

Table 7.2 The 1,4-addition to ester type sugar-auxiliaries.

	Compound	Auxiliary	Method	Yield	dr	Stereochemistry
1	**41a**	**Aux**[5a]	A	51–90%	87 : 13; 88 : 12	*R*
2	**41b**	**Aux**[5b]		61–77%	96 : 4–98 : 2	*R*
3	**41c**	**Aux**[5c]		58–85%	92 : 8–98 : 2	*R*
4	**41c**	**Aux**[5c]	B	92–95%	4 : 96–2 : 98	*S*
5	**42a**	**Aux**[6a]	A	84%	88 : 12	76% ee (*S*)
6	**43a**	**Aux**[7a]		82%	84 : 16	68% ee (*S*)
7	**42b**	**Aux**[6b]	A	78%	89 : 11	82% ee (*R*)
8	**43b**	**Aux**[7b]		94%	n.d.	96% ee (*R*)

n.d., not determined.

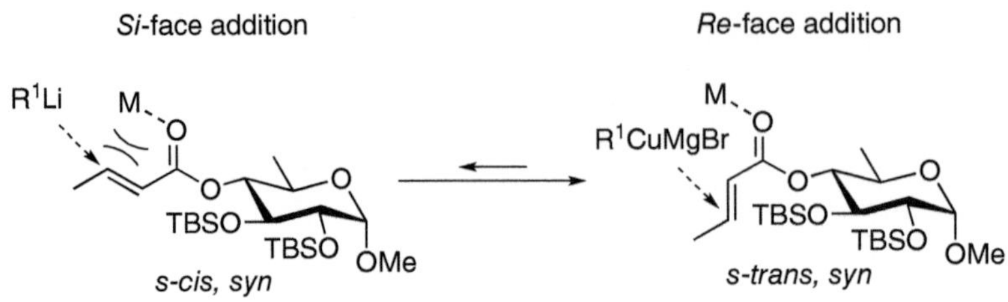

Scheme 7.7 The conformational explanation of observed stereoselectivity difference in 1,4-addition of different metalloorganic species.

The stereoselective radical 1,4-addition can be successfully accomplished by introduction of **Aux**^{5d} where hydroxyl groups at C2 and C3 position were protected by 2,4,6-trimethylbenzoyl (MesCO). Among the Lewis acids investigated, Et_2AlCl was the most effective catalyst under common radical condition involving alkyl iodide, tributyltin hydride, and triethyl borane with oxygen (Scheme 7.8). Four to five equivalents of Et_2AlCl were needed to provide the quantitative yield and diastereoselectivity (up to 95 : 5) [37]. As a continuation of this research, Tadano developed additional auxiliaries **Aux**^{6} and **Aux**^{7} (see Figure 7.3) for diastereoselective 1,4-addition based on D-mannopyranose and D-galactopyranose, respectively [35]. In general, both compounds exhibited similar reactivity in the addition of vinylmagnesium bromide (R^1 =vinyl) with a catalytic amount of copper and provided a high stereoselectivity (Table 7.2, entries 5–8). It is worth mentioning that when benzyl moiety serves as the R protecting group (**Aux**^{6a}, **Aux**^{7a}), the predominating products **42a** and **43a** have (*S*)-configuration (Table 7.2, entries 5, 6, the enantioselectivity refers to compounds after auxiliary detachment), while the derivatives **Aux**^{6b} and **Aux**^{7b} bearing pivaloyl group result in the formation of (*R*)-enantiomers of **42b** and **43b**(Table 7.2, entries 7, 8).

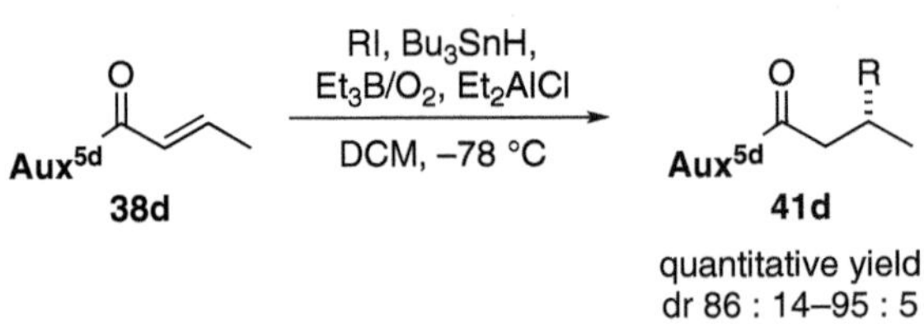

Scheme 7.8 Stereoselective 1,4-addition of a radical to **Aux**^{5d} crotonate (**38d**).

The auxiliary **Aux**^{5c} has been also examined by Tadano in asymmetric dihydroxylation with osmium tetraoxide and *N*-methylmorpholine (Scheme 7.9). A dihydroxylation product was synthesized with yields exceeding 90%, but diastereoselectivity up to 89 : 11 [34].

Another group of chiral ester bearing sugar-derived auxiliaries are acrylates, widely used as dienophiles or dipolarofiles in [4+2] and [3+2] cycloadditions.

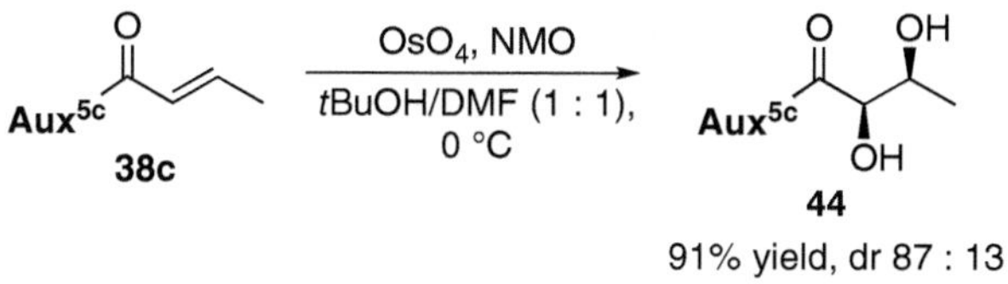

Scheme 7.9 Stereoselective dihydroxylation of double bond of crotonate **38c**.

Acrylic acid can be simply combined with sugar moieties by esterification of selected positions like anomeric (**Aux**[8]), C-2 (**Aux**[9–11]), or C-4 (**Aux**[5b,c]). The L-arabionpyranoside derivative **45** and dienes (i.e. **19**) in the presence of Et_2AlCl undergo [4+2] cycloaddition with good yields (60–90%) and high diastereoselectivity (dr 85 : 15–96 : 4). Within the absence of chromium complex responsible for steric interaction blocking the *Re*-side of the acrylate the selectivity dramatically decreases (Table 7.3, entry 1) [38].

Table 7.3 The Diels–Alder reaction of crotonic acid esters bearing carbohydrate-auxiliaries **45–49**.

	Compound	Auxiliary	Diene	Method	Yield	*endo/exo*	dr
1	**45**	**Aux**[9]	**19**	A	77% (60%)a	—	95 : 5 (78 : 22)a (*R*)
2	**46**	**Aux**[8]	**50**	B	95%	96 : 4	96 : 4 (*R*)
3	**47a**	**Aux**[5b]	**50**	B	79%	>95 : 5	90 : 10 (*R*)
4	**47a**	**Aux**[5b]	**50**	C	96%	89 : 11	87 : 13 (*S*)
5	**47b**	**Aux**[5c]	**50**	C	97%	80 : 20	>95 : 5 (*S*)
6	**48**	**Aux**[10]	**51**	A	88%	91 : 9	98 : 2 (*S*)
7	**49**	**Aux**[11]	**51**	A	89%	>95 : 5	99 : 1 (*R*)

a) Values in brackets refer to **Aux**[9] in the absence of Cr complex.
Source: Totani et al. [34] and Shing et al. [38].

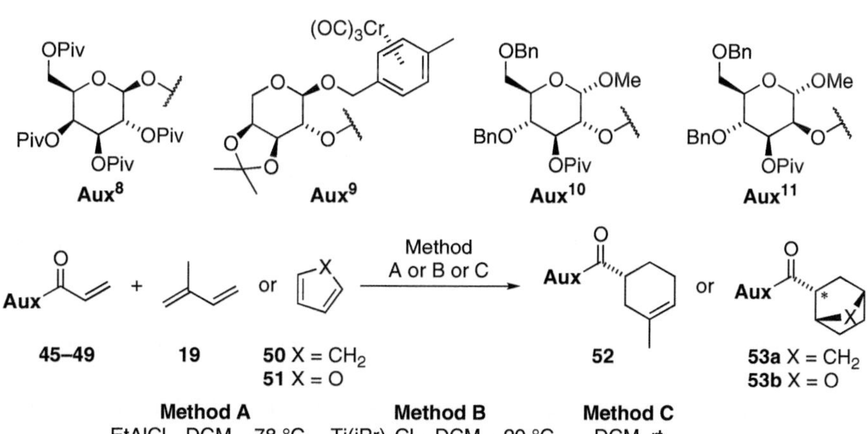

Satisfying results have been also obtained for tetra-*O*-pivaloyl-β-galacopyranose derivative **Aux⁸** (Table 7.3, entry 2). In reactions activated by 5 equiv of $Ti(OiPr)_2Cl_2$ the cycloadduct was formed with 96 : 4 *endo/exo* selectivity and preferred formation of the (*R*)-configured diastereoisomer of *endo*-products (dr 96 : 4). The observed stereodifferentiation is an outcome of the *exo*-anomeric effect responsible for the acrylate orientation and shielding by the 2-pivaloyl group strengthened by the coordination of titanium metal. Indeed, the corresponding α-anomer of **Aux⁸**, which lacks this *exo*-anomeric effect, reacts more slowly and with lower diastereoselectivity (7 : 3), but with the reversed (*S*)-isomer preference [4]. Tadano et al. investigated the application of 6-deoxy-α-D-glucopyranoside **Aux⁵** as a chiral helper. In a presence of Lewis Acid compound, **Aux⁵ᶜ** reacts with cyclopentadiene with excellent *endo*-selectivity and good diastereoselectivity (Table 7.3, entry 3). However, the same reaction also takes place under thermal conditions and provides lower yields but inverts the π-facial selectivity (Table 7.3, entries 4, 5) [38]. The same group reported on a Diels–Alder reaction of **48** and **49**, differing from each other by inverted configuration at C-2, with furane in the presence of Et_2AlCl. Under the same reaction conditions both compounds react similarly, but with a reversed selectivity, leading to the formation of enantiomeric products (Table 7.3, entries 6 vs. 7) [34].

The selective 1,3-dipolar cycloaddition was achieved by use of **47b** with nitrile oxides (Scheme 7.10). Prepared from benzonitrile oxide in high yield (above 90%) isoxazolines were characterized by almost complete regio- and stereoselectivity [39].

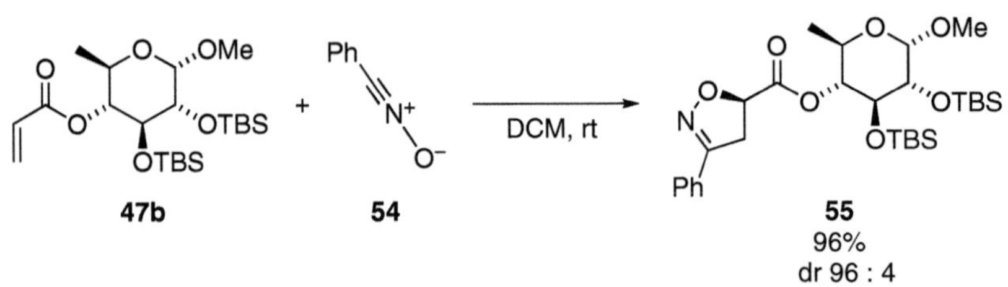

Scheme 7.10 1,3-dipolar cycloaddition involving the sugar-auxiliary developed by Tadano.

The application of chiral auxiliaries is a widespread stereochemistry controlling strategy for reactions involving ester or amide enolates as nucleophiles. However, early attempts of sugar auxiliary incorporation to enolate chemistry by Heathcock et al. [40], Costa et al. [41], and Kunz and Mohr [42] were rather unsuccessful. Described methods suffer from low yield and poor selectivity, caused by multiple possible sites for metal ion coordination to the auxiliary and reactant decomposition under the reaction conditions. The first satisfactory attempt has been reported for 3-*epi*-D-allofuranose derivative **56**. The (*E*)-enolate **56a** generated at −90 °C by treatment with LDA reacted with methyl iodide to afford methylated products **57** with a yield in a range of 27–100% and diastereoselectivity 1 : 4 to 1 : 10 (Scheme 7.11) [42].

Pioneering work on the use of sugar-derived auxiliaries in aldol reactions was performed by Brandänge between **58** and acetophenone that in optimized conditions provides dr 75 : 25 in favor of the (*S*)-isomer. The complexation of metal ion has huge

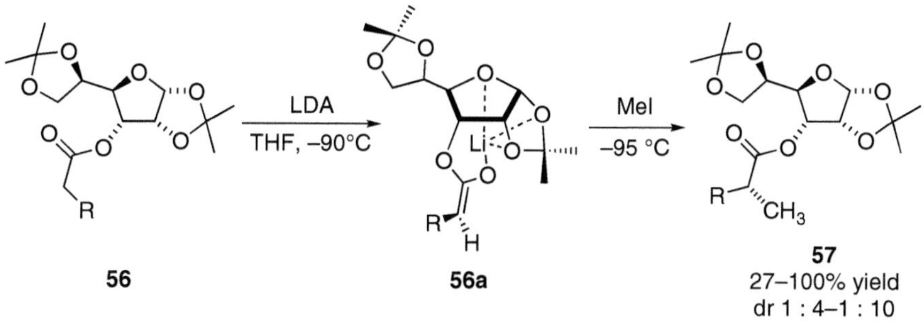

Scheme 7.11 Application of *allo*-furanose derivative **56** as a chiral auxiliary in enolate alkylation. Source: Based on Kunz and Mohr [42].

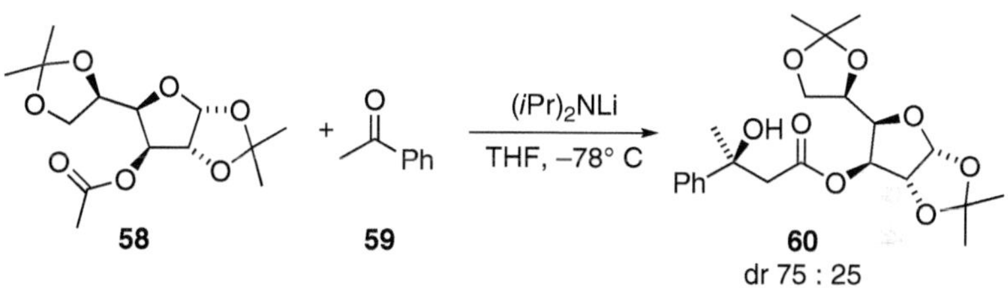

Scheme 7.12 The aldol reaction supported by the sugar-derived auxiliary **58** developed by Brandänge. Source: Based on Brandänge et al. [43].

influence on the process, since addition of $MgCl_2$ significantly reduces the selectivity and promotes the (*R*)-isomer formation (Scheme 7.12). [43]

Another successful approach has been reported by Kakinuma, who showed that the modified dicyclohexylidene-D-glucose **61** creates the appropriate chiral environment for asymmetric alkylation of enolate [44]. The lithium enolate in a presence of excess LiCl provides the product of benzylation **61a** in 60% yield and dr 96 : 4, whereas silyl-enol ether **61b** in the presence of HMPA exhibit reversed selectivity in favor of (*S*)-isomer with a similar diastereoselectivity. Notably, the addition of HMPA is crucial to access the high selectivity (Scheme 7.13).

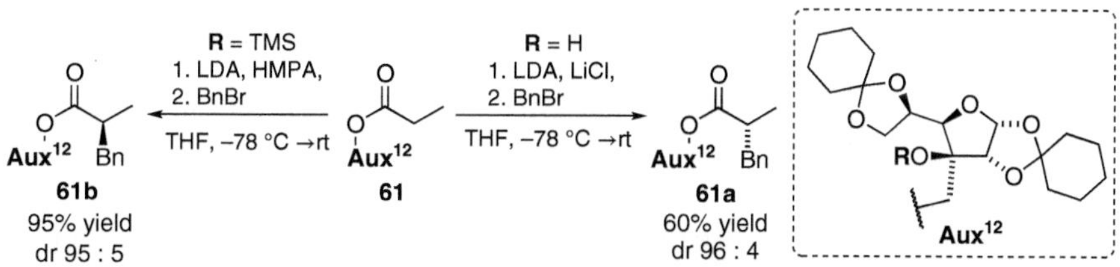

Scheme 7.13 The stereoselective benzylation of ester **61** reported by Kakinuma.

Tadano et al. have introduced their concept based on *tert*-butyldimethylsilyl (TBS)-*O*-protected-D-glucopyranoside (**Aux**^{5d}) into diastereoselective enolate chemistry [45]. Compound **62** was transformed to (*Z*)-enolate by treatment with NaHDMS. The enolate preference is a consequence of steric interaction between the R substituent and the carbohydrate framework, while the *Si*-side of enolate is shielded by the TBS group protecting hydroxyl at the C-3 position. High yields of 97% and 80% along with excellent diastereoselectivity of 95 : 5 and 98 : 2 have been obtained for BnBr and MeI, respectively (Scheme 7.14). The same strategy has been introduced to enolate Michael reaction between **62** and methyl crotonate that became a vital step in the total synthesis of (−)-lasiol [46]. The target product was obtained in 75% yield with *syn*/*anti* ratio > 5 : 95 and diastereoselectivity exceeding 95 : 5 for the *anti*-isomer.

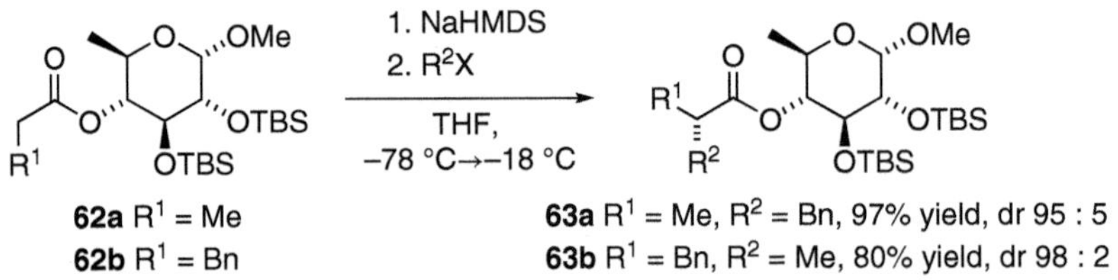

Scheme 7.14 The Tadano's approach for alkylation of esters with auxiliary **Aux5d**.

This concept has been also extended for dialkylation of β-ketoesters **64** and **67** resulting in complementary selectivity of enantiomeric products containing the quaternary stereogenic center (Scheme 7.15).

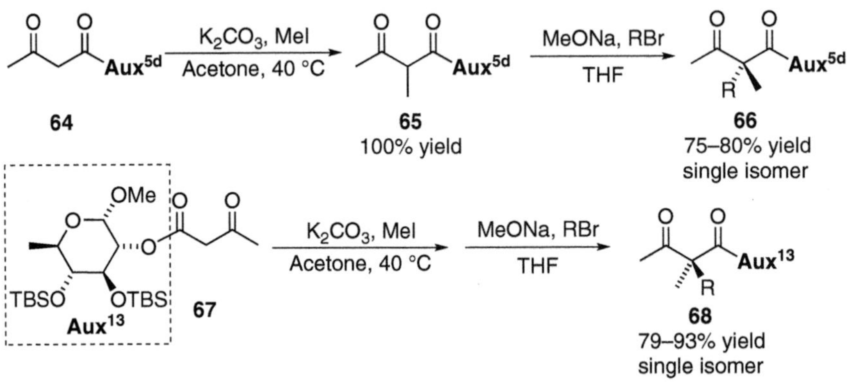

Scheme 7.15 The Tadano's approach for dialkylation of β-ketoesters with auxiliaries **Aux5d** and **Aux13**.

The product **66** was obtained as a single diastereoisomer in 75% yield, however, the order of alkylation steps (methylation followed by benzylation) is essential to achieve high selectivity [47]. Analogously, the same synthetic steps sequence for **67** resulted in good yields of **68** as an apparently single diastereomer with inverted stereochemistry [48]. An original application of sugar auxiliary bonded by the ester group to ferrocene was reported by Krishna and Mongin [49]. The various types of carbohydrate moieties were investigated in a selective deprotonation of the ferrocene framework **69**. The ferrocene iodides **70** formed after an interception of carbanion by iodide were

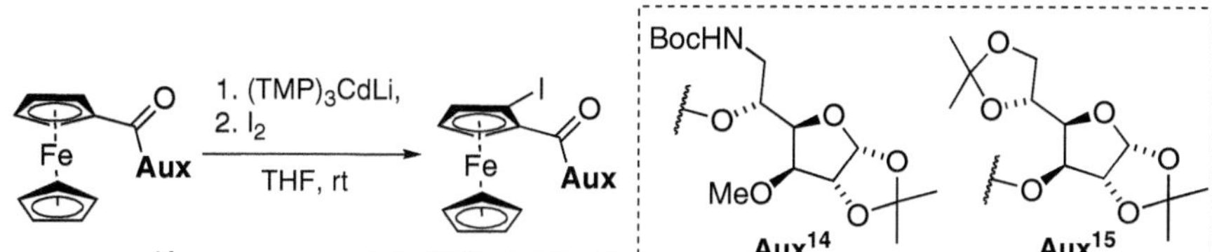

69a - Aux^{14}
69b - Aux^{15}
70a 85%, dr 87 : 13
70b 34% isomer R$_P$

Scheme 7.16 Ferrocene selective iodination with application of sugar-based auxiliaries. Source: Based on Sreeshailam et al. [49].

obtained with good yields and moderate stereoselectivity. Among tested auxiliaries derivatives, **Aux**^{14} and **Aux**^{15} provided the best results (Scheme 7.16) [49].

7.1.4 Chiral Ethers

The most prominent representatives of this type of carbohydrate-derived auxiliaries are sugar allyl ethers commonly used in cyclopropanation and epoxidation reactions. Currently, there are several comprehensive reviews dedicated to this topic [50, 51]. Herein, we just briefly disclose this concept, which is based on the coordination of a Simmons–Smith reactant or an oxidizing agent by the carbohydrate moiety to enforce the interaction with double bond from one side. The first described example of allylic-D-glucopranose **71** provides the broad scope of cyclopropane derivatives **72** with a diasteromeric ratio exceeding 1 : 50. The free hydroxyl at C-2 position is essential for such high selectivity by coordination of the Simmons-Smith reagent through zinc cation (**TS.2**, Scheme 7.17) [52]. When *m*-CBPA (*m*-chloroperbenzoic acid) is used instead, the formation of epoxide **73** takes place with good stereoselectivity (only one example) [53]. In both cases the unprotected hydroxyl group at C-2 position is required to afford the high

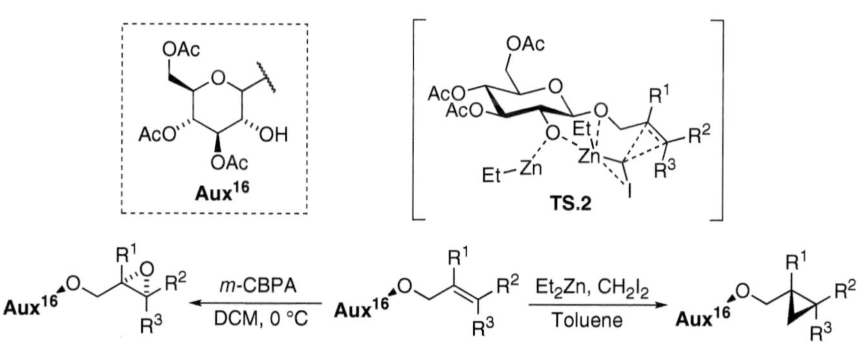

73
R^1 = R^3 = H, R^2 = Me
>95% yield, dr 4.1 : 1

71

72
>97% yield
dr > 50 : 1

Scheme 7.17 Sugar-based ether auxiliaries application in [2+1] cycloaddition. Source: Based on Charette et al. [52].

asymmetric induction level. Protection of this part of sugar scaffold (**Aux**[16]) results in significant drop of the selectivity. Worthy of note is the fact that compound **71** exhibits inverted face-selectivity for the oxygen implantation than methylene group from the Simmons–Smith reagent. This observation indicates that oxygen delivery from the *m*-CBPA takes place from the opposite side of the double bond in comparison to methylene group from the Simmons–Smith reaction.

The use of sugar aromatic ethers in asymmetric synthesis has been also reported, for example compound **75** was successfully introduced into a Diels-Alder reaction with cyclopentadiene and Danishefski's diene. In both cases excellent regio- and diastereoselectivities were observed since in each case only one cycloadduct was isolated in 92% and 83% yield, respectively (Scheme 7.18) [54]. High diastereoselectivity was rationalized by the *exo*-anomeric effect that oriented the naphthoquinone in a fashion in which the front side of this substituent is efficiently blocked by 2-*O*-acetyl group.

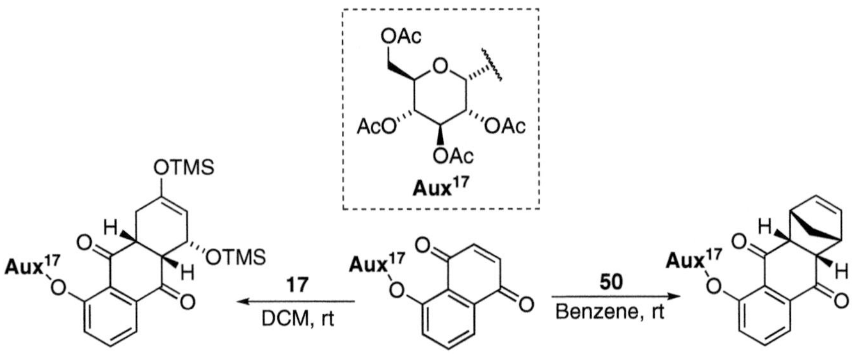

75
92% yield
single isomer

74

76
83% yield
single isomer

Scheme 7.18 Sugar-based ether auxiliaries application in [2 + 4] cycloaddition. Source: Based on Beagley et al. [54].

Quite recently, Linker and coworkers presented an interesting application of sugar auxiliary **Aux**[16] in a stereoselective single oxygen addition to naphthalenes **77** (Scheme 7.19) [55]. The corresponding endoperoxides **78a**/**78b** were obtained with

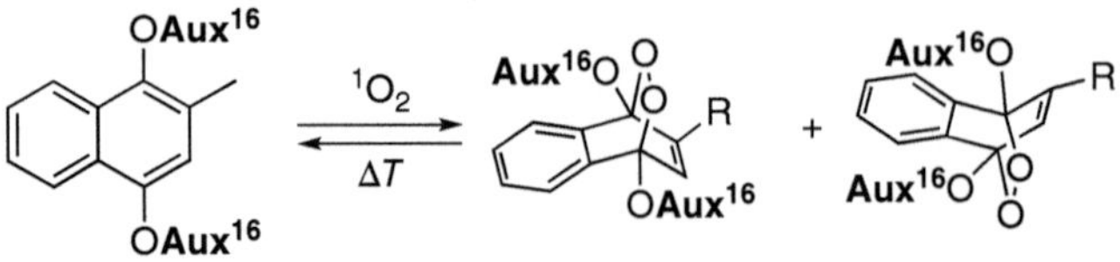

77

(1*S*,4*R*)-**78a**
Major

(1*R*,4*S*)-**78b**
Minor

Scheme 7.19 Stereoselective singlet oxygen addition to naphthalene ring. Source: Based on Bauch et al. [55].

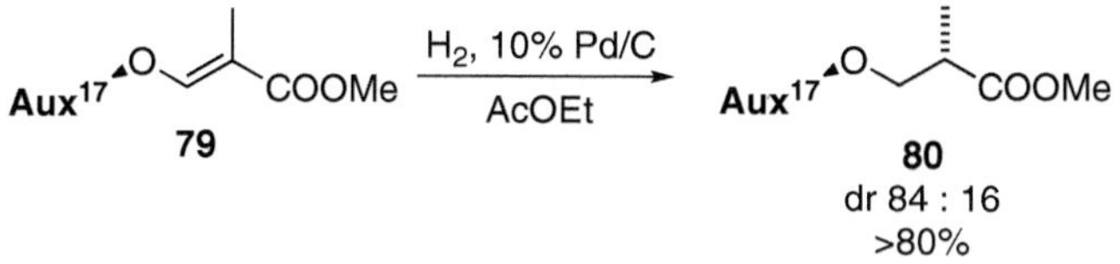

Scheme 7.20 Stereoselective hydrogenation of the double bond. Source: Based on Idris et al. [56] and Lersen et al. [57].

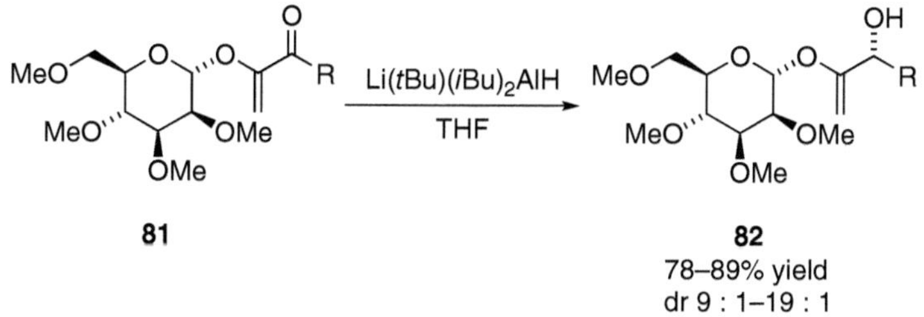

Scheme 7.21 Stereoselective reduction of the carbonyl group. Source: Akashi et al. [48].

diastereoselectivity ranging from 51 : 49 to 91 : 9 dr. Most of those products occurred to be unstable and retro-reaction was observed; only two compounds (R = H or *t*-Bu) were stable enough to be subjected for further analysis. The observed results were rationalized by the selective shielding of one face of naphthalene ring by carbohydrate in predominating rotamer structure that is stabilized by steric and electrostatic interactions.

The sugar-enol ethers constitute another group of useful reactants with versatile applications in various cycloaddition reactions like cyclopropanation, [2+2] addition of ketenes and isocyanides, 1,3-dipolar cycloadditions as well as Diels–Alder and hetero-Diels–Alder reactions. Detailed information about aforementioned reactions can be found in comprehensive reviews and monographs available in the literature [4, 50, 51]. The less common but worthy of mention applications of carbohydrate enol-ethers are asymmetric reductions. The auxiliary-supported heterogeneous hydrogenation of double bond of D-glucoside **79** proceeded in good yield and moderate diastereoselectivity (Scheme 7.20) [56, 57].

Several examples of stereodifferentiating carbonyl reduction of conjugated ketones have been also reported. DiCesare developed the D-mannose derivative **81** that in a presence of bulky reducing agent undergoes a selective reduction to the corresponding (*S*)-allylic alcohols **82**. The size of the substituent (R) has a strong impact on the reaction diastereoselectivity, however the *exo*-anomeric effect is proposed as a main factor affecting observed diastereomer preference (Scheme 7.21) [50].

7.2 Chiral Catalysts

Asymmetric catalysis is another emerging synthetic approach to access enantiomerically pure compounds. Stereoselective, catalytic transformations are commonly used in contemporary synthesis since they are attractive from time-saving and cost-reducing viewpoints. Carbohydrates are an ideal source of chirality for a chiral catalyst design and have already proven to be useful scaffolds for such purposes. This chapter is dedicated to present the most prominent examples of sugar-based catalysts together with their versatile applications in organic chemistry. It is possible to distinguish two types of asymmetric catalysts. Chiral complexes in which sugar-bearing ligand creates the chiral pocket around a metal-containing active center of a catalyst. The second type is organocatalysis where the functional groups attached to carbohydrate skeleton, arranged in a defined three-dimensional orientation, activate substrate molecules for further transformation.

7.2.1 Chiral Complexes

The homogeneous catalysis based on metal-chiral ligand complexes attracts considerable attention from academics and industry thanks to features like high activity, selectivity, and turnover number. The development of a successful catalyst requires the proper design of the chiral ligand with a precisely defined three-dimensional structure. Indeed, the ligand structure is responsible for the reactants arrangement during the reaction thus inducing the stereochemistry control, whereas metal is responsible for an initiation of the process. On this basis, sugars containing polyfunctional, stereo-defined structures seem to be perfect candidates for chiral ligand precursors either for the introduction of donor centers for metal coordination or the attachment of steric repulsion by bulky groups. This chapter is dedicated to highlighting the most prominent and general examples of carbohydrates introduction into asymmetric metal catalysis since several comprehensive reviews were already dedicated to this topic [51, 58–61]. The number of sugar-based analogs of popular chiral ligands have already been reported [58]. Among them, chiral phosphorus compounds (**83–85**) are the most widespread ligands bearing sugar motifs inside their structures. A notable example of the utility of these ligands is the asymmetric rhodium-catalyzed hydrogenation of α,β-unsaturated carboxylic acid derivatives (**86**, **88**, Scheme 7.22).

Several scientific groups of Cullen [62], Descotes [63], Selke [64], and Thompson [65] independently reported a successful application of diphosphinite derivatives of glucopyranose in the asymmetric reduction of a double bond (**86**). Moreover, Selke investigated the effect of the pyranose configuration on enantioselectivity [66]. It revealed that equatorial arrangement in vicinal diphosphinite hexapyranoside (**83**) has a fundamental influence on the enantiomeric discrimination of the process. As an example, D-manno-configured-**85** gave a racemic product, the D-galacto-configured ligand (**84**) afforded only 46% ee, and the D-glucose configuration of ligand (**83**) provided product **87** with 96% ee (Scheme 7.22). Further improvement was achieved by RajanBabu, who investigated the electronic effects of the phosphinite motifs on reaction stereoselectivity [67–69]. Ligands bearing electron-deficient aryl groups (**83b**) resulted in low enantioselectivity, whereas electron-rich substituents (**83c**) led to improved enantiomeric excess up to 99% ee. The versatility of the catalyst **83c** was proven by a broad substrate scope, giving the

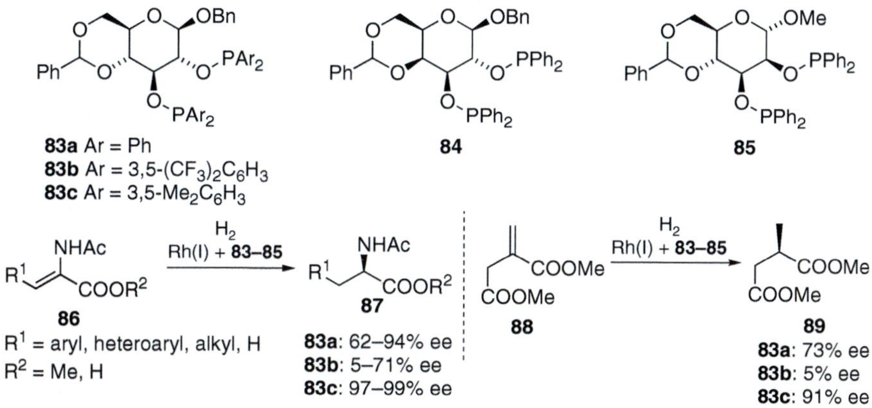

Scheme 7.22 Sugar-based diphosphinite derivatives utilized in the asymmetric hydrogenation.

(*S*)-configured products (**87**, 91–99% ee). The (*R*)-enantiomer of **87** with excellent selectivity could be synthetized through application of catalysts **90a** (96–97% ee) and **90b** (95–98% ee).

Notably, both types of *pseudo*-enantiomeric catalysts **90a,b**, which relationship to **83** have been shown on Scheme 7.23 were prepared from the same starting material – D-glucose. Nickel-**83** complex was used by RajanBabu and coworkers in stereocontrolled hydrocyanation of vinylnaphthalene **91** with HCN (Scheme 7.24) [70]. In this case electronic effects of aryl residues at diphosphinite groups of **83** also significantly affected enantioselectivity. However, an opposite influence was observed, as the electron-deficient ligand **83c** provided high enantiomeric excess (up to 91% ee at 0 °C for **91b**) in comparison to electron rich ligand **83b** (16% ee) [71–73]. Different aromatic alkenes and dienes were also implemented resulting in formation of enantioenriched products with 55–84% ee [74].

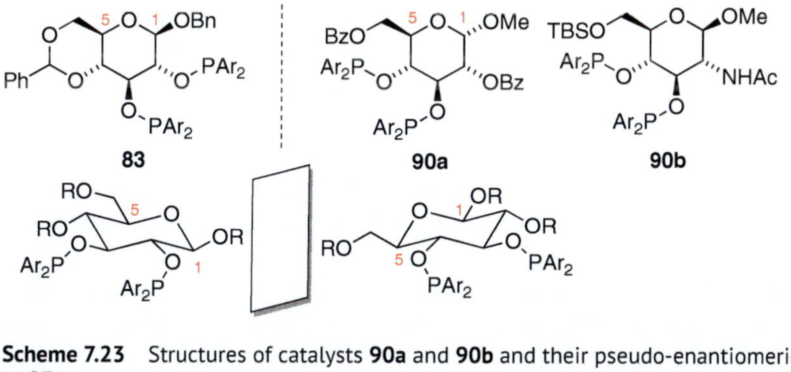

Scheme 7.23 Structures of catalysts **90a** and **90b** and their pseudo-enantiomeric relation to **83**.

Trost ligands are another group of chiral diphosphine compounds used in complexation of palladium to perform asymmetric allylic alkylation [75]. Ruffo

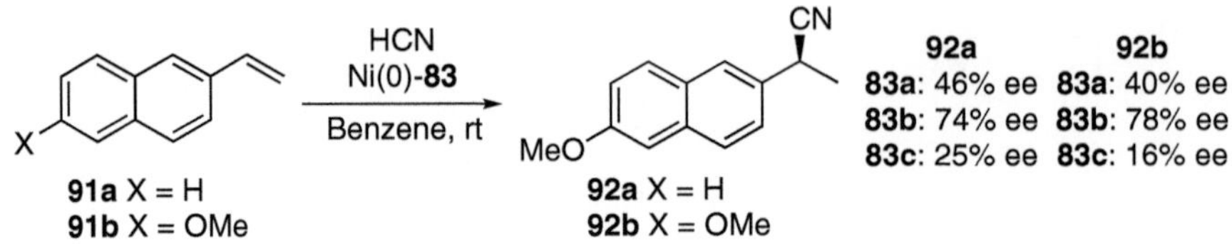

Scheme 7.24 The asymmetric addition of cyanide to a vinylnaphthalene promoted by Ni-**83** complexes. Source: RajanBabu et al. [69].

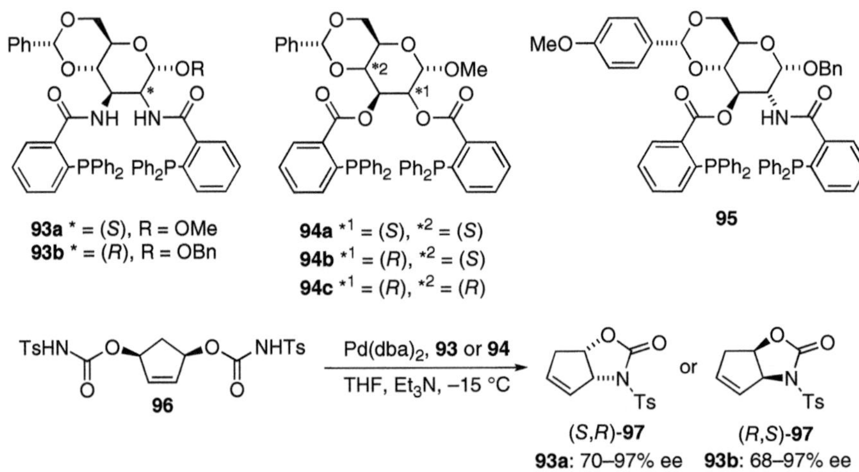

Scheme 7.25 Structures of carbohydrate-based Trost ligands **93–95** and their application in the desymmetrization reaction of **96.** Source: Based on Ruffo et al. [76].

developed the Trost-type ligands containing sugar scaffolds (**93–95**) which were successfully applied in the desymmetrization of *meso*-diols **96** with high stereoselectivity (Scheme 7.25) [76]. Depending on the nature of sugar skeleton (mannose-type **93a** or glucose-type **93b**) opposite enantioselectivity can be accessed. Unfortunately, the utility of catalysts **93a** and **93b** is quite limited due to the multi-step synthesis of these molecules. The more accessible ligands **94a–94c** have been also reported resulting in similar efficiency (>99% yield) to **93**, but lower enantioselectivity (up to 82% ee) [77].

In addition, ligand **95** was used by Woodward to create a chiral copper(II) complex that catalyzed the 1,4 addition of Et_2Zn to **98** with excellent enantioselectivity (Scheme 7.26) [78].

The stereoselective cyanosililation reaction can be accomplished with D-glucose-based ligands **100a–100c** developed by Shibasaki's group [79, 80]. Complexes of $Ti(OiPr)_4$ with **100a** in ratio 1 : 1 [79] and $Gd(OiPr)_3$ in ratio 1 : 2 with **100b** [80] afforded products of cyanosiliation of ketones in high yield and opposite enantioselectivities up to 97% ee (Scheme 7.27). The best results were obtained for aryl-alkyl ketones, whereas dialkyl ketones exhibited significantly lower stereoselectivity.

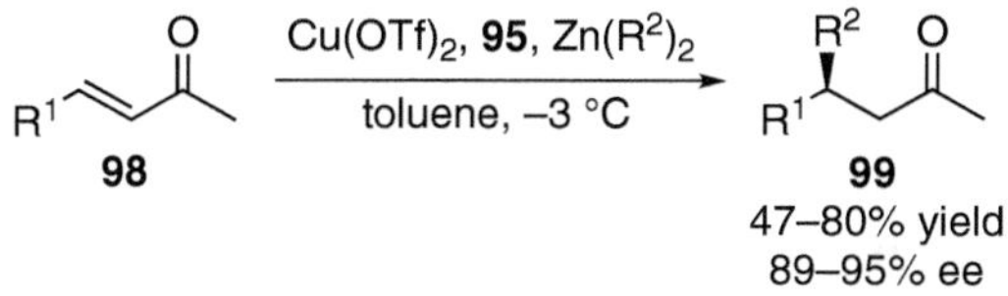

Scheme 7.26 The Woodward application of ligand **95** in the copper initiated Michael addition. Source: Based on De Roma et al. [78].

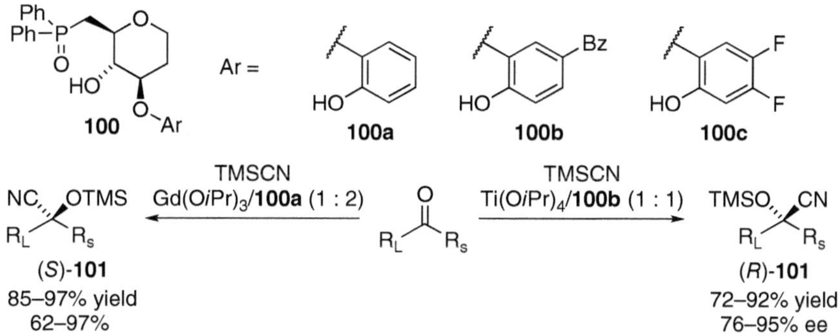

Scheme 7.27 The asymmetric cyanosiliation of ketones promoted by chiral complexes of **100.**

The class of sugar-derived *trans*-1,2-diaminocyclohexane (DACH) ligands were prepared from D-gluco (**102a**) or D-manno-diamines (**102b**). Molybdenum complexes of **102a** were used as catalysts of allylic alkylation with methyl malonate of both **103** and **104** initiated by microwave irradiation (Scheme 7.28). High enantioselectivities (99% ee for **103** and 96% ee for **104**) and regioselectivity in favor of branched product **105** were observed for this reaction. The mannose-derived **102b** resulted in formation of enantiomeric (*S*)-**105** with reduced efficiency [81].

C_2-Symmetric bisoxazoline (BOX) derivatives are undoubtedly one of the most common and versatile examples of chiral ligands with a proven utility in a variety of Lewis acid-catalyzed asymmetric transformations. The usefulness of glucosamine as a component of the BOX structure has been confirmed by **107a–107d** and **108a–108h** developed by Boysen and coworkers (Scheme 7.29) [82–84]. The conformational analysis revealed that ligands **107** adopt the twisted-boat conformation of the sugar moiety, while **108** containing a benzylidene moiety adopts the chair-like geometry. Moreover, the proper functionalization of the C-3 hydroxyl, accessible even at late-stage functionalization, has a strong impact on the catalyst's asymmetric induction. The applicability of designed ligands was demonstrated in the indole cyclopropanation with diazoacetates promoted by copper(I) (Scheme 7.29). The ligands efficiency in asymmetric induction depends highly on the carbohydrate conformation and steric hindrance of the functional group attached to C-3 position. The methodology was further used in the challenging cyclopropanation of styrene and aliphatic alkenes in good yields (40–75%) and 90% ee for the *trans*-isomer (*cis*/*trans* 73 : 27) [85].

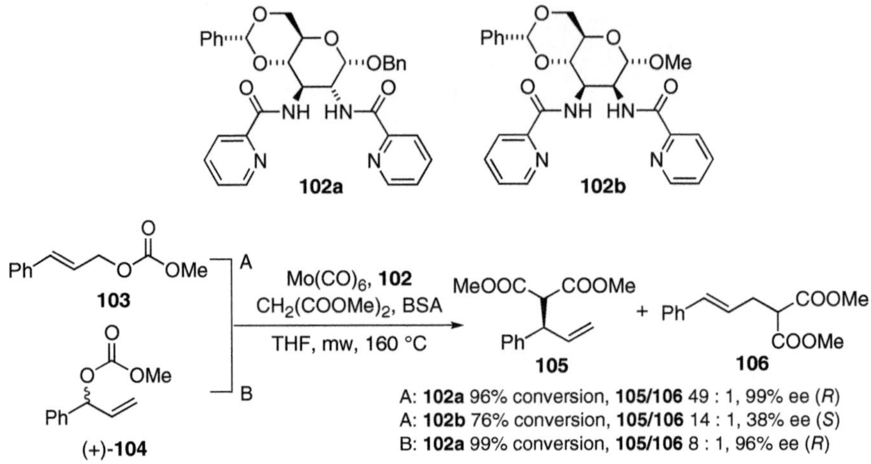

Scheme 7.28 The asymmetric allylic alkylation of **103** and **104** catalyzed by sugar-based DACH ligands **102**.

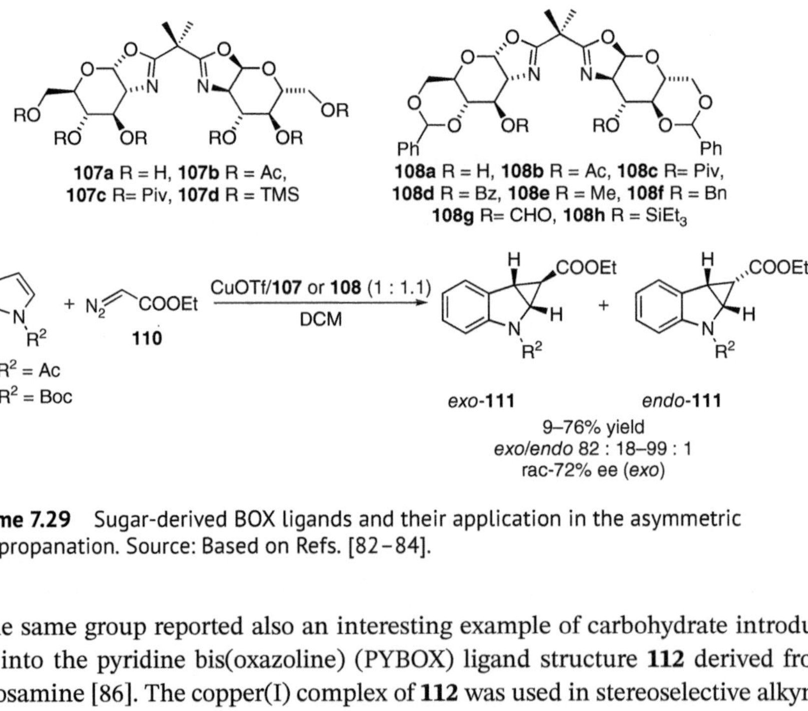

Scheme 7.29 Sugar-derived BOX ligands and their application in the asymmetric cyclopropanation. Source: Based on Refs. [82–84].

The same group reported also an interesting example of carbohydrate introduction into the pyridine bis(oxazoline) (PYBOX) ligand structure **112** derived from glucosamine [86]. The copper(I) complex of **112** was used in stereoselective alkynylation of imines **113** with moderate yields and excellent enantioselectivity in most cases (Scheme 7.30).

Carbohydrate-derived ligands **116** combining the functionality of oxazolines and phosphorus groups are another emerging group of ligands successfully used in several asymmetric transformations. Diéguez and Andersson performed asymmetric hydrogenation of olefins **117** in the presence of iridium-**116** complex. The broad

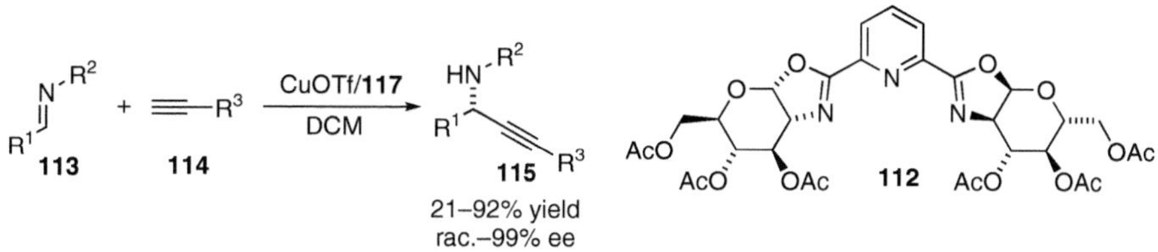

Scheme 7.30 The asymmetric alkynylation of imines catalyzed by Cu/**113** complex.

spectrum of di- and trisubstituted alkenes were hydrogenated with an excellent enantioselectivity (Scheme 7.31) [87].

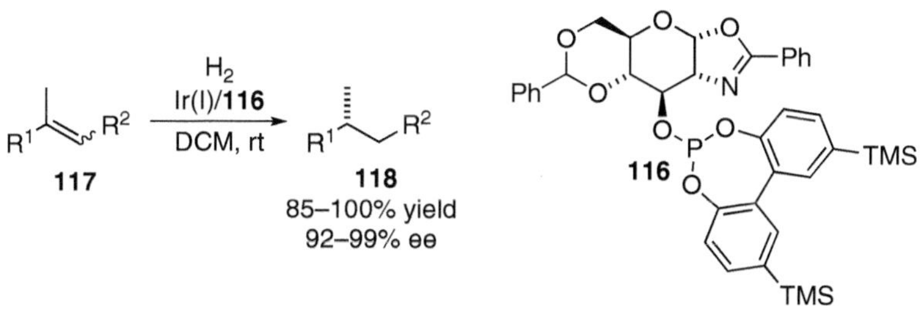

Scheme 7.31 The asymmetric hydrogenation of olefins catalyzed by Ir/**116** complex.
Source: Based on Diéguez et al. [87].

The palladium complex with phosphine-(**119**) [88], phosphinites-(**120a–120c**) [89, 90], and phosphites-oxazoline(**121**) [91, 92] ligands were investigated in the asymmetric allylation reaction of malonate ester or benzyl amine. The sugar-oxazoline palladium complexes were also used in the Heck reaction (Scheme 7.32) [93–95]. The catalysts **120** resulted in high yield (**120a** – 88%, **120b** – 24%, **120c** – 100%) and enantioselectivity (**120a** – 91% ee, **120b** – 92% ee, **120c** – 96% ee), but the substrate scope was limited to a single alkene [93]. Another example was provided by Diéguez, who assessed ligand **121** previously proven to be efficient in asymmetric allylation in the Heck reaction [94, 95]. Excellent results were obtained for a broad spectrum of olefins and triflate reactants.

The chiral variants of NHC catalysts are of high demand since carbene-catalysis has a wide range of applications in modern organic synthesis. On this basis, carbohydrates were also examined as chiral scaffolds for the NHC-catalysts development (Scheme 7.33) [96]. The conjugations of imidazole with gluco- (**125**) and galactopyranose (**126**) at the anomeric position were evaluated in a variety of types of metathesis [98]. The sugar moiety has been proved to affect the olefin geometry by steric interactions and provide asymmetric induction on observable level.

Another approach to sugar-NHC catalysts was reported for rhodium complexes of C_2-symmetric, sugar-disubstituted imidazolium ligands. Henderson et al. performed asymmetric hydrosilylation of ketones with **127** and reported interesting results (Scheme 7.33) [97]. The carbohydrate structure had a strong effect on catalyst efficiency since NHC synthetized from mannosamine was far less effective than glucosamine-derivative.

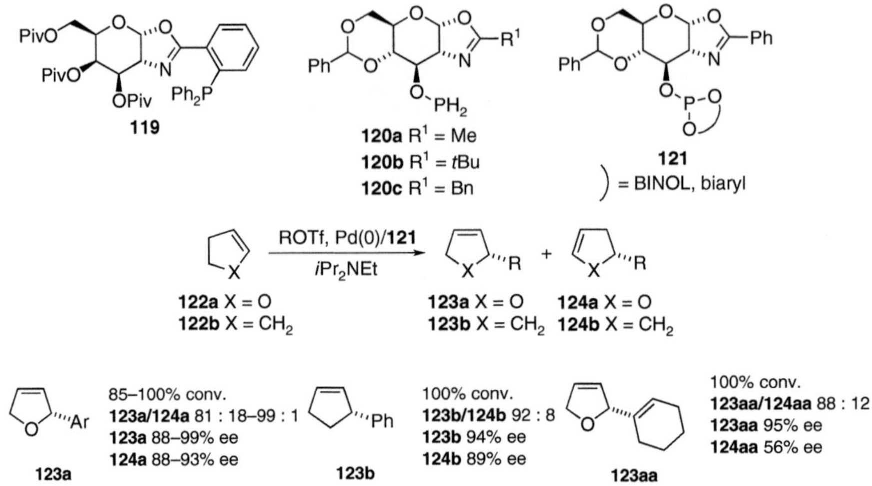

Scheme 7.32 The asymmetric Heck reaction catalyzed by Pd/**121** complexes. Source: Refs. [93–95].

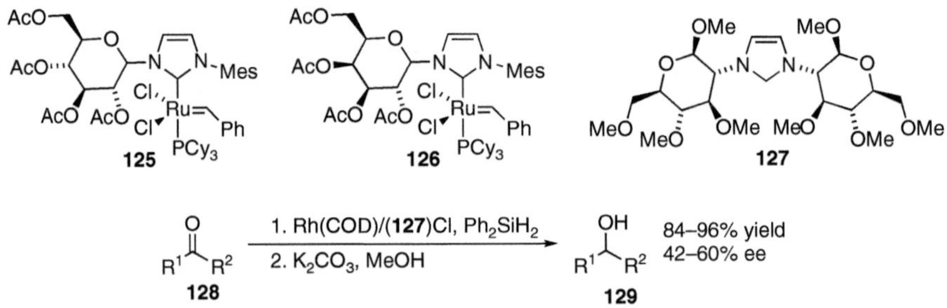

Scheme 7.33 The NHC induced hydrosilylation of ketones. Source: Based on Tewes et al. [96] and Henderson et al. [97].

7.2.2 Organocatalysts

Organocatalysis as an alternative to metal-complex catalysis has arisen over the past two decades to become a well-recognized methodology of asymmetric synthesis. The concept of organocatalysis assumes application of simple organic molecules as direct catalysts working throughout covalent or non-covalent interaction with activated substrates. To address the necessity of the efficient stereochemistry control by organocatalysts the development of molecules easily-available in optical pure form is of high demand. Sugars with their well-defined stereochemistry, high diversity, low cost, insensitivity to water and air are ideal candidates for this purpose. Carbohydrate moieties have been successfully introduced to organocatalyst structures used in several asymmetric transformations [61, 99]. The most recognizable organocatalytic approach involving sugar-catalysis is Shi epoxidation of olefins using chiral ketone derived from D-fructose (Scheme 7.34) [100]. The concept of this process is based on the oxidation of carbonyl group by Oxone® to

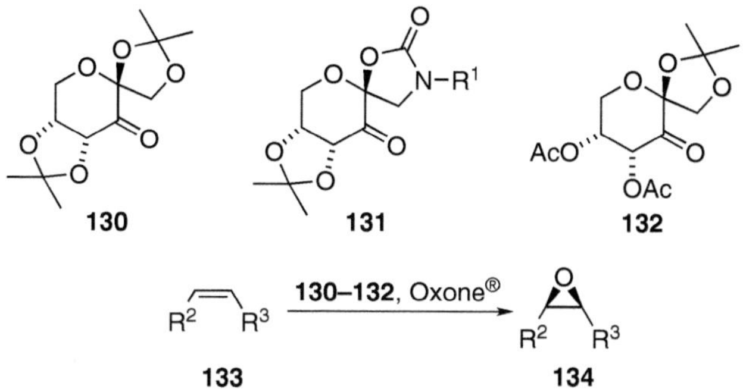

Scheme 7.34 The Shi epoxidation of olefins initiated by chiral ketones **130-132**. Source: Based on Tu et al. [100].

the chiral dioxirane which can further react with a double bond in an asymmetric fashion. Both enantiomers of Shi catalyst **130** can be accessed from D-fructose in two steps and L-sorbose in a five-step procedure [101]. The substrate scope for the classic Shi epoxidation is already impressive, however the best results were obtained for *trans*-disubstituted and trisubstituted olefins [50, 61, 102]. Subsequent investigations of Shi catalyst structure have led to development of novel derivatives that extend this scope. Ketones **130** and **131** are suitable for epoxidation of *cis*-disubstituted alkenes [103], whereas **132** has been proven to efficiently epoxidize *trans*-disubstituted and trisubstituted α,β-unsaturated carbonyl compounds recognized as challenging substrates [104]. The more details about Shi epoxidation can be found in comprehensive reviews dedicated to this subject [102].

The combination of carbohydrates with proline, which is already a well-recognized organocatalyst itself, significantly improved the efficiency and selectivity of aldol reactions. Since 2007 when Machinami and coworkers reported the first example of such a combination **135** [105], several sugar derived prolinamides (**136–139**) have been reported (Figure 7.4) [99]. The main advantage of this methodology lies in maintaining of the high efficiency in water [106] or brine [107] as reaction media. In general, for the standard aldol reaction substrates like cyclic ketones and

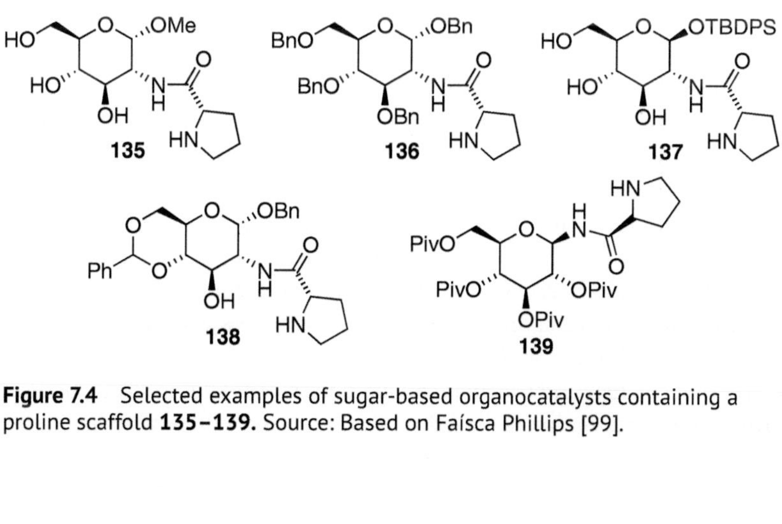

Figure 7.4 Selected examples of sugar-based organocatalysts containing a proline scaffold **135–139.** Source: Based on Faísca Phillips [99].

acetone with aromatic aldehydes, high yields with excellent diastereoselectivities and enantioselectivities were reported (Table 7.4) [105–111]. In particular cases the catalyst loading can be reduced to 0.1 mol% maintaining similar selectivity, but with a significant decrease of yield [108].

Table 7.4 The asymmetric aldol reaction catalyzed by **135–139**.

	Catalyst	Ketone	Solvent	Yield	*anti/syn*	ee
1	**135**	Acetone	Neat	69%	—	61%
2	**136**	Cyclohexanone	Neat	82–99%	71 : 29–98 : 2	91–99%
	136	Acetone	Neat	86–98%	—	93–95%
3	**137**	Cyclohexanone	Brine	86–99%	87 : 13–98 : 2	82–99%
4	**137**	Acetone	Brine	21–54%	—	66–75%
	138	Cyclohexanone	H_2O	62–99%	75 : 25–99 : 1	85–99%
	139	Cyclohexanone	DMSO:H_2O (1 : 1)	82–99%	91 : 9–97 : 3	88–97%

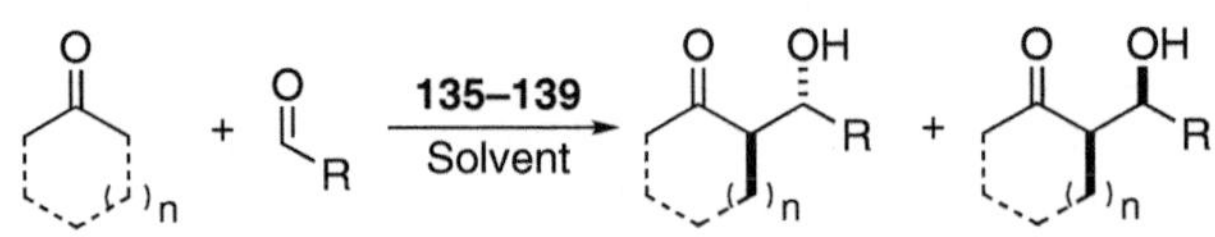

140 **141** *anti*-**142** *syn*-**142**

Optically pure thiourea derivatives are prominent examples of organocatalysts that often exhibit the dual mode of catalytic activity depending on the functionalities located on both sides of the thiourea group. The assembly of carbohydrate scaffold and optically pure primary (**143**) [112, 113], secondary (**144**) [114], or tertiary amines (**145**) [115] to thiourea group are prominent examples of this class of carbohydrate-derived organocatalysts (Scheme 7.35) [99]. The most common application of thiourea-carbohydrate hybrid catalysts is the asymmetric 1,4-addition of ketones or β-dicarbonyl compounds to nitro-olefins (Scheme 7.35). In general, high yield and excellent enantioselectivity were obtained for reactions activated by carbohydrate-derived thioureas [112–114, 116]. The match and mismatch effects are often observed between carbohydrate moiety and optically pure amines attached to the other side of thiourea leading to inversion or reduction of the selectivity.

Bifunctional thiourea **143** was also implemented into the three-component Biginelli reaction resulting in heterocycle formation with good yields and moderate to high enantioselectivities [117]. The presence of additives of *t*BuNH_2·TFA salt and 2,4,6-trichlorobenzoic acid is crucial for the reaction efficiency. The urea **145** was also used for the addition of malonate esters to *N*-Boc imines with high selectivity, albeit in a low yield (Scheme 7.36) [116].

The combination of sugar and *Cinchona* alkaloid in thiourea catalyst **146** has been found to promote cyanosilylation of dimethyl benzylophosphonates with high levels

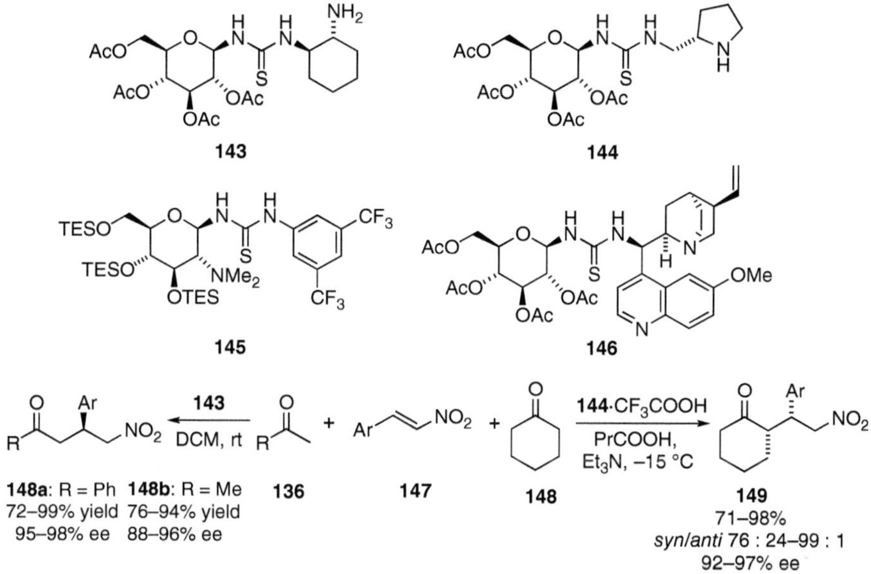

Scheme 7.35 The asymmetric Michael addition to nitroolefins **147** catalyzed by thioureas **143** and **144.** Source: Based on Faísca Phillips [99].

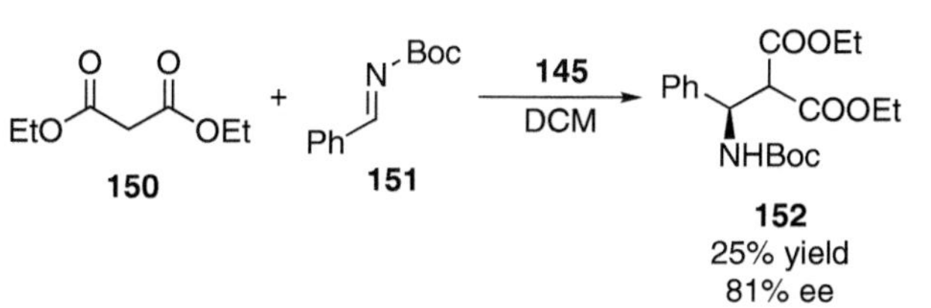

Scheme 7.36 The asymmetric addition of **150** to imine **151** promoted by urea **145.** Source: Based on Imrich et al. [118].

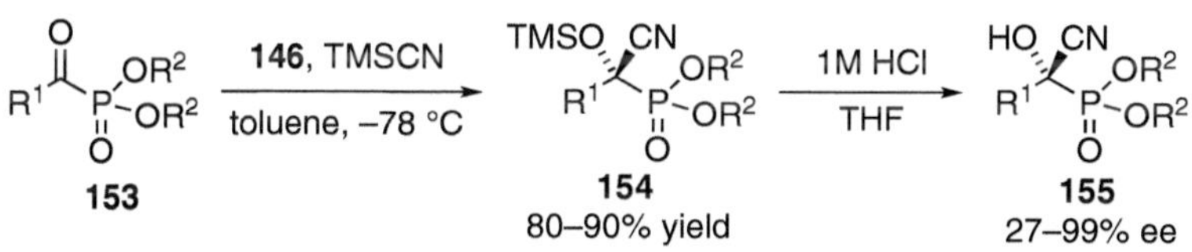

Scheme 7.37 The asymmetric cyanosilylation of benzylophosphonates **153** catalyzed by **146.** Source: Based on Kong et al. [115].

of conversion and enantiomeric induction. High yield, but with lower enantioselectivity was reported for aliphatic acylphosphonates (Scheme 7.37) [115].

The carbohydrate iodoarenes **156** were recently investigated by Ziegler and coworker as activators of asymmetric, oxidative spirolactonization of **157** (Scheme 7.38) [119]. The first developed catalysts **156a–156d** afforded the corresponding spirolactone **158** in yields up to 77% and moderate enantioselectivity

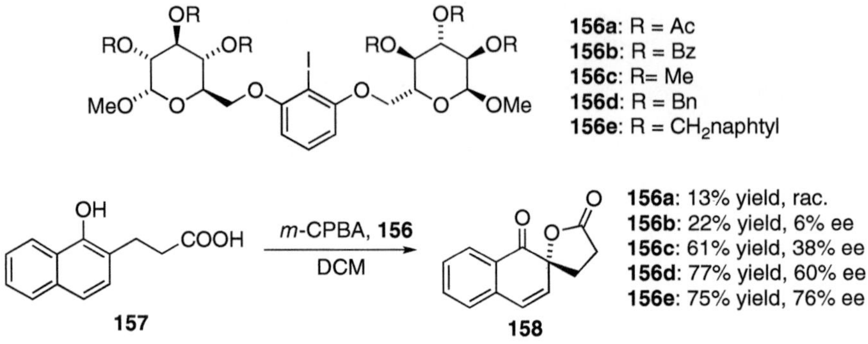

Scheme 7.38 The oxidative spirolactonization of **157** promoted by sugar-derived iodoarenes **156.** Source: Based on Imrich and Ziegler [119].

up to 60% ee. Further investigation revealed that the configuration and functionalization of carbohydrate C-4 position has a significant effect on reaction selectivity [118]. The D-galactose derivative **156e** with bulky substituents at C-4 position improves the enantioselectivity up to 76% ee. It is worth mentioning that the phenolic connection to C-6 position is crucial for catalyst efficiency.

Bakó et al. developed sugar-based phase transfer catalysts **159** for asymmetric epoxidation of chalcones **160** with *t*-butylhydroperoxide (TBHP) [120, 121]. The yield (33–82%) and stereoselectivity (8–94% ee) of monosacharide-based crown ethers **159** depended highly on the type of sugar moiety and nitrogen substituent (Scheme 7.39). The obtained results supported with density functional theory (DFT) calculation revealed that glucopyranose-derived **159a** and mannopyranose-based **159b** behave like pseudo-enantiomers and initiate the formation of the epoxides **161** with the opposite configuration [122, 123].

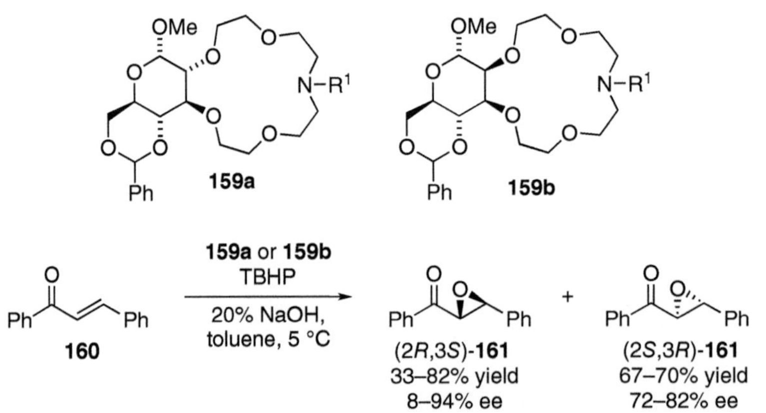

Scheme 7.39 The phase transfer catalyzed epoxidation of chalcone **160** with sugar-based crown ethers **159.**

Recently, the same research group reported D-glucose-based quaternary ammonium salts **162–163** and triazolium salts **164–165** as novel class of phase transfer

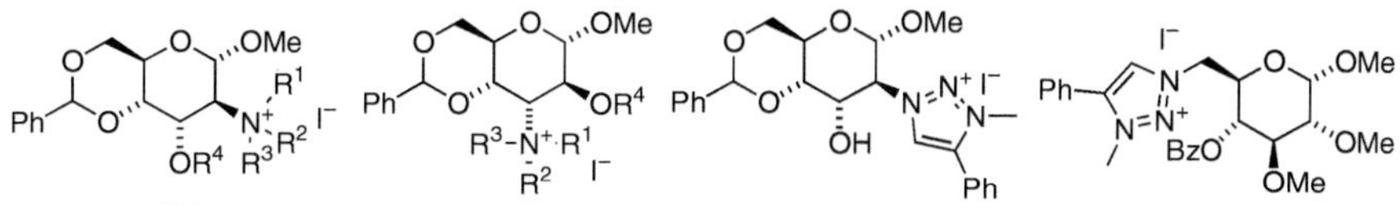

161 162 163 164

Figure 7.5 The examples of phase transfer catalysts **161–165** derived from carbohydrates. Source: Based on Bakó et al. [124].

catalysts (Figure 7.5) [124]. The synthetized catalysts were evaluated in the phase-transfer catalyst (PTC) promoted alkylation of glycine-derived Schiff base. Unfortunately, the reported enantioselectivities did not exceed 21% ee and the yield ranged from 67% to 88%.

It is remarkable that structures of such common compounds as monosaccharides play an important role in the enantioselective catalysis or can be used as chiral auxiliaries. In particular, enantioselective organocatalysis certainly will attract more attention from the scientific community in future investigations. Considering the widespread availability and low costs monosaccharides are frequently the chiral building blocks of choice for the synthesis of preparative amounts of enantiomerically pure compounds.

References

1 Corey, E.J. and Ensley, H.E. (1975). Preparation of an optically active prostaglandin intermediate via asymmetric induction. *J. Am. Chem. Soc.* 97 (23): 6908–6909.

2 Diaz-Muñoz, G., Miranda, I.L., Sartori, S.K. et al. (2019). Use of chiral auxiliaries in the asymmetric synthesis of biologically active compounds: a review. *Chirality* 31 (10): 776–812.

3 Vasella, A. (1977). Stereoselektivität und Reaktivität bei der 1,3-dipolaren cycloaddition chiraler *N*-(Alkoxyalkyl)nitrone. *Helv. Chim. Acta* 60 (4): 1273–1295.

4 Kunz, H. and Stoye, A. (2017). Carbohydrates as stereodifferentiating auxiliaries. In: *Heterocycles as Chiral Auxiliaries in Asymmetric Synthesis* (ed. M. Braun), 1–72. Cham: Springer International Publishing.

5 Kunz, H. and Sager, W. (1987). Diastereoselective Strecker synthesis of α-aminonitriles on carbohydrate templates. *Angew. Chem. Int. Ed. Engl.* 26 (6): 557–559.

6 Kunz, H., Sager, W., Schanzenbach, D., and Decker, M. (1991). Carbohydrates as chiral templates: stereoselective Strecker synthesis of D-α-amino nitriles and acids using *O*-pivaloylated D-galactosylamine as the auxiliary. *Liebigs Ann. Chem.* 1991 (7): 649–654.

7 Kunz, H., Sager, W., Pfrengle, W., and Schanzenbach, D. (1988). Reversal of asymmetric induction in stereoselective Strecker synthesis on galactosyl amine as the chiral matrix. *Tetrahedron Lett.* 29 (35): 4397–4400.

8 Kunz, H., Pfrengle, W., Rück, K., and Sager, W. (1991). Stereoselective synthesis of L-amino acids via Strecker and Ugi Reaktions on carbohydrate templates. *Synthesis* 1991 (11): 1039–1042.

9 Kunz, H. and Pfrengle, W. (1988). Carbohydrates as chiral templates: asymmetric ugi-synthesis of alpha-amino acids using galactosylamines as the chiral matrices. *Tetrahedron* 44 (17): 5487–5494.

10 Kunz, H., Pfrengle, W., and Sager, W. (1989). Carbohydrates as chiral templates: diastereoselective Ugi synthesis of (*S*)-amino acids using *O*-acylated D-arabinopyranosylamine as the auxiliary. *Tetrahedron Lett.* 30 (31): 4109–4110.

11 Oertel, K., Zech, G., and Kunz, H. (2000). Stereoselective combinatorial Ugi-multicomponent synthesis on solid phase. *Angew. Chem. Int. Ed.* 39 (8): 1431–1433.

12 Kunz, H. and Schanzenbach, D. (1989). Carbohydrates as chiral templates: stereoselective synthesis of β-amino acids. *Angew. Chem. Int. Ed.* 28 (8): 1068–1069.

13 Kunz, H., Burgard, A., and Schanzenbach, D. (1997). Asymmetric Mannich synthesis of β-amino acids with two new stereogenic centers at the α and β positions. *Angew. Chem. Int. Ed. Engl.* 36 (4): 386–387.

14 Cui, B., Kong, S., Wu, G. et al. (2012). Lewis acid catalyzed diastereoselective vinylogous Mannich reaction induced by *O*-pivaloylated d-galactosylamine as the chiral auxiliary: stereoselective synthesis of 6-arylpiperidin-2-ones. *Synthesis* 44 (1): 111–119.

15 Laschat, S. and Kunz, H. (1990). Carbohydrates as chiral templates: stereoselective synthesis of chiral homoallyl amines and β-amino acids. *Synlett* 1990 (01): 51–52.

16 Laschat, S. and Kunz, H. (1990, 1990). Carbohydrates as chiral templates: stereoselective synthesis of (*R*)-homoallyl amines using L-fucose as the auxiliary formally enantiomeric to D-galactose. *Synlett* 10: 629–630.

17 Li, B.Y., Li, Z.J., and Meng, X.B. (2010). $InCl_3$-catalyzed asymmetric aza-Friedel–Crafts reaction of indoles with imines generated from *O*-pivaloylated β-D-galactosylamine. *Carbohydr. Res.* 345 (12): 1708–1712.

18 Yang, H., Cui, B., Wu, G. et al. (2012). Asymmetric aza-Friedel–Crafts reaction of indoles induced by *O*-pivaloylated D-galactosylamine as the chiral auxiliary. *Tetrahedron* 68 (24): 4830–4837.

19 Laschat, S. and Kunz, H. (1991). Carbohydrates as chiral templates: diastereoselective synthesis of *N*-glycosyl-*N*-homoallylamines and β-amino acids from imines. *J. Org. Chem.* 56 (20): 5883–5889.

20 Wang, G., Li, B., Lou, Q. et al. (2013). Diastereoselective Povarov-like reaction involving *O*-pivaloylated D-galactosylimine. *Adv. Synth. Catal.* 355 (2, 3): 303–307.

21 Laschat, S. and Kunz, H. (1992). Carbohydrates as chiral templates: stereoselective synthesis of (*R*)- and (*S*)-α-aminophosphonic acid derivatives. *Synthesis* 1992 (1/2): 90–95.

22 Kunz, H. and Pfrengle, W. (1989). Carbohydrates as chiral templates: stereoselective tandem Mannich–Michael reactions for the synthesis of piperidine alkaloids. *Angew. Chem. Int. Ed. Engl.* 28 (8): 1067–1068.

23 Weymann, M., Pfrengle, W., Schollmeyer, D., and Kunz, H. (1997). Enantioselective syntheses of 2-alkyl-, 2,6-dialkylpiperidines and indolizidine alkaloids through diastereoselective Mannich–Michael reactions. *Synthesis* 1997 (10): 1151–1160.

24 Pfrengle, W. and Kunz, H. (1989). Hetero-Diels–Alder reactions on a carbohydrate template: stereoselective synthesis of (*S*)-anabasin. *J. Org. Chem.* 54 (18): 4261–4263.

25 Zhao, Y., Wang, G., Zhou, S. et al. (2014). Diastereoselective formation of aziridines from diazocarbonyl compounds and *N*-(*O*-pivaloylated D-galactosyl)benzylideneamines and ring-opening reactions with *p*-toluenethiol. *Org. Biomol. Chem.* 12 (21): 3362–3365.

26 Kunz, H. and Pees, K.J. (1989). 1,4-Addition of dialkylaluminium chlorides to α,β-unsaturated carboxylic acid derivatives: a new synthesis of β-branched carboxylic acids. *J. Chem. Soc., Perkin Trans. 1* (6): 1168–1169.

27 Elzner, S., Maas, S., Engel, S., and Kunz, H. (2004). Stereoselective conjugate addition of mixed organoaluminum reagents to α,β-unsaturated *N*-acyloxazolidinones derived from carbohydrates. *Synthesis* (13): 2153–2164.

28 Rück-Braun, K., Stamm, A., Engel, S., and Kunz, H. (1997). β-Branched α-halo carboxylic acid derivatives via stereoselective 1,4 addition of dialkylaluminum chlorides to α,β-unsaturated *N*-acyloxazolidinones. *J. Org. Chem.* 62 (4): 967–975.

29 Rück, K. and Kunz, H. (1992). A bicyclic carbohydrate oxazolidinone template for stereoselective 1,4-additions of organoaluminium chlorides to unsaturated carboxylic acid derivatives. 1992 (Synlett, 04): 343–344.

30 Vega-Pérez, J.M., Vega, M., Blanco, E., and Iglesias-Guerra, F. (2001). Stereoselective synthesis of oxiranes using oxazolidines derived from 2-amino-2-deoxy-D-allose as chiral auxiliaries. *Tetrahedron: Asymmetry* 12 (22): 3189–3203.

31 Köll, P. and Lützen, A. (1996). D-xylose derived oxazolidin-2-ones as chiral auxiliaries in stereoselective alkylations. *Tetrahedron: Asymmetry* 7 (3): 637–640.

32 Lützen, A. and Köll, P. (1997). D-Xylose derived oxazolidin-2-ones as chiral auxiliaries in stereoselective acylations and halogenations. *Tetrahedron: Asymmetry* 8 (1): 29–32.

33 Lützen, A. and Köll, P. (1997). D-Xylose derived oxazolidin-2-ones as chiral auxiliaries in stereoselective aldol reactions. *Tetrahedron: Asymmetry* 8 (8): 1193–1206.

34 Totani, K., Takao, K.I., and Tadano, K.I. (2004). Sugar as a tool for asymmetric synthesis: some effective approaches. *Synlett* (12): 2066–2080.

35 Totani, K., Nagatsuka, T., Yamaguchi, S. et al. (2001). Highly diastereoselective 1,4-addition of an organocuprate to methyl α-D-gluco-, α-D-manno-, or α-D-galactopyranosides tethering an α,β-unsaturated ester. Novel asymmetric access to β-C-substituted butanoic acids. *J. Org. Chem.* 66 (18): 5965–5975.

36 Totani, K., Asano, S., Takao, K., and Tadano, K. (2001). Highly stereoselective α-alkylations, 1,4-additions, and one-pot 1,4-addition/α-methylations achieved on 4-*O*-acyl and 4-*O*-crotonyl derivatives of methyl 6-deoxy-2,3-di-*O*-(*t*-butyldimethylsilyl)-α-D-glucopyranoside. *Synlett* 2001 (11): 1772–1776.

37 Munakata, R., Totani, K., Takao, K.I., and Tadano, K.I. (2000). Highly stereoselective Lewis acid mediated conjugate radical additions to methyl α-D-glucopyranoside derivatives tethering an unsaturated ester moiety at C-4. *Synlett* (7): 979–982.

38 Shing, T.K.M., Chow, H.F., and Chung, I.H.F. (1996). Sugarometallic chemistry: aglycone-chromium complex as chiral auxiliary in asymmetric Diels–Alder reaction. *Tetrahedron Lett.* 37 (21): 3713–3716.

39 Tamai, T., Asano, S., Totani, K. et al. (2003). Highly stereoselective [3+2] cycloadditions of nitrile oxides to methyl 4-*O*-acryloyl-6-deoxy-2,3-*O*-(*t*-butyldimethylsilyl)-α-D-glucopyranoside. *Synlett* (12): 1865–1867.

40 Heathcock, C.H., White, C.T., Morrison, J.J., and VanDerveer, D. (1981). Acyclic stereoselection. 11. Double stereodifferentiation as a method for achieving superior Cram's rule selectivity in aldol condensations with chiral aldehydes. *J. Organ. Chem.* 46 (7): 1296–1309.

41 Costa, P.R.R., Ferreira, V.F., Alencar, K.G. et al. (1996). 2,3:4,5-DI-*O*-isopropylidene-β-D-fructopyranose as chiral auxiliary in asymmetric α-alkylation of ester enolates. *J. Carbohydr. Chem.* 15 (6): 691–699.

42 Kunz, H. and Mohr, J. (1988). Carbohydrates as chiral templates: reactivity and stereoselectivity of carbohydrate ester enolates. *J. Chem. Soc., Chem. Commun.* (19): 1315–1317.

43 Brandänge, S., Josephson, S., Mörch, L. et al. (1981). Asymmetric synthesis of β-hydroxy esters by Reformatsky reactions and amide base mediated condensations. *Acta Chem. Scand.* 35b: 273–277.

44 Kishida, M., Eguchi, T., and Kakinuma, K. (1996). Remote asymmetric induction observed in the alkylation of propionate attached to a carbohydrate template. *Tetrahedron Lett.* 37 (12): 2061–2062.

45 Totani, K., Asano, S., Takao, K.I., and Tadano, K.I. (2001). Highly stereoselective α-alkylations, 1,4-additions, and one-pot 1,4-addition/α-methylations achieved on 4-*O*-acyl and 4-*O*-crotonyl derivatives of methyl 6-deoxy-2,3-di-*O*-(*t*-butyldimethylsilyl)-α-D-glucopyranoside. *Synlett* (11): 1772–1776.

46 Asano, S., Tamai, T., Totani, K. et al. (2003). Highly stereoselective 1,4-addition of the enolate generated from 6-deoxy-D-glucopyranoside-derived propionyl ester to methyl crotonate: application to total synthesis of (−)-lasiol. *Synlett* (14): 2252–2254.

47 Kozawa, I., Akashi, Y., Takiguchi, K. et al. (2007). Stereoselective double alkylation of the acetoacetate ester α-carbon on a D-glucose-derived template: application to the synthesis of enantiopure cycloalkenones bearing an asymmetric quaternary carbon. *Synlett* (3): 399–402.

48 Akashi, Y., Takao, K., and Tadano, K. (2009). Stereoselective α-alkylation of methyl 6-deoxy-3,4-di-*O*-(tert-butyldimethylsilyl)-2-*O*-(2-methyl-3-oxobutanoyl)-α-D-glucopyranoside. *Tetrahedron Lett.* 50 (10): 1139–1142.

49 Sreeshailam, A., Dayaker, G., Chevallier, F. et al. (2011). Diastereoselective deprotonative metalation of sugar-derived ferrocene esters using mixed lithium-cadmium combinations. *Eur. J. Org. Chem.* (20-21): 3715–3718.

50 Furman, B. and Stecko, S. (2013). Carbohydrate derived auxiliaries: mono (and disaccharide) derivatives. *ChemInform* 44: https://doi.org/10.1002/chin.201329253.

51 Boysen, M.M.K. (ed.) (2013). *Carbohydrates – Tools for Stereoselective Synthesis.* Weinheim: Wiley-VCH Verlag GmbH & Co. KGaA.

52 Charette, A.B., Côé, B., and Marcoux, J.F. (1991). Carbohydrates as chiral auxiliaries: asymmetric cyclopropanation reaction of acyclic olefins. *J. Am. Chem. Soc.* 113 (21): 8166–8167.

53 Charette, A.B. and Côté, B. (1993). Carbohydrates as chiral auxiliaries: the asymmetric epoxidation reaction of olefins. *Tetrahedron: Asymmetry* 4 (11): 2283–2286.

54 Beagley, B., Curtis, A.D.M., Pritchard, R.G., and Stoodley, R.J. (1992). Asymmetric Diels–Alder reactions. Part 6. Regio- and stereo-selective cycloadditions of 5-(2′,3′,4′,6′-*tetra-O*-acetyl-β-D-glucopyranosyloxy)-1,4-naphthoquinone. *J. Chem. Soc., Perkin Trans. 1* (15): 1981–1991.

55 Bauch, M., Fudickar, W., and Linker, T. (2021). Stereoselective [4+2] cycloaddition of singlet oxygen to naphthalenes controlled by carbohydrates. *Molecules* 26: 804–821.

56 Idris, M.S., Larsen, D.S., Schofield, A. et al. (1995). Regio- and stereoselective bromoalkoxylations of (*E*)-α-(2′,3′,4′,6′,-*tetra-O*-acetyl-β-D-glucopyranosyloxymethylene) carbonyl compounds: a route to stereopure α-bromo-α-dioxymethyl carbonyl compounds. *Tetrahedron Lett.* 36 (18): 3251–3254.

57 Larsen, D.S., Schofield, A., Stoodley, R.J., and Tiffin, P.D. (1994). Diastereoselective hydrogenations of α-(2′,3′,4′,6′-*tetra-O*-acetyl-β-D-glucopyranosyloxy methylene) carboxylic esters: a route to stereopure aldol derivatives. *Tetrahedron Lett.* 35 (49): 9285–9288.

58 Benessere, V., Del Litto, R., De Roma, A., and Ruffo, F. (2010). Carbohydrates as building blocks of privileged ligands. *Coord. Chem. Rev.* 254: 390–401.

59 Castillón, S., Claver, C., and Díaz, Y. (2005). C_1 and C_2-symmetric carbohydrate phosphorus ligands in asymmetric catalysis. *Chem. Soc. Rev.* 34 (8): 702–713.

60 Henderson, A.S., Bower, J.F., and Galan, M.C. (2016). Carbohydrates as enantioinduction components in stereoselective catalysis. *Org. Biomol. Chem.* 14 (17): 4008–4017.

61 Lehnert, T., Özüduru, G., Grugel, H. et al. (2011). More than just sweet – sugar-derived stereodifferentiating agents for asymmetric synthesis. *Synthesis* (17): 2685–2708.

62 Cullen, W.R. and Sugi, Y. (1978). Asymmetric hydrogenation catalyzed by diphosphinite rhodium complexes derived from a sugar. *Tetrahedron Lett.* 19 (19): 1635–1636.

63 Sinou, D. and Descotes, G. (1980). Asymmetric reduction of unsaturated acids catalyzed by diphosphinite rhodium complexes derived from sugars. *React. Kinet. Catal. Lett.* 14 (4): 463–466.

64 Selke, R. (1979). Phosphinites of carbohydrates as chiral ligands for asymmetric synthesis catalyzed by complexes, I. Hydrogenation of aminoacid precursors by rhodium(I) chelates of methyl 4,6-*O*-benzylidente-2,3-bis-*O*-(diphenylphosphino)-α-D-glucopyranosides. *React. Kinet. Catal. Lett.* 10 (2): 135–138.

65 Jackson, R. and Thompson, D.J. (1978). Asymmetric hydrogenation using readily prepared chiral sugar diphosphiniterhodium complexes. *J. Organomet. Chem.* 159 (4): 29–31.

66 Selke, R. (1987). Phosphinites of carbohydrates as chiral ligands for asymmetric synthesis catalysed by complexes. IV. Influence of the solvent and the structure of cationic 4,6-*O*-benzylidene-2,3-*O*-bis(diphenylphosphino)-D-hexopyranoside-rhodium(I) chelates on enantioselec. *J. Prakt. Chem.* 329 (4): 717–724.

67 RajanBabu, T.V., Casalnuovo, A.L., and Ayers, T.A. (1994). Electronic amplification of selectivity in Rh-catalyzed hydrogenations: D-glucose-derived ligands for the synthesis of D-or L-amino acids. *J. Am. Chem. Soc.* 116 (9): 4101–4102.

68 RajanBabu, T.V., Ayers, T.A., Halliday, G.A. et al. (1997). Carbohydrate phosphinites as practical ligands in asymmetric catalysis: electronic effects and dependence of backbone chirality in Rh-catalyzed asymmetric hydrogenations. Synthesis of *R*- or *S*-amino acids using natural sugars as ligand precursors. *J. Org. Chem.* 62 (17): 6012–6025.

69 RajanBabu, T.V., Radetich, B., You, K.K. et al. (1999). Electronic effects in asymmetric catalysis: structural studies of precatalysts and intermediates in Rh-catalyzed hydrogenation of dimethyl itaconate and acetamidocinnamic acid derivatives using C_2-symmetric diarylphosphinite ligands. *J. Org. Chem.* 64 (10): 3429–3447.

70 Nugent, W.A., RajanBabu, T.V., and Burk, M.J. (1993). Beyond nature's chiral pool: enantioselective catalysis in industry. *Science* 259 (5094): 479–483.

71 Casalnuovo, A.L., RajanBabu, T.V., Ayers, T.A., and Warren, T.H. (1994). Ligand electronic effects in asymmetric catalysis: enhanced enantioselectivity in the asymmetric hydrocyanation of vinylarenes. *J. Am. Chem. Soc.* 116 (22): 9869–9882.

72 RajanBahu, T.V. and Casalnuovo, A.L. (1994). Electronic effects in asymmetric catalysis: enantioselective carbon–carbon bond forming processes. *Pure Appl. Chem.* 66 (7): 1535–1542.

73 RajanBabu, T.V. and Casalnuovo, A.L. (1992). Tailored ligands for asymmetric catalysis: the hydrocyanation of vinyl arenes. *J. Am. Chem. Soc.* 114 (15): 6265–6266.

74 Saha, B. and RajanBabu, T.V. (2006). Nickel(0)-catalyzed asymmetric hydrocyanation of 1,3-dienes. *Org. Lett.* 8 (20): 4657–4659.

75 Holz, J., Stürmer, R., Schmidt, U. et al. (2001). Synthesis of chiral 2,5-bis(oxymethyl)-functionalized bis(phospholanes) and their application in

Rh- and Ru-catalyzed enantioselective hydrogenations. *Eur. J. Org. Chem.* 24: 4615–4624.

76 Ruffo, F., Del Litto, R., De Roma, A. et al. (2006). Bis(phosphinoamides) based on sugars for highly enantioselective allylic substitution: inversion of stereocontrol by switching from glucose to mannose. *Tetrahedron: Asymmetry* 17 (15): 2265–2269.

77 Del Litto, R., De Roma, A., and Ruffo, F. (2007). A convenient route from simple sugars to new chiral bis(phosphinoesters) for asymmetric catalysis. *Inorg. Chem. Commun.* 10 (5): 618–622.

78 De Roma, A., Ruffo, F., and Woodward, S. (2008). Amino-sugar modular ligands – useful cores for the formation of asymmetric copper 1,4-addition catalysts. *Chem. Commun.* (42): 5384–5386.

79 Hamashima, Y., Kanai, M., and Shibasaki, M. (2000). Catalytic enantioselective cyanosilylation of ketones [14]. *J. Am. Chem. Soc.* 122 (30): 7412–7413.

80 Yabu, K., Masumoto, S., Yamasaki, S. et al. (2001). Switching enantiofacial selectivities using one chiral source: catalytic enantioselective synthesis of the key intermediate for (20*S*)-camptothecin family by (*S*)-selective cyanosilylation of ketones [7]. *J. Am. Chem. Soc.* 123 (40): 9908–9909.

81 Del Litto, R., Benessere, V., Ruffo, F., and Moberg, C. (2009). Carbohydrate-based pyridine-2-carboxamides for Mo-catalyzed asymmetric allylic alkylations. *Eur. J. Org. Chem.* 9: 1352–1356.

82 Irmak, M., Groschner, A., and Boysen, M.M.K. (2007). glucoBox – A new carbohydrate-based bis(oxazoline) ligand. Synthesis and first application. *Chem. Commun.* (2): 177–179.

83 Minuth, T. and Boysen, M.M.K. (2010). Bis(oxazolines) based on glycopyranosides – steric, configurational and conformational influences on stereoselectivity. *Beilstein J. Org. Chem.* 6: 1–7.

84 Minuth, T. and Boysen, M.M.K. (2008). A general and efficient route to 3-*O*-modified carbohydrate bis(oxazoline) ligands. *Synlett* 10: 1483–1486.

85 Minuth, T. and Boysen, M.M.K. (2010). Carbohydrate-derived bis(oxazoline) ligand in the total synthesis of grenadamide. *Synthesis* 16: 2799–2803.

86 Irmak, M. and Boysen, M.M.K. (2008). A new pyridyl bis(oxazoline) ligand prepared from D-glucosamine for asymmetric alkynylation of imines. *Adv. Synth. Catal.* 350 (3): 403–405.

87 Diéguez, M., Mazuela, J., Pàmies, O. et al. (2008). Chiral pyranoside phosphite-oxazolines: a new class of ligand for asymmetric catalytic hydrogenation of alkenes. *J. Am. Chem. Soc.* 130 (23): 7208–7209.

88 Gläser, B. and Kunz, H. (1998). Enantioselective allylic substitution using a novel (phosphino-α-D-glucopyrano-oxazoline)palladium catalyst. *Synlett* (1): 53–54.

89 Yonehara, K., Hashizume, T., Mori, K. et al. (1999). Palladium-catalyzed asymmetric allylic substitution reactions using new chiral phosphinite-oxazoline ligands derived from D-glucosamine. *J. Org. Chem.* 64 (26): 9374–9380.

90 Yonehara, K., Hashizume, T., Mori, K. et al. (1999). Palladium-catalysed asymmetric allylic alkylation using new chiral phosphinite-nitrogen ligands derived from D-glucosamine. *Chem. Commun.* 2 (5): 415–416.

91 Mata, Y., Pámies, O., and Diéguez, M. (2009). Pyranoside phosphite-oxazoline ligand library: highly efficient modular P,N ligands for palladium-catalyzed allylic substitution reactions. A study of the key palladium allyl intermediates. *Adv. Synth. Catal.* 351 (18): 3217–3234.

92 Mata, Y., Diéguez, M., Pàmies, O., and Claver, C. (2005). New carbohydrate-based phosphite-oxazoline ligands as highly versatile ligands for palladium-catalyzed allylic substitution reactions. *Adv. Synth. Catal.* 347 (15): 1943–1947.

93 Mata, Y., Diéguez, M., Pàmies, O., and Claver, C. (2005). Chiral phosphite-oxazolines: a new class of ligands for asymmetric heck reactions. *Org. Lett.* 7 (25): 5597–5599.

94 Mata, Y., Pàmies, O., and Diéguez, M. (2007). Screening of a modular sugar-based phosphite-oxazoline ligand library in asymmetric Pd-catalyzed Heck reactions. *Chem. Eur. J.* 13 (12): 3296–3304.

95 Yonehara, K., Mori, K., Hashizume, T. et al. (2000). Palladium-catalyzed asymmetric intermolecular arylation of cyclic or acyclic alkenes using phosphinite-oxazoline ligands derived from D-glucosamine. *J. Organomet. Chem.* 603 (1): 40–49.

96 Tewes, F., Schlecker, A., Harms, K., and Glorius, F. (2007). Carbohydrate-containing N-heterocyclic carbene complexes. *J. Organomet. Chem.* 692 (21): 4593–4602.

97 Henderson, A.S., Bower, J.F., and Galan, M.C. (2014). Carbohydrate-based N-heterocyclic carbenes for enantioselective catalysis. *Org. Biomol. Chem.* 12 (45): 9180–9183.

98 Keitz, B.K. and Grubbs, R.H. (2010). Ruthenium olefin metathesis catalysts bearing carbohydrate-based N-heterocyclic carbenes. *Organometallics* 29 (2): 403–408.

99 Faísca Phillips, A.M. (2014). Applications of carbohydrate-based organocatalysts in enantioselective synthesis. *Eur. J. Org. Chem.* 2014 (33): 7291–7303.

100 Tu, Y., Wang, Z., Shi, Y. et al. (1996). An efficient asymmetric epoxidation method for trans-olefins mediated by a fructose-derived ketone. *J. Am. Chem. Soc.* 118 (11): 9806–9807.

101 Zhao, M.X. and Shi, Y. (2006). Practical synthesis of an L-fructose-derived ketone catalyst for asymmetric epoxidation of olefins. *J. Org. Chem.* 71 (14): 5377–5379.

102 Zhu, Y., Wang, Q., Cornwall, R.G., and Shi, Y. (2014). Organocatalytic asymmetric epoxidation and aziridination of olefins and their synthetic applications. *Chem. Rev.* 114 (16): 8199–8256.

103 Shu, L., Wang, P., Gan, Y., and Shi, Y. (2003). Asymmetric epoxidation catalyzed by N-aryl-substituted oxazolidinone-containing ketones: further evidence for electronic effects. *Org. Lett.* 5 (3): 293–296.

104 Wu, X.Y., She, X., and Shi, Y. (2002). Highly enantioselective epoxidation of α,β-unsaturated esters by chiral dioxirane. *J. Am. Chem. Soc.* 124 (30): 8792–8793.

105 Tsutsui, A., Takeda, H., Kimura, M. et al. (2007). Novel enantiocontrol system with aminoacyl derivatives of glucoside as enamine-based organocatalysts for aldol reaction in aqueous media. *Tetrahedron Lett.* 48 (30): 5213–5217.

106 Shen, C., Shen, F., Zhou, G. et al. (2012). Novel carbohydrate-derived prolinamide as a highly efficient, recoverable catalyst for direct aldol reactions in water. *Catal. Commun.* 26: 6–10.

107 De Nisco, M., Pedatella, S., Bektaş, S. et al. (2012). D-Glucosamine in a chimeric prolinamide organocatalyst for direct asymmetric aldol addition. *Carbohydr. Res.* 356: 273–277.

108 Shen, C., Liao, H., Shen, F., and Zhang, P. (2013). Novel synthesis of carbohydrate-derived organocatalysts and their application in asymmetric aldol reactions. *Catal. Commun.* 41: 106–109.

109 Agarwal, J. and Peddinti, R.K. (2012). Synthesis and characterization of monosaccharide derivatives and application of sugar-based prolinamides in asymmetric synthesis. *Eur. J. Org. Chem.* 32: 6390–6406.

110 Agarwal, J. and Peddinti, R.K. (2010). Highly efficient and solvent-free direct aldol reaction catalyzed by glucosamine-derived prolinamide. *Tetrahedron: Asymmetry* 21 (15): 1906–1909.

111 Pedatella, S., De Nisco, M., Mastroianni, D. et al. (2011). Diastereo- and enantioselective direct aldol reactions in aqueous medium: a new highly efficient proline-sugar chimeric catalyst. *Adv. Synth. Catal.* 353 (9): 1443–1446.

112 Li, X.J., Liu, K., Ma, H. et al. (2008). Highly enantioselective Michael addition of malonates to nitroolefins catalyzed by chiral bifunctional tertiary amine-thioureas based on saccharides. *Synlett* 20: 3242–3246.

113 Gu, Q., Guo, X.T., and Wu, X.Y. (2009). Highly enantioselective Michael addition of acetone to nitroolefins catalyzed by chiral bifunctional primary amine-thiourea catalysts with acetic acid. *Tetrahedron* 65 (27): 5265–5270.

114 Lu, A., Gao, P., Wu, Y. et al. (2009). Highly enantio- and diastereoselective Michael addition of cyclohexanone to nitroolefins catalyzed by a chiral glucose-based bifunctional secondary amine-thiourea catalyst. *Org. Biomol. Chem.* 7 (15): 3141–3147.

115 Kong, S., Fan, W., Wu, G., and Miao, Z. (2012). Enantioselective synthesis of tertiary α-hydroxy phosphonates catalyzed by carbohydrate/cinchona alkaloid thiourea organocatalysts. *Angew. Chem. Int. Ed.* 51 (35): 8864–8867.

116 Puglisi, A., Benaglia, M., Raimondi, L. et al. (2011). Novel carbohydrate-based bifunctional organocatalysts for nucleophilic addition to nitroolefins and imines. *Org. Biomol. Chem.* 9 (9): 3295–3302.

117 Wang, Y., Yang, H., Yu, J. et al. (2009). Highly enantioselective Biginelli reaction promoted by chiral bifunctional primary amine-thiourea catalysts: asymmetric synthesis of dihydropyrimidines. *Adv. Synth. Catal.* 351 (18): 3057–3062.

118 Imrich, M.R., Biehler, L.E., Maichle-Mössmer, C., and Ziegler, T. (2019). Carbohydrate-based chiral iodoarene catalysts: a survey through the development of an improved catalyst design. *Molecules* 24 (21): 3883.

119 Imrich, M.R. and Ziegler, T. (2019). Carbohydrate based chiral iodoarene catalysts for enantioselective dearomative spirocyclization. *Tetrahedron Lett.* 60 (38): 15–16.

120 Bakó, P., Bakó, T., Mészáros, A. et al. (2004). Phase transfer catalysed asymmetric epoxidation of chalcones using chiral crown ethers derived from D-glucose and D-mannose. *Synlett* 4: 643–646.

121 Bakó, T., Bakó, P., Keglevich, G. et al. (2004). Phase-transfer catalyzed asymmetric epoxidation of chalcones using chiral crown ethers derived from D-glucose, D-galactose, and D-mannitol. *Tetrahedron: Asymmetry* 15 (10): 1589–1595.

122 Makó, A., Menyhárd, D.K., Bakó, P. et al. (2008). Theoretical study of the asymmetric phase-transfer mediated epoxidation of chalcone catalyzed by chiral crown ethers derived from monosaccharides. *J. Mol. Struct.* 892 (1–3): 336–342.

123 Makó, A., Rapi, Z., Keglevich, G. et al. (2010). Asymmetric epoxidation of substituted chalcones and chalcone analogues catalyzed by α-D-glucose- and α-D-mannose-based crown ethers. *Tetrahedron: Asymmetry* 21 (8): 919–925.

124 Nemcsok, T., Rapi, Z., Bagi, P., and Bakó, P. (2019). Synthesis and application of novel carbohydrate-based ammonium and triazolium salts. *Synth. Commun.* 49 (18): 2388–2400.

8

Monoterpenes as Chiral Building Blocks

Agata J. Pacuła-Miszewska, Magdalena Obieziurska-Fabisiak, and Jacek Ścianowski

Nicolaus Copernicus University, Department of Organic Chemistry, Faculty of Chemistry, 7 Gagarin Street, 87-100 Torun, Poland

8.1 Introduction

Enantiopure compounds found in the natural kingdom create a collection called "*The chiral pool.*" This entity includes amino acids, carbohydrates, and terpenes [1, 2]. Given the complexity of the carbon skeleton and the broad set of derivatives (about 55 000 structures), terpenes can be considered the dominant chiral pool contributors [3]. Their uniqueness is striking in their specific biosynthesis in plant cells [4]. Many complex hydrocarbon frameworks are obtained from simple non-chiral substrates by just a few enzyme-catalyzed reactions [5–8]. This extremely well-coordinated process creates products with a perfectly designed length and form of the carbon skeleton accompanied by a fully maintained stereochemistry. For selected monoterpenes (pinene **4**, verbenone **5**, limonene **6**, and menthol **7**), sesquiterpenes (zingiberene **11** and turmerone **12**), and diterpenes (kaurene **18** and steviol **19**), a simplified biosynthetic process is presented in Scheme 8.1 [9–13]. For all isoprenoids, the biosynthesis consists of four stages and starts from synthesizing two C_5-precursors: dimethylallyl diphosphate (DMAPP) and isopenthenyl diphosphate (IPP). Next, specific prenyltransferase enzymes catalyze the head-to-tail condensation of these building blocks (Stage 2). Depending on the number of used IPP molecules, precursors of mono- (geranyl diphosphate, GPP), sesqui- (farnesyl diphosphate, FPP), and diterpenes (geranylgeranyl diphosphate, GGPP) are formed. When the length of the chain is fixed, terpene synthases facilitate the isoprenoid core formation (Stage 3). The key step is eliminating a diphosphate group (structures **1**, **8**, and **13**), then generating a reactive carbocation **2**, **9**, and **14**. These species can be variously rearranged (transformation **2** → **3**), they can undergo cyclization (**14** → **15**), proton transfer (**9** → **10**), and alkyl shift (**16** → **17**), until they are trapped by an external nucleophile or deprotonated to form alkenes (**4**, **6**, **11**, and **18**). The created carbohydrate frameworks are equipped with various functional groups. The last stage of the biosynthetic process involves further modifications of

Chiral Building Blocks in Asymmetric Synthesis: Synthesis and Applications, First Edition.
Edited by Elżbieta Wojaczyńska and Jacek Wojaczyński.

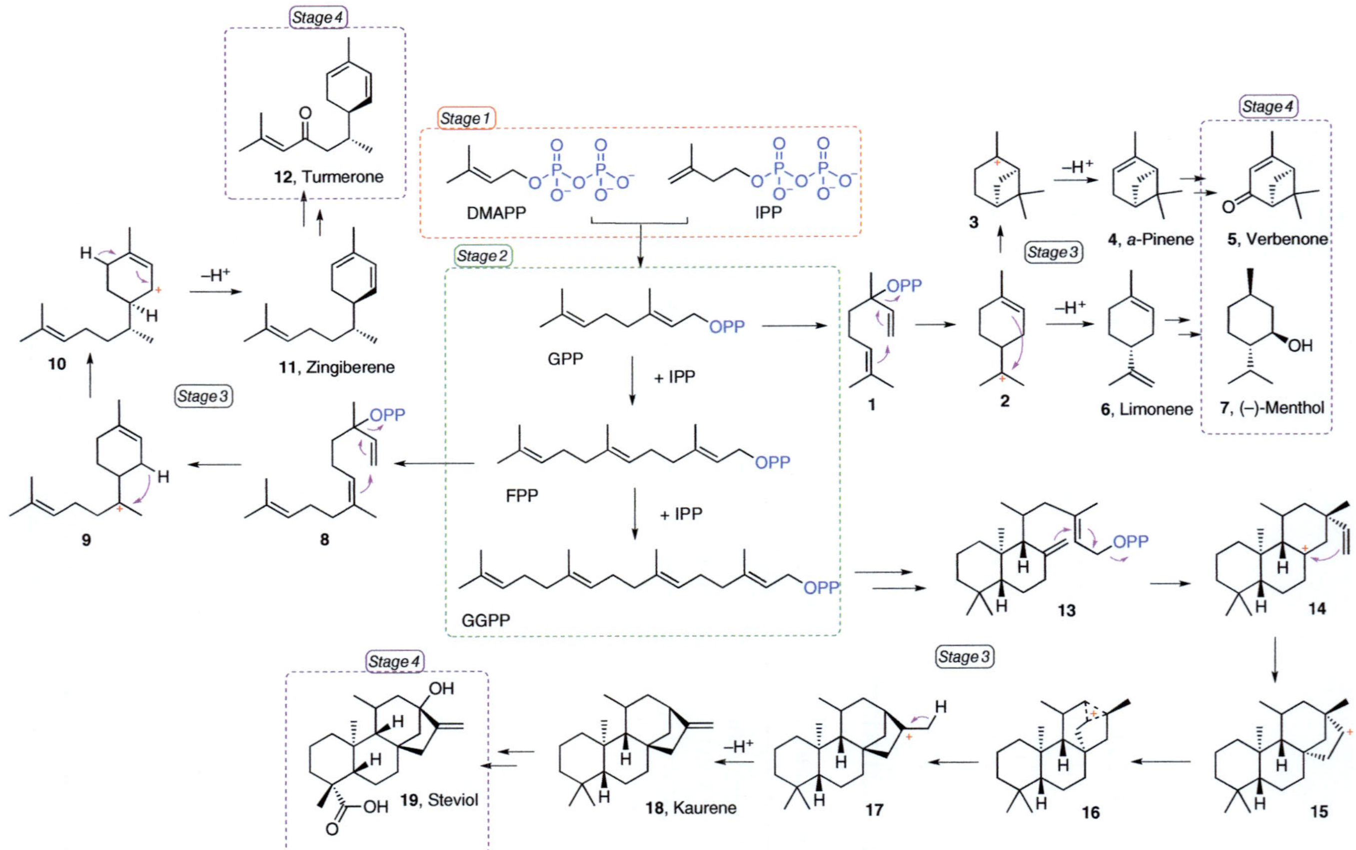

Scheme 8.1 Biosynthesis of terpenes shown on the example of verbenone **5**, menthol **7**, turmerone **12**, and steviol **19**.

the primary core by substitution or redox reactions that lead to the formation of alcohols **7**, ketones **5**, **12**, and carboxylic acids **19** as the final products.

The biosynthesis of isoprenoids, an example of a complex and unique total synthesis, presents the significant advantages of monoterpenes as precursors in constructing more complex natural products – structural diversity, feasibility of various functionalizations, and stereocontrol through all reaction steps. High abundance and general commercial availability also justify their broad applicability. Monoterpenes are volatile and easily extracted from natural sources by steam distillation. Turpentine oil from pine trees, also a bio-waste from the paper industry, is the most crucial source. The isoprenoid composition of turpentine depends on its origin, but the main ingredients are α- and β-pinene, which can be isomerized to camphene, 3-carene, or limonene. The production of citrus juice is another industrial source of terpene-containing bio-wastes. *R*-Limonene, which can be further oxidized to isocarveol, carvone, and limonene oxide, can be recycled from orange peel by extraction. Terpene feedstocks are easily accessible and, once isolated, can be readily transformed to one another, making them a cheap and diversified source of chiral building blocks for the total synthesis of natural products [14]. Many regio- and stereoisomers possessing various functional groups can be easily modified and further incorporated into the structure of the final product.

Several reviews devoted to the applicability of isoprenoids in the total synthesis have been published [15–19]. This chapter will summarize specific transformations of solely monoterpene scaffolds in the total syntheses of natural products presented in the last 10 years. The review will be divided according to the used isoprenoid framework as acyclic, monocyclic, and bicyclic monoterpene substrates.

8.2 Acyclic Monoterpene Building Blocks

Acyclic monoterpenes are formed from two isoprene units that create a linear hydrocarbon molecule based on the 2,6-dimethyloctane framework. Chiral pool terpenes from this isoprenoid group that has been used in the synthesis of natural products include derivatives with various functional groups – alkenes (*R*-(+)-citronellene), alcohols (*S*-(−)-citronellol and (*R*)-(−)-linalool), and aldehydes (*S*-(−)-citronellal) (Figure 8.1).

Fortunately, (*S*)-(−)-citronellol and (*R*)-(−)-linalool are inexpensive and commercially available. In addition, (*S*)-(−)-citronellol can be easily oxidized to the corresponding aldehyde – *S*-(−)-citronellal. The most recent total syntheses with the utility of these monoterpenes as building blocks are presented.

In 2015, (*R*)-(−)-linalool **20** was used as a starting material in the synthesis of (+)-artemone **22**, a tetrahydrofuran-based sesquiterpene [20, 21], and Plinol A **25** [22], a cyclopentanoid precursor of several natural products. In both cases, the presence of Δ1,2, Δ6,7 double bonds enabled an oxidative transformation of monoterpene **20**. Under microwave irradiation, allylic oxidation with SeO_2/*t*BuOOH system leads to the aldehyde **21** in 52% yield. The synthesis of Plinol A involved

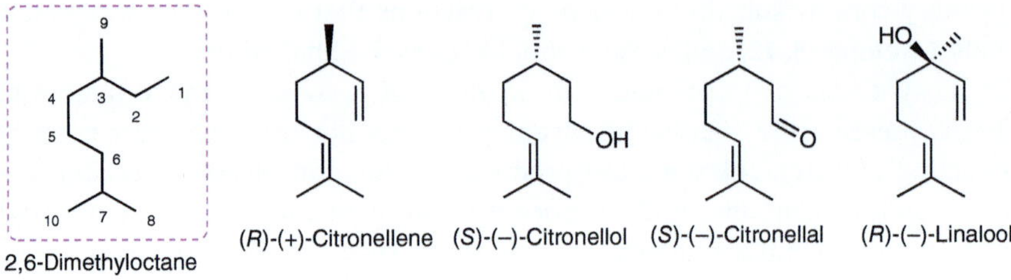

Figure 8.1 Acyclic monoterpenes most frequently used as chiral pool.

a Sharpless-type epoxidation with vanadyl acetylacetonate and *t*BuOOH, which furnished an equimolar mixture of two diastereomers **23**, directly used for further transformation. Next, an intramolecular radical cyclization of **23**, catalyzed by Ti(III) species, yielded hydroxyplinol **24**, further reduced to Plinol A **25**. Finally, compound **25** served as a substrate for the synthesis of (−)-cyclonerodiol **26**, chokol G **27**, and piperitone **28** [22] (Scheme 8.2).

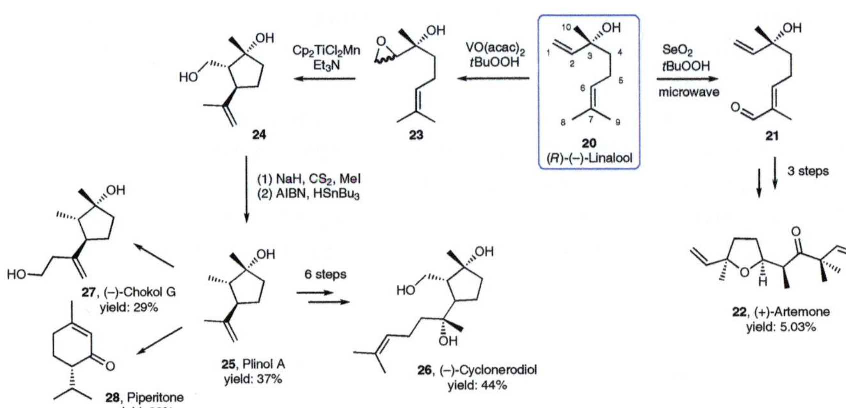

Scheme 8.2 Synthesis of (+)-artemone **22** and (−)-cyclonerodiol **26**, chokol G **27** and piperitone **28** starting from (*R*)-(−)-linalool **20**.

(*R*)-(−)-Linalool also served as a chiral building block in the total synthesis of ophiobolins, 5,8,5-fused polycyclic sesterterpenes, currently evaluated as anticancer agents [23, 24]. A ring-closing metathesis of isoprenoid **20**, followed by silylation (alkene **29**) and ruthenium-catalyzed allylic oxidation, afforded a cyclic enone **30**. Next, derivative **30** was coupled with cyclopropane iodides **34** and **39**. Compounds **33** and **38** were obtained by zinc-mediated cyclopropanation in the presence of enantiopure dioxaborolane ligand **32** and further transformation to iodides by an Appel reaction (iodide **34**) or mesylation and NaI substitution (iodide **39**). Depending on the used terpene, farnesol **31** or geraniol **37**, different ophiobolins could be obtained, making the linalool-derived enanone **30** a versatile sesterpene precursor (Scheme 8.3).

In acyclic monoterpenes, (*S*)-(−)-citronellol **43** also holds an important place in the terpene chiral pool. In the total synthesis of cytotoxic macrolactone

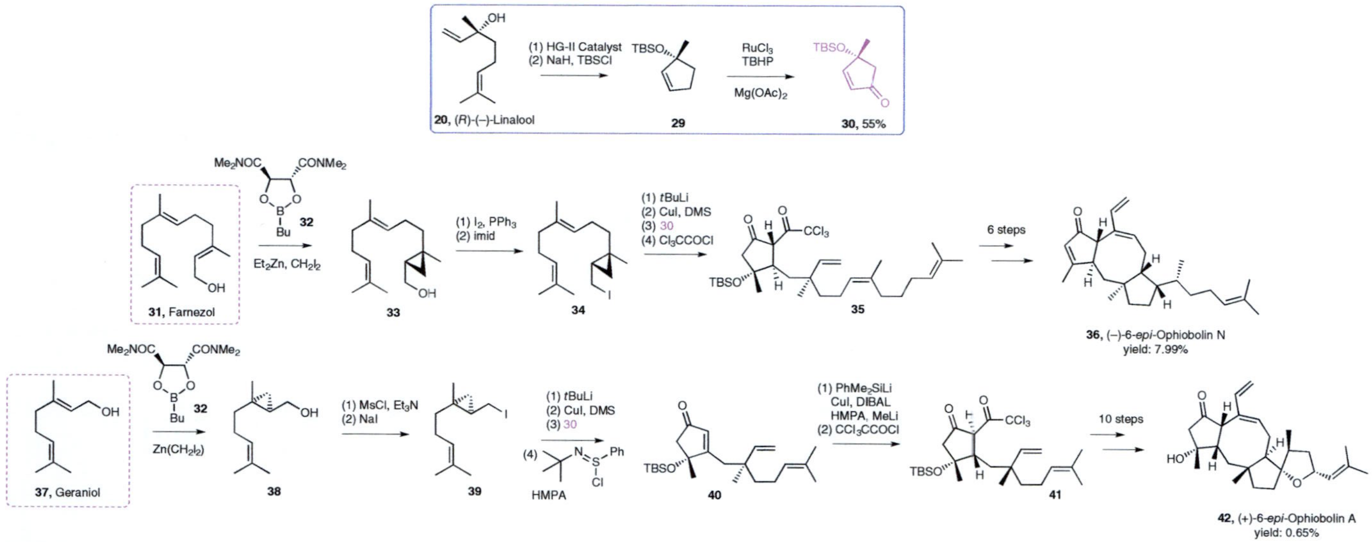

Scheme 8.3 Synthesis of ophiobolins A and N.

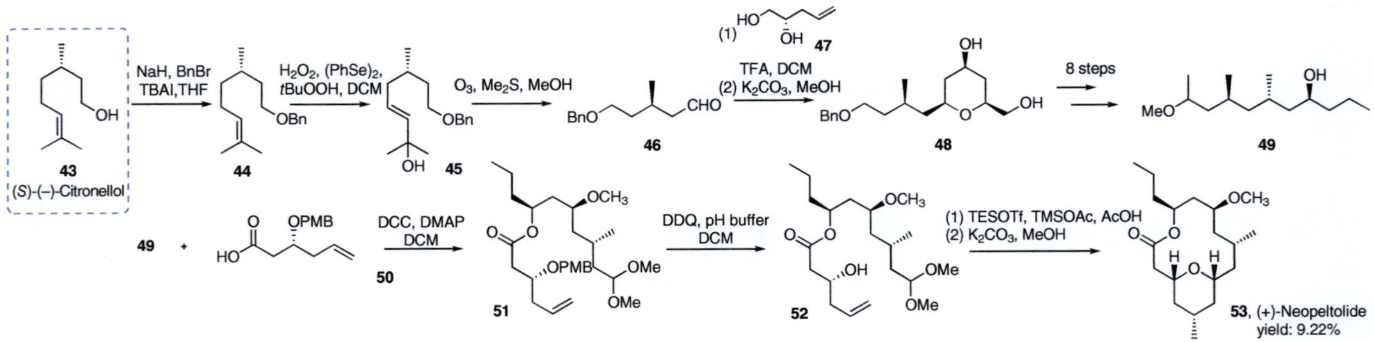

Scheme 8.4 Synthesis of (+)-neopeltolide **53** employing citronellol **43**.

(+)-neopeltolide **53** [25], compound **43** was utilized to obtain the chiral intermediate alcohol **49**. First, (*S*)-(−)-citronellol was protected with a benzyl group and transformed into alcohol **45** through PhSeOH/*t*BuOOH allylic oxidation. Further ozonation and reduction with Me_2S produced aldehyde **46**. In TFA, an intermolecular Prins cyclization of compound **46** with diol **47** resulted in pyran **48**, which later gave the building block **49**. The DCC-coupling of **49** with carboxylic acid **50** led to compound **51**. Deprotection with DDQ yielded alcohol **52**, which was subjected to a second Prins cyclization to give the final macrolide **53** (Scheme 8.4).

(*S*)-(−)-Citronellol **43** and the corresponding alkene (*R*)-(+)-citronellene **57** were also used to form chiral-substituted cyclopentane building blocks **54** and **58** that were installed in the structure of complex sesterterpenes leucosceptroid A **56**, B **55** [26], and nitidasin **59** [27] (Scheme 8.5).

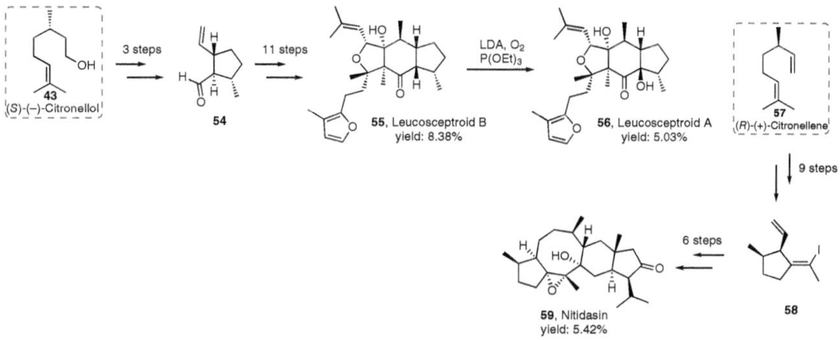

Scheme 8.5 Citronellol and citronellene as starting material in the synthesis of sesterterpenes **55**, **56**, and **59**.

In Mulzer's total synthesis of kendomycin, a macrocyclic anticancer drug, (*R*)-(+)-citronellene **57**, was incorporated to define two stereogenic centers of the final product [28]. The presence of two double bonds enabled the transformation of compound **57** to a benzofuran scaffold, an essential component of the synthesized macrocycle.

Aldehyde (*S*)-(−)-citronellal **60**, from the same subgroup of monoterpenes, served as starting material in the total synthesis of pseudopteroxazole [29], diterpene isolated from marine corals with tuberculostatic activity. The intermediate **64** had to be obtained to form the tricyclic core of the final product. Two alternative protocols commenced from the aforementioned citronellal **60** or the monocyclic terpene (*S*)-(+)-carvone **65**. In both cases, isolation of the keto-alcohol **63** as a single diastereoisomer was essential. This could not be accomplished when (*S*)-(+)-carvone **65** was used as the chiral building block. The isomers **63**/**63a** were inseparable. The enantiopure procedure started from the ZnB_2-mediated cyclization affording a mixture of diastereomeric alcohols **60a** and **60b**. Conveniently, compound **60a** could be inverted to the desirable isomer **60b**. Next, double bond hydroboration, TBSCl protection, and PCC-oxidation gave the ketone **61**, which was further transformed to the α,β-unsaurated ester **62**. Finally, compound **62** was transformed to a Weinreb salt, then converted to a ketone with MeMgI and desilylated with TBAF to afford the keto-alcohol **63**, which yielded the key intermediate **64** in the pseudoptheroxazole synthesis after six steps (Scheme 8.6).

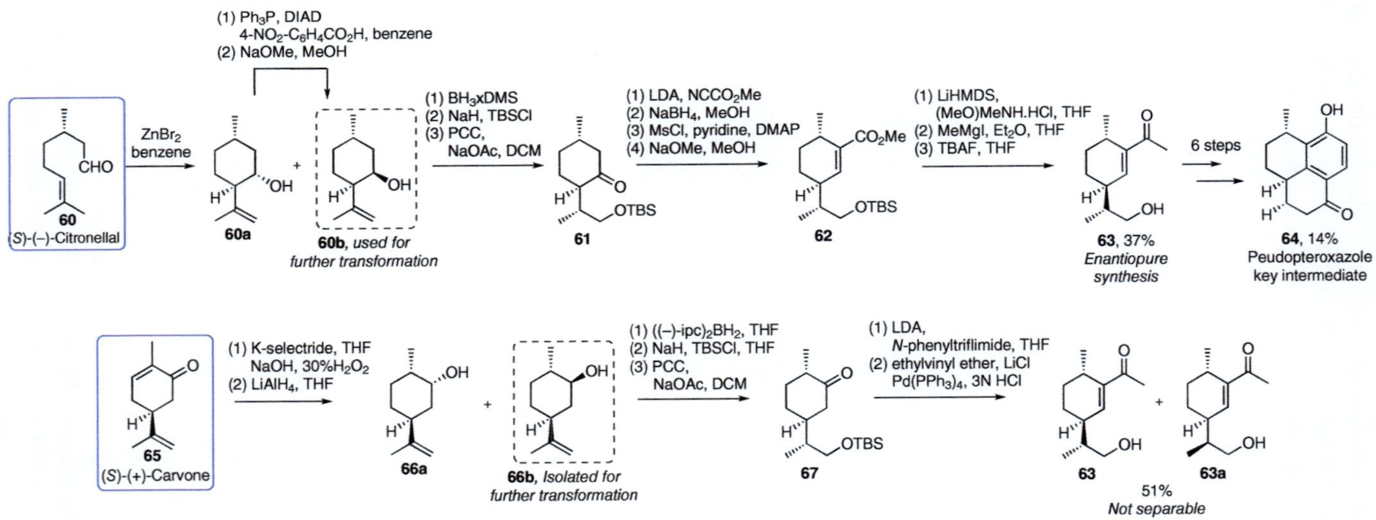

Scheme 8.6 Synthesis of the tricyclic core **64** of the pseudopteroxazole.

Additionally, the chiral pool aldehyde (*R*)-(+)-citronellal was utilized in the total synthesis of two sesquiterpenoids: englerin a (from the bark of a South African plant *Phyllanthus engleri*) [30] and jiadifenolide (from the pericarp of *Illicium jiadifengp*) [31].

8.3 Monocyclic Terpene Building Blocks

The next group of monoterpenes used as natural chiral building blocks possess one ring and are the derivatives of the *p*-menthane system. Compounds include ketones ((*S*)-(+)-/(*R*)-(−)-carvone, (*S*)-(−)-/(*R*)-(+)-pulegone and (1*R*,4*S*)-(−)-menthone), alcohols ((*S*)-(−)-/(*R*)-(+)-perillyl alcohol and (1*R*,3*R*,4*S*)-(−)-/(1*S*,3*S*,4*R*)-(+)-isopulegol), alkenes ((*R*)-(+)-limonene), and epoxides ((*R*)-(−)-epoxycarvone and (*R*)-(+)-limonene oxide) with chiral centers located on the C1 or C4 carbon atoms of the *p*-menthane skeleton (Figure 8.2).

In a recent literature, carvone seems to be the most exploited, especially as the (*R*)-(−)-enantiomer. This ketone is the major constituent of the caraway seeds oil ((*S*)-(+)-enantiomer, 50–70%) and spearmint oil ((*R*)-(−)-enantiomer, 51%). Examples of natural products that have been obtained with the incorporation of (*S*)-(+)- or (*R*)-(−)-carvone backbone are presented in Scheme 8.7 [32–42]. Selected total syntheses will be described in detail below.

Lindel et al. used (*S*)-(+)-carvone **65** to construct the cubitane skeleton **80** of the diterpene cubitanoids family. This was achieved in the total synthesis of (+)-cubitene **85** [43], isolated from East African termites *Chlorolestes umbratus*. The route to the specific 12-membered ring included an interesting intramolecular SmI_2-induced cyclization of the carvone-derived intermediate **83**. The phosphate **83** synthesis commenced from aldol condensation of **65**-enolate with geranylaldehyde **81** affording enone **82**. Next, silylation of secondary –OH, followed by LiOH-ester hydrolysis and phosphate formation, yielded the cyclization precursor **83**. To eliminate side reactions like dimerization, the solution of SmI_2 was cooled to 0 °C, and compound **83** was slowly added overnight to keep its concentration low (Scheme 8.8).

Samarium iodide was also later applied to initiate a radical cascade that enabled the macrocyclic backbone of (+)-pleuromutilin [44], a fungal metabolite and currently an antibiotic drug. In the synthesis of (+)-cubitene, the SmI_2-promoted cyclization precursor was constructed starting from a 1-isopropyl-4-methylcyclohexane core, the commercially available dihydrocarvone.

Another application of (*S*)-(+)-enantiomer of carvone was the synthesis of *Abies* sequiterpenoids, including abiespiroside A **93** [45], obtained from endangered Chinese fir trees. The route included a Pd-catalyzed lactonization giving a diastereomeric mixture of lactones **91a**/**91b**, separated after the hydroxyl group protection with TBSCl. The major diastereoisomer **91a** was used as a precursor of oxaspirolactone terpenoids analogs to product **93**. In the procedure, the chiral pool compound **65** was first transformed to *trans*-diol **86**. In the catalytic cycle, the $PdCl_2L_2$ reacted with $SnCl_2$ to form the $PdCl(SnCl_3)L_2$ complex, which combined with compound

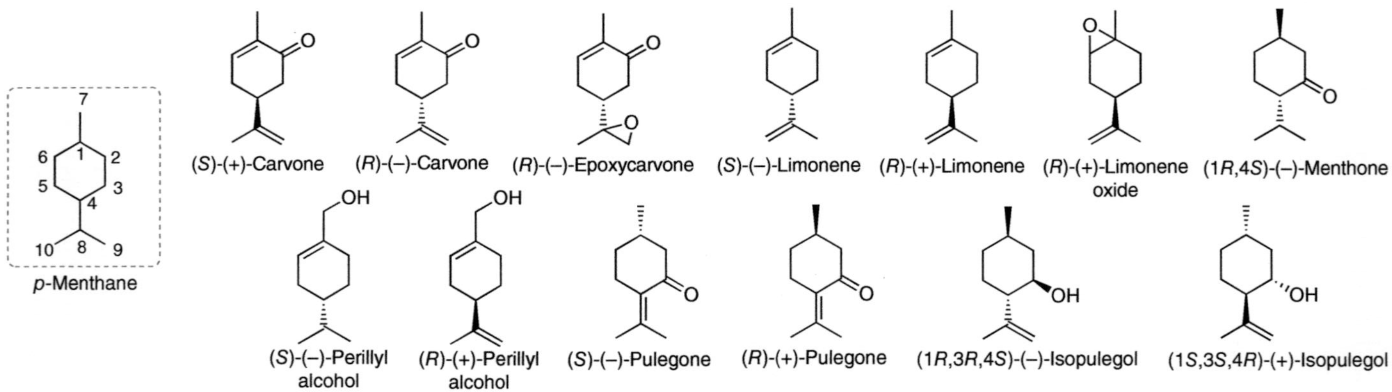

Figure 8.2 Monocyclic monoterpenes most frequently used as chiral pool.

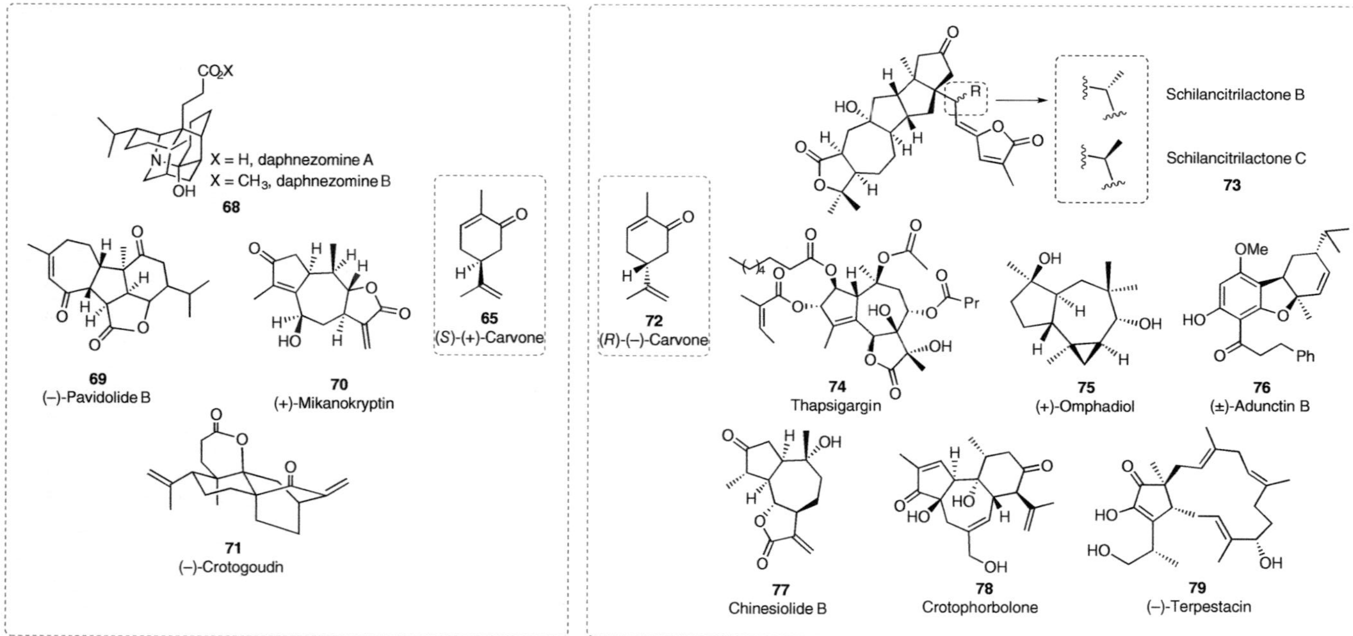

Scheme 8.7 Examples of natural products obtained using carvone-derived chiral building blocks.

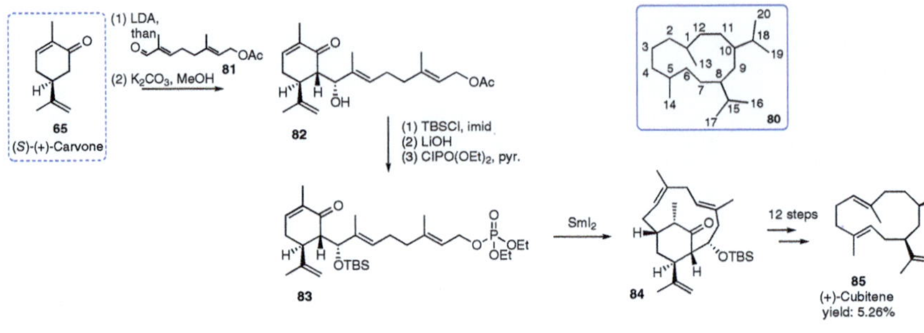

Scheme 8.8 Synthesis of (+)-cubitene **85** through the SmI_2-induced cyclization of carvone-derived precursor **83**.

86 formed the reactive $PdH(SnCl_3)L_2$ **92** and C^9-chlorinated **87** as the by-product. Coordination of the Pd-complex to the allylic alcohol **86** and the subsequent carbon monoxide migratory insertion leads to the formation of the metal species **90**, which after reductive elimination, gives the final lactone **91a/91b** (Scheme 8.9).

In 2012, (*R*)-(−)-carvone was used as starting material in the total synthesis of (+)-cyperolone **97** [46], sesquiterpenoid isolated from Japanese nutgrass. The procedure provided a universal method to obtain various bicyclic cyperane (**98**)-like compounds starting from chiral building block **72**. To construct the bicyclic core, (*R*)-(−)-carvone was first transformed to diol **94** through the formation of chlorohydrine and basic hydrolysis. Next, silylation of the 1° hydroxyl group, PCC-induced oxidative rearrangement, followed by addition of propargyl magnesium bromide and final silylation of 3° hydroxyl group, gave the 3-siloxy-1,5-enyne **95** as the key intermediate. Finally, a Pt-catalyzed cycloisomerization furnished the bicyclic compound **96**, which yielded the (+)-cycloperolone **97** after eight steps (Scheme 8.10).

Another interesting transformation of (*R*)-(−)-carvone **72**, primarily oxidized to epoxycarvone **99**, was presented by Sarpong and coworkers [47]. Using the procedure established by Bajero [48], compound **99** was first transformed by a Ti-induced radical cyclization to a bicyclic pinene diol **100**. Next, an efficient Pd-catalyzed cross-coupling of compound **100** with vinyl iodide **101**, through the cleavage of the cyclobutanol ring, gave the adduct **102**. Subsequent bromination and acylation of hydroxyl groups, followed by radical cyclization, yielded the [3.3.1]bicyclic product **103**, which was later reduced to alcohol **104**. The synthesis of compound **104** obtained xishacorene B **105**, diterpene isolated from Chinese coral with immunomodulating properties (Scheme 8.11).

Limonene, another member of the monocyclic terpene family, has also been used as a building block in several organic syntheses. Naturally occurring in the oil of *Citrus* fruits, it is known to exist in two enantiomeric forms: the (*S*)-(−)-enantiomer, a major component of the lemon essential oil, and (*R*)-(+)-isomer, contributing up to 95% of the orange peel composition.

In 2010, two synthetic protocols were published where the (*R*)-(+)-limonene scaffold was converted to an unsaturated carbonyl derivative that could undergo an intra- or intermolecular pericyclic reaction. To synthesize the guaiane sesquiterpene aciphyllene **111** (isolated in 1983 from the roots of *Lindera glauca*), Srikrishna and

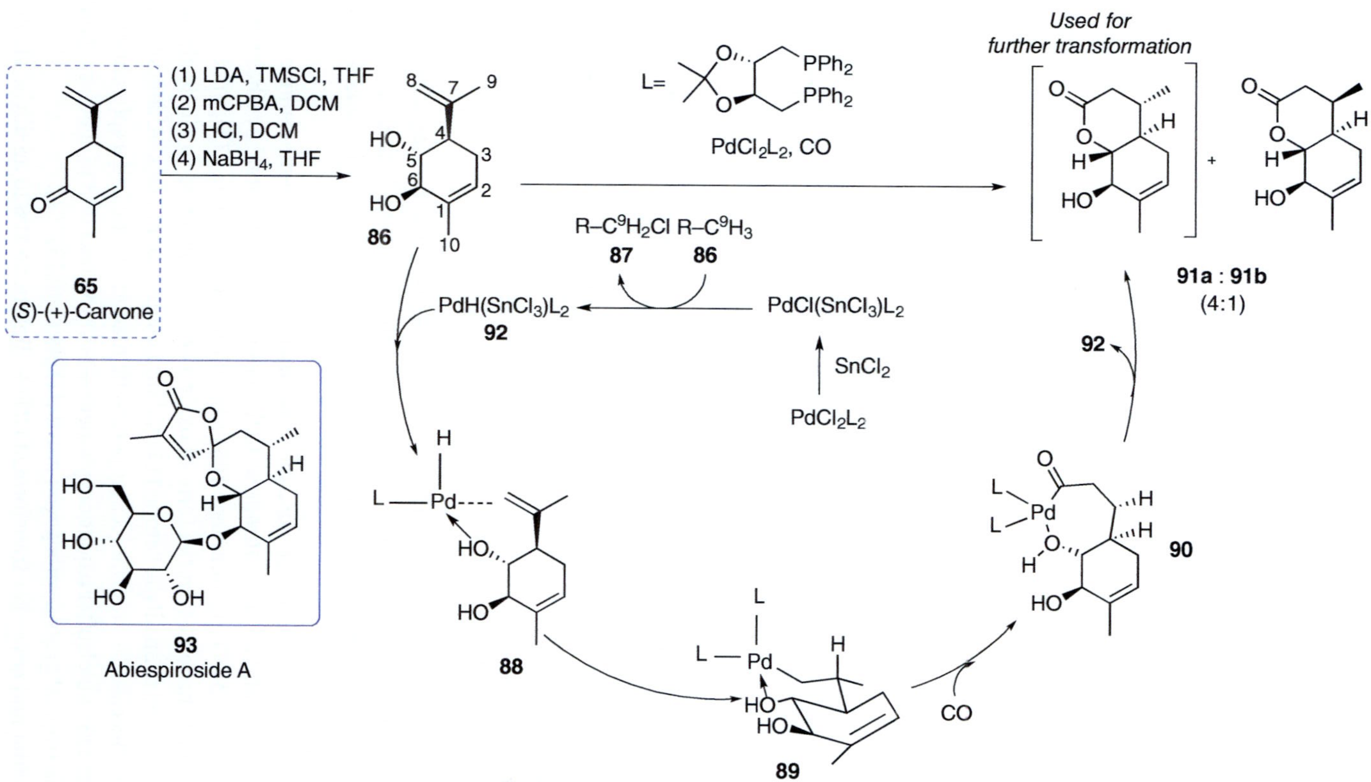

Scheme 8.9 Synthesis of precursor **91** as the precursor of *Abies* sesquiterpenoids.

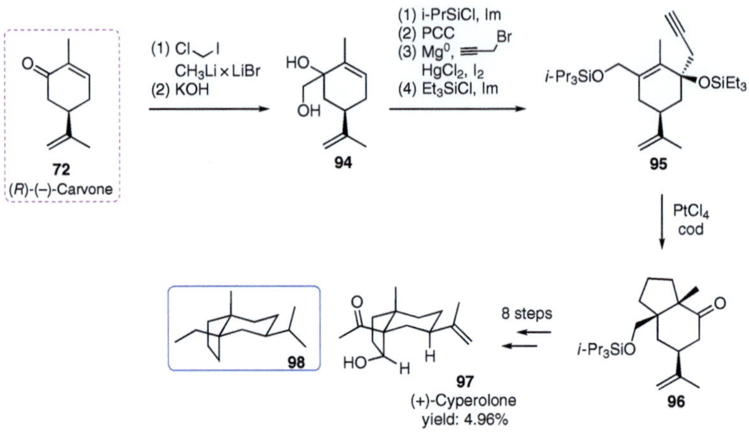

Scheme 8.10 Synthesis of (+)-cyperolone **97** utilizing (*R*)-(−)-carvone **72** as the starting material.

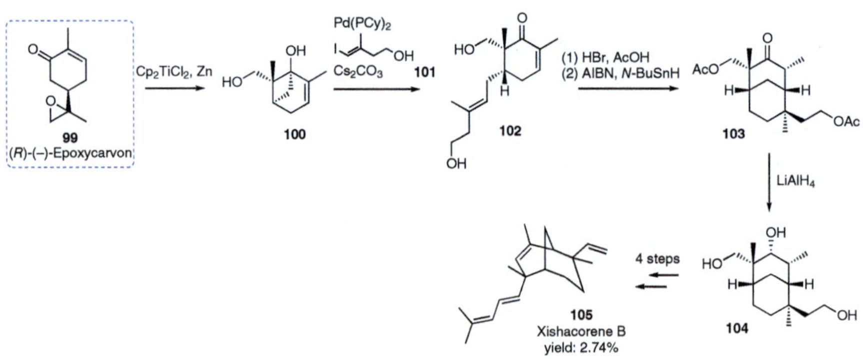

Scheme 8.11 Transformation of (*R*)-(−)-carvone epoxide **99** to [3.3.1]bicyclic compund **103**, precursor of xishacorene B.

coworker transformed limonene **106** to aldehyde **107** [49] and further converted it to a ketal derivative **108**. Treatment with boron trifluoride diethyl etherate initiated the intramolecular pericyclic ene reaction, yielding the final bicyclic backbone **110** [50] (Scheme 8.12).

In 2012 and 2020, other guaiane sesquiterpenoides, (+)-indicanone [51], (−)-rotundone, and (−)-*epi*-rotundone [52] were obtained by a total synthesis that commenced from (*R*)-(+)-limonene **106** as the chiral building block (Figure 8.3).

(*R*)-(+)-Limonene **106**, in the form of oxide **112**, was converted to allylic alcohol **113** and oxidized to an allylic ketone **114** enabling a subsequent Diels–Alder cycloaddition. In this case, an intermolecular pericyclic reaction furnished a spirochroman precursor **115**, which after oxidation, yielded (+)-cymbodiacetal **116**, a natural product of the tropical grass *Cymbopogon martini* (Scheme 8.13) [53].

Srikrishna and coworker transformed (*R*)-(+)-limonene **106** to an allylic alcohol **117**, which was used as a substrate in the total synthesis of linear type triquinane cucumin H **121** [54] and angular triquinane sesquiterpene silphiperfolan-6-ol **122** [55]. Terpene **106** was subjected to ozonolysis to form a ketoaldehyde. In piperidine

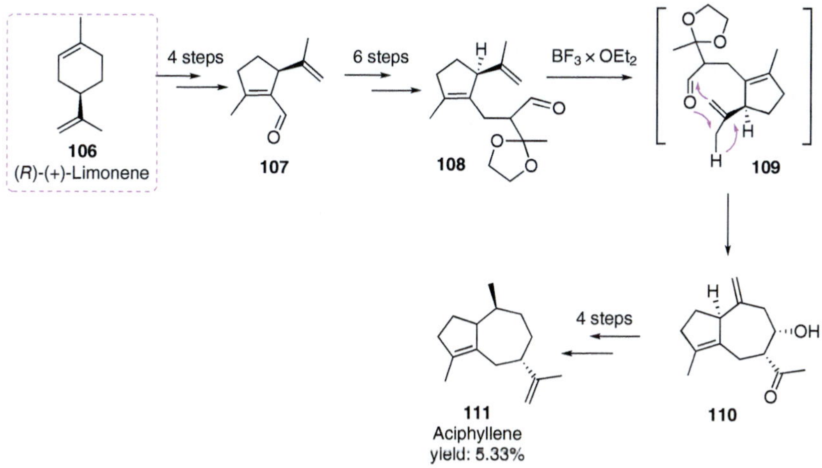

Scheme 8.12 Synthesis of aciphyllene **111** based on the (*R*)-(+)-limonene **106** chiral backbone.

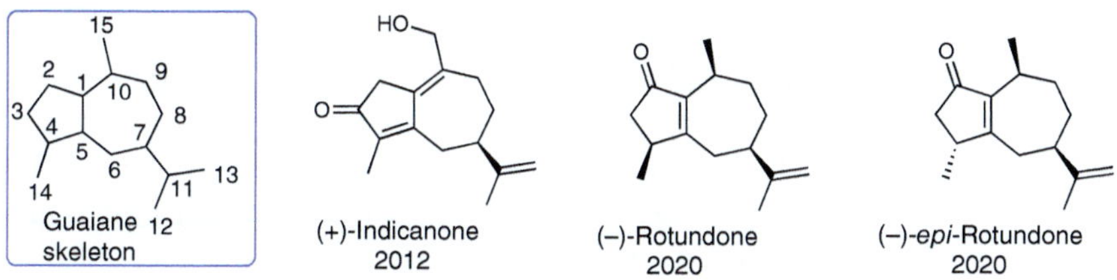

Figure 8.3 Guaiane-type sesquiterpenoides obtained from (*R*)-(+)-limonene **106**.

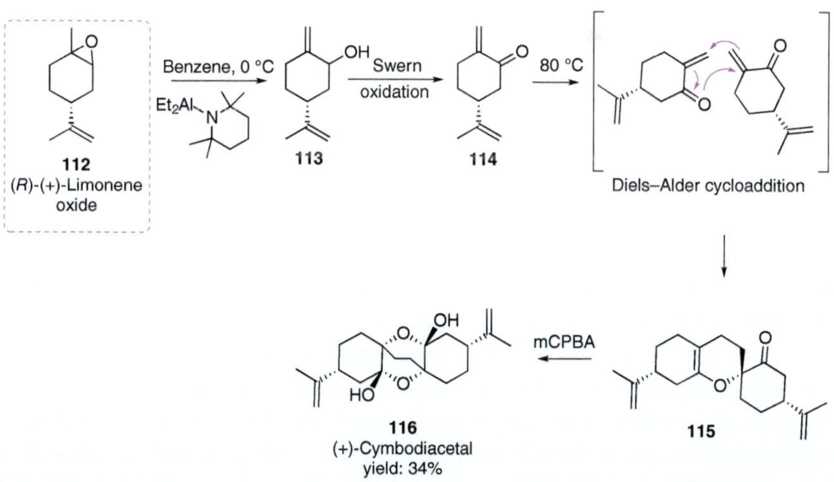

Scheme 8.13 Diels–Alder cycloaddition of limonene-derived ketone **114** leading to (+)-cymbodiacetal **116**. Source: Uroos et al. [53]/American Chemical Society.

and acetic acid, it underwent an intramolecular aldol condensation to a cyclopentane aldehyde, which after $NaBH_4/CeCl_3 \times 6H_2O$ Luche reduction, furnished compound **117**. Further transformations to the unsaturated ester **118**, carboxylic acid **119**, and diazoketone **120** were the starting point for the 12-step synthesis of cucumin H **121** and silphiperfolan-6-ol **122** (Scheme 8.14).

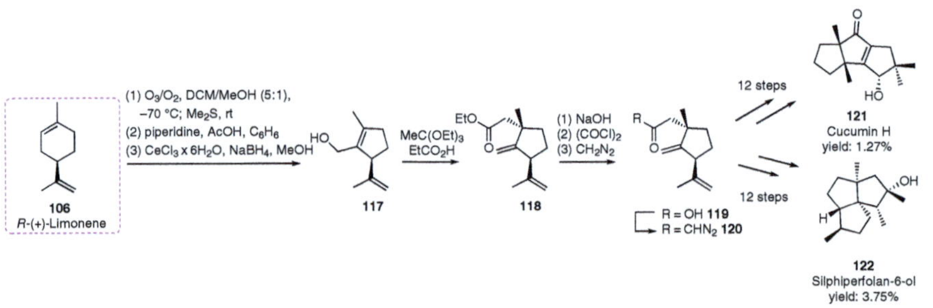

Scheme 8.14 Limonene-derived allylic alcohol **117** as a precursor of sesquiterpenes **121** and **122** obtained **132**.

As the (*S*)-(−)-enantiomer **123**, limonene was efficiently utilized by Pitsinos et al. in the total synthesis of (−)-laurenditerpenol **125** [56], a cytotoxic agent from marine alga (*Laurencia intricate*). Monoterpene **123** was transformed to lactone **124**, responsible for the intense sweet smell of several wines. The procedure was previously established by Guth in 1996 [57]. Later on, the precursor **124** was subjected to nine reaction steps affording the final natural product **125** (Scheme 8.15).

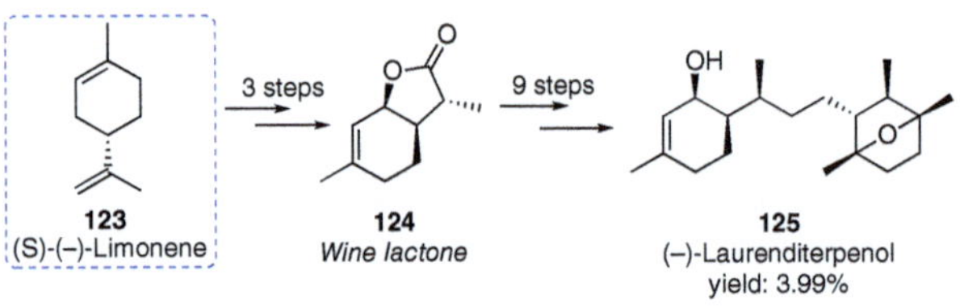

Scheme 8.15 Transformation of (*S*)-(−)-limonene **123** to wine lactone **124** and (−)-laurenditerpenol **125**.

Through the mevalonic pathway of terpene biosynthesis, limonene is transformed to perillyl alcohol, the component of essential oils of several plants, including lavender, cherries, and spearmint. In the total synthesis of natural products, it has been used as a building block in both enantiomeric forms. In 2016, Li's procedure to obtain rubiflordilactone B **128**, a seven-membered ring triterpenoid from the Schisandraceae family, was to apply (*S*)-(−)-enantiomer **126** to build the cycloheptane ring, which served as the core to construct the complete polycyclic structure of compound **128** [58]. To achieve this, terpene **126** was treated with Zn(II) iodide and triethylphosphite to form a phosphonate ester. Next, ozonolysis to a keto-aldehyde and subsequent Horner–Wadsworth–Emmons olefination in potassium carbonate furnished the enone **127** (Scheme 8.16).

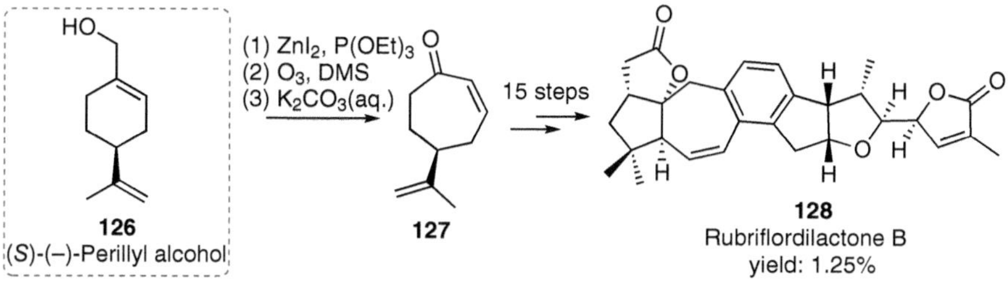

Scheme 8.16 (*S*)-(−)-perillyl alcohol **126** as starting material in the synthesis of rubriflordilactone B **128**.

The (*R*)-(+)-perillyl alcohol **129** was used in 2017 by Clark and coworker to synthesize furanocembranoid 7-*epi*-pukalide **133** [59]. Terpene **129** was first converted to the β-ketoester **130**, coupling partner of aldehyde **131**, to obtain epoxyfuran **132**. In tetrahydrothiophene, a rapid furan formation, through a sulfur ylide, yielded **132** by a one-pot Knoveagel condensation. A subsequent ring-closing metathesis furnished the final product **133** (Scheme 8.17).

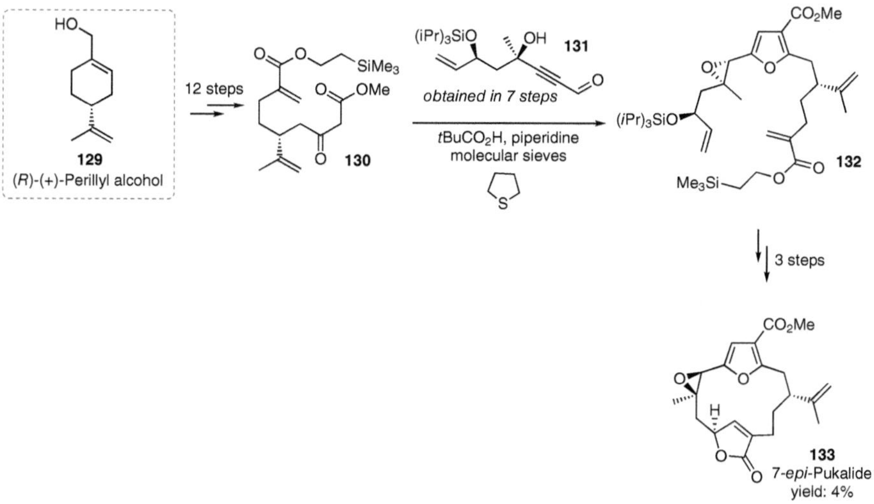

Scheme 8.17 Utility of (*R*)-(+)-perillyl alcohol-derived β-ketoester **130** in the one-pot Knovenagel condensation.

Later, pulegone, a menthane-type ketone, was applied to synthesize sesquiterpene jiadifenolide **138** and diterpene (+)-ryanodol **145**. Compound **138** was also obtained by Schenvi's group starting from citronellal [31]. In this procedure, pulegone as the (*R*)-(+)-isomer was transformed to a five-membered ring precursor, which was further used to obtain **138** by two protocols: through an 18-step reaction by Sorrensen in 2014 [60] and 13-step by Zhang and coworkers in 2015 [61]. In both procedures, (*R*)-(+)-pulegone was first converted to a cyclopentane α-ketoester **137** (Scheme 8.18).

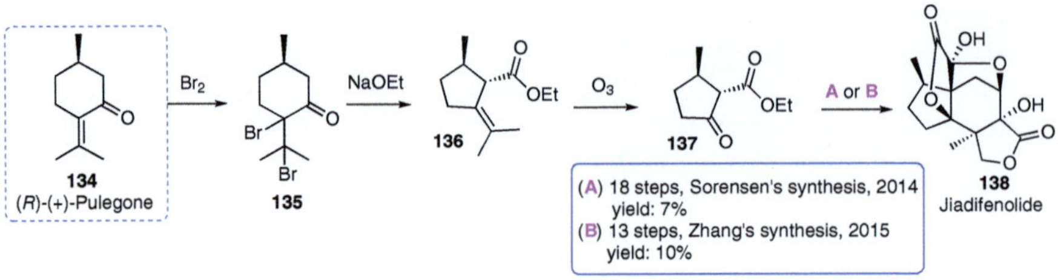

Scheme 8.18 (*R*)-(+)-Pulegone **134** as substrate in the synthesis of jiadifenolide **138**.

(*S*)-(−)-Enantiomer of pulegone **139** begins the synthesis of a highly oxidized compound (+)-ryanodol **145** [62]. First, oxidation to diol and further protection as benzyloxymethyl ethers led to compound **140**. Subsequent multiple functionalizations by Grignard reagents furnished enyne **143**. Using a ruthenium catalyst, Pauson–Khand cyclization constructed the final backbone, and after the additional seven-step reaction yielded the diterpene **145** (Scheme 8.19).

A series of total syntheses were also performed with the C3-functionalized *p*-menthane-type alcohol (1*R*,3*R*,4*S*)-(−)-isopulegol **146** as a chiral building block. To synthesize (−)-englerin A **151**, initially obtained starting from citronellal (similar to jiadifenolide **138**), a different approach was designed, beginning with (1*R*,3*R*,4*S*)-(−)-isopulegol **146** [63]. Through the cleavage of the cyclohexyl ring and a re-closing with a Pd-catalyst, a cyclopentane aldehyde **147** was obtained. Further addition of α-bromoester and ring-closing metathesis with a second-generation Grubbs catalyst furnished the guaiane core **150**, that was used to build-up the final structure of (−)-englerin A **151** (Scheme 8.20).

In 2013, the (1*R*,3*R*,4*S*)-(−)-isopulegol structure **146** was efficiently modified to construct leubethanol **152** through 13 reaction steps. This diterpene was isolated from the bark of *Leucophyllum frutescens*, used in traditional Mexican medicine, and belongs to the family of serrulatanes possessing significant antimicrobial properties. The use of monoterpene **139** provided two stereocenters at C1 and C4 of the final natural product **152** [64, 65] (Scheme 8.21).

Similarly, the stereocontrol of the synthesis of three benzoxazole alkaloids (+)-*sec*-pseudopteroxazole **158**, (+)-pseudopteroxazole **159**, and (+)-ileabethoxazole **160** was achieved using (1*R*,3*R*,4*S*)-(−)-isopulegol **146** as a starting material [66]. Also, the stereochemistry at the C1 and C4 of the monoterpene skeleton was transferred to the structure of the final products. The procedure commenced from Mitsunobu inversion at C3, followed by silylation (alkyne **153**), Dess–Martin oxidation, and triflation to afford triflate **154**. Next, an efficient Stille coupling in $Pd(PPh_3)_4$ formed two new C—C bonds (compound **156**). Finally, direct heating induced an aromatization sequence to create the benzoxazole precursor **157** (Scheme 8.22).

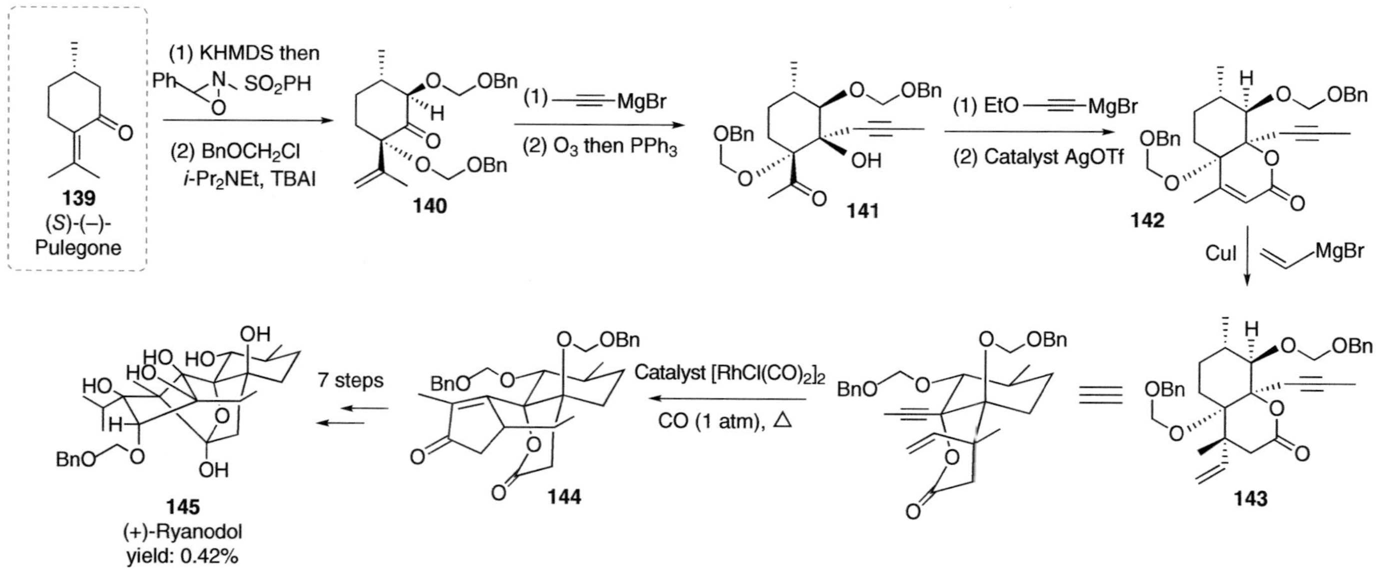

Scheme 8.19 (*S*)-(−)-Pulegone **139** as substrate in the synthesis of (+)-ryanodol **145**.

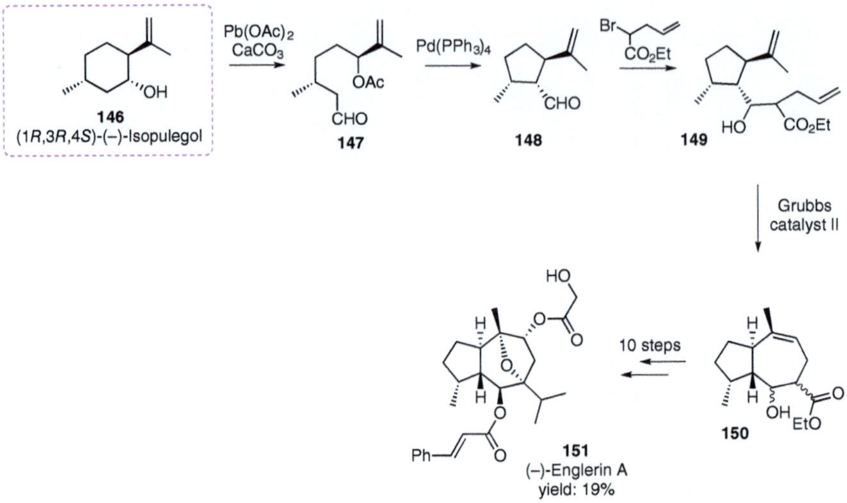

Scheme 8.20 Synthesis of (−)-englerin A **151** starting from (1*R*,3*R*,4*S*)-(−)-isopulegol **146**.

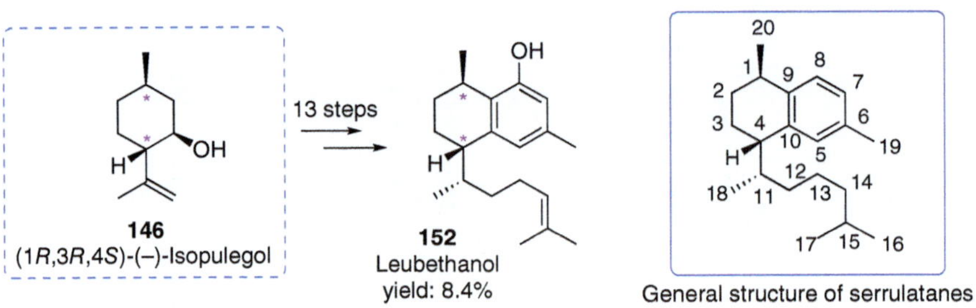

Scheme 8.21 (1*R*,3*R*,4*S*)-(−)-Isopulegol **139** as starting material in the synthesis of leubethanol **152**.

While using isopulegol produced several natural compounds, a similar monoterpene ketone, (1*R*,4*S*)-(−)-menthone, lacking an unsaturated C=C bond, was rarely used as a building block. One example is the synthesis of *ent*-nanolobalolide **166** performed by Chen and coworkers in 2011, where the cyclohexyl ring of the monoterpene is expanded to a cycloheptane-core [67]. In this case, silylation and cyclopropanation under Simmons–Smith conditions afforded compound **163**, rapidly transforming into a cycloheptanone **165** through $FeCl_3$ and sodium acetate treatment. After additional eight reaction steps, the precursor **165** was transformed to *ent*-nanolobalolide **166** (Scheme 8.23).

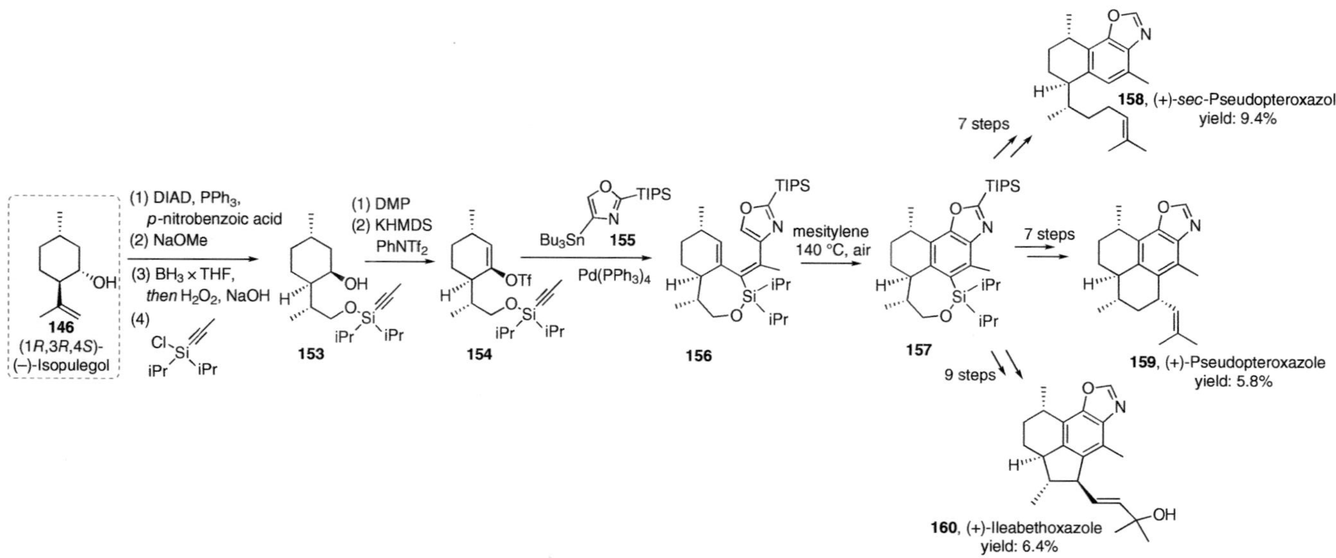

Scheme 8.22 (1*R*,3*R*,4*S*)-(–)-Isopulegol **146** transformation to an alkaloid benzoxazole precursor **157**.

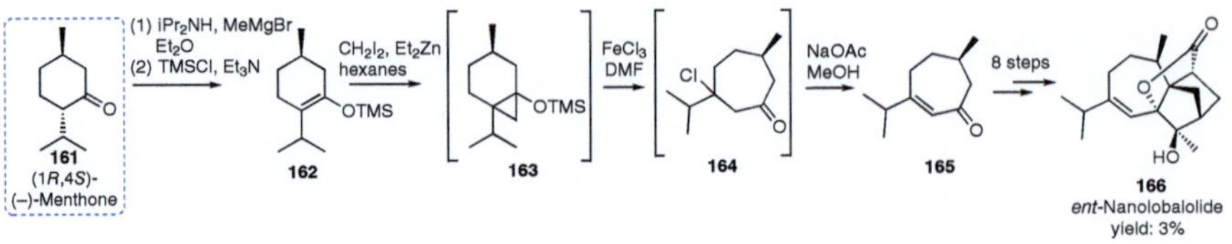

Scheme 8.23 Utility of (−)-menthone **161** as a chiral reactant.

8.4 Bicyclic Terpene Building Blocks

The final group of building blocks presented are the bicyclic monoterpenes based on different carbon skeletons: a carane bicyclo[4.1.0] – (1*S*,6*R*)-(+)-3-carene and (1*S*,6*R*)-(+)-2-carene; bornane bicyclo[2.2.1] – (1*S*,4*R*)-(+)-fenchone; and pinane bicyclo[3.1.1] – (1*R*,5*R*)-(+)-α-pinene, (1*S*,5*S*)-(−)-β-pinene, (1*R*,5*R*)-(+)-verbenone and (1*R*,5*S*)-(−)-myrtenal (Figure 8.4).

In the last 10 years, several total syntheses have been developed utilizing (1*S*,6*R*)-(+)-3-carene. In 2012, Baran and coworkers performed a 14-step synthesis of anticancer ingenane-type diterpene (+)-ingenol **173** starting from (1*S*,6*R*)-(+)-3-carene **167** [68, 69]. Terpene **167** was first transformed to chloroketone **168** by chlorination with NCS and subsequent ozonolysis. Next, treatment with lithium naphthalenide, HMPA, and methyl iodide, followed by adding LiHMDS and aldehyde **169**, forming aldol **170**. Addition of ethynylmagnesium bromide to the carbonyl and protection of both hydroxy groups furnished the silylated product **171**. A Pauson–Khand cyclization formed the dienone **172**. The additional seven reaction steps led to (+)-ingenol **173**. In 2016, the authors successfully used the same precursor **172** to design the total synthesis of tigliane-type diterpene (+)-phorbol **174** [70], an anticancer drug in phase II clinical trial for the treatment of acute myeloid leukemia. The utility of (1*S*,6*R*)-(+)-3-carene as starting material shortened the synthesis from 40–52 to only 19 steps (Scheme 8.24).

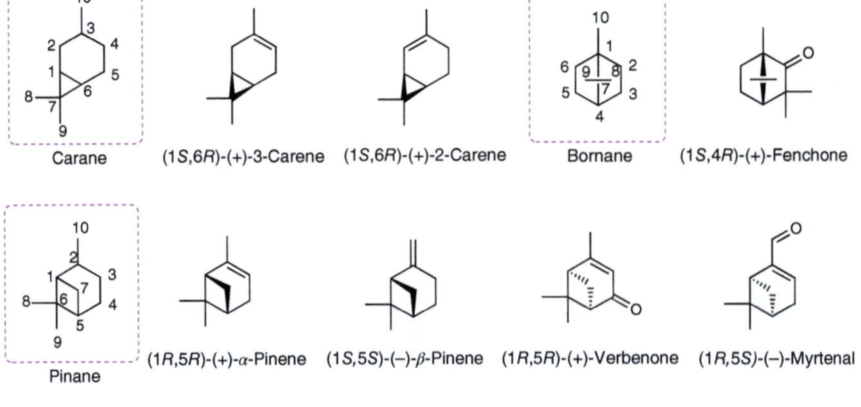

Figure 8.4 Bicyclic monoterpenes most frequently used as chiral pool.

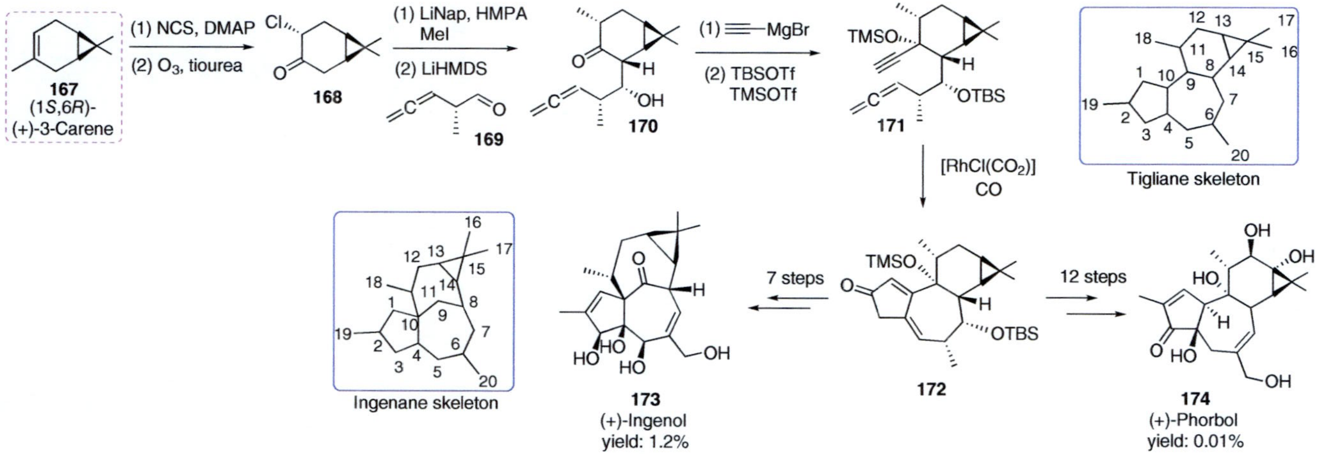

Scheme 8.24 Transformation of (1*S*,6*R*)-(+)-3-carene **167** to ingenane and tigliane diterpenes precursor **172**.

Regioisomer of **167** – (1*S*,6*R*)-(+)-2-carene **174** was also used as a chiral reactant. Williams et al. proposed a seven-step synthesis of *seco*-casbane **181**, a diterpene isolated from the Australian plant *Croton insularis* [71]. The casbane family of terpenes possesses a characteristic *gem*-dimethylcyclopropyl moiety. In the presented procedure, 2-carene was used to construct this particular cyclopropyl backbone with a fixed stereochemistry on both chiral centers. First, dihydroxylation with osmium tetroxide and oxidative cleavage led to ketoaldehyde **175**. Subsequent Wittig olefination with phosphorane **176** furnished the ketoester **177**. Next, **178** was attached to form enyne **179**. Finally, incorporation of γ-alkylidenebutenolide was achieved using iodoacrylic acid **180** to afford the *seco*-casbane **181** (Scheme 8.25).

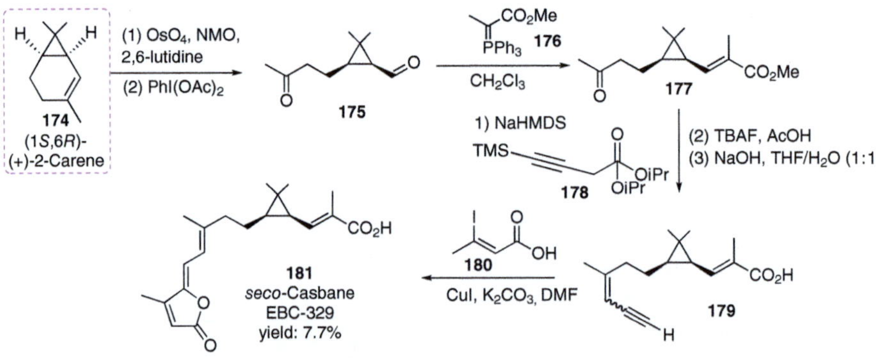

Scheme 8.25 Utility of (1*S*,6*R*)-(+)-2-carene **174** to construct a *gem*-dimethylcyclopropyl moiety of the *seco*-casbane **181**.

Among the bicyclic monoterpenes possessing a pinane skeleton, alkenes (1*R*,5*R*)-(+)-α-pinene **184** and (1*S*,5*S*)-(−)-β-pinene **185** were used to synthesize chromane-type meroterpenoids guadial B **186** and C **187** [72]. First, compound **183** was obtained from commercially available derivative **182** by the treatment with benzaldehyde and TMEDA that led to a methide **182b**, which after passing through silica and methanol, formed the final ether **183**. Then, heating of compound **183** directly with (1*R*,5*R*)-(+)-α- or (1*S*,5*S*)-(−)-β-pinene afforded the final natural products **186** and **187** (Scheme 8.26).

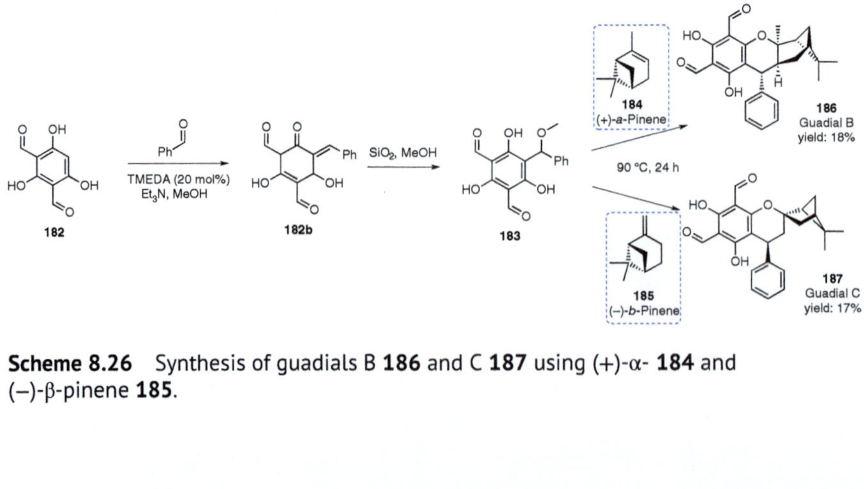

Scheme 8.26 Synthesis of guadials B **186** and C **187** using (+)-α- **184** and (−)-β-pinene **185**.

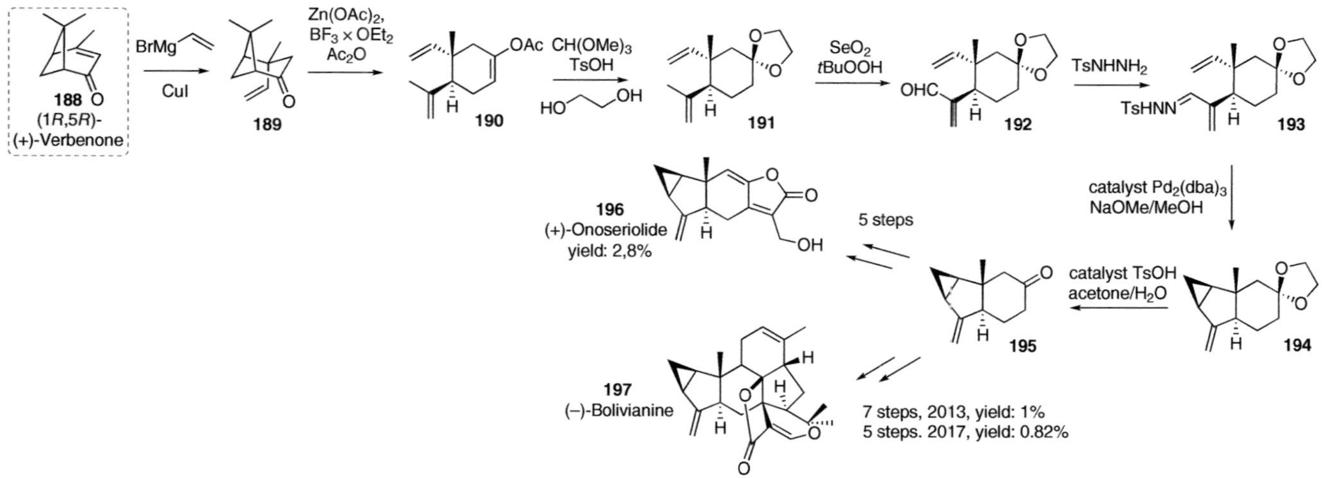

Scheme 8.27 Utility of (1*R*,5*R*)-(+)-verbenone **188** in the synthesis of (+)-onoseriolide **196** and (–)-bolivianine **197**.

Meanwhile, a pinane-derived ketone (1*R*,5*R*)-(+)-verbenone **188** was used by Liu et al. as a chiral building block in the total synthesis of terpenoids (+)-onoseriolide **196** and (−)-bolivianine **197**, from the Chloranthaceae plant family [73, 74]. The procedure commenced from the Cu-catalyzed addition of vinyl magnesium bromide to obtain compound **189** and cyclobutene ring-opening to afford acetate **190**. By ketal **191** formation and selenium dioxide allylic oxidation, aldehyde **192** was formed. Conversion to tosylhydrazone **193** followed by palladium-catalyzed cyclopropanation (**194**) and deketalization with TsOH (catalyst) furnished ketone **195**, a precursor of (+)-onoseriolide **196** (five steps) and (−)-bolivianine **197** (seven steps). In 2017, the authors improved the synthesis of **197**, shortening the process to five reaction steps starting from ketone **195** [75] (Scheme 8.27).

A different carbonyl pinane-derivative (1*R*,5*S*)-(−)-myrtenal **198** was used by Hu and Maimone in the synthesis of (+)-cardamon peroxide **203** [76]. The presence of an —O—O— bond arms the molecule with significant antimalarial activity. The synthetic procedure arose from the hypothesis that the biosynthesis of **203** is based on the coupling of two pinene and three oxygen molecules. As the pinane skeleton donor, (1*R*,5*S*)-(−)-myrtenal **198** first underwent dimerization to afford the triene **199**. Subsequent [4+2] cycloaddition reaction induced by O_2, treatment with DBU, and Dess–Martin periodinate oxidation led to enone **200**. Exposure to the catalytic amount of $Mn(dpm)_3$, oxygen, and $PhSiH_3$ formed the intermediate **201**. Peroxyradical cyclization, O_2-trapping, and PPh_3-reduction furnished the final peroxide **203** (Scheme 8.28).

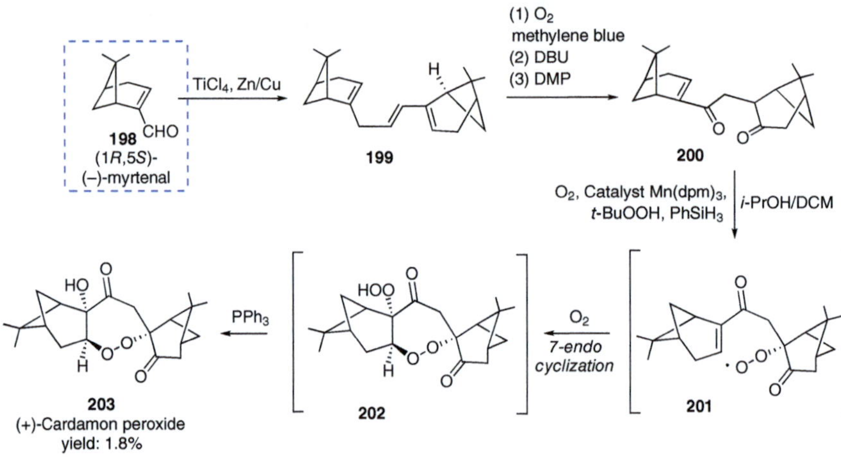

Scheme 8.28 Synthesis of (+)-cardamon peroxide **203** commenced from (1*R*,5*S*)-(−)-myrtenal **198**.

The last bicyclic monoterpenes presented are those derived from the bornane skeleton. The ketone (1*S*,4*R*)-(+)-fenchone **204** represents this group of chiral building blocks in the total synthesis of (−)-aplyviolene **210**, isolated from marine sponge *Chelonaplysilla violacea* [77]. In this procedure, the bicyclic terpene skeleton was transformed to a 5,7-fused ring system, the core of the final natural product. First, (1*S*,4*R*)-(+)-fenchone **204** was converted to oxime and subjected to Beckmann fragmentation affording nitrile **205**. Reduction to aldehyde and further Wittig olefination yielded alcohol **206**. Subsequent Appel reaction and silver nitrite treatment furnished nitroalkene **207**. Treatment with phenylisocyanate and triethylamine induced an intramolecular cycloaddition to form isoxazoline **208**, which was reduced to ketoalcohol **209** bearing the 5,7-fused ring system building up the backbone of (−)-aplyviolene **210** (Scheme 8.29).

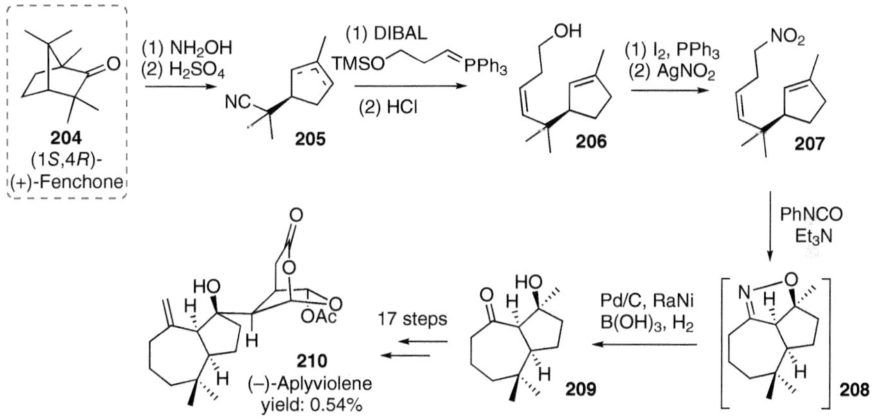

Scheme 8.29 Transformation of (1*S*,4*R*)-(+)-fenchone **204** to a 5,7-fused ring precursor of (−)-aplyviolene **210**.

8.5 Conclusions

The presence of double bonds and oxidized carbon atoms (hydroxy, epoxy, and carbonyl groups) in the structurally and stereochemically diversified monoterpene skeletons enables variously functionalizing the primary structure and expanding it to a highly complex molecule with precisely defined chiral centers. The presented review highlighted that monoterpenes could be transformed by allylic oxidation, epoxidation, intra- or intermolecular cyclization, ring-closing metathesis, ozonolysis, and reduction to various aldehydes, Wittig olefination, functionalization with Grignard reagents, ring expansion, or contraction to form cyclopentanes/cyclopropanes. In many cases, the total synthesis from terpenes is the fastest and most efficient route to obtain a natural product. Monoterpenes held their position in modern organic syntheses as cheap and versatile building blocks and will presumably continue to inspire and resolve various synthetic problems.

References

1 Casiraghi, G., Zanardi, F., Rassu, G., and Spanu, P. (1995). Stereoselective approaches to bioactive carbohydrates and alkaloids—with a focus on recent syntheses drawing from the chiral pool. *Chem. Rev.* 95 (6): 1677–1716.

2 Blaser, H.-U. (1992). The chiral pool as a source of enantioselective catalysts and auxiliaries. *Chem. Rev.* 92 (5): 935–952.

3 Brill, Z.G., Condakes, M.L., Ting, C.P., and Maimone, T.J. (2017). Navigating the chiral pool in the total synthesis of complex terpene natural products. *Chem. Rev.* 117 (18): 11753–11795.

4 Jaegera, R. and Cuny, E. (2016). Terpenoids with special pharmacological significance: a review. *Nat. Prod. Commun.* 11 (9): 1373–1390.

5 Wedler, H.B., Pemberton, R.P., and Tantillo, D.J. (2015). Carbocations and the complex flavor and bouquet of wine: mechanistic aspects of terpene biosynthesis in wine grapes. *Molecules* 20 (6): 10781–10792.

6 Bohlmann, J., Meyer-Gauen, G., and Croteau, R. (1998). Plant terpenoid synthases: molecular biology and phylogenetic analysis. *Proc. Natl. Acad. Sci. U.S.A.* 95 (8): 4126–4133.

7 Lange, B.M. and Ahkami, A. (2013). Metabolic engineering of plant monoterpenes, sesquiterpenes and diterpenes—current status and future opportunities. *Plant Biotechnol. J.* 11 (2): 169–196.

8 Cho, K.S., Lim, Y., Lee, K. et al. (2017). Terpenes from forests and human health. *Toxicol. Res.* 33 (2): 97–106.

9 Christianson, D.W. (2008). Unearthing the roots of the terpenome. *Curr. Opin. Chem. Biol.* 12 (2): 141–150.

10 Rani, K. (1999). Cyclisation of farnesyl pyrophosphate into sesquiterpenoids in ginger rhizomes. *Fitoterapia* 70 (6): 568–574.

11 Bergman, M.E. and Phillips, M.P. (2021). Structural diversity and biosynthesis of plant derived *p*-menthane monoterpenes. *Phytochem. Rev.* 20 (2): 433–459.

12 Moon, J.H., Lee, K., Lee, J.H., and Lee, P.C. (2020). Redesign and reconstruction of a steviol-biosynthetic pathway for enhanced production of steviol in *Escherichia coli*. *Microb. Cell Fact.* 19 (1): 20–22.

13 Brandle, J.E. and Telmer, P.G. (2007). Steviol glycoside biosynthesis. *Phytochemistry* 68 (14): 1855–1863.

14 Rubulotta, G. and Quadrelli, E.A. (2019). Chapter 11: Terpenes: a valuable family of compounds for the production of fine chemicals. In: *Horizons in Sustainable Industrial Chemistry and Catalysis*, Studies in Surface Science and Catalysis, vol. 178 (ed. S. Albonetti, S. Perathoner and E.A. Quadrelli), 215–229. London UK: Elsevier.

15 Urabe, D., Asaba, T., and Inoue, M. (2015). Convergent strategies in total syntheses of complex terpenoids. *Chem. Rev.* 115 (17): 9207–9231.

16 Gaich, T. and Mulzer, J. (2012). Chiral pool synthesis. Starting from terpenes. In: *Comprehensive Chirality*, vol. 2 (ed. E.M. Carreira and H. Yamamoto), 163–206. London, UK: Elsevier.

17 Hanson, J.R. (2007). Steroids: partial synthesis in medicinal chemistry. *Nat. Prod. Rep.* 24 (6): 1342–1349.

18 Lednicer, D. (2011). *Steroid Chemistry at a Glance*. Hoboken, NJ, USA: Wiley.

19 Hung, K., Hu, X., and Maimone, T.J. (2018). Total synthesis of complex terpenoids employing radical cascade processes. *Nat. Prod. Rep.* 35 (2): 174–202.

20 Wan, K.K., Evans-Klock, C.D., Fielder, B.C., and Vosburg, D.A. (2013). Synthesis of *cis*- and *trans*-davanoids: artemone, hydroxydavanone, isodavanone, and nordavanone. *Synthesis* 45 (11): 1541–1545.

21 Nacsa, E.D., Fielder, B.C., Wetzler, S.P. et al. (2015). Biomimetic synthesis of (+)-artemone via a stereoselective, organocatalytic cyclization. *Synthesis* 47 (17): 2599–2602.

22 Morales, C.P., Herrador, M.M., Quílez del Moral, J.F., and Barrero, A.F. (2015). Expedient access to enantiopure cyclopentanic natural products: total synthesis of (−)-cyclonerodiol. *Nat. Prod. Commun.* 10 (1): 1–4.

23 Thach, D.Q., Brill, Z.G., Grover, H.K. et al. (2020). Total synthesis of (+)-6-*epi*-ophiobolin A. *Angew. Chem. Int. Ed.* 59 (4): 1532–1536.

24 Brill, Z.G., Grover, H.K., and Maimone, T.J. (2016). Enantioselective synthesis of an ophiobolin sesterterpene via a programmed radical cascade. *Science* 352 (6289): 1078–1082.

25 Yadav, J.S. and Narayana Kumar, G.G.K.S. (2010). A concise stereoselective formal total synthesis of the cytotoxic macrolide (þ)-neopeltolide via Prins cyclization. *Tetrahedron* 66 (2): 480–487.

26 Guo, S., Liu, J., and Ma, D. (2015). Total synthesis of leucosceptroids A and B. *Angew. Chem. Int. Ed.* 54 (4): 1298–1301.

27 Hog, D.T., Huber, F.M.E., Mayer, P., and Trauner, D. (2014). The total synthesis of (−)-nitidasin. *Angew. Chem. Int. Ed.* 53 (32): 8513–8517.

28 (a) Martin, H.J., Magauer, T., and Mulzer, J. (2010). In pursuit of a competitive target: total synthesis of the antibiotic kendomycin. *Angew. Chem. Int. Ed.* 49 (33): 5614–5626. (b) Magauer, T., Martin, H.J., and Mulzer, J. (2010). Ring-closing metathesis and photo-fries reaction for the construction of the ansamycin antibiotic kendomycin: development of a protecting group free oxidative endgame. *Chem. Eur. J.* 16 (2): 507–519.

29 Yadav, J.S., Vijaya Bhasker, E., and Srihari, P. (2010). Synthesis of a key intermediate for the total synthesis of pseudopteroxazole. *Tetrahedron* 66 (11): 1997–2004.

30 Li, Z., Nakashige, M., and Chain, W.J. (2011). A brief synthesis of (−)-englerin A. *J. Am. Chem. Soc.* 133 (17): 6553–6556.

31 Lu, H.-H., Martinez, M.D., and Shenvi, R.A. (2015). An eight-step gram-scale synthesis of (−)-jiadifenolide. *Nat. Chem.* 7 (7): 604–607.

32 Zhang, P.-P., Yan, Z.-M., Li, Y.-H. et al. (2017). Enantioselective total synthesis of (−)-pavidolide B. *J. Am. Chem. Soc.* 139 (40): 13989–13992.

33 Dethe, D.H. and Dherange, B.D. (2018). Total synthesis of adunctin B. *J. Org. Chem.* 83 (6): 3392–3396.

34 Liu, G. and Romo, D. (2011). Total synthesis of (+)-omphadiol. *Angew. Chem. Int. Ed.* 50 (33): 7537–7540.

35 Asaba, T., Katoh, Y., Urabe, D., and Inoue, M. (2015). Total synthesis of crotophorbolone. *Angew. Chem. Int. Ed.* 54 (48): 14457–14461.

36 Hu, X., Xu, S., and Maimone, T.J. (2017). A double allylation strategy for gram-scale guaianolide production: total synthesis of (+)-mikanokryptin. *Angew. Chem. Int. Ed.* 56 (6): 1624–1628.

37 Jin, Y. and Qiu, F.G. (2012). A convergent stereocontrolled total synthesis of (−)-terpestacin. *Org. Biomol. Chem.* 10 (28): 5452–5455.

38 Wang, L., Wang, H., Li, Y., and Tang, P. (2015). Total synthesis of schilancitrilactones B and C. *Angew. Chem. Int. Ed.* 54 (19): 5732–5735.

39 Finkbeiner, P., Murai, K., Rö, M., and Sarpong, R. (2017). Total synthesis of terpenoids employing a "benzannulation of carvone" strategy: synthesis of (−)-crotogoudin. *J. Am. Chem. Soc.* 139 (33): 11349–11352.

40 Chen, D. and Evans, P.A. (2017). A concise, efficient and scalable total synthesis of thapsigargin and nortrilobolide from (*R*)-(−)-carvone. *J. Am. Chem. Soc.* 139 (17): 6046–6049.

41 Finkbeiner, P., Murai, K., Röpke, M., and Sarpong, R. (2020). Total synthesis of (−)-daphnezomines A and B. *J. Am. Chem. Soc.* 142 (36): 15240–15245.

42 Elford, T.G. and Hall, D.G. (2010). Total synthesis of (+)-chinensiolide B *via* tandem allylboration/lactonization. *J. Am. Chem. Soc.* 132 (5): 1488–1489.

43 Simon, K., Wefer, J., Schöttner, E., and Lindel, T. (2012). Enantioselective total synthesis of the diterpene (+)-cubitene. *Angew. Chem. Int. Ed.* 51 (43): 10889–10892.

44 Fazakerley, N.J., Helm, M.D., and Procter, D.J. (2013). Total synthesis of (+)-pleuromutilin. *Chem. Eur. J.* 19 (21): 6718–6723.

45 Davis, D.C., Hoch, D.G., Wu, L. et al. (2018). Total synthesis, biological evaluation, and target identification of rare *Abies* sesquiterpenoids. *J. Am. Chem. Soc.* 140 (50): 17465–17473.

46 Klahn, P., Duschek, A., Liebert, C., and Kirsch, S.F. (2012). Total synthesis of (þ)-cyperolone. *Org. Lett.* 14 (5): 1250–1253.

47 Kerschgens, I., Rovira, A.R., and Sarpong, R. (2018). Total synthesis of (−)-xishacorene B from (*R*)-carvone using a C–C activation strategy. *J. Am. Chem. Soc.* 140 (31): 9810–9813.

48 Bermejo, F.A., Mateos, A.F., Escribano, A.M. et al. (2006). Ti(III)-promoted cyclizations. Application to the synthesis of (*E*)-*endo*-bergamoten-12-oic acids. Moth oviposition stimulants isolated from *Lycopersicon hirsutum*. *Tetrahedron* 62 (38): 8933–8942.

49 Srikrishna, A. and Babu, N.C. (2001). An enantiospecific formal total synthesis of (−)-aplysin and (−)-debromoaplysin. *Tetrahedron Lett.* 42 (29): 4913–4914.

50 Srikrishna, A. and Pardeshi, V.H. (2010). Enantiospecific total synthesis of aciphyllene. *Tetrahedron* 66 (41): 8160–8168.

51 Hayashi, Y., Ogawa, K., Inagaki, F., and Mukai, C. (2012). First total synthesis of (+)-indicanone. *Org. Biomol. Chem.* 10 (24): 4747–4751.

52 Rüthia, F. and Schröde, F. (2020). Total synthesis of (−)-rotundone and (−)-*epi*-rotundone from monoterpene precursors. *Helv. Chim. Acta* 103 (11): e2000129.

53 Uroos, M., Lewis, W., Blake, A.J., and Hayes, C.J. (2010). Total synthesis of (+)-cymbodiacetal: a re-evaluation of the biomimetic route. *J. Org. Chem.* 75 (24): 8465–8470.

54 Srikrishna, A. and Dethe, D.H. (2011). Enantiospecific total synthesis of cucumin H. *Indian J. Org. Chem.* 50B (8): 1092–1106.

55 Srikrishn, A., Nagaraju, G., and Sheth, V.M. (2012). Enantiospecific first total synthesis of (6*S*,7*R*)-silphiperfolan-6-ol. *Tetrahedron* 68 (12): 2650–2656.

56 Pitsinos, E.N., Athinaios, N., and Vidali, V.P. (2012). Enantioselective total synthesis of (−)-laurenditerpenol. *Org. Lett.* 14 (17): 4666–4669.

57 Guth, H. (1996). Determination of the configuration of wine lactone. *Helv. Chim. Acta* 79 (6): 1559–1571.

58 Yang, P., Yao, M., Li, J. et al. (2016). Total synthesis of rubriflordilactone B. *Angew. Chem. Int. Ed.* 55 (24): 6964–6968.

59 McAulay, K. and Clark, J.S. (2017). Total synthesis of 7-*epi*-pukalide and 7-acetylsinumaximol B. *Chem. Eur. J.* 23 (41): 9761–9765.

60 Siler, D.A., Mighion, J.D., and Sorensen, E.J. (2014). An enantiospecific synthesis of jiadifenolide. *Angew. Chem. Int. Ed.* 53 (21): 5332–5335.

61 Shen, Y., Li, L., Pan, Z. et al. (2015). Protecting-group-free total synthesis of (−)-jiadifenolide: development of a [4+1] annulation toward multisubstituted tetrahydrofurans. *Org. Lett.* 17 (21): 5480–5483.

62 Chuang, K.V., Xu, C., and Reisman, S.E. (2016). A 15-step synthesis of (+)-ryanodol. *Science* 353 (6302): 912–915.

63 Zahel, M., Keßberg, A., and Metz, P. (2013). A short enantioselective total synthesis of (−)-englerin A. *Angew. Chem. Int. Ed.* 52 (20): 5390–5392.

64 Anakok, O.F., Ndi, C.P., Barton, M.D. et al. (2011). Antibacterial spectrum and cytotoxic activities of serrulatane compounds from the Australian medicinal plant *Eremophila neglecta*. *J. Appl. Microbiol.* 112 (1): 197–204.

65 Lu, J.M.H., Perkins, M.V., and Griesser, H.J. (2013). Total synthesis and structural confirmation of the antibacterial diterpene leubethanol. *Tetrahedron* 69 (31): 6468–6473.

66 Yang, M., Yang, X., Sun, H., and Li, A. (2016). Total synthesis of ileabethoxazole, pseudopteroxazole and *seco*-pseudopteroxazole. *Angew. Chem. Int. Ed.* 55 (8): 2851–2855.

67 Cheng, M.H., Tian, W., Peixoto, P.A. et al. (2011). Synthesis of *ent*-nanolobatolide. *Angew. Chem. Int. Ed.* 55 (8): 2851–2855.

68 Jørgensen, L., McKerrall, S.J., Kuttruff, C.A. et al. (2013). 14-Step synthesis of (+)-ingenol from (+)-3-carene. *Science* 341 (6148): 878–882.

69 McKerrall, S.J., Jørgensen, L., Kuttruff, C.A. et al. (2014). Development of a concise synthesis of (+)-ingenol. *J. Am. Chem. Soc.* 36 (15): 5799–5810.

70 Kawamura, S., Chu, H., Felding, J., and Baran, P.S. (2016). Nineteen-step total synthesis of (+)-phorbol. *Nature* 532 (7597): 90–93.

71 Vanden Berg, T.J., Pinkerton, D.M., and Williams, C.M. (2017). 7-step total synthesis of (+)-EBC-329: photoisomerisation reveals new *seco*-casbane family member. *Org. Biomol. Chem.* 15 (34): 7102–7105.

72 Dethe, D.H., Kumar, V.B., and Maiti, R. (2018). Biomimetic total syntheses of chromane meroterpenoids, guadials B and C, guapsidial A and psiguajadial D. *Org. Biomol. Chem.* 16 (26): 4793–4796.

73 Du, B., Yuan, C., Yu, T. et al. (2014). Asymmetric total synthesis of onoseriolide, bolivianine, and isobolivianine. *Chem. Eur. J.* 20 (9): 2613–2622.

74 Yuan, C., Du, B., Yang, L., and Liu, B. (2013). Bioinspired total synthesis of bolivianine: a Diels–Alder/ intramolecular hetero-Diels–Alder cascade approach. *J. Am. Chem. Soc.* 135 (42): 9291–9294.

75 Li, J.-P., Yuan, C.-C., Du, B., and Liu, B. (2017). An alternative total synthesis of bolivianine. *Chin. Chem. Lett.* 28 (1): 113–116.

76 Hu, X. and Maimone, T.J. (2014). Four-step synthesis of the antimalarial cardamom peroxide via an oxygen stitching strategy. *J. Am. Chem. Soc.* 136 (14): 5287–5290.

77 Schnermann, M.J. and Overman, L.E. (2012). A concise synthesis of (−)-aplyviolene facilitated by a strategic tertiary radical conjugate. *Angew. Chem. Int. Ed.* 51 (38): 9576–9580.

9

Diterpene Acids as Starting Materials for the Synthesis of Biologically Active Compounds

Ignacio E. Tobal, Alejandro M. Roncero, Rosalina F. Moro, David Díez, and Isidro S. Marcos

Universidad de Salamanca, Facultad de Ciencias Químicas, Departamento de Química Orgánica, Plaza de los Caídos 1-5, 37008 Salamanca, Spain

One of the main challenges of the modern organic chemistry is the effective synthesis of enantiomerically pure molecules (homochirals). In general, the biologically active compounds are chiral molecules, and usually their activity is bound to only one of the enantiomers, although sometimes it happens that both enantiomers show a different activity [1, 2].

These enantiopure compounds are obtained by resolution of racemic mixtures [3, 4], synthesis starting from homochiral natural products ("chiral pool") [5, 6], and asymmetric synthesis using chiral auxiliaries [7, 8], reagents [9, 10], or catalysts [11–19].

Due to the special complexity of many biologically active compounds, to obtain sufficient quantities for testing, the total synthesis is not advised since it typically involves multistep protocols resulting in a low overall yield. The possibility to achieve these synthetic targets by transformations of abundant natural products is widely used in organic chemistry [20, 21]. This approach has an advantage with respect to the total synthesis, making the process cheaper by decreasing the number of synthetic steps [22–24].

As an example of this kind of approach, the Taxol® synthesis is paradigmatic. Transformation of baccatin III, taxane abundant in nature, leads to Taxol [25, 26] that is used in the antitumoral therapy. Motika and Hergenrother have highlighted the use of natural products for the synthesis of new compounds with unexpected biological activities and cellular targets different from those of the parent compound [27]. Recently, several articles have been published in the use and understanding of the role of natural products for chemistry, biology, and medicine, thus demonstrating the continuing importance of these compounds [28]. These articles contextualize this chapter in the use of diterpenes for the synthesis of not only biologically active compounds but also new frameworks. Hereby, we reviewed the

Ignacio E. Tobal and Alejandro M. Roncero have contributed equally to this work.

Chiral Building Blocks in Asymmetric Synthesis: Synthesis and Applications, First Edition.
Edited by Elżbieta Wojaczyńska and Jacek Wojaczyński.

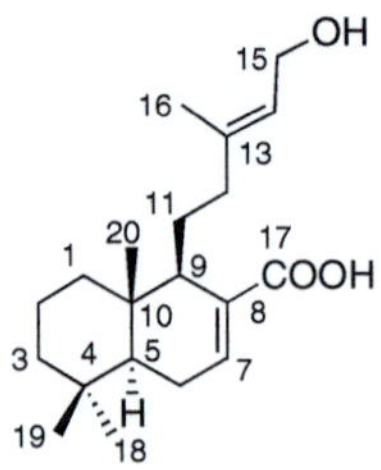

Zamoranic acid, 1
H.viscosum
(Valparaiso)

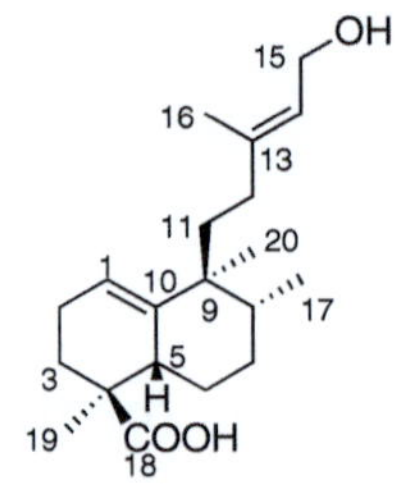

***Ent*-halimic acid, 2**
H.viscosum
(Villarino de los Aires)

Halimium viscosum
(Villarino de los Aires chemotype)

Figure 9.1 Natural products zamoranic acid, **1**, and *ent*-halimic acid, **2**, isolated from *Halimium viscosum*. Source: David Díez.

use of bicyclic diterpenes with labdane or halimane backbone for the synthesis of a high variety of carbon skeletons of compounds with biological interest.

Halimium viscosum is a plant of the *Cistaceae* family that is abundant in the west of the Iberian Peninsula. It is located in several places such as Villarino de los Aires (Salamanca, Spain) [29], La Fregeneda (Salamanca, Spain) [30, 31], and Valparaiso (Zamora, Spain) [32], not far away from each other. Surprisingly, in the first location, rearranged *ent*-labdanes (*ent*-halimanes) with unsaturated side chain were mainly isolated while, in the second, *ent*-halimanes with a saturated side chain dominated and, in the third, labdanes with structurally strange carboxyl function at C-17. At the time of these studies, finding different chemotypes of the same plant was a novelty. Further, other two chemotypes of *H. viscosum* have been localized in Portugal near the border with Salamanca, Celorico da Beira [33], and San João da Pesqueira [34].

This chapter accounts for the use of two diterpenes such as zamoranic acid, **1** [32], and *ent*-halimic acid, **2** [29] (Figure 9.1), isolated from *H. viscosum*, and their methyl esters **3** and **4** as starting materials for the synthesis of compelling compounds. These compounds are present largely in *H. viscosum* plants of the Iberian Peninsula (Spain and Portugal). Recently, the extraction and isolation technics for these natural products have been improved, not only increasing the isolated yield of these starting materials but also optimizing time and cost [35].

Due to the high number of rearranged *ent*-labdane derivatives isolated from *H. viscosum* and to simplify their nomenclature and to facilitate their classification, the name of *ent*-halimane was proposed for that carbon skeleton. Recently, a review on natural halimanes has been published [35].

The use of zamoranic, **1** [32], and *ent*-halimic, **2** [29], acids as starting materials has made possible to obtain a large number of active compounds such as antibiotics, antifeedants, antitumorals, antiadhesives, antivirals, etc. In this chapter, many of the molecules synthesized from these acids will be shown that account for their versatility. First, compounds synthesized from zamoranic acid, **1**, will be studied, followed by compounds obtained from *ent*-halimic acid, **2**.

9.1 Zamoranic Acid 1 as a Precursor of Interesting Compounds

Zamoranic acid **1** is isolated from *H. viscosum*, Valparaíso (Zamora) chemotype, being its main terpenoid component (Figure 9.1). Due to its unique functionalization, an allylic hydroxyl group in the side chain, a carboxylic acid at C-17, and a ring B unsaturated in Δ^7, this labdane is an excellent starting material for the synthesis of bioactive compounds.

Zamoranic acid **1** [32] derivatives with an allylic functionalization at C-17, as other labdanes with the same type of functionality in the ring B, exhibit the same kind of rearrangement in acidic media to give compounds with a isofregenedane skeleton [36]. This reactivity was used in the synthesis of (−)-chrysolic acid and (+)-isofregenedol, establishing in this manner the absolute configuration for both compounds [37] (Figure 9.2). Zamoranic acid has been also used for the synthesis of a series of natural labdanes, diols and triols, which permitted the corroboration of their structures and absolute configuration [38, 39]. Compound **1** has been the starting material for the synthesis of sesquiterpenes (Figure 9.2), such as the antifeedants like isodrimeninol [40], poligodial, and warburganal [41] and the antibiotics like pereniporins A and B [42–45]. The global yields in the synthesis of these compounds by a natural product transformation are the best known until date. Labdanolides as dilactone (+)-limonidilactone [46] have been obtained, too, confirming their structure and absolute configuration. The functionalization at C-17 and at the side chain permitted to achieve the synthesis of tricyclic diterpenes with different skeletons such as podocarpanes, dinorpimaranes, abietanes, totaranes, or cleistantanes and tetracyclic ones such as hibaenes [47].

The following classifies the syntheses of all the compounds mentioned above:

1) Synthesis of chrysolic acid and isofregenedol.
2) Synthesis of new labdane diterpenoids.
3) Synthesis of drimanes:
   3.1 Synthesis of polygodial, warburganal, pereniporin A, and pereniporin B.
   3.2 Synthesis of isodrimeninol.
4) Synthesis of labdanolides and furolabdanes:
   4.1 Synthesis of (+)-limonidilactone.
   4.2 Synthesis of gutierrezianolic acid.
5) Synthesis of tri- and tetracyclic diterpenes.

9.1.1 Synthesis of Chrysolic Acid and Isofregenedol

Chrysolic acid [48], isolated from *Chrysothamnus paniculatus*, and 14-isofregeneden-13-ol [49] and isofregenedol, isolated from *Haplopappus parvifolius*, are isofregenedane diterpenes in which their structures were spectroscopically determined but their proposed absolute configuration was based on biogenetic grounds and not proven (Scheme 9.1).

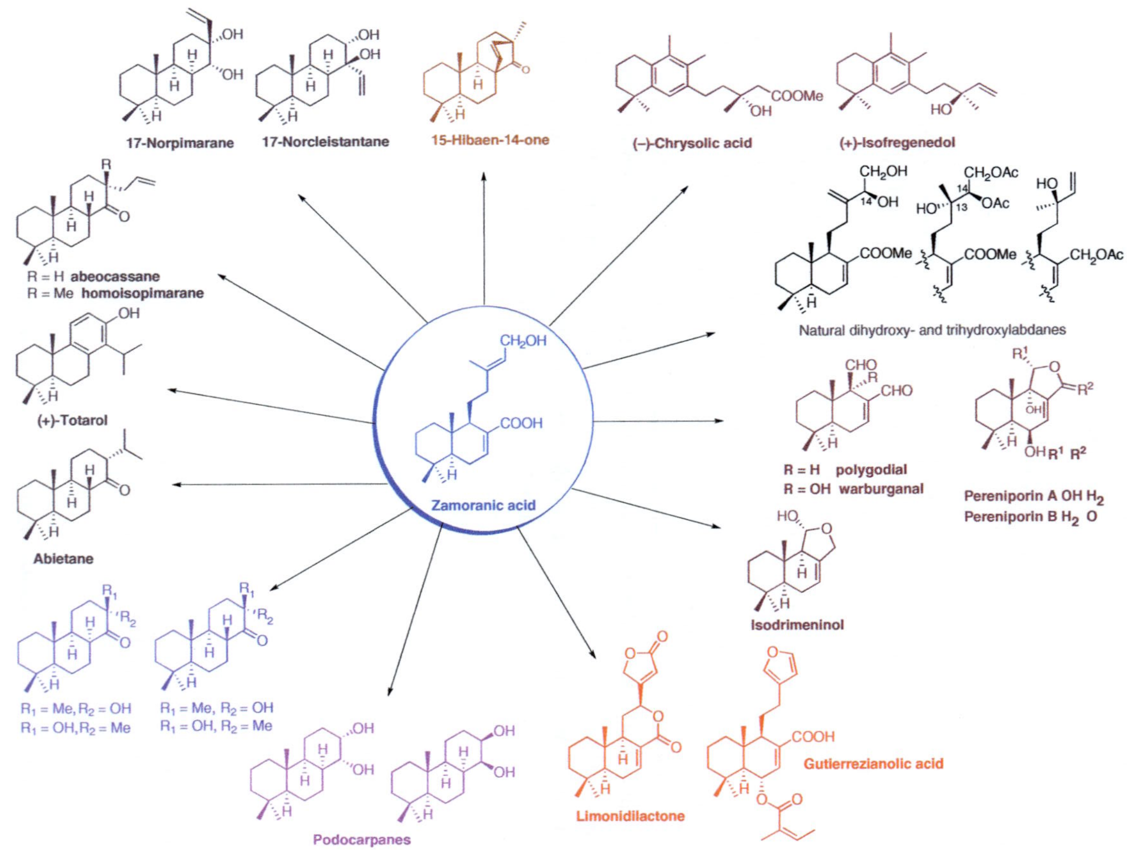

Figure 9.2 Synthesized compounds using zamoranic acid **1** as starting material.

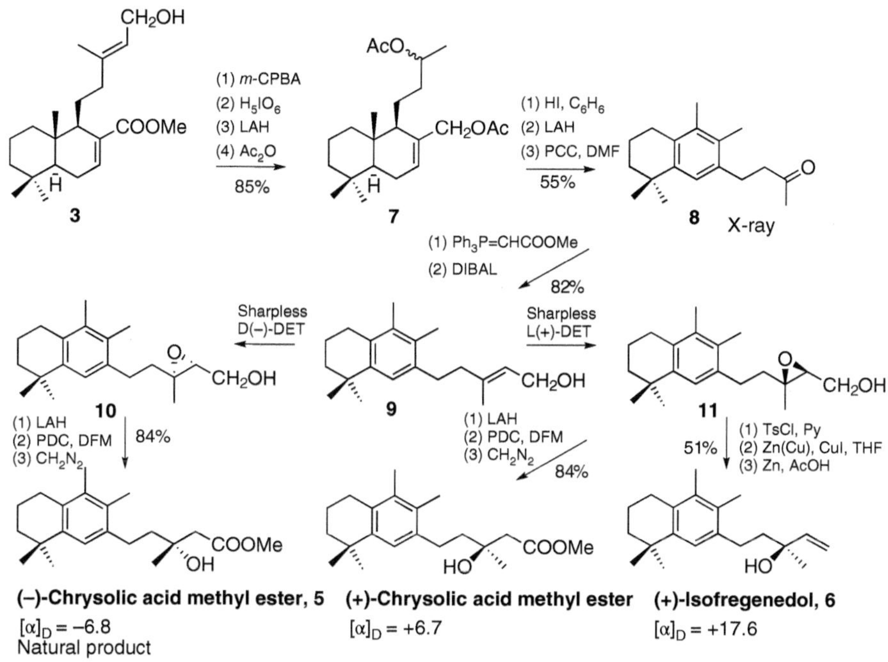

Scheme 9.1 Synthesis of chrysolic acid methyl esters **5** and isofregenedol **6**.

The synthesis of chrysolic acid methyl ester **5** and isofregenedol **6** from ketone **8**, easily obtained from zamoranic acid methyl ester, **3**, established the absolute configuration of the natural products (Scheme 9.1). The synthetic key step of the tetrahydronaphtalenic (THN) system characteristic of these compounds is a cationic rearrangement of adequately functionalized bicyclic systems, previously described by us [36].

These synthetic procedures, with a total control of the absolute configuration and high yields, have permitted to corroborate the structures of new natural compounds, such as chrysolic acid methyl ester **5** and isofregenedol **6**, and to establish their absolute configuration.

9.1.2 Synthesis of New Labdane Diterpenoids

Zamoranic acid has been used as starting material for the synthesis of six new labdane diterpenoids; four of them are diterpene acids isolated as their methyl esters from *H. viscosum*, corroborating in this manner the structure and establishing the absolute configuration of these natural products. Two are allylic epimeric diols **12** and **13**, other two are the diacetyl derivates of the side chain triols **14** and **15**, and the remaining two are the allylic alcohols, **16** and **17**: methyl 14*R*,15-dihydroxy-7,13(16)-labdadien-17-oate, **12**; methyl 14*S*,15dihydroxy-7,13(16)-labdadien-17-oate, **13** [38]; methyl 13*R*-hydroxy-14*S*,15-diacetoxy-7-labden-17-oate, **14**; methyl 13*S*-hydroxy-14*R*,15-diacetoxy-7-labden-17-oate, **15**; and 17-acetoxy-7,14-labdadien-13*R*- or 13*S*-ol, **16** and **17** [39].

The stereochemistry of the side chain stereogenic centers was determined by asymmetric synthesis of **12**, **13**, **14**, **15**, **16**, and **17** from zamoranic acid methyl ester **3** (Scheme 9.2).

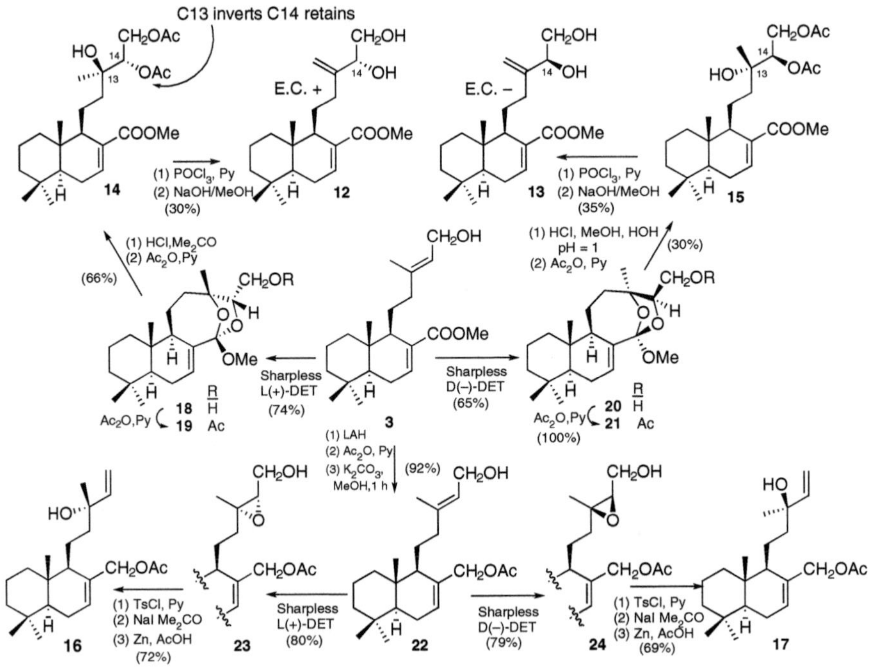

Scheme 9.2 Synthesis of labdane diterpenoids from zamoranic acid, **1**.

The absolute configuration at C14 as *R* of **12** was confirmed by circular dichroism ($\Delta\varepsilon_{313} = +1.2$). Once the C14 configuration is resolved, the natural product could be established as methyl 13*R*-hydroxy-14*S*,15-diacetoxy-7-labden-17-oate whose physical properties are identical to these of **14**.

Compound **13** was obtained following a similar synthesis to that for **12**. By this synthetic route, the absolute configuration of the natural products **15** and **13** was established; methyl 13*S*-hydroxy-14*R*,15-diacetoxy-7-labden-17-oate for the former (**15**) and methyl 14*S*,15-dihydroxy-7,13(16)-labdadien-17-oate for the latter (**13**).

The structure and stereochemistry of 17-acetoxy-7,14-labdadien-13*R*- or 13*S*-ol for **16** and **17** are confirmed by enantioselective synthesis starting with **3** (Scheme 9.2).

9.1.3 Synthesis of Drimanes

Drimanes are sesquiterpenes that have attracted the attention of the organic chemists due to their biological activities such as antifeedant, antimicrobial, cytotoxic, growth regulator, or molluscicide [42–45].

Functionalization of the bicyclic system as in the side chain of zamoranic acid, **1**, is very adequate for the synthesis of antifeedant drimanes such as poligodial and warburganal isolated from *Polygonum hydropiper* and *Warburgia ugandensis*, respectively, and for the preparation of antibiotics like pereniporins A and B isolated from culture of basideomycetes *Perennipora medullaepanis*. A different strategy was used for the transformation of zamoranic acid into isodrimeninol, as will be seen later, providing the best results known until date.

9.1.3.1 Synthesis of Poligodial, Warburganal, Pereniporin A, and Pereniporin B

The key step for the synthesis of poligodial **32**, warburganal **33**, pereniporin A **34**, and pereniporin B **35** from zamoranic acid, **1**, consists in the side chain degradation to obtain the methylketones **25** or **26** by a Norrish II-type photochemical reaction (Scheme 9.3). This transformation leads to drimane dienes **27** or **28**, from which epoxides **29** or **30** can be obtained and from them diol **31**, a key intermediate in the synthesis of the biological active drimanes.

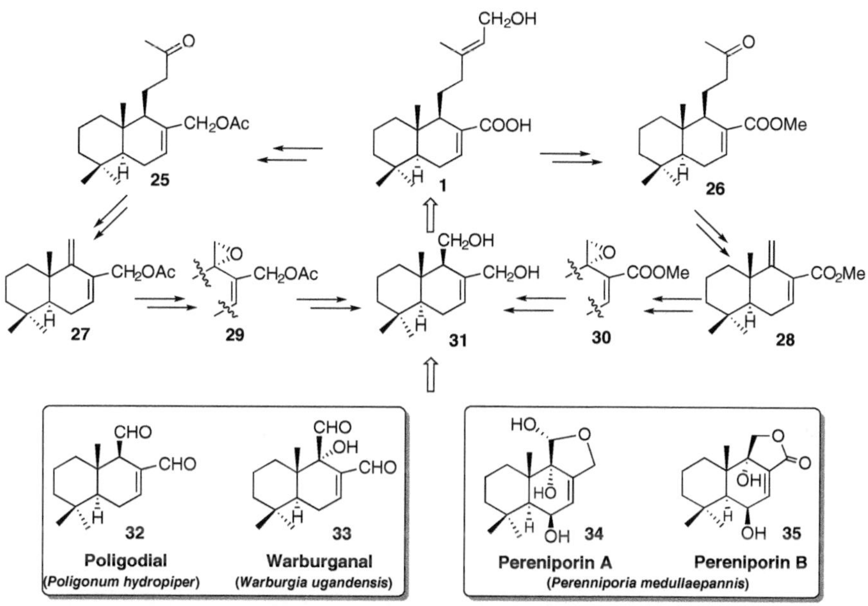

Scheme 9.3 Synthesis and synthetic plan of drimanes from zamoranic acid.

9.1.3.2 Synthesis of Isodrimeninol

Isodrimeninol synthesis from zamoranic acid, **1**, demands C11 functionalization (Scheme 9.4). Treatment of acetyl zamoranic acid methyl ester with $SnCl_4$ at room temperature resulted in excellent yield triene **36** as a single isomer, with C11 functionalization in place. Isodrimeninol was obtained following the sequence shown in Scheme 9.4.

This synthesis is the most efficient to date (56% global yield) allowing access to multigram quantities of the natural product Isodrimeninol **42**.

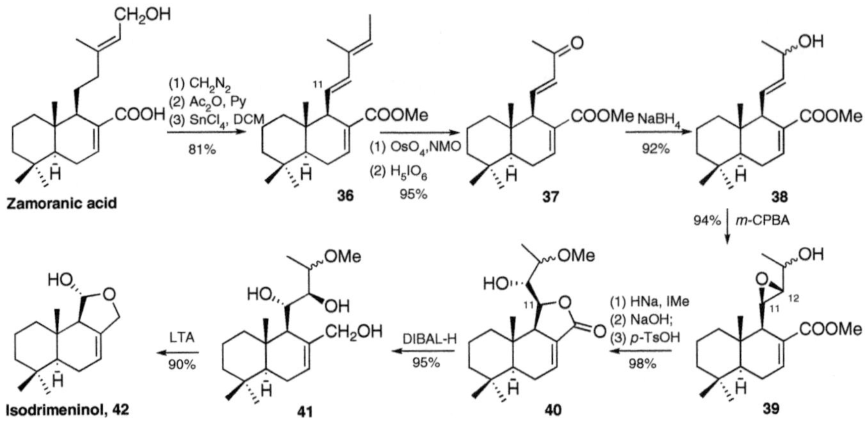

Scheme 9.4 Isodrimeninol synthesis from zamoranic acid.

9.1.4 Synthesis of Labdanolides and Furolabdanes

Diterpenes of the labdane type belong to a class of compounds with increasing pharmacological interest [1]. Antimutagenic, antibacterial, antifungal, and blood coagulation influencing effects have been reported [2]. Some of them are present in small quantities in nature. The synthesis of the minor components from other abundant natural products is of longstanding interest. For this reason, the syntheses of a labdanolide (+)-limonidilactone and of a furolabdane gutierrezianolic acid were undertaken as described afterward.

9.1.4.1 Synthesis of (+)-Limonidilactone

Limonidilactone isolated from *Vitex limonifolia*es is a labdanolide that contains a γ-butenolide and a δ-lactone. The structure of (−)-limonidilactone was established by spectroscopic methods and confirmed by X-ray, but the absolute configuration was unknown [46].

(+)-Limonidilactone **49** (Scheme 9.5) was afforded from zamoranic acid **1** by a six-step sequence in 25% overall yield, establishing in this manner the absolute configuration of the natural (−)-limonidilactone. The key steps of the synthesis are functionalization at C12, controlling the stereochemistry of that center, and functionalization at C16 to finally elaborate the γ-lactone fragment as shown in Scheme 9.5.

The spectroscopic properties of **49** are identical to these of natural limonidilactone. As the natural product shows a negative rotation, $[\alpha]_D = -23.8$ (c 1.4, $CHCl_3$), this implies that the natural (−)-limonidilactone belongs to the labdane antipodal series.

9.1.4.2 Synthesis of Gutierrezianolic Acid

Angeloyl-gutierrezianolic acid methyl ester **52** (Scheme 9.5) is a furolabdane isolated by Timmermann et al. from *Baccharis pingraea* DC and before by Bohlmann et al. from *Gutierrezia espinosae*. These plants are used in South America in folk medicine due to their antirheumatic, antiinflammatory, antitumoral, and antibiotic properties. The absolute configuration of **52** was proposed on the basis of biogenetic

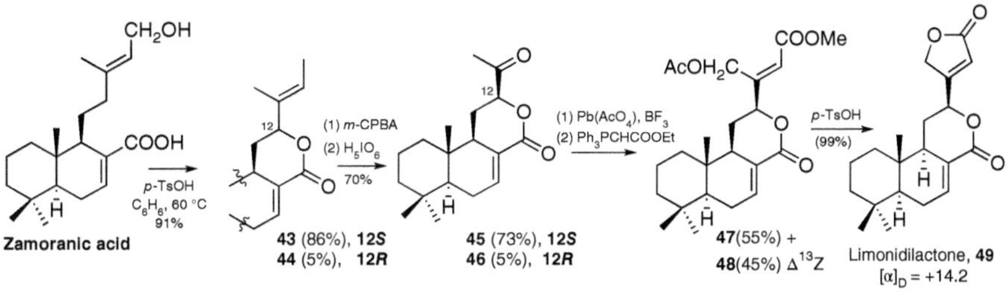

Scheme 9.5 Limonidilactone, **49**, synthesis from zamoranic acid.

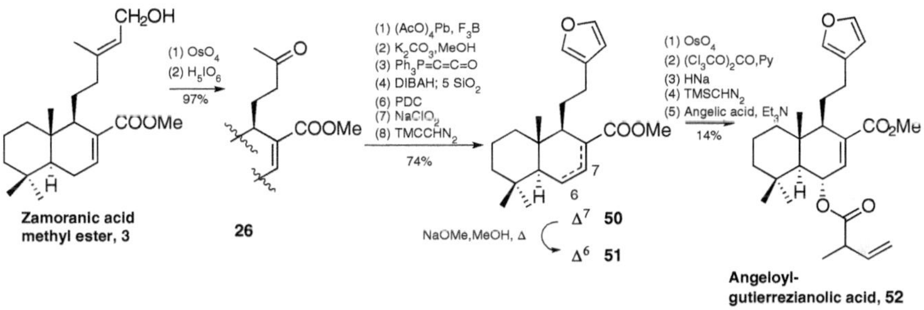

Scheme 9.6 Synthesis of angeloyl-gutierrezianolic acid methyl ester from zamoranic acid methyl ester.

and taxonomic considerations, but until now it has not been rigorously established [50].

As a continuation of our studies on the synthesis of furolabdane and labdanolide terpenoids, we decided to synthesize angeloyl-gutierrezianolic acid, **52**, from zamoranic acid methyl ester, **3**. The synthesis of the target compound, the methyl ester of the angeloyl-gutierrezianolic acid, **52**, can be carried out following the synthetic route of Scheme 9.6. The first goal is to build the furan ring at the side chain of the methylketone **26** to obtain **50** followed by the development of the allylic system at ring B and ulterior esterification with angelic acid to obtain **52**.

The optical rotation of **52**, $[\alpha]_D^{22} = +62.0$ (c 1.0, $CHCl_3$), is close to the natural product, $[\alpha]_D^{22} = +72.0$ (c 1.0, $CHCl_3$), corroborating the absolute configuration of the natural product.

9.1.5 Synthesis of Tri- and Tetracyclic Diterpenes

Tri- and tetracyclic diterpenes (Figure 9.3) are very common compounds in nature and show a great variety of carbon skeletons. Among tricyclic terpenes, abietanes, pimaranes, isopimaranes, cassanes, cleistanthanes, totaranes, and the trinor-derived podocarpanes can be found. In the tetracyclic series, there are kauranes, beyeranes, or hibaenes and isohibaenes. These compounds exhibit antifungal, antiinflammatory, antioxidant, antibiotic, antitumor, antimalarial, cardioactive, antiviral, and nitric oxide biological activities, among other [47, 51, 52].

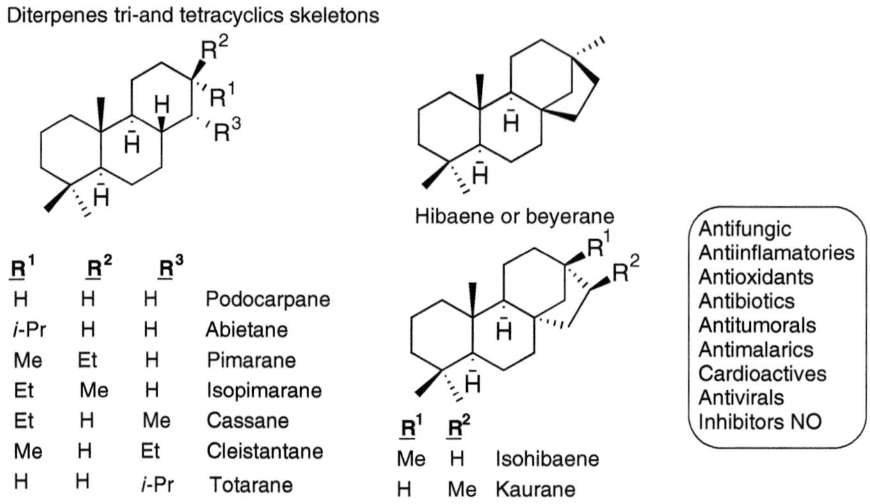

Figure 9.3 Tri- and tetracyclic diterpene skeletons.

Zamoranic acid, **1**, has been used in the synthesis of this kind of natural products. The access to the tri- and tetracyclic skeletons **53**, **54**, **55**, or **56**, from zamoranic acid, **1** (Scheme 9.7), can be achieved using the tricyclic diols **60**, **61**, and **62** as intermediates obtained from the dicarbonyl compounds **57**, **58**, and **59** by a SmI_2-induced cyclization.

Hibaenone **53**, totarol **54**, and other tricyclic diterpenes with a range of different skeletons as **55** and **56** (Scheme 9.7) were obtained using dicarbonyl intermediates **57**, **58**, and **59**, respectively, as well as tricyclic diols **60**, **61**, and **62**, precursors of the natural compounds.

The third ring was obtained by synthesis of the dicarbonyl compounds and their treatment with samarium iodide. Dicarbonyl compounds **57**, **58**, and **59** were obtained as shown in Scheme 9.8. Cyclization with SmI_2 of the corresponding dicarbonyl compounds afforded the key intermediate tricyclic diols **60**, **61**, and **62**, respectively, in the synthesis of 15-hibaene-14-one, (−)-totarol, and other mentioned natural products.

The SmI_2-promoted cyclization results in a *cis* disposition of the hydroxyl groups, probably due to chelation of the oxygen to samarium in the coupling reaction.

In conclusion, several dicarbonyl compounds were obtained from zamoranic acid. These advanced intermediates undergo an efficient stereoselective cyclization promoted by SmI_2, leading to tri- and tetracyclic diterpenes with different carbon skeletons.

9.2 *Ent*-halimic Acid as a Precursor of Biologically Active Compounds and Other Interesting Derivatives

Ent-halimic acid, **2**, is the main component of the room temperature ethyl acetate extract of *H. viscosum* (Villarino de los Aires chemotype). It can be obtained in

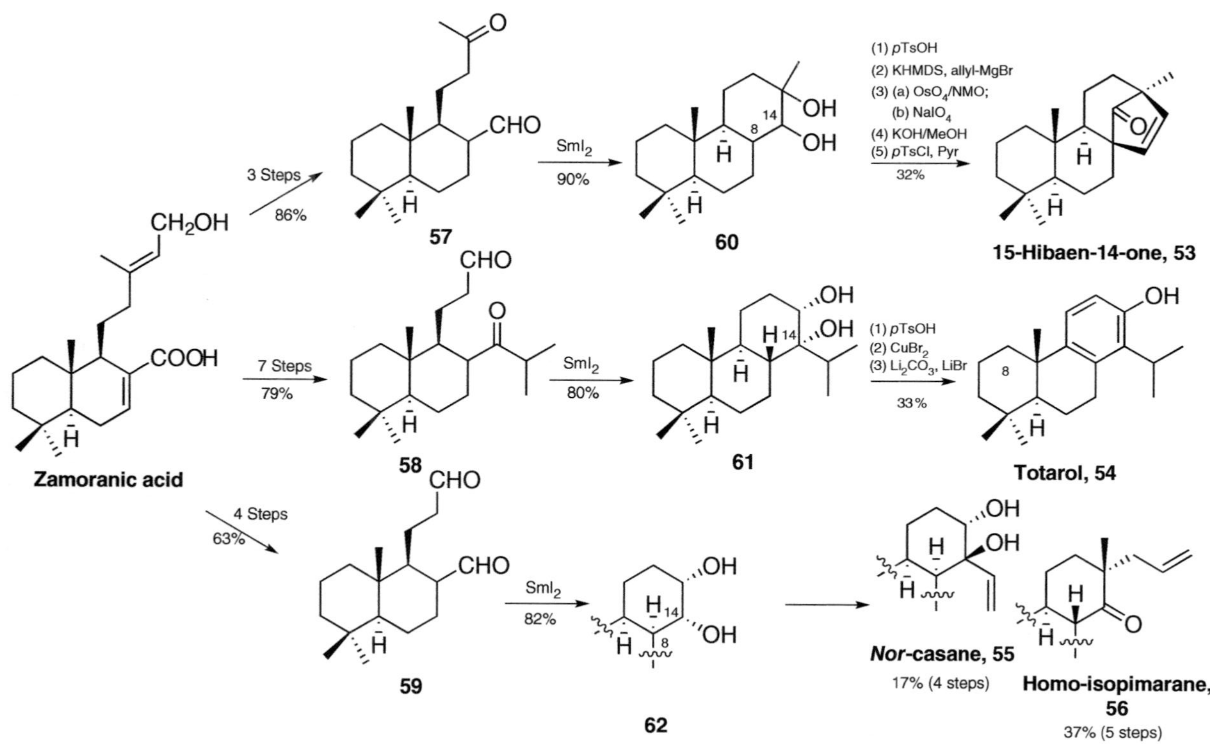

Scheme 9.7 Synthesis for tri- and tetracyclic diterpenes from zamoranic acid, **1**.

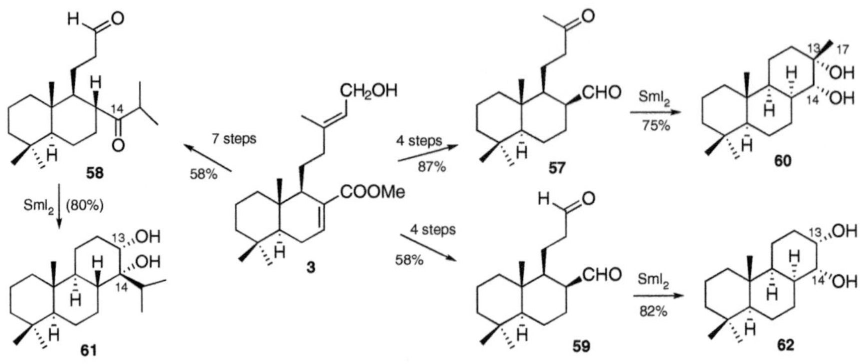

Scheme 9.8 Synthesis of tricyclic key intermediates from zamoranic acid.

multigram quantities (3.4% of the dry plant weight) by a quick and efficient method. It bears an unsaturated side chain, Δ^{13}, functionalized at C-15, a carboxyl group at C18, and a double bond $\Delta^{1(10)}$. This functionalization permits its use for the synthesis of high added value biologically active compounds.

In Figure 9.4, the structures of the following derivatives obtained from *ent*-halimic acid, **2**, are shown:

1) *Ent*-halimanolides.
2) Chettaphanins I and II.
3) Sesterterpenolides.
4) Bioconjugate sesterterpenolides and glycerophospholipids.
5) Rearranged derivatives: *ent*-labdanes, abeopicrasanes, and propellanes.
6) Sesquiterpene quinone/hydroquinones.
7) Sesqui- and diterpene alkaloids.

In the following paragraphs, the synthesis of the referred compounds will be discussed. This permitted the corroboration of their structure along with the determination of some of their absolute configuration, as well as testing their biological activities.

9.2.1 Synthesis of *Ent*-halimanolides

A series of natural *ent*-halimanolides along with α- and β-hydroxybutanolides and a natural furo-*ent*-halimanolide have been achieved using *ent*-halimic acid as starting material. The synthetic methodology will be described in the following paragraphs.

9.2.1.1 Synthesis of Natural *Ent*-halimanolide Synthesis

The first three natural *ent*-halimanolides **63**, **64**, and **65** were synthesized from *ent*-halimic acid **2**, using methyl ketone **66** as intermediate [53, 54] (Scheme 9.9). In this manner, their structures have been confirmed, and their absolute configurations established. Bestmann methodology has been used for the synthesis of butenolides [55], and for the preparation of γ-hydroxybutenolides, the Boukouvalas method [56] has been employed. Biological testing has been carried out on these

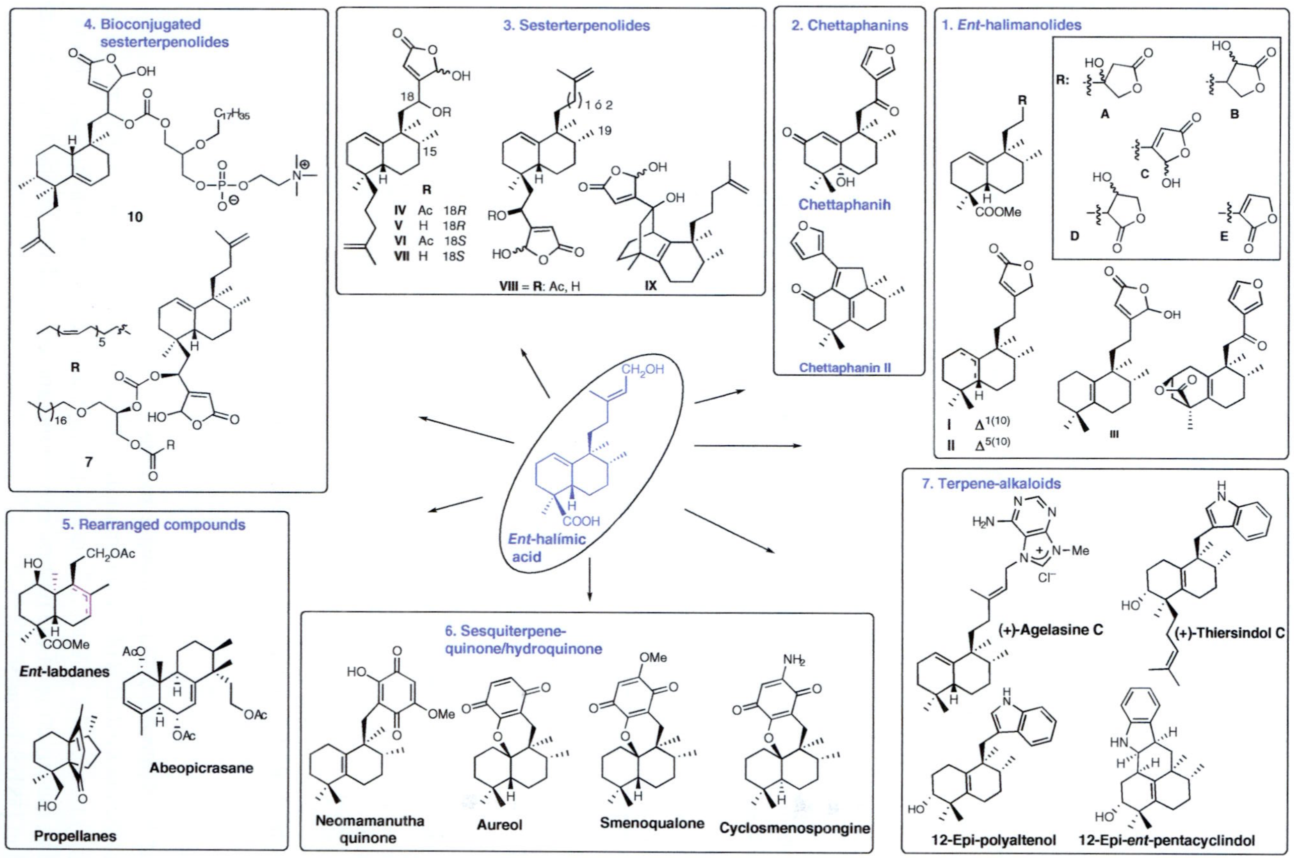

Figure 9.4 Synthesis of interesting compounds from *ent*-halimic acid.

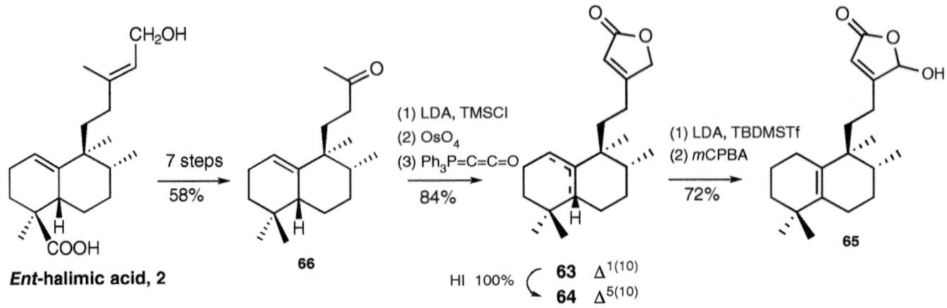

Scheme 9.9 Synthesis of natural *ent*-halimanolide from *ent*-halimic acid, **2**.

compounds. Compound **63** possesses cytotoxic and antiviral activity [HELAM cells (IC_{50} = 5.0), MDCK (IC_{50} = 5.1) and influenza virus (IC_{50} = 6.8)] [53].

9.2.1.2 Synthesis of α and β-Hydroxyhalimanolides

In the same way, two series of *ent*-halimanolides (15,16-olides **67–70** and 16,15-olides **71–74**), analogues of biologically active diterpene lactones, were synthesized starting from *ent*-halimic acid methyl ester **4** (Scheme 9.10). The absolute configuration was determined by CD spectroscopy and corroborated by X-ray analysis. The intermediate triol **75** can be accessed from **2** by singlet oxygen oxidation and ulterior hydroboration–oxidation. For the synthesis of the γ-hydroxybutanolide **76**, the Reformatski methodology was used. Antifeedant testing has been carried out on these compounds [57].

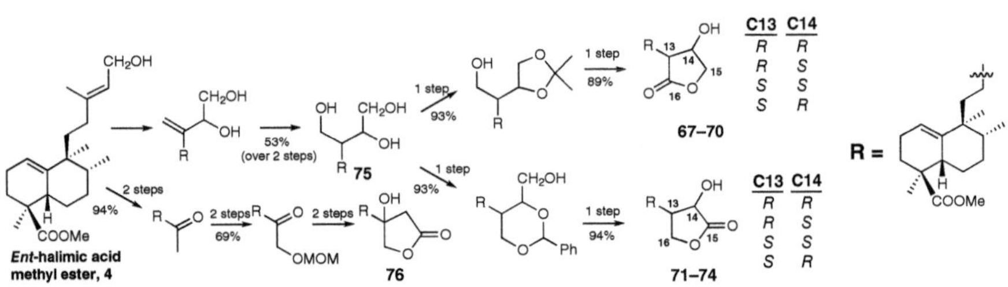

Scheme 9.10 Synthesis of α and β-hydroxyhalimanolides.

9.2.1.3 Synthesis of Furo-*ent*-halimanolide

An efficient synthesis of *ent*-halimanolide **77** (15,16-epoxy-12-oxo-*ent*-halima-5(10), 13(16),14-trien-18,2β-olide), from *ent*-halimic acid, **2**, has been achieved, corroborating the structure of the natural compound and establishing its absolute configuration [58] (Scheme 9.11).

Using the dinorderivatives **80** and **79** as intermediates gave access to the tetranor derivative **78** by a route in multigram scale. From **78**, the natural lactone **77** was obtained with a moderately activity against HeLa (human cervix cancer).

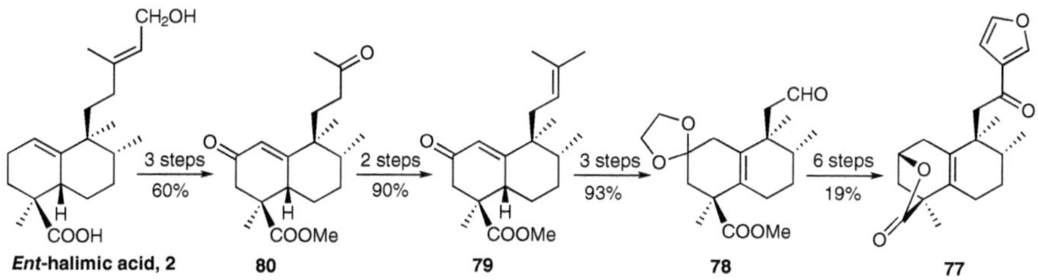

Scheme 9.11 Synthesis of furo-*ent*-halimanolides.

9.2.2 Synthesis of Chettaphanins

An efficient synthesis of the first known *ent*-halimanes, chettaphanin I **81** and II **82**, has been achieved from *ent*-halimic acid **2** (Scheme 9.12). The structure of the natural products was established by NMR measurements and by X-ray analysis of chettaphanin II, **82**, for which the absolute configuration remained unknown.

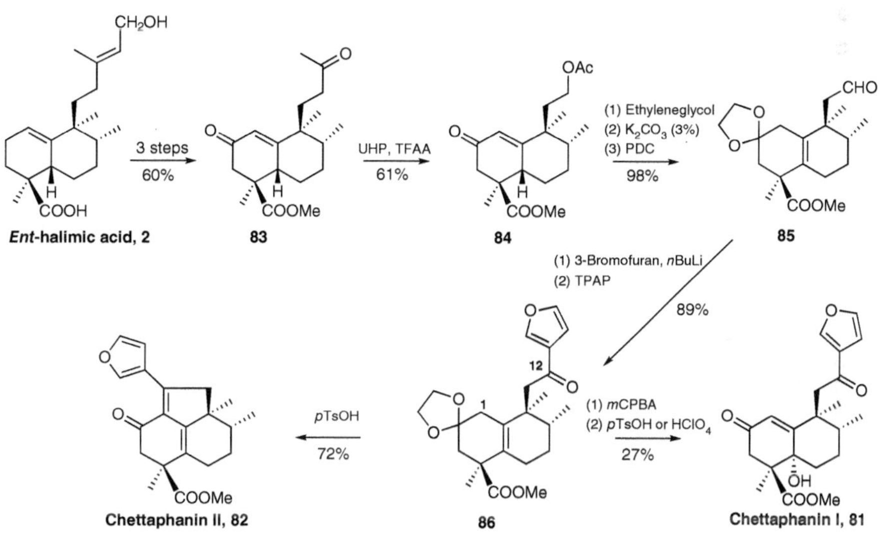

Scheme 9.12 Synthesis of chettaphanins from *ent*-halimic acid.

To synthesize chettaphanin I **81** and chettaphanin II **82** from *ent*-halimic acid **2**, it was necessary to functionalize C2 and C12 and to introduce a furyl group in the side chain; furthermore, for the synthesis of **81**, it was necessary to introduce a hydroxyl group on C5. The synthesis was developed according to Scheme 9.12 [59, 60].

9.2.3 Synthesis of Sesterterpenolides

Using *ent*-halimic acid, **2**, as starting material, the synthesis of bioactive sestert-erpenoid γ-hydroxybutenolides 15,18-bisepi-*ent*-cladocoran A and B, **87** and **88**

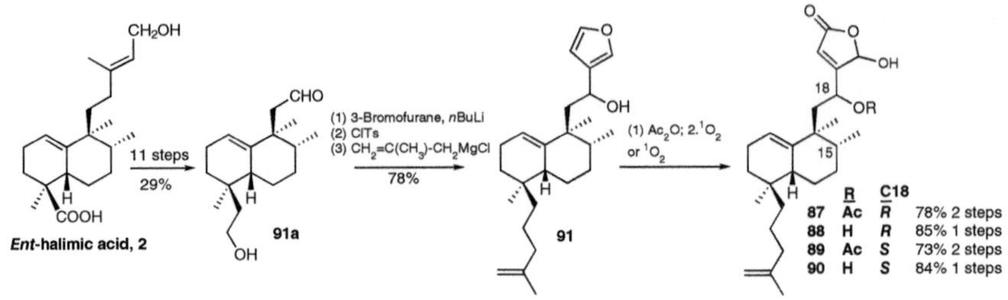

Scheme 9.13 Synthesis of sesterterpenolides.

(Scheme 9.13), and the epimers of these compounds at C18, 15-epi-*ent*-cladocoran A and B, **89** and **90**, has been carried out, showing that the structures proposed by Fontana et al. [61] for cladocorans A and B (**87** and **88**) should be revised [62, 63]. This task was achieved by Miyaoka et al. [64].

9.2.3.1 Synthesis of Sesterterpenolides, Analogues of Dysidiolide

Sesterterpenolides **87–90** are structural analogues of the natural sesterterpenolide dysidiolide [65–67], an inhibitor of protein phosphatases cdc25A (IC_{50} = 9.4 μM) and cdc25B (IC_{50} = 87 μM), which are essential for cell proliferation. Dysidiolide inhibits the growth of A-549 human lung carcinoma and P388 murine leukemia cell lines at low micromolar concentrations [68–70]. The encouraging biological activities shown by sesterterpenolides **87–90**, analogues of dysidiolide and in some cases more active than the dysidiolide itself, boost the search for the synthesis of this kind of analogues. In this manner, several sesterterpenolide analogues of dysidiolide **92–96** (Scheme 9.14) have been synthesized from *ent*-halimic acid **2** showing *in vitro* antitumoral activity against human HeLa, A549, HT-29, and HL-60 carcinoma cells. The proliferation inhibition data showed a significant antitumor activity of the compounds **92–96** inhibiting proliferation of distinct cancer cell types with an IC_{50} in the low micromolar range [65].

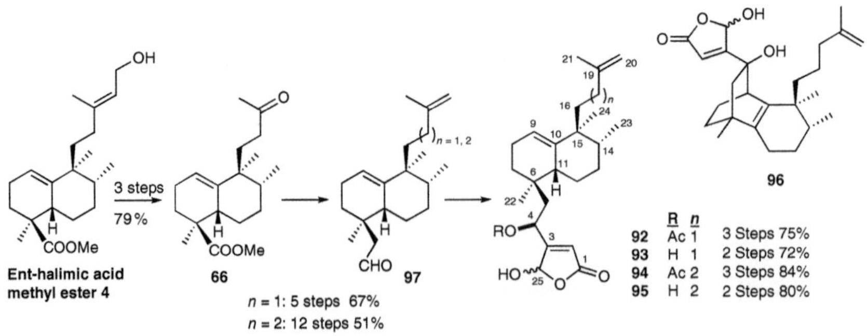

Scheme 9.14 Synthesis of sesterterpenolide analogues of dysidioide.

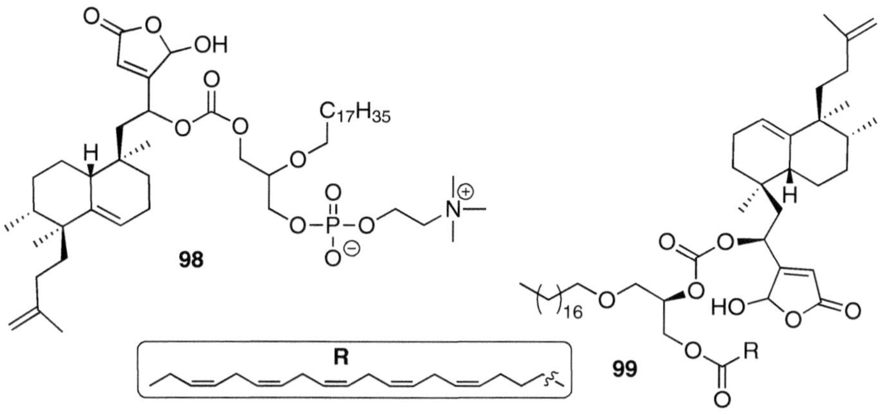

Figure 9.5 Sesterterpenolide and glycerol hybrids.

9.2.4 Synthesis of Sesterterpenolide and Glycerol Hybrids

One of the sesterterpenolides, **92**, derived from *ent*-halimic acid, **2**, has been used in the synthesis of a series of bioconjugated sesterterpenoids with phospholipids and polyunsaturated fatty acids (PUFAs) as **98** and **99** (Figure 9.5) for biological activity testing as antiproliferative agents against several cancer cell lines. Different substitution analogues of the original lipidic ether edelfosine (1-*O*-octadecyl-2-*O*-methyl-*rac*-glycero-3-phosphocholine) were obtained varying the sesterterpenoid in position 1 or 2 of the glycerol or a phosphocholine or PUFA unit in position 3. Simple bioconjugates of sesterterpenoids and eicosapentaenoic acid (EPA) have been obtained, too. All synthetic derivatives were tested against the human tumor cell lines HeLa (cervix) and MCF-7 (breast). Some compounds showed good IC_{50} (0.3 and 0.2 μM) values against these cell lines [71].

9.2.5 Synthesis of Rearranged Compounds

Due to its functionality, *ent*-halimic acid **2** could be also considered an excellent synthon for the synthesis of new rearranged compounds with interest in perfumery.

9.2.5.1 Synthesis of *Ent*-labdanes from *Ent*-halimanes

Scheme 9.15 shows the 1,2-shift of Me-20 of *ent*-halimanes leading to *ent*-labdanes. This rearrangement is opposite to the biosynthetic route in which the *ent*-labdanes are the precursors of *ent*-halimanes. For the first time, *ent*-labdanes have been synthesized starting from *ent*-halimic acid, **2** (Scheme 9.15) [72]. Effectively, the *ent*-halimane epoxyderivative **100** (Scheme 9.15), used previously by the group, leads to the *ent*-labdane tetranorderivatives **101** and **102**. This rearrangement enables the 1(10)-halimanes as starting material for the synthesis of biologically active labdanes.

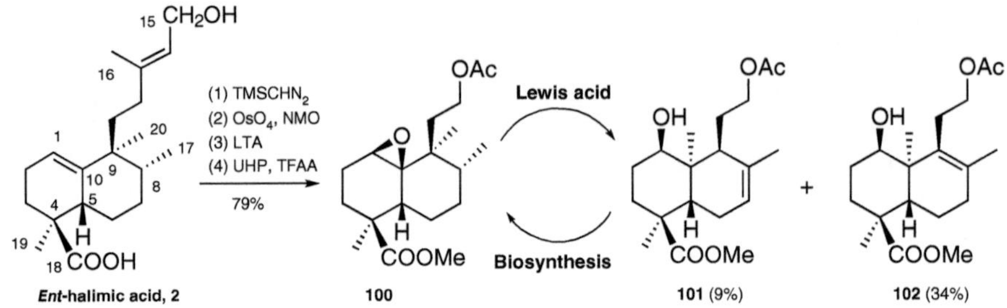

Scheme 9.15 Synthesis of *ent*-labdanes from *ent*-halimanes. Source: Marcos et al. [72]/American Chemical Society.

9.2.5.2 Synthesis of Abeopicrasanes from *Ent*-halimanes

Picrasanes are quassinoids, degraded triterpenes, interesting for their antitumoral properties [73]. One of their components is bruceantin [74]. An advanced intermediate **105** (Scheme 9.16) with the ABC ring of the picrasane quassinoid skeleton **106** has been synthesized from *ent*-halimic acid, **2** [73]. The bicyclic system of the starting material has been incorporated as the BC part of the ABC system. Until date, no diterpenes of the antipode series have been used in this kind of approach of triterpenes of quassinoids with picrasane skeleton. The tricyclic system was elaborated using the tetranorderivative **103** as intermediate by allylic oxidation incorporating on C18 the necessary four carbons to access to the 4,5-*seco*-abeopicrasane **104**. Carbons C-4 and C-5 of the halimane are incorporated as C-10 and C-9 of the new seco-abeopicrasane, respectively, with the right configuration. Dione **104** annulation leads to the corresponding abeopicrasane **105**.

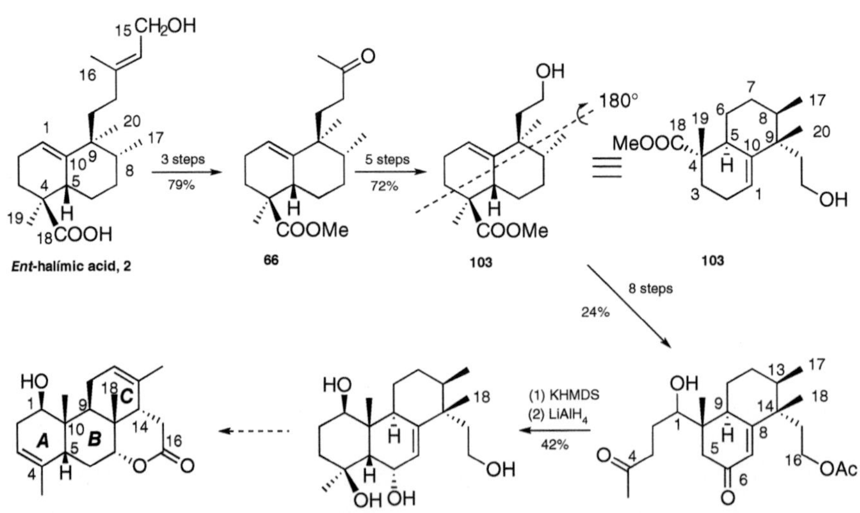

Scheme 9.16 Synthesis of pricrasane and abeopicrasane from *ent*-halimic acid, **2**.

9.2.5.3 Synthesis of [4.3.3]propellanes from *Ent*-halimanes

Ent-halimic acid **2** was also used as the starting material for the efficient synthesis of a pair of tetranor derivatives **107** and **108**, functionalized at C-18 (Scheme 9.17). These compounds could be used to synthesize [4.3.3]propellanes **110**, oxides **111**, and lactones **112** similar to ambreinolide, all of which may be of interest to the perfume industry [75].

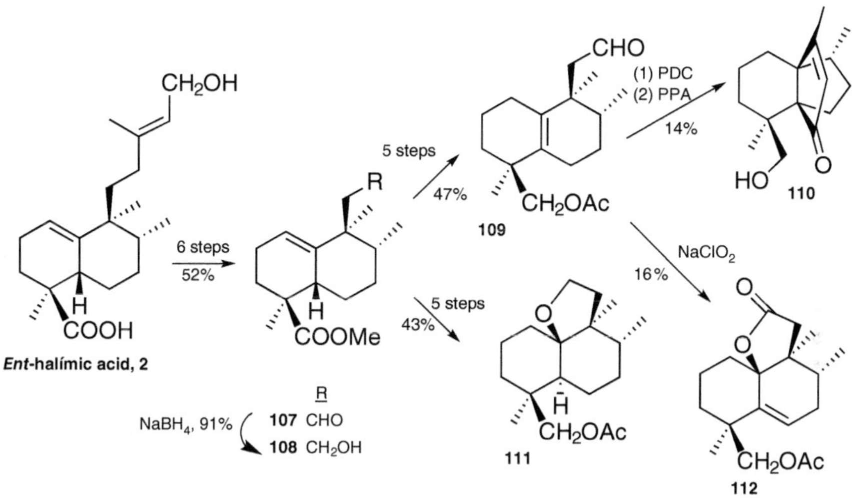

Scheme 9.17 Synthesis of [4.3.3]propellanes from *ent*-halimic acid, **2**.

9.2.6 Synthesis of Quinone/Hydroquinone Sesquiterpenes

Quinone/hydroquinone sesquiterpenes [76, 77] with drimane or rearranged drimane skeleton are a very wide and diverse group of secondary metabolites of mixed biogenesis with the terpene unit associated to a quinone or quinol. Those compounds are very interesting not only for their biological activities but also for the diverse structures that they possess.

Ent-halimic acid **2** has been used in the synthesis of the quinone/hydroquinone sesquiterpenes (−)-aureol **113**, (−)-smenoqualone **114**, and (−)-neomamanuthaquinone **116** and in the formal synthesis of (−)-cyclosmenospongine **115** [76] (Figure 9.6).

The syntheses of those quinone/hydroquinone sesquiterpenes **113–116** from *ent*-halimic acid, **2**, were planned according to the following synthesis (Scheme 9.18), which is developed by the approach AB/ABD/ABCD.

The synthesis of compounds with a quinoid or hydroquinoid moiety **113–116** can be achieved through an intermediate with the ring D incorporated, for example, compound **118** or an analogue (Scheme 9.18). This intermediate could be obtained from the tetranorderivative **117**, which would be previously synthesized from *ent*-halimic acid **2**. The key step in this sequence is a Barton decarboxylation reaction in the presence of benzoquinone [78–80]. With **118** in hand, (−)

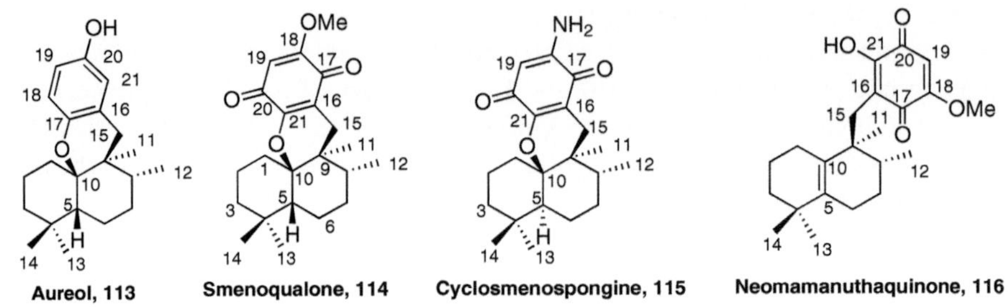

Figure 9.6 Natural quinone/hydroquinone sesquiterpenes.

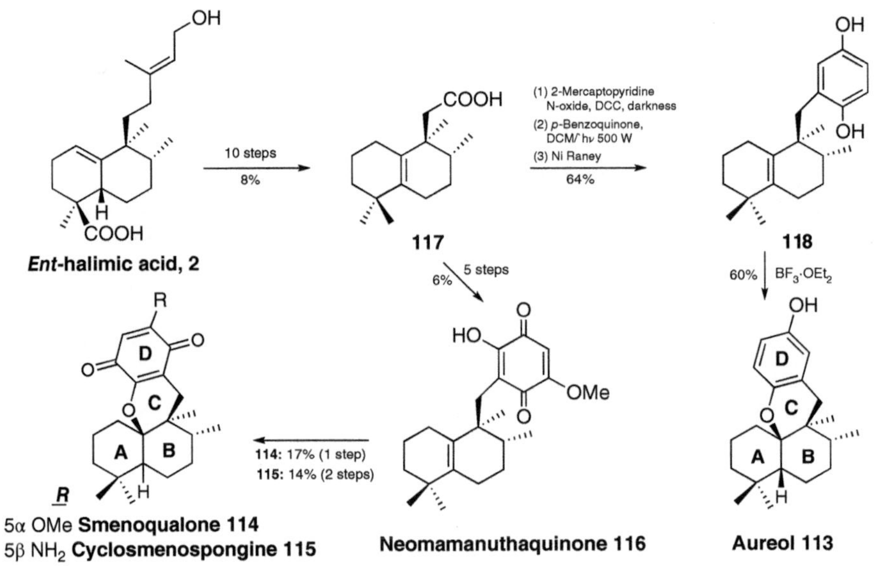

Scheme 9.18 Synthesis of sesquiterpenes quinone/hydroquinone.

neomamanuthaquinone **116** (Scheme 9.18) was obtained by addition of NaOMe to the quinone ring. The tricyclic sesquiterpene quinone, **116**, would give access to (−)-**114** and (−)-cyclosmenospongine **115**. On the other hand, the ring cyclization of intermediate **118** provided (−)-aureol **113**. The formation of the tetracyclic compound was achieved with *p*-TsOH, in a stereoselective manner with $BF_3 \cdot Et_2O$.

9.2.7 Synthesis of Sesqui- and Diterpene Alkaloids

Ent-halimic acid, **2**, has been also used for the synthesis of terpene-alkaloids, specifically in the preparation of 7,9-dialkylpurines (agelasine C) and diterpene and sesquiterpene indoles.

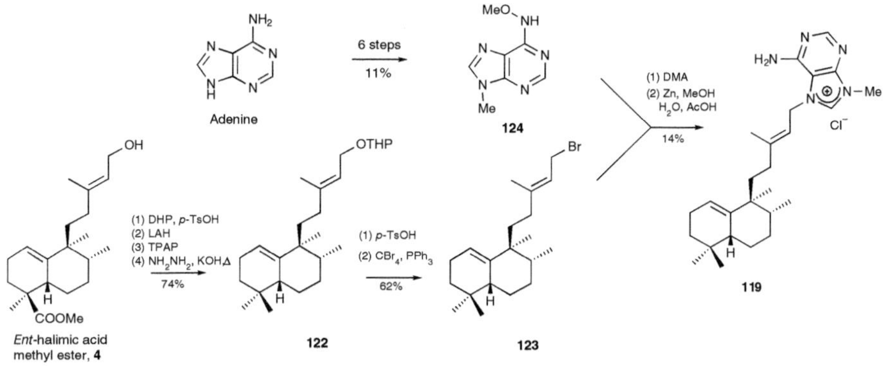

Scheme 9.19 Synthesis of (+)-agelasine C.

9.2.7.1 Synthesis of the Diterpene Alkaloid, (+)-Agelasine C

Ent-halimic acid methyl ester **4** has been used for the synthesis of different terpene alkaloids. Here, the preparation of 7,9-dialkylpurine (+)-agelasine C **119** will be presented (Scheme 9.19) [81].

Agelasines are a diterpene alkaloid group, 7,9-dialkylpurine salts, isolated from marine sponges of the genus *Agelas* [82]. (−)-Agelasine C [83] is one of the first four known agelasines isolated by Nakamura and coworkers who proposed the structural formula **120**. (−)-Agelasine C showed powerful inhibitory effects on Na, K-ATPase, and antimicrobial activities (figure 7). Epi-agelasine C **121** was isolated by Hattori et al. [84] from the marine sponge *Agelas mauritiana* as an antifouling substance active against macroalgae. Due to the interest of epi-agelasine C as antifouling agent [84] and having the possibility of establishing the absolute configuration of this compound, the synthesis of **119** (Scheme 9.14) was carried out [81].

The synthesis of **119** was planned following a synthetic pathway analogue to other agelasines, which consists of coupling the terpenic fragment **123** with a purine derivative **124** [81] (Scheme 9.19).

The physical properties of the synthesized product **119** were very different from the natural product epi-agelasine C; thus its proposed structure **121** (Figure 9.7) should be revised. On the other hand, when the ^{1}H and ^{13}C NMR spectra of **119** were compared with the ones of (−)-agelasine C proposed structure **120**, the two pairs of spectra were identical. However, the optical rotation of **119** and natural (−)-agelasine C were similar in absolute value but different sign. Thus, it should be concluded that the structure of the natural product (−)-agelasine C should be corrected to structure **125** (Figure 9.7), which is the enantiomer of the synthesized product **119** (+)-agelasine C (Figure 9.7, Scheme 9.19).

Spectroscopic considerations made when comparing the spectra of **119** with the ones of epi-agelasine C and their specific rotations permitted to suggest the structure **126** for the natural epi-agelasine C, as it appears in Figure 9.7.

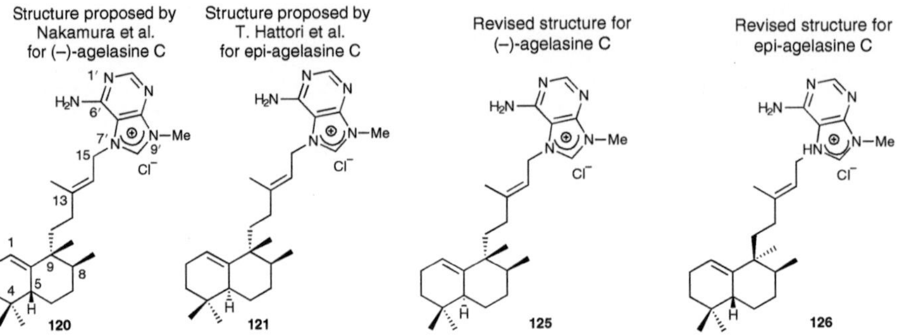

Figure 9.7 Structures proposed and revised for (−)-agelasine C and epi-agelasine C.

9.2.7.2 Synthesis of Diterpene Alkaloid, (+)-Thiersindole C

The indole diterpene alkaloid (+)-thiersindole C **131** has been synthesized from *ent*-halimic acid, **2** [85] (Scheme 9.20). Firstly, the bicyclic system, functionalized in C3, was elaborated by intermediates **127** and **128** synthesis. Secondly, a Fischer indolization was used to obtain the north side chain, leading to the intermediate **129,** and finally elongation of the south side chain with an isoprene unit in two steps through **130** led to (+)-thiersindole C, **131**.This synthesis corroborated the absolute configuration for the natural product (−)-thiersindole C. The synthesized (+)-thiersindole C **131** showed antitumor activity against several human tumor cell lines with an IC_{50} in the range of 10^{-5} M.

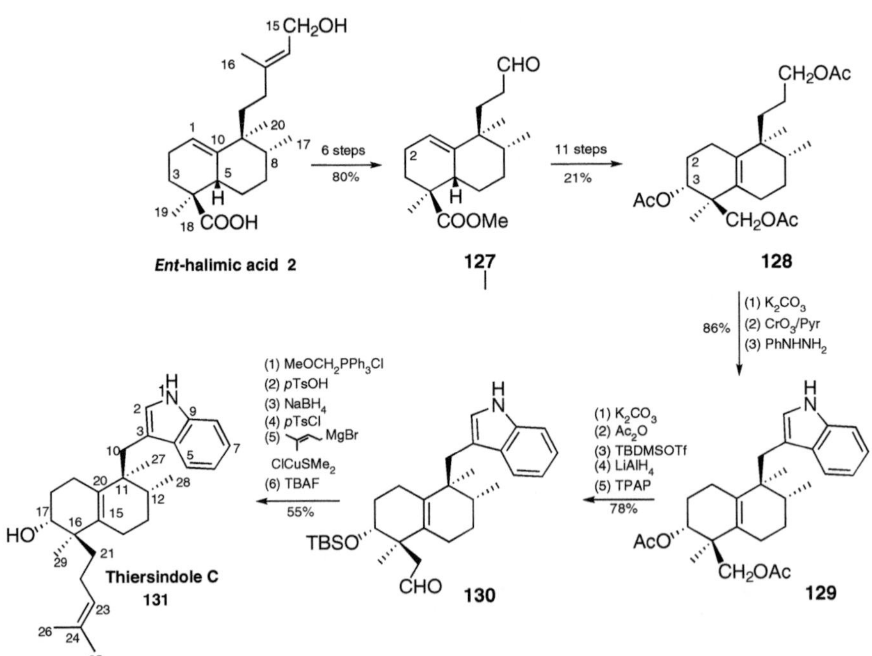

Scheme 9.20 Synthesis of (+)-thiersindole C, **131**, from *ent*-halimic acid, **2**.

9.2.7.3 Synthesis of Sesquiterpenyl Indoles

The synthesis of 12-epi-*ent*-polyalthenol **135** and 12-epi-*ent*-pentacyclindole **137** from *ent*-halimic acid **2** has been carried out using the trinorderivatives **132** and the C3-functionalized intermediates **133–134** [86, 87] (Scheme 9.21). The synthesis of pentacyclic compounds can be considered biomimetic, using cyclization of 3-(but-3-enyl) indole derivatives that produced polycyclic compounds with a hexahydrocarbazole structure. In this reaction, three new stereogenic centers are generated in one step [88]. The structure of the natural product polyalthenol **138** and pentacyclindole **139** (Scheme 9.21) are confirmed this way. Several of the sesquiterpenyl indoles synthesized show cellular proliferation inhibition of a number of human leukemic and solid tumor cell lines [86, 87].

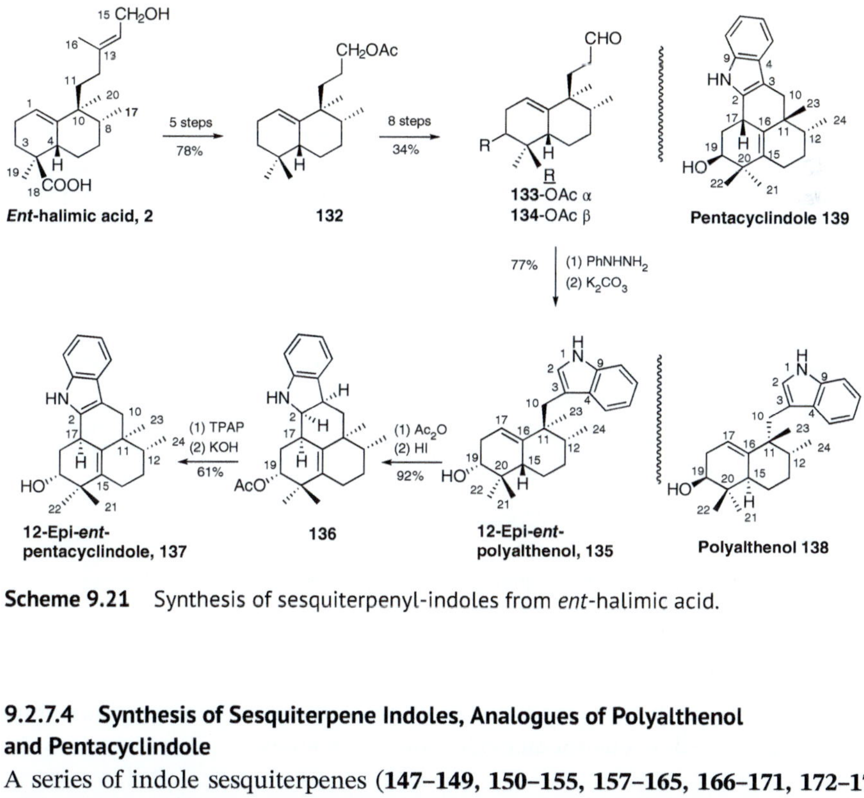

Scheme 9.21 Synthesis of sesquiterpenyl-indoles from *ent*-halimic acid.

9.2.7.4 Synthesis of Sesquiterpene Indoles, Analogues of Polyalthenol and Pentacyclindole

A series of indole sesquiterpenes (**147–149, 150–155, 157–165, 166–171, 172–175**; Figure 9.8), analogues of polyalthenol **138** and pentacyclindole **139**, have been synthesized, following similar procedures as the ones described in the last section, starting from *ent*-halimic acid, **2**, to test their biological activity. These 42 analogues include diverse oxidation levels at the sesquiterpenyl moiety and different functionalization on the indole ring. All synthetic derivatives were tested against a representative panel of Gram-positive and Gram-negative bacterial strains and human solid tumor cell lines A549 (non-small cell lung), HBL-100 (breast), HeLa (cervix), SW1573 (non-small cell lung), T-47D (breast), and WiDr (colon). Overall, the compounds presented activity against the cancer cell lines. The result, displaying

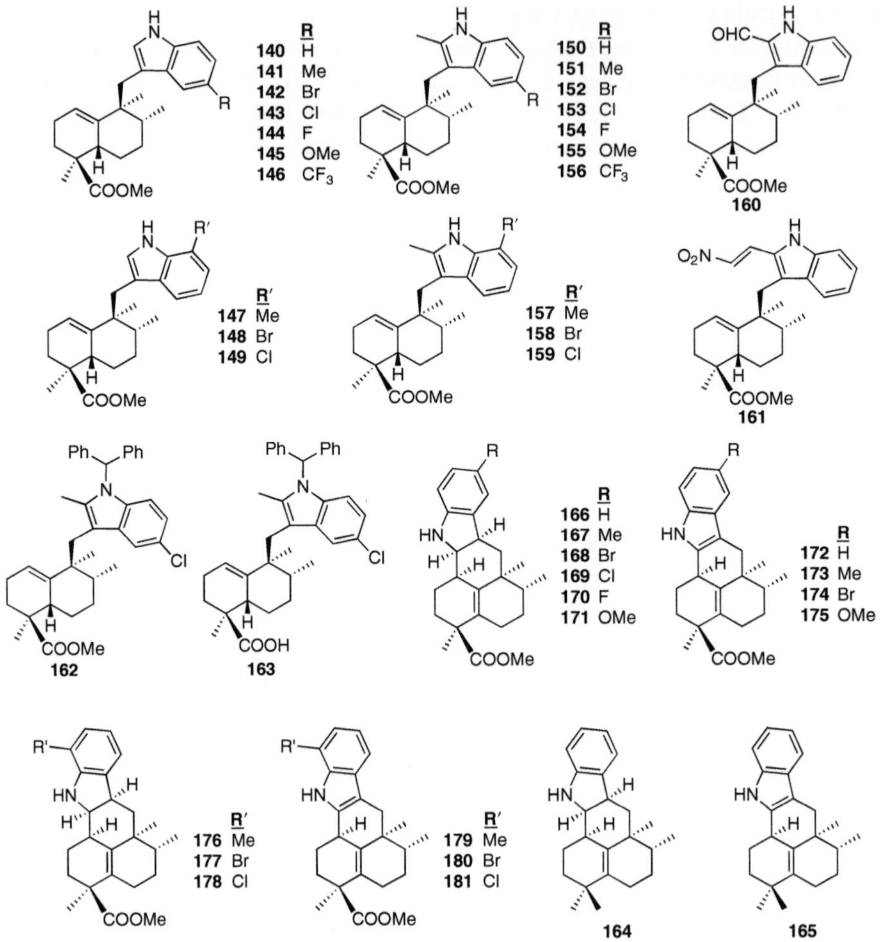

Figure 9.8 Sesquiterpene indoles, analogues of polyalthenol and pentacyclindole.

a polyalthenol scaffold, showed GI50 values in the range of 1.2–5.7 μM against all cell lines tested [89].

In this work, we summarized the high utility of diterpene halimanes as a chiral pool in the synthesis of enantiomerically pure compounds.

Acknowledgments

European Regional Development Fund (FEDER), the Ministerio de Economía y Competitividad of Spain (MINECO) (SAF 2017-89672, CTQ 2015-68175-R; AGL 2016-79813-C2-2-R), Junta de Castilla y León (UIC 21), and the University of Salamanca (Programa Propio I, 2019) are acknowledged for their support. IET thanks FSE (European Social Fund) and Junta de Castilla y León, and AMR thanks the Ministerio de Educación of Spain for their grants.

References

1 Brown, H. (1990). *Chirality in Drug Design and Synthesis*. New York: Academic Press.

2 Kotha, S. (1994). Opportunities in asymmetric synthesis: an industrial prospect. *Tetrahedron* 50 (12): 3639.

3 Jacques, J. and Collet, A. (1981). *Enantiomers Racemates and Resolutions*. New York: Wiley.

4 Kagan, H. and Fiand, J. (1988). *Topics in Stereochemistry*, vol. 18. Wiley.

5 Scout, J. (1984). *Asymmetric Synthesis*, vol. 4. New York: Academic Press.

6 Ho, T.-L. (1992). *Enantioselective Synthesis: Natural Products from Chiral Terpenes*. New York: Wiley.

7 Evans, D. and Gage, J. (1990). Diastereoselective aldol condensation using chiral oxazolidinone auxiliary: (2s, 3s)-3-hydroxy-3-phenyl-2-methylpropanoic acid. *Org. Synth.* 68: 83.

8 Bull, S.D., Davies, G.S., Jones, S. et al. (1999). Asymmetric alkylations using superquat auxiliaries – an investigation into the synthesis and stability of enolates derived from 5,5-disubstituted oxazolidin-2-ones. *J. Chem. Soc. Perkin Trans.* 4: 387.

9 Brown, H.C. and Ramachandran, P.V. (1992). Asymmetric reduction with chiral organoboranes based on .alpha.-pinene. *Acc. Chem. Res.* 25 (1): 16.

10 Ramachandran, H.C.B.P.V. (1991). The boron approach to asymmetric synthesis. *Pure Appl. Chem.* 63 (3): 307.

11 Uozumi, Y. and Hayashi, T. (1991). Catalytic asymmetric synthesis of optically active 2-alkanols via hydrosilylation of 1-alkenes with a chiral monophosphine-palladium catalyst. *J. Am. Chem. Soc.* 113 (26): 9887.

12 Knowles, W.S. and Sabacky, M.J. (1968). Catalytic asymmetric hydrogenation employing a soluble, optically active, rhodium complex. *Chem. Commun.* 22: 1445.

13 Sheldon, R.A. (1996). *Asymmetric Catalysis in Organic Synthesis*, vol. 115. Chichester: Wiley.

14 Noyori, R. (1990). Chiral metal complexes as discriminating molecular catalysts. *Science* 248 (4960): 1194.

15 Johnson, R.A. and Barry Sharpless, K. (1991). *3.2: Addition Reactions with Formation of Carbon–Oxygen Bonds: (II) Asymmetric Methods of Epoxidation*, 389. Oxford: Pergamon.

16 Moon Kim, B. and Sharpless, K.B. (1990). Heterogeneous catalytic asymmetric dihydroxylation: use of a polymer-bound alkaloid. *Tetrahedron Lett.* 31 (21): 3003.

17 Katsuki, T. and Martin, V. (1996). Asymmetric epoxidation of allylic alcohols: the Katsuki–Sharpless epoxidation reaction. *Org. React.* 48: 1–300.

18 Crispino, G.A., Ho, P.T., and Sharpless, K.B. (1993). Selective perhydroxylation of squalene: taming the arithmetic demon. *Science* 259 (5091): 64.

19 Noyori, R. (1994). *Asymmetric Catalysis in Organic Synthesis*, vol. 2. Wiley.

20 Boukouvalas, J., Wang, J.-X., Marion, O. et al. (2006). Synthesis and stereochemistry of the antitumor diterpenoid (+)-zerumin B. *J. Organomet. Chem.* 71 (17): 6670.

21 Pathak, A., Aslaoui, J., and Morin, C. (2005). Synthesis of (+)-6β–isovaleryloxylabda-8,13-diene-7α,15-diol, a metabolite from *Trismusculus reticulatus. J. Organomet. Chem.* 70 (10): 4184.

22 Marcos, I.S., Moro, R.F., Santiago Carballares, M. et al. (1999). Synthesis of (+)-limonidilactone: absolute configuration of (−)-limonidilactone. *Tetrahedron Lett.* 40 (13): 2615.

23 Hagiwara, H., Takeuchi, F., Hoshi, T. et al. (2003). First synthesis of 1,9-dideoxyforskolin from ptychantin A. *Tetrahedron Lett.* 44 (11): 2305.

24 Jung, M., Lee, S., and Yoon, B. (1997). Conversion of sclareol into (+)-galanolactone and (+)-labdienedial. *Tetrahedron Lett.* 38 (16): 2871.

25 Kingston, D.G., Jagtap, P.G., Yuan, H. et al. (2002). The chemistry of taxol and related taxoids. *Fortschr. Chem. Org. Naturst.* 84: 53.

26 Gennari, C., Vulpetti, A., Donghi, M. et al. (1996). Semisynthesis of taxol: a highly enantio- and diastereoselective synthesis of the side chain and a new method for ester formation at C13 using thioesters. *Angew. Chem. Int. Ed.* 35 (15): 1723.

27 Motika, S.E. and Hergenrother, P.J. (2020). Re-engineering natural products to engage new biological targets. *Nat. Prod. Rep.* 37 (11): 1395.

28 Romo, D. and Vanderwal, C.D. (2020). Synthetic strategies for mining the information-rich content of natural products for biology and medicine. *Nat. Prod. Rep.* 37 (11): 1393.

29 Urones, L.G., De Pascual Teresa, J., Marcos, L.S. et al. (1987). Diterpenoids from *Halimium viscosum. Phytochemistry* 26 (4): 1077.

30 De Pascual Teresa, J., Urones, J.G., Basabe, P. et al. (1985). Diterpenoids of *Halimium viscosum. Phytochemistry* 24 (4): 791.

31 De Pascual Teresa, J., Urones, J.G., Carrillo, H. et al. (1979). Acidos Diterpénicos Del *Halimium viscosum* (Wilk) P. Silva. Ácido Hidrohalímico (Ácido *Ent*-9,1-Friedolabd-1(10)-En-15-Hidroxi-18-Oico) Y Derivados. *An. Quim.* 75: 140.

32 De Pascual Teresa, J., Urones, J.G., Marcos, I.S. et al. (1986). Labdane diterpenoids from *Halimium viscosum. Phytochemistry* 25 (3): 711.

33 Rodilla, J.M.L., De Mendonça, D.I.M., Urones, J.G. et al. (1998). Hydroxylated diterpenoids from *Halimium viscosum. Phytochemistry* 49 (3): 817.

34 Rodilla, J.M.L., Mendonça, D.I.D., Ismael, M.I.G. et al. (2001). Bi- and tricyclic diterpenoids from *Halimium viscosum. Nat. Prod. Lett.* 15 (6): 401.

35 Roncero, A.M., Tobal, I.E., Moro, R.F. et al. (2018). Halimane diterpenoids: sources, structures, nomenclature and biological activities. *Nat. Prod. Rep.* 35 (9): 955.

36 Marcos, I.S., Basabe, P., Laderas, M. et al. (2003). Side-chain migration reactions and ring B aromatization in labdanes: scope and limitations. Synthesis of isofregenedane type tetrahydronaphthalenic diterpenes. *Tetrahedron* 59 (13): 2333.

37 Marcos, I.S., Laderas, M., Dı́ ez, D. et al. (2003). Synthesis and absolute configuration of (−)-chrysolic acid and (+)-isofregenedol. *Tetrahedron Lett.* 44 (29): 5419.

38 Urones, J.G., Marcos, I.S., Basabe, P. et al. (1990). Formation of orthoesters in the sharpless asymmetric epoxidation: hemisynthesis of labdanes. *Tetrahedron* 46 (7): 2495.

39 Urones, J.G., Marcos, I.S., Basabe, P. et al. (1994). Compounds with the labdane skeleton from *Halimium viscosum*. *Phytochemistry* 35 (3): 713.

40 Marcos, I.S., Moro, R.F., Carballares, M.S. et al. (2000). An efficient total synthesis of isodrimeninol from zamoranic acid. *Synlett* 2000 (04): 541.

41 Urones, J.G., Marcos, I.S., Perez, B.G. et al. (1994). Transformation of labdanes into drimanes. Semisyntheses of poligodial and warburganal from zamoranic acid. *Nat. Prod. Lett.* 5 (1): 21.

42 Urones, J.G., Marcos, I.S., Gómez-Pérez, B. et al. (1994). Diastereoselective ring – opening of 12-acetoxy-9α and 9β(11)-epoxy-7-drimene: homochiral semisynthesis of poligodial and warburganal. *Tetrahedron Lett.* 35 (22): 3781.

43 Urones, J.G., Marcos, I.S., Gómez Pérez, B. et al. (1994). Chemistry of zamoranic acid. Part V homochiral semisyntheses of active drimanes: pereniporin B, polygodial and warburganal. *Tetrahedron* 50 (37): 10995.

44 Urones, J.G., Marcos, I.S., Gómez Pérez, B. et al. (1995). Chemistry of zamoranic acid. Part IX homochiral synthesis of polygodial and warburganal from 17-acetoxy-7-labden-15-ol. *Tetrahedron* 51 (6): 1845.

45 Urones, G.J., Díez, D., Gómez, M.P. et al. (1997). Chemistry of zamoranic acid. Part 10. Homochiral hemisynthesis of Pereniporin A. *J. Chem. Soc. Perkin Trans.* 12: 1815.

46 Marcos, I.S., Moro, R.F., Carballares, S. et al. (2001). Synthesis of (+)-limonidilactone and 12-epi-limonidilactone. *Tetrahedron* 57 (4): 713.

47 Marcos, I.S., Cubillo, M.A., Moro, R.F. et al. (2005). Synthesis of tri- and tetracyclic diterpenes. Cyclisations promoted by SmI_2. *Tetrahedron* 61 (4): 977.

48 Timmermann, B.N., Hoffmann, J.J., Jolad, S.D. et al. (1982). Constituents of *Chrysothamnus paniculatus* (compositae). 2. Chrysolic acid, a new labdane-derived diterpene with an aromatic B ring. *J. Organomet. Chem.* 47 (21): 4114.

49 Zdero, C., Bohlmann, F., and Niemeyer, H.M. (1991). Seco-, nor-, normal and rearranged labdanes from *Haplopappus parvifolius*. *Phytochemistry* 30 (11): 3683.

50 Marcos, I.S., Benéitez, A., Moro, R.F. et al. (2010). Semisynthesis of (+)-angeloyl-gutierrezianolic acid methyl ester diterpenoid. *Tetrahedron* 66 (45): 8605.

51 Marcos, I.S., Moro, R.F., Carballares, M.S. et al. (2002). Cyclization of bicyclic diterpenes promoted by SmI_2. Synthesis of tri- and tetracyclic diterpenes. *Synlett* 2002 (03): 0458.

52 van der Vlugt, J.I., Grutters, M.M.P., Ackerstaff, J. et al. (2003). Possphites–Monophosphites derived from incompletely condensed silsesquioxanes. *Tetrahedron Lett.* 44 (45): 8301.

53 Marcos, I.S., Pedrero, A.B., Sexmero, M.J. et al. (2003). Synthesis and absolute configuration of three natural *ent*-halimanolides with biological activity. *Tetrahedron Lett.* 44 (2): 369.

54 Marcos, I.S., Pedrero, A.B., Sexmero, M.J. et al. (2005). Synthesis of *ent*-halimanolides from *ent*-halimic acid. *Synthesis* 2005 (19): 3301.

55 Bestmann, H.J. (1977). Phosphacumulene ylides and phosphaallene ylides [new synthetic methods (19)]. *Angew. Chem. Int. Ed.* 16 (6): 349.

56 Boukouvalas, J. and Lachance, N. (1998). A mild, efficient and general method for the synthesis of trialkylsilyl (Z)-4-oxo-2-alkenoates and Γ-hydroxybutenolides. *Synlett* 1998 (01): 31.

57 Marcos, I.S., Gonzalez, J.L., Sexmero, M.J. et al. (2000). Diterpenic α- and β-hydroxybutanolides with antifeedant activity: semisynthesis and absolute configuration. *Tetrahedron Lett.* 41 (15): 2553.

58 Marcos, I., Conde, A., Moro, R. et al. (2008). Synthesis of an *ent*-halimanolide from *ent*-halimic acid. *Molecules* 13 (5): 1120.

59 Marcos, I.S., Hernández, F.A., Sexmero, M.J. et al. (2002). Synthesis and absolute configuration of (−)-chettaphanin II. *Tetrahedron Lett.* 43 (7): 1243.

60 Marcos, I.S., Hernández, F.A., Sexmero, M.J. et al. (2003). Synthesis and absolute configuration of (−)-chettaphanin I and (−)-chettaphanin II. *Tetrahedron* 59 (5): 685.

61 Fontana, A., Ciavatta, M.L., and Cimino, G. (1998). Cladocoran A and B: two novel Γ-hydroxybutenolide sesterterpenes from the Mediterranean coral *Cladocora cespitosa*. *J. Organomet. Chem.* 63 (9): 2845.

62 Marcos, I.S., Pedrero, A.B., Sexmero, M.J. et al. (2003). Synthesis of bioactive sesterterpenolides from *ent*-halimic acid. 15-epi-*ent*-cladocoran A and B. *J. Organomet. Chem.* 68 (19): 7496.

63 Marcos, I.S., Pedrero, A.B., Sexmero, M.J. et al. (2002). Synthesis and absolute configuration of the supposed structure of cladocoran A and B. *Synlett* 2002 (01): 0105.

64 Miyaoka, H., Yamanishi, M., Kajiwara, Y. et al. (2003). Total synthesis of cladocorans A and B: a structural revision. *J. Organomet. Chem.* 68 (9): 3476.

65 Marcos, I.S., Escola, M.A., Moro, R.F. et al. (2007). Synthesis of novel antitumoural analogues of dysidiolide from *ent*-halimic acid. *Bioorg. Med. Chem.* 15 (17): 5719.

66 Gunasekera, S.P., McCarthy, P.J., Kelly-Borges, M. et al. (1996). Dysidiolide: a novel protein phosphatase inhibitor from the Caribbean sponge *Dysidea etheria* de Laubenfels. *J. Am. Chem. Soc.* 118 (36): 8759.

67 Corey, E.J. and Roberts, B.E. (1997). Total synthesis of dysidiolide. *J. Am. Chem. Soc.* 119 (51): 12425.

68 Takahashi, M., Dodo, K., Sugimoto, Y. et al. (2000). Synthesis of the novel analogues of dysidiolide and their structure–activity relationship. *Bioorg. Med. Chem. Lett.* 10 (22): 2571.

69 Brohm, D., Metzger, S., Bhargava, A. et al. (2002). Natural products are biologically validated starting points in structural space for compound library development: solid-phase synthesis of dysidiolide-derived phosphatase inhibitors. *Angew. Chem. Int. Ed.* 41 (2): 307.

70 Draetta, G. and Eckstein, J. (1997). Cdc25 protein phosphatases in cell proliferation. *Biochim. Biophys. Acta, Rev. Cancer* 1332 (2): M53.

71 Gil-Mesón, A., Roncero, A., Tobal, I. et al. (2016). Synthesis of bioconjugate sesterterpenoids with phospholipids and polyunsaturated fatty acids. *Molecules* 21 (1): 47.

72 Marcos, I., Sexmero, M., Hernández, F. et al. (2006). Synthetic studies towards the *ent*-labdane diterpenoids: rearrangement of *ent*-halimanes. *Molecules* 11 (10): 792.

73 Marcos, I.S., García, N., Sexmero, M.J. et al. (2007). Synthetic studies towards Picrasane quassinoids. *Tetrahedron* 63 (10): 2335.

74 VanderRoest, J.M. and Grieco, P.A. (1993). Total synthesis of (.+−.)-bruceantin. *J. Am. Chem. Soc.* 115 (13): 5841.

75 Marcos, I.S., Martínez, B., Sexmero, M.J. et al. (2006). Chemistry of *ent*-halimic acid: synthesis of [4.3.3]propellanes. *Synthesis* 2006 (22): 3865.

76 Marcos, I.S., Conde, A., Moro, R.F. et al. (2010). Synthesis of quinone/hydroquinone sesquiterpenes. *Tetrahedron* 66 (42): 8280.

77 Marcos, I.S., Conde, A., Moro, R.F. et al. (2010). Quinone/hydroquinone sesquiterpenes. *Mini Rev. Org. Chem.* 7 (3): 230.

78 Barton, D.H.R., Bridon, D., and Zardb, S.Z. (1987). The invention of radical chain reactions: part XIV. A decarboxylative radical addition to quinones. *Tetrahedron* 43 (22): 5307.

79 Barton, D.H.R. and Sas, W. (1990). The invention of radical reactions. Part XIX. The synthesis of very hindered quinones. *Tetrahedron* 46 (10): 3419.

80 Barton, D.H.R., Lacher, B., and Zard, S.Z. (1987). The invention of radical reactions: part XVI. Radical decarboxylative bromination and iodination of aromatic acids. *Tetrahedron* 43 (19): 4321.

81 Marcos, I.S., García, N., Sexmero, M.J. et al. (2005). Synthesis of (+)-agelasine C. a structural revision. *Tetrahedron* 61 (49): 11672.

82 Rosemeyer, H. (2004). The chemodiversity of purine as a constituent of natural products. *Chem. Biodivers.* 1 (3): 361.

83 Nakamura, H., Wu, H., Ohizumi, Y. et al. (1984). Agelasine-A, -B, -C and -D, novel bicyclic diterpenoids with a 9-methyladeninium unit possessing inhibitory effects on Na,K-Atpase from the Okinawa Sea sponge *Agelas* sp.1. *Tetrahedron Lett.* 25 (28): 2989.

84 Hattori, T., Adachi, K., and Shizuri, Y. (1997). New agelasine compound from the marine sponge *Agelas mauritiana* as an antifouling substance against macroalgae. *J. Nat. Prod.* 60 (4): 411.

85 Marcos, I.S., Escola, M.A., Moro, R.F. et al. (2007). Synthesis of (+)-thiersindole C. *Synlett* 2007 (13): 2017.

86 Marcos, I.S., Moro, R.F., Costales, I. et al. (2012). Synthesis of 12-epi-*ent*-polyalthenol an antitumour indole sesquiterpene alkaloid. *Tetrahedron* 68 (38): 7932.

87 Marcos, I.S., Moro, R.F., Costales, I. et al. (2013). Biomimetic synthesis of an antitumour indole sesquiterpene alkaloid, 12-epi-*ent*-pentacyclindole. *Tetrahedron* 69 (35): 7285.

88 Marcos, I.S., Moro, R.F., Costales, I. et al. (2009). Synthesis of hexahydrocarbazoles by cyclisation of 3-(but-3-enyl) indole derivatives. *Tetrahedron* 65 (49): 10235.

89 Marcos, I.S., Moro, R.F., Costales, I. et al. (2014). Synthesis and biological activity of polyalthenol and pentacyclindole analogues. *Eur. J. Med. Chem.* 73: 265.

10

Alkaloids as Chiral Building Blocks, Auxiliaries, Ligands, and Molecular Diversity

Karol Kacprzak[1], *Elżbieta Wojaczyńska*[2], *Andrzej Trochimczuk*[2], *Franz Steppeler*[2], *and Jacek Wojaczyński*[3]

[1]*Adam Mickiewicz University in Poznań, Faculty of Chemistry, Ul. Uniwersytetu Poznańskiego 8, 61-614 Poznań, Poland*
[2]*Wrocław University of Science and Technology, Faculty of Chemistry, Wybrzeże Wyspiańskiego 27, 50 370 Wrocław, Poland*
[3]*University of Wrocław, Faculty of Chemistry, F. Joliot-Curie St. 14, 50 383 Wrocław, Poland*

10.1 Introduction

Alkaloids constitute a large, heterogeneous family of natural products with c. 20 000 members of great importance for asymmetric synthesis and medicine [1–5]. Most of them are chiral, having often a unique scaffold and additional functionalities, e.g. hydroxyl group, or multiple nitrogen atoms offering different coordinating sites. These structural features make many of them able to transfer chirality, both in molecular recognition and enantioselective catalytic processes, where they work as selectors, organocatalysts, auxiliaries, or ligands for metal ions. For more than a century, numerous alkaloids, such as four major members of Cinchona family (quinine, quinidine, cinchonidine, and cinchonine) ephedrines, brucine, or strychnine, have been used as chiral bases for resolution of racemic carboxylic acids by diastereomeric salts formation and their subsequent crystallization [6, 7]. Alkaloids have also been studied as selectors for various chiral recognition and separation processes. Worth to note are chiral stationary phases based on pseudoenantiomeric quinine and quinidine 9-*O*-carbamates that were commercialized (ChiraPak QN-AX) and are available for an enantioseparation of many types of acidic analytes [8, 9]. Most important studies employing alkaloids were however devoted to stereoselective synthesis [10]. In this field various alkaloids and their derivatives were investigated as useful chiral reagents and auxiliaries [11], in particular ephedrines [12–14] or as stoichiometric chiral ligands (mainly sparteine and its surrogate [15–18]). Asymmetric catalysis uses many alkaloids and their derivatives, mostly those having 1,2-aminoalcohol motif such as ephedrines [13, 14], and Cinchona alkaloids. The latter, in line with numerous designed derivatives, have been recognized as privileged organocatalysts or ligands for metal-catalyzed reactions. Cinchona alkaloids have been shown to efficiently catalyze more than 100 types of

Chiral Building Blocks in Asymmetric Synthesis: Synthesis and Applications, First Edition.
Edited by Elżbieta Wojaczyńska and Jacek Wojaczyński.

stereoselective transformations in all available catalysis formats – homogeneous, biphasic (phase transfer) heterogeneous, and as solid phase support [10, 19–23]. Several of Cinchona-catalyzed reactions have been performed on scale in various stages of process development [24–27], and one has also been commercialized in the 1980s by Lonza for an enantioselective citramalic acid manufacturing [25]. Catalytic asymmetric dihydroxylation (AD) of olefins into *syn*-1,2-diols performed with the rationally developed bis-dihydroquinine or bis-dihydroquinidine-phthalazine ligands ((DHQ)$_2$PHAL and (DHQD)$_2$PHAL) in water–organic media with *in situ* regeneration of Os(VIII) resembles biocatalysis in a performance [28]. Less common alkaloids and their derivatives such as (−)-brucine, brucine-21,22-diol [29], (−)-strychnine, and (−)-nicotine are also occasionally employed in stereoselective synthesis and are a subject of a devoted review [11].

Last but not least certain native alkaloids are still in use in medicine and other industries (morphine, codeine, quinine, cytisine, pseudoephedrine, nicotine, camptothecin) [1–5], and some alkaloids serve as precursors for large-scale production of valuable semisynthetic drugs that cannot be prepared by total synthesis or biotechnologically by cost-effective manner. This is the case of anticancer camptothecin and its derivatives [30–33], vinca alkaloids [34–36], or a number of opioid-based analgesics, such as morphine [37, 38] and many others (*vide infra*) [2–4].

Limited supply, as well as typically a single enantiomer availability, high cost, and legal issues are limiting work with some alkaloids and make their stoichiometric chemistry difficult. Compared with other chiral pool chemicals, such as α-amino- and α-hydroxyacids or monoterpenes, alkaloids with their rather complex structure (framework + numerous stereogenic centers, see examples in Figure 10.1) are not an obvious source of basic and versatile building blocks for asymmetric synthesis. Beside numerous catalytic applications (not presented here, see [10, 11, 13–29]) they are rather used *per se* for derivatization or conjugation toward novel semisynthetic drugs, identification of novel bioactivities, or for construction of designed catalysts, ligands, or enantioselective separation systems. Just a very few simple alkaloids, such as ephedrines, quinine/quinidine, but also cocaine, can however be easily converted into a number of valuable chiral building blocks that were used for stereoselective synthesis.

Nevertheless, due to the unique biological activities of alkaloids and the constant pressure from pharma industry expecting novel drug leads, candidates, and screening libraries there is remarkable progress in their chemistry. Alkaloids are also a polygon for testing of novel synthetic tools, either for their total synthesis or for late-stage derivatization [39]. In this context transition-metal-catalyzed cross-couplings [40–42], as well as a plethora of recently expanded heteroarene CH-functionalization methods [43–45] involving electro- and photochemical activation [46–48] and click chemistry (Cu(I)-catalyzed Huisgen cycloaddition of azides and alkynes) [49] have been widely explored.

Surprisingly, more complex alkaloids have recently been considered as an advantageous source of chiral molecular diversity through ring-distortion approaches (formally belonging to the diversity-oriented synthesis, DOS) [50–53]. This subset of DOS uses readily available natural products, e.g. alkaloids or terpenoids, whose cyclic molecular frameworks can readily be converted into defined different cyclic scaffolds (contracted, expanded, rearranged, annulated, or cleaved to the linear

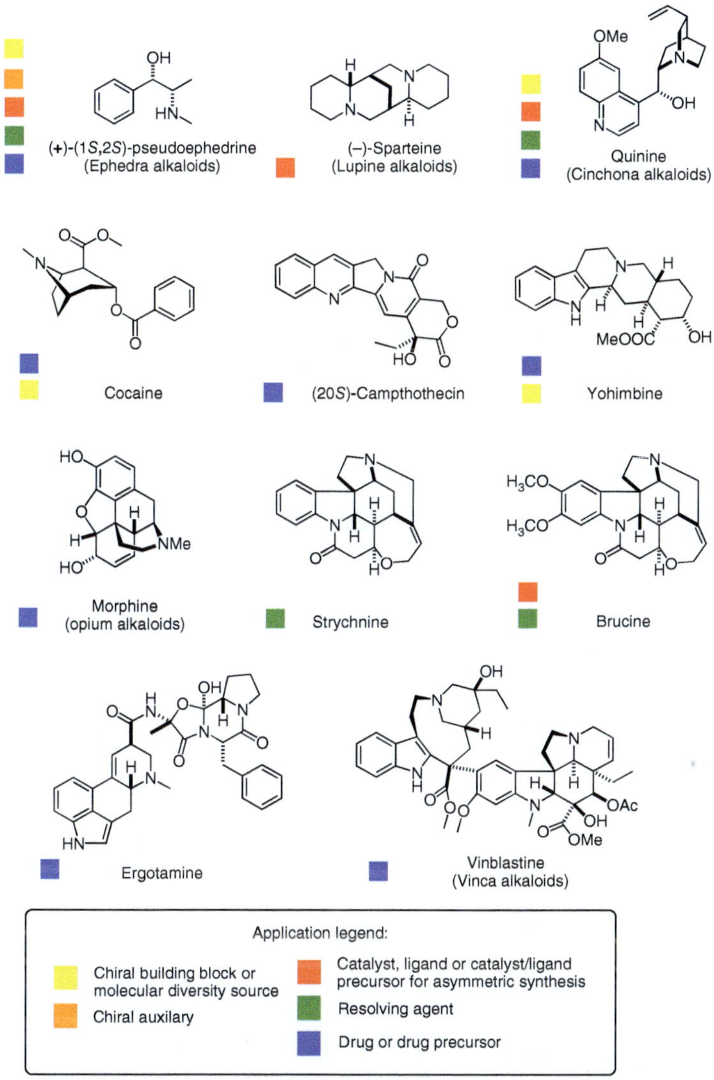

Figure 10.1 Selected alkaloids presented in this chapter and their applications. Note that all are biologically active, but their applicability in stereoselective synthesis drastically decreases with increasing structural and stereochemical complexity and decreasing availability.

products) while retaining a complex and defined stereochemistry. Their further modification led to highly diverse libraries of pseudo-natural products suitable for bioactivity screening.

A representative example of ring-distortion approach applied to tryptoline ring containing commercially available and relatively inexpensive indole alkaloid yohimbine has been reported by Huigens et al. [51] A set of properly identified reactions enabled synthesis of 70 complex and diverse compounds **1–6** that were both architecturally and stereochemically unique as compared to the parent alkaloid (Scheme 10.1). These products were subjected to both phenotypic screenings and reporter gene assays that resulted in an identification of new hits having antiproliferative activities against cancer cells with functional hypoxia-inducible factors, nitric oxide inhibition, or inhibition and activation of the antioxidant response element (ARE) [52]. It is worth to note that there are several examples of similar ring-cleavage or rearrangement reaction applied to alkaloids in literature, identified by their structure elucidation, synthesis, or probing the reactivity. Other examples of ring-distortion approach involving quinine [53], (−)-cytisine [54] and other natural products [53] are also available.

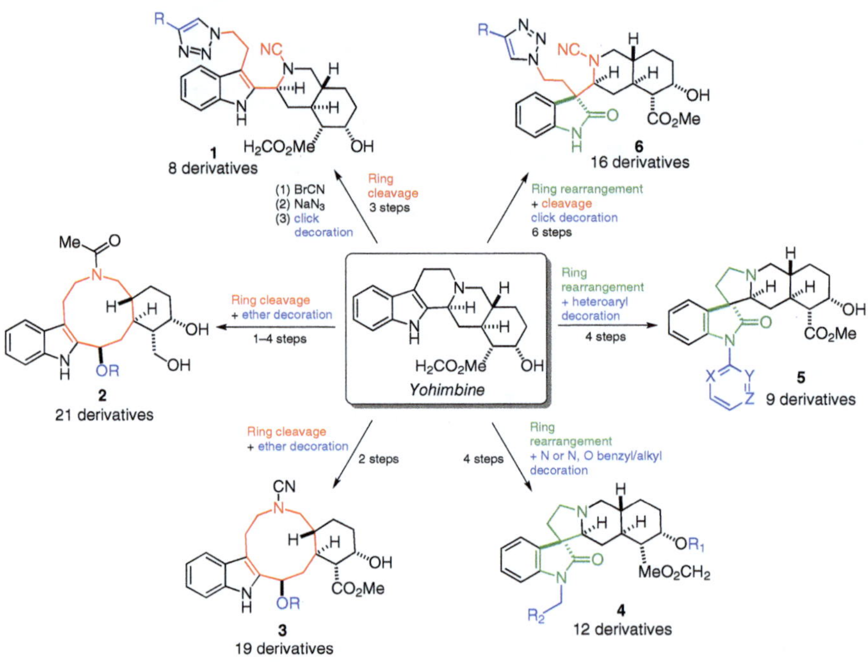

Scheme 10.1 Ring distortion of yohimbine with a subsequent decoration of fragments toward highly diverse chiral library. Source: Paciaroni et al. [52]/John Wiley & Sons.

Before the systematic presentation of material, it is worth to highlight some general limitations hampering the work with alkaloids. Natural origin of alkaloids and their biological role results in typically low abundance of these products over plant and animal world. Several medicinally relevant alkaloids are available on kilogram to ton scale (ephedrines, Cinchona alkaloids, sparteine, nicotine, strychnine, brucine, opium alkaloids, colchicine, and anticancer camptothecin, vinca as well as few others). The isolation of other alkaloids requires access to the biological material and uses a rather laborious blend of extractive, chromatographic, or crystallization steps to get the alkaloid in pure form [2, 55]. The concentration of alkaloids in biological material varies and may be affected by sensitivity of crops to external factors (weather conditions, pathogens presence, cultivation culture). This may result in limited market availability of some alkaloids, as it was highlighted recently for widely used (−)-sparteine, which is relatively abundant in plants [56].

The quest for the rise of the accessibility of several important alkaloids for medicine and industry resulted in a dynamic development of their biochemistry and biotechnology as well as total syntheses. The understanding of the biosynthesis of alkaloids, both on enzymatic and genetic levels, and the use of current biotechnological techniques offer some general approaches for optimizing/increasing their production. In particular, the use of plant cell cultures in line with plant hormones and specific elicitors enhances the secondary metabolites production [57–61]. On the other hand, the use of genetic engineering techniques such as overexpression of key enzymes involved in the biosynthesis of desired alkaloids or even an installation of a complete biochemical plant pathway into microorganisms has enabled the production of plant alkaloids for simple precursors [62–66].

The total synthesis of complex, enantiomerically pure alkaloids was, since their isolation and structure elucidation, a challenge and a driving force for conceptual and technical inventions in organic chemistry. Despite the successful preparation of vast numbers of complex alkaloids, their total synthesis is not expected to provide soon large quantities of products by cost-effective manner. For example, attempts for the enantioselective synthesis of rather structurally simple antimalarial quinine initiated by Dering/Woodward in the period of WW2 failed and have resulted in presenting the formal synthesis. Curiously, quinine has only been prepared in enantioselective fashion by Stork in 2001, in a route involving 13 steps and giving <10% yield (for an retrospective account referring this topic see an excellent review [67]). Although even unnatural enantiomer of (+)-quinine has recently been prepared by 15-step synthesis (longest linear sequence [LLS]) with 16% overall yield [68], it is not expected that the total synthesis of alkaloids replace their isolation from natural sources or biotechnological production.

So far relatively few simple alkaloids are produced synthetically or by combined chemical/biotechnological synthesis, and these include all diastereomeric ephedrines or synthetic (*S*)- and racemic nicotine.

An interesting alternative to a canonical chemical synthesis of alkaloids is the use of dedicated enzymes participating in their biosynthesis for their chemoenzymatic preparation. Enzymes are able to construct sophisticated carbon framework by various C—C bond formations or may be used for functionalization of synthetic or natural alkaloids [69–71]. The use of enzymes also assures the chemo-, regio-, and enantioselective course of reaction and results often in a remarkable shortening and simplifying of the synthetic sequence. This growing field of research involves, for example, the application of peptide synthetase SfmC that catalyzes Pictet–Spengler reactions as key steps for the preparation of the tetrahydroisoquinoline skeleton in antitumor antibiotics such as saframycin, use of strictosidine synthase, norcoclaurine synthase, prenylation–oxidative modification enzyme participating in the biosynthesis of indole mycotoxin such as penitrem or versatile Diels–Alderases identified in plants producing monoterpene indole alkaloids [69–71].

The use of enzymes is not restricted to those participating in a particular alkaloid biosynthesis. For example, Turner and coworkers demonstrated a "toolbox" of designed monoamine oxidase variants from *Aspergillus niger* having broad substrate scope and tolerance for sterically demanding motifs for an efficient asymmetric synthesis of (*R*)-coniine, (*R*)-eleagnine, (*R*)-leptaflorine, and (*R*)-harmicine [72]. Further development of the chemoenzymatic approach combining advantages of either enzymes or modern chemistries may constitute a practical and commercially viable alternative to the classical total synthesis [69, 70].

Last but not least working with most common alkaloids is limited by legal regulation. Many alkaloids exert a strong physiological effect on human organism and were classified as narcotics. For example, chiral cocaine, morphine, ergine, or ibogaine, having fascinating molecular architecture and bioactivity, belong to strictly controlled compounds. The access to them is therefore heavily restricted which hampers their wider use in science. Interestingly, cocaine that is illegally produced in multiton quantities might be a convenient source of chiral pyrrolidines (*vide infra*) but is rather legally destroyed than utilized as valuable chiral product for research or medicine. Ephedrines and pseudoephedrines are treated as precursors for methamphetamine and other illegal designer drugs [73], so handling them also undergoes controlling with special permission varying from country to country. An interesting alternative to the restricted ephedrines has been demonstrated by Myers and coworkers, who designed their substitutes, namely (1*S*,2*S*)- and (1*R*,2*R*)-pseudoephenamines **8**. They remain regulatory free and are easily accessible from 1,2-aminoalcohol **7** in a three-step synthesis (Scheme 10.2). These synthetic replacements of ephedra alkaloids are often superior to parent pseudoephedrines in asymmetric synthesis. In particular, they give equal or greater diastereoselectivities in asymmetric alkylation reactions, with notable improvements in the selectivity of formation of quaternary stereocenters. The pseudoephenamine-derived amides tend to be crystalline substances as compared with the corresponding pseudoephedrine derivatives and provide sharp, well-defined signals in NMR spectra facilitating the isolation and analytics during synthesis [74, 75].

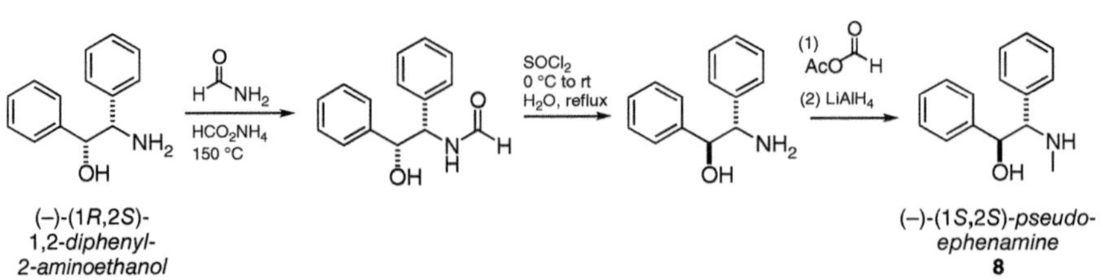

Scheme 10.2 Synthesis of regulatory free pseudoephenamine substitute of pseudoephedrine having similar outcome in asymmetric synthesis. Source: Based on Morales et al. [74] and Mellem and Myers [75].

Working under a strict regime, with fully controlled material, usually in pharmaceutical industry, discourages most academic organic chemists and is a reason for a slower progress of alkaloids chemistry in the last decades. An inspection of databases confirmed that apart from medicinal use, isolation, analytics, or total synthesis many alkaloids have not generated a chemical output since 2000 (e.g. scopolamine, ajmaline, aconitine).

Consequently, among c. 20 000 identified alkaloids just a very few are commercially available, in particular, those of relatively simple structure with a smaller number of stereogenic centers (apart from their catalytic application) are used *per se* as chiral building blocks or are prone to undergo a well-defined, high-yielding, and stereoselective degradation to provide an access to smaller chiral unique fragments.

This chapter presents an overview of the stoichiometric chemistry of the common alkaloids belonging to different classes with a focus on their transformations into chiral building blocks for further synthesis, application as chiral auxiliaries or direct use in stereoselective synthesis of natural products, as well as a source of chiral molecular architectures. Occasionally, simple derivatization or conjugation of alkaloids will also be mentioned when they led to useful products. The chapter is organized according to the rising structural complexity of alkaloids. Small molecular entities such as ephedra and tobacco alkaloids are presented first followed by lupine, Cinchona, and tropane families. Application of alkaloids for the construction of chiral, functional polymers is covered in Section 10.7.

The chapter covers recent examples from the literature published in the twenty-first century although some older examples will occasionally be presented to give a better overall picture of alkaloids significance in stereoselective synthesis and medicinal chemistry. Use of alkaloids as catalysts and ligands in stereoselective synthesis as well as their biological activities are a dominant part of published work and were not referred to here (for general reviews see [1–23]).

10.2 Ephedra Alkaloids

These simple alkaloids **9–12** (Figure 10.2) contain chiral 1,2-aminoalcohol functionality that is especially favored as a motif in catalysts, ligands, and resolving agents used in stereoselective synthesis [12–14]. Ephedrine **9** is still isolated in ton quantities from *Ephedra sinica*; its main current source is, however, chemical or biochemical synthesis, which provides all four diastereomers of the products (two enantiomeric ephedrines and two pseudoephedrines). A detailed review by Bautista

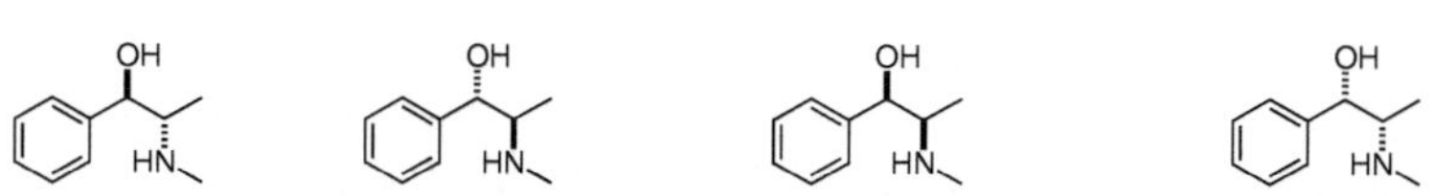

(–)-(1*R*,2*S*)ephedrine **9** (+)-(1*S*,2*R*)-ephedrine **10** (–)-(1*R*,2*R*)-pseudoephedrine **11** (+)-(1*S*,2*S*)-pseudoephedrine **12**

Figure 10.2 Ephedrine and its stereoisomers.

and coworkers provides a complete list of 26 alkaloids, mainly based on ephedrine skeleton, isolated from plants belonging to *Ephedra* genus, and their applications in organic synthesis, biological activity, and medical use [76].

Majority of papers, however, focus on the applications of the natural (−)-(1*R*,2*S*)-ephedrine **9**, due to its common availability and reasonable price. Ephedra alkaloids early found a broad use in asymmetric synthesis – as resolution agents for chiral organic acids, as chiral auxiliaries, and, after appropriate modifications, as ligands for metal-based catalysts or organocatalysts for a number of stereoselective reactions. The most remarkable early achievements include Evans' oxazolidinones or Abiko–Masamune auxiliaries for stereoselective aldol reaction or bis-oxazolines for various additions and are a subject of reviews by Cruz and coworkers [13, 14] as well as Hitchcock and Dore [12]. In particular, ephedrine can be easily transformed into a number of useful chiral heterocycles of special interest for the subject of the current book [77, 78].

Ephedra alkaloids have been relatively rarely used as starting chiral pool reactants in asymmetric synthesis (Scheme 10.3). However, a series of other 1,2-aminoalcohols **13** were prepared by Page et al. by reductive ring opening of oxazolidines **14** formed from (1*S*,2*R*)-ephedrine **10** or (1*S*,2*S*)-pseudoephedrine **12** [79]. The latter was used by Pramanik et al. in the scalable route to benzphetamine hydrochloride **15**, a drug against anorexia [80]. More frequently, ephedrines were applied to the stereoselective preparation of chiral heterocycles **16–21**. For example, substituted tetrahydroisoquinolines **16** and oxazilidines **17** were synthetized in a stereospecific manner as described by Davis and coworkers [81]. Henry and O'Neill obtained morpholine *N*-oxides **18** [82], whereas Terán's group morpholin-3-ones **20** via epoxyderivatives **19** [83], in both cases with defined configurations of the four stereogenic centers. Ephedrine derivatives served as chiral building blocks in the diastereoselective synthesis of optically pure benzosultams **21** that were obtained by superelectrophilic activation by Michelet et al. [84] (Scheme 10.3).

An active field of ephedrines research involves their use for preparation of chiral ligands and catalysts [12–14]. Simple *N*-methylephedrine **22** and *N*-pyrrolidinylnorephedrine **24** are effective sub- and stoichiometric ligands for the preparation of enantiomerically pure propargylic alcohols **23** by selective addition of terminal acetylenes to aldehydes or ketones. The reaction reported by Carreira and coworkers involves *in situ* formation of the zinc alkynylide intermediate by the reaction of a base and zinc trifluoromethanesulfonate, and provides a direct access to wide range of propargylic alcohols **23** in good-to-excellent yields and enantioselectivities (Scheme 10.4) [85]. Catalytic version of this addition was also reported [86]. Asymmetric alkynylation of a trifluoromethylketone **25** to Efavirenz precursor **26** mediated by the chiral zinc aminoalkoxide complex prepared from *N*-pyrrolidinylnorephedrine **24** was accomplished on a 50-g scale with a high yield of 95% and asymmetric induction level of 99% ee (Scheme 10.4) [87].

Substoichiometric amount (0.45 equiv) of *N*-methylephedrine **22** in combination with zinc has been used by Palomo et al. as a chiral ligand for enantioselective aza-Henry reaction, providing the corresponding 1,2-nitroamines in high enantioselectivities and yields [88]. Similarly, substoichiometric loading of the same

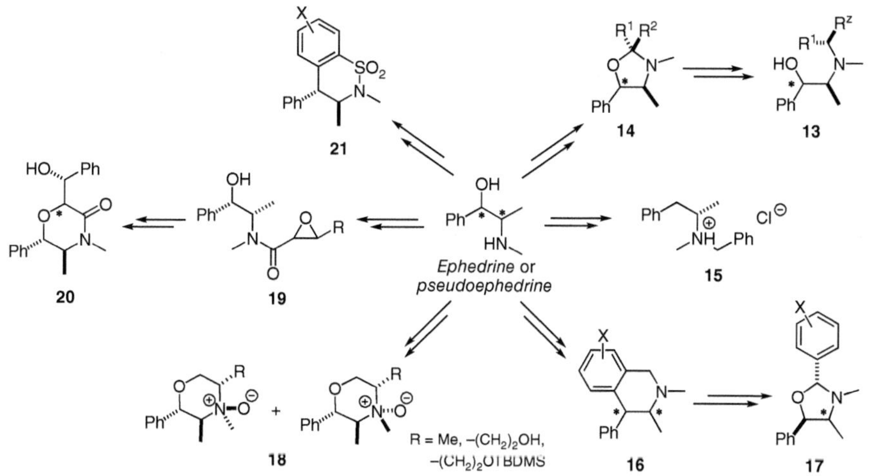

Scheme 10.3 Ephedrine isomers as chiral building blocks. Source: Based on Refs. [79–84].

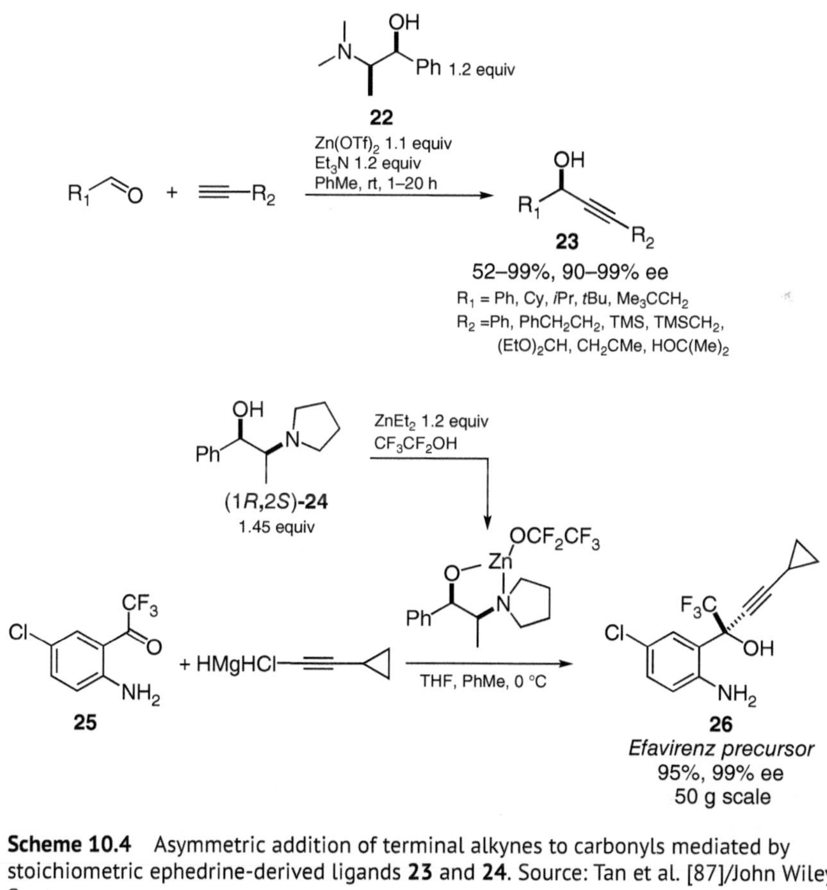

Scheme 10.4 Asymmetric addition of terminal alkynes to carbonyls mediated by stoichiometric ephedrine-derived ligands **23** and **24**. Source: Tan et al. [87]/John Wiley & Sons.

ligand was employed in an asymmetric imino-Reformatsky reaction by Cozzi and coworkers. The three-component one-pot process combining an aldehyde, 2-methoxyaniline, α-bromoester, and dimethylzinc and a catalytic amount of a nickel salt led to the corresponding 2-aminoester with high enantiomeric excess up to 92% ee and a good yield [89, 90].

P- and C-stereogenic aminophosphines **27–29** were prepared as ligands for Pd- and Rh-catalyzed asymmetric: allylation [91], hydroformylation [92], and hydrogenation [93]. A series of oxazolidines **30** (Scheme 10.5) derived from (1*R*,2*S*)-ephedrine **9** or (1*S*,2*S*)-pseudoephedrine **12** and salicylaldehydes were also tested in addition of diethylzinc to aryl aldehydes, which proceeded with good yields, but moderate enantioselectivities [94]. Braga and coworkers obtained a series of selenides **31** and diselenides **32** based on ephedrine and used them as catalysts for enantioselective addition of boronic acids, and highly enantioselective diethylzinc addition to aryl aldehydes [95]. Ephedrine-based oxazolidine bearing phosphines **33** were tested in palladium-catalyzed asymmetric allylic alkylation by Jin et al. [96], whereas Page and co-workers used ephedra-based 1,2-aminothioethers **34** in the same reaction [97]. Chiral thioureas **35** derived from ephedrine and pseudoephedrine introduced by Bolm's group turned out to be a good organocatalyst for the asymmetric Michael addition [98].

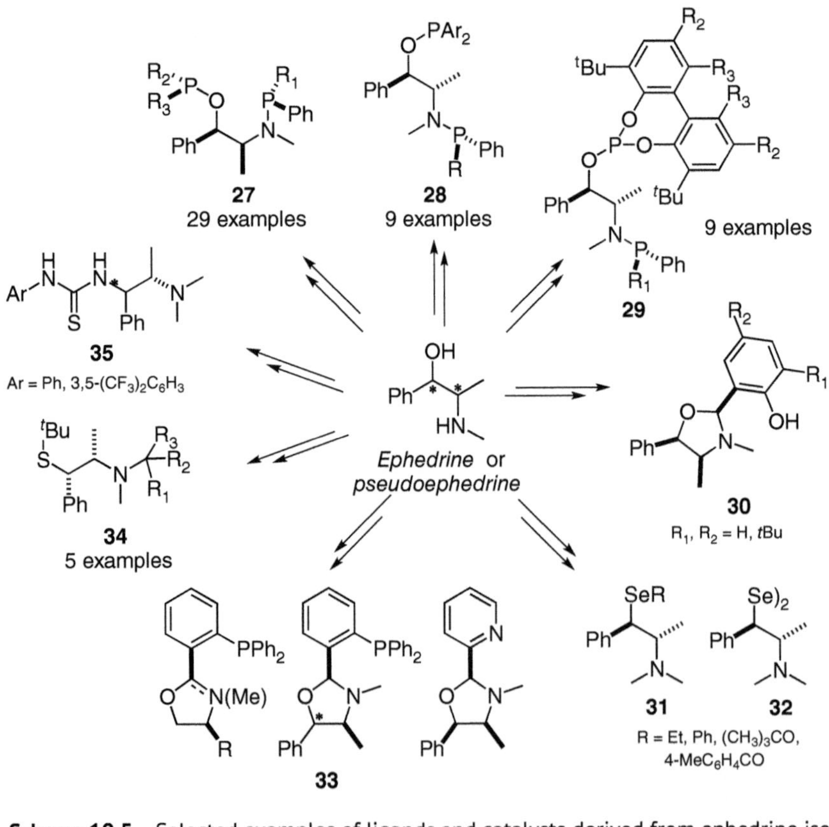

Scheme 10.5 Selected examples of ligands and catalysts derived from ephedrine isomers.

The most fundamental significance of ephedrines in stereoselective synthesis is however their performance as versatile and efficient chiral auxiliaries. In the mid-1990s Myers developed [99] a general and highly diastereoselective alkylation of ephedrine amides **36** that are easily accessible by the reaction of ephedrine with acyl chloride or anhydride [100, 101]. These amides are transformed *in situ* into the corresponding enolates **37** (prepared at low temperature in the presence of lithium chloride) reacting further with a variety of electrophiles in a highly diastereoselective fashion (Scheme 10.6). The resultant products **38** can be converted into various chiral compounds such as acids [102], aldehydes, alcohols [103], ketones, amides [104], or a variety of amino acids [105-107], among others (Scheme 10.6) [12–14, 99–101]. The chiral auxiliary is removed by using specific hydrolysis (cleavage) protocol of the amide adduct **38** and can be recycled in most syntheses. Myers' alkylation has quickly become very popular due to its generality and high chiral induction level. Its numerous, further extensions studied mainly by Badia and coworkers involve amination [107–109], Mannich [110–113], and aldol reaction [114–117] (Scheme 10.6; for a dedicated review see also [13]). These routes give convenient access to the more sophisticated building blocks, such as various β-aminoesters [110–112, 118], β-lactams [111], and many chiral heterocycles: 7,8-disubstituted 3-aryl-1,2,3,4-tetrahydroisoquinolin-4-ols [107], 5-substituted-3-methyl-pyrrolidin-2-ones [108], piperidines and fully substituted piperidinones [113], pyrrolidines [119], as well as piperidine alkaloids [120]. Diastereoselective conjugate addition to unsaturated amides **39** was employed for the synthesis of β-aminoesters and their derivatives [121], α,β-branched

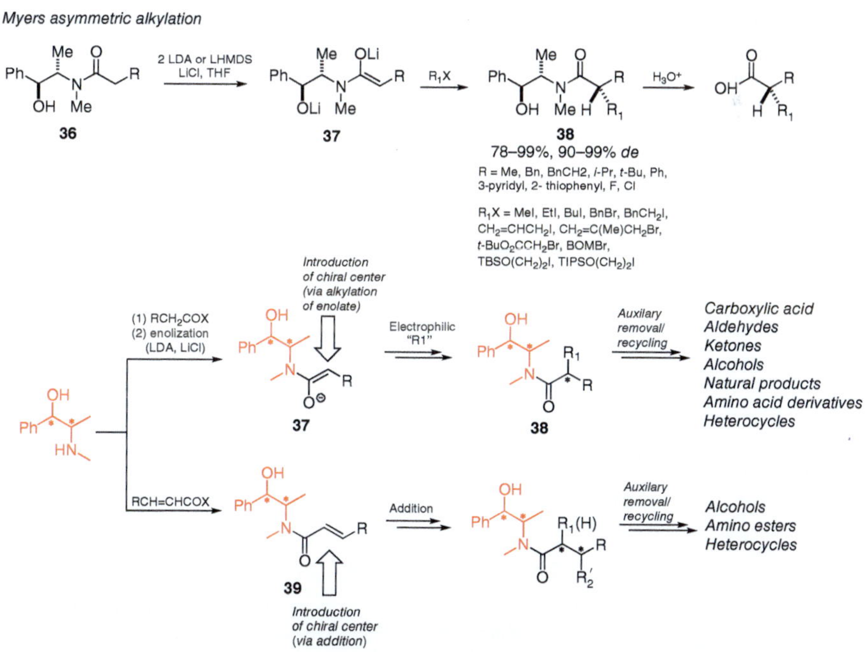

Scheme 10.6 Ephedrine-derived amides **36** and **38** for enantio- or diastereoselective alkylation/additions. Source: Based on Refs. [102–105].

alcohols [122], 3-aryl-δ-lactones and tetrahydropyran NK_1 receptor antagonist [123, 124], α-chiral-β,γ-unsaturated carboxylic acids [125], enantiomerically enriched α-chiral β-trifluroboratoamides and their Suzuki–Miyaura cross-coupling [126], or γ-butyrolactones [127–129].

The amide alkylation process has also been widely employed as key stereoselective step in complex natural products syntheses, for example for the preparation of (−)-terpestacin **40**, a fungal metabolite used in AIDS infection and (−)-fusaproliferin **41** (25 g scale of preparation of the key chiral propionamide) [130] or (−)-quinocarcin **42**, a pentacyclic alkaloid with antiproliferative properties [131] by the group of Myers, (−)-platensimycin **43** – bacterial fatty acids biosynthesis inhibitor by Nicolaou and coworkers [132] or C1–C19 polyol fragment of the brasilinolides **44** by Paterson et al. [133] (Figure 10.3) as well as homophymine A [134] or chaetochalasin A [135].

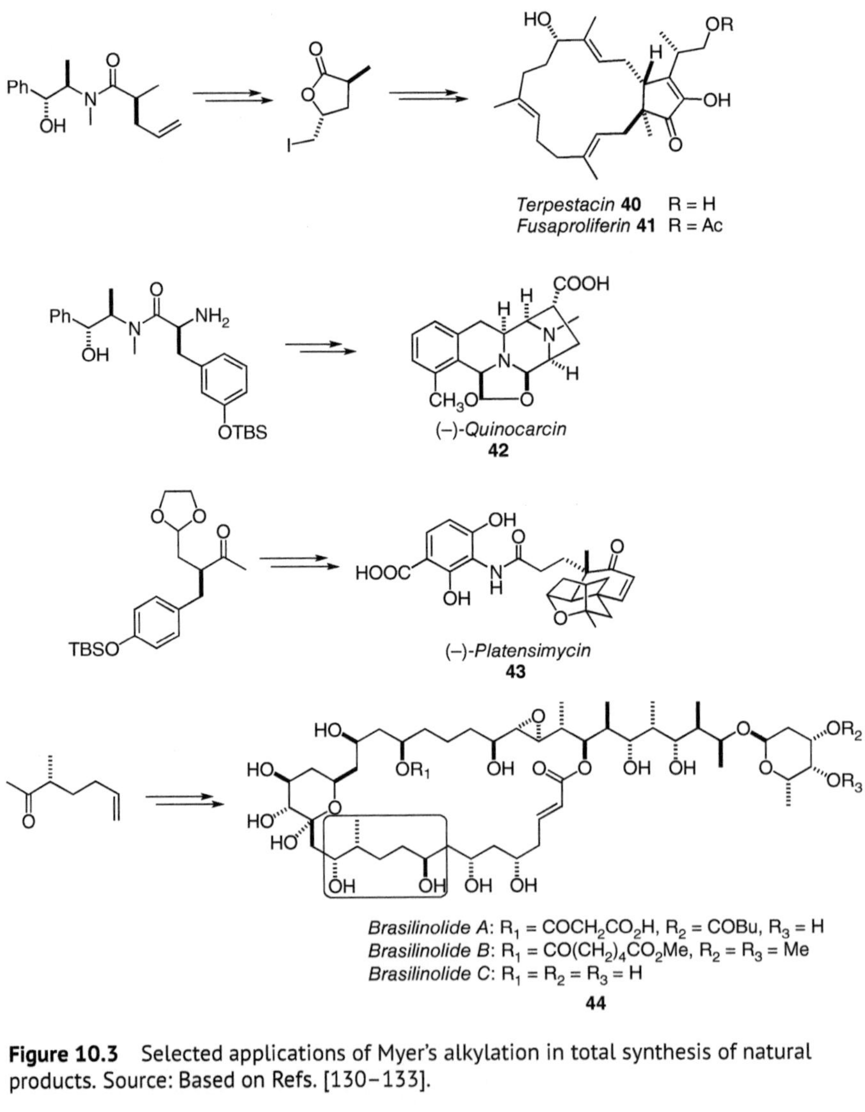

Terpestacin **40** R = H
Fusaproliferin **41** R = Ac

(−)-*Quinocarcin*
42

(−)-*Platensimycin*
43

Brasilinolide A: $R_1 = COCH_2CO_2H$, $R_2 = COBu$, $R_3 = H$
Brasilinolide B: $R_1 = CO(CH_2)_4CO_2Me$, $R_2 = R_3 = Me$
Brasilinolide C: $R_1 = R_2 = R_3 = H$
44

Figure 10.3 Selected applications of Myer's alkylation in total synthesis of natural products. Source: Based on Refs. [130–133].

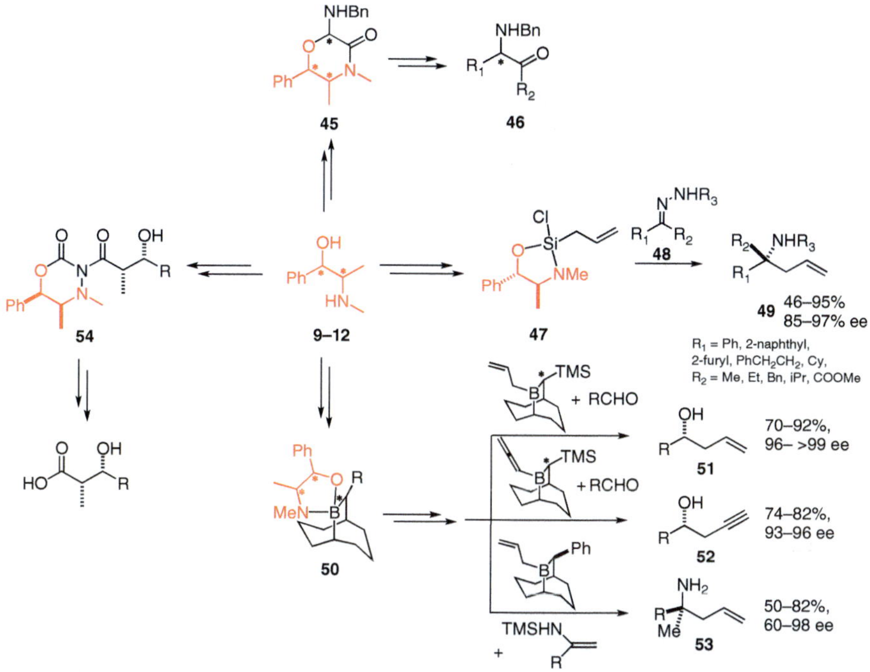

Scheme 10.7 Selected applications of ephedrines-derived heterocyclic chiral auxiliaries **45**, **47**, **50**, and **54**. Source: Based on Refs. [136–143].

Many heterocyclic auxiliaries, such as **45**, **47**, **50**, and **54** that are easily accessible from ephedra alkaloids (Scheme 10.7), were also successfully used in number of stereoselective syntheses. For example, morpholin-3-ones **45** prepared from ephedrine by the reaction with glyoxalic acid hydrate were shown to be precursors of enantioenriched aminoalcohols or aminoketones **46** and have also been employed in the total synthesis of (−)-β-conhydrine and (+)-α-conhydrine [136].

Practical, scalable, and highly enantioselective synthesis of tertiary carbinamines **49** was reported by Leighton and coworkers. It was based on the addition of a (*S*,*S*)-pseudoephedrine-derived allylsilane reagent **47** to a diverse set of ketone-derived benzoylhydrazones **48**. The chiral auxiliary (pseudoephedrine) was recovered by simple extraction (Scheme 10.7) [137].

Hitchcock and coworkers developed a synthetic route to another ephedrine-derived chiral auxiliary, namely 3,4,5,6-tetrahydro-2*H*-1,3,4-oxadiazin-2-one **54**, used for stereoselective aldol reaction [138, 139]. This heterocycle can be obtained in gram quantities from the alkaloid or its (1*S*,2*S*) epimer and after *N*-acylation provided a substrate undergoing a highly diastereoselective aldol reaction. Recovery of the chiral auxiliary was possible as well [138, 139].

Soderquist and coworkers reported a series of protocols for highly enantioselective synthesis of homoallylic **51** and homopropargylic alcohols **52** as well as *N*-propargylamides or tertiary carbamines **53** by asymmetric allyl- or allenylboration. To this end, they used an air-stable crystalline pseudoephedrine borinic ester complexes of 10-(trimethylsilyl)-9-borabicyclo[3.3.2]decanes **50** which after reacting with appropriate Grignard reagents led to respective chiral boranes. Their reaction with aldehydes or *N*-acylimines provided the desired products **51–53** (Scheme 10.7) [140–143].

Readily available (*R*,*S*)-*N*-tosylnorephedrine has been conveniently used as a chiral substrate for the synthesis of enantiomerically pure sulfoxides [144] also on a larger scale [145]. The key reactant, namely oxathiazolidine, can be prepared in kilogram quantities in 95% isolated yield and can be easily recycled after the reaction. Ephedrines are highly efficient chiral auxiliaries for the synthesis of *P*-chiral phosphines (Jugé–Stephan method) that unlike their tertiary amine counterparts are configurationally stable at room temperature [146–150]. *P*-Chiral phosphines, such as DiPAMP, BisP*, MiniPhos, and many others, are an important class of ligands for asymmetric catalysis used in enantioselective hydrogenations, conjugate addition, hydrosilylation, hydroacylation, or ring opening (to name the most explored), and typically work with efficiencies comparable to *C*-chiral phosphines while being superior for selected applications. *P*-Chiral phosphines are also convenient precursors of other *P*-chiral compounds [146, 148–150]. The method is based on a one-pot formation of 1,3,2-oxazaphospholidine boranes **56** and **58** from bis(diethylamino)aryl or alkyl phosphine **55** and (−)-ephedrine or (+)-pseudoephedrine, followed by protection with BH_3 [151]. The cyclization reaction is stereoselective and the chirality transmission is controlled by the ephedrine enantiomer: (−)-ephedrine **9** gives preferentially the (S_P)-diastereoisomer of **57** in 90% de, whereas (R_P)-**59** can be synthetized using (+)-pseudoephedrine **12** (Scheme 10.8). Oxazaphospholidines **56, 58** react with a variety of electrophiles or nucleophiles giving access to linear diverse *P*-chiral compounds [148–152]. This

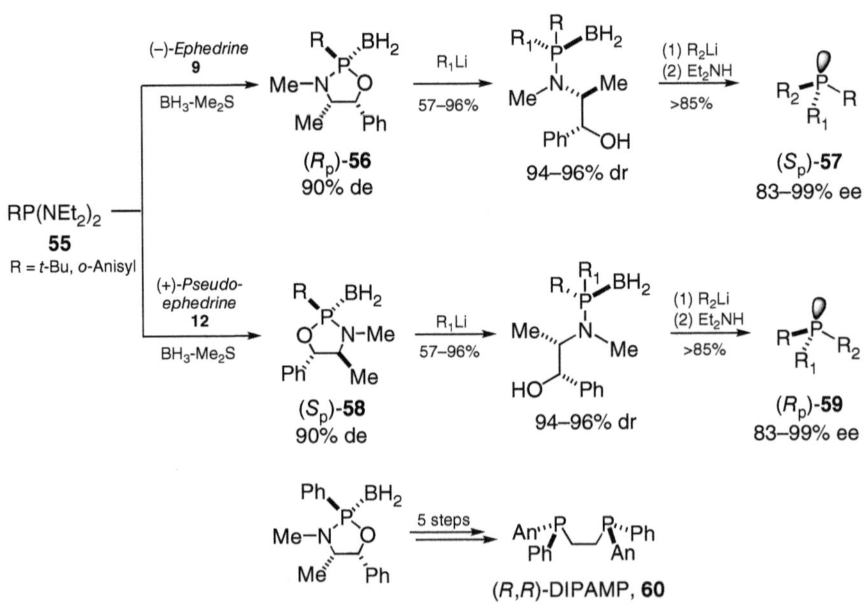

Scheme 10.8 Synthesis of enantiomers of (S_P)-**57** or (R_P)-**59** from (−)-ephedrine or (+)-pseudoephedrine.

methodology has been also used for the preparation of various bis-*P*-chiral ligands, such as enantiomerically pure (*R*,*R*)-DIPAMP **60**, its analogs and others [151, 153].

Application of *P*-chiral ligands containing one or two stereogenic phosphorus atoms in asymmetric synthesis is a subject of several dedicated reviews [146–150].

Ephedrines have been also used as building blocks of chiral ionic liquids [154, 155]. Vo-Thanh and coworkers converted ephedrine to the quaternary ammonium salts which were used for chiral recognition, and as solvent and sole chirality inductor in the stereoselective Baylis–Hillman and Michael addition giving products with moderate asymmetric induction level [154]. Bica and coworkers used ephedrine and pseudoephedrine among other amino alcohols for preparation of coordinating chiral ionic liquids [155, 156]. The utility of ephedrine and dimethylephedrinium cation as structure-directing agents for the construction of chiral nanoporous aluminophosphates has also been reported [157, 158].

10.3 Tobacco Alkaloids (Nicotine and Anabasine)

Nicotiana tabacum (cultivated tobacco) and *Nicotiana rustica* (Aztec or strong tobacco) originating from North and South Americas constitute the most widely grown non-food crop producing relatively simple pyridine–pyrrolidine type of alkaloids, called collectively as tobacco alkaloids. Leaves of *N. tabacum*, which is commonly used by tobacco industry, contain a mixture of alkaloids, consisting of 93–95% of (*S*)-nicotine (3-(1-methyl-2-pyrrolidinyl)pyridine) **61**, followed by anatabine **65**, nornicotine **62**, anabasine **64**, and many others (Figure 10.4 includes also cotinine **63**, one of the minor components, but is the major product of nicotine metabolism) [159]. (*S*)-Enantiomer is a predominant form of nicotine, with 99.5% of the total amount. Tobacco contains typically 1–3% of nicotine but in *N. rustica* the level is much higher, up to 9%.

Nicotine can serve as a source of chiral-substituted pyrrolidine moiety and has been employed by several groups as a substrate for the synthesis of various related derivatives, including other, less abundant tobacco alkaloids. However, one can notice that it is much less successfully used as organocatalyst or ligand compared to the frequently used another natural chiral pyrrolidine containing product – proline. The reactivity of nicotine is specific due to the presence of two heterocyclic moieties: a combined experimental and theoretical studies confirmed a deactivating effect of pyridyl ring on pyrrolidine subunit, decreasing Lewis basicity and nucleophilicity of nitrogen atom [160].

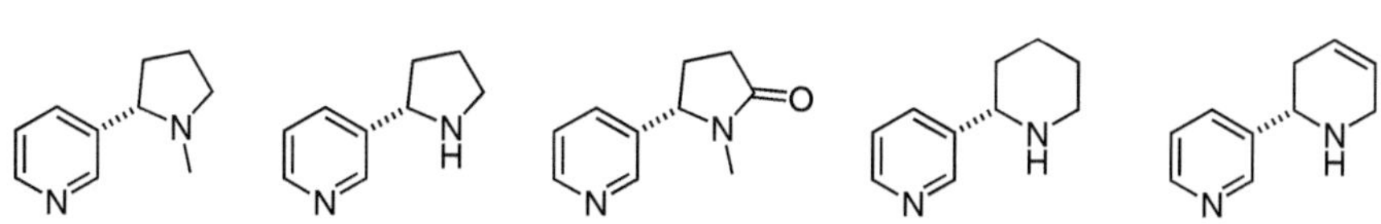

(*S*)-(–)-*Nicotine* **61** | *Nornicotine* **62** | *Cotinine* **63** | *Anabasine* **64** | *Anatabine* **65**

Figure 10.4 Most common tobacco alkaloids.

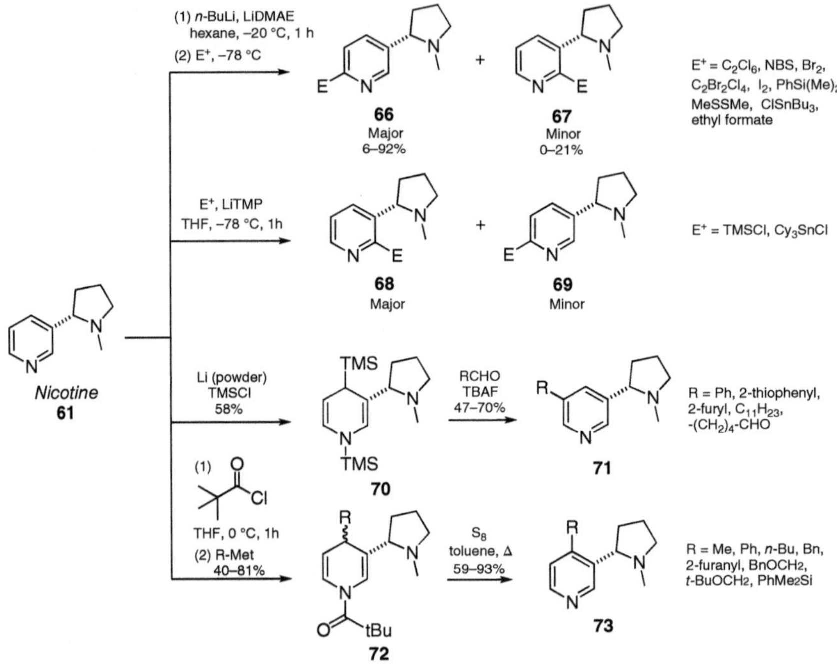

Scheme 10.9 Regioselective substitution of pyridine ring in (*S*)-nicotine **61**. Source: Comins et al. [161]/American Chemical Society.

Comins and his group are recently most active in this field. They used nicotine as chiral scaffold for synthesis of related natural products and developed a number of protocols for regioselective substitution of pyridine subunit of nicotine in all positions (Scheme 10.9). Regioselective *ortho*-lithiation is completed with *n*BuLi-LiDMAE system followed by addition of electrophile and led to various 6-substituted derivatives **66** without racemization in 55–92% yield [163]. The product **68** substituted in position 2 appeared more challenging (due to steric hindrance exerted by pyrrolidine unit), but could be prepared by changing the base to lithium tetramethylpiperidide (LiTMP). C-5-Substituted derivatives **71** were prepared via a reductive disilylation of (*S*)-nicotine by a reaction with lithium and chlorotrimethylsilane, and addition of carbonyl nucleophiles to the resulting 1,4-bis(trimethylsilyl)-1,4-dihydronicotine **70** in the presence of TBAF [162]. Its treatment with aromatic and aliphatic aldehydes led to the introduction of the desired groups in 5-position in 47–70% yield. A two-step reaction sequence involving acylation of pyridine nitrogen atom with pivaloyl chloride and addition of cuprate reagents gave 1,4-dihydronicotine products **72** in 40–79% yield and a complete diastereoselection. Their rearomatization with sulfur afforded 4-substituted nicotines **73** (59–93%, >99% ee; Scheme 10.9) [161].

Further exploration of nicotine chemistry by Comins and coworker led to development of protocols for regioselective iodination of 6-chloronicotine at position 2′, 4′, and 5′ [164]. These derivatives served later for synthesis of other nicotine derivatives with the aid of cross-coupling chemistry [165, 166].

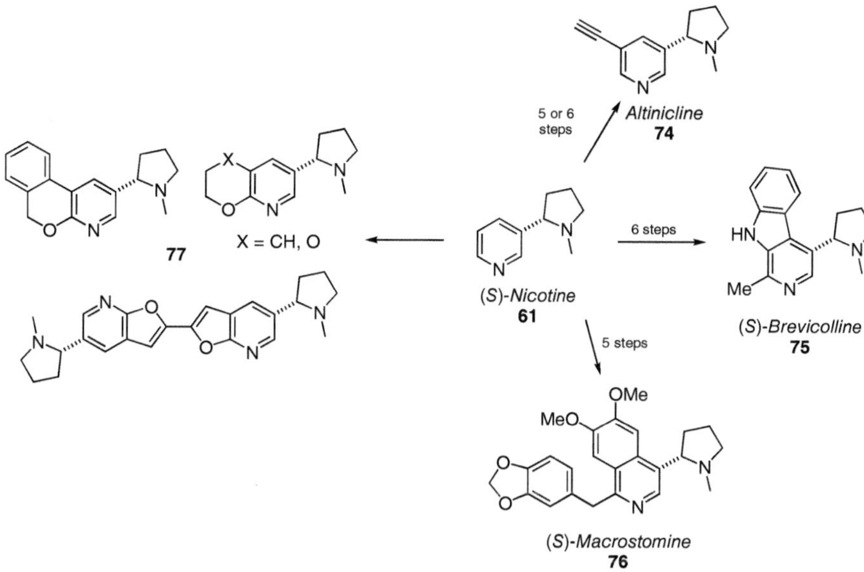

Scheme 10.10 (*S*)-Nicotine as building block for related alkaloids: altinicline **74**, (*S*)-brevicolline **75**, (*S*)-macrostomine **76**, and a library of bicyclic derivatives **77** (selected examples shown). Source: Based on Refs. [167–171].

Based on their expertise, Comins and coworkers demonstrated syntheses of various nicotine analogs dedicated for bioactivity screening using a set of modern synthetic chemistry tools such as various cross-coupling reactions or Buchwald amination. Thus, nicotine served as substrate for syntheses of altinicline **74** tested in a treatment of Parkinson's disease [167, 168], fused ring (*S*)-brevicolline **75** with a phototoxic effect on bacteria and fungi [169] and (*S*)-macrostomine **76**, a bioactive isoquinoline alkaloid found in plants of the genus *Papaver* (Scheme 10.10) [170]. A library of chiral heterocycles **77** including bicyclic dioxino-, dihydropyranol-, dihydrofuro-, and tricyclic benzofuro- and benzopyranonicotines in the enantiomerically pure form has also been reported (Scheme 10.10) [171].

A progress in current nicotine chemistry is a subject of authoritative review by Wagner and Comins published in 2007 [172]. A kilogram-scale nicotine chemistry involving iridium-catalyzed borylation and bromination has been employed for design of a novel therapeutic agent to treat tobacco addiction by Singer and coworkers [173]. Conceptually similar vaccine using cotinine **63** (a metabolite of nicotine, widely used as a biomarker of tobacco consumption and drug candidate for psychiatric disorders [174]) has also been developed by Gallagher and coworkers [175].

Pyridine–piperidine alkaloid anabasine **64** (Figure 10.4) found in Asian plant *Anabasis aphylla* and in tree tobacco (*Nicotiana glauca*) gained attention in life science as nicotinic receptor modulator. In chemistry, however, it has been used as a platform for synthesis of some "drug-like" heterocyclic derivatives. For example, Krasavin and Sandulenko used anabasine in two isocyanide-based multicomponent processes. A conversion to imidazo[1,2-*a*]pyridines **78** was achieved in the course of the Groebke–Blackburn reaction; unfortunately the products were racemic – most probably due to the initial harsh condition of pyridine amination. A small series of dipeptoid products **79** were accessible by Ugi reaction of anabasine with a complete diastereoselectivity albeit in low yields (Scheme 10.11) [176]. Brel with coworkers showed that *N*-propargyl and 2-azidoethyl anabasines **80**, **83** could be used as substrates for click chemistry reactions, such as Cu(I)-catalyzed Huisgen 1,3-dipolar cycloaddition of alkynes and azides or cycloaddition of alkynes to nitrile oxides. The respective products, namely, 1,4-regioisomeric 1,2,3-triazoles **81**, **84** and 3,5-regioisomeric isoxazoles **82**, were successfully prepared, but further screening tests were not reported (Scheme 10.11) [177, 178]. *N*-Phenacyl salts of anabasine with *N*-protected piperidine ring underwent another 1,3-dipolar cycloaddition with alkynes giving a series of isomeric indolizines with 2-piperidyl substituent in either 6- or 8-position **85** (Scheme 10.11) [179].

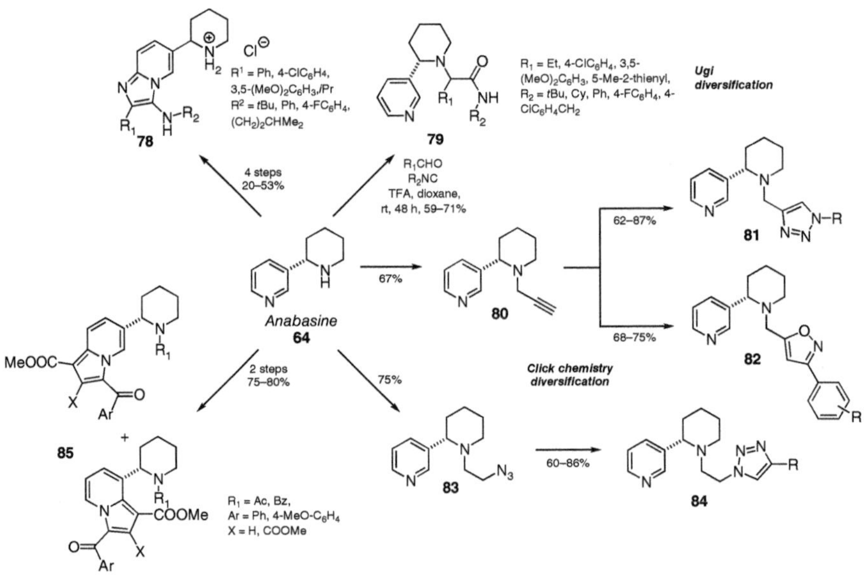

Scheme 10.11 Anabasine used as a chiral scaffold in various syntheses. Source: Based on Refs. [176–179].

10.4 Lupin Alkaloids

Around 200 alkaloids have been found in lupin and related species belonging to *Leguminosae/Fabaceae* family. They contain a quinolizidine core **86**, in many cases built into a tricyclic or tetracyclic structure bearing a central [3,7]-diazabicyclo[3.3.1]nonane (bispidine) **87** nucleus (Figure 10.5). The most common members of this class are tetracyclic (−)-sparteine **88** isolated from Scotch broom (*Cytisus scoparius*) or synthetized, (−)-cytisine **90** isolated from Golden Rain tree (*Laburnum anagyroids*) seeds, (−)-lupanine **92**, as well as tricyclic (+)-sparteine surrogate **91** (prepared from (−)-cytisine or obtained synthetically) [16, 180]. Among them (−)-sparteine **88** and (+)-sparteine surrogate **91** have been

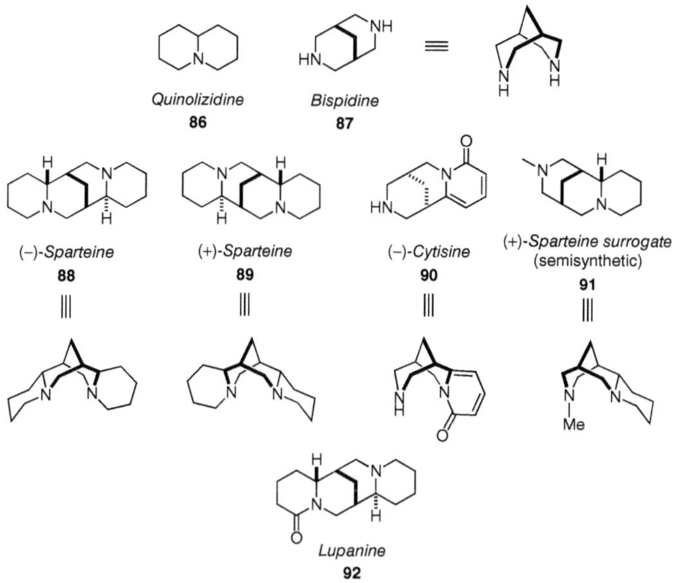

Figure 10.5 Structural features and common representatives of lupin alkaloids and derivatives.

recognized as efficient stereoorienting stoichiometric chiral ligands (without prior covalent attachment) in a number of processes involving mainly enantioselective deprotonation–electrophile trapping [15, 16, 181, 182], but there is an increasing number of different reactivities [183–185].

These alkaloids possess also distinct biological activities stimulating their medicinal chemistry as reviewed by Sacchetti and Rossetti [180]. For example, (−)-sparteine is antiarrhythmic and antimicrobial. (−)-Cytisine is, due to being a partial agonist of the nicotinic acetylcholine receptors (nAChRs), used as a smoking cessation drug (marketed in many countries under the name Tabex®) and approved in Eastern Europe and Canada. It acts also as analgesic, antihypertensive, inotropic, antioxidant, and antidepressant [186, 187].

Major application of sparteine and sparteine surrogate in stereoselective synthesis involves a large pool of asymmetric deprotonation–substitution reactions. They are promoted by *in situ* prepared chiral bases through butyllithium (BuLi) or *sec*-butyllithium (*s*-BuLi) complexation with sparteine or sparteine surrogates (in general, any chiral suitable diamine ligand) that are able to distinguish enantiotopic protons and effect enantioselective deprotonation [15, 180, 182]. In contrast to the chiral auxiliaries these diamine ligands do not require prior installation and removal steps, and can be recycled after the reaction.

The origin of the chiral induction in the enantioselective deprotonation is substrate dependent and has been recognized as a simple asymmetric deprotonation, a dynamic kinetic resolution, or a dynamic thermodynamic resolution (mechanistic aspects are discussed in reviews [182, 183]). Interestingly, among numerous tested chiral diamines of both natural and synthetic origin typically

(−)-sparteine, O'Brien's (+)-sparteine surrogate, and complementary non-alkaloidal (*R*,*R*)-cyclohexanediamine derivatives were usually found to be the most efficient chiral ligands. It is worth to note that despite intense research aiming to switch this methodology into a catalytic fashion, these diamines work best in slightly overstoichiometric quantities. The use of lower amounts is sometimes possible (see [15, 188]), but usually compromises substantially the chiral induction. In catalysis, better results were obtained using copper complexes of (−)-sparteine, and (+)-sparteine surrogate [15, 185]

Organolithium-(−)-sparteine complexes assembled *in situ* have been quickly recognized as valuable chiral bases for highly enantioselective deprotonation (Hoppe–Beak deprotonation) of α-protons to an oxygen, nitrogen, sulfur, or a phosphorus and benzylic lithiation-trapping, *ortho*-, and carbolithiation. The subsequent substitution of lithiated intermediates by a number of electrophiles led to an array of enantiomerically pure or enriched products.

A seminal work in this field was published in 1971 by Hoppe, who reported an enantioselective deprotonation–substitution of a *O*-alkylcarbamates [189, 190], and Beak on *N*-Boc pyrrolidine [191, 192]. Carbamates **93** favorably serve as activating groups for the deprotonation and as protecting groups for the alcohol. In general, carbamates are pro-*S* hydrogen selectively deprotonated by the chiral base prepared from *s*-BuLi/(−)-sparteine **88**. This leads to a configurationally stable lithiated intermediate **94** which is further trapped by an electrophile with a complete stereoretention, giving diverse oxygen-containing products **95–97** with a high degree of enantioselectivity (Scheme 10.12). In 2004 Hoppe reported on a preparation of chiral boronic acids **95** by a sparteine-mediated asymmetric deprotonation of *N*,*N*-diisopropylcarbamates (OCb) and a subsequent trapping of organolithium with triisopropyl borate. The products can easily be oxidized to respective chiral alcohols **96** (Scheme 10.12) [193].

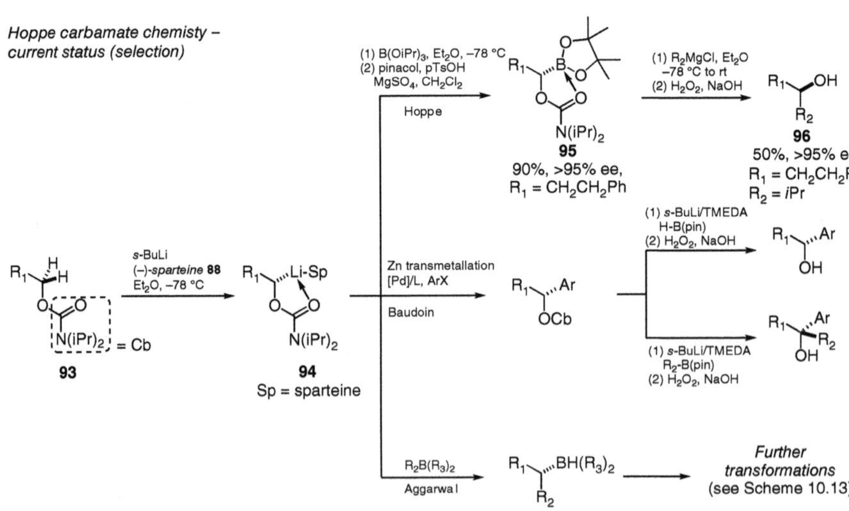

Scheme 10.12 Sparteine-mediated asymmetric deprotonation of Hoppe carbamates **93** and their selected stereoselective transformations. Source: Based on Refs [193–195].

On the other hand, Baudoin and coworkers reported a general and highly enantioselective α-arylation of such carbamates derived from primary alcohols by combining Hoppe's asymmetric deprotonation with Negishi cross-coupling (Scheme 10.12) [194]. Carbamates **93** were also initially used by Aggarwal to develop an extraordinary powerful lithiation–borylation sequence that offered an access to a variety of enantioenriched boronic esters **97** serving as versatile chiral intermediates (Schemes 10.12 and 10.13) [195–198].

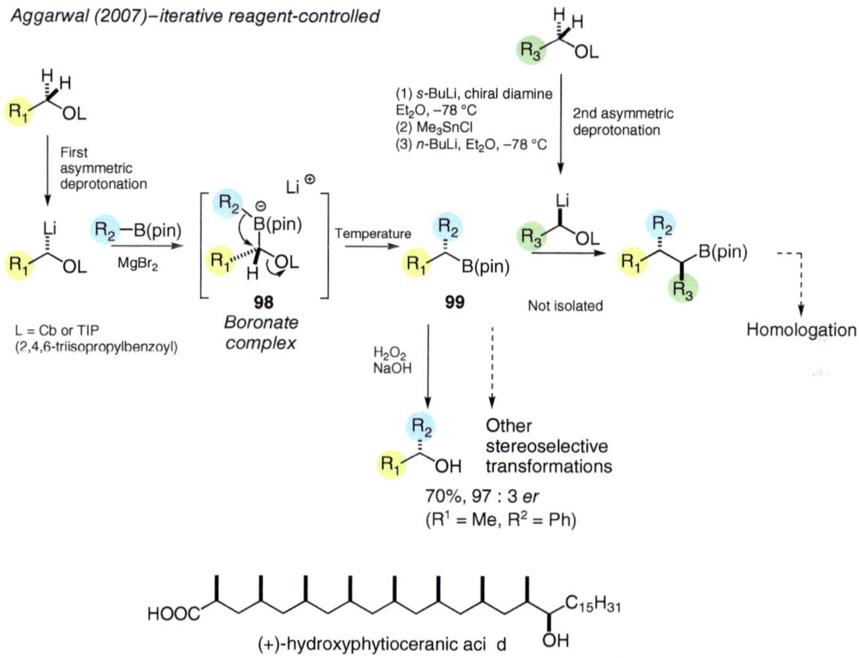

Scheme 10.13 Aggarwal iterative asymmetric deprotonation mediated by sparteine coupled with borylation approach and their selected applications. Source: Based on Refs. [195–198].

The lithiation–borylation route to chiral boronic esters **97** and their stereoselective transformations has been extensively studied by Aggarwal's group. It is worth to note that secondary and tertiary chiral boron esters are more stable than organolithium species and can be used for a formation of a new C—C, C—O, C—N, C—X, or C—H bonds at stereogenic centers as well as for stereoselective preparation of amines, halides, arenes, and alkynes [195–198].

For example, enantioenriched secondary alcohols flanked by similar side chains can be conveniently prepared using a lithiation–borylation route. It involves the standard preparation of the α-lithiated carbamates by stereoselective deprotonation in the presence of (−)-sparteine or (+)-sparteine surrogate, depending on the planned stereochemistry of product. Next, the organolithium intermediate reacts with a borane or boronic ester to form the respective boron-ate complex **98** with a retention of the stereochemistry. This complex (Scheme 10.13) undergoes a 1,2-metallate rearrangement accompanied with a migration of the R group and expulsion of the carbamate leaving group. This eventually leads to a secondary homologated boronic ester **99** which can easily be oxidized to secondary alcohols in very high enantiomeric ratios (Scheme 10.13) [193]. An extension of this approach for a one-pot synthesis of chiral *anti*-β-hydroxy-*E*-allylsilanes and popular in natural products 2-ene-*anti*-1,4-diol motifs was reported by the same group, in line with a demonstration of stereoselective synthesis of (−)-decarestrictine D nanomolar inhibitor of liver cell cholesterol biosynthesis [199] or solandelactone F [200].

A replacement of the carbamates that give sometimes insufficient rates of the 1,2-metallate rearrangement, for hindered 2,4,6-triisopropylbenzoates as alternative leaving groups brought substantial improvement, as demonstrated by Aggarwal. This modification allows for the exploitation of much wider arrays of boronic esters, including those that were inactive with the carbamates [201].

The iterative lithiation–borylation (Scheme 10.13) developed by Aggarwal and coworkers is an especially convenient tool for a straightforward installation of multiple contiguous stereogenic centers on "growing" carbon chains. Proofs of concept have widely been demonstrated by this group in the stereoselective synthesis (+)-kalkitoxin and (+)-hydroxyphthioceranic acid (Scheme 10.13). Six iterative homologations using chiral lithiated benzoate esters (>99% ee) obtained by Hoppe–Beak deprotonation of ethyl 2,4,6-triisopropyl-benzoate with *s*-BuLi in the presence of (+)- or (−)-sparteine were performed to build up the core of the (+)-kalkitoxin. In the case of (+)-hydroxyphthioceranic acid, 16 iterative homologations were conducted to reach the target. Purification of intermediates between several homologations could be largely avoided, providing an additional benefit to this protocol [202]. A similar iterative approach used a key C—C bond forming step in a stereoselective synthesis of Dysoxylactam A, a 17-membered macrocyclic lipid, highly potent in reverse multidrug resistance in cancer cells [203] and antiparasitic (−)-Stemaphylline [204]. It is worth to highlight that lithiation–borylation method can be implemented also as a late-stage operation as shown by Aggarwal in stereoselective synthesis of anticancer (+)-Giganin [205].

An interesting recent example of Hoppe–Beak chemistry reported by Olofsson, Pace, and coworkers involves a novel enantioselective acylation of carbamate-based organolithiums generated in the presence of (−)-sparteine with Weinreb amides for the preparation of highly enantioenriched α-pyrrolydinyl ketones and other derivatives. The authors provided a complete technology that enabled full recovery of the solvent (cyclopentyl methyl ether), sparteine, and the released Weinreb amine (HNMe(OMe)) by a simple work-up [206]. Chemistry and application of the lithiation–borylation approach in complex natural product synthesis is a subject of authoritative reviews by Aggarwal [195, 196, 207, 208].

Many *N*-Boc heterocycles, in particular *N*-Boc pyrrolidine **100**, are especially well-suited substrates for sparteine-mediated enantioselective deprotonation as first demonstrated by Kerrick and Beak [191, 192]. These organolithium intermediates can be easily transformed into a number of useful enantiomerically enriched α-substituted products **101–105** with the use of diverse electrophiles (Scheme 10.14). In the seminal work, pro-*S* selective lithiation of *N*-Boc pyrrolidine mediated by *s*-BuLi and (−)-sparteine **88** in Et_2O at −78 °C, followed by trapping of the chiral lithiated intermediate with Me_3SiCl, gave α-substituted (*S*)-configurated

N-Boc pyrrolidine **102** with a high enantioselectivity. This lithiation has been later successfully applied to other heterocycles, such as *N*-Boc: piperidines, piperazines, indolines, imidazolidines, oxazolidines [18], and recently for 1,3-oxazinanes [209]. The scope of the lithiation–substitution of *N*-Boc pyrrolidine could be largely extended by transmetallating the sensitive organolithium to organocuprates or configurationally and chemically more stable organozinc species. The organocuprates allow trapping with alkyl, vinyl, allyl, and propargyl halides, and α,β-unsaturated carbonyl compounds provide access to α-functionalized pyrrolidines **104** or non-natural alkaloid **105** [210, 211]. Organozinc intermediates undergo efficient Pd-catalyzed Negishi cross-coupling with aryl bromides, iodides, or chlorides giving a wide spectrum of α-arylated *N*-Boc pyrrolidines **103** (Scheme 10.14) [212].

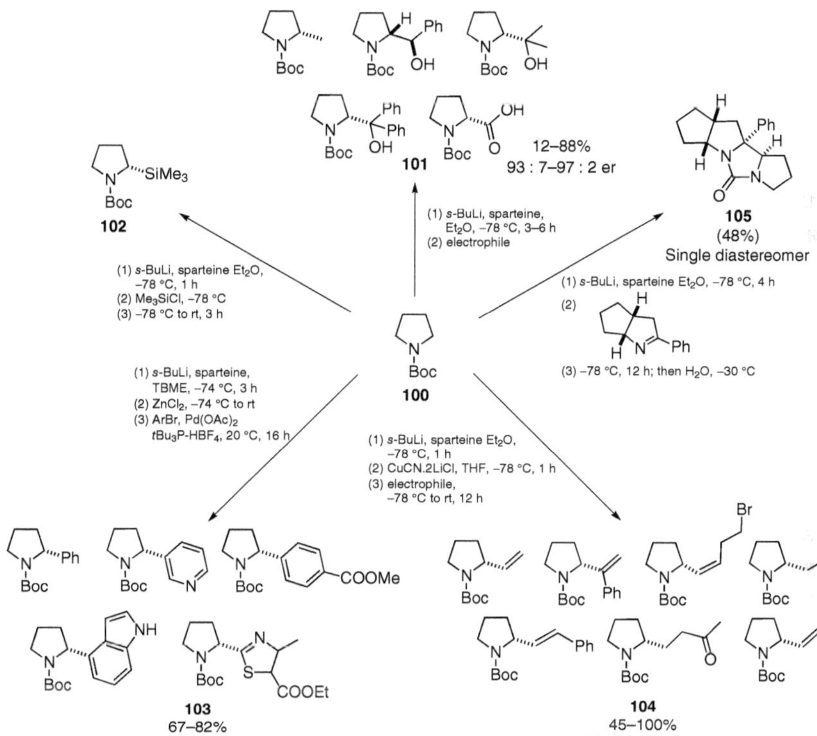

Scheme 10.14 Selected examples of (−)-sparteine **88**-mediated enantioselective deprotonation/substitution of *N*-Boc pyrrolidine **100** (comparable results are obtained for *N*-Boc-piperidine; the use of (+)-sparteine surrogate **91** gives an opposite enantiomeric induction). Source: Based on Refs. [209–212].

The enantioselective lithiation–substitution of nitrogen heterocycles was often used for preparation of diverse biologically active compounds as well as natural products (Figure 10.6). For example, enantioselective lithiation of *N*-Boc pyrrolidine **100** using *s*-BuLi/(−)-sparteine, subsequent transmetallation with $ZnCl_2$ and Pd-catalyzed Negishi coupling was a key step in the 100-g scale synthesis of the glucokinase activator (*R*)-**106** [213]. A similar approach, employing both

(−)-sparteine **88** and the (+)-sparteine surrogate **91**, was used for enantioselective synthesis of (*R*)-crispine A **107**, (*S*)-nicotine, and its derivative (*S*)-SIB-1508Y and (+)-maackiamine **108** with a convenient *in situ* IR spectroscopic monitoring of the process [214], as well as (−)-hygroline **109** [215].

(−)-Sparteine-mediated enantioselective Pd-catalyzed α-arylation of *N*-Boc-pyrrolidine was also successfully used for a combinatorial synthesis of a library of chiral 2-arylpyrrolidines [216]. Neurokinin-1 receptor antagonist (+)-L-733,060 **110** was synthetized employing the lithiation of *N*-Boc pyrrolidine **100** as a key step in a modification involving the use of substoichiometric amount of sparteine (0.3 equiv) together with an additional achiral amino alcohol ligand [217]. Two enantioselective syntheses of (−)-kainic acid **111** based on lithiation–substitution of the chiral 3,4-disubstituted *N*-Boc pyrrolidine were also reported [218, 219]. Bicyclic proline analog motif of Telaprevir **114**, registered drug for treating the hepatitis C virus, was prepared during process-scale development by enantioselective lithiation–carboxylation of *N*-Boc bicyclic pyrrolidine precursor mediated by (+)-sparteine surrogate. Although the desired intermediate was obtained with high diastereo- and enantioselectivity, concerns regarding large-scale supply of the ligand finally made the classical resolution approach the preferred method of kg-scale synthesis (Figure 10.6) [220]. Intermediate organocuprates were also applied to the total synthesis of (−)-pyrrolam A **112** [221] and (+)-elaeokanine A **113** [222].

An elegant extension of this approach toward regio- and enantiodivergent synthesis of 2- and 3-substituted β-aminoacids was reported by Baudoin and coworkers [209]. It was based on a one-pot (−)-sparteine or (+)-sparteine surrogate-mediated enantioselective lithiation of a Boc-1,3-oxazinane, a subsequent transmetallation to organozinc followed by a direct or migratory Negishi coupling with an electrophile.

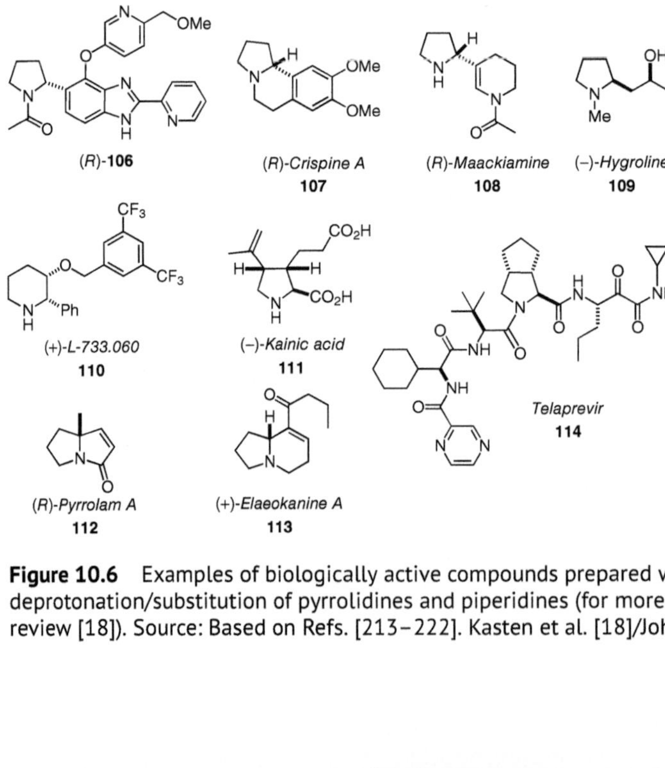

Figure 10.6 Examples of biologically active compounds prepared with the use of deprotonation/substitution of pyrrolidines and piperidines (for more applications see review [18]). Source: Based on Refs. [213–222]. Kasten et al. [18]/John Wiley & Sons.

The regioselectivity of the Negishi coupling in Boc-1,3-oxazinanes could be controlled using an external ligand. Both C4- and C5-functionalized oxazinanes were then transformed into highly enantioenriched β2- and β3-amino acids with the (*R*) or (*S*) configuration, depending on the chiral ligand used.

For the most recent review dedicated to enantioselective lithiation of hetereocycles and their applications, see the excellent work of O'Brian and coworkers [18].

Asymmetric C–H functionalization at benzylic position is particularly attractive as many drugs contain multiply-substituted benzylic (or heterobenzylic) carbon. Benzylic protons show increased acidity making a lithiation substitution relatively easy and popular due to wide access to achiral or racemic starting benzylic substrates. Regioselectivity of lithiation is typically controlled by the use of an adjacent directing group appended on the benzylic heteroatom or the aromatic *ortho*-position. This process allows for the selection of diverse types of electrophiles, thus giving access to a wide range of substituted products [223].

For example, Schlosser showed deprotonation of *N*-Boc-*N*-methylbenzylamine with *s*-BuLi-(−)-sparteine with subsequent trapping with electrophiles that led to products with high enantioselectivities. Interestingly, the enantioselectivity direction of this process is highly dependent on the solvent selection. In hexane or diethyl ether pro-*S* proton was substituted by electrophiles, whereas the use of THF activated the pro-*R* proton [224]. Park and coworkers reported (−)- or (+)-sparteine-mediated asymmetric lateral substitutions of 2-benzyl-*N*-pivaloyl aniline that reacted with various electrophiles, such as ketones, aldehydes, alkyl halides, or α,β-unsaturated esters affording diverse 2-alkyl-substituted anilines with high enantioselectivities [225]. Later the same group adopted this methodology for the synthesis of highly enantioenriched 3,4-disubstituted tetrahydroquinolines [226] and *trans*-2,3-disubstituted indolines [227, 228].

The intramolecular asymmetric addition of organolithiums to simple, unactivated alkenes (carbolithiation) is an attractive route for functionalized carbo- and heterocyclic compounds having two contiguous stereogenic centers [229]. An enantioselective carbolithiation mediated by (−)-sparteine has been initiated by Normant and Marek [230]. The reaction of cinnamyl alcohol with BuLi and (−)-sparteine led to chiral benzyl lithium intermediate that reacted with electrophiles in a highly diastereoselective manner. For an enantioselective carbolithiation mediated by sparteine see review [229].

An interesting application of (−)-sparteine/*s*-BuLi mediated *ortho*-lithiation and dilithiation is the concise and highly enantioselective synthesis of diverse planar chiral mono- and disubstituted 1,*n*-dioxa[*n*]paracyclophanes **115** and **116** reported by Shibata (Scheme 10.16) [231]. This method was later used for the synthesis of planar-chiral [*n*]paracyclophane mono- and C_2-symmetric diphosphines. Their screening as chiral ligands in Ag-catalyzed allylation of imines and Pd-catalyzed couplings gave moderate results. In contrast, rare asymmetric Sonogashira coupling of diiodoparacyclophanes possessing a flippable ansa chain and asymmetric desymmetrization of tricarbonyl(η^6-*ortho*-dichlorobenzene)chromium by Suzuki–Miyaura gave high enantioselectivities of up to 90% and 85% ee, respectively [232].

In a similar research field O'Shea demonstrated a first asymmetric synthesis of planar chiral [2.2]metacyclophanes **119**. The three-step method began with an initial formation of the 1,2-diarylethanes **117** equipped with directing methoxy- or *N*,*N*-diisopropylamide groups that underwent a subsequent ring closure by a dimetallation–oxidation step affording achiral C-5 substituted cyclophanes **118**. The final key step introducing chirality involved a (−)-sparteine mediated aryl lithiation, followed by an electrophile trapping to give the desired planar chiral cyclophanes **119** with enantiomeric ratios above 90 : 10 (Scheme 10.15) [233].

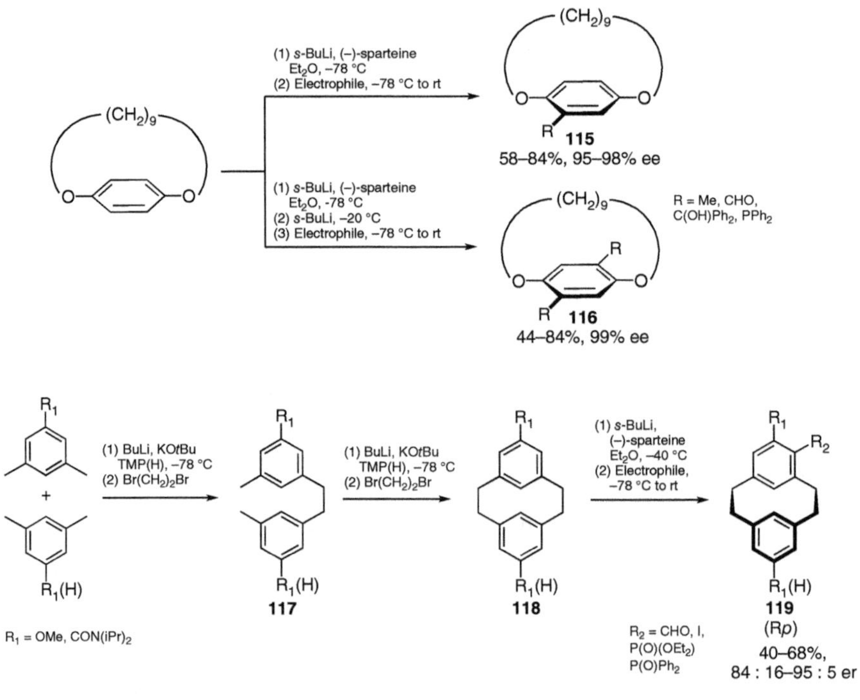

Scheme 10.15 Sparteine-mediated *ortho*-lithiation for the synthesis of planar chiral cyclophanes **115**, **116**, and **119**. Source: Based on Refs. [231–233].

An enantioselective deprotonation of phosphine boranes with sparteine with a subsequent substitution is also applicable for the synthesis of P-chiral compounds as pioneered by Evans and coworkers [234]. This widely explored approach became one of the most useful and general methods for the synthesis of P-chiral compounds [146–150]. Thus, an asymmetric deprotonation of prochiral dimethylarylphosphine–borane complex **120** mediated by *s*-BuLi-(−)-sparteine led to chiral alkyl lithium intermediates **121** and **122**. They were reacted with a number of electrophiles to form diverse P-chiral compounds **123–125**, **128** [235], whereas direct oxidative coupling produced chiral bis-phosphines **126** with high enantioselectivities, and *meso* bis-phosphines **127** (Scheme 10.16). Borane could be easily removed from phosphine–borane complex by the treatment with diethylamine.

This methodology was adopted for the synthesis of a number of important P-chiral phosphine ligands either with the use of (−)-sparteine **88** or (+)-sparteine surrogate **91**, such as tris-phosphine MT-Siliphos [236] or recent preparation of trichickenfootphos, BisP*, Mini-PHOS, and precursor of QuinoxP* with a substoichiometric

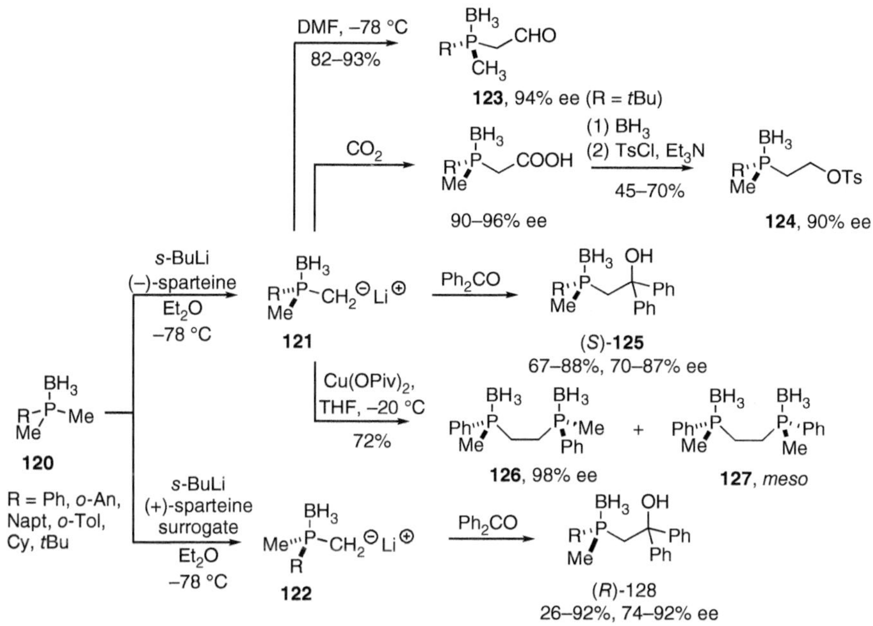

Scheme 10.16 (−)-Sparteine **88** and (+)-sparteine surrogate **91**-mediated asymmetric deprotonation–substitution of phosphines–borane complex toward P-chiral compounds **123–128**. Source: Johansson et al. [235]/American Chemical Society.

loading (0.3–0.4 equiv) of the chiral ligands [188]. Mechanistic aspects, stereochemical outcome, chemistry, and application in asymmetric catalysis of these complex reactions are discussed in detail in two excellent reviews by Kolodiazhnyi [148, 149].

Lack of larger quantities of enantiomeric (+)-sparteine isolated from natural sources or prepared by total synthesis limits the use of the presented reactions toward synthesis of the single enantiomer (obtained with (−)-sparteine). For this reason, an intense research was undertaken in the last two decades to construct simplified (+)-sparteine surrogates with the same chiral induction as (+)-sparteine.

O'Brien and coworkers were especially active in this field carrying out systematic structure activity relationship (SAR) studies of diverse sparteine analogs in model enantioselective deprotonation–silylation of *N*-Boc pyrrolidine. They revealed that the A–C ring-scaffold of (+)-sparteine was essential for high stereoinduction and reactivity, whereas the D ring was not and could be omitted. These findings eventually led to development of (+)-sparteine surrogate **91** (Figure 10.5) acting as highly efficient replacement of the hardly available alkaloid enantiomer [237]. Interestingly, it became soon evident that (+)-sparteine surrogate **91** has a greatly enhanced reactivity in a complex with *s*-BuLi, thus offering even broader synthetic applicability in comparison to natural (−)-sparteine **88** [237]. Higher reactivity of the *s*-BuLi/**91** system was conveniently exploited in a total synthesis of (−)-stemaphylline by Aggarwal and coworkers [204]. A detailed (+)-sparteine surrogate chemistry, its development, and application is a subject of a review by O'Brien [237].

Beside problems with an access to (+)-sparteine, over the last two decades also availability of more common (−)-sparteine has varied dramatically (especially on a larger scale) [56], which together with their high prices hamper the wider use of these valuable ligands (note the case of the development of hepatitis C drug, Telaprevir, *vide supra*). Similarly, commercially available (+)-sparteine surrogate **91** is also obtained from another lupine alkaloid (−)-cytisine **90** [238] making it expensive and dependent on the natural supply.

Solution of the problem with a convenient access to both sparteines and sparteine surrogate is sought in the development of a more efficient biotechnological production of lupine alkaloids [239], optimization of existing isolation technologies [240], or their total synthesis [180, 241, 242]. It is worth to note that despite intense screening their replacement for other more accessible chiral 1,2-diamines, such as *trans*-1,2-diaminocyclohexane derivatives (Aleksiakis ligand), is typically not superior.

Attempts to make synthetic sparteine surrogates more accessible came mainly from the group of O'Brien. Initially (−)-sparteine surrogate was prepared by the racemic sparteine surrogate resolution via diastereomeric salt formation with (−)-*O*,*O*′-di-*p*-toluoyl-L-tartaric acid [241]. Later on, O'Brien's group showed an optimized multigram route for the preparation of both (−)-sparteine (10 steps) and (−)-sparteine surrogate (8 steps). These syntheses started from ethyl 2-pyridyl acetate and used the *Burkholderia cepacia* lipase kinetic resolution of racemic ester as a key step. The kinetic resolution led to the respective enantiopure acid (49%, 98 : 2 er) and ester (46%, >99 : 1 er) which were utilized in further steps. These intermediates were conveniently isolated by simple filtration and aqueous work-up without chromatography. Finally, 3.5 g of (−)-sparteine surrogate was obtained from 30 g of racemic intermediate with overall 22% yield, with three chromatographic purification steps. 1 g of (−)-sparteine was received from 6.5 g of two key chiral intermediates in 31% yield [242].

(−)-Cytisine, found in plants belonging to the Leguminosae family, in particular *Laburnum anagyroides* (golden rain tree) seeds, can be efficiently extracted and is produced on relatively large scale due to the medical use [186, 187]. This alkaloid is directly used in treatment of nicotine addiction. In chemistry, however, (−)-cytisine **90** is a convenient substrate for scalable synthesis of the (+)-sparteine surrogate possessing the requisite absolute stereochemistry [238]. The three-step route involved acylation, hydrogenation of pyridine ring, and reduction of the carbonyl group with $LiAlH_4$ (Scheme 10.17) leading to a single diastereomer of the final product **91**, conveniently prepared in multigram quantity.

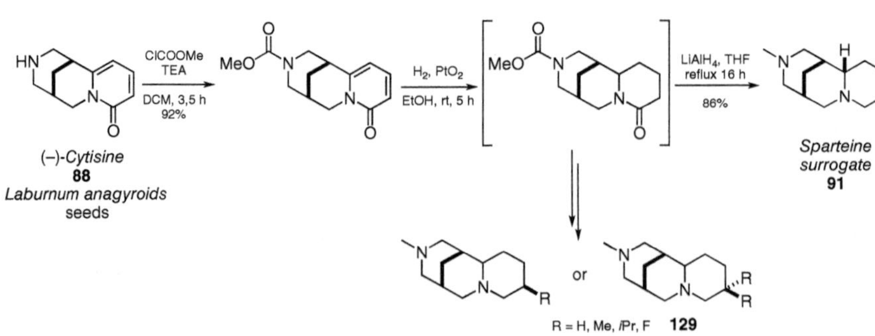

Scheme 10.17 Synthesis of (+)-sparteine surrogate **91** and other bispidines from (−)-cytisine **90**. Source: Based on Dixon et al. [238] and Breuning and Hein [243].

Another attempt to a series of bispidines starting from (−)-cytisine was reported by Breuning and Hein as well (Scheme 10.17) [243]. Among them, 5-*exo*-methyl-substituted compound **129**, tested in an enantioselective lithiation/stannylation of *O*-alkyl carbamate, resulted in stereoselectivity induction at the level comparable to (+)-sparteine surrogate.

Cytisine is also used for the preparation of other natural products. Ohmiya and coworkers reported a three-step route to (+)-hupeol **131** with a preserved configuration of the bridgehead atoms (Scheme 10.18) [244]. Gallagher's group demonstrated a preparation of (+)-kuraramine using *N*-methylcytisine **130** as a starting material [245]. Further comprehensive studies in this chemistry by Aggarwal and coworkers resulted in the transformation that *N*-benzyl cytisine **130** underwent: an efficient ring cleavage via directed C(6) lithiation, borylation, and oxidation sequence provided highly functionalized, enantiopure *cis*-3,5-disubstituted piperidines **132** belonging to a "privileged" scaffold in medicinal chemistry (Scheme 10.18) [54].

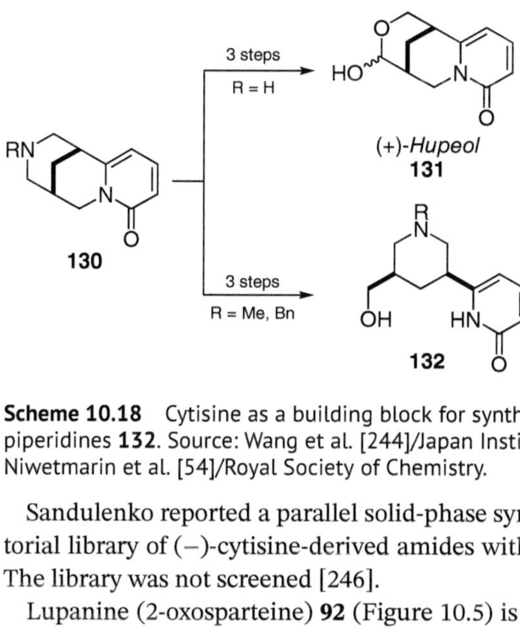

Scheme 10.18 Cytisine as a building block for synthesis of (+)-hupeol **131** and chiral piperidines **132**. Source: Wang et al. [244]/Japan Institute of Heterocyclic Chemistry and Niwetmarin et al. [54]/Royal Society of Chemistry.

Sandulenko reported a parallel solid-phase synthesis of 436-membered combinatorial library of (−)-cytisine-derived amides with two and three points of diversity. The library was not screened [246].

Lupanine (2-oxosparteine) **92** (Figure 10.5) is a naturally occurring lactam alkaloid of *Lupinus albus* typically isolated as a racemate or enriched (+)-enantiomer [247]. An efficient method of lupanine extraction and enantioseparation was reported by Przybył and Kubicki [247] and Maulide et al. [248] who has also used it for the preparation of (+)-sparteine. Easily accessible 2,2-dialkylsparteines obtained by the reaction of racemic lupanine (activated using triflic anhydride) with Grignard compounds were evaluated as possible long-lasting anesthetics acting as irreversible sodium channel blockers [249]. Arylsparteines were also prepared and tested toward their antiproliferative and autophagic flux inhibitory properties [250].

10.5 *Cinchona* Alkaloids

Cinchona alkaloids, with quinine **133**, quinidine **135**, cinchonidine **134**, and cinchonine **136** as major members (Figure 10.7), occupy an exceptional position in

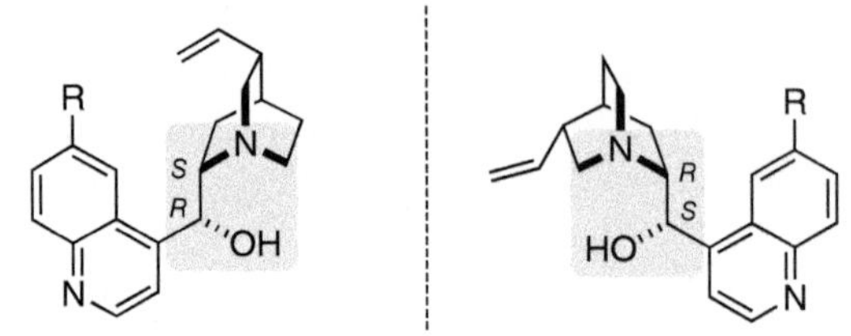

Quinine **133** (QN, R = OMe)
Cinchonidine **134** (CD, R = H)

Quinidine **135** (QD, R = OMe)
Cinchonine **136** (CN, R = H)

Figure 10.7 Four major members of Cinchona alkaloids. Highlighted in gray are enantiomeric 1,2-aminoalcohol parts of alkaloids allowing calling them pseudoenantiomers.

stereoselective synthesis and chirotechnology (*vide supra*) [19–23]. Wide use of quinine in medicine, beverage, and chemical industry resulted in its stable production by extraction from the bark of various genera of *Cinchona* trees, at the level of several hundred tons/year. Lower abundant members of this family such as quinidine, cinchonine, and cinchonidine are isolated in the process of quinine purification, and additional larger amount of quinidine is also produced directly from quinine [251]. Due to their relatively high abundance in the bark of *Cinchona* genus tree (5–14%) they are relatively inexpensive natural products (c. 300 € per 1 kg of quinine) with a high potential to be even more extensively explored as chiral substrate, both in stereoselective synthesis and in medicinal chemistry.

Although quinine and quinidine, as well as cinchonidine and cinchonine are diastereoisomers, they are often called pseudoenantiomers, because the 1,2-aminoalcohol functionality of key importance for catalysis or chiral discrimination remains enantiomeric in these pairs (Figure 10.7).

The chiral 1,2-aminoalcohol structural motif, including the highly basic and bulky quinuclidine nitrogen, is primarily responsible for the enantiodiscriminative supramolecular interactions during catalytic or chiral recognition processes. The major area of a dynamic exploration of Cinchona alkaloids remains asymmetric catalysis with a number of spectacular successes, both in academia and industry [19–28]. However, these catalytic applications are out of scope of this chapter and will not be presented (for fundamental reviews see [19–23]). This group of natural products are frequently used for the construction of diverse chiral molecular systems for an enantiodiscrimination and enantioseparation dedicated to liquid chromatography [8, 9], capillary electrophoresis, membrane separation, and other applications (for a review see [252]). A number of those Cinchona alkaloid-based systems constitute polymer-supported architectures described in Section 10.7 of this chapter.

The most important landmarks in Cinchona alkaloids-driven catalysis and enantioseparation include (Scheme 10.19):

- catalytic asymmetric dihydroxylation (AD) of olefins into *syn*-1,2-diols with the use of bis-dihydroquinine or dihydroquinidine-phthalazine ligands ($(DHQ)_2$PHAL **137** and $(DHQD)_2$PHAL) [28];
- number of biphasic reactions, such as alkylation of O'Donnell Schiff base, Michael addition, epoxidation catalyzed by quaternary salts of Cinchona alkaloids **138** [253];
- organocatalytic application of cupreines **139** [254] 9-amino-9-epialkaloids **142** [255], thioureas **140** [256], squaramides **141** [257];

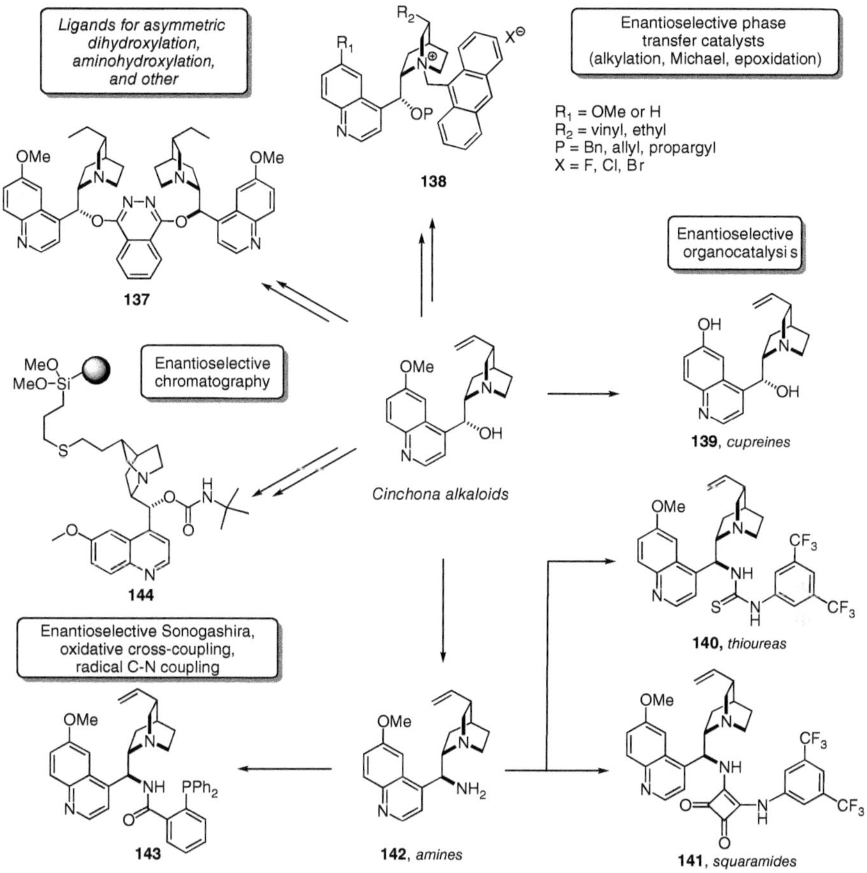

Scheme 10.19 Major applications of Cinchona alkaloid derivatives **137–144** (illustrated with quinine as an example) used as efficient organocatalysts, ligands, PTC catalysts, and chiral stationary phases in catalytic stereoselective synthesis and enantioseparation.

- use of Cinchona ligands, such as **143** in enantioselective cross-coupling reactions [258, 259];
- development of silica-immobilized Cinchona 9-*O*-carbamates **144** as enantioselective stationary phases for chromatography [8, 9].

In contrast to hundreds of catalytic applications, the use of the parent unmodified Cinchona alkaloids as stoichiometric ligands is relatively rare [20] (for older examples see [23]). In this context, quinine and quinidine were used successfully by Senanayake and coworkers as stereodiscriminating ligands in a practical multigram sulfinylation reaction leading to chiral sulfinates and sulfoxides (Scheme 10.20). Thus, reaction of quinine **133** or quinidine **135** with thionyl chloride and triethylamine gave intermediate pseudooxathiazolidine-2-oxide "ate" complex **145**. It reacted further with Grignard reagent providing the *tert*-butanesulfinate **146** with 96% de. Second addition of a variety of Grignard reagents resulted in the formation of highly enantiopure *tert*-butylalkylsulfoxides **147** [260].

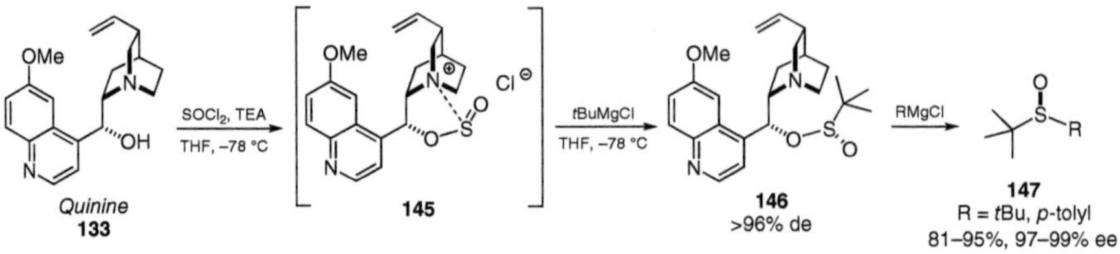

Scheme 10.20 Stoichiometric chiral sulfinyl transfer complex derived from quinine and quinidine. Source: Lu et al. [260]/American Chemical Society.

Quinidine has been used for highly enantioselective stoichiometric desymmetrization of *meso*-anhydrides [261], but many Cinchona alkaloid derivatives perform it well in a catalytic fashion with low loadings of organocatalyst [19, 20]. Cinchonine was also reported to be an efficient ligand in a highly enantioselective Reformatsky reaction of ketones having the sp^2-nitrogen atom adjacent to the reactive carbonyl center. Such substrates are capable of chelation contributing to the enantioface discrimination [262]. Both pseudoenantiomeric cinchonine and cinchonidine promoted a highly enantioselective addition of diethylzinc to *N*-diphenylphosphinoylbenzylimine giving corresponding protected amines in high enantiomeric excess up to 93%. Nevertheless, a catalytic version of this reaction (with 0.2 equiv of alkaloid) in the presence of methanol as an additive exhibited a comparable level of stereoinduction [263].

The multifunctional nature of Cinchona alkaloids makes them especially prone for complex reactivities that occur with degradation or rearrangement of the carbon moiety, often controlled by the stereochemistry of the particular alkaloid. These transformations, characteristic for Cinchona alkaloids, include in particular an acid-catalyzed cleavage of the N1—C8 bond in both quinine/quinidine and the cinchonidine/cinchonine pairs with the formation of cinchotoxines **148** (quinicine and cinchonicine, respectively) described as early as in 1853 by Pasteur (Scheme 10.21) [21, 264]. These easily accessible chiral 3,4-*syn*-difunctionalized piperidines having three stereogenic centers at C9, C8, and N1, can be easily further diversified. Not surprisingly, this simple reaction in a slightly modified version has recently served as a tool for ring distortion of quinine for the generation of chiral molecular diversity (*vide infra*) [51].

Other acidic rearrangement reaction of Cinchona alkaloids led to the synthesis of diverse cyclic ethers, e.g. β-isoquinidine **149** or β-isocupreidine **150** bearing oxazatwistane moiety [19, 21, 265, 266] (Scheme 10.22), which were used as efficient catalysts of stereoselective Baylis–Hilmann reactions [267] or scaffolds for medicinal chemistry [268]. There are also other distinct cage rearrangements of Cinchona 9-*O*-mesylates or 9-halides studied in detail by Hoffmann that provide cage-expanded [3.2.2]-azabicyclic α-aminoethers **151** or β-functionalized [3.2.2]-azabicyclic derivatives **152** (depending on conditions and the alkaloid used) [19, 265, 269].

Cinchona alkaloids undergo also relatively simple oxidative degradations that provide a few attractive heterocyclic chiral building blocks containing pharmacologically favored motifs such as quinuclidine **153** and **154** or piperidine derivatives

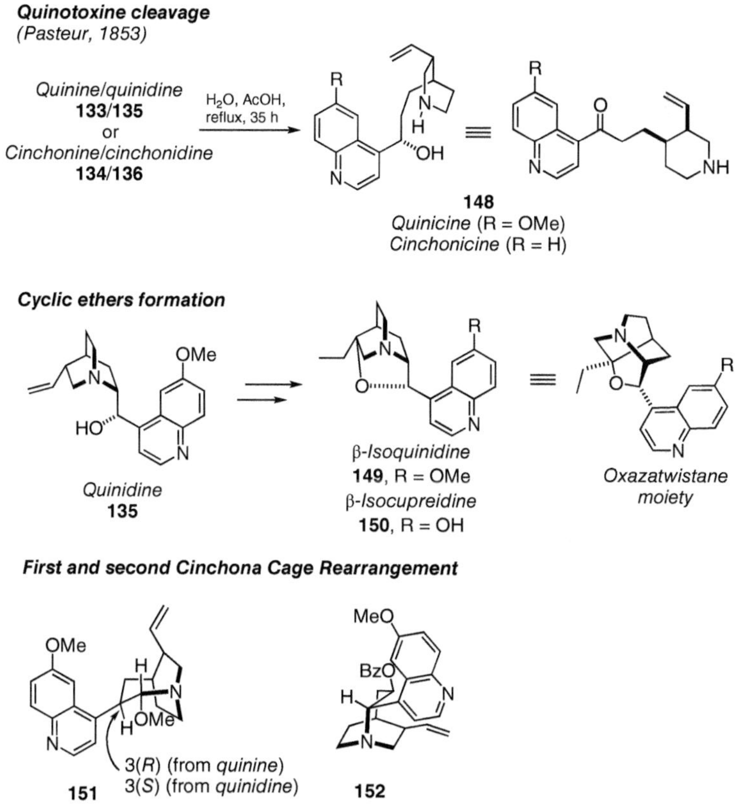

Scheme 10.21 Ring cleavages and rearrangements of Cinchona alkaloids. Source: Based on Boratyński et al. [21] and Biddle [264].

(meroquinene ester **156**). Hoffmann demonstrated an oxidative cleavage of C-9-4′ bond in quinine or quinidine by $LiAlH_4$ and oxygen leading to chiral quincorine **153** or quincoridine **154**, respectively, with a preserved configuration at the C-8 stereogenic center. After technical improvements facilitating the isolation of products with the aid of acylation by di-*tert*-butyl dicarbonate, this process has been further commercialized by Buchler GmbH giving an access to larger quantities of these versatile building blocks (Scheme 10.22) [19, 265]. Similar to the parent Cinchona alkaloids, quincorines **153**, **154** exhibit analogous reactivities at hydroxyl, vinyl, or quinuclidine nitrogen summarized in a review of Boratyński [21].

On the other hand, practical and scalable oxidative degradation of quininone **155** (Oppenauer oxidation product of quinine or quinidine) carried out in the presence of potassium butoxide and oxygen resulted in a cleavage of the N1—C8 bond of the alkaloid. This led to the meroquinene ester **156** that can efficiently be cyclized to *N*-acylated *cis*-decahydroisoquinolones **157** with a maintenance of the *cis* stereochemistry at the ring juncture [270]. This easily accessible chiral building block has been used in a stereoselective synthesis of *cis*-isoquinolinone and *cis*-isoquinolonone derivatives **157** [270], AB ring system of manzamine alkaloids [271], antihistaminic (+)-mequitazine **158** [272], and recently also for the

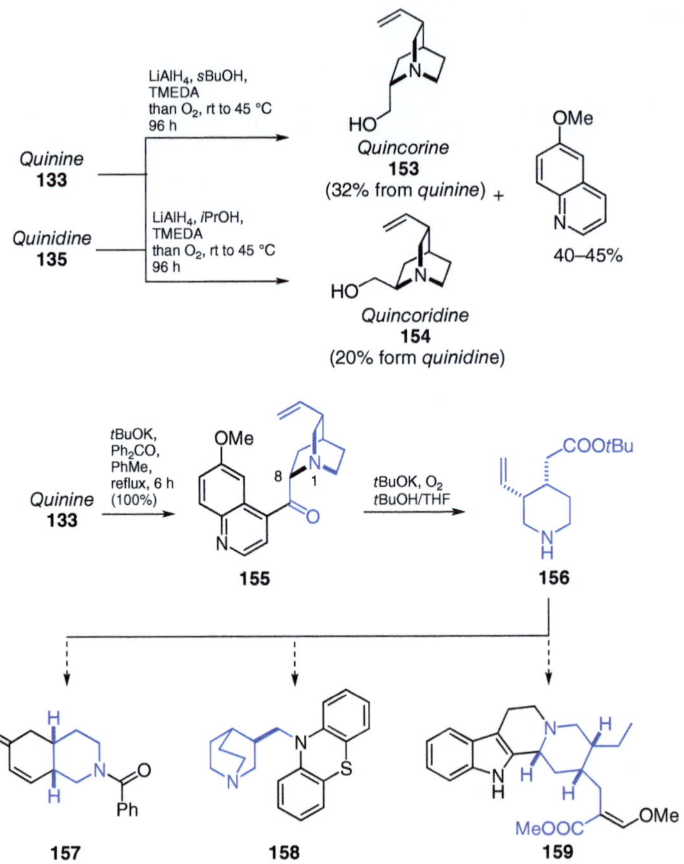

Scheme 10.22 Chiral building blocks from Cinchona alkaloids through an oxidative cleavage of: (top) C4′—C9 bond of quinine and quinidine leading to quincorine **153** and quincoridine **154**, and (bottom) C8—N1 bond of quininone **155** to meroquinene ester **156** and its subsequent transformations. Source: Based on Refs. [19, 21, 265, 266].

preparation of monoterpenoid indole alkaloid uncarialin A **159** – a potent serotonin 5-HT1A receptor agonist (Scheme 10.22) [273].

Ring-distortion strategy applied to quinine combining the presented above reactivity has been reported by Huigens et al. [51]. Thus, a relatively simple transformation involving one to five steps led to five distinct structures **160–164** (Scheme 10.23). For example, a reaction of quinine with *O*-phenyl thionochloroformate resulted in a novel tandem ring cleavage of quinuclidine occurring selectively at N1—C2 bond, and a ring fusion by rearrangement of the free alcohol and thiocarbamate to form the five-membered *S*-alkyl thiocarbamate **160** as a single stereoisomer. On the other hand, an acid-catalyzed Hoffmann-type elimination of quinine (modification of the process discovered by Pasteur) proceeded selectively at N1—C8 bond, and further addition of benzyl chloroformate gave ketone **165**. Petasis olefination of ketone **165** followed by 1,2-ring fusion installed by ring-closing metathesis with Grubbs-2 catalyst resulted in a formation of [4.4.0]-bicyclic system

161. A three-step sequence applied to **165** involving quinoline *N*-oxidation with *m*CPBA, chlorination with oxalyl chloride and nucleophilic displacement of the chloride by (*S*)-2-(methoxymethyl)pyrrolidine led to amine **162**. A single-step ring distortion of the quinoline ring was accomplished by the addition of isoamylmagnesium bromide to quinine that led to product **163** as a single diastereomer. Variation of this route involved the reduction of hemiaminal ether formed through the addition of phenylmagnesium chloride to quinine with $NaBH_3CN$ to give tetrahydroquinoline **166**. Further elaboration of **166** by the reaction with *O*-phenyl thionochloroformate resulted in bis-acylation and gave **164** as the major product without chlorine incorporation observed (Scheme 10.23) [51].

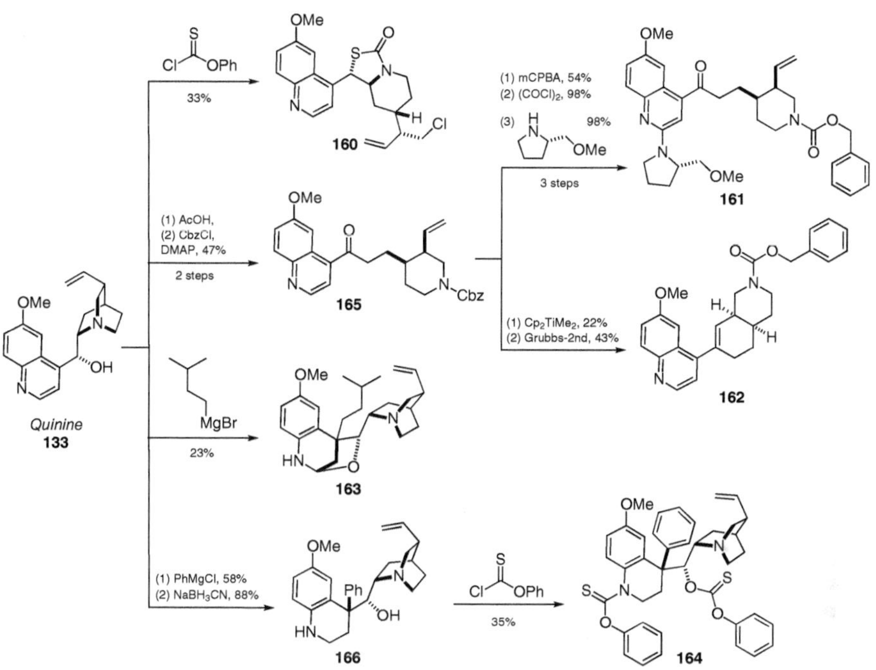

Scheme 10.23 Ring distortion of quinine toward chiral molecular diversity. Source: Huigens et al. [51]/Springer Nature.

If other stereoisomeric Cinchona alkaloids or their derivatives will be applied in these exemplary reaction set, the number of stereochemically and structurally distinct products may significantly rise.

Vast numbers of Cinchona alkaloid derivatives and their applications are a consequence of dynamic studies of their chemistry, in particular in last three decades. Apart from reactivity presented above they involved also stereoselective conversion of 9-hydroxy group, numerous modification of the peripheral vinyl moiety, and transformation of aromatic quinoline ring which are reviewed in detail by Boratyński et al. [21] and Kacprzak [251].

Below, selected examples of modern Cinchona alkaloids chemistry are presented, especially reactions allowing their further diversification relevant for stereoselective catalysis and growing medicinal chemistry of that class of alkaloids. Nucleophilic substitution of the aromatic quinoline ring of Cinchona alkaloids typically gives 2′-substituted products **167** [41, 274], but sterically less demanding Grignard reagents react selectively at 3-carbon yielding **168**. (Scheme 10.24) [274]. An interesting variant of a selective functionalization of quinoline ring of quinine either in 2′ or in 3′ position by metallation has been reported by Knochel group. Typically, the metallation occurs at position 3′, but the increasing bulkiness of the C-9 hydroxy group by formation of silyl ether selectively directs the metallation to the position 2′. Sometimes the addition leads to the loss of aromaticity and to unexpected formation of bicyclic *N*,*O*-acetals **169** with a unique highly diversified Cinchona alkaloids architecture [41, 274].

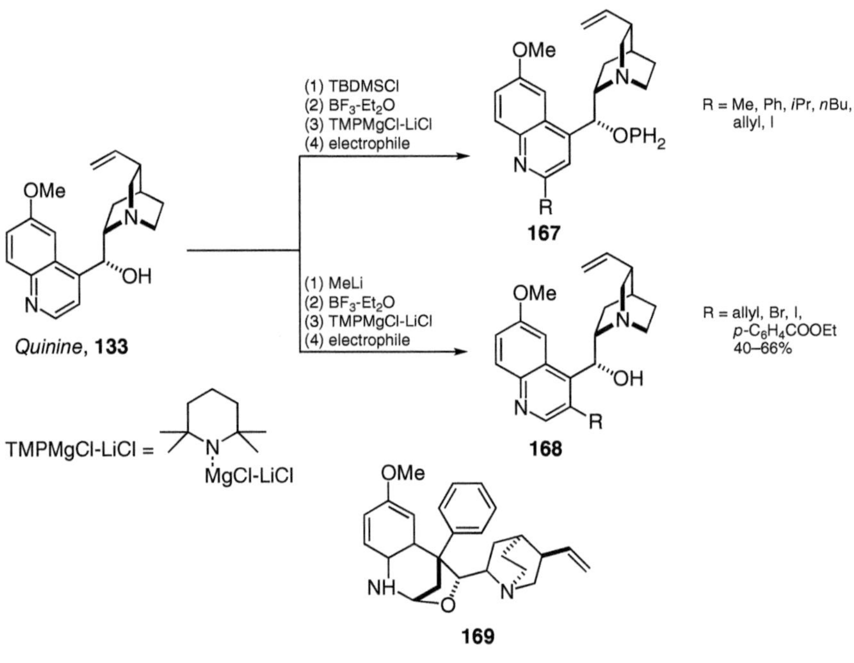

Scheme 10.24 Nucleophilic substitution of Cinchona alkaloids. Source: Hintermann et al. [274]/John Wiley & Sons.

Complementary to nucleophilic substitution but operationally much simpler is a radical CH-functionalization of heteroaryl ring. This pool of reactions became recently an important research field in organic synthesis, offering easy access to a number of privileged structures sought in medicinal chemistry also as a convenient late-stage diversification [39, 275]. An elegant general approach to CH functionalization of heteroarenes based on zinc sulfinates as a source of radicals has been developed by Baran and coworkers [43–45]. This protocol was successfully employed for functionalization of many complex molecules including 10,11-dihydroquinine **170** (Scheme 10.25). Thus, depending on the radical type, substitution at the positions 2′ and 7′ in quinoline ring of the alkaloid is possible. Highly electrophilic radicals, such as trifluoromethyl, undergo regioselective addition to the position 7′ – **171** (with the use of either sodium or zinc trifluoromethanesulfinate as radical

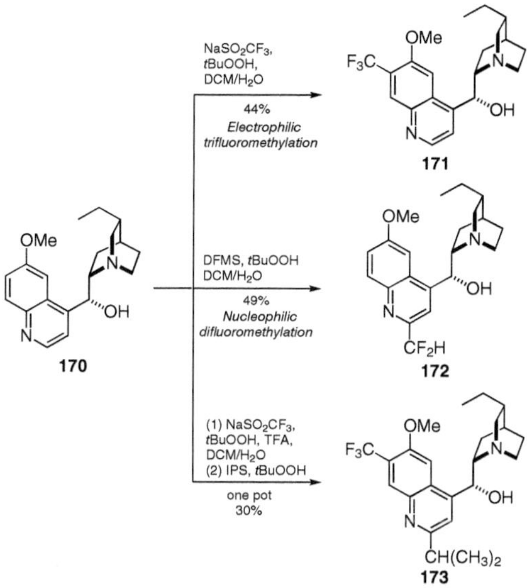

Scheme 10.25 Baran's zinc sulfinates for radical regioselective modification of quinoline ring of dihydroquinine **170**. Source: Based on Fujiwara et al. [43, 44].

precursor). In contrast, less electrophilic radicals, for example, difluoromethyl preferentially add to the position 2′ – **172**. One-pot sequential addition using zinc trifluoromethanesulfinate and zinc isopropylsulfinate led to the installation of both trifluoromethyl and isopropyl groups selectively at the positions 7′ and 2′, respectively, yielding product **173** (Scheme 10.25) [43, 44].

There are also several photomediated CH-heteroarene alkylation protocols with the use of different sources of alkyl radicals and activation modes (Scheme 10.26) [47, 48, 276–278]. These reactions are frequently tested on Cinchona alkaloids and

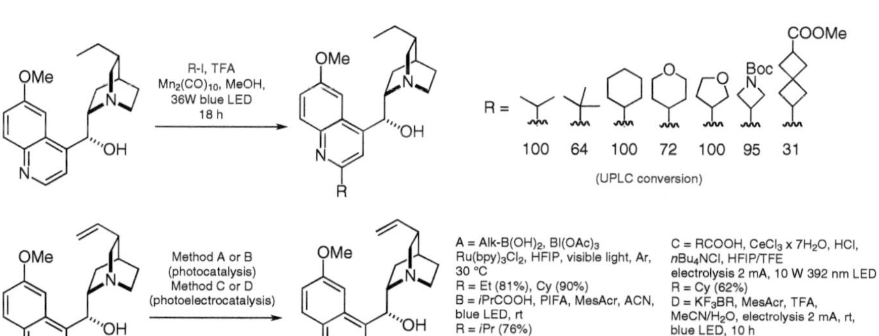

Scheme 10.26 Selected examples of CH-alkylation of Cinchona alkaloids using photocatalytic activation. Source: Based on Refs. [47, 48, 276–278].

gave various 2′-alkylated products in good yields. It is worth to note that some of the protocols are compatible with the vinyl group of quinine, expanding the chemical space accessible to Cinchona alkaloids chemistry [47, 48, 276–278].

Expanding area of Cinchona alkaloids chemistry is powered by click chemistry. These alkaloids are especially well suited for Cu(I)-catalyzed Huisgen 1,3-dipolar cycloaddition of azides and alkynes (CuAAC) because installation of either azide or alkyne functionalities at various positions in Cinchona alkaloids is straightforward and typically does not exceed more than two to three simple steps from the commercially available parent alkaloids [21, 49, 279, 280].

CuAAC has been used for immobilization of alkaloids for preparation of chiral stationary phases, such as **174** [281, 282] and polymer-supported catalysts **175** [283]. Several other 1,2,3-triazoles such as **176–179** derived from Cinchona alkaloids were shown to exhibit remarkable cytotoxic or antiprotozoal activities [284–288]. For example, Cinchona alkaloid-5-floxuridine conjugates such as **176** linked by 1,2,3-triazole have potent cytotoxic activity *in vitro* (Figure 10.8) [287].

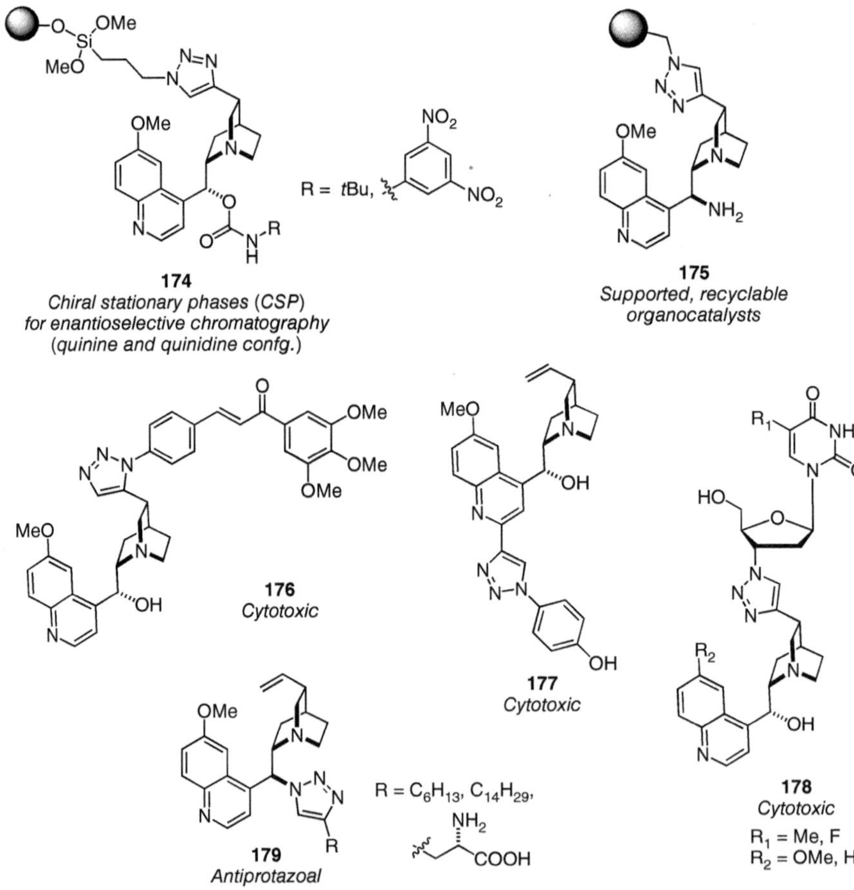

Figure 10.8 Click chemistry Cinchona alkaloids-derived 1,2,3-triazoles and their applications. Source: Celewicz et al. [287]/Canadian Intellectual Property Office.

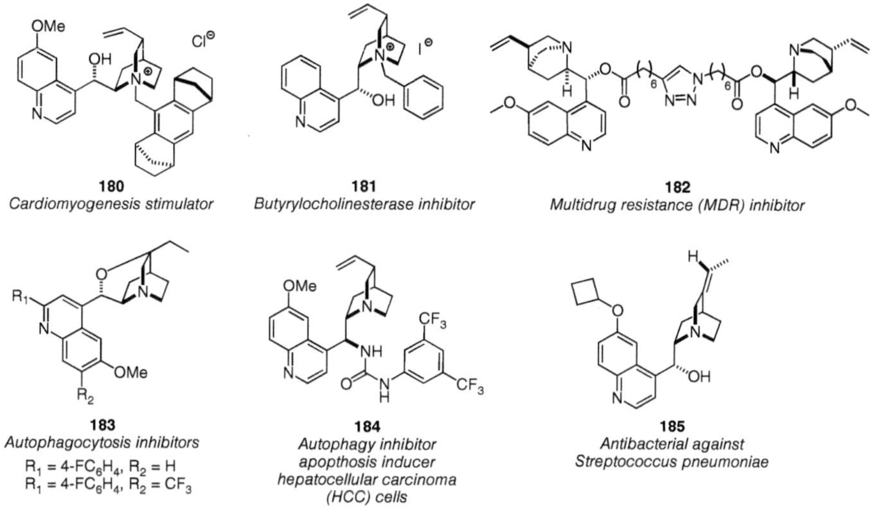

Figure 10.9 Selected examples of biologically active diversified Cinchona alkaloids. Source: Wang et al. [293]/American Chemical Society [289–293].

A growing area of Cinchona alkaloids research is medicinal chemistry [251]. Screening of numerous derivatives often intended as organocatalysts led to the identification of a pool of diverse bioactivities. For example, quaternary salts of Cinchona such as **180** were identified as cardiomyogenesis stimulators of transgenic murine embryonic stem [289]. Other simple quaternary salts derived from cinchonidine **181** were shown to be inhibitors of butyrylocholinesterase exhibiting 250-fold selectivity increase as compared with acetylcholinesterase [290]. Dimeric quinine **182** linked by an ester bond was shown to be highly active in multidrug resistance (MDR) completely reversing the P-glycoprotein (P-gp)-mediated paclitaxel resistance phenotype as well as inhibiting its transport in MCF-7/DX1 cell in *in vitro* studies [291]. Recently, modified β-isocupreidines **183** were found to be potent autophagy inhibitors [268] similar to urea derivative **184** that inhibited autophagy and induced apoptosis in hepatocellular carcinoma (HCC) cells [292]. The cyclobutyl ether of apocupreine **185** was found during systematic SAR studies to be active against *Streptococcus pneumoniae* (Figure 10.9) [293].

10.6 Tropane Alkaloids

The group of c. 200 known alkaloids containing a bicyclic tropane (*N*-methyl-8-azabicyclo[3.2.1]octane) core are found mainly in two plant families, *Solanaceae* and *Erythroxylaceae* (coca) [294–296] with hyoscyamine **186**, scopolamine **187**, and cocaine **188** as common members (Figure 10.10). Tropane alkaloids exhibit various potent pharmacological activities on nervous system, acting as cholinergic agents or analgesics. Consequently, c. 20 derivatives are currently used in medicine as mydriatics, antiemetics, antispasmodics, anesthetics, and bronchodilators [295, 296]. Most of them are derived from scopolamine having the status of the Essential

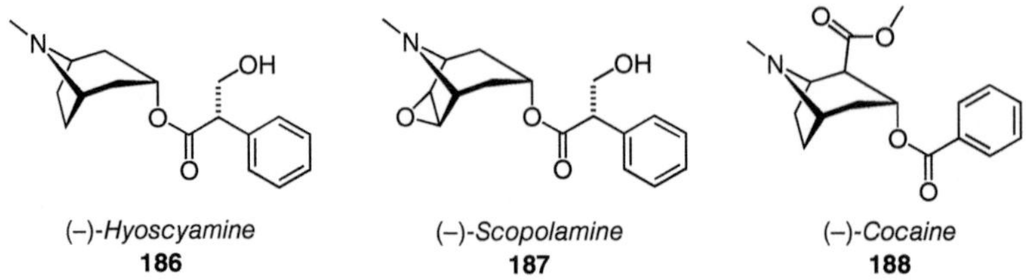

Figure 10.10 The most common members of tropane alkaloids.

Medicine of the World Health Organization. Scopolamine is currently produced on scale by an extraction from the corkwood tree genus *Duboisia* cultivated industrially in Australia [294].

In living systems scopolamine is biosynthesized from hyoscyamine [294]. To satisfy the higher demand of the pharmaceutical industry for the former many biotechnological attempts (but not the total synthesis) are studied. These include overexpression of a gene encoding hyoscyamine-6β-hydroxylase, responsible for hydroxylation of hyoscyamine and its further epoxidation to scopolamine. With the use of transgenic henbane (*Hyoscyamus niger*) hairy root cultures in bioreactors increased production of the desired alkaloid was confirmed [296, 297]. Alternatively, the same gene was also expressed in *N. tabacum* and *Saccharomyces cerevisiae* strains, which were able to convert hyoscyamine into scopolamine [298].

Although several derivatives obtained by modifications of scopolamine were introduced into the pharmaceutical market in the past, including scopolamine-*N*-methyl bromide, scopolamine-*N*-butyl bromide, and tiotropium bromide [295, 296], the most recent research on scopolamine is rather modest. A few simple derivatives, including the product of its basic degradation to scopoline **189** (Scheme 10.27), were studied by Sharwar et al. as inhibitors of acetylcholine esterase [299]. In another study, α- and β-scopodonnines, minor tropane alkaloids found in plants from Solanaceae family, were prepared by thermal dimerization of scopolamine and converted to quaternary salts that were evaluated as myorelaxants [300].

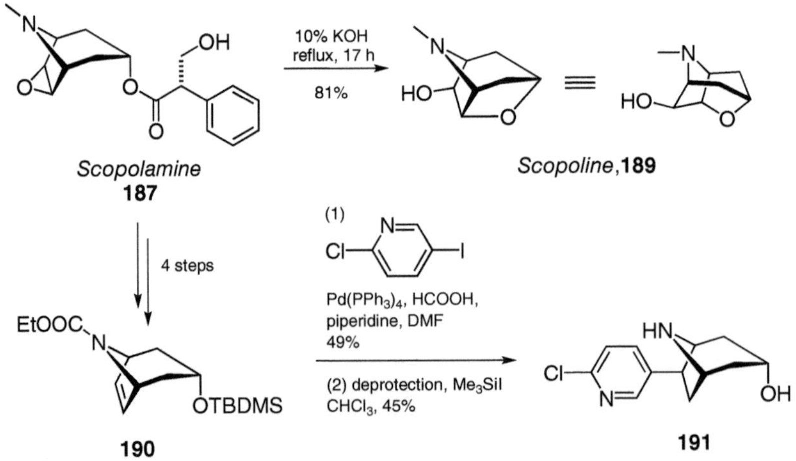

Scheme 10.27 Formation of scopoline and epibatidine analog form scopolamine.

Scopolamine **187** was also used as a starting material for the synthesis of new amine-based, bridgehead-hydroxylated glycosidase inhibitors by Bremner et al. [301]. Later, the same group applied this alkaloid for preparation of a seven-membered analog of epibatidine (Scheme 10.27). This powerful analgesic isolated from the skin of Ecuadorian frogs is characterized by a low therapeutic index; therefore, safer derivatives are sought. The synthetic route involved initial reduction of epoxide moiety to alkene and its base hydrolysis to tropenol. After considerable elaboration an introduction of the chloropyridyl substituent was completed using a reductive Heck coupling. Deprotection yielded the target compound **191**, unfortunately as a racemate, albeit with desired *exo*-stereochemistry of the aryl group [302].

Cocaine **188**, being the second most frequently consumed (thus produced on tons scale) illicit drug globally, bears a relatively simple tropane skeleton (itself achiral due to the symmetry) with four stereogenic centers. Surprisingly, cocaine can serve as an attractive chiral building block for asymmetric synthesis. An interesting attempt toward this goal was demonstrated by Trudell and coworkers [303, 304]. The starting material described as "confiscated grade cocaine" **188**·HCl obtained from the National Institute of Drug Abuse has been efficiently converted into (+)-2-tropinone **192** in 68% yield on multigram scale [303]. **192** was transformed to the corresponding enol ether **193** and immediately subjected to ozonolysis which resulted in ring opening with the preserved configurations of both stereocenters. This eventually led to 2,5-*cis*-disubstituted pyrrolidine **194**, having a common structural motif found in certain alkaloids, for which relatively few synthetic methods

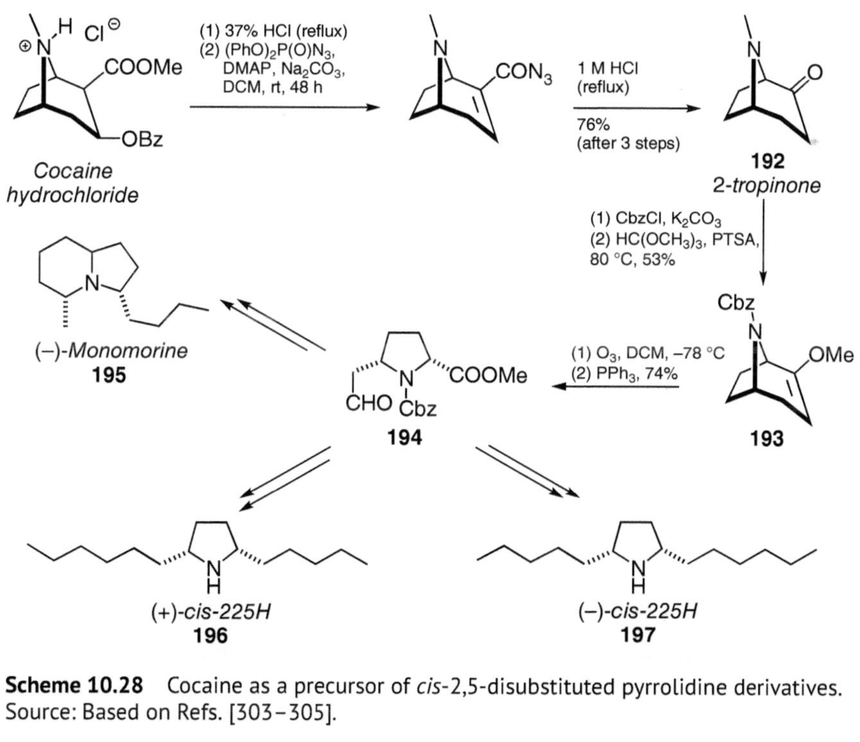

Scheme 10.28 Cocaine as a precursor of *cis*-2,5-disubstituted pyrrolidine derivatives. Source: Based on Refs. [303–305].

were available (Scheme 10.28) [304]. The pyrrolidine derivative **194** was then used in the preparation of (−)-monomorine **195**, an alkaloid found in ants and frog skin, bearing the indolizidine ring system. The target compound **195** was obtained in 37% yield over six steps as a single enantiomer, as a proof of concept of the proposed methodology. Starting from 2,5-disubstituted pyrrolidine **194**, two enantiomers of other amphibian alkaloid, namely *cis*-225*H* **196**, **197**, were synthesized, in line with their absolute configuration assignment [305].

In their study on the selectivity of cocaine transporters, Kozikowski and coworkers prepared constrained, N,C3-bridged tricyclic tropaquinuclidines **201**, **202**, starting from norcocaine formed by demethylation of the parent alkaloid **188** with α-chloroethyl chloroformate [306, 307]. Alkylation with propargyl bromide, hydrolysis, dehydration, and re-esterification provided an unsaturated ester **198** which was subjected to the radical 6-*endo-trig* cyclization. Two diastereomeric *Z* and *E* vinylstannanes **199**, **200** were separated, and subsequent Stille coupling with a series of iodoarenes yielded a set of 10 products (Scheme 10.29) serving for further modification [308]. Later, the synthetic methodology was extended to the preparation of various derivatives bearing biaryl or heteroaromatic ring in the formed ring [309].

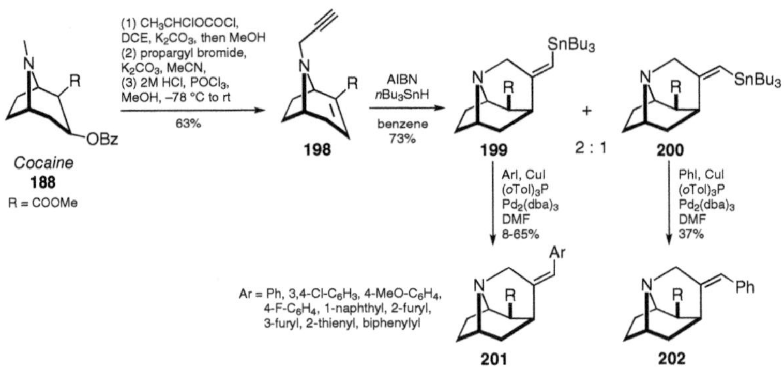

Scheme 10.29 Preparation of tropaquiniclidines **199–202** from cocaine by Kozikowski and coworkers. Source: Hoepping et al. [306]/American Chemical Society.

10.7 Alkaloids as Building Blocks in the Syntheses of Chiral Polymers and Their Application

Immobilization of functionalized alkaloids onto the surfaces of solid supports or their direct installation into a dedicated polymeric network is a frequent requirement in the development of reusable catalysts, chromatographic materials and sensors, or drug-delivery systems. An insolubility of immobilized or polymerized systems in the reaction medium allows for an easy separation of catalyst and the product from a mixture as well as to perform the process with an excess of the substrate to shift the equilibrium toward product. The heterogeneous format promises also a facile catalyst re-use and compatibility with automated flow chemistry systems. To secure the catalytic or biological activity of immobilized or polymerized alkaloids as well as their stability, dedicated immobilization and polymerization techniques must preserve the chemical integrity, allowing the convenient control of site and the density of the ligand binding. Alkaloids, in particular Cinchona derivatives, are frequently used for construction of such immobilized or heterogenized systems for enantioselective catalysis [310], enantiomer separation [8, 9], and for drug delivery [311]. Presented here is a selection of recent Cinchona alkaloids-based polymers intended for stereoselective synthesis. For earlier attempts toward polymer-supported Cinchona systems see review [23], those dedicated to asymmetric dihydroxylation see [312], and sparteine-based polymers are a subject of review by Okamoto [313].

There are two major strategies for the preparation of Cinchona alkaloid-based polymers. The first approach uses the alkaloid or their derivatives for the polymerization resulting in a formation of a chiral polymeric main chain. In the second, alkaloids are introduced into pendant arms of an existing polymer.

In the most recent work Itsuno reported Cinchona bis-urea-type polymers **204** and their use as catalysts in the asymmetric Michael addition. Polymers **204** were obtained in a two-step procedure involving an initial reaction of Cinchona alkaloids with either mono- or bifunctional isocyanates resulting in a formation of urea-type monomers **203** with single or double vinyl groups, respectively. The use of structurally diverse diisocyanates OCN-L-NCO allowed for controlling the spacer length between two Cinchona moieties that may work independently or cooperatively. Finally, the polymerization was carried out using Mizoroki–Heck coupling reaction of the respective dimers **203** having two vinyl groups with 1,4-diiodobenzene (Scheme 10.30) [314].

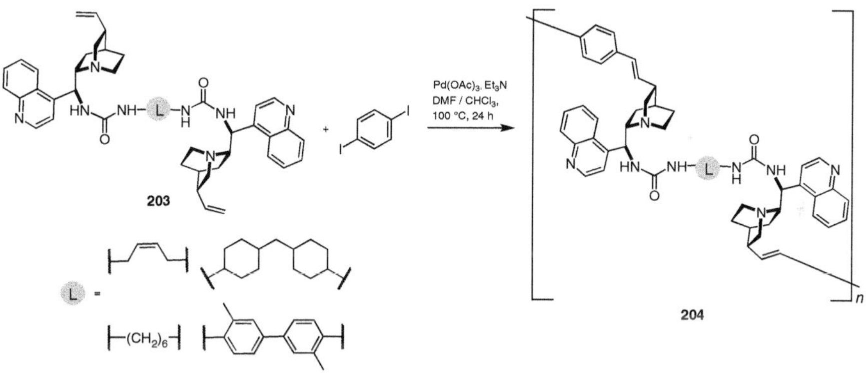

Scheme 10.30 Mizoroki–Heck polymerization of Cinchona alkaloid dimers with the aid of vinyl groups. Source: Abdelkawy et al. [314]/John Wiley & Sons.

Gel permeation chromatography (GPC) analysis of the obtained polymers **204** revealed low average molecular weight in the range of 3400–6700 Da and weight average molecular weight in the range of 6800–9900 Da, which makes them oligomers rather than polymers. Since the molecular weight of the repeating unit is slightly above 800, the average polymeric chain contains 4–8 or maximum of 12 such units.

The use of polymeric catalysts **204** in the model Michael addition of methyl 2-oxocyclopentanecarboxylate and *trans*-β-nitrostyrene gave good results only when

L was an aromatic moiety. In the best case polymer **204** (L = 3,3′-dimethylbiphenyl) allowed for the formation of product in 85% yield, dr = 7 : 1, and ee > 99%. Under the same conditions the respective soluble dimer (used for comparison) gave the product in 98% yield, dr = 4.7 : 1, and ee > 99%. The authors examined the recyclability of the polymeric catalyst and found no changes in its performance in five consecutive repetitions [314].

Quaternization polymerization has also been used for the preparation of another type of polymeric Cinchona-based catalysts **206**. Conceptually similar approach to the synthesis involved an initial formation of bis-9-*O*-ethers **205** from cinchonidine **134** and dihydrocinchonidine (not shown) followed by their quaternization with a series of benzyl-type dichlorides or dibromides. This resulted in the formation of intramolecular bonds with quaternized nitrogen in quinuclidine parts (Scheme 10.31) [315].

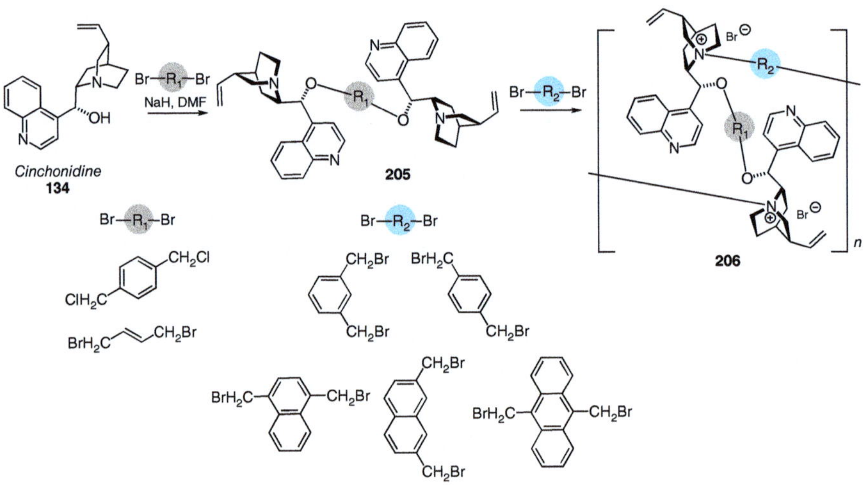

Scheme 10.31 Quaternization of Cinchona bis 9-*O*-ethers toward polymers **206**. Source: Ahamed et al. [315]/Elsevier.

Again, in this case low-molecular-weight <10 kDa was determined, indicating an oligomeric nature of products **206**. Their catalytic activity and re-usability was confirmed in model asymmetric benzylation of O'Donnell Schiff base (*N*-diphenylmethylene glycine *tert*-butyl ester). Both high yields of alkylation and enantioselectivities up to 95% were achieved in most cases [315].

The same group presented later other quaternary polymers having Cinchona alkaloids in the main chain. In this case alkaloid dimers were prepared by quaternization with naphthalene disulfonic acid sodium salt (ionic salts) and with α,α′-dibromoxylene or 4,4′-dichloromethylbiphenyl (covalently linked dimers), whereas Mizoroki–Heck reaction were employed for polymerization. The resultant polymers were obtained in good yield (70–85%) and remarkably higher molecular weight in the range 20–37 kDa. A similar level of performance of these polymers in alkylation of O'Donnell Schiff base was reported [316].

Squaramides having Cinchona 9-amino-9-epialkaloids are powerful organocatalysts [257]; not surprisingly, such a motif was also used for the design of catalytic polymers. Itsuno showed that bis-squaramides with two Cinchona alkaloids underwent the Mizoroki–Heck polymerization with a series of structurally diverse diiodoarenes. The oligomeric rather than polymeric products with M_n in the range of 4000–8400 Da were obtained. They proved to be efficient catalysts in a test Michael addition of β-ketoesters to nitroolefins giving the products with a very high level of enantio- and diastereoselectivity and good yields. The recovered polymeric catalysts maintained fully their catalytic activity over three consecutive runs [317]. Similar polymers having an extra linker between Cinchona-squaramide units were also prepared and tested in the asymmetric Michael addition with a comparable output [318].

A remedy for a low degree of polymerization was demonstrated in 2019 by the group of Itsuno who designed a series of Cinchona squaramide-based hyperbranched polymers. Such polymeric architectures **207** and **208** (Figure 10.11) were similarly obtained by the Mizoroki–Heck reaction of Cinchona dimers with aryl di-, tri-, and tetraiodides [319, 320].

The polymers **207** had much higher M_n than those linear ones obtained formerly [314] lying in the range of 16–37 kDa and M_w in the range of 27–72 kDa, (polydispersity index ~2). Polymers **207** and **208** tested in the model Michael addition (reaction of methyl 2-oxocyclopentanecarboxylate with *trans*-β-nitrostyrene) showed a very high enantio- and diastereoselectivity induction (ee > 99%. dr up to >100 : 1) and chemical yield 90%. Recyclability experiments for **207** over six cycles revealed a constant asymmetric induction level, but a minor drop in the yield from initial 90% to 80–86% [319].

Conceptually similar polymers having Cinchona sulfonamide moieties in the main chain have also been synthetized by the same group [321]. Again, the Mizoroki–Heck reaction of Cinchona monomers bearing vinyl and iodo groups led to the polymeric products of moderate molecular weights. The polymers tested in desymmetrization of cyclic anhydrides gave high enantioselectivities (>90% in most cases) and 99% yield [321]. Other examples of Cinchona sulfonamide-based polymers obtained by polycondensation were also reported [322]. The same group described a synthesis of chiral ionic polymers prepared from 10,11-didehydrocinchonidinium salt. The produced polymers of relatively low-molecular weight (M_n 2500–7400) performed well in benzylation of O'Donnell Schiff base, being slightly superior to monomeric counterparts [323]. Other examples of low-molecular-weight Cinchona polyesters [324] and Cinchona polyurethanes are also available [325].

A different approach for the preparation of functional polymers is an introduction of ligand or drug as pendant groups into a preformed polymer matrix. Studies toward such molecular systems based on the polymerization of Cinchona alkynes were presented by Yashima and coworkers [326]. Thus, synthesis of chiral poly(phenylacetylenes) **209–212** (Figure 10.12) having a pendant alkaloid was achieved directly from alkyne-bearing Cinchona derivatives by rhodium-catalyzed polymerization. Interestingly, the centered chirality of Cinchona alkaloid at

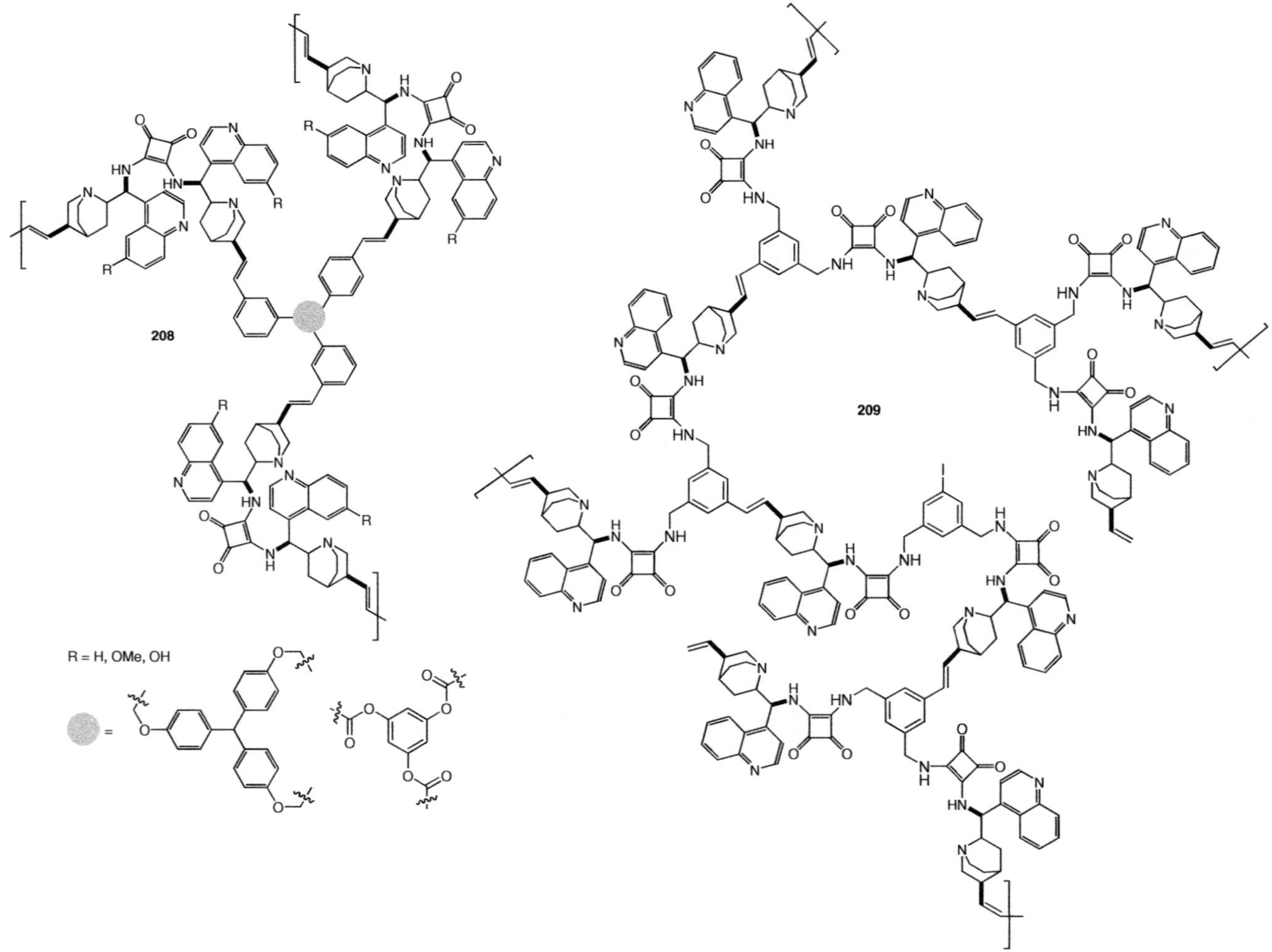

Figure 10.11 Examples of hyperbranched polymers with Cinchona squaramide moieties. Source: Based on Chhanda and Itsuno [319, 320].

Figure 10.12 Helical poly(phenylacetylene)s **209–212** bearing Cinchona alkaloid pendant groups. Source: Based on Refs. [326–328].

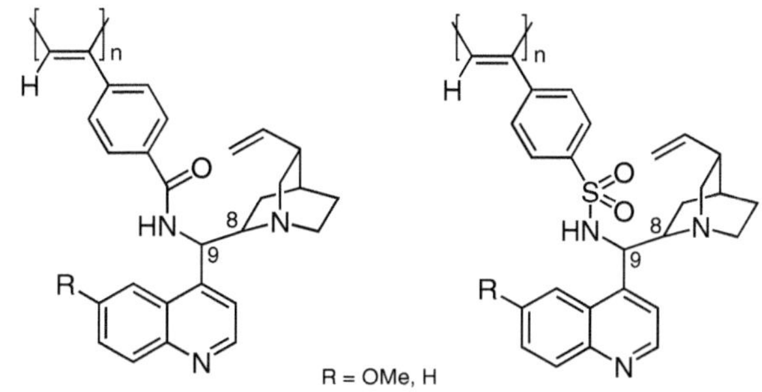

9-position is responsible for an induction of the helically chiral conformation of poly(phenylacetylene) chain. The helical poly(acetylene)s proved to be a highly efficient catalyst in the Henry reaction of various aldehydes with nitromethane [327]. An extension of these studies involving synthesis and catalytic application of poly(phenylacetylenes) bearing Cinchona alkaloid pendants linked by sulfonamide group was published [328]. In this case the respective polymer of molecular weight c. 15 000 Da catalyzed the enantioselective aza-Michael addition of aniline to chalcone with a high asymmetric induction level which exceeded that obtained with either nonhelical polymers or a corresponding monomer [328].

Cinchona alkaloids served also as chiral building blocks for the synthesis of advanced coordination polymers. For example, Xue reported that 4-vinylbenzylcinchonidinium chloride solvothermally reacted with CuCl to produce 2D homochiral copper(I)–olefin coordination polymer capable to selectively intercalating (*R*)-2-butanol with low enantioselection (ee ~ 25%). [329]. Another approach to the synthesis of chiral polymers was published by Lewiński and coworkers. The authors demonstrated that dinuclear metal–Cinchona alkaloid complexes are effective chiral and rigid building blocks that could be used for the preparation of novel heterometallic coordination polymers of various topologies [330]. Polymers that contained pendant quinine molecules were synthesized by ring-opening polymerization of substituted cyclic carbonates by Kiesewetter. The monomer, namely cyclic carbonate with quinine side group, was polymerized using thiourea catalysts to give an oligomer with M_n 2400–7100 Da in a moderate yield of 40–50% [331].

An important class of Cinchona alkaloids-supported polymers and materials is that prepared by direct immobilization of the alkaloid onto polymers [23, 252, 312]. This strategy is commonly used for manufacture of chiral stationary phases based on Cinchona selectors [8, 9, 281, 282] and for the preparation of most of polymer-supported catalysts and reagents [283, 332, 333].

Cinchona squaramides **213–215** bearing free amino groups (Scheme 10.32) were immobilized onto poly(glycidyl methacrylate) solid support. Reaction of those primary amino groups with epoxide-functionalized solid support resulted in an immobilization of catalysts with loading of squaramides **213–215** ranging from 0.04 to 0.2 mmol g^{-1} of resulting material. The polymer-supported catalysts were tested in the model Michael addition of pentane-2,4-dione to *trans* β-nitrostyrene and

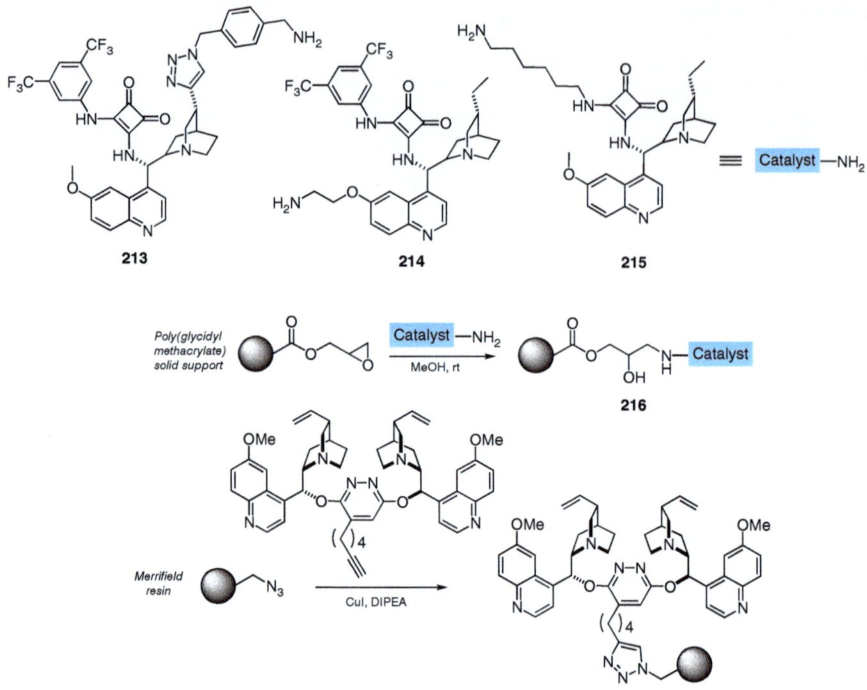

Scheme 10.32 Immobilization of Cinchona organocatalysts onto poly(glycidyl methacrylate) and Merrifield resin using click chemistry. Source: Based on Nagy et al. [332] and Jumde and Mandoli [333].

performed well both regarding yield (up to 96%) and enantiomeric induction (up to 96% ee). The polymeric catalysts were recycled and reused up to five times without any loss of yield and ee [332].

Polystyrene is a common support for immobilization of catalysts. 9-Amino(9-deoxy)*epi*-quinine (**175**, Figure 10.8) having alkyne functionality was grafted to azido-modified polystyrene using click chemistry by Pericas and coworkers [283]. The resultant polymer-supported catalysts **216** were found to efficiently catalyze the Michael addition with high levels of conversion and enantioselectivity with the use of different nucleophiles and structurally diverse enones. Moreover, these easily recyclable catalysts have been successfully employed to single-pass, continuous-flow process (with residence time 40 minutes) that was operated for 21 hours without significant decrease in conversion and even with improved enantioselectivity comparing to batch operation [283].

Last but not least, Mandoli demonstrated an exceptionally efficient performance of a polystyrene-supported bis-Cinchona ether **217** with 1,2,3-triazole linkage in the enantioselective α-amination of 2-oxindoles with diethyl azodicarboxylate (Scheme 10.32). The supported catalysts provided a product with nearly quantitative yields and high enantiomeric purity (89–95% ee), and their stability was confirmed during the course of 100 reaction cycles taking >5300 hours continuous operation time over eight months [333].

Acknowledgments

Support of this work by National Science Centre (contract grant no. UMO-2016/21/B/ST5/00297) is kindly acknowledged.

References

1 Buckingham, J., Baggaley, K.H., Roberts, A.D. et al. (ed.) (2010). *Dictionary of Alkaloids with CD-ROM*, 2e. London: CRC Press.

2 Fattorusso, E. and Taglialatela-Scafati, O. (ed.) (2007). *Modern Alkaloids: Structure, Isolation, Synthesis and Biology*. Weinheim: Wiley.

3 Heinrich, M., Mah, J., and Amirkia, V. (2021). Alkaloids used as medicines: structural phytochemistry meets biodiversity—an update and forward look. *Molecules* 26 (7): 1836.

4 Bhambhani, S., Kondhare, K.R., and Giri, A.P. (2021). Diversity in chemical structures and biological properties of plant alkaloids. *Molecules* 26 (11): 3374.

5 Adamski, Z., Blythe, L.L., Milella, L. et al. (2020). Biological activities of alkaloids: from toxicology to pharmacology. *Toxins (Basel)* 12 (4): 210.

6 Jacques, J., Collet, A., and Willen, S.H. (1981). *Enantiomers, Racemates, and Resolutions*. New York: Wiley-Interscience.

7 Kozma, D. (ed.) (2002). *Optical Resolutions via Diastereomeric Salt Formation*. London: CRC Press.

8 Ilisz, I., Bajtai, A., Péter, A. et al. (1985). Cinchona alkaloid-based zwitterionic chiral stationary phases applied for liquid chromatographic enantiomer separations: an overview. In: *Chiral Separations. Methods in Molecular Biology* (ed. G.K.E. Scriba), 251–277. New York, NY: Humana Press.

9 Ilisz, I., Bajtai, A., Lindner, W. et al. (2018). Liquid chromatographic enantiomer separations applying chiral ion-exchangers based on Cinchona alkaloids. *J. Pharm. Biomed. Anal.* 159: 127–152.

10 Song, C.E. (2000). Alkaloids as chirality transmitters in asymmetric catalysis. In: *The Alkaloids. Chemistry and Biology*, vol. 53 (ed. G.A. Cordell), 1–56. San Diego/London: Academic Press.

11 Hitchcock, S.R., Wolfe, J.A., and Dore, D.D. (2012). Alkaloid derived auxiliaries: miscellaneous alkaloids, Chapter 3.11. In: *Comprehensive Chirality*, vol. 3 (ed. E.M. Carreira and H. Yamamoto), 284–296. Elsevier.

12 Hitchcock, S.R. and Dore, D.D. (2012). Alkaloid-derived auxiliaries: ephedra alkaloids, Chapter 3.10. In: *Comprehensive Chirality*, vol. 3 (ed. E.M. Carreira and H. Yamamoto), 248–283. Elsevier.

13 Cruz, A., Padilla-Martínez, I.I., and Bautista-Ramirez, M.E. (2017). Ephedrines as chiral auxiliaries in enantioselective alkylation reactions of acyl ephedrine amides and esters: a review. *Curr. Org. Synth.* 15 (1): 38–83.

14 Cruz, A. and Esther Bautista Ramirez, M. (2011). Ephedrines and their acyclic derivatives. *Curr. Org. Synth.* 8 (6): 901–928.

15 Chuzel, O. and Riant, O. (2005). Sparteine as a chiral ligand for asymmetric Catalysis. In: *Chiral Diazaligands for Asymmetric Synthesis. Topics in*

Organometallic Chemistry, vol. 15 (ed. M. Lemaire and P. Mangeney), 59–92. Berlin, Heidelberg: Springer.

16 Kizirian, J.C. (2007). Chiral tertiary diamines in asymmetric synthesis. *Chem. Rev.* 108 (1): 140–205.

17 Tomioka, K., Yamamoto, Y., and Yamada, K.I. (2012). Stoichiometric auxiliary ligands for metals and main group elements: ligands for lithium. In: *Comprehensive Chirality*, vol. 3 (ed. E.M. Carreira and H. Yamamoto), 626–654. Elsevier.

18 Kasten, K., Seling, N., and O'Brien, P. (2019). Enantioselective lithiation–substitution of nitrogen-containing heterocycles. In: *Organic Reactions*, vol. 100 (ed. S.E. Denmark), 255–328. New York: Wiley.

19 Song, C.E. (ed.) (2009). *Cinchona Alkaloids in Synthesis and Catalysis: Ligands, Immobilization and Organocatalysis*. Weinheim: Wiley-VCH.

20 Ager, D. (2012). Alkaloid derived auxiliaries: Cinchona alkaloids and derivatives. In: *Comprehensive Chirality*, vol. 3 (ed. E.M. Carreira and H. Yamamoto), 223–247. Elsevier.

21 Boratyński, P.J., Zielińska-Błajet, M., and Skarżewski, J. (2019). Cinchona alkaloids—derivatives and applications. In: *Alkaloids: Chemistry and Biology*, vol. 82 (ed. H.-J. Knölker), 29–145. Academic Press.

22 Marcelli, T. and Hiemstra, H. (2010). Cinchona alkaloids in asymmetric organocatalysis. *Synthesis* 8: 1229–1279.

23 Kacprzak, K. and Gawroński, J. (2001). Cinchona alkaloids and their derivatives: versatile catalysts and ligands in asymmetric synthesis. *Synthesis* 7: 961–998.

24 Carlone, A. and Bernardi, L. (2019). Enantioselective organocatalytic approaches to active pharmaceutical ingredients – selected industrial examples. *Phys. Sci. Rev.* 4 (8): 20180097.

25 Stutte, P. (1994). The preparation and uses of enantiomerically pure β-lactones. In: *Chirality in Industry: The Commercial Manufacture and Applications of Optically Active Compounds*, vol. 1 (ed. A.N. Collins, G.N. Sheldrake and J. Crosby), 341–348. Chichester: Wiley.

26 Scott, J.P., Ashwood, M.S., Brands, K.M.J. et al. (2007). Development of a phase transfer catalyzed asymmetric synthesis for an estrogen receptor beta selective agonist. *Org. Process Res. Dev.* 12 (4): 723–730.

27 Xu, F., Corley, E., Zacuto, M. et al. (2010). Asymmetrie synthesis of a potent, aminopiperidine-fused imidazopyridine dipeptidyl peptidase IV inhibitor. *J. Org. Chem.* 75 (5): 1343–1353.

28 Kolb, H.C., VanNieuwenhze, M.S., and Sharpless, K.B. (1994). Catalytic asymmetric dihydroxylation. *Chem. Rev.* 94 (8): 2483–2547.

29 Kim, H.Y. and Oh, K. (2015). Stereodivergent asymmetric reactions catalyzed by brucine diol. *Synlett* 26 (15): 2067–2087.

30 Khaiwa, N., Maarouf, N.R., Darwish, M.H. et al. (2021). Camptothecin's journey from discovery to WHO essential medicine: fifty years of promise. *Eur. J. Med. Chem.* 223: 113639.

31 Naeem, M., Aftab, T., and Khan, M.M.A. (ed.) (2017). *Catharanthus roseus: Current Research and Future Prospects*. Cham: Springer.

32 Martino, E., Della Volpe, S., Terribile, E. et al. (2017). The long story of camptothecin: from traditional medicine to drugs. *Bioorg. Med. Chem. Lett.* 27 (4): 701–707.

33 Kacprzak, K.M. (2013). Chemistry and biology of camptothecin and its derivatives. In: *Natural Products: Phytochemistry, Botany and Metabolism of Alkaloids, Phenolics and Terpenes* (ed. K. Ramawat and J.M. Mérillon), 643–682. Berlin, Heidelberg: Springer.

34 Mayer, S., Keglevich, P., Keglevich, A. et al. (2021). New anticancer Vinca alkaloids in the last decade – A mini-review. *Curr. Org. Chem.* 25 (10): 1224–1234.

35 Haque, A., Rahman, M.A., Faizi, M.S.H. et al. (2018). Next generation antineoplastic agents: a review on structurally modified vinblastine (VBL) analogues. *Curr. Med. Chem.* 25 (14): 1650–1662.

36 Martino, E., Casamassima, G., Castiglione, S. et al. (2018). Vinca alkaloids and analogues as anti-cancer agents: looking back, peering ahead. *Bioorg. Med. Chem. Lett.* 28 (17): 2816–2826.

37 Kaboudin, B. and Sohrabi, M. (2021). Chemistry and synthesis of major opium alkaloids: a comprehensive review. *J. Iran. Chem. Soc.* 18 (12): 3177–3218.

38 Reed, J.W. and Hudlicky, T. (2015). The quest for a practical synthesis of morphine alkaloids and their derivatives by chemoenzymatic methods. *Acc. Chem. Res.* 48 (3): 674–687.

39 Moir, M., Danon, J.J., Reekie, T.A. et al. (2019). An overview of late-stage functionalization in today's drug discovery. *Expert Opin. Drug Discovery* 14 (11): 1137–1149.

40 Sánchez-Velasco, O.A., Saavedra-Olavarría, J., Araya-Santelices, D.A.A. et al. (2021). Synthesis of *N*-arylcytisine derivatives using the copper-catalyzed Chan-Lam coupling. *J. Nat. Prod.* 84 (7): 1985–1992.

41 Jaric, M., Haag, B.A., Manolikakes, S.M. et al. (2011). Selective and multiple functionalization of pyridines and alkaloids via Mg- and Zn-organometallic intermediates. *Org. Lett.* 13 (9): 2306–2309.

42 Frackenpohl, J., Braje, W.M., and Hoffmann, H.M.R. (2001). Cross-coupling reactions in Cinchona alkaloid chemistry: aryl-substituted and dimeric quinine, quinidine, as well as quincorine and quincoridine derivatives. *Perkin Trans.* 1 (1): 47–65.

43 Fujiwara, Y., Dixon, J.A., Rodriguez, R.A. et al. (2012). A new reagent for direct difluoromethylation. *J. Am. Chem. Soc.* 134 (3): 1494–1497.

44 Fujiwara, Y., Dixon, J.A., O'hara, F. et al. (2012). Practical and innate carbon–hydrogen functionalization of heterocycles. *Nature* 492 (7427): 95–99.

45 Smith, J.M., Dixon, J.A., Degruyter, J.N. et al. (2019). Alkyl sulfinates: radical precursors enabling drug discovery. *J. Med. Chem.* 62 (5): 2256–2264.

46 Dong, J., Lyu, X., Wang, Z. et al. (2019). Visible-light-mediated Minisci C–H alkylation of heteroarenes with unactivated alkyl halides using O_2 as an oxidant. *Chem. Sci.* 10 (4): 976–982.

47 Nuhant, P., Oderinde, M.S., Genovino, J. et al. (2017). Visible-light-initiated manganese catalysis for C–H alkylation of heteroarenes: applications and mechanistic studies. *Angew. Chem. Int. Ed.* 56 (48): 15309–15313.

48 Li, G.X., Morales-Rivera, C.A., Wang, Y. et al. (2016). Photoredox-mediated Minisci C–H alkylation of *N*-heteroarenes using boronic acids and hypervalent iodine. *Chem. Sci.* 7 (10): 6407–6412.

49 Kacprzak, K., Skiera, I., Piasecka, M. et al. (2016). Alkaloids and isoprenoids modification by copper(I)-catalyzed Huisgen 1,3-dipolar cycloaddition (click chemistry): toward new functions and molecular architectures. *Chem. Rev.* 116 (10): 5689–5743.

50 Yi, S., Varun, B.V., Choi, Y. et al. (2018). A brief overview of two major strategies in diversity-oriented synthesis: build/couple/pair and ring-distortion. *Front. Chem.* 6: 507.

51 Huigens, R.W. III, Morrison, K.C., Hicklin, R.W. et al. (2013). A ring-distortion strategy to construct stereochemically complex and structurally diverse compounds from natural products. *Nat. Chem.* 5 (3): 195–202.

52 Paciaroni, N.G., Ratnayake, R., Matthews, J.H. et al. (2017). A tryptoline ring-distortion strategy leads to complex and diverse biologically active molecules from the indole alkaloid yohimbine. *Chem. Eur. J.* 23 (18): 4327–4335.

53 Liu, J., Flegel, J., Otte, F. et al. (2021). Combination of pseudo-natural product design and formal natural product ring distortion yields stereochemically and biologically diverse pseudo-sesquiterpenoid alkaloids. *Angew. Chem. Int. Ed.* 60 (39): 21384–21395.

54 Niwetmarin, W., Campello, H.R., Sparkes, H.A. et al. (2018). (−)-Cytisine: access to a stereochemically defined and functionally flexible piperidine scaffold. *Org. Biomol. Chem.* 16 (32): 5823–5832.

55 Verpoorte, R. and Schripsema, J. (1994). Isolation, identification, and structure elucidation of alkaloids. A general overview. In: *Alkaloids. Modern Methods of Plant Analysis*, vol. 15 (ed. H.F. Linskens and J.F. Jackson), 1–24. Berlin, Heidelberg: Springer.

56 Ritter, S.K. (2017). Where has all the sparteine gone? *C&EN Glob. Enterp.* 95 (17): 18–20.

57 Ehrenworth, A.M. and Peralta-Yahya, P. (2017). Accelerating the semisynthesis of alkaloid-based drugs through metabolic engineering. *Nat. Chem. Biol.* 13 (3): 249–258.

58 Lalaleo, L., Khojasteh, A., Fattahi, M. et al. (2016). Plant anti-cancer agents and their biotechnological production in plant cell biofactories. *Curr. Med. Chem.* 23 (39): 4418–4441.

59 Leonard, E., Runguphan, W., O'Connor, S. et al. (2009). Opportunities in metabolic engineering to facilitate scalable alkaloid production. *Nat. Chem. Biol.* 5 (5): 292–300.

60 Wawrosch, C. and Zotchev, S.B. (2021). Production of bioactive plant secondary metabolites through in vitro technologies—status and outlook. *Appl. Microbiol. Biotechnol.* 105 (18): 6649–6668.

61 Ziegler, J. and Facchini, P.J. (2008). Alkaloid biosynthesis: metabolism and trafficking. *Annu. Rev. Plant Biol.* 59: 735–769.

62 Narcross, L., Fossati, E., Bourgeois, L. et al. (2016). Microbial factories for the production of benzylisoquinoline alkaloids. *Trends Biotechnol.* 34 (3): 228–241.

63 Dehghan, E., Hosseini, B., Naghdi Badi, H. et al. (2010). Application of conventional and new biotechnological approaches for improving of morphinane alkaloids production. *J. Med. Plants* 9 (35): 33–50.

64 Brown, S., Clastre, M., Courdavault, V. et al. (2015). De novo production of the plant-derived alkaloid strictosidine in yeast. *Proc. Natl. Acad. Sci. U.S.A.* 112 (11): 3205–3210.

65 Galanie, S., Thodey, K., Trenchard, I.J. et al. (2015). Complete biosynthesis of opioids in yeast. *Science* 349 (6252): 1095–1100.

66 Nakagawa, A., Matsumura, E., Koyanagi, T. et al. (2016). Total biosynthesis of opiates by stepwise fermentation using engineered *Escherichia coli. Nat. Commun.* 7: 10390.

67 Kaufman, T.S., and Rúveda, E.A. (2005). The quest for quinine: those who won the battles and those who won the war. *Angew. Chem. Int. Ed.* 44 (6): 854–885.

68 Shiomi, S., Misaka, R., Kaneko, M. et al. (2019). Enantioselective total synthesis of the unnatural enantiomer of quinine. *Chem. Sci.* 10 (41): 9433–9437.

69 Tanifuji, R., Minami, A., Oguri, H. et al. (2020). Total synthesis of alkaloids using both chemical and biochemical methods. *Nat. Prod. Rep.* 37 (8): 1098–1121.

70 Patil, M.D., Grogan, G., and Yun, H. (2018). Biocatalyzed C—C bond formation for the production of alkaloids. *ChemCatChem* 10 (21): 4783–4804.

71 Gober, C.M. and Joullié, M.M. (2016). Joining forces: fermentation and organic synthesis for the production of complex heterocycles. *J. Org. Chem.* 81 (21): 10136–10144.

72 Ghislieri, D., Green, A.P., Pontini, M. et al. (2013). Engineering an enantioselective amine oxidase for the synthesis of pharmaceutical building blocks and alkaloid natural products. *J. Am. Chem. Soc.* 135 (29): 10863–10869.

73 Presley, B., Bianchi, B., Coleman, J. et al. (2018). Efficiency of extraction and conversion of pseudoephedrine to methamphetamine from tamper-resistant and non-tamper-resistant formulations. *J. Pharm. Biomed. Anal.* 156: 16–22.

74 Morales, M.R., Mellem, K.T., and Myers, A.G. (2012). Pseudoephenamine: a practical chiral auxiliary for asymmetric synthesis. *Angew. Chem. Int. Ed.* 51 (19): 4568–4571.

75 Mellem, K.T. and Myers, A.G. (2013). A simple, scalable synthetic route to (+)- and (−)-pseudoephenamine. *Org. Lett.* 15 (21): 5594–5597.

76 González-Juárez, D.E., Escobedo-Moratilla, A., Flores, J. et al. (2020). A review of the *Ephedra* genus: distribution, ecology, ethnobotany, phytochemistry and pharmacological properties. *Molecules* 25 (14): 3283.

77 Cruz, A. and Juarez-Juarez, M. (2004). Heterocyclic compounds derived from ephedrines. *Curr. Org. Chem.* 8 (8): 671–693.

78 Cruz, A., Padilla-Martínez, I.I., and Bautista-Ramírez, M.E. (2017). Synthesis, structure and biological activity of ephedra heterocycles. In: *Alkaloids – Alternatives in Synthesis, Modification and Application* (ed. V. Georgiev and A. Pavlov), 25–56. London: IntechOpen.

79 Bulman Page, P.C., Buckley, B.R., Elsegood, M.R.J. et al. (2007). Synthesis of enantiomerically pure tertiary 1,2-aminoalcohols by the highly diastereoselective reductive ring opening of oxazolidines. *Tetrahedron* 63 (45): 10991–10999.

80 Pramanik, C., Bapat, K., Chaudhari, A. et al. (2014). An efficient, scalable process for benzphetamine hydrochloride. *Org. Process Res. Dev.* 18 (4): 495–500.

81 Coote, S.J., Davies, S.G., Fletcher, A.M. et al. (2010). Enantiospecific stereodivergent synthesis of *trans*- and *cis*-*N*(2),3-dimethyl-4-phenyl-1,2,3,4-tetrahydroisoquinolines. *Chem. Asian J.* 5 (3): 589–604.

82 Henry, N. and O'Neil, I.A. (2007). The synthesis of highly functionalized morpholine *N*-oxides from ephedrine and pseudoephedrine utilizing a tandem Cope elimination/reverse Cope elimination protocol. *Tetrahedron Lett.* 48 (10): 1691–1694.

83 Aparicio, D.M., Gnecco, D., Juárez, J.R. et al. (2012). Diastereoselective synthesis of aryl and alkyl *trans*-glycidic amides from pseudoephedrine-derived sulfonium salt. Chemospecific *exo-tet* ring closure for morpholin-3-ones. *Tetrahedron* 68 (49): 10252–10256.

84 Michelet, B., Castelli, U., Appert, E. et al. (2020). Access to optically pure benzosultams by superelectrophilic activation. *Org. Lett.* 22 (13): 4944–4948.

85 Frantz, D.E., Fässler, R., and Carreira, E.M. (2000). Facile enantioselective synthesis of propargylic alcohols by direct addition of terminal alkynes to aldehydes. *J. Am. Chem. Soc.* 122 (8): 1806–1807.

86 Anand, N.K. and Carreira, E.M. (2001). A simple, mild, catalytic, enantioselective addition of terminal acetylenes to aldehydes. *J. Am. Chem. Soc.* 123 (39): 9687–9688.

87 Tan, L., Chen, C.-Y., Tillyer, R.D. et al. (1999). A novel, highly enantioselective ketone alkynylation reaction mediated by chiral zinc aminoalkoxides. *Angew. Chem. Int. Ed* 38 (5): 711–713.

88 Palomo, C., Oiarbide, M., Halder, R. et al. (2006). Enantioselective aza-Henry reactions assisted by ZnII and *N*-methylephedrine. *Angew. Chem. Int. Ed.* 45 (1): 117–120.

89 Cozzi, P.G., Rivalta, E., Cozzi, P.G. et al. (2005). Highly enantioselective one-pot, three-component imino-Reformatsky reaction. *Angew. Chem. Int. Ed.* 44 (23): 3600–3603.

90 Cozzi, P.G. (2006). A catalytic enantioselective imino-Reformatsky reaction. *Adv. Synth. Catal.* 348 (15): 2075–2079.

91 Jaillet, A., Darcel, C., Bayardon, J. et al. (2020). Design of P-chirogenic aminophosphine-phosphinite ligands at both phosphorus centers: origin of enantioselectivities in Pd-catalyzed allylic reactions. *J. Org. Chem.* 85 (22): 14391–14410.

92 Ewalds, R., Eggeling, E.B., Hewat, A.C. et al. (2000). Application of P-stereogenic aminophosphine phosphinite ligands in asymmetric hydroformylation. *Chem. Eur. J.* 6 (8): 1496–1504.

93 Biosca, M., De La Cruz-Sánchez, P., Pàmies, O. et al. (2020). P-Stereogenic *N*-phosphine-phosphite ligands for the Rh-catalyzed hydrogenation of olefins. *J. Org. Chem.* 85 (7): 4730–4739.

94 Parrott, R.W. and Hitchcock, S.R. (2007). Intramolecular chiral relay at stereogenic nitrogen: oxazolidine catalysts derived from ephedra alkaloids. *Tetrahedron: Asymmetry* 18 (3): 377–382.

95 Soares, L.C., Alberto, E.E., Schwab, R.S. et al. (2012). Ephedrine-based diselenide: a promiscuous catalyst suitable to mimic the enzyme glutathione peroxidase (GPx) and to promote enantioselective C–C coupling reactions. *Org. Biomol. Chem.* 10 (32): 6595–6599.

96 Jin, M.J., Sarkar, M.S., Jung, J.Y. et al. (2006). Palladium-catalyzed asymmetric allylic alkylation using ephedrine-derived phosphinooxazolidines. *Bull. Korean Chem. Soc.* 27 (5): 773–775.

97 Bulman Page, P.C., Heaney, H., Reignier, S. et al. (2003). 1,2-Aminothioethers derived from ephedrine and pseudoephedrine: heterobidentate ligands for the palladium-catalysed asymmetric allylic substitution reaction. *Synlett* 1: 22–28.

98 Flock, A.M., Krebs, A., and Bolm, C. (2010). Ephedrine- and pseudoephedrine-derived thioureas in asymmetric Michael additions of keto esters and diketones to nitroalkenes. *Synlett* 8: 1219–1222.

99 Myers, A.G., Yang, B.H., Chen, H. et al. (1994). Use of pseudoephedrine as a practical chiral auxiliary for asymmetric synthesis. *J. Am. Chem. Soc.* 116 (20): 9361–9362.

100 Davies, S.G., Fletcher, A.M., and Thomson, J.E. (2013). Direct asymmetric syntheses of chiral aldehydes and ketones via *N*-acyl chiral auxiliary derivatives including chiral Weinreb amide equivalents. *Chem. Commun.* 49 (77): 8586–8598.

101 Minko, Y. and Marek, I. (2014). Stereodefined acyclic trisubstituted metal enolates towards the asymmetric formation of quaternary carbon stereocentres. *Chem. Commun.* 50 (84): 12597–12611.

102 Muñoz, L., Bosch, M.P., Rosell, G. et al. (2009). Asymmetric synthesis of (*R*)- and (*S*)-4-methyloctanoic acids. A new route to chiral fatty acids with remote stereocenters. *Tetrahedron: Asymmetry* 20 (4): 420–424.

103 Tietze, L.F., Raith, C., Brazel, C.C. et al. (2007). Enantioselective synthesis of 2-substituted alcohols using (+)-(1*S*,2*S*)-pseudoephedrine as chiral auxiliary. *Synthesis* 2: 229–236.

104 Li, B.F., Hughes, R.M., Le, J. et al. (2009). Efficient synthesis of (2*S*,3*S*)-2-ethyl-3-methylvaleramide using (1*S*,2*S*)-pseudoephedrine as a chiral auxiliary. *Org. Process Res. Dev.* 13 (3): 463–467.

105 Nagula, G., Huber, V.J., Lum, C. et al. (2000). Synthesis of α-substituted β-amino acids using pseudoephedrine as a chiral auxiliary. *Org. Lett.* 2 (22): 3527–3539.

106 Krebs, A., Ludwig, V., Pfizer, J. et al. (2004). Enantioselective synthesis of non-natural aromatic α-amino acids. *Chem. Eur. J.* 10 (2): 544–553.

107 Anakabe, E., Vicario, J.L., Badía, D. et al. (2001). Asymmetric synthesis of arylglycines and their use as chiral templates for the stereocontrolled synthesis of 7,8-disubstituted 3-aryl-1,2,3,4-tetrahydroisoquinolin-4-ols. *Eur. J. Org. Chem.* 22: 4343–4352.

108 Vicario, J.L., Badía, D., and Carrillo, L. (2001). Aziridine ring-opening reactions with chiral enolates. Stereocontrolled synthesis of 5-substituted-3-methyl-pyrrolidin-2-ones. *J. Org. Chem.* 66 (17): 5801–5807.

109 Vicario, J.L., Badía, D., Carrillo, L. et al. (2002). Asymmetric synthesis of 1,2-diaryl-2-amino ethanols. *Tetrahedron: Asymmetry* 13 (7): 745–751.

110 Vicario, J.L., Badía, D., and Carrillo, L. (2001). Asymmetric synthesis of β-substituted α-methyl-β-amino esters by Mannich reaction of (*S*,*S*)-(+)-pseudoephedrine acetamide derived enolate with imines. *Org. Lett.* 3 (5): 773–776.

111 Vicario, J.L., Badía, D., and Carrillo, L. (2001). Stereocontrolled Mannich reaction with enolizable imines using (*S*,*S*)-(+)-pseudoephedrine as chiral auxiliary. Asymmetric synthesis of α,β-disubstituted β-aminoesters and β-lactams. *J. Org. Chem.* 66 (26): 9030–9032.

112 Iza, A., Vicario, J.L., Carrillo, L. et al. (2006). (+)-(*S*,*S*)-Pseudoephedrine as a chiral auxiliary in asymmetric Mannich Reactions: Scope and Limitations. *Synthesis* 23: 4065–4074.

113 Iza, A., Uria, U., Reyes, E. et al. (2013). A general approach for the asymmetric synthesis of densely substituted piperidines and fully substituted piperidinones employing the asymmetric Mannich reaction as key step. *RSC Adv.* 3 (48): 25800–25811.

114 Vicario, J.L., Badía, D., and Carrillo, L. (2003). An easy and straightforward approach to the asymmetric synthesis of isoflavanones. *Tetrahedron: Asymmetry* 14 (4): 489–495.

115 Vicario, J.L., Rodriguez, M., Badía, D. et al. (2004). Double stereodifferentiation in the "acetate-type" aldol reaction with Garner's aldehyde. Stereocontrolled synthesis of polyhydroxylated γ-amino carbonyl compounds. *Org. Lett.* 6 (18): 3171–3174.

116 Rodríguez, M., Vicario, J.L., Badía, D. et al. (2005). (*S*,*S*)-(+)-Pseudoephedrine as chiral auxiliary in asymmetric acetate aldol reactions. *Org. Biomol. Chem.* 3 (10): 2026–2030.

117 Ocejo, M., Carrillo, L., Vicario, J.L. et al. (2011). Role of pseudoephedrine as chiral auxiliary in the "acetate-type" aldol reaction with chiral aldehydes; asymmetric synthesis of highly functionalized chiral building blocks. *J. Org. Chem.* 76 (2): 460–470.

118 Etxebarria, J., Vicario, J.L., Badia, D. et al. (2004). Asymmetric synthesis of β-amino esters by aza-Michael reaction of α,β-unsaturated amides using (*S*,*S*)-(+)-pseudoephedrine as chiral auxiliary. *J. Org. Chem.* 69 (7): 2588–2590.

119 Alonso, B., Ocejo, M., Carrillo, L. et al. (2013). Using heteroaryl-lithium reagents as hydroxycarbonyl anion equivalents in conjugate addition reactions with (*S*,*S*)-(+)-pseudoephedrine as chiral auxiliary; enantioselective synthesis of 3-substituted pyrrolidines. *J. Org. Chem.* 78 (2): 614–627.

120 Etxebarria, J., Vicario, J.L., Badía, D. et al. (2007). A general and enantiodivergent method for the asymmetric synthesis of piperidine alkaloids: concise synthesis of (*R*)-pipecoline, (*S*)-coniine and other 2-alkylpiperidines. *Tetrahedron* 63 (46): 11421–11428.

121 Etxebarria, J., Vicario, J.L., Badia, D. et al. (2005). (*S*,*S*)-(+)-Pseudoephedrine as chiral auxiliary in asymmetric aza-Michael reactions. Unexpected selectivity change when manipulating the structure of the auxiliary. *J. Org. Chem.* 70 (22): 8790–8800.

122 Reyes, E., Vicario, J.L., Carrillo, L. et al. (2006). Tandem asymmetric conjugate addition/α-alkylation using (*S*,*S*)-(+)-pseudoephedrine as chiral auxiliary. *Org. Lett.* 8 (12): 2535–2538.

123 Smitrovich, J.H., Boice, G.N., Qu, C. et al. (2002). Pseudoephedrine as a chiral auxiliary for asymmetric Michael reactions: synthesis of 3-aryl-δ-lactones. *Org. Lett.* 4 (11): 1963–1966.

124 Huffman, M.A., Smitrovich, J.H., Rosen, J.D. et al. (2005). Synthesis of a tetrahydropyran NK1 receptor antagonist via asymmetric conjugate addition. *J. Org. Chem.* 70 (11): 4409–4413.

125 Poremba, K.E., Lee, V.A., and Sculimbrene, B.R. (2014). Synthesis of α-chiral-β,γ-unsaturated carboxylic acid derivatives using chiral auxiliaries. *Tetrahedron* 70 (35): 5463–5467.

126 Molander, G.A., Shin, I., and Jean-Gérard, L. (2010). Palladium-catalyzed Suzuki–Miyaura cross-coupling reactions of enantiomerically enriched potassium β-trifluoroboratoamides with various aryl- and hetaryl chlorides. *Org. Lett.* 12 (19): 4384–4387.

127 Kerrigan, N.J., Hutchison, P.C., Heightman, T.D. et al. (2003). Application of an ephedrine chiral linker in a solid-phase, 'asymmetric catch-release' approach to γ-butyrolactones. *Chem. Commun.* 3 (12): 1402–1403.

128 Kerrigan, N.J., Hutchison, P.C., Heightman, T.D. et al. (2004). Development of a solid-phase 'asymmetric resin-capture–release' process: application of an ephedrine chiral resin in an approach to γ-butyrolactones. *Org. Biomol. Chem.* 2 (17): 2476–2482.

129 Fukuzawa, S., Seki, K., Tatsuzawa, M. et al. (1997). A facile synthesis of chiral γ-butyrolactones in extremely high enantioselectivity mediated by samarium(II) iodide. *J. Am. Chem. Soc.* 119 (6): 1482–1483.

130 Myers, A.G., Siu, M., and Ren, F. (2002). Enantioselective synthesis of (−)-terpestacin and (−)-fusaproliferin: clarification of optical rotational measurements and absolute configurational assignments establishes a homochiral structural series. *J. Am. Chem. Soc.* 124 (16): 4230–4232.

131 Kwon, S. and Myers, A.G. (2005). Synthesis of (−)-quinocarcin by directed condensation of α-amino aldehydes. *J. Am. Chem. Soc.* 127 (48): 16796–16797.

132 Nicolaou, K.C., Edmonds, D.J., Li, A. et al. (2007). Asymmetric total syntheses of platensimycin. *Angew. Chem. Int. Ed.* 46 (21): 3942–3945.

133 Paterson, I., Mühlthau, F.A., Cordier, C.J. et al. (2009). Toward the total synthesis of the brasilinolides: stereocontrolled assembly of a C1–C19 polyol segment. *Org. Lett.* 11 (2): 353–356.

134 Bellotta, F., D'Auria, M.V., Sepe, V. et al. (2009). Synthetic studies on homophymine A: stereoselective synthesis of (2*R*,3*R*,4*R*,6*R*)-3-hydroxy-2,4,6-trimethyloctanoic acid. *Tetrahedron* 65 (18): 3659–3663.

135 Thomas, E.J. and Willis, M. (2014). Approaches to the total synthesis of chaetochalasin A. *Org. Biomol. Chem.* 12 (38): 7537–7550.

136 Powell, W.C. and Walczak, M.A. (2018). Asymmetric synthesis of chiral 1,2-amino alcohols and morpholin-2-ones from arylglyoxals. *J. Org. Chem.* 83 (17): 10487–10500.

137 Berger, R., Duff, K., and Leighton, J.L. (2004). Enantioselective allylation of ketone-derived benzoylhydrazones: practical synthesis of tertiary carbinamines. *J. Am. Chem. Soc.* 126 (18): 5686–5687.

138 Casper, D.M., Burgeson, J.R., Esken, J.M. et al. (2002). Toward the development of a structurally novel class of chiral auxiliaries: diastereoselective aldol reactions of a (1*R*,2*S*)-ephedrine-based 3,4,5,6-tetrahydro-2*H*-1,3,4-oxadiazin-2-one. *Org. Lett.* 4 (21): 3739–3742.

139 Casper, D.M. and Hitchcock, S.R. (2003). An improved procedure for the asymmetric aldol reaction of the titanium enolate of an N_3-propionyl-3,4,5,6-tetrahydro-2*H*-1,3,4-oxadiazin-2-one. *Tetrahedron: Asymmetry* 14 (4): 517–521.

140 Burgos, C.H., Canales, E., Matos, K. et al. (2005). Asymmetric allyl- and crotylboration with the robust, versatile, and recyclable 10-TMS-9-borabicyclo[3.3.2]decanes. *J. Am. Chem. Soc.* 127 (22): 8044–8049.

141 Lai, C. and Soderquist, J.A. (2005). Nonracemic homopropargylic alcohols via asymmetric allenylboration with the robust and versatile 10-TMS-9-borabicyclo[3.3.2]decanes. *Org. Lett.* 7 (5): 799–802.

142 Canales, E., Hernandez, E., and Soderquist, J.A. (2006). Nonracemic 3°-carbamines from the asymmetric allylboration of *N*-trimethylsilyl ketimines with *B*-allyl-10-phenyl-9-borabicyclo[3.3.2]decanes. *J. Am. Chem. Soc.* 128 (27): 8712–8713.

143 Gonzalez, A.Z., Canales, E., and Soderquist, J.A. (2006). N-propargylamides via the asymmetric Michael addition of *B*-alkynyl-10-TMS-9-borabicyclo[3.3.2]decanes to *N*-acylimines. *Org. Lett.* 8 (15): 3331–3334.

144 Han, Z., Krishnamurthy, D., Grover, P. et al. (2003). A highly selective and practical method for enantiopure sulfoxides utilizing activated and functionally differentiated *N*-sulfonyl-1,2,3-oxathiazolidine-2-oxide derivatives. *Angew. Chem. Int. Ed.* 42 (18): 2032–2035.

145 Han, Z., Song, J.J., Yee, N.K. et al. (2007). Development of a preparative-scale asymmetric synthesis of (*R*)-*p*-tolyl methyl sulfoxide for use in a one-pot synthesis of a drug intermediate containing a trifluoromethyl-substituted alcohol functionality. *Org. Process Res. Dev.* 11 (3): 605–608.

146 Dutartre, M., Bayardon, J., and Jugé, S. (2016). Applications and stereoselective syntheses of P-chirogenic phosphorus compounds. *Chem. Soc. Rev.* 45 (20): 5771–5794.

147 Huber, R., Passera, A., and Mezzetti, A. (2019). Which future for stereogenic phosphorus? Lessons from P* pincer complexes of iron(II). *Chem. Commun.* 55 (63): 9251–9266.

148 Kolodiazhnyi, O.I. (2014). Recent advances in asymmetric synthesis of P-stereogenic phosphorus compounds. In: *Phosphorus Chemistry I. Topics in Current Chemistry*, vol. 360 (ed. J.L. Montchamp), 161–236. Cham: Springer.

149 Kolodiazhnyi, O.I. (2012). Recent developments in the asymmetric synthesis of P-chiral phosphorus compounds. *Tetrahedron: Asymmetry* 23 (1): 1–46.

150 Harvey, J.S. and Gouverneur, V. (2010). Catalytic enantioselective synthesis of P-stereogenic compounds. *Chem. Commun.* 46 (40): 7477–7485.

151 Jugé, S., Stephan, M., Laffitte, J.A. et al. (1990). Efficient asymmetric synthesis of optically pure tertiary mono and diphosphine ligands. *Tetrahedron Lett.* 31 (44): 6357–6360.

152 Jugé, S. (2008). Enantioselective synthesis of P-chirogenic phosphorus compounds via the ephedrine-borane complex methodology. *Phosphorous, Sulfur, Silicon Relat. Elem.* 183 (2, 3): 233–248.

153 Stephan, M., Šterk, D., and Mohar, B. (2009). DiPAMP's big brother "i-Pr-SMS-Phos" exhibits exceptional features enhancing rhodium(I)-catalyzed hydrogenation of olefins. *Adv. Synth. Catal.* 351 (17): 2779–2786.

154 Truong, T.-K.-T., Nguyen Van Buu, O., Aupoix, A. et al. (2012). Chiral ionic liquids derived from (−)-ephedrine and carbohydrates: synthesis, properties and applications to asymmetric synthesis and catalysis. *Curr. Org. Synth.* 9 (1): 53–64.

155 Vasiloiu, M., Leder, S., Gaertner, P. et al. (2013). Coordinating chiral ionic liquids. *Org. Biomol. Chem.* 11 (46): 8092–8102.

156 Vasiloiu, M., Cervenka, I., Gaertner, P. et al. (2015). Amino alcohol-derived chiral ionic liquids: structural investigations toward chiral recognition. *Tetrahedron: Asymmetry* 26 (18, 19): 1069–1082.

157 Álvaro-Muñoz, T., López-Arbeloa, F., Pérez-Pariente, J. et al. (2014). (1*R*,2*S*)-Ephedrine: a new self-assembling chiral template for the synthesis of aluminophosphate frameworks. *J. Phys. Chem. C* 118 (6): 3069–3077.

158 Bernardo-Maestro, B., Gálvez, P., González, D. et al. (2018). Conformational space of (1*R*,2*S*)-dimethyl-ephedrinium and (1*S*,2*S*)-dimethyl-pseudoephedrinium in the synthesis of nanoporous aluminophosphates. *J. Phys. Chem. C* 122 (35): 20377–20390.

159 Benowitz, N.L., Hukkanen, J., and Jacob, P. (2009). Nicotine chemistry, metabolism, kinetics and biomarkers. In: *Nicotine Psychopharmacology. Handbook of Experimental Pharmacology*, vol. 192 (ed. J.E. Henningfield, E.D. London and S. Pogun), 29–60. Berlin, Heidelberg: Springer.

160 Byrne, P.A., Kobayashi, S., Breugst, M. et al. (2016). Quantification of the nucleophilic reactivity of nicotine. *J. Phys. Org. Chem.* 29 (12): 759–767.

161 Février, F.C., Smith, E.D., and Comins, D.L. (2005). Regioselective C-2 and C-6 substitution of (S)-nicotine and nicotine derivatives. Org. Lett. 7 (24): 5457–5460.

162 Smith, E.D., Février, F.C., and Comins, D.L. (2005). Synthesis of nicotine derivatives via reductive disilylation of (*S*)-nicotine. *Org. Lett.* 8 (2): 179–182.

163 Comins, D.L., King, L.S., Smith, E.D. et al. (2005). Synthesis of C-4 substituted nicotine derivatives via an *N*-acylpyridinium salt of (*S*)-nicotine. *Org. Lett.* 7 (22): 5059–5062.

164 Wagner, F.F. and Comins, D.L. (2006). Regioselective 5-, 4-, and 2-substitution of (*S*)-6-chloronicotine and 4-substitution of (*S*)-5-chloronicotine. *Eur. J. Org. Chem.* 2006 (16): 3562–3565.

165 Ondachi, P.W. and Comins, D.L. (2008). Synthesis and regioselective substitution of C-6 alkoxy derivatives of (*S*)-nicotine. *Tetrahedron Lett.* 49 (3): 569–572.

166 Zhu, J., Enamorado, M.F., and Comins, D.L. (2016). Synthesis of C-4 substituted amido nicotine derivatives via copper(I)- and (II)-catalyzed cross-coupling reactions. *J. Org. Chem.* 81 (22): 11529–11534.

167 Comins, D.L. and Smith, E.D. (2006). A six-step synthesis of (*S*)-5-ethenyl-3-(1-methyl-2-pyrrolidinyl)pyridine (SIB-1508Y) from (*S*)-nicotine. *Tetrahedron Lett.* 47 (9): 1449–1451.

168 Wagner, F.F. and Comins, D.L. (2006). Expedient five-step synthesis of SIB-1508Y from natural nicotine. *J. Org. Chem.* 71 (22): 8673–8675.

169 Wagner, F.F. and Comins, D.L. (2006). Six-step synthesis of (*S*)-brevicolline from (*S*)-nicotine. *Org. Lett.* 8 (16): 3549–3552.

170 Enamorado, M.F., Ondachi, P.W., and Comins, D.L. (2010). A five-step synthesis of (*S*)-macrostomine from (*S*)-nicotine. *Org. Lett.* 12 (20): 4513–4515.

171 Ondachi, P.W. and Comins, D.L. (2010). Synthesis of fused-ring nicotine derivatives from (*S*)-nicotine. *J. Org. Chem.* 75 (5): 1706–1716.

172 Wagner, F.F. and Comins, D.L. (2007). Recent advances in the synthesis of nicotine and its derivatives. *Tetrahedron* 63 (34): 8065–8082.

173 Sieser, J.E., Maloney, M.T., Chisowa, E. et al. (2018). Ir-catalyzed borylation as an efficient route to a nicotine hapten. *Org. Process Res. Dev.* 22 (4): 527–534.

174 Moran, V.E. (2012). Cotinine: beyond that expected, more than a biomarker of tobacco consumption. *Front. Pharmacol.* 3: 173.

175 Oliver, J.L., Pashmi, G., Barnett, P. et al. (2007). Development of an anti-cotinine vaccine to potentiate nicotine-based smoking cessation strategies. *Vaccine* 25 (42): 7354–7362.

176 Sandulenko, Y. and Krasavin, M. (2012). Natural products as templates for bioactive compound libraries. 3*. Novel heterocycles and peptidomimetics generated from anabasine by isocyanide-based multicomponent reactions. *Chem. Heterocycl. Compd.* 48 (4): 606–612.

177 Artyushin, O.I., Vinogradova, N.M., Sharova, E.V. et al. (2016). Novel approach to the design of potential bioactive alkaloid anabasine conjugates using click chemistry methodology. *Heteroat. Chem* 27 (5): 245–252.

178 Brel, V.K. (2016). Click chemistry methodology in the synthesis of anabasine and cytisine conjugates with isoxazole derivatives. *Russ. J. Org. Chem.* 52 (1): 54–60.

179 Musina, L.A., Shults, E.E., Bagryanskaya, I.Y. et al. (2009). Syntheses based on anabasine. *Russ. Chem. Bull.* 57 (1): 140–150.

180 Sacchetti, A. and Rossetti, A. (2021). Synthesis of natural compounds based on the [3,7]-diazabicyclo[3.3.1]nonane (bispidine) core. *Eur. J. Org. Chem.* 2021 (10): 1491–1507.

181 Kizirian, J.C. (2010). Mechanism and stereochemical features in asymmetric deprotonation using RLi/(−)-sparteine bases. In: *Stereochemical Aspects of Organolithium Compounds: Topics in Stereochemistry*, vol. 26 (ed. J.S. Siegel and R.E. Gawley), 189–251. Zürich: Helvetica Chimica Acta.

182 Gawley, R.E. (2010). Overview of carbanion dynamics and electrophilic substitutions in chiral organolithium compounds. In: *Stereochemical Aspects of Organolithium Compounds: Topics in Stereochemistry*, vol. 26 (ed. J.S. Siegel and R.E. Gawley), 93–133. Zürich: Helvetica Chimica Acta.

183 Gu, Q.S. and Yang, D. (2017). Enantioselective synthesis of (+)-mitomycin K by a palladium-catalyzed oxidative tandem cyclization. *Angew. Chem. Int. Ed.* 56 (21): 5886–5889.

184 Shintani, R. and Fu, G.C. (2002). Highly enantioselective desymmetrization of anhydrides by carbon nucleophiles: reactions of Grignard reagents in the presence of (−)-sparteine. *Angew. Chem. Int. Ed.* 41 (6): 1057–1059.

185 Canipa, S.J., Stute, A., and O'Brien, P. (2014). Use of copper(II)/diamine catalysts in the desymmetrisation of meso-diols and asymmetric Henry reactions: comparison of (−)-sparteine and (+)-sparteine surrogates. *Tetrahedron* 70 (40): 7395–7403.

186 Rouden, J., Lasne, M.C., Blanchet, J. et al. (2013). (−)-Cytisine and derivatives: synthesis, reactivity, and applications. *Chem. Rev.* 114 (1): 712–778.

187 Pérez, E.G., Méndez-Gálvez, C., and Cassels, B.K. (2012). Cytisine: a natural product lead for the development of drugs acting at nicotinic acetylcholine receptors. *Nat. Prod. Rep.* 29 (5): 555–567.

188 Granander, J., Secci, F., Canipa, S.J. et al. (2011). One-ligand catalytic asymmetric deprotonation of a phosphine borane: synthesis of P-stereogenic bisphosphine ligands. *J. Org. Chem.* 76 (11): 4794–4799.

189 Hoppe, D., Hintze, F., and Tebben, P. (1990). Chiral lithium-1-oxyalkanides by asymmetric deprotonation; enantioselective synthesis of 2-hydroxyalkanoic acids and secondary alkanols. *Angew. Chem. Int. Ed.* 29 (12): 1422–1424.

190 Hoppe, D. and Hense, T. (1997). Enantioselective synthesis with lithium/(−)-sparteine carbanion pairs. *Angew. Chem. Int. Ed.* 36 (21): 2282–2316.

191 Kerrick, S.T. and Beak, P. (1991). Asymmetric deprotonations: enantioselective syntheses of 2-substituted (*tert*-butoxycarbonyl)pyrrolidines. *J. Am. Chem. Soc.* 113 (25): 9708–9710.

192 Beak, P., Basu, A., Gallagher, D.J. et al. (1996). Regioselective, diastereoselective, and enantioselective lithiation–substitution sequences: reaction pathways and synthetic applications. *Acc. Chem. Res.* 29 (11): 552–560.

193 Stymiest, J.L., Dutheuil, G., Mahmood, A. et al. (2007). Lithiated carbamates: chiral carbenoids for iterative homologation of boranes and boronic esters. *Angew. Chem. Int. Ed.* 46 (39): 7491–7494.

194 Royal, T., Baumgartner, Y., and Baudoin, O. (2017). Enantioselective α-arylation of *O*-carbamates via sparteine-mediated lithiation and Negishi cross-coupling. *Org. Lett.* 19 (1): 166–169.

195 Sandford, C. and Aggarwal, V.K. (2017). Stereospecific functionalizations and transformations of secondary and tertiary boronic esters. *Chem. Commun.* 53 (40): 5481–5494.

196 Collins, B.S.L., Wilson, C.M., Myers, E.L. et al. (2017). Asymmetric synthesis of secondary and tertiary boronic esters. *Angew. Chem. Int. Ed.* 56 (39): 11700–11733.

197 Leonori, D. and Aggarwal, V.K. (2015). Stereospecific couplings of secondary and tertiary boronic esters. *Angew. Chem. Int. Ed.* 54 (4): 1082–1096.

198 Leonori, D. and Aggarwal, V.K. (2015). Reagent-controlled lithiation–borylation. In: *Topics in Organometallic Chemistry*, vol. 49 (ed. E. Fernández and A. Whiting), 271–295. Cham: Springer.

199 Binanzer, M., Yu Fang, G., Aggarwal, V.K. et al. (2010). Asymmetric synthesis of allylsilanes by the borylation of lithiated carbamates: formal total synthesis of (−)-decarestrictine D. *Angew. Chem. Int. Ed.* 49 (25): 4264–4268.

200 Robinson, A. and Aggarwal, V.K. (2012). Stereocontrolled asymmetric synthesis of *syn-E*-1,4-diol-2-enes using allyl boronates and its application in the total synthesis of solandelactone F. *Org. Biomol. Chem.* 10 (9): 1795–1801.

201 Larouche-Gauthier, R., Fletcher, C.J., Couto, I. et al. (2011). Use of alkyl 2,4,6-triisopropylbenzoates in the asymmetric homologation of challenging boronic esters. *Chem. Commun.* 47 (47): 12592–12594.

202 Balieu, S., Hallett, G.E., Burns, M. et al. (2015). Toward ideality: the synthesis of (+)-kalkitoxin and (+)-hydroxyphthioceranic acid by assembly-line synthesis. *J. Am. Chem. Soc.* 137 (13): 4398–4403.

203 Rogers, J.J., Aggarwal, V.K., Rogers, J. et al. (2021). Synthesis of dysoxylactam A using iterative homologation of boronic esters. *Asian J. Org. Chem.* 10 (9): 2338–2341.

204 Varela, A., Garve, L.K.B., Leonori, D. et al. (2017). Stereocontrolled total synthesis of (−)-stemaphylline. *Angew. Chem. Int. Ed.* 56 (8): 2127–2131.

205 Fletcher, C.J., Wheelhouse, K.M.P., Aggarwal, V.K. et al. (2013). Stereoselective total synthesis of (+)-giganin and its C10 epimer by using late-stage lithiation–borylation methodology. *Angew. Chem. Int. Ed.* 52 (9): 2503–2506.

206 Monticelli, S., Holzer, W., Langer, T. et al. (2019). Sustainable asymmetric organolithium chemistry: enantio- and chemoselective acylations through recycling of solvent, sparteine, and Weinreb "Amine". *ChemSusChem* 12 (6): 1147–1154.

207 Leonori, D. and Aggarwal, V.K. (2014). Lithiation–borylation methodology and its application in synthesis. *Acc. Chem. Res.* 47 (10): 3174–3183.

208 Aiken, S.G., Bateman, J.M., and Aggarwal, V.K. (2019). Boron "ate" complexes for asymmetric synthesis. In: *Science of Synthesis 2019/6: Advances in Organoboron Chemistry towards Organic Synthesis* (ed. E. Fernández), 393–458. Stuttgart: Georg Thieme Verlag.

209 Lin, W., Zhang, K.F., and Baudoin, O. (2019). Regiodivergent enantioselective C–H functionalization of Boc-1,3-oxazinanes for the synthesis of β2- and β3-amino acids. *Nat. Catal.* 2 (10): 882–888.

210 Dieter, R.K., Topping, C.M., Chandupatla, K.R. et al. (2001). Enantioselectivity in the reactions of chiral α-(*N*-carbamoyl)alkylcuprates. *J. Am. Chem. Soc.* 123 (21): 5132–5133.

211 Dieter, R.K., Oba, G., Chandupatla, K.R. et al. (2004). Reactivity and enantioselectivity in the reactions of scalemic stereogenic α-(*N*-carbamoyl)alkylcuprates. *J. Org. Chem.* 69 (9): 3076–3086.

212 Campos, K.R., Klapars, A., Waldman, J.H. et al. (2006). Enantioselective, palladium-catalyzed α-arylation of *N*-Boc-pyrrolidine. *J. Am. Chem. Soc.* 128 (11): 3538–3539.

213 Klapars, A., Campos, K.R., Waldman, J.H. et al. (2008). Enantioselective Pd-catalyzed α-arylation of *N*-Boc-pyrrolidine: the key to an efficient and practical synthesis of a glucokinase activator. *J. Org. Chem.* 73 (13): 4986–4993.

214 Barker, G., McGrath, J.L., Klapars, A. et al. (2011). Enantioselective, palladium-catalyzed α-arylation of *N*-Boc pyrrolidine: in situ react IR spectroscopic monitoring, scope, and synthetic applications. *J. Org. Chem.* 76 (15): 5936–5953.

215 Liniger, M., Estermann, K., and Altmann, K.H. (2013). Total synthesis of hygrolines and pseudohygrolines. *J. Org. Chem.* 78 (21): 11066–11070.

216 Cho-Schultz, S., Patten, M.J., Huang, B. et al. (2009). Solution-phase parallel synthesis of Hsp90 inhibitors. *J. Comb. Chem.* 11 (5): 860–874.

217 Bilke, J.L., Moore, S.P., O'Brien, P. et al. (2009). Catalytic asymmetric synthesis of piperidines from pyrrolidine: concise synthesis of L-733,060. *Org. Lett.* 11 (9): 1935–1938.

218 Morita, Y., Tokuyama, H., and Fukuyama, T. (2005). Stereocontrolled total synthesis of (−)-kainic acid. Regio- and stereoselective lithiation of pyrrolidine ring with the (+)-sparteine surrogate. *Org. Lett.* 7 (20): 4337–4340.

219 Tomooka, K., Akiyama, T., Man, P. et al. (2008). Asymmetric synthesis of (−)- and (+)-kainic acid using a planar chiral amide as a chiral building block. *Tetrahedron Lett.* 49 (44): 6327–6329.

220 Tanoury, G.J., Chen, M., Dong, Y. et al. (2014). Stereoselective lithiation and carboxylation of Boc-protected bicyclopyrrolidine: synthesis of a key building block for HCV protease inhibitor telaprevir. *Org. Process Res. Dev.* 18 (10): 1234–1244.

221 Watson, R.T., Gore, V.K., Chandupatla, K.R. et al. (2004). Synthesis of (−)-(*R*)-pyrrolam A and studies on its stability: a caveat on computational methods. *J. Org. Chem.* 69 (18): 6105–6114.

222 Dieter, R.K. and Chen, N. (2006). Asymmetric synthesis of enantioenriched (+)-elaeokanine A. *J. Org. Chem.* 71 (15): 5674–5678.

223 Wong, J.Y.F. and Barker, G. (2020). Recent advances in benzylic and heterobenzylic lithiation. *Tetrahedron* 76 (50): 131704.

224 Schlosser, M. and Limat, D. (2002). Sparteine-mediated α-lithiation of *N*-Boc-*N*-methylbenzylamine: rapid racemization and subsequent deracemization. *J. Am. Chem. Soc.* 117 (49): 12342–12343.

225 Ko, Y.K., Im, C., Do, J. et al. (2014). (+)-Sparteine-mediated substitution of *o*-benzyl-*N*-pivaloylaniline with ketones. *Eur. J. Org. Chem.* 2014 (16): 3460–3467.

226 Choi, Y.S., Kang, K.H., and Park, Y.S. (2015). Asymmetric syntheses of 3,4-disubstituted tetrahydroquinoline derivatives using (+)-sparteine-mediated electrophilic substitution. *Bull. Korean Chem. Soc.* 36 (5): 1500–1503.

227 Kang, K.H., Kim, Y., Im, C. et al. (2013). Mitsunobu cyclodehydration of *N*-pivaloyl-2-aminophenethyl alcohol for asymmetric synthesis of *trans*-2,3-disubstituted indolines. *Tetrahedron* 69 (12): 2542–2549.

228 Kang, K.H., Do, J., and Park, Y.S. (2012). Asymmetric synthesis of *trans*-2,3-disubstituted indoline derivatives. *J. Org. Chem.* 77 (1): 808–812.

229 Gómez-SanJuan, A., Sotomayor, N., and Lete, E. (2013). Inter- and intramolecular enantioselective carbolithiation reactions. *Beilstein J. Org. Chem.* 9 (1): 313–322.

230 Norsikian, S., Marek, I., Klein, S. et al. (1999). Enantioselective carbometalation of cinnamyl derivatives: new access to chiral disubstituted cyclopropanes – configurational stability of benzylic organozinc halides. *Chem. Eur. J.* 5 (7): 2055–2068.

231 Kanda, K., Endo, K., and Shibata, T. (2010). Enantioselective synthesis of planar-chiral 1,*n*-dioxa[*n*]paracyclophanes via catalytic asymmetric *ortho*-lithiation. *Org. Lett.* 12 (9): 1980–1983.

232 Kanda, K., Oshima, S., Shizuno, T. et al. (2014). Enantioselective synthesis of planar-chiral phosphines with 1,*n*-dioxa[*n*]paracyclophane scaffold and their application as chiral ligands. *Heterocycles* 8 (2): 1355–1370.

233 Blangetti, M., Müller-Bunz, H., and O'Shea, D.F. (2013). First asymmetric synthesis of planar chiral [2.2]metacyclophanes. *Chem. Commun.* 49 (55): 6125–6127.

234 Muci, A.R., Campos, K.R., and Evans, D.A. (1995). Enantioselective deprotonation as a vehicle for the asymmetric synthesis of C_2-symmetric *P*-chiral diphosphines. *J. Am. Chem. Soc.* 117 (35): 9075–9076.

235 Johansson, M.J., Andersson, K.H.O., and Kann, N. (2008). Modular asymmetric synthesis of *P*-chirogenic β-amino phosphine boranes. *J. Org. Chem.* 73 (12): 4458–4463.

236 Cain, M.F., Glueck, D.S., Golen, J.A. et al. (2012). Asymmetric synthesis and metal complexes of a C_3-symmetric P-stereogenic triphosphine, (*R*)-MeSi(CH_2PMe(*t*-Bu))$_3$ (MT-Siliphos). *Organometallics* 31 (3): 775–778.

237 O'Brien, P. (2008). Basic instinct: design, synthesis and evaluation of (+)-sparteine surrogates for asymmetric synthesis. *Chem. Commun.* 6: 655–667.

238 Dixon, A.J., McGrath, M.J., and O'Brien, P. (2006). Synthesis of (+)-(1*R*,2*S*,9*S*)-11-methyl-7,11-diazatricyclo[7.3.1.02,7]tridecane, A (+)-sparteine surrogate. *Org. Synth.* 83: 141–154.

239 Parmaki, S., Tsipa, A., Vasquez, M.I. et al. (2020). Resolution of alkaloid racemate: a novel microbial approach for the production of enantiopure lupanine via industrial wastewater valorization. *Microb. Cell Fact.* 19 (1): 67.

240 Regla, I., López-Ortiz, M., Jiménez, O. et al. (2019). Sustainable process for sparteine sulfate preparation. *Org. Process Res. Dev.* 23 (11): 2567–2570.

241 Firth, J.D., O'Brien, P., and Ferris, L. (2014). Revisiting the sparteine surrogate: development of a resolution route to the (−)-sparteine surrogate. *Org. Biomol. Chem.* 12 (46): 9357–9365.

242 Firth, J.D., Canipa, S.J., Ferris, L. et al. (2018). Gram-scale synthesis of the (−)-sparteine surrogate and (−)-sparteine. *Angew. Chem. Int. Ed.* 57 (1): 223–226.

243 Breuning, M. and Hein, D. (2013). 5-Substituted derivatives of the tricyclic (+)-sparteine surrogate in the enantioselective lithiation/stannylation of an *O*-alkyl carbamate. *Eur. J. Org. Chem.* 33: 7575–7582.

244 Wang, Y.H., Higashiyama, K., Kubo, H. et al. (2000). Synthesis of (+)-hupeol from (−)-cytisine: transformation of a cyclic imine into a cyclic hemiacetal. *Heterocycles* 53 (3): 545–548.

245 Frigerio, F., Haseler, C.A., and Gallagher, T. (2010). Synthesis of (+)-kuraramine. *Synlett* 5: 729–730.

246 Sandulenko, Y.B., Ivashchenko, A.V., Kravchenko, D.V. et al. (2009). Parallel solid-phase synthesis of novel 3,7-diazabicyclo[3.3.1]nonane derivatives starting from natural alkaloid cytisine. *Russ. Chem. Bull.* 57 (9): 1999–2004.

247 Przybył, A.K. and Kubicki, M. (2011). Simple and highly efficient preparation and characterization of (−)-lupanine and (+)-sparteine. *Tetrahedron* 67 (40): 7787–7793.

248 Maulide, N., Peng, B., Mateus Afonso, C.A., et al. (2014). Process for converting lupanine into sparteine. WO2014191261A1, filed 20 May 2014.

249 Gawali, V.S., Simeonov, S., Drescher, M. et al. (2017). C_2-Modified sparteine derivatives are a new class of potentially long-acting sodium channel blockers. *ChemMedChem* 12 (22): 1819–1822.

250 Gabr, M.T. and Abdel-Raziq, M.S. (2018). Synthesis, antiproliferative activity and autophagic flux inhibition of new arylsparteine derivatives. *Phytochem. Lett.* 27: 203–207.

251 Kacprzak, K.M. (2013). Chemistry and biology of Cinchona alkaloids. In: *Natural Products* (ed. K. Ramawat and J.M. Mérillon), 605–641. Berlin, Heidelberg: Springer.

252 Kacprzak, K. and Gawronski, J. (2009). Resolution of racemates and enantioselective analytics by Cinchona alkaloids and their derivatives. In: *Cinchona Alkaloids in Synthesis and Catalysis: Ligands, Immobilization and Organocatalysis* (ed. C.E. Song), 421–469. Weinheim: Wiley-VCH.

253 Majdecki, M., Niedbała, P., Tyszka-Gumkowska, A. et al. (2021). Assisted by hydrogen-bond donors: Cinchona quaternary salts as privileged chiral catalysts for phase-transfer reactions. *Synthesis* 53 (16): 2777–2786.

254 Bryant, L.A., Fanelli, R., and Cobb, A.J.A. (2016). Cupreines and cupreidines: an established class of bifunctional Cinchona organocatalysts. *Beilstein J. Org. Chem.* 12 (1): 429–443.

255 Melchiorre, P. (2012). Cinchona-based primary amine catalysis in the asymmetric functionalization of carbonyl compounds. *Angew. Chem. Int. Ed.* 51 (39): 9748–9770.

256 Connon, S.J. (2008). Asymmetric catalysis with bifunctional Cinchona alkaloid-based urea and thiourea organocatalysts. *Chem. Commun.* 22: 2499–2510.

257 Rouf, A. and Tanyeli, C. (2016). Squaramide based organocatalysts in organic transformations. *Curr. Org. Chem.* 20 (28): 2996–3013.

258 Zhang, Z.-H., Dong, X.-Y., Du, X.-Y. et al. (2019). Copper-catalyzed enantioselective Sonogashira-type oxidative cross-coupling of unactivated C(sp^3)–H bonds with alkynes. *Nat. Commun.* 10 (1): 5689.

259 Jiang, S.P., Dong, X.Y., Gu, Q.S. et al. (2020). Copper-catalyzed enantioconvergent radical Suzuki–Miyaura C(sp^3)–C(sp^2) cross-coupling. *J. Am. Chem. Soc.* 142 (46): 19652–19659.

260 Lu, B.Z., Jin, F., Zhang, Y. et al. (2005). New general sulfinylating process for asymmetric synthesis of enantiopure sulfinates and sulfoxides. *Org. Lett.* 7 (8): 1465–1468.

261 Bolm, C., Schiffers, I., Dinter, C.L. et al. (2000). Practical and highly enantioselective ring opening of cyclic meso-anhydrides mediated by Cinchona alkaloids. *J. Org. Chem.* 65 (21): 6984–6991.

262 Ojida, A., Yamano, T., Taya, N. et al. (2002). Highly enantioselective Reformatsky reaction of ketones: chelation-assisted enantioface discrimination. *Org. Lett.* 4 (18): 3051–3054.

263 Beresford, K.J.M. (2004). Enantioselective addition of diethylzinc to *N*-diphenylphosphinoylimines employing cinchonidine and cinchonine as chiral ligands. *Tetrahedron Lett.* 45 (31): 6041–6044.

264 Biddle, H.C. (1912). The conversion of cinchonine and quinine into their poisonous isomers, cinchotoxine and quinotoxine, and the relation of this conversion to the toxicity of the Cinchona alkaloids. *J. Am. Chem. Soc.* 34 (4): 500–515.

265 Hoffmann, H.M.R. and Frackenpohl, J. (2004). Recent advances in Cinchona alkaloid chemistry. *Eur. J. Org. Chem.* 21: 4293–4312.

266 Nakamoto, Y., Urabe, F., Takahashi, K. et al. (2013). α-Isocupreine, an enantiocomplementary catalyst of β-isocupreidine. *Chem. Eur. J.* 19 (38): 12653–12656.

267 Zhong, N.J., Wang, Y.Z., Cheng, L. et al. (2018). Recent advances in the annulation of Morita–Baylis–Hillman adducts. *Org. Biomol. Chem.* 16 (29): 5214–5227.

268 Laraia, L., Ohsawa, K., Konstantinidis, G. et al. (2017). Discovery of novel Cinchona-alkaloid-inspired oxazatwistane autophagy inhibitors. *Angew. Chem. Int. Ed.* 56 (8): 2145–2150.

269 Franz, M.H., Röper, S., Wartchow, R. et al. (2004). The first and second Cinchona rearrangement. Two fundamental transformations of alkaloid chemistry. *J. Org. Chem.* 69 (9): 2983–2991.

270 Hutchison, D.R., Khau, V.V., Martinelli, M.J. et al. (1998). Synthesis of *cis*-4*a*(*S*),8*a*(*R*)-perhydro-6(2*H*)-isoquinolinones from quinine: 4*a*(*S*), 8*a*(*R*)-2-benzoyloctahydro-6(2*H*)-isoquinolinone. *Org. Synth.* 75: 223.

271 Clark, J.S., Townsend, R.J., Blake, A.J. et al. (2001). A concise enantioselective synthesis of the AB ring system of the manzamine alkaloids by ring-closing enyne metathesis. *Tetrahedron Lett.* 42 (18): 3235–3238.

272 Leroux, S., Larquetoux, L., Nicolas, M. et al. (2011). Asymmetric synthesis of (+)-mequitazine from quinine. *Org. Lett.* 13 (13): 3549–3551.

273 Ūdris, N., Jaudzems, K., and Smits, G. (2021). Total synthesis of the proposed structure of uncarialin A. *J. Org. Chem.* 86 (9): 6927–6930.

274 Hintermann, L., Schmitz, M., and Englert, U. (2007). Nucleophilic addition of organometallic reagents to Cinchona alkaloids: simple access to diverse architectures. *Angew. Chem. Int. Ed.* 46 (27): 5164–5167.

275 Bera, A., Kabadwal, L.M., Bera, S. et al. (2021). Recent advances on non-precious metal-catalyzed C–H functionalization of *N*-heteroarenes. *Chem. Commun.* 58 (1): 10–28.

276 Genovino, J., Lian, Y., Zhang, Y. et al. (2018). Metal-free-visible light C–H alkylation of heteroaromatics via hypervalent iodine-promoted decarboxylation. *Org. Lett.* 20 (11): 3229–3232.

277 Lai, X.L., Shu, X.M., Song, J. et al. (2020). Electrophotocatalytic decarboxylative C–H functionalization of heteroarenes. *Angew. Chem. Int. Ed.* 59 (26): 10626–10632.

278 Yan, H., Hou, Z.W., and Xu, H.C. (2019). Photoelectrochemical C–H alkylation of heteroarenes with organotrifluoroborates. *Angew. Chem. Int. Ed.* 58 (14): 4592–4595.

279 Kacprzak, K.M., Lindner, W., and Maier, N.M. (2008). An improved synthesis of 10,11-didehydro Cinchona alkaloids. *Chirality* 20 (3, 4): 441–445.

280 Kacprzak, K. and Gierczyk, B. (2010). Clickable 9-azido-(9-deoxy)-Cinchona alkaloids: synthesis and conformation. *Tetrahedron: Asymmetry* 21 (21, 22): 2740–2745.

281 Kacprzak, K.M., Maier, N.M., and Lindner, W. (2006). Highly efficient immobilization of Cinchona alkaloid derivatives to silica gel via click chemistry. *Tetrahedron Lett.* 47 (49): 8721–8726.

282 Kacprzak, K.M., Maier, N.M., and Lindner, W. (2011). Triazolo-linked cinchona alkaloid carbamate anion exchange-type chiral stationary phases: synthesis by click chemistry and evaluation. *J. Chromatogr. A* 1218 (11): 1452–1460.

283 Izquierdo, J., Ayats, C., Henseler, A.H. et al. (2015). A polystyrene-supported 9-amino(9-deoxy)epi quinine derivative for continuous flow asymmetric Michael reactions. *Org. Biomol. Chem.* 13 (14): 4204–4209.

284 Jernei, T., Duró, C., Dembo, A. et al. (2019). Synthesis, structure and in vitro cytotoxic activity of novel cinchona—chalcone hybrids with 1,4-disubstituted- and 1,5-disubstituted 1,2,3-triazole linkers. *Molecules* 24 (22): 4077.

285 Boratyński, P.J., Gałęzowska, J., Turkowiak, K. et al. (2018). Triazole biheterocycles from Cinchona alkaloids: coordination and antiproliferative properties. *ChemistrySelect* 3 (32): 9368–9373.

286 Baraniak, D., Kacprzak, K., and Celewicz, L. (2011). Synthesis of 3′-azido-3′-deoxythymidine (AZT)—cinchona alkaloid conjugates via click chemistry: toward novel fluorescent markers and cytostatic agents. *Bioorg. Med. Chem. Lett.* 21 (2): 723–726.

287 Celewicz, L., Kacprzak, K., Baraniak, D., et al. (2015). 2′,3′-Dideoxy-5-fluorouridine derivatives, a process for the manufacture thereof and application thereof. WO2015050467A1, filed 04 March 2014.

288 Sahu, A., Agrawal, R.K., and Pandey, R.K. (2019). Synthesis and systemic toxicity assessment of quinine-triazole scaffold with antiprotozoal potency. *Bioorg. Chem.* 88: 102939.

289 Berkessel, A., Seelig, B., Schwengberg, S. et al. (2010). Chemically induced cardiomyogenesis of mouse embryonic stem cells. *ChemBioChem* 11 (2): 208–217.

290 Nawaz, S.A., Ayaz, M., Brandt, W. et al. (2011). Cation–π and π–π stacking interactions allow selective inhibition of butyrylcholinesterase by modified quinine and cinchonidine alkaloids. *Biochem. Biophys. Res. Commun.* 404 (4): 935–940.

291 Pires, M.M., Emmert, D., Hrycyna, C.A. et al. (2009). Inhibition of P-glycoprotein-mediated paclitaxel resistance by reversibly linked quinine homodimers. *Mol. Pharmacol.* 75 (1): 92–100.

292 Jin, P.R., Ta, Y.N.N., Chen, I.T. et al. (2021). Cinchona alkaloid-inspired urea-containing autophagy inhibitor shows single-agent anticancer efficacy. *J. Med. Chem.* 64 (19): 14513–14525.

293 Wang, X., Zeng, Y., Sheng, L. et al. (2019). A cinchona alkaloid antibiotic that appears to target ATP synthase in *Streptococcus pneumoniae*. *J. Med. Chem.* 62 (5): 2305–2332.

294 Kohnen-Johannsen, K.L. and Kayser, O. (2019). Tropane alkaloids: chemistry, pharmacology, biosynthesis and production. *Molecules* 24 (4): 796.

295 Grynkiewicz, G. and Gadzikowska, M. (2008). Tropane alkaloids as medicinally useful natural products and their synthetic derivatives as new drugs. *Pharmacol. Rep.* 60 (4): 439–463.

296 Ullrich, S.F., Hagels, H., and Kayser, O. (2016). Scopolamine: a journey from the field to clinics. *Phytochem. Rev.* 16 (2): 333–353.

297 Zhang, L., Ding, R., Chai, Y. et al. (2004). Engineering tropane biosynthetic pathway in *Hyoscyamus niger* hairy root cultures. *Proc. Natl. Acad. Sci.* 101 (17): 6786–6791.

298 Häkkinen, S.T., Moyano, E., Cusidó, R.M. et al. (2005). Enhanced secretion of tropane alkaloids in *Nicotiana tabacum* hairy roots expressing heterologous hyoscyamine-6β-hydroxylase. *J. Exp. Bot.* 56 (420): 2611–2618.

299 Shahwar, D., Raza, M.A., and Khan, T. (2012). Synthesis and evaluation of enzyme inhibitory potential of some derivatives of scopolamine. *Turk. J. Chem.* 36 (2): 257–266.

300 Razzakov, N.A., Vakhabov, A., and Aripova, S.F. (2003). Quaternary derivatives of α- and β-scopodonnines and their pharmacologic activity. *Chem. Nat. Compd.* 39 (2): 215–217.

301 Bremner, J.B. (2000). Alkaloids as starting materials for synthesis: the use of scopolamine and castanospermine. *ACGC Chem. Res. Commun.* 11: 20–28.

302 Bremner, J.B., Godfrey, C.A., Jensen, A.A. et al. (2004). Synthesis and nicotinic receptor activity of a hydroxylated tropane. *Bioorg. Med. Chem. Lett.* 14 (1): 271–273.

303 Zhang, C., Lomenzo, S.A., Ballay, C.J. II, et al. (1997). An improved synthesis of (+)-2-tropinone. *J. Org. Chem.* 62 (22): 7888–7889.

304 Zhang, S., Xu, L., Miao, L. et al. (2007). General strategy for the construction of enantiopure pyrrolidine-based alkaloids. Total synthesis of (−)-monomorine. *J. Org. Chem.* 72 (8): 3133–3136.

305 Shu, H., Noble, A.R., Zhang, S. et al. (2010). Enantioselective syntheses of both enantiomers of *cis*-pyrrolidine 225*H*. *Tetrahedron* 66 (25): 4428–4433.

306 Hoepping, A., Johnson, K.M., George, C. et al. (2000). Novel conformationally constrained tropane analogues by 6-*endo-trig* radical cyclization and Stille coupling – switch of activity toward the serotonin and/or norepinephrine transporter. *J. Med. Chem.* 43 (10): 2064–2071.

307 Hoepping, A., George, C., Flippen-Anderson, J. et al. (2000). Radical cyclization strategies for the formation of ring constrained tricyclic tropane analogues. *Tetrahedron Lett.* 41 (39): 7427–7432.

308 Zhang, A., Zhou, G., Hoepping, A. et al. (2002). Further studies on conformationally constrained tricyclic tropane analogues and their uptake inhibition at monoamine transporter sites: synthesis of (*Z*)-9-(substituted arylmethylene)-7-azatricyclo[4.3.1.03,7]decanes as a novel class of serotonin transporter inhibitors. *J. Med. Chem.* 45 (9): 1930–1941.

309 Zhou, J., Zhang, A., Kläss, T. et al. (2003). Biaryl analogues of conformationally constrained tricyclic tropanes as potent and selective norepinephrine reuptake inhibitors: synthesis and evaluation of their uptake inhibition at monoamine transporter sites. *J. Med. Chem.* 46 (10): 1997–2007.

310 Kristensen, T.E. and Hansen, T. (2010). Polymer-supported chiral organocatalysts: synthetic strategies for the road towards affordable polymeric immobilization. *Eur. J. Org. Chem.* 2010 (17): 3179–3204.

311 Guo, X., Wang, L., Wei, X. et al. (2016). Polymer-based drug delivery systems for cancer treatment. *J. Polym. Sci., Part A: Polym. Chem.* 54 (22): 3525–3550.

312 Mandoli, A., Pini, D., Fiori, M. et al. (2005). Asymmetric dihydroxylation with recoverable Cinchona alkaloid derivatives: a warning note and an improved, insoluble polymer-bound ligand architecture. *Eur. J. Org. Chem.* 7: 1271–1282.

313 Okamoto, Y. (2004). (−)-Sparteine: the compound that most significantly influenced my research. *J. Polym. Sci., Part A: Polym. Chem.* 42 (18): 4480–4491.

314 Abdelkawy, M.A., Aly, E.S.A., El-Badawi, M.A. et al. (2020). Synthesis of cinchona urea polymers and their evaluation as catalyst in the asymmetric reactions. *ChemistrySelect* 5 (21): 6294–6298.

315 Ahamed, P., Haque, M.A., Ishimoto, M. et al. (2013). Synthesis of chiral quaternary ammonium polymers for asymmetric organocatalysis application. *Tetrahedron* 69 (19): 3978–3983.

316 Parvez, M.M., Haraguchi, N., and Itsuno, S. (2014). Synthesis of Cinchona alkaloid-derived chiral polymers by Mizoroki–Heck polymerization and their application to asymmetric catalysis. *Macromolecules* 47 (6): 1922–1928.

317 Ullah, M.S. and Itsuno, S. (2017). Synthesis of Cinchona alkaloid squaramide polymers as bifunctional chiral organocatalysts for the enantioselective Michael addition of β-ketoesters to nitroolefins. *Mol. Catal.* 438: 239–244.

318 Ullah, M.S. and Itsuno, S. (2018). Cinchona squaramide-based chiral polymers as highly efficient catalysts in asymmetric Michael addition reaction. *ACS Omega* 3 (4): 4573–4582.

319 Chhanda, S.A. and Itsuno, S. (2019). Design and synthesis of chiral hyperbranched polymers containing cinchona squaramide moieties and their catalytic activity in the asymmetric Michael addition reaction. *J. Catal.* 377: 543–549.

320 Chhanda, S.A. and Itsuno, S. (2020). Design and synthesis of Cinchona-based chiral hyperbranched polymers and their application in asymmetric reactions. *Tetrahedron* 76 (24): 131247.

321 Takata, S., Endo, Y., Shahid Ullah, M. et al. (2016). Synthesis of cinchona alkaloid sulfonamide polymers as sustainable catalysts for the enantioselective desymmetrization of cyclic anhydrides. *RSC Adv.* 6 (76): 72300–72305.

322 Endo, Y., Takata, S., Kumpuga, B.T. et al. (2017). Synthesis of cinchona alkaloid sulfonamide polymers as enantioselective catalysts for the Michael addition reaction of β-ketoester and nitrostyrene. *ChemistrySelect* 2 (31): 10107–10111.

323 Hassan, M.M., Haraguchi, N., and Itsuno, S. (2016). Highly active polymeric organocatalyst: chiral ionic polymers prepared from 10,11-didehydrogenated cinchonidinium salt. *J. Polym. Sci., Part A: Polym. Chem.* 54 (5): 621–627.

324 Kumpuga, B.T. and Itsuno, S. (2018). Synthesis of chiral polyesters of cinchona alkaloid catalysts for enantioselective Michael addition of anthrone to nitroalkenes. *J. Catal.* 361: 398–406.

325 Kumpuga, B.T. and Itsuno, S. (2019). Synthesis of chiral polyurethanes of Cinchona alkaloids for the enantioselective synthesis in asymmetric catalysis. *Catal. Commun.* 118: 5–9.

326 Miyake, G.M., Iida, H., Hu, H.Y. et al. (2011). Synthesis of helical poly(phenylacetylene)s bearing Cinchona alkaloid pendants and their application to asymmetric organocatalysis. *J. Polym. Sci., Part A: Polym. Chem.* 49 (24): 5192–5198.

327 Tang, Z., Iida, H., Hu, H.Y. et al. (2012). Remarkable enhancement of the enantioselectivity of an organocatalyzed asymmetric Henry reaction assisted by helical poly(phenylacetylene)s bearing Cinchona alkaloid pendants via an amide linkage. *ACS Macro Lett.* 1 (2): 261–265.

328 Iida, H., Tang, Z., and Yashima, E. (2013). Synthesis and bifunctional asymmetric organocatalysis of helical poly(phenylacetylene)s bearing Cinchona alkaloid pendants via a sulfonamide linkage. *J. Polym. Sci., Part A: Polym. Chem.* 51 (13): 2869–2879.

329 Xie, Y.R., Wang, X.S., Zhao, H. et al. (2003). Unprecedented homochiral olefin-copper(I) 2D coordination polymer grid based on chiral ammonium salts as building blocks. *Organometallics* 22 (22): 4396–4398.

330 Lewiński, J., Kaczorowski, T., Justyniak, I. et al. (2011). Development of chiral *N,N*-ditopic metalloligands based on a Cinchona alkaloids' backbone for constructing homochiral coordination polymers. *Chem. Commun.* 47 (3): 950–952.

331 Edward, J.A., Kiesewetter, M.K., Kim, H. et al. (2012). Organocatalytic synthesis of quinine-functionalized poly(carbonate)s. *Biomacromolecules* 13 (8): 2483–2489.

332 Nagy, S., Fehér, Z., Kárpáti, L. et al. (2020). Synthesis and applications of Cinchona squaramide-modified poly(glycidyl methacrylate) microspheres as recyclable polymer-grafted enantioselective organocatalysts. *Chem. Eur. J.* 26 (59): 13513–13522.

333 Jumde, R.P. and Mandoli, A. (2016). Long-lived polymer-supported dimeric cinchona alkaloid organocatalyst in the asymmetric α-amination of 2-oxindoles. *ACS Catal.* 6 (7): 4281–4285.

11

Chiral Building Blocks for Total Steroid Synthesis and the Use of Steroids as Chiral Building Blocks in Organic Synthesis

Izabella Jastrzębska[1] *and Douglas F. Covey*[2,3,4]

[1] *University of Białystok, Faculty of Chemistry, ul. Ciołkowskiego 1K, 15-245 Białystok, Poland*
[2] *Washington University in St. Louis, School of Medicine, Department of Developmental Biology, 660 South Euclid Avenue, St. Louis, MO 63110, USA*
[3] *Washington University in St. Louis, School of Medicine, Department of Psychiatry, 660 South Euclid Avenue, St. Louis, MO 63110, USA*
[4] *Washington University in St. Louis, Taylor Family Institute for Innovative Psychiatric Research, School of Medicine, Department of Anesthesiology, 660 South Euclid Avenue, St. Louis, MO 63110, USA*

11.1 Introduction

Steroids, naturally occurring small molecules, are interesting due to their biological and pharmacological activities. Their core structure is a *cyclopenta[a]perhydrophenanthrene* that consists of rings A–D (Figure 11.1a). Naturally occurring steroids have a strictly-defined configuration of several stereogenic centers (Figure 11.1b). Despite having a common backbone, steroids with different substituents and their various possible stereochemical relationships can induce an extensive range of valuable physiological effects.

This chapter reviews chiral building blocks (CBBs) used to construct steroids published in 2010–2021. Since steroids can also be applied as CBB to synthesize other classes of compounds and other more sophisticated steroids, their use as CBB will be the subject of the second part of this chapter.

The configurations at the stereocenters in steroidal compounds indicate their biological actions. For this reason, natural products that have required configurations at initial stereocenters as well as directing those created during steroid total synthesis have been established [1]. Substrates from the chiral pool shown in Figure 11.2 serve as examples of molecules used for this purpose.

The classical CBBs for steroid synthesis are the *Hajos–Parish ketone* and *Wieland–Miescher ketone* [2, 3] (Figure 11.3).

Chiral Building Blocks in Asymmetric Synthesis: Synthesis and Applications, First Edition.
Edited by Elżbieta Wojaczyńska and Jacek Wojaczyński.

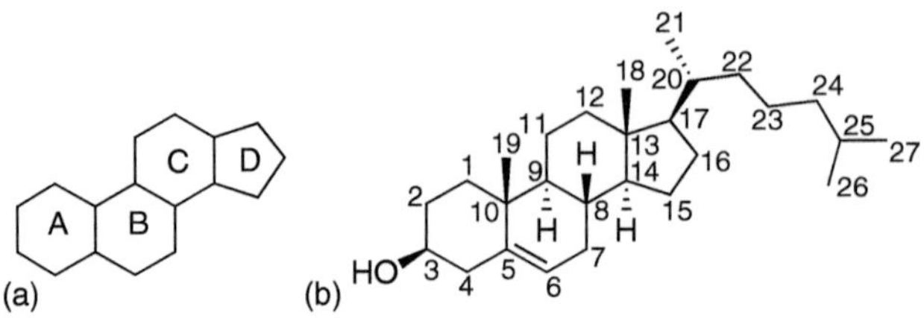

Figure 11.1 (a) The steroid skeleton; (b) cholesterol structure with carbon numbering; eight stereogenic centers (3, 8, 9, 10, 13, 14, 17, and 21) are visible.

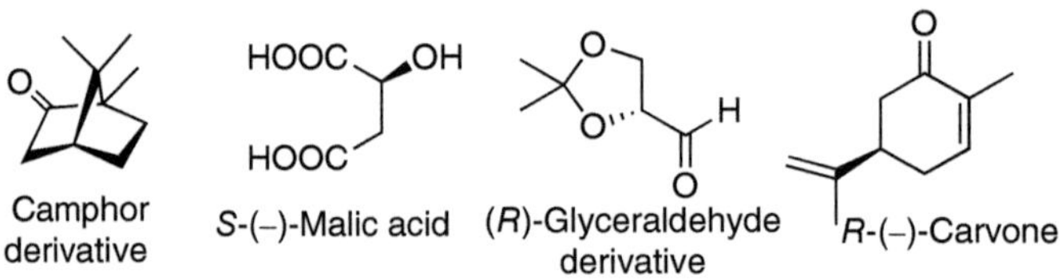

Figure 11.2 Examples of substrates from the chiral pool used to synthesize steroids.

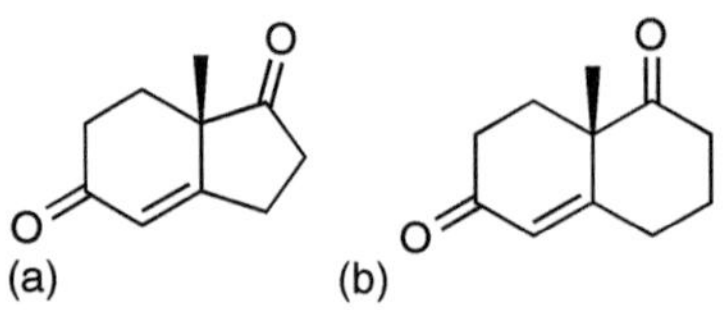

Figure 11.3 The structure of (a) Hajos–Parish ketone and (b) Wieland–Miescher ketone.

11.2 Chiral Building Blocks for the Construction of Steroids

Recent CBBs used to synthesize the steroid system are more complex than those used in classical steroid syntheses. These chiral molecules are often systems containing two or three rings.

11.2.1 Chiral Building Block 1 (CBB 1)

In **CBB 1**, three stereogenic centers: C-8, C-13, and C-14, with the D-ring template are established. The synthetic strategy for preparing **CBB 1** is shown in Scheme 11.1 [4]. The synthesis started from the bromination of 3-methyl anisole (**1**) followed by benzylic displacement with allenyl magnesium bromide and then a one-carbon extension to give the derivative of propargylic alcohol **3**. The *Z*-isomer was attained by reduction with Zn/Br$(CH_2)_2$Br. This was converted into a phosphorus derivative **6** formed as a reaction product with hexachloroacetone and then an Arbuzov reaction with phospholane **5**. Deprotonation of the phosphorus derivative **6** followed by the addition of 2-methyl-2-cyclopentanone **7** and then an alkylation reaction gave chiral building block **CBB 1**. As noted by the authors, conjugate addition reactions of allylic phosphonates are known to be very diastereoselective, with the allylic double

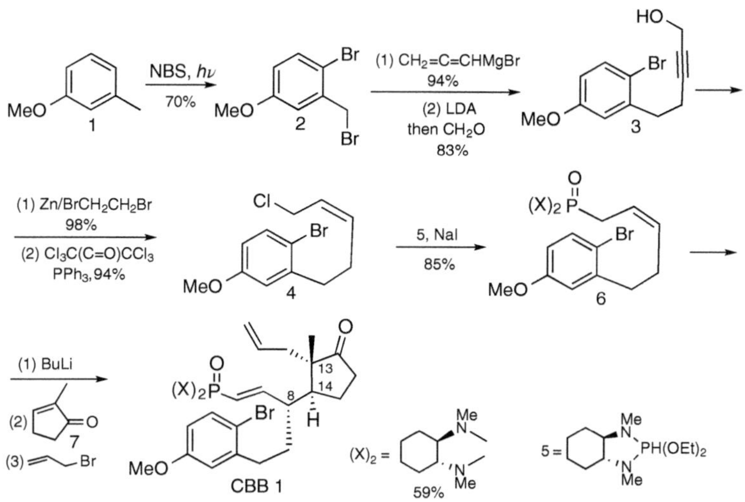

Scheme 11.1 The synthesis of **CBB 1**. Abbreviations: NBS = *N*-bromosuccinimide, LDA = lithium diisopropylamide. Source: Foucher et al. [4]/American Chemical Society.

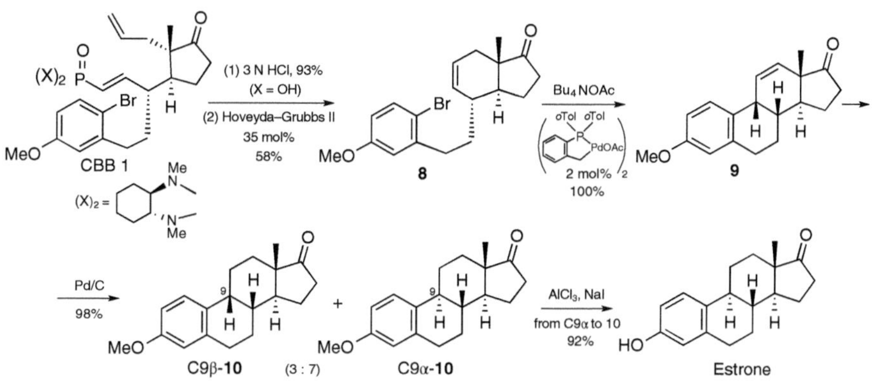

Scheme 11.2 The synthesis of estrone. Source: Katritzky et al. [5]/American Chemical Society.

bond configuration being translated into the relative configuration of the sp^3-centers of the formed C—C bond [5].

Using chiral molecule **CBB 1**, the *trans*-fused CD ring junction formation was achieved. This avoided the problem of cis-isomer contamination, which is a common complication in total steroid synthesis. Scheme 11.2 illustrates a synthesis of estrone that started with **CBB 1**.

11.2.2 Chiral Building Block 2 (CBB 2)

Chiral building block **2** (**CBB 2**) was obtained via a one-pot Michael–Michael-aldol reaction sequence (Scheme 11.3) [6]. Double Michael reaction of starting aldehyde

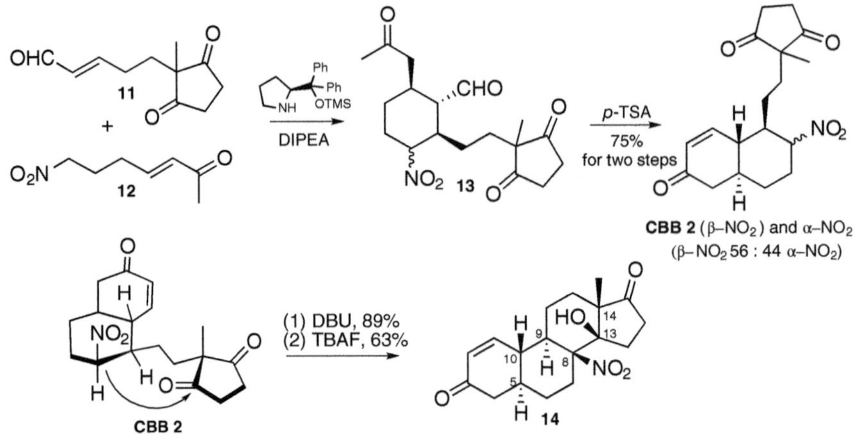

Scheme 11.3 The synthesis of steroid system via **CBB 2**. Abbreviations: DIPEA = *N,N*-diisopropyl-ethylamine, *p*-TSA = *p*-toluenesulfonic acid, DBU = 1,8-diazabicyclo(5.4.0)undec-7-ene, TBAF = tetra-*n*-butylammonium fluoride. Source: Jhuo et al. [6]/American Chemical Society.

11 and nitro compound **12**, followed by aldol condensation, gave **CBB 2**. Using Henry reaction, compound **CBB 2** was transformed into a steroid. The synthesis via **CBB 2** presents a concise methodology to achieve a steroidal molecule with six quaternary carbon stereogenic centers (at C-5, 8, 9,10, 13, and C-14) with high enantioselectivity (99% ee).

11.2.3 Chiral Building Block 3 (CBB 3)

The enantioselective synthesis of oxy-functionalized steroids was performed using **CBB 3** [7]. The key step was the asymmetric variant of the Michael reaction with copper(II) bis-oxazoline hexafluoroantimonate complex (Scheme 11.4). The steroidal compound **18** was achieved by a double aldol cyclization strategy. The preparation of the suitable steroid epimer (natural diastereoisomer) was obtained using appropriate aldolization reaction conditions (LiHMDS, THF, −78 to 60 °C). Under proper conditions, it was possible to control the configuration at C-8, C-13, and C-14 in the product.

11.2.4 Chiral Building Block 4 (CBB 4)

Structurally similar to **CBB 3,** Michael adduct **4** (**CBB 4**) was achieved as the result of a reaction between a chloro ester **19** and an α,β-unsaturated ketone **16** using chiral copper(II) bis-oxazoline hexafluoroantimonate complex as a catalyst (Scheme 11.5). After intramolecular condensation followed by an epimerization/reductive transposition reaction sequence, the intermediate of a cardiotonic steroid backbone (compound **20**) was obtained [8].

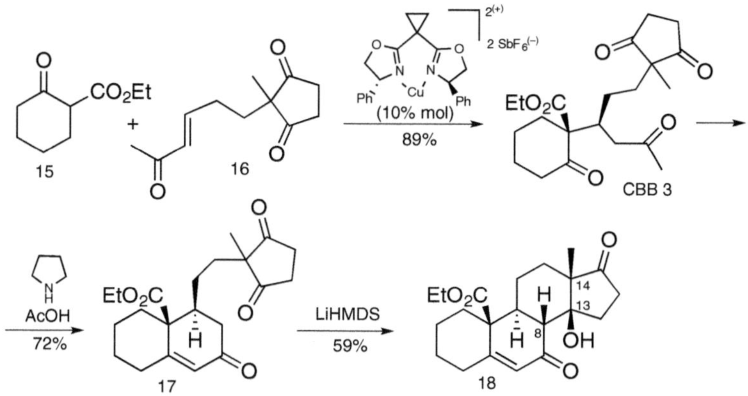

Scheme 11.4 The enantioselective synthesis of oxy-functionalized steroids. Abbreviation: LiHMDS = lithium bis(trimethylsilyl)amide.

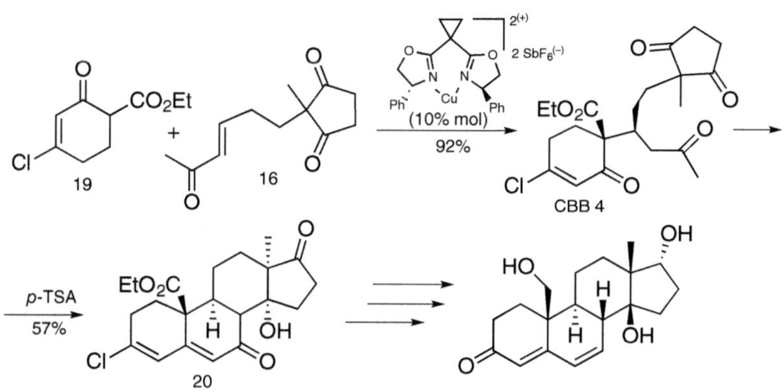

Scheme 11.5 The synthesis of cardiotonic steroids intermediate via **CBB 4**.

11.2.5 Chiral Building Block 5 (CBB 5)

Chiral enone **5** (**CBB 5**) is easily available from the 2-methyl-1,3-cyclopentanedione **21** and acrolein via Michael addition. The stereochemistry of the benzyloxy group in **CBB 5** was established using the chiral catalyst (*S*)-2-(diphenyl((trimethylsilyl)oxy)methyl)pyrrolidine in step 2. This chiral enone will establish the C-9 and C-11 configurations in subsequently formed steroids. A diastereoselective Michael reaction was carried out with a Cu(II) catalyst followed by *p*-TSA treatment to give the steroid core [9]. The synthesis is outlined in Scheme 11.6.

Using **CBB 5**, syntheses of the highly oxygenated cardiotonic steroid ouabagenin and other cardiotonic steroids with lower oxidation states were performed [10]. As a result, ouabagenin (oxygenated in positions: C-1, C-3, C-5, C-11, C-14, and C-19) was obtained by a 22-step synthesis (Scheme 11.7). This strategy, starting from **CBB 5**, is convergent and concise, considering the high oxidation state of the obtained steroid molecule.

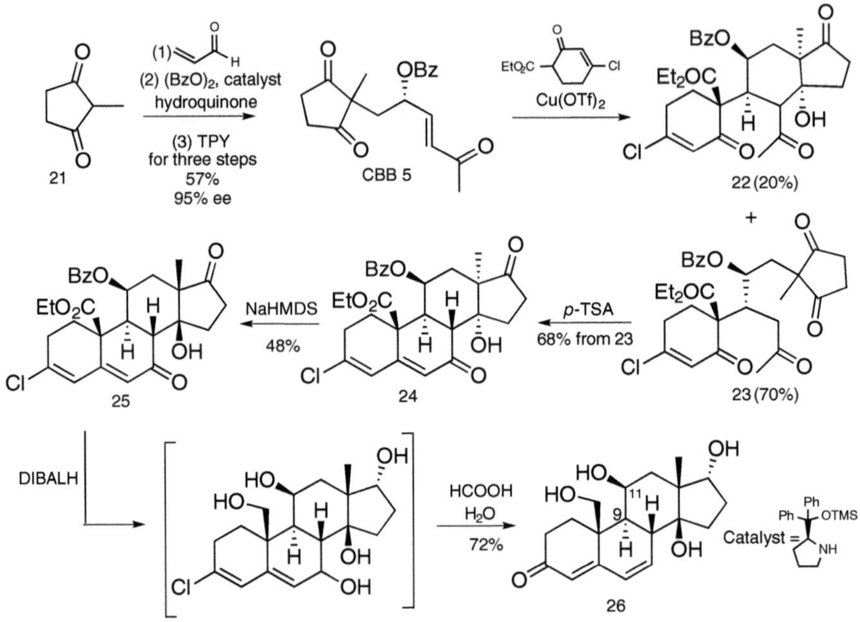

Scheme 11.6 The synthesis of cardenolide via **CBB 5**. Abbreviations: TPY = (triphenylphosphoranylidene)-2-propanone, NaHMDS = sodium bis(trimethylsilyl)amide, *p*-TSA = *p*-toluenesulfonic acid, DIBALH = diisobutylaluminium hydride.

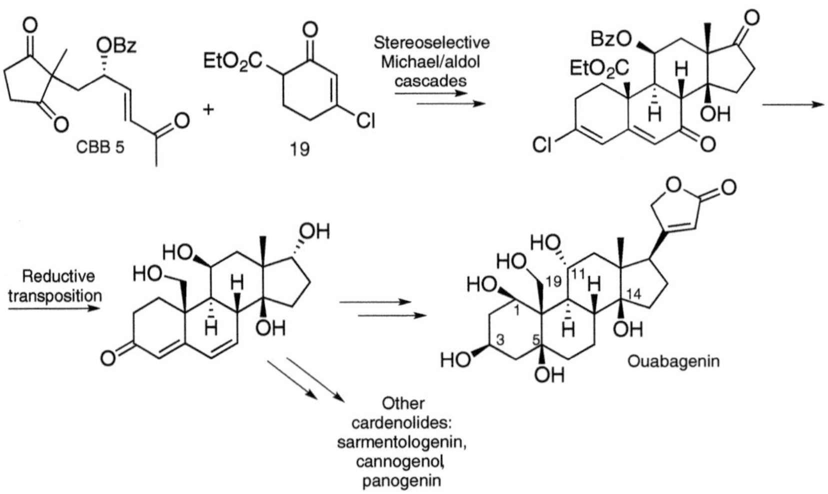

Scheme 11.7 The synthetic approach to ouabagenin from **CBB 5**.

11.2.6 Chiral Building Block 6 (CBB 6)

Cardiotonic steroids are characterized by the presence of the β-OH group in position C-14 and cis-fused AB and CD rings. Therefore, a tricyclic lactone **CBB 6** was suitable for synthesizing this type of compound. The synthesis of **CBB 6**

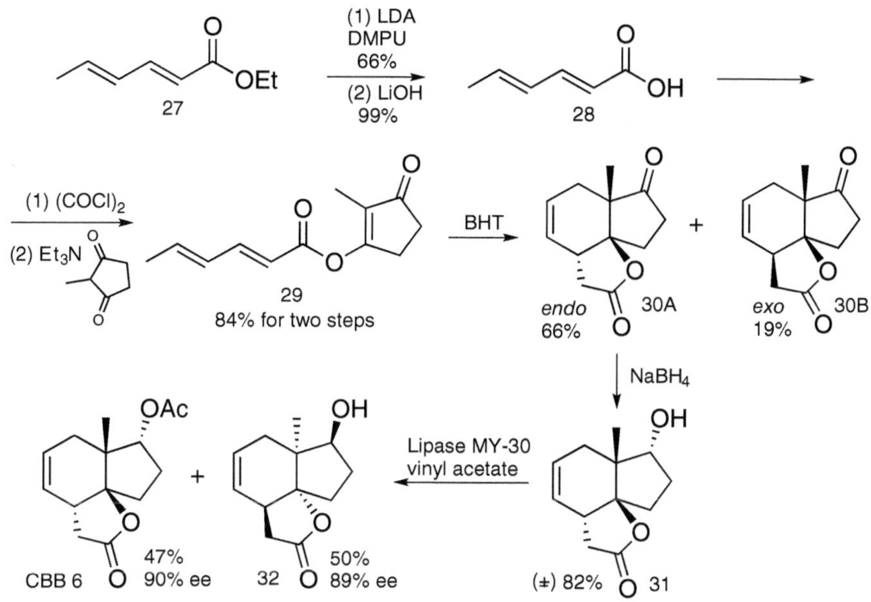

Scheme 11.8 The synthesis of **CBB 6**. Abbreviations: LDA = lithium diisopropylamide, DMPU = *N,N′*-dimethylpropyleneurea, BHT = butylated hydroxytoluene.

commenced from ethyl sorbate (**27**), which has been isomerized then hydrolyzed. The resulting acid **28** was treated with oxalyl chloride followed by a reaction with 2-methyl-1,3-cyclopantanedione to give the enol ester 29. The intramolecular Diels–Alder reaction provided *endo*- and *exo*-tricyclic ketones **30A–30B** (Scheme 11.8). After reduction with $NaBH_4$, the obtained mixture of alcohols **31** was separated by optical resolution using lipase MY-30 (Meito Sangyo Co. Ltd.) in the presence of vinyl acetate. The absolute configuration of **CBB 6** was established by X-ray analysis [11].

The steroidal skeleton was created from **CBB 6** and a bromoarene via Mizoroki–Heck reaction using a palladacycle catalyst (Hermans catalyst) followed by reduction with DIBALH (Scheme 11.9). The subsequent Friedel–Crafts reaction provided a tetracyclic compound that was readily converted to cardiotonic steroids.

11.2.7 Chiral Building Block 7 (CBB 7)

A derivative of **CBB 6** was used to prepare chiral building block **7** (**CBB 7**), as shown in Scheme 11.10 [12]. **CBB 7** was produced from the Mizoroki–Heck reaction of 2-pyrone triflate and a tricyclic alkene **36**. The resultant compound was obtained as a single diastereomer and its regioisomer **37**. Treatment of **CBB 7** with lithium trimethylsilylacetylide in the presence of a Lewis acid caused the opening of the lactone. Reduction of the obtained ketone **38** using Luche conditions ($NaBH_4/CeCl_3$) provided the alcohol **39**, which was then subjected to an intramolecular Diels–Alder reaction to get the precursor of cardiotonic steroids (compound **40**).

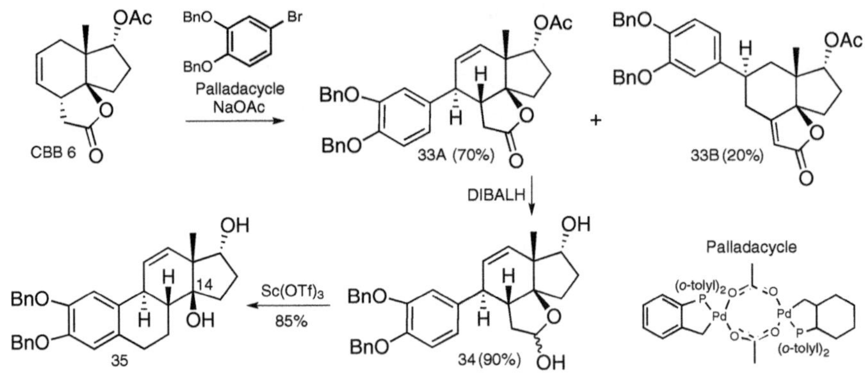

Scheme 11.9 The synthesis of a tetracyclic steroidal core from **CBB 6**. Abbreviation: DIBALH = diisobutylaluminium hydride.

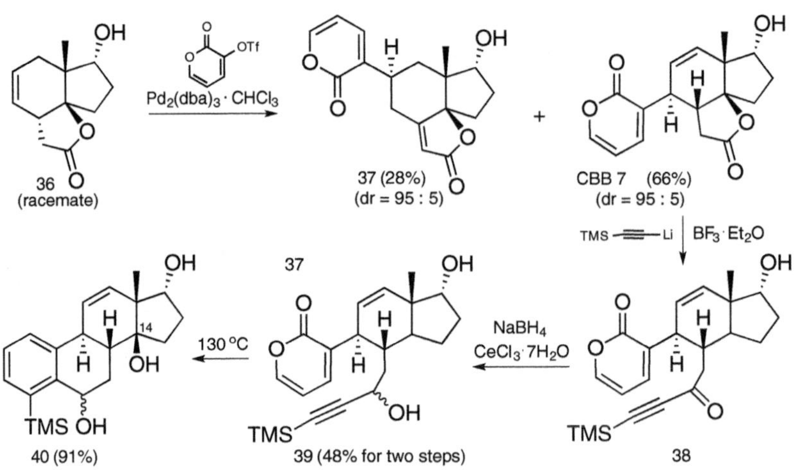

Scheme 11.10 The synthesis of an estrogenic skeleton from **CBB 7**. Source: Watanabe et al. [12]/American Chemical Society.

11.2.8 Chiral Building Block 8 (CBB 8)

A different approach to the synthesis of steroids, through **CBB 8**, which allows one to obtain partially aromatic steroidal compounds from the chiral starting material (*R*)-epichlorohydrin (**41**) is shown in Scheme 11.11 [13]. The methodology uses as the critical step a metallacycle-mediated annulative cross-coupling reaction of chiral enyne **CBB 8** and (TMS)-phenylacetylene. The product hydrindane **42A** is transformed into a vinylcyclopropane intermediate **43** followed by intramolecular Friedel–Crafts alkylation.

ent-**CBB 8**, an optical antipode of **CBB 8**, readily available from (*S*)-epichlorohydrin, is a convenient building block for preparing a whole range of *ent*-steroids with an aromatic ring A (Scheme 11.12). *ent*-Steroids are a class of drug-like

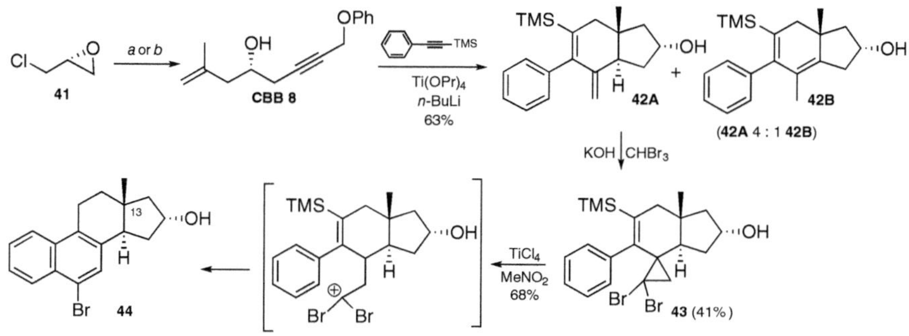

Scheme 11.11 The synthesis of a steroid analogue via **CBB 8**; (a) (i) Phenylpropargyl ether, *n*-BuLi (17%), (ii) 2-propenylmagnesium bromide, CuI (70%); (b) (i) phenylpropargyl ether, *n*-BuLi, $BF_3 \cdot OEt_2$, (ii) KO*t*-Bu, (iii) 2-propenylmagnesium bromide, CuI (63% over three steps). Source: Kim et al. [13]/Springer Nature.

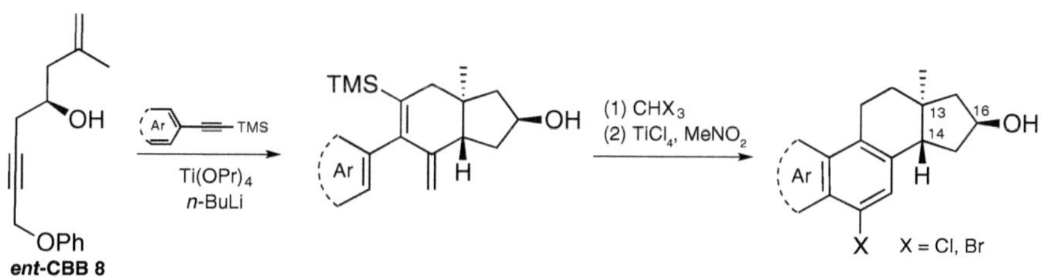

Scheme 11.12 The synthesis of *ent*-steroids from *ent*-**CBB 8**. Source: Kim et al. [13]/Springer Nature.

molecules that remain largely marginalized in biology and medicine as they are unobtainable from natural sources.

11.2.9 Chiral Building Block 9 (CBB 9)

(*R*)-Epichlorohydrin was a starting material for chiral building block **9** (**CBB 9**) [14]. Although **CBB 9**, which contains a *p*-OMe on the aromatic ring, is an analog of **CBB 8**, a different synthetic path to the steroid skeleton was applied, as depicted in Scheme 11.13. Using this methodology, it is possible to introduce the C-9 quaternary center. TMS alkyne reacted with enyne **CBB 9** in an alkoxide-directed metallacycle-mediated annulative cross-coupling reaction. Hydrindane **45** obtained this way was subjected to a double asymmetric Brønsted acid-mediated Friedel–Crafts ring closure to give a tetracyclic skeleton (compound **46**) with a relatively good yield.

Subsequent oxidative rearrangement resulted in the 1,2-methyl group shift from C-9 to C-10, crucial for generating steroids with a C-10 methyl group. The oxidative reaction proceeded with (diacetoxyiodo)benzene (phenyliodine(III) diacetate [PIDA]) as presented in Scheme 11.14.

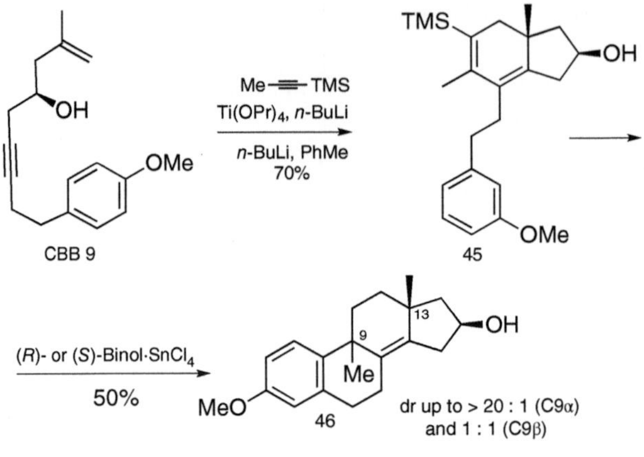

Scheme 11.13 The synthesis of a steroidal precursor from **CBB 9** [15].

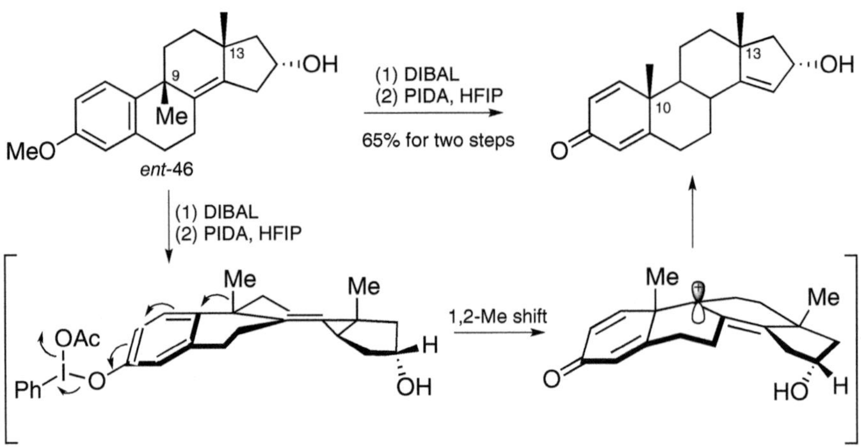

Scheme 11.14 Possible mechanism of the oxidative rearrangement [14]. Abbreviation: DIBALH = diisobutylaluminium hydride, PIDA = phenyliodine(III) diacetate, HFIP = hexafluoroisopropanol.

11.3 Steroids as Chiral Building Blocks

Steroids as natural compounds with strictly defined stereogenic centers are a useful tool for constructing various biologically active substances. This subsection presents selected examples of isoprenoid synthesis using steroids as CBBs.

11.3.1 Synthesis of C-*nor*-D-*homo*-Steroids

C-*nor*-D-*homo*-Steroids possess a steroid core with a different 6–6–5–6 pattern of rings. In this class of natural compounds, the *Veratrum* steroidal alkaloid cyclopamine is the most known example (Scheme 11.15a) [16]. For the construction of C-*nor*-D-*homo*-steroids, the Comins reagent (*N*-(5-chloro-2-pyridyl)triflimide)

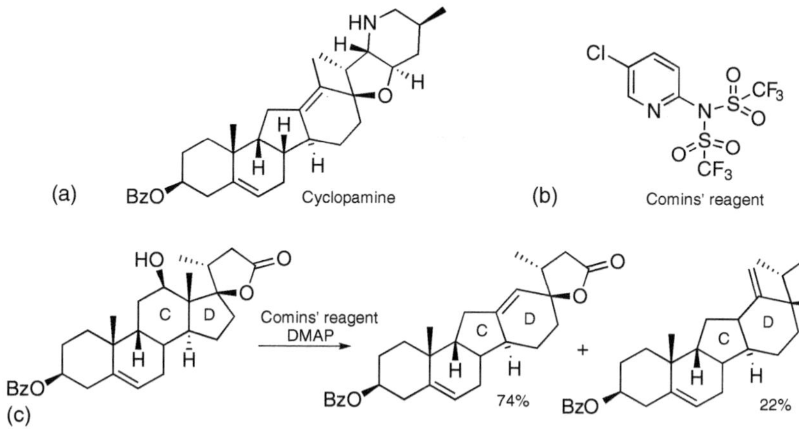

Scheme 11.15 (a) The structure of cyclopamine; (b) the structure of Comins reagent; (c) a biomimetic synthesis of C-nor-D-homo-steroids. Source: Li et al. [16]/Royal Society of Chemistry.

(Scheme 11.15b) in the presence of 4-dimethylaminopyridine (DMAP) was used [17]. This combination of reagents resulted in the rearrangement of 12β-hydroxy steroids as presented in Scheme 11.15c.

11.3.2 Pregnenolone as a Chiral Building Block: Synthesis of Cyclocitrinol

Cyclocitrinol is a member of the natural steroids family with an uncommon bicyclo[4.4.1] A/B ring scaffold. The compound was extracted from the fungus *Penicillium citrinum* in 2000 [18]. Afterward, many other representatives of this type of steroidal compound have been isolated [19–23]. Studies have shown that some cyclocitrinols show potential activity in treating various neurological disorders [20]. Pregnenolone, which contains five of the required correct stereocenters (C-3,9,13,14,17) of cyclocitrinol, was selected as a CBB. Cyclocitrinol was obtained in 10 steps (Scheme 11.16) [24]. For the construction of the bicyclo[4.4.1] A/B ring system, a biomimetic cascade rearrangement was used. Activation of the C-19 methyl group engendered an electrophilic center which forced the attack by the C5–C6 alkene to give a cyclopropyl ring. The fragmentation of the cyclopropane via deprotonation at C-1 formed the bicyclo[4.4.1] A/B ring scaffold.

11.3.3 4-Androstene-3,17-Dione as a Chiral Building Block: Synthesis of Vulgarobufotoxin

Bufadienolides are a large family of steroidal compounds found in animals and plants [25]. These steroids possess attractive biological activities, including cardiotonic, anticancer, and anti-inflammatory effects [26]. For example, bufadienolides have nontypical cis-fused A/B and C/D ring cores and a heterocyclic ring substituent in the C-17β-position. The synthesis of vulgarobufotoxin was

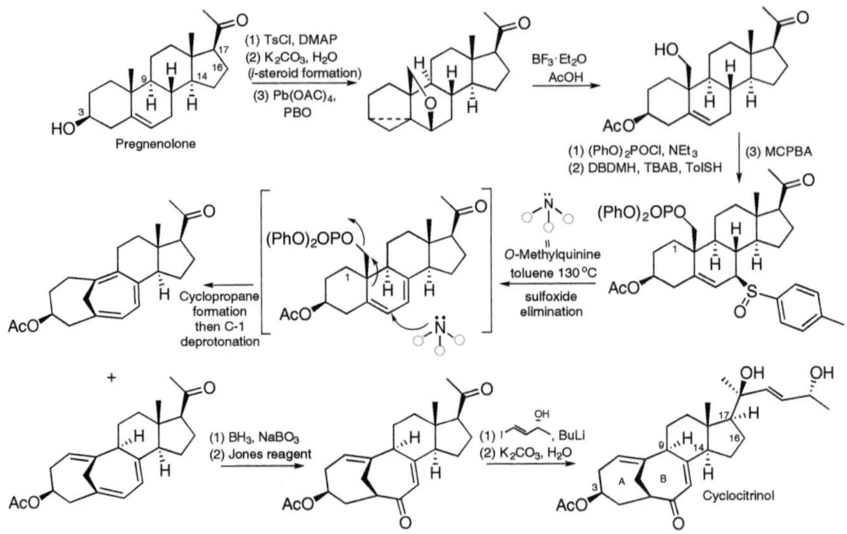

Scheme 11.16 The synthesis of cyclocitrinol. Abbreviation: Ts = tosyl, DMAP = 4-dimethylaminopyridine, BPO = benzoyl peroxide, DBDMH = 1,3-dibromo-5,5-dimethylhydantoin, TBAB = *tetra*-*n*-butylammonium bromide, MCPBA = *meta*-chloroperoxybenzoic acid. Source: Wang et al. [24]/American Chemical Society.

commenced with 4-androstene-3,17-dione (Scheme 11.17). The starting material possesses four correct stereogenic centers of bufadienolides (C-8,9,10,13). After the crucial C=C double bond reduction (A/B *cis* ring fusion) and the protection of the C-3 hydroxyl group, the desired C-14 configuration was achieved (C/D *cis* ring fusion). The Stille reaction and a subsequent epoxidation introduced the functionalization at the C-16β position. Rearrangement of the epoxide with Lewis acid gave the desired 16-oxo steroid, which was transformed into vulgarobufotoxin and other bufadienolides [27]. The synthesis was accomplished in 13–16 steps starting from 4-androstene-3,17-dione.

11.3.4 Ergosterol as a Chiral Building Block

11.3.4.1 Synthesis of Pleurocin A/Matsutakone

Pleurocin A/matsutakone and pleurocin B represent the unusual 11(9 → 7)*abeo*-steroid family. These fungal metabolites were isolated by two separate groups from two different fungal species. The steroidal molecule matsutakone, named after the mushrooms *Tricholoma matsutake* and pleurocin A named after *Pleurotus eryngii*, show structurally highly oxygenated cyclopenta[*a*]anthracene systems [28, 29]. The biological tests revealed inhibitory effects on acetylcholine levels (IC_{50} = 20.9 μM) and endothelial NO-synthase (25 μM) inhibition, and no cytotoxicity at the reported concentration. Starting from ergosterol, the synthesis of pleurocin A/matsutakone was accomplished with 17 steps [30]. Ergosterol provided five stereocenters (C-13,14,17,20,24) for the fungal metabolites that remained intact during the synthetic steps. Configurations of two additional centers (at C-3 and

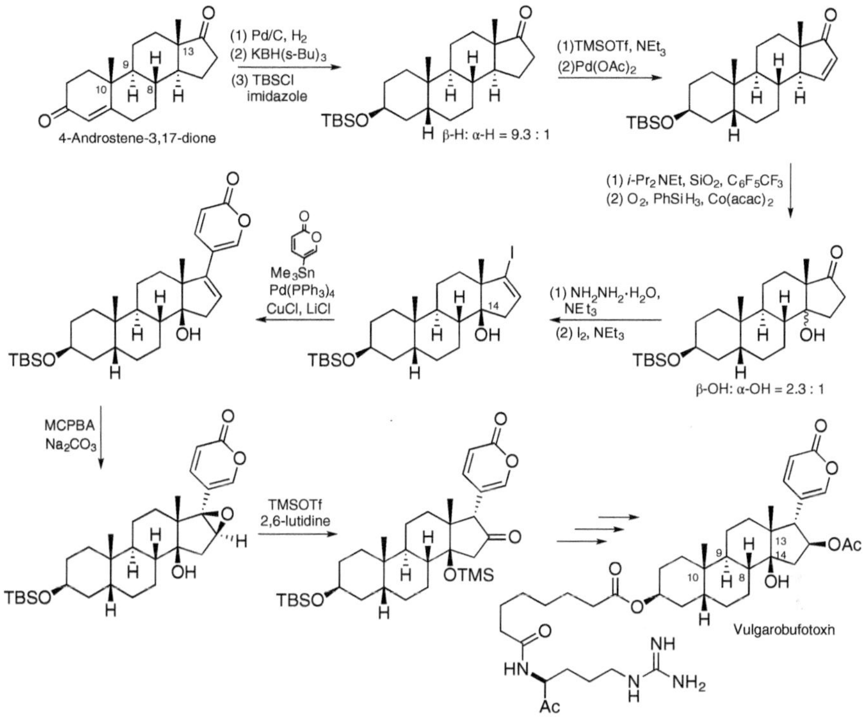

Scheme 11.17 The synthesis of vulgarobufotoxin. Abbreviations: TBS = *tert*-butyldimethylsilyl, TMSOTf = trimethylsilyl trifluoromethanesulfonate, $Co(acac)_2$ = bis(acetyloacetonato)cobalt(II).

C-10) were inverted to obtain the desired product. A significant transformation in the presented synthesis is the complicated 9,11-*seco* iodo-compound radical cyclization (Scheme 11.18).

11.3.4.2 Synthesis of Strophasterol A

Four secosterols, named strophasterols A–D with an uncommon steroid skeleton, have been extracted from the mushroom *Stropharia rugosoannulata* [31]. Ergosterol was selected as a starting material because it has six correct stereocenters (C-3,10,13,17,20,24) of strophasterol A [32]. A key step was obtaining the chloro-diketone, which was transformed into a tricyclic diketoacid under basic conditions. This decisive ring scission resulted from vinylogous α-ketol rearrangement followed by loss of chloride. The additional ring of the secosterol was obtained from the iodide as a result of the radical process ($BEt_3 \cdot O_2/nBu_3SnH$). Summarizing, strophasterol A was achieved in 18 steps from commercially available ergosterol (Scheme 11.19).

11.3.4.3 Synthesis of Herbarulide

Herbarulide was isolated from the fungus *Pleospora herbarum* in 1999, the sponge *Biemna fortis* in 2006, the fungi *Antrodia camphorata* in 2017, and *Stereum hirsutum*

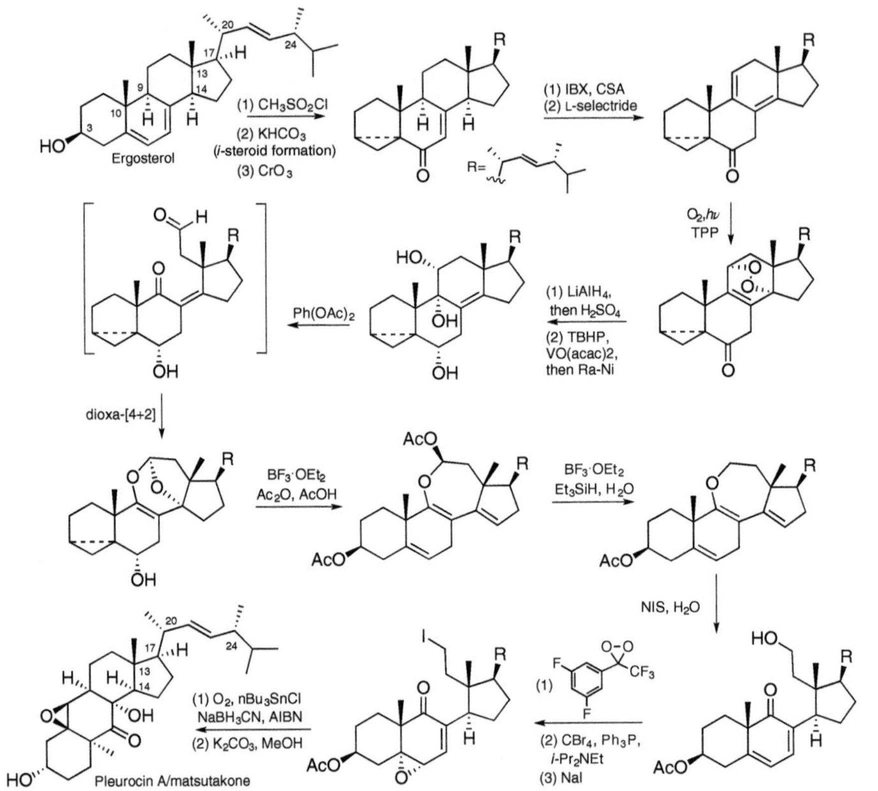

Scheme 11.18 The synthesis of pleurocin A/matsutakone. Abbreviations: CSA = camphorsulfonic acid, DMSO = dimethyl sulfoxide, IBX = 2-iodoxybenzoic acid, Ra-Ni = Raney nickel, TBHP = *tert*-butyl hydroperoxide.

in 2019 [33–36]. Herbarulide is a 5,6-epoxy-5,6-secosteroid. The synthesis of the 5,6-secosteroid started from ergosterol which possesses seven correct stereogenic centers of herbarulide (C-9,10,13,14,17,20,24) and the side chain. It was accomplished in four steps (Scheme 11.20). The decisive ring B expansion was achieved by 5-alkoxy radical scission of the C5—C6 bond followed by recombination of the 5-oxo moiety with acyl radicals [37]. In the presented work, the structural revision of herbarulide was simultaneously determined (configuration a C-24).

11.3.4.4 Synthesis of Pinnigorgiol E and Pinnigorgiol B

Two compounds, pinnigorgiol E and pinnigorgiol B, are natural 9,11-secosterols with synthetically challenging structures. Pinnigorgiols, characterized by an unusual tricyclo[5.2.1.1]decane core with γ-diketone moiety, have been isolated from a *Pinnigorgia* coral species in 2016 [38]. Starting with ergosterol via semipinacol rearrangement of a 5,6-diol ergosterol analog, the 7,5-bicyclo framework was prepared. After C ring-opening, the seven-membered A ring was severed. Then the acyl radical precursor was obtained using iodothiol. The resulting thioester was transformed into pinnigorgiol E by an acyl radical cyclization/hemiketalization

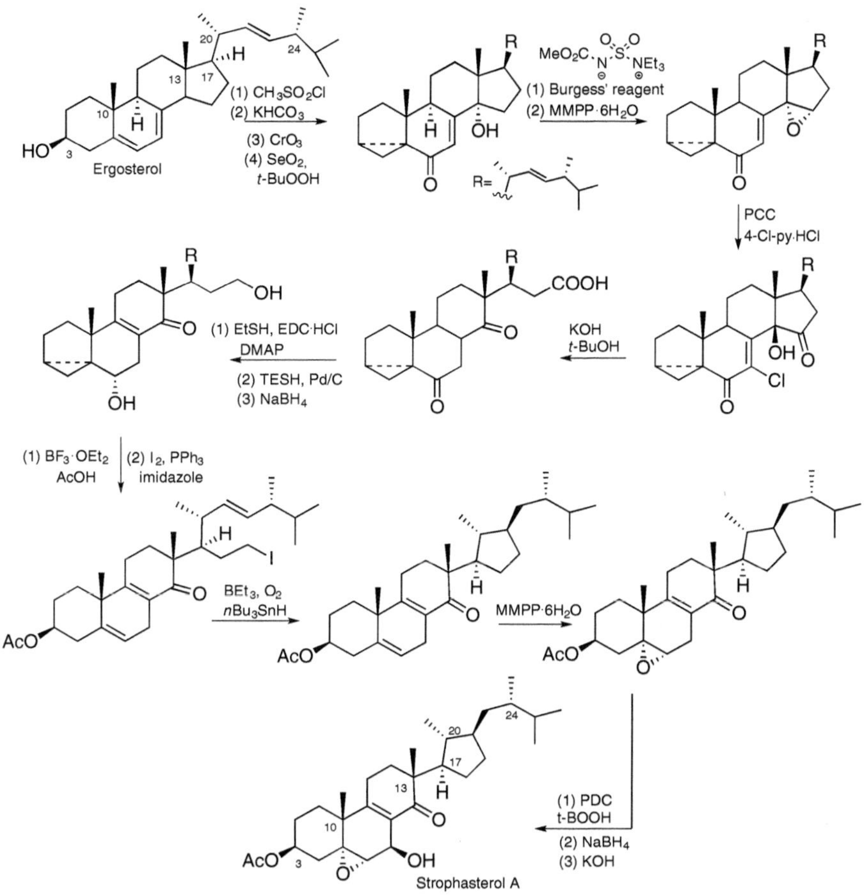

Scheme 11.19 The synthesis of strophasterol. Abbreviations: DMAP = 4-dimethylamino-pyridine, EDC = 1-ethyl-3-(3-dimethylaminopropyl)carbodiimide, MMPP = magnesium bis(monoperoxyphthalate), PCC = pyridinium chlorochromate, PDC = pyridinium dichromate, TESH = triethyl silane.

cascade (Scheme 11.21). Pinnigorgiol B was readily obtained from pinnigorgiol E due to basic hydrolysis [39].

11.3.5 Tigogenin as a Chiral Building Block

11.3.5.1 Synthesis of Clathsterol

Polyoxygenated steroids are widely distributed compounds in marine and terrestrial organisms. Clathsterol, a sterol sulfate, was isolated from RedSea sponge *Clathria* sp. [40]. The compound was shown to inhibit human virus type 1 (HIV-1) reverse transcriptase at a concentration of 10 μM. The synthesis of clathsterol was accomplished in 24 steps starting from tigogenin, a naturally existing steroidal sapogenin [41]. Tigogenin was chosen as a starting material because it has nine stereogenic centers (C-5,8,9,10,13,14,16,17) consistent with those present in clathsterol. Scheme 11.22

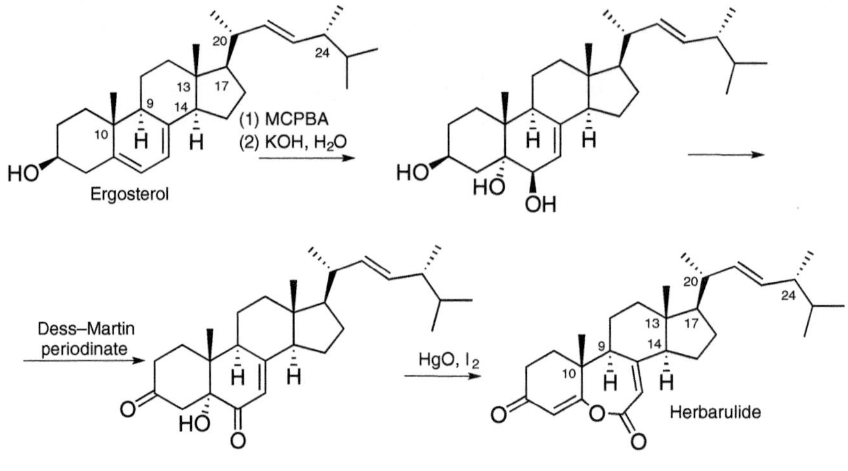

Scheme 11.20 The synthesis of herbarulide. Abbreviation: MCPBA = *meta*-chloroperoxybenzoic acid.

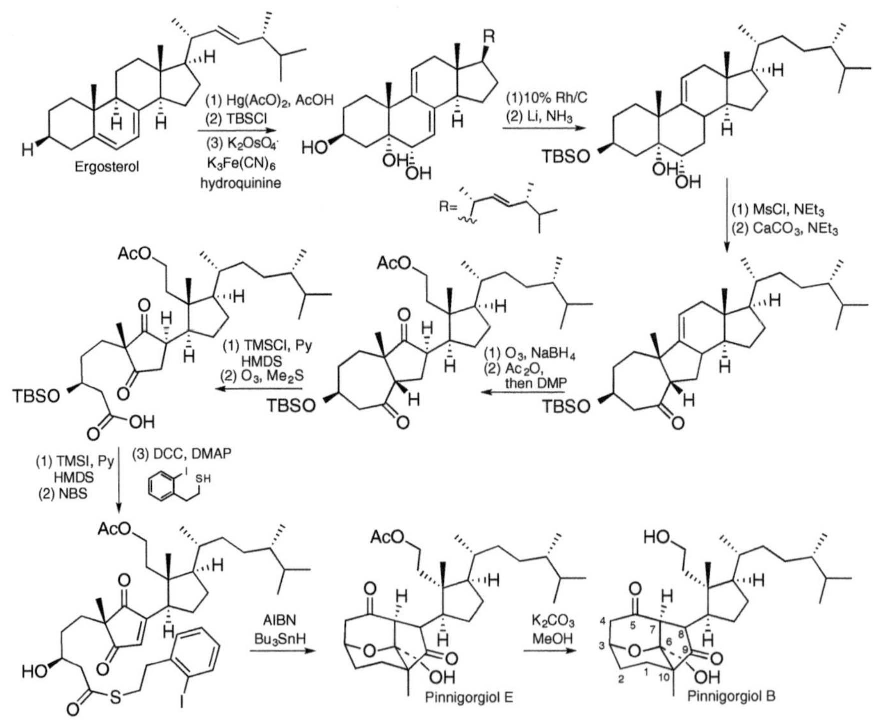

Scheme 11.21 The synthesis of pinnigorgiol E and pinnigorgiol B. Abbreviations: TBS = *t*-butyldimethylsilyl, Ms = mesyl, DMP = Dess–Martin periodinane, TMS = trimethylsilyl, HMDS = 1,1,1,3,3,3-hexamethyl-disilazane, NBS = *N*-bromosuccinimide, DCC = dicyclohexylcarbodiimide, DMAP = 4-dimethylaminopyridine, AIBN = 2,2′-azobis(isobutyro-nitrile).

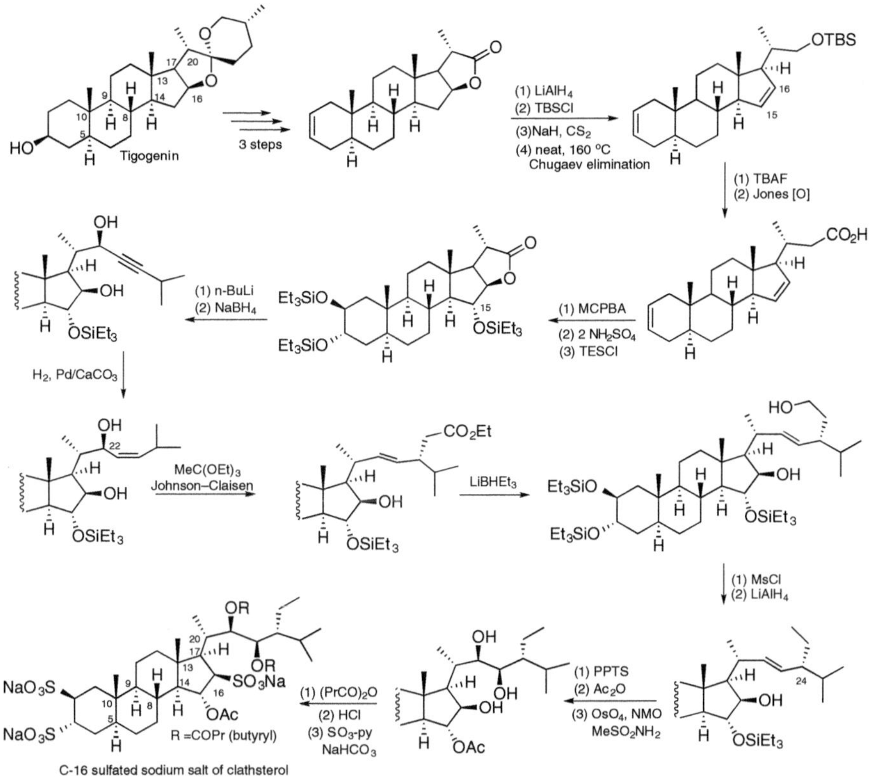

Scheme 11.22 The synthesis of clathsterol. Abbreviations: TBS = *tert*-butyldimethylsilyl, MCPBA = *meta*-chloroperoxybenzoic acid, TES = trimethylsilyl, TBAF = *tetra-n*-butylammonium fluoride, Ms = mesyl, PPTS = pyridinium *p*-toluenesulfonate, NMO = 4-methylmorpholine *N*-oxide.

presents the synthesis of the C-16 sulfated sodium salt of clathsterol. Crucial steps are the Chugaev reaction to introduce the required C15=C16 double bond followed by desylilation and Jones oxidation to obtain the acid. Epoxidation with MCPBA then lactone ring closure in acidic conditions has allowed the installation of a hydroxyl group in C-15. Lactone addition/reduction protocol provided C-22 with a suitable configuration (*R*). Jonson–Claisen rearrangement followed by reduction with $LiBHEt_3$, mesylation, and subsequent reduction furnished the required C-24 configuration. Dihydroxylation with OsO_4, acylation with butyrate anhydride generated the C-22 and C-23 hydroxyl groups, and a final treatment with SO_3-py and $NaHCO_3$ of the intermediate gave the sulfated analog of clathsterol.

11.3.5.2 Synthesis of Nortriterpenoid Propindilactone G

Schinortriterpenoid propindilactone G belongs to a *Schisandra* nortriterpenoid family of compounds, a group of highly oxygenated polycyclic nortriterpenoids. Propindilactone G possesses a unique 5/5/7/6/5 pentacyclic skeleton and 10 stereogenic centers. Moreover, it was isolated from the plant *Schisandraceae propinqua* var. *propinqua* in 2008 [42]. The synthesis of propindilactone G was inspired by a

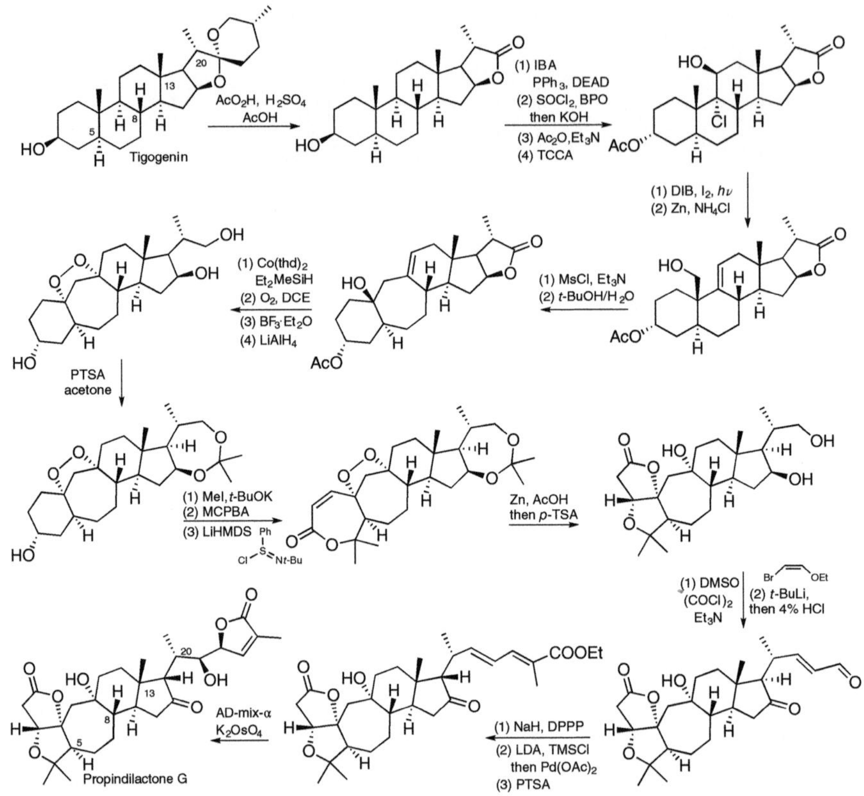

Scheme 11.23 The synthesis of propindilactone G. Abbreviations: IBA = 3-iodobenzoic acid, DEAD = diethyl azodicarboxylate, BPO = benzoyl peroxide, TCCA = trichloroisocyanuric acid, MsCl = methanesulfonyl chloride, DIB = (diacetoxyiodo)benzene, thd = 2,2,6,6-tetramethyl-3,5-heptanedionato, DCE = 1,2 dichloroethane, *p*-TSA = *p*-toluenesulfonic acid, DMP = Dess–Martin periodinane, MCPBA = *meta*-chloroperoxybenzoic acid, LHMDS = lithium bis(trimethylsilyl)amide, LDA = lithium diisopropylamide; DPPP = 2-(diphenoxyphosphoryl)propanoate.

biosynthetic pathway. The synthesis commenced with tigogenin and was achieved in 20 steps (Scheme 11.23). A nortriterpenoid and tigogenin share stereogenic centers at C-5, C-8, C-13, and C-20, as well as a common C/D-ring junction. Crucial conversions include Breslow's functionalization at positions C-9 and C-11, Suarez radical process, a ring enlargement achieved by Wagner–Meerwein rearrangement, inversion of configuration at C-10, and transesterification/oxa-Michael cascade [43].

References

1 Chapelon, A.-S., Moraléda, D., Rodriguez, R. et al. (2007). Enantioselective synthesis of steroids. *Tetrahedron* 63 (47): 11511–11616.

2 Hajos, Z.G. (1974). Asymmetric synthesis of bicyclic intermediates of natural product chemistry. *J. Org. Chem.* 39 (12): 1615–1621.

3 Wieland, P. and Miescher, K. (1950). Über die Herstellung mehrkerniger ketone. *Helv. Chim. Acta* 33 (7): 2215–2228.

4 Foucher, V., Guizzardi, B., Groen, M.B. et al. (2010). A novel, versatile D → BCD steroid construction strategy, illustrated by the enantioselective total synthesis of estrone. *Org. Lett.* 12 (4): 680–683.

5 Katritzky, A.R., Piffl, M., Lang, H., and Anders, E. (1999). Regioselectivity of the reactions of heteroatom-stabilized allyl anions with electrophiles. *Chem. Rev.* 99 (3): 665–722.

6 Jhuo, D.-H., Hong, B.-C., Chang, C.-W., and Lee, G.-H. (2014). One-pot organocatalytic enantioselective Michael–Michael–Aldol–Henry reaction cascade. A facile entry to the steroid system with six contiguous stereogenic centers. *Org. Lett.* 16 (10): 2724–2727.

7 Cichowicz, N.R., Kaplan, W., Khomutnyk, Y. et al. (2015). Concise enantioselective synthesis of oxygenated steroids via sequential copper(II)-catalyzed Michael addition/intramolecular Aldol cyclization reactions. *J. Am. Chem. Soc.* 137 (45): 14341–14348.

8 Bhattarai, B. and Nagorny, P. (2018). Enantioselective total synthesis of cannogenol-3-*O*-α-L-rhamnoside via sequential Cu(II)-catalyzed Michael addition/intramolecular Aldol cyclization reactions. *Org. Lett.* 20 (1): 154–157.

9 Kaplan, W., Khatri, H.R., and Nagorny, P. (2016). Concise enantioselective total synthesis of cardiotonic steroids 19-hydroxysarmentogenin and trewianin aglycone. *J. Am. Chem. Soc.* 138 (22): 7194–7198.

10 Khatri, H.R., Bhattarai, B., Kaplan, W. et al. (2019). Modular total synthesis and cell-based anticancer activity evaluation of ouabagenin and other cardiotonic steroids with varying degrees of oxygenation. *J. Am. Chem. Soc.* 141 (12): 4849–4860.

11 Nakazaki, A., Hashimoto, K., Ikeda, A. et al. (2017). De novo synthesis of possible candidates for the Inagami–Tamura endogenous digitalis-like factor. *J. Org. Chem.* 82 (17): 9097–9111.

12 Watanabe, S., Nishikawa, T., and Nakazaki, A. (2019). Synthesis of oxy-functionalized steroidal skeletons via Mizoroki–Heck and intramolecular Diels–Alder reactions. *Org. Lett.* 21 (18): 7410–7414.

13 Kim, W.S., Du, K., Eastman, A. et al. (2018). Synthetic *nat*- or *ent*-steroids in as few as five chemical steps from epichlorohydrin. *Nat. Chem.* 10 (1): 70–77.

14 Kim, W.S., Shalit, Z.A., Nguyen, S.M. et al. (2019). A synthesis strategy for tetracyclic terpenoids leads to agonists of ERβ. *Nat. Commun.* 10 (1): 2448.

15 Millham, A.B., Bhatt, C.P., and Micalizio, G.C. (2020). From metallacycle-mediated annulative cross-coupling to steroidal tetracycles through intramolecular C9–C10 bond formation. *Org. Lett.* 22 (16): 6595–6599.

16 Li, H.-J., Jiang, Y., and Li, P. (2006). Chemistry, bioactivity and geographical diversity of steroidal alkaloids from the Liliaceae family. *Nat. Prod. Rep.* 23 (5): 735.

17 Heretsch, P., Rabe, S., and Giannis, A. (2010). A biomimetic approach to C-*nor*-D-*homo*-steroids. *J. Am. Chem. Soc.* 132 (29): 9968–9969.

18 Kozlovsky, A.G., Zhelifonova, V.P., Ozerskaya, S.M. et al. (2000). Cyclocitrinol, a new fungal metabolite from *Penicillium citrinum*. *Pharmazie* 55 (6): 470–471.

19 do Marinho, A.M., R., Rodrigues-Filho, E., Ferreira, A.G., and Santos, L.S. (2005). C25 steroid epimers produced by *Penicillium janthinellum*, a fungus isolated from fruits *Melia azedarach*. *J. Braz. Chem. Soc.* 16 (6b): 1342–1346.

20 Du, L., Zhu, T., Fang, Y. et al. (2008). Unusual C25 steroid isomers with bicyclo[4.4.1]A/B rings from a volcano ash-derived fungus *Penicillium citrinum*. *J. Nat. Prod.* 71 (8): 1343–1351.

21 Xia, M.-W., Cui, C.-B., Li, C.-W., and Wu, C.-J. (2014). Three new and eleven known unusual C25 steroids: activated production of silent metabolites in a marine-derived fungus by chemical mutagenesis strategy using diethyl sulphate. *Marine Drugs* 12 (3): 1545–1568.

22 Ying, Y.-M., Zheng, Z.-Z., Zhang, L.-W. et al. (2014). Rare C_{25} steroids produced by *Penicillium chrysogenum* P1X, a fungal endophyte of *Huperzia serrata*. *HCA* 97 (1): 95–101.

23 Yu, F.-X., Li, Z., Chen, Y. et al. (2017). Four new steroids from the endophytic fungus *Chaetomium* sp. M453 derived of Chinese herbal medicine *Huperzia serrata*. *Fitoterapia* 117: 41–46.

24 Wang, Y., Ju, W., Tian, H. et al. (2018). Scalable synthesis of cyclocitrinol. *J. Am. Chem. Soc.* 140 (30): 9413–9416.

25 Steyn, P.S. and van Heerden, F.R. (1998). Bufadienolides of plant and animal origin. *Nat. Prod. Rep.* 15 (4): 397.

26 Cheng, C., Wang, J., Chen, J. et al. (2019). New therapeutic aspects of steroidal cardiac glycosides: the anticancer properties of Huachansu and its main active constituent Bufalin. *Cancer Cell Int.* 19 (1): 92.

27 Shimizu, S., Hagiwara, K., Itoh, H., and Inoue, M. (2020). Unified total synthesis of five bufadienolides. *Org. Lett.* 22 (21): 8652–8657.

28 Zhao, Z.-Z., Chen, H.-P., Wu, B. et al. (2017). Matsutakone and matsutoic acid, two (nor)steroids with unusual skeletons from the edible mushroom *Tricholoma matsutake*. *J. Org. Chem.* 82 (15): 7974–7979.

29 Kikuchi, T., Horii, Y., Maekawa, Y. et al. (2017). Pleurocins A and B: unusual 11(9 → 7)-*abeo*-ergostanes and eringiacetal B: a 13,14-*seco*-13,14-epoxyergostane from fruiting bodies of *Pleurotus eryngii* and their inhibitory effects on nitric oxide production. *J. Org. Chem.* 82 (19): 10611–10616.

30 Heinze, R.C. and Heretsch, P. (2019). Translation of a polar biogenesis proposal into a radical synthetic approach: synthesis of pleurocin A/matsutakone and pleurocin B. *J. Am. Chem. Soc.* 141 (3): 1222–1226.

31 Wu, J., Tokuyama, S., Nagai, K. et al. (2012). Strophasterols A to D with an unprecedented steroid skeleton: from the mushroom *Stropharia rugosoannulata*. *Angew. Chem. Int. Ed.* 124 (43): 10978–10980.

32 Heinze, R.C., Lentz, D., and Heretsch, P. (2016). Synthesis of strophasterol A guided by a proposed biosynthesis and innate reactivity. *Angew. Chem. Int. Ed.* 55 (38): 11656–11659.

33 Krohn, K., Biele, C., Aust, H.-J. et al. (1999). Herbarulide, a ketodivinyllactone steroid with an unprecedented homo-6-oxaergostane skeleton from the endophytic fungus *Pleospora herbarum*. *J. Nat. Prod.* 62 (4): 629–630.

34 Huang, X.-C., Guo, Y.-W., and Song, G.-Q. (2006). Fortisterol, a novel steroid with an unusual seven-membered lactone ring B from the Chinese marine sponge *Biemna fortis* Topsent. *J. Asian Nat. Prod. Res.* 8 (6): 485–489.

35 Tong, Z.-B., Cui, X.-H., Wang, J. et al. (2017). Constituents from solid-cultured *Antrodia camphorata*. *Nat. Prod. Res.* 31 (21): 2564–2567.

36 Zhao, Z.-Z., Han, K.-Y., Li, Z.-H. et al. (2019). Cytotoxic ergosteroids from the fungus *Stereum hirsutum*. *Phytochem. Lett.* 30: 143–149.

37 Duecker, F.L., Heinze, R.C., Mueller, M. et al. (2020). Synthesis of the alleged structures of fortisterol and herbarulide and structural revision of herbarulide. *Org. Lett.* 22 (4): 1585–1588.

38 Chang, Y.-C., Kuo, L.-M., Su, J.-H. et al. (2016). Pinnigorgiols A–C, 9,11-secosterols with a rare ring arrangement from a gorgonian coral *Pinnigorgia* sp. *Tetrahedron* 72 (7): 999–1004.

39 Li, X., Zhang, Z., Fan, H. et al. (2021). Concise synthesis of 9,11-secosteroids pinnigorgiols B and E. *J. Am. Chem. Soc.* 143 (13): 4886–4890.

40 Rudi, A., Yosief, T., Loya, S. et al. (2001). Clathsterol, a novel anti-HIV-1 RT sulfated sterol from the sponge *Clathria* species. *J. Nat. Prod.* 64 (11): 1451–1453.

41 Zhou, T., Feng, F., Shi, Y., and Tian, W.-S. (2016). Synthesis toward and stereochemical assignment of clathsterol: exploring diverse strategies to polyoxygenated sterols. *Org. Lett.* 18 (9): 2308–2311.

42 Lei, C., Huang, S.-X., Chen, J.-J. et al. (2008). Propindilactones E–J, schiartane nortriterpenoids from *Schisandra propinqua* var. *propinqua*. *J. Nat. Prod.* 71 (7): 1228–1232.

43 Wang, Y., Chen, B., He, X., and Gui, J. (2020). Bioinspired synthesis of nortriterpenoid propindilactone G. *J. Am. Chem. Soc.* 142 (11): 5007–5012.

12

Chiral Organophosphorus Compounds in Asymmetric Synthesis

Elżbieta Łastawiecka, Sylwia Sowa, Katarzyna Szwaczko, Kamil Dziuba, Marek Stankevič, and Adam Włodarczyk

Maria Curie-Skłodowska University, Institute of Chemical Sciences, Department of Organic Chemistry, Faculty of Chemistry, 33 Gliniana Street, 20-614 Lublin, Poland

12.1 Introduction

Chirality is one of the most unusual phenomena in organic compounds. Simply, it means the existence of two non-imposable isomers possessing the same substitution pattern at a specified position but differing in their mutual arrangement in space. Chiral compounds play a major role in living organisms. Amino acids and sugars are involved in the most fundamental processes like tissue formation, energy transfer, signaling, biosynthesis, etc. Considering this, the synthesis of organic compounds in nonracemic applications in pharmacy, medicine, or material chemistry is one of the most emerging goals in synthetic organic chemistry. This is also true for organophosphorus compounds, which play an important role in many branches of chemistry. This chapter will be devoted to discussing recent developments in chiral organophosphorus compounds and their applications in asymmetric synthesis. The chapter has been divided into five parts in which different applications of *P*- and *C*-chiral organophosphorus compounds will be discussed.

12.2 Organophosphorus Compounds with Incorporated Chiral Terpene Moieties

In the chemistry of organophosphorus compounds, mono or bicyclic monoterpene hydrocarbons are interesting small-molecule building blocks mainly due to their availability and occurrence in an enantiomerically pure form (Figure 12.1). Terpenic alcohols like (−)-menthol are commonly used chiral auxiliaries to synthesize *P*-stereogenic compounds, e.g. chiral *H*-phosphonates, *H*-phosphinates, and secondary phosphine oxides (SPOs). The diastereomerically pure *H*-menthyl phosphinate, obtained by a resolution from the diastereomeric mixture, can

Chiral Building Blocks in Asymmetric Synthesis: Synthesis and Applications, First Edition.
Edited by Elżbieta Wojaczyńska and Jacek Wojaczyński.

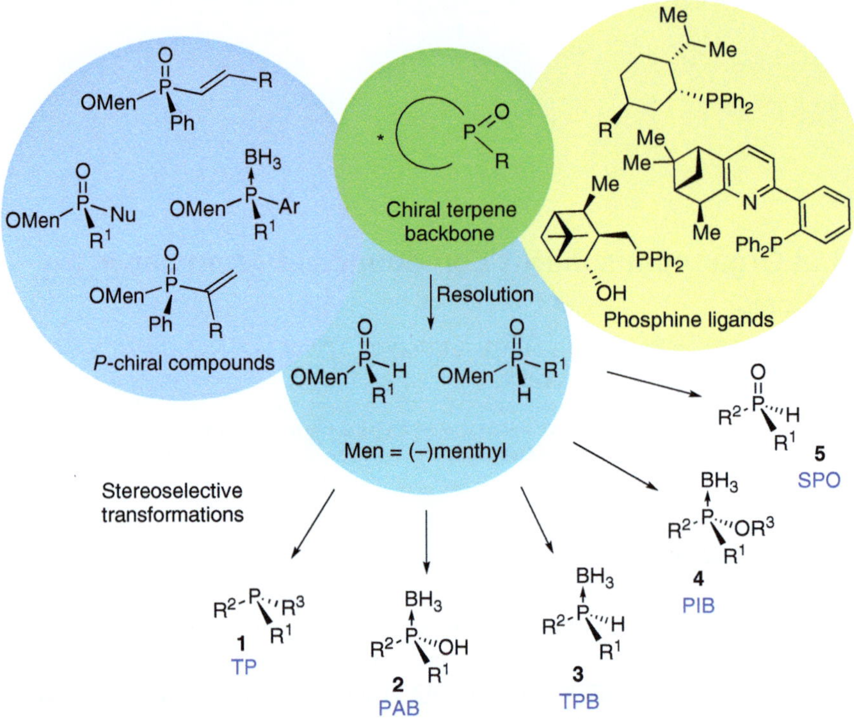

Figure 12.1 (–)-L-Menthol in the synthesis of *P*-stereogenic compounds.

be used in the preparation of optically active tertiary phosphines (TPs), SPOs, compounds with dative phosphorus–boron bond (PAB, phosphinous acid–borane; PIB, phosphinite–borane; TPB, tertiary phosphine–borane), etc. Further development of these menthol-based synthetic procedures reported initially by Horner and Knowles (Nobel Prize in 2001) and others led to many optically pure phosphines [1–3]. As part of the chiral pool strategy, terpenes have also been used as substrates for synthesizing phosphorus-based ligands for organometallic chemistry and asymmetric catalysis. Below, the key terpene building blocks with phosphorus functionality will be summarized.

One of the hardest tasks in organic chemistry is the preparation of optically active organophosphorus compounds, especially those possessing a chirality center at a phosphorus atom [4–9]. Various methods have been reported to synthesize *P*-stereogenic building blocks based on kinetic resolution or chiral auxiliary approaches. The last method is a compelling approach to control chirality and has become a widely preferred methodology for synthesizing chiral organophosphorus compounds. Commonly used chiral pool auxiliaries are the terpene alcohols, which transfer the chiral information to the phosphorus atom through the formation of diastereomeric compounds. The latter can be separated by chromatography or recrystallization, and the chiral auxiliary can be recovered for reuse. This well-known methodology gives access to both enantiomers of some phosphorus(V) species. Naturally occurring (1*R*,2*S*,5*R*)-2-isopropyl-5-methylcyclohexanol,

(−)-menthol (L-menthol) was first used as a chiral auxiliary in the late 1960s by Mislow [10–14] and others [15–17]. An example is L-menthyl *H*-phenylphosphinate (*R*_p-**7**), which has been used as a substrate for the synthesis of other classes of organophosphorus compounds (Scheme 12.1) (for recent reviews, see [18–20]).

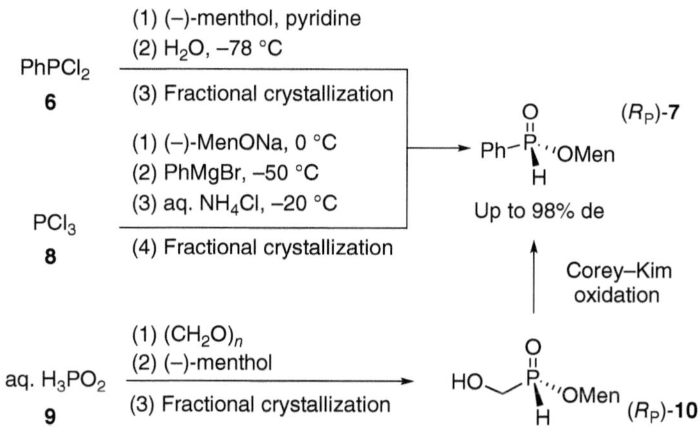

Scheme 12.1 Synthesis of *H*-Phosphinate (R_P)-**7**.

H-Phosphinate (R_P)-**7** was easily obtained by a reaction of $PhPCl_2$ **6** with pyridine and (−)-menthol and subsequent hydrolysis. The two diastereomers of menthyl phosphinates were separated by fractional crystallization. This methodology has been well-described on a multi-gram scale in the literature, initially by Mislow et al. [10]. Recently, a more efficient protocol (yield 58%, de >98%) was presented by Han and coworkers [21]. In 2011, Buono and coworkers reported a detailed preparation of (R_P)-**7** using $MenOPCl_2$ and phenylmagnesium chloride in 16% overall yield and (R_P)-**7**/(S_P)-**7** ratio 99/1 after two recrystallizations [22]. The authors also prepared *o*-tolyl and 1-naphthyl derivatives in comparable yields. Further, the versatile and inexpensive synthesis of *P*-stereogenic phosphinate building blocks from (−)-menthol was presented by Berger and Montchamp [23]. The reaction of hypophosphorous acid **9**, paraformaldehyde, and (−)-menthol afforded menthyl(hydroxymethyl)phosphinate (R_P)-**10** with 98% de after recrystallization. Despite the low yield (6%), this process can be regarded as a general method, and the obtained (R_P)-**10** was converted in a highly stereoselective manner by Corey–Kim oxidation to the corresponding *H*-phosphinate (R_P)-**7** [24]. Since 1970, several valuable optically pure pairs of L-menthyl phosphinates were synthesized and separated using L-menthol as a chiral auxiliary (Figure 12.2) [25].

Over the last 10 years, many research groups have shown the incredible synthetic potential of L-menthyl *H*-phosphinates in constructing *P*-stereogenic organophosphorus compounds. For example, Han and coworkers used optically active phosphinate (R_P)-**7** for reaction with ketones or aldehydes in a base (the Pudovik reaction), forming the corresponding *P*,*C*-stereogenic α-hydroxyphosphinates **17** with high diastereomeric ratio and retention of the configuration at the phosphorus

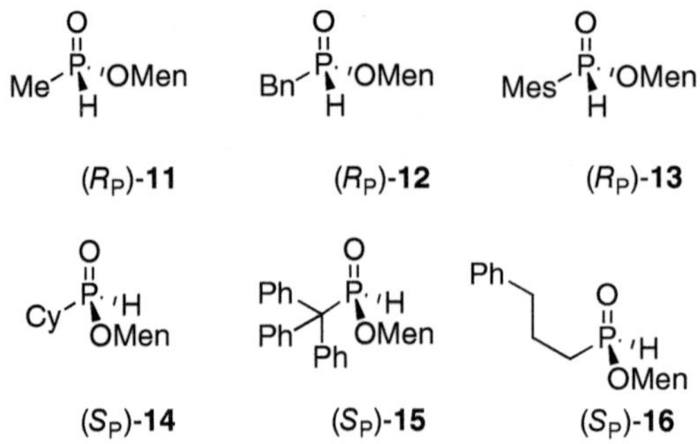

Figure 12.2 Optically pure L-menthyl phosphinates.

atom [26]. Palladium-catalyzed stereospecific addition of alkynes to phenylphosphinate (R_P)-**7** resulted in the corresponding Markovnikov-type adducts **18** [27]. In turn, radical or base-catalyzed addition of various electron-rich alkenes opens access to anti-Markovnikov-type optically pure β-adducts **19** [28]. Copper(II) chloride was also applied as a chlorination reagent to transform *P*-chiral *H*-phosphinates into the corresponding phosphoryl chlorides **20**. Han and coworkers also showed that nucleophiles like amines, alcohols, or thiols served as good substrates for Atherton–Todd reaction, providing a general method for synthesizing optically active organophosphorus derivatives **21** [29–31]. In 2012, Zhao and coworkers described the hydrophosphinylation of a Schiff base with phenylphosphinate (R_P)-**7**, followed by crystallization of the product, which afforded the optically pure phosphinate **22** [32]. The same substrate has been used to synthesize *P*-chiral phosphonothioates and phosphonoselenoates **23** as described by Wang et al. [33]. In stereoselectivity, the synthesis of terpene derivative **21** proceeded with inversion of configuration at the phosphorus atom, whereas for compounds **17–20** and **22**, retention was observed.

Structurally similar (R_P)-**10**, possessing hydroxymethyl group at the phosphorus atom, has been developed and extensively studied by Montchamp's group. The reaction with alkyl bromides resulted in the formation of an intermediate, which in the presence of *s*-BuLi afforded the rearranged product **24**. In turn, palladium-catalyzed cross-coupling reaction of (R_P)-**10**, manganese-catalyzed arylation, and sila-Arbuzov alkylation opened the door to various *P*-stereogenic menthyl derivatives such as **25** (Scheme 12.2) [26].

The use of L-menthol as a source of chirality can be expanded over different organophosphorus compounds. (–)-*O*,*O*-Di-(1*R*,2*S*,5*R*)-menthyl *H*-phosphonate **26** can be easily obtained in the reaction of (–)-menthol with $MeOPCl_2$, followed by hydrolysis [34]. The authors demonstrated the utility of dimenthyl phosphite **26** in synthesizing optically active NH-iminophosphonates **27** and their further functionalization. Mannich reaction of **27** with acetone and proline as a catalyst gave the α-amino-α-trifluoromentyl-γ-oxobutylphosphonate **28**, and the aza-Henry reaction with electron-rich heterocycles afforded the diastereomerically enriched products. Base-catalyzed hydrophosphonylations of aldehydes with **26** were realized by Schmutzler and coworkers and afforded the appropriate phosphonic acid diesters

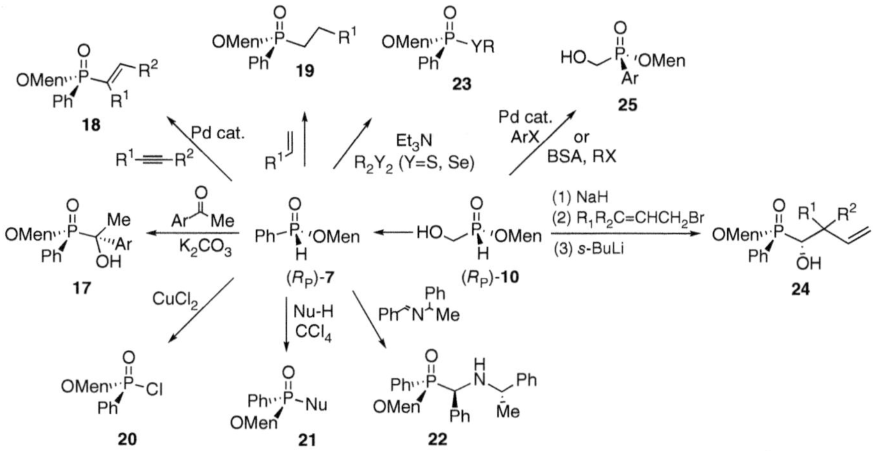

Scheme 12.2 L-Menthyl *H*-phosphinates in the construction of *P*-stereogenic compounds. BSA, *N,O*-bis(trimethylsilyl)acetamide.

29 [35]. These compounds were obtained with moderate selectivity, and pure major diastereomers were separated by crystallization (Scheme 12.3).

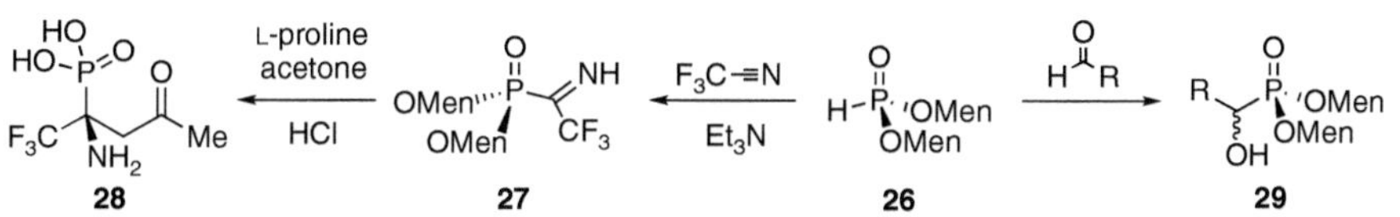

Scheme 12.3 The utility of dimenthyl phosphite **26** in the synthesis of **28** and **29**.

(1*R*,2*S*,5*R*)-Menthyl phosphinate **30** was prepared by the reaction of hypophosphorous acid with (−)-menthol in the presence of trimethyl orthoformate. It was used in palladium-catalyzed Heck reaction with iodobenzene to yield the diastereomeric (−)-menthyl phenylphosphinate (R_P)-**7**. The Atherton–Todd reaction of **30** proceeded with the formation of amide **31**, whereas a reaction with Schiff bases resulted in bis-amidophosphinates **32** (Scheme 12.4) [36].

Terpene fragments can also be attached via a phosphorus–carbon bond, not via a phosphorus–oxygen bond. Treatment of dichloro(phenyl)phosphine with menthylmagnesium chloride and then fractional crystallization resulted in optically pure menthyl phenylphosphine oxide (R_P)-**33** as reported by Zhao and coworkers [37–39] Addition of (R_P)-**33** to α,β-unsaturated aldehydes resulted in a series of 1,3-bisphosphinylpropanes **34** with multiple stereogenic centers. The authors assigned high stereoselectivity of the reaction to the unusual reaction mechanism. In turn, the addition of chiral phosphine oxide (R_P)-**33** to unsaturated ketones led to the appropriate β-phosphinoylketones **35** (Scheme 12.5).

Another application of (R_P)-**33** came from Loh and coworkers. It was based on allylic carbonates in the presence of the catalytic amount of quinine, which afforded **36**, possessing menthyl substituent at phosphorus atom [40]. The phospha-Mannich

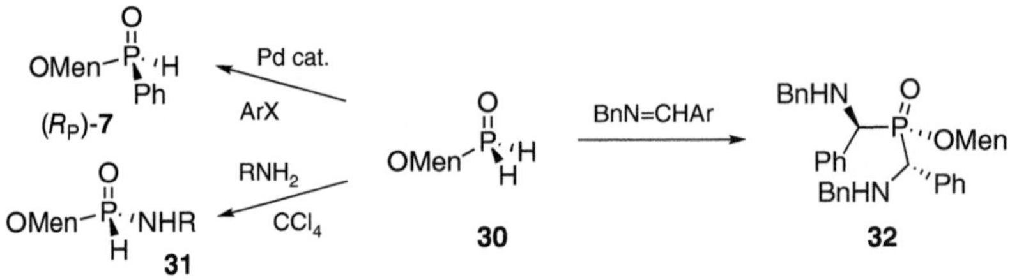

Scheme 12.4 The transformation of **30** leading to (−)-menthyl phenylphosphinate and **31–32**.

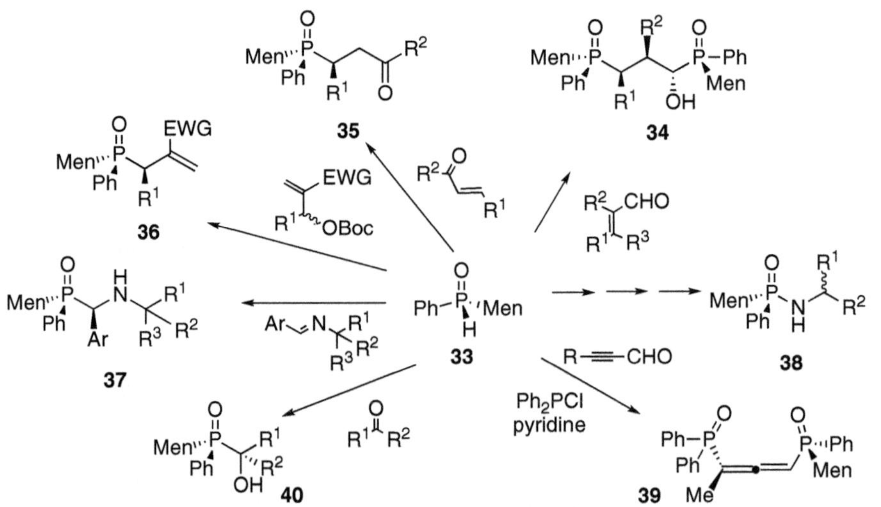

Scheme 12.5 The synthesis of numerous chiral organophosphorus compounds from **33**.

reactions with chiral and non-chiral imines were evaluated by Zhao's group [41]. The reaction resulted in **37** as two pairs of diastereoisomers, easily separated with preparative thin layer chromatography (TLC). Finally, the same authors described the use of (R_P)-**33** in the Atherton–Todd coupling reaction with ammonia to give (−)-menthyl phenylphosphinamide, which was subsequently converted to *N*-phosphonyl imines and then in a reaction with the Grignard reagent to **38** [42]. Recently, Li and coworkers demonstrated that menthylphenylphosphine oxide (R_P)-**33** reacted with alkenyl aldehydes to yield α-(phosphinoyl)propargyl alcohols, which in the presence of pyridine and Ph_2PCl could be subsequently converted to various *P*-axial stereogenic allenyl phosphine oxides **39** [43]. Base-catalyzed nucleophilic addition of (R_P)-**33** to aldehydes and ketones was also evaluated. The desired α-hydroxyphosphine oxides **40** were obtained with good yields and high diastereoselectivity up to 98% de [44].

Using L-menthol as a chiral alkoxy group to synthesize the corresponding phosphoric acid esters is not the only possibility in organophosphorus chemistry. Another application includes using L-menthyl chloroacetate or L-menthyl bromoacetate as a removable chiral auxiliary. When treated with tertiary phosphines, the

Figure 12.3 *P*-Stereogenic phosphonium salt and phosphine oxides.

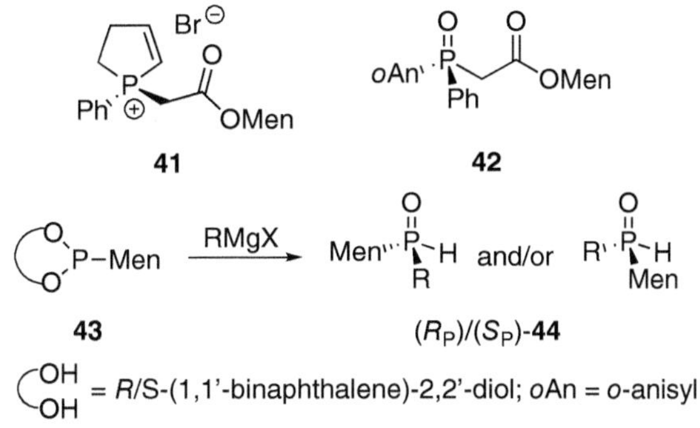

mixtures of diastereomeric phosphonium salts, such as compound **41**, or treated in a substitution reaction with SPOs, yield a mixture of diastereoisomers, which can be separated to obtain pure menthyl ester **42**. Pietrusiewicz and coworkers recently used this methodology in synthesizing functionalized *P*-stereogenic phosphine oxides (Figure 12.3) [45, 46].

Recently, Hey-Hawkins and coworkers applied chloromethyl (1*R*,2*S*,5*R*)-(−)-menthyl ether for the synthesis of chiral 1,2-diphosphole derivative, followed by [4+2]-cycloaddition with non-chiral maleic anhydride, which afforded the appropriate *P*-chiral phosphines [47].

In 2020, Zhao's group disclosed the elegant procedure for synthesizing chiral SPO's with menthyl substituent using L-menthyl group as a chiral auxiliary [48]. Menthyl group was introduced to a cyclic P–Cl species with a binaphthol moiety, forming the compound **43** with C—P bond as a single stereoisomer. When **43** was treated with an organometallic reagent, chiral menthyl-substituted SPO's (R_P)/(S_P)-**44** were formed with >90% yields and dr of c. 99 : 1. The authors suggested that the menthyl group acted not only as an auxiliary but also stabilized the configuration at the phosphorus atom, ensuring the stereoselective formation of the C—P bond.

The discussed compounds were built around the pentavalent phosphorus atom that possessed the phosphoryl (P=O) group. A different class of organophosphorus compounds includes phosphine-boranes. The phosphorus atom is at a lower oxidation state despite an additional coordination bond with the BH_3 group. Diastereoisomeric menthylphosphinite–boranes **46** are popular substrates for the preparation of chiral organophosphorus compounds. A typical synthetic procedure of menthylphosphine-boranes was presented by Imamoto et al. [49, 50].

Today, L-menthyl *H*-PIBs are used to prepare several *P*-stereogenic compounds. Some examples are provided in Scheme 12.6. L-Menthyl *H*-PIBs **45** undergo reaction with methyl iodide and sodium hydride to provide L-menthyl methylphenylphosphinite-borane with 100% de [51]. Palladium-catalyzed coupling of **46** with haloarenes proceeds with an inversion or retention of configuration at the phosphorus atom, depending on the used solvent [12, 52].

Recently, Stankevič and coworkers have developed another method based on diastereomerically pure *O*-(−)-menthyl phosphinous acid monoester monoamide-borane [53]. Treatment of dichlorophenylphosphine with (−)-menthol and triethylamine, followed by the reaction with diethylamine, resulted in an intermediate, which was

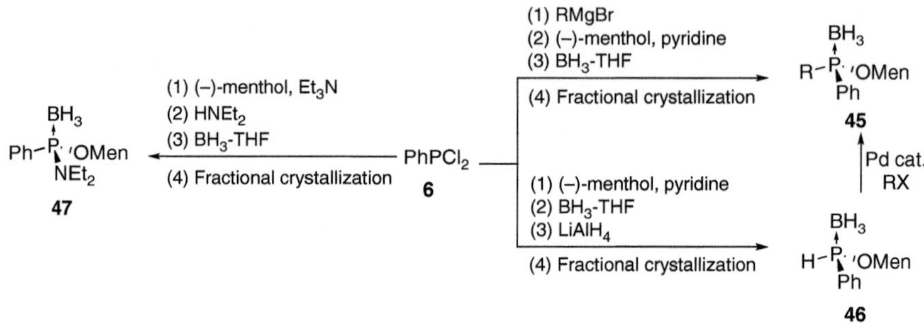

Scheme 12.6 Synthesis of L-menthyl *H*-phosphinite-boranes and L-menthyl *H*-phosphinates.

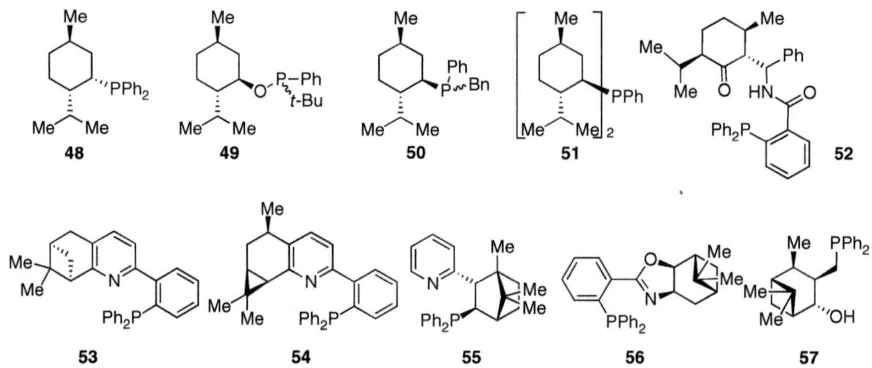

Figure 12.4 (–)-Menthol-derived catalysts **48–57**.

reacted with BH_3–THF (tetrahydrofuran) complex. The product of this three-step synthesis, phenylphosphonous acid-borane L-menthyl ester *N*,*N*-diethylamide **47**, was obtained in diastereomerically pure form after fractional crystallization. This compound was used in the synthesis of various enantiomerically enriched PAB amides and secondary or TPBs.

The diversity of chemical structure and availability make terpenes attractive starting materials in synthesizing optically pure ligands. In addition to the previous examples where terpenes such as (–)-menthol have been used as a removable chiral auxiliary, other transformations are also known where they become an integral part of the molecule. Among them simple monodentate menthol-derived phosphines **48–51** presented in Figure 12.4 can be found [54–56]. (+)-Isomenthone was also an attractive molecule for synthesizing a phosphine ligand. A slight modification of its structure led to polydendate phosphine **52** [57]. A series of bidentate *N*,*P*-donor pyridine ligands were synthesized from (–)-β-pinene (compound **53**), (+)-3-carene (compound **54**), (+)-2-carene, and (–)-α-pinene and were investigated in the asymmetric palladium-catalyzed Heck reaction [58]. At the end of the 1990s, several imine-type ligands and their palladium complexes were also obtained from terpenes: (+)-fenchone, (+)-2-hydroxypinan-3-one, (–)-menthone, (+)-camphor, and (+)-ketopinic acid [59, 60]. In turn, a library of optically active terpene-derived phosphite-type compounds were developed by Gavrilov and coworkers [61]. Particularly noteworthy are the pyridine *P,N*-ligands obtained by Knochel's group, from (+)-camphor (**55**) or (+)-nopinone; their iridium(II) complexes were successfully applied as catalysts for asymmetric hydrogenation of non-symmetrically substituted styrenes [62]. Similarly, monoterpene PHOX ligands **56** and phosphine **57** reported by Krzemiński were derived from pinene [63, 64].

It can be concluded that terpenes are a perfect source of chirality if organophosphorus compounds in a nonracemic form are required. Furthermore, the use of terpenes allows the preparation of different types of organophosphorus compounds, and terpene scaffold may be incorporated into the structure of the target molecule.

12.3 Organophosphorus Compounds with Axial Chirality

Nowadays, atropoisomeric phosphines such as BINAP **58**, MeO-BIPHEP **59**, and monodentate BINAPINE **60** are among the most widely used chiral ligands in diverse asymmetric catalytic reactions (Figure 12.5). Moreover, BINAP is a chiral ligand produced on an industrial scale and available commercially as single enantiomers.

Considering the importance of the above compounds in organic synthesis, it is not surprising that synthetic methodologies for the modifications of binaphthyl-based phosphorus ligands to increase efficiency (TON and TOF), selectivity (ee), and facilitate separation have been frequently reported in the past two decades. The chemistry described in this progress report provides a broad view of the opportunities to use axially chiral phosphine compounds as nonracemic building blocks in organic synthesis.

Structural diversity in the BINAP library can be afforded by introducing different substituents in almost all positions of binaphthyl backbone. Phosphine oxide analogs of BINAP are typically used in scaffold functionalization by electrophilic reaction or deprotonation-functionalization sequence.

Modifying the position adjacent to the phosphorus atom is used to increase the steric hindrance around the ligating center. In 2002, Zhang reported a versatile synthetic approach to prepare BINAP-type ligands bearing different aryl or alkyl substituents at the 3,3′-positions of the binaphthyl core (Scheme 12.7) [65].

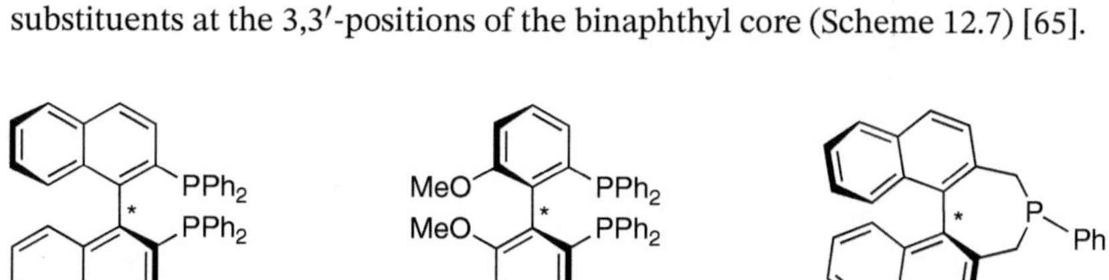

(S_a)-BINAP **58** | (S_a)-MeO-BIPHEP **59** | (S_a)-BINAPINE **60**

Figure 12.5 Structures of atropisomeric phosphine ligands.

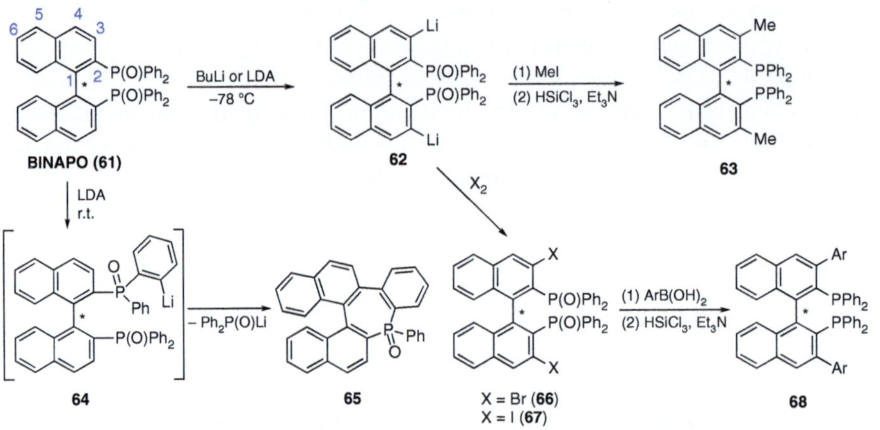

Scheme 12.7 Functionalization in the 3,3′-positions. Source: Zhang [65] / Google Patents.

The synthesis of these compounds was accomplished via selective 3,3′-lithiation with butyllithium or LDA of chiral BINAP dioxide **61**, followed by alkylation with alkyl iodide or halogenation with bromine or iodine. Suzuki coupling of iodide **67** with arylboronic acids gives the arylated BINAP derivatives **68**. Similarly, Widhalm and Mereiter reported an interesting intermolecular cyclization of BINAPO to give phosphepine oxide **65** through C–P cleavage [66].

In 2001, Köckritz and coworkers reported a simple and extremely regioselective bromination of 4,4′-positions of BINAP oxide using Br_2 in the presence of pyridine in dichloromethane [67]. The main drawback of this protocol is the low reaction rate, which could be improved by using NBS in room-temperature ionic liquids (RTILs) [68]. In this case, the product was obtained in quantitative yield. 4,4′-Dibromo-BINAPO **69** is a versatile precursor for further transformations presented below (Scheme 12.8).

Using 4,4′-dibromo intermediate **69**, Lin and coworkers introduced various groups, such as chloride, methyl, phenyl, etc., using halogen metathesis or Suzuki coupling (**70–76**) [69]. In another paper, the same authors effectively obtained compounds **77–81** by lithation of reduced **69**, followed by treatment with various electrophiles. Suzuki–Miyaura coupling of **77–81** with boronic acids allowed the preparation of 4,4′-bis(4-benzonitrile)-BINAPO **73** [70], 4,4′-bis(4-methoxycarbonyl)-BINAPO **74**, 4,4′-bis(3-pyridyl)-BINAPO **75**, 4,4′-bis(4-pyridyl)-BINAPO **76**, [71], or a more complex derivative **82** [72]. Coupling of **69** with copper cyanide led to the formation of 4,4′-dicyano-BINAPO **86**, which was subsequently reduced to 4,4′-diamBINAP **87** [73]. Köckritz and coworkers described also a preparation of phosphonate BINAP derivatives **89** through Pd-catalyzed reaction with diethyl phosphate [67]. Coupling of **69** with potassium vinyltrifluoroborate gave access to 4,4′-divinyl-BINAP **83** as described by Ding and coworkers [74], and Lemaire and coworkers obtained the perfluoroalkylated BINAP's **84–85** directly via a copper-mediated cross-coupling reaction between perfluoroalkyl iodide and **69** [75].

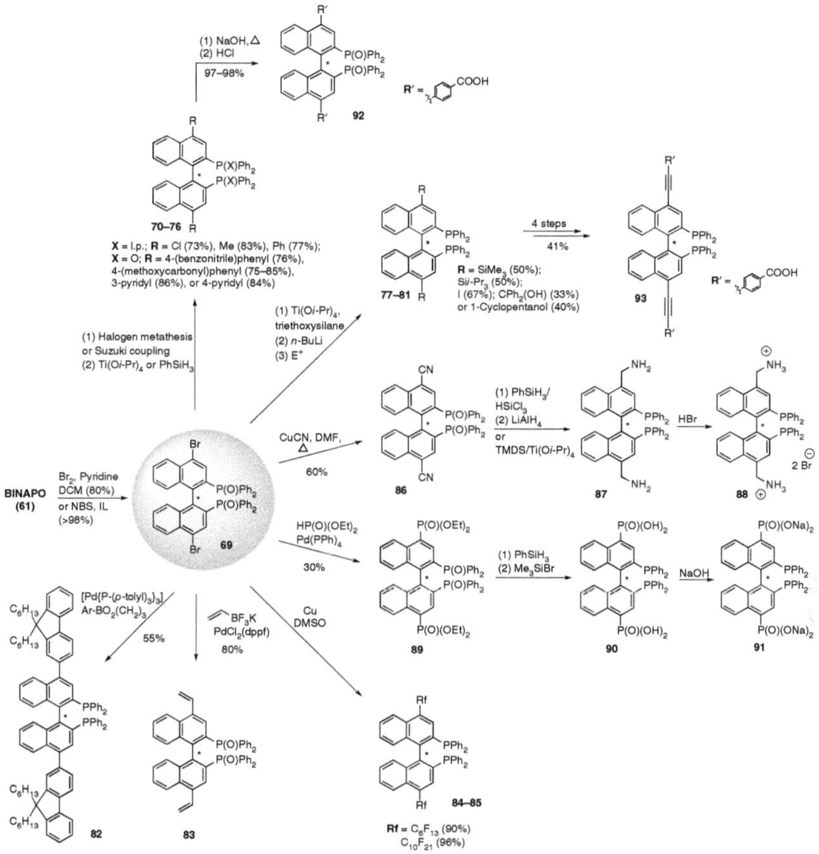

Scheme 12.8 Functionalization in the 4,4′-positions.

Subsequent transformations of 4,4′-substituted-BINAP derivatives (**73**, **74**, **79**, **87**, and **89**) were also reported. The development of catalytic systems for water or ionic liquids as alternative solvents requires the modification of ligands. Lemaire and coworkers demonstrated that quaternization of 4,4′-diamBINAP **87** provided a water-soluble BINAP analog suitable for asymmetric biphasic catalytic hydrogenation [73]. Hydrolysis of phosphorylated BINAP **89** afforded the appropriate 4,4′-diphosphonic acid-substituted BINAP, which was further transformed into its sodium salt **91** [67, 76]. Furthermore, treatment of chiral **90** with $[Ru(benzene)Cl_2]_2$, (*R*,*R*)-DPEN, and $Zr(O\text{-}t\text{-}Bu)_4$ yielded porous zirconium phosphonates of the approximate formula $Zr[Ru(\mathbf{L90})(DPEN)Cl_2]\text{-}4H_2O$ for heterogeneous asymmetric hydrogenation of prochiral ketones [77].

Carboxylic acid-substituted BINAPO **92** derivative was obtained by hydrolysis of the corresponding nitrile **73** or esters **74** with NaOH [70, 71]. Shortly after, Lin and coworkers reported synthesizing BINAP-derived dicarboxylic acid **93** from 4,4′-I_2-BINAP **79** with a 41% overall yield. This compound has been used to prepare the first example of a BINAP-based **MOF-1** (Scheme 12.9) [78]. The authors also reported the metalation of **BINAP-MOF-1** to afford catalysts for heterogeneous catalysis. Another example of a heterogeneous catalyst based on 4,4′-substituted BINAP is **BINAP-A-SBA-15**. This catalyst has been obtained via covalent anchoring to the inner walls of mesoporous SBA-15 [79].

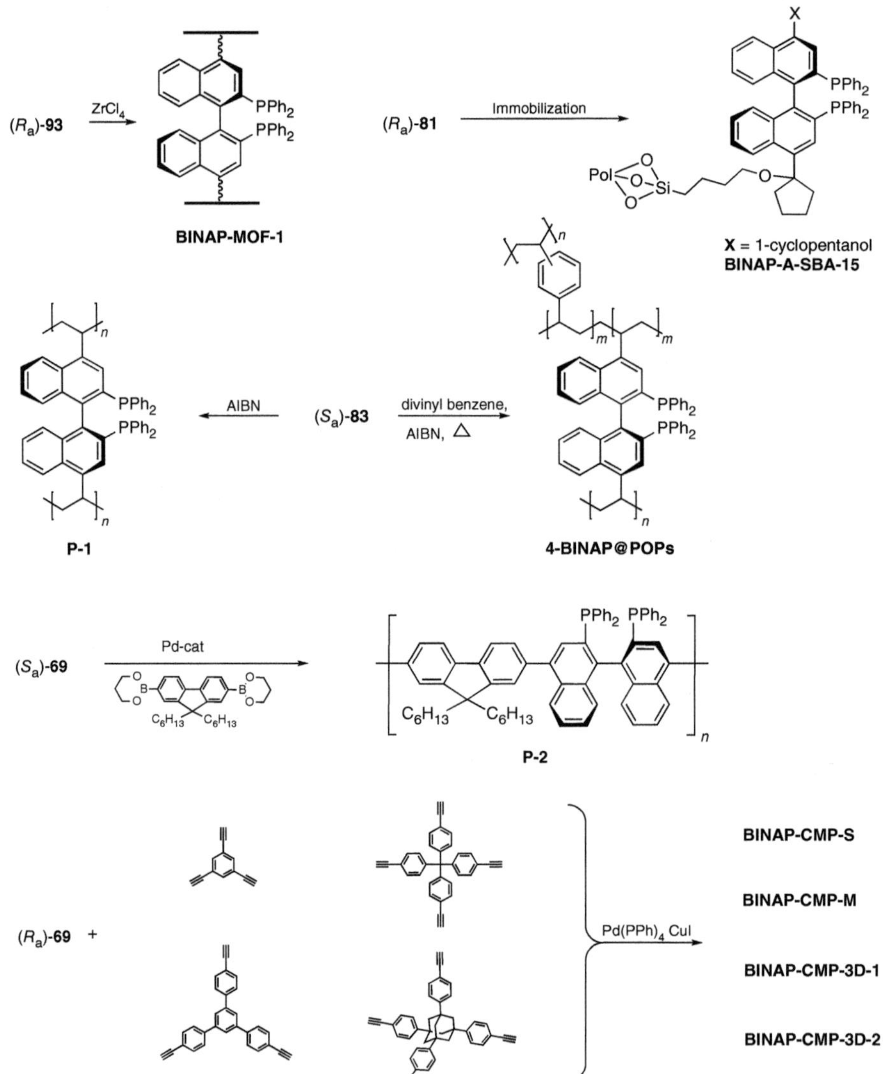

Scheme 12.9 Polymer supported 4,4′-substituted-BINAP-based ligands. DVB, divinylbenzene. Source: Falkowski et al. [78] / Royal Society of Chemistry.

Preparation of polymer-supported 4,4′-substituted-BINAP-based ligands has been described as well (Scheme 12.9). Both **83** and **69** were successfully used as monomers to prepare chiral polymers such as **P-1** [80], **4-BINAP@POPs**

[74], **P-2**, and **BINAP-CMP** [81]. Chiral BINAP-based porous organic polymers (**4-BINAP@POPs**) were efficiently prepared via copolymerization of **83** with divinylbenzene. A new kind of chiral conjugated copolymer **P-2** with (S_a)-BINAP and fluorene as the building blocks was synthesized by the Pd-catalyzed Suzuki coupling reaction of **69** with boronic acids. Recently, Li and coworkers reported synthesizing a series of conjugated microporous polymers CMP's by embedding the (R_a)-BINAP ligand into CMP networks [81].

Functionalization of 5 and 5′ positions of binaphthyl-based diphosphines can be achieved by two strategies: nitration or halogenation of BINAP dioxide. In 1985, Kumobayashi patented a procedure involving treatment of BINAPO with a nitrating mixture, followed by reducing N=O and P=O bonds, yielding 5,5′-diamino-BINAP **94** (Scheme 12.10) [82]. The compound **94** was converted to 5,5′-diacetamido-BINAP **95** and has shown to be a versatile precursor for the preparation of polymeric and dendritic supramolecular structures.

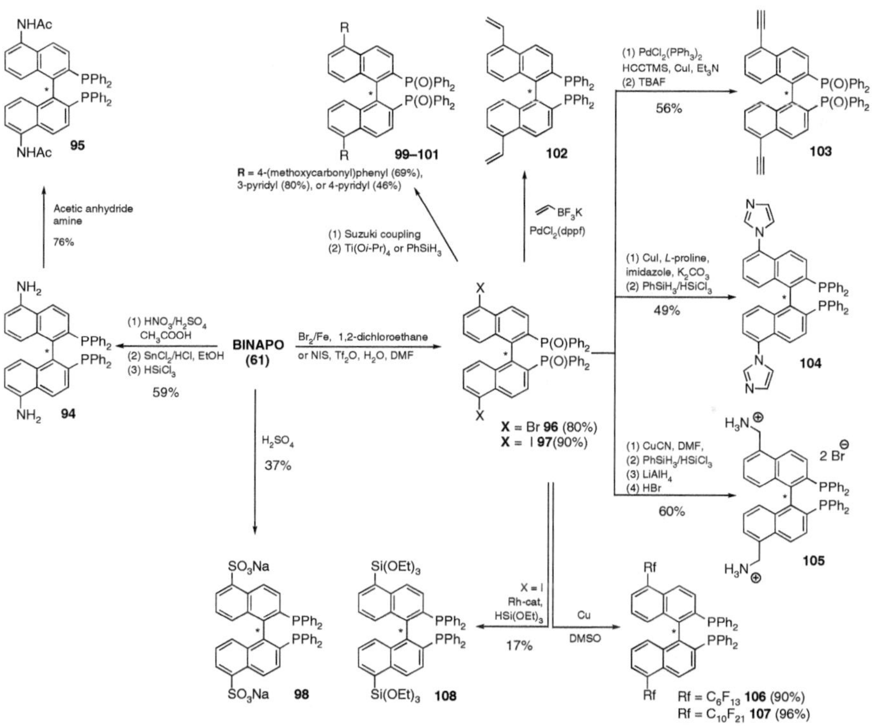

Scheme 12.10 Functionalization in the 5,5′-positions. DMSO, dimethyl sulfoxide; TBAF, tetrabutylammonium fluoride. Source: Okano et al. [82] / Google Patents.

Lemaire et al. [83] reported a regioselective dibromination at the 5 and 5′ positions to give 5,5′-dibromo-BINAPO **96** in 80% yield with no traces of **69**. Additionally, bis(pyridine)iodonium tetrafluoroborate in the presence of trifluoroacetic acid [84] or NIS in the presence of rhodium complex [85] were reported to be successfully used for regioselective iodination of BINAP dioxide to give

5,5′-I_2-BINAP dioxide **97**. Furthermore, sulfonylation of BINAPO yielding **98** has been reported by Kumobayashi [86].

Thus obtained 4,4′-I_2-BINAPO, 5,5′-Br_2-BINAPO, and 5,5′-I_2-BINAPO could be transformed into various derivatives via coupling reactions, e.g. **99–101** [71], **102** [74], **103** [84], **106**, and **107** [75]. Using **96** as the starting material and Cu-catalyzed C—N bond formation as the crucial synthetic step, Jin et al. were able to synthesize (*S*)-5,5′-diimidazole BINAP **104** [87], and Lemaire and coworkers obtained 5,5′-dibamBINAP **105** using the same methodology [73]. Additionally, 5,5′-diiodoBINAP dioxide **97** was successfully used in Rh-catalyzed Masuda reaction, which afforded bis(triethoxysilyl)-BINAP **108** [86].

Some of the 5,5′-substituted BINAP derivatives have appeared to be useful precursors for preparing polymeric and dendrimeric molecules (Figure 12.6). 5,5′-Diamino-BINAP **94** has been used for the first time by Chan and coworkers to prepare soluble polymer-supported catalysts ***polyEster*-BINAP** for asymmetric hydrogenation [88]. Compound **94** was condensed with various copolymers (e.g. **P-3** [89] and **CPC-BINAP** [90]) and was used as a linking point for attachment to supports, e.g. **P-4** (Figure 12.6) [91]. The first dendritic BINAP derivative obtained from **94** was **Den-1** [92]. Later, Fan and coworkers prepared the soluble dendritic BINAP-based organometallic catalysts **Gn-Den-BINAP** [93, 94], **Den-2**, and **Den-3** [95]. The same group also reported the synthesis of a soluble bifunctional polymeric ligand **polyBINOL-BINAP-B** [96]. Starting from vinyl-functionalized BINAP **102**, Ding and coworkers synthesized polymeric **P-5**, copolymeric **5-BINAP@POPs**, and embedded it into three different porous organic polymers (**Poly-1**, **Poly-2**, and **Poly-3**) (Figure 12.6) [74, 97]. Furthermore, Shimada's group prepared a series of molecular BINAP-based building blocks containing allylsilyl groups, which could be incorporated into the appropriate sol–gel precursors using 5,5′-diethynyl-BINAP dioxide **103** [98].

In 2010, Shimada and coworkers patented the preparation of copolymer possessing multiple (meth)acryloyl groups (Figure 12.6) [99]. The method for manufacturing the polymer includes a copolymer reaction with 5,5′-diiodo-BINAP dioxide **97** by a cross-coupling reaction, then polymerization and reduction of P=O bonds in the polymer **P-7–P-11**.

As opposed to modifying the mentioned positions in the BINAP scaffold, the direct synthesis of 6,6′-substituted BINAP derivatives appears impossible. These positions cannot be modified via electrophilic substitution directly. In contrast, 6 and 6′ positions are more reactive in BINOL and protected BINOL and extensive modifications of these positions can be achieved. For that reason, 6,6′-substituted BINOLs were used to prepare the 6,6′-disubstituted BINAP derivatives.

Compared to the binaphthyl core, the biphenyl scaffold-like MeO-BIPHEP **59** offers a tremendous advantage: positions 6 and 6′ can be easily functionalized. By changing the substituent pattern at positions 6 and 6′, the dihedral angle of the biphenyl core can be controlled. Zhang et al. first reported the preparation of chiral bisphosphines, called TunaPhos ligands, from enantiomerically pure MeO-BIPHEP **59**. MeO-BIPHEP was demethylated to provide HO-BIPHEP **109** and

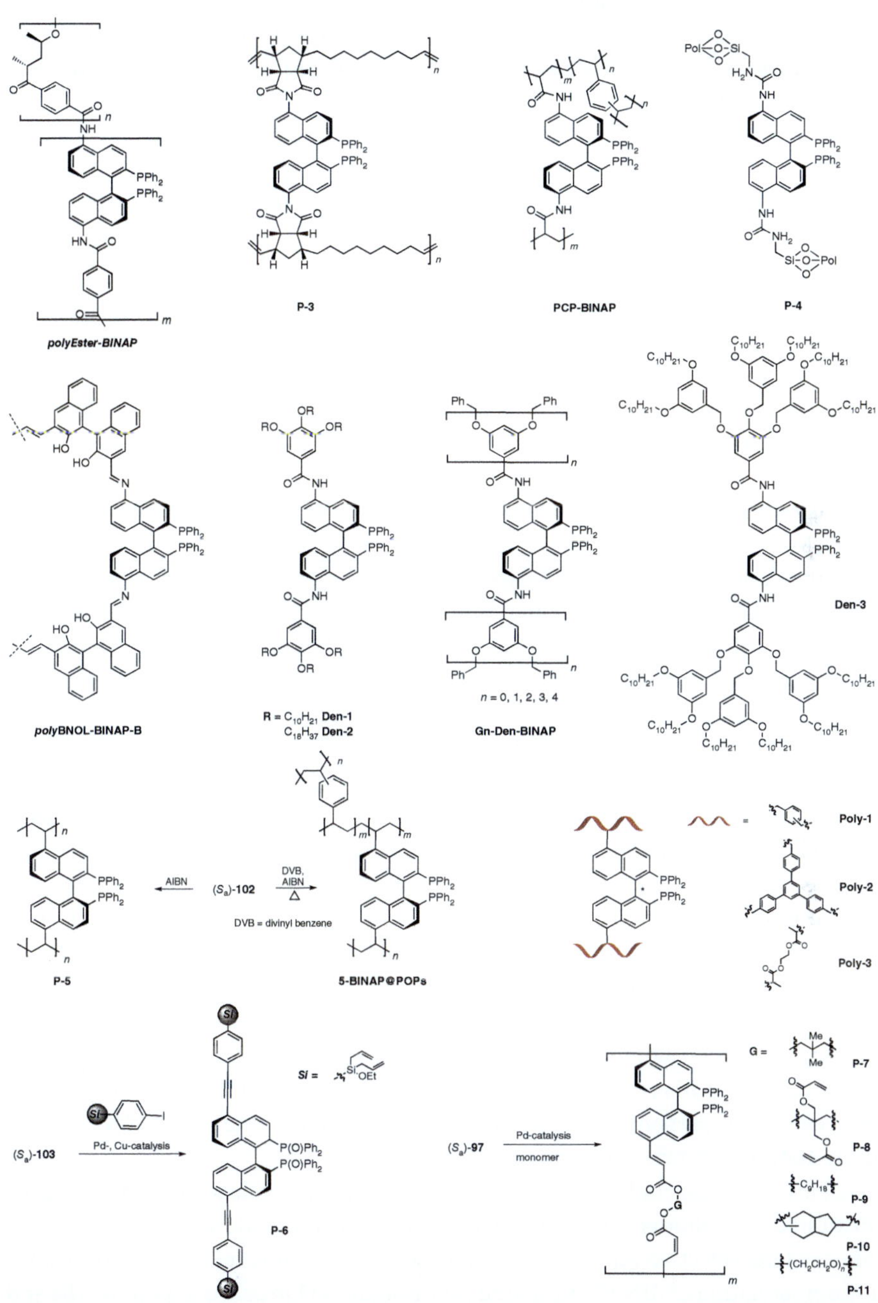

Figure 12.6 Polymeric and dendrimeric of 5,5′-disubstituted BINAP derivatives. DVB, divinylbenzene.

was subsequently alkylated with alkyl dihalides in the presence of a base in DMF to afford C1–C6-TunaPhos ligands **110–115** (Scheme 12.11) [100].

Through cyclization of the key starting material **109** with an array of bis-tosylated electrophiles, a series of chiral bisphosphines **116–122** with fused and tunable dihedral angles have been obtained (Scheme 12.11) [101–103].

Similarly, Wang and Zhou described the preparation of a series of tunable chiral bisphosphine ligands **123–139**. The reaction of (S_a)-**109** with alkyl halides or sulfonates in the presence of Cs_2CO_3 afforded products with 27–91% yields (Scheme 12.12) [104]. Compound **109** was shown to be a versatile precursor for the preparation of dendritic BIPHEP **140–142** [105].

Alexakis and coworkers reported that chiral 6,6′-dibromodiphenyldiphosphines ((R_a)-**143** and (R_a)-**145**) can be applied to the synthesis of chiral ligands bearing various substituents at positions 6 and 6′ of the biphenyl core (Scheme 12.13) [106].

The synthesis of diphosphine **144** was conducted through the Suzuki–Miyaura coupling of (R_a)-**143** with phenylboronic acid. A different approach was applied for a bis(dicyclohexylphosphine) **145**. In this case, Br–Li exchange was performed in toluene with *s*-BuLi, followed by quenching with different electrophiles such as I, Cl, or TMS.

Another important class of phosphines containing a binaphthyl backbone are derivatives of 5-dehydro-3*H*-dinaphtho[2,1-c;1′,2′-*e*]phosphepine (BINAPINE) **60,** which, unlike BINAP ligands, possess a phosphorus atom incorporated in a seven-membered ring. This BINAPINE is a valuable building block for synthesizing various derivatives by simple modification of benzylic position.

In 2003, Zhang and cowokers reported the first example of a chiral bisphosphepine ligand (R_a,R_a)-**152** with stereogenic C and P atoms. The employed synthetic protocol consisted of metallation of enantiopure (R_a)-*t*-Bu-BINAPHINE, followed by dimerization via oxidative coupling in the presence of copper(II) chloride [107]. Two years later, Widhalm and coworkers showed the synthesis of α-chiral monophosphine ligands (**153–162**) from phosphine sulfide by one or two subsequent diastereoselective α-alkylations, followed by desulfuration with Raney nickel [108]. The phosphine-olefin hybrid ligand **163** was obtained by elaborating the parent BINAPINE-borane via lithation next to phosphorus and treatment with cinnamyl bromide. Borane deprotection with diethylamine furnished a desired phosphine ligand **163** [109]. Furthermore, using the same synthetic strategy, Mazet and coworkers prepared another monosubstituted α-chiral BINAPINE ligand family **164–183**, shown in Figure 12.7 [110–112].

Recently, Mikami and coworkers have reported the synthesis of perfluoroaryl-substituted phosphorus compounds via nucleophilic aromatic substitution (S_NAr). This diastereoselective S_NAr reaction was successfully utilized for stereoselective perfluoroarylation of BINAPINE-based phosphinate **151** to give compounds **184** and **185** [113].

Moreover, Fu and coworkers have developed methodologies to modify the axially chiral biphenyl-derived phosphepines via Suzuki–Miyaura coupling with *m*-disubstituted phenylboronic acids (Scheme 12.14) [114].

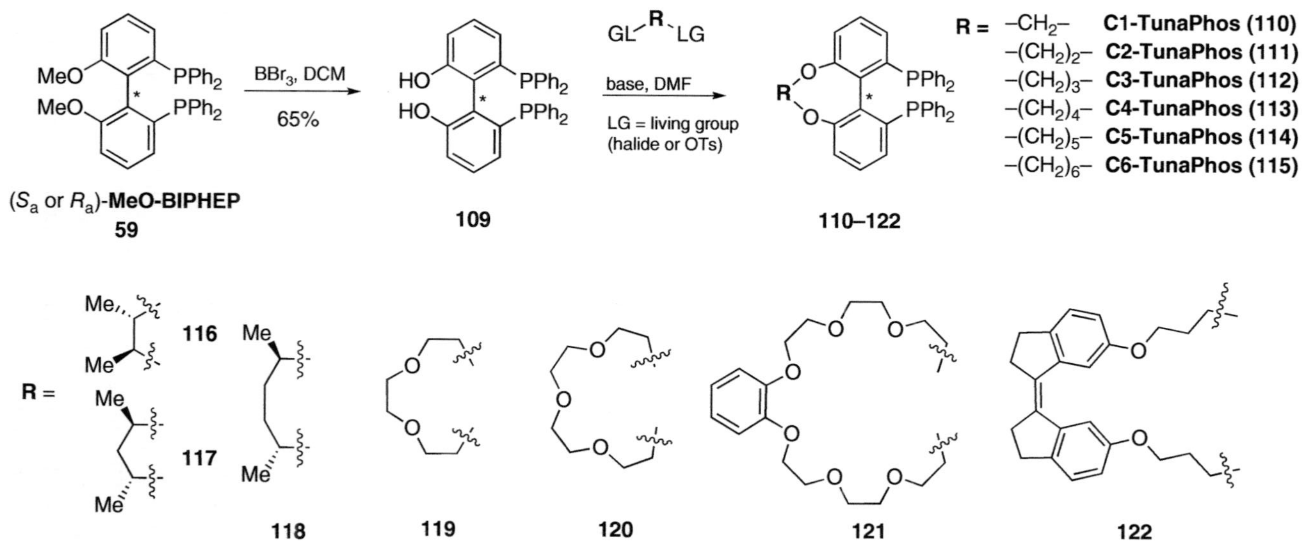

Scheme 12.11 Modification of MeO-BIPHEP. Source: Zhang et al. [100] / American Chemical Society.

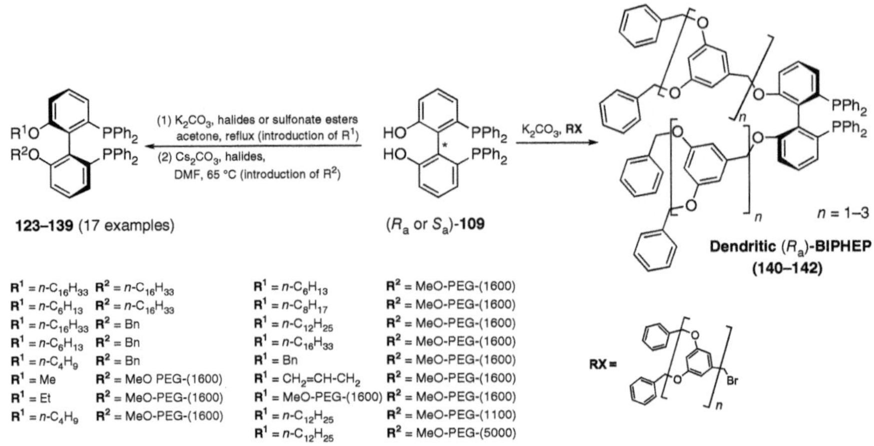

Scheme 12.12 Modification of HO-BIPHEP. Source: Wang et al. [104] / John Wiley & Sons.

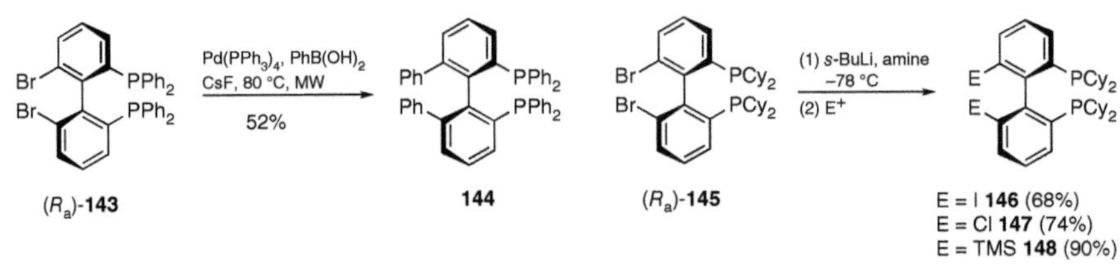

Scheme 12.13 Modification of positions 6 and 6′ of biphenyl core. MW, microwave irradiation. Source: Graff et al. [106] / John Wiley & Sons.

All examples discussed above indicate that organophosphorus compounds possessing axial chirality, and, sometimes, chirality at the phosphorus atom, are highly important in organic synthesis and asymmetric catalysis. Furthermore, axially chiral mono- and diphosphines in conjunction with transition metals are very potent tools in preparing chiral nonracemic organic compounds for applications in many industries. Therefore, it is evident that extensive modification of ligand precursors will occur to achieve the optimal performance of catalytic systems.

12.4 Chiral Aminophosphonic Acids and Their Analogs

Chiral aminophosphonic acids, the analogs of amino acids, are among the most important phosphorus-containing chiral building blocks. Their bioactive properties make them valuable intermediates targeting natural products, pharmaceuticals, and agrochemicals [115–120]. In general, aminophosphonic acids can be divided into two classes: the first results from a formal isosteric substitution of the planar and less bulky carboxylic group (CO_2H) by a tetrahedral phosphonic acid functionality (PO_3H_2); and the second possesses both carboxylic and phosphonic acid groups [120]. Further structural recognition is associated with the proximity

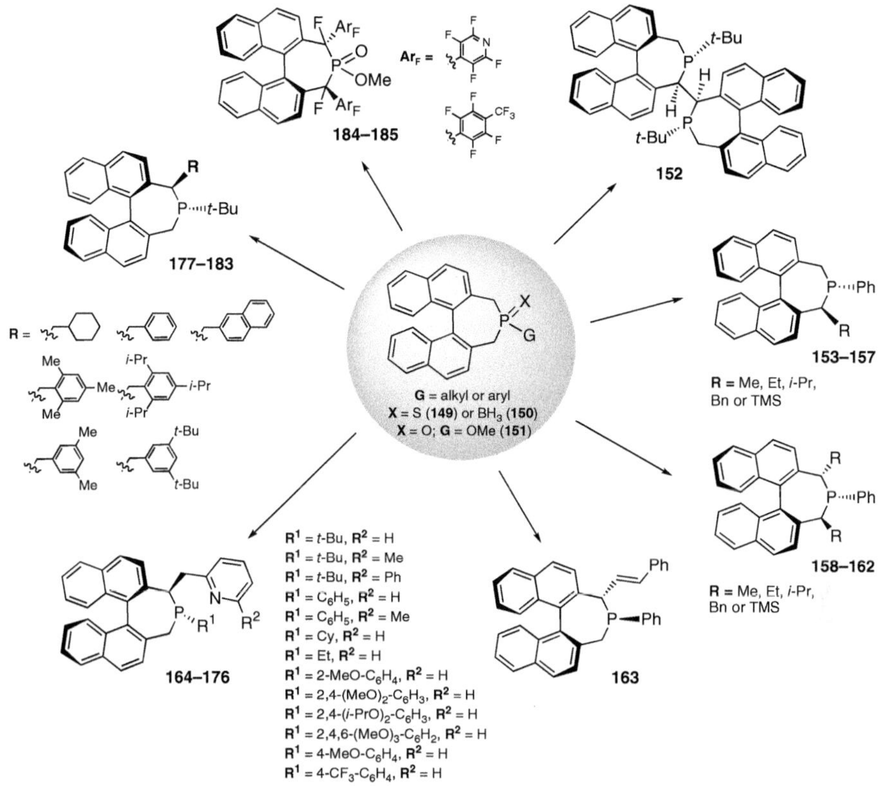

Figure 12.7 Modification of BINAPINE.

of the amino moiety to the phosphorus center and other functional groups. Nevertheless, aminophosphonic acids are a frequent choice as substrates in asymmetric synthesis [121].

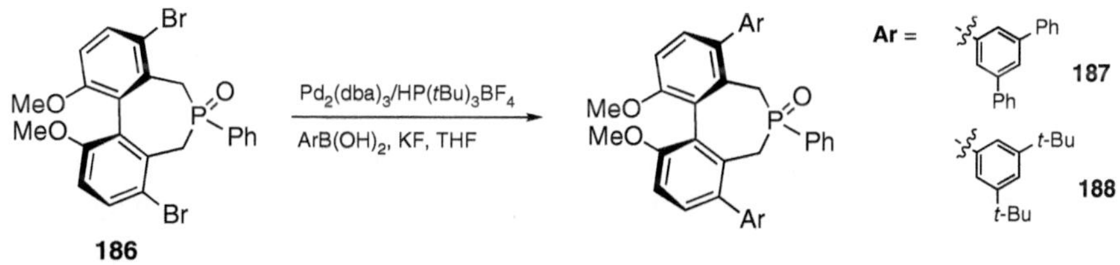

Scheme 12.14 Modification of biphenyl-derived phosphepines. Source: Ziegler et al. [114] / John Wiley & Sons.

In 2008, Carraro and Bonchio employed phosphonoalanine and phosphonovaline as ligands in preparation of pentamolybdate units $[(R^*PO_3)_2Mo_5O_{15}]^{2-}$ **189** and **190** used in the formation of polyoxometalate (POM)-based soft material with self-assembling properties (Figure 12.8) [122].

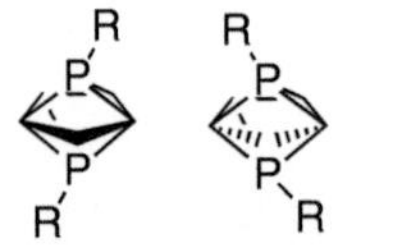

$[(RPO_3)_2Mo_5O_{15}]^{2-}$

(*S*,*S*) or (*R*,*R*) R = $CH_3CH(NH_3)^+$ **189**
(*S*,*S*) or (*R*,*R*) R = $CH_3CH(CH_2)CH(NH_3)^+$ **190**

Figure 12.8 "Envelope" representation of aminophosphonate pentamolybdate units.

Later, the same group built up the protected phosphonoalanine into the structure of Keggin-type polyoxotungstates, which were further used as catalysts for enantioselective oxidation of sulfides to sulfoxides [123]. More recently, the same authors have described the cell internalization properties of fluoresceine-functionalized polyoxotungstate scaffold (*R*,*R*)-FITC-PW_9 based on phosphoalanine [124]. The investigated fluorescent-tagged POM has shown a preferential nuclear accumulation in HEK cells, further applied in nano-theranostics.

L-Ala(P) **191** has been applied as a starting material in the three-step enzymatic transformation, leading to dehydrophos (DHP) **193**, an antibiotic produced naturally in *Streptomyces luridus* (Scheme 12.15) [125, 126].

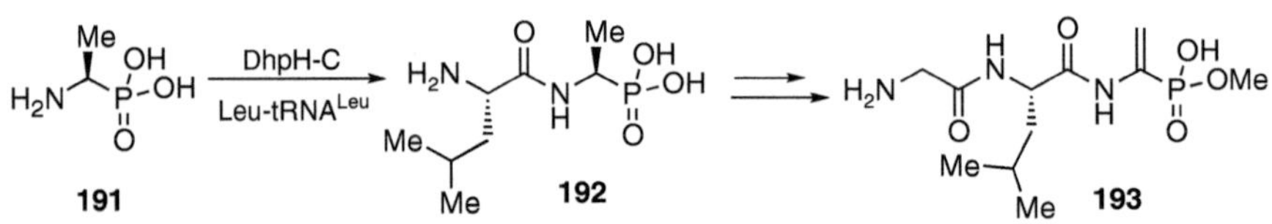

191 **192** **193**

Scheme 12.15 The enzymatic transformation of **191** leading to dehydrophos (DHP) **193**.

The formation of an amide bond in **192** was catalyzed by the C-terminal domain of DhpH enzyme (DhpH-C) using Leu-tRNA^{Leu} as a source of L-Leu. Later, the same authors analyzed both DhpH-C and tRNA to determine the catalytic residues and investigate the mechanism for the tRNA substrate recognition [126]. Recently, the same group made efforts to obtain antimicrobial phosphonodipeptides derived from **191** (analogs of alaphosphin) with other amino acids (Ile, Ala, Val, Met, norvaline, and norleucine) [127].

In 2016, Osborn and coworkers applied L-*R*-aminoethylphosphonic acid diethyl ester **196** to synthesize glucosides and galactosides that could exhibit activity in the bacterial cell via active carbohydrate transport mechanisms (Scheme 12.16) [128]. Compound **196** was linked to sugar molecules via ester, amide, or a carbamate bond. Among all tested derivatives, only carbamates **199** and **200** revealed inhibitory activity for the examined organisms. β-4-Nitrobenzylcarbonates **194** and **195**, obtained from acylated glucose and galactose derivatives, were reacted with **196** to form carbamates **197** and **198**, which were submitted to subsequent hydrolysis affording the target phosphonic acids **199** and **200**.

Ordónez and Cativiela applied a phosphophenylalanine scaffold to synthesize *R* and *S*,2,3,4-tetrahydroisoquinoline-3-phosphonic acids **203** (Scheme 12.17) [129].

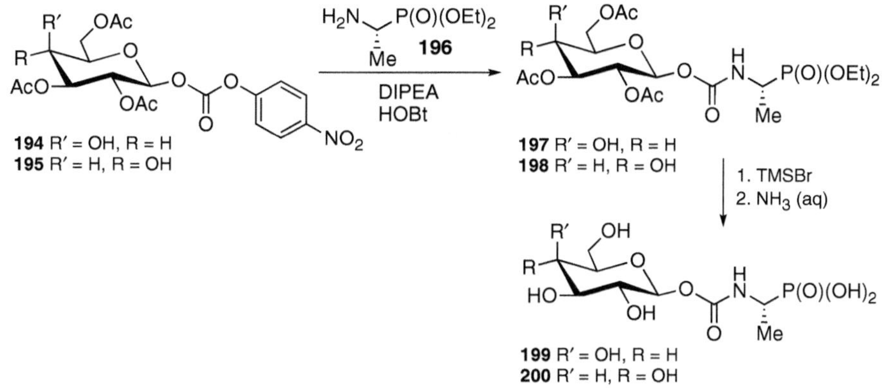

Scheme 12.16 The synthesis of glucosides and galactosides **199–200**. Source: Bovill et al. [128] / Elsevier.

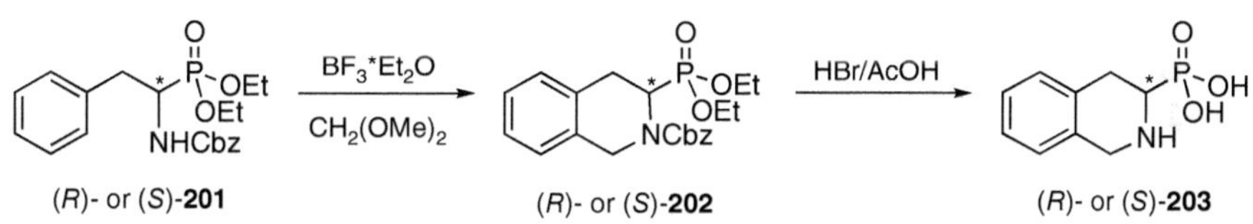

Scheme 12.17 The synthesis of *R* and *S* 1,2,3,4-tetrahydroisoquinoline-3-phosphonic acids **203**. Source: Viveros-Ceballos et al. [129] / John Wiley & Sons.

Protected (*R*)- and (*S*)-phosphophenylalanine diethyl esters **201**, obtained in the enantiomerically pure form via preparative high-performance liquid chromatography (HPLC) resolution with chiral stationary phase have provided both enantiomers of **203** in a two-step process including Pictet–Spengler cyclization and deprotection of amine.

Both enantiomers of phosphonoalanine **204** and their fluorinated analogs **205** were used to synthesize a library of α-aminophosphonic acids **207** and **208**, which are the allosteric inhibitors of the human farnesyl pyrophosphate synthase (FPPS) (Scheme 12.18) [130].

The authors have found that the aryl group in the benzylic substituent played a significant role in the inhibitor/enzyme complex formation. They have also demonstrated the impact of chirality on the binding mode of these compounds to the enzyme's active sites.

Herdewijn and coworkers employed 3-phosphono-L-alanine **210** to synthesize a series of deoxynucleotide derivatives **211–213** (Scheme 12.19) [131]. The configuration of the phosphorus atom was not determined. However, single peaks were reported in the ^{31}P nuclear magnetic resonance spectroscopy (NMR) spectra of **211–213**.

The synthesized compounds showed excellent properties as a mimic of the pyrophosphate (PPi) moiety of deoxyadenosine triphosphate as potential substrates for HIV-1 reverse transcriptase.

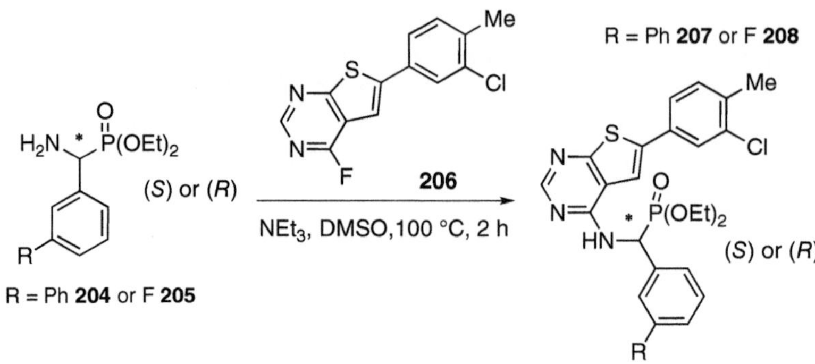

Scheme 12.18 The synthesis of a library of α-aminophosphonic acids **207** and **208**. DMSO, dimethyl sulfoxide. Source: Feng et al. [130] / American Chemical Society.

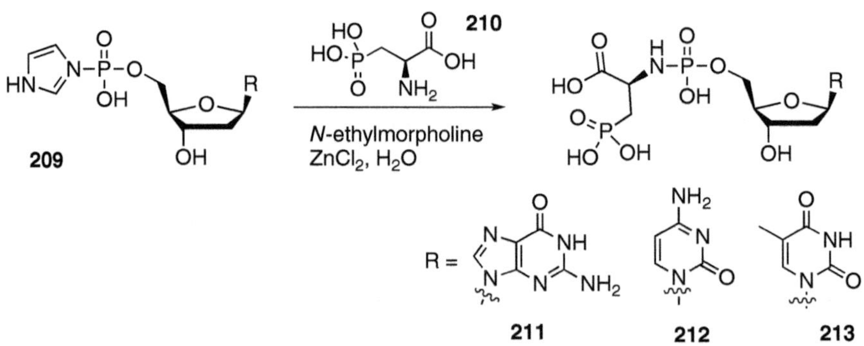

Scheme 12.19 The formation of deoxynucleotide derivatives **211–213**. Source: Yang et al. [131] / Royal Society of Chemistry.

Watanabe et al. performed the synthesis of 2-amino-4-{[3-(carboxymethyl) phenoxy](methoxy)phosphoryl}butanoic acid (GGsTop) **217** recognized as an irreversible inhibitor of glutamyl transpeptidase (GGT). Starting separately from both enantiomers of phosphonoglutamic acid **214**, the authors accessed four stereoisomers of **217** (Scheme 12.20) [132].

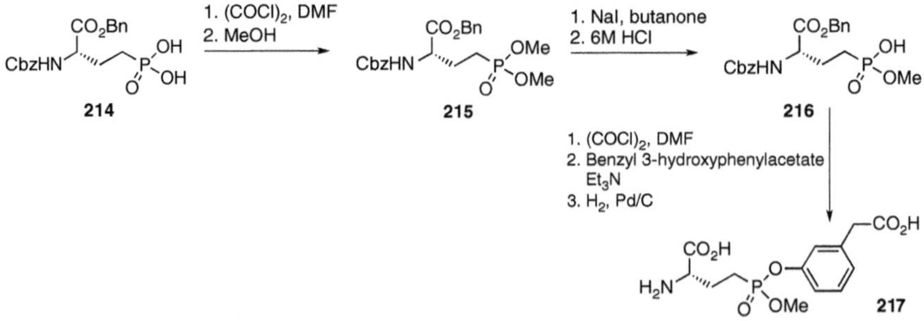

Scheme 12.20 The synthetic route to GGsTop **217**. Source: Watanabe et al. [132] / Elsevier.

The synthetic route to **217** started from the esterification of phosphonic acid **214** to form diester **215**. Next, deprotection of one ester group enabled the formation of monoester monoacid **216**, which after reaction with benzyl 3-hydroxyphenylacetate and benzyl hydrogenation afforded a diester as a diastereomeric mixture. Finally, the separation of isomers **217**, obtained from L- and D-phosphonoamino acid, was achieved via HPLC using a column with a chiral stationary phase. The studies of the inhibitory activity against human GGT have revealed that the enzyme is highly sensitive toward S_P isomer. In contrast, the recognition level concerning the configuration of the α-carbon atom is not relevant.

Lecorvey and coworkers intensively explored a library of bifunctional organocatalysts **222–223**, combining amino catalysis and aminophosphonic acid (Scheme 12.21) [133, 134].

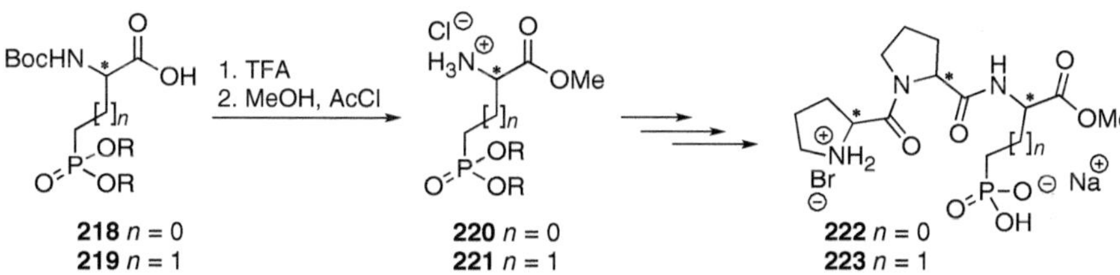

Scheme 12.21 The synthesis of aminophosphonic acids **222–223**.

These catalysts based on aspartic **218** or glutamic **219** acid scaffold have been evaluated in Michael reaction. In many cases, compounds **222** and **223** efficiently promoted an addition of propanal to *trans*-β-nitrostyrene, affording products in very high yields, high *syn*/*anti* selectivities, and enantiomeric excesses.

In 2013, Micura and coworkers reported the synthesis of the solid-supported phosphorylated 3′-aminoacyl-tRNASec mimic **228** (Scheme 12.22) [135].

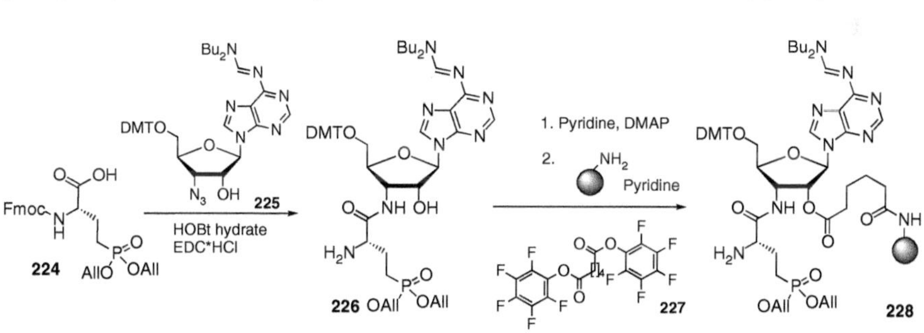

Scheme 12.22 The synthesis of tRNASec mimic **228**. Source: Rigger et al. [135] / John Wiley & Sons.

Part of the mimic was prepared using (*S*)-2-amino-4-(allyloxyphosphoryl)butanoic acid **224**. Solid-supported compound **228** was obtained in a three-step procedure, including the coupling of acid **224** with 3′-azidoadenosine **225**, followed by esterifying the 2′ position in **225** with adipic acid bis(pentafluorophenyl) ester **227**.

In the last step, a transamidation reaction using amino-functionalized support allowed immobilization on a solid phase.

Caroff et al. reported the synthesis of P2Y12 antagonists, which structure consisted of a piperazine carbamate, a central amino acid core, and a pyrimidine moiety (Scheme 12.23) [136]. Piperazine carbamate fragment **229** was coupled to the *N*-protected amino acid **230–232** using EDC-Cl/HOBt, HATU, or PyBOP as activating agents. After the deprotection step, amino moiety from aminophosphonic acids **233–235** was coupled to functionalized carboxylic acid **239,** leading to phosphonic acids **240–242** after final deprotection. The authors found compound **240** as the most effective among all tested phosphonic acids **240–242** in terms of a wide therapeutic window for rats and dogs. However, due to the low bioavailability of receptor antagonist **240**, it was necessary to transform it into a diester prodrug **243**. Therefore, both **240** and **243** were selected for further preclinical development.

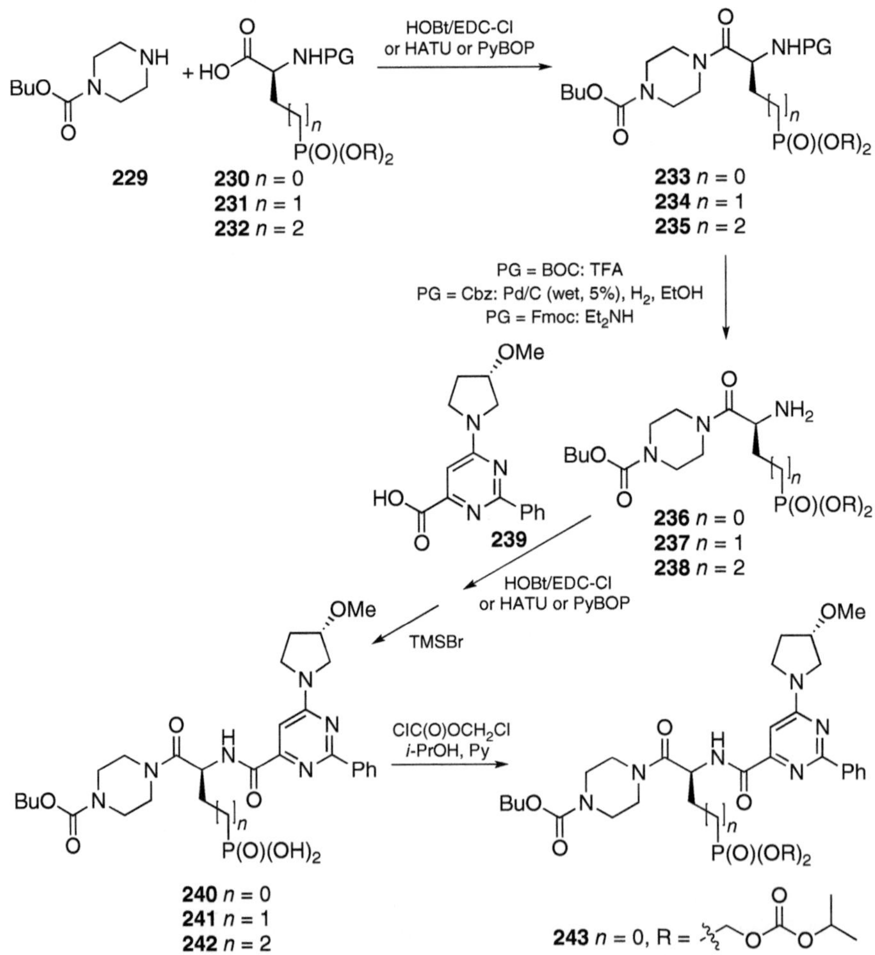

Scheme 12.23 The synthetic route to receptor antagonists **240–242** and prodrug **243**.

In nature, *O*-phosphorylated side chains of serine, threonine, or tyrosine are responsible for protein–protein interactions [137]. These interactions proceed through phosphorylation sites that interact with phosphotyrosine or phosphoserine/threonine motifs. However, their delivery to a cell is connected with many difficulties like transport via lipophilic cell membrane and dephosphorylation. Therefore, studies on the synthesis and application of mimics of naturally occurring *O*-phosphorylated amino acids are an important part of the chemistry of phosphonoamino acids.

Although the synthesis of α,α-difluoromethylenephosphoserine analogs was pioneered by Berkowitz et al. at the beginning of the 1990s, [138] these compounds are still important motifs in developing peptides and new potential drugs. Arrendale et al. reported the synthesis of phosphoserine phosphonamidite prodrug **246** starting from α,α-difluoromethylenephosphoserine precursor **244**. In the next few steps, **244** was successfully transformed into the inhibitor prodrug **246**, tested as a mixture of diastereomers against human DG75 leukemia cells (Scheme 12.24) [139].

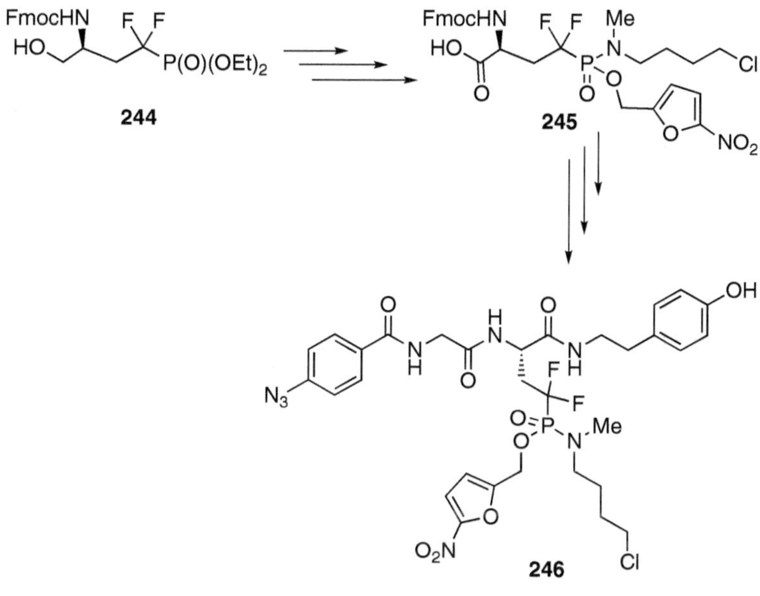

Scheme 12.24 The steps of the synthesis of the inhibitor prodrug **246**. Source: Arrendale et al. [139] / Springer Nature.

Later, Han and coworkers obtained a series of β-amino-γ,γ-difluoro-ω-phosphonoglutamic acid derivatives with the structure analogous to compound **247**. The aminophosphonic acid **247** was used to synthesize dipeptide **251**, achieved in the sequence of NH_2 protection with tosyl moiety and deprotection of both amino groups in **248**, leading to compound **249** (Scheme 12.25) [140]. The coupling of diamine **249** with Cbz-protected alanine **250** resulted in the formation of **251** with a 51% overall yield starting from **247**.

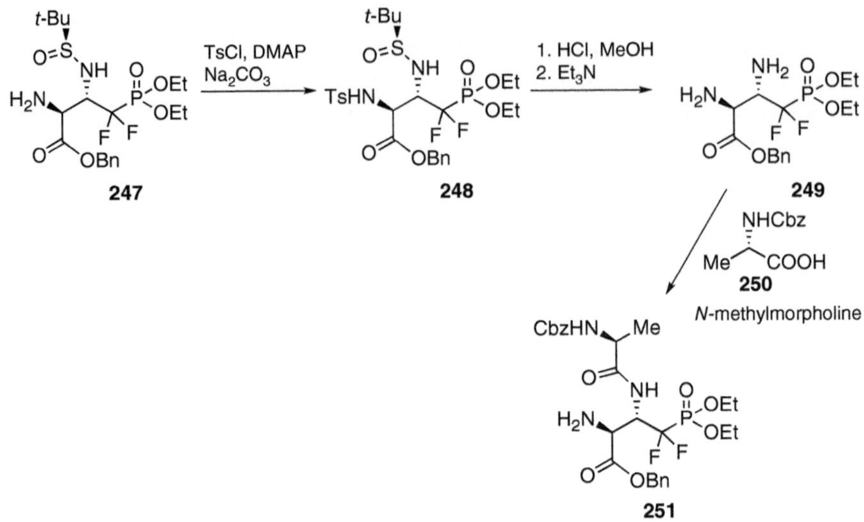

Scheme 12.25 The synthesis of dipeptide **251** starting from **247**. Source: Zhang et al. [140] / John Wiley & Sons.

Compared to difluorophosphonates, their monofluoro analogs have been-less explored due to higher synthetic challenges. Moreover, the stereoselective formation of the new stereogenic center at the α-carbon atom is even more challenging. In 2000, Otaka and coworkers showed the application of a diastereomeric mixture of monofluorinated pSer mimetic as a building block in synthesizing peptide H–Gly–FPab–Val–Pro–Met–Leu for evaluation in kinase and phosphatase inhibition [141]. Later, Macdonald and coworkers described the use of phosphoserine mimetic FPab for the synthesis of *N*-arylamide phosphonates, subtype-selective agonists, and antagonists of sphingosine L-phosphate receptors (S1P1–5) [142].

Hymel and Burke have prepared (2*S*,3*R*)-2-amino-3-methyl-4-phosphonobutanoic acid (Pmab) **252**, a phosphatase-stable analog of phosphothreonine (pThr) [143]. Besides this compound, access to a few of its analogs from **252** to **255** was also established. Aminophosphonic acids **252–255** were transformed into solid phase peptide synthesis (SPPS)-compatible derivatives and used as building blocks to synthesize a library of peptides by solid-phase coupling methods (Scheme 12.26). Peptidomimetic ligands **256–259** obtained by this route were found to enhance binding affinities with the polo-box domain (PBD) of polo-like kinase 1 (Plk1). They approached similar affinities to that of the corresponding pThr containing peptides. Recently, another group presented a protocol leading to the Fmoc-protected **245**, further subjected to SPPS to access two phosphono-peptide inhibitors [144]. The same authors also reported the synthesis of α,α-difluorinated phosphonate pSer/pThr mimetics via rhodium-catalyzed asymmetric hydrogenation of α-(acylamino)acrylates [145].

Recently, a highly enantioselective approach to L-phosphonomethylphenylalanine (L-Pmp) di-*t*-butyl ester **260** has been demonstrated [146]. The authors incorporated

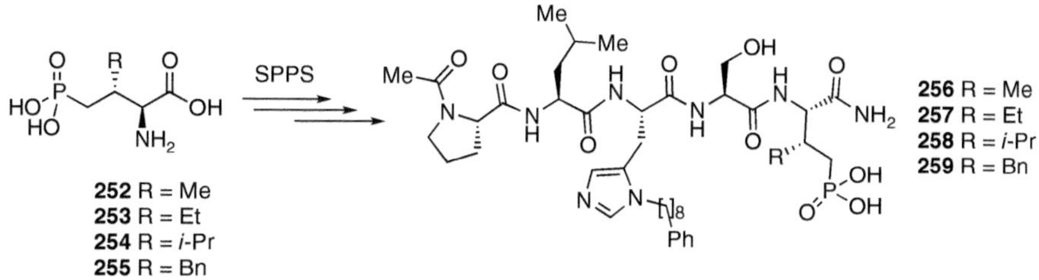

Scheme 12.26 The synthesis of a library of peptides **256–259**.

the L-Pmp residue into peptides and peptidomimetics using standard Fmoc protocol. Compound **260** was used to build a series of Pmp-containing Grb2-SH2 domain antagonists, which can be used as chemotherapeutic leads to treat breast cancer like compound **261** (Scheme 12.27).

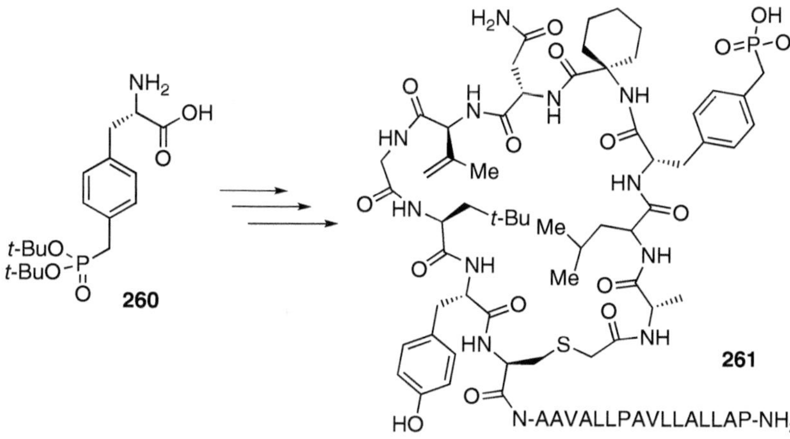

Scheme 12.27 Aminophosphonic acid **253** as a building block for peptide **261**.

Cairo and coworkers presented the first synthesis of Fmoc-L-BrPmp(OMe_2)-OH **262**, the brominated analog of Tyr(P) [147]. This molecule was incorporated into a short peptide sequence Asp–BrPmp–Leu **263** using solid-phase Fmoc protecting group strategy (Scheme 12.28). The comparison of the inhibitory activities of **262** and **263** against the protein tyrosine phosphatase CD4 revealed that tripeptide **263** was fourfold more potent than **262** alone.

Fluorinated phosphatyrosine analog *N*-Fmoc-F_2Pmp-OH **264** has become an important structural motif in the synthesis of peptidomimetics. Compound **264** was built into the structure of the PTP-1B inhibitor **265** (Scheme 12.29) [148]. The authors investigated structural factors crucial for the inhibitory activity, selectivity, and peptide-enzyme complex structure. Later, Meyer and Köhn employed **264** in the multi-step synthesis of the known peptide Ac–Asp–Ala–Asp–Glu–F_2Pmp–Leu–NH_2 **266** (Scheme 12.29) and its modified analog Ac–Ala–Ser–Gly–Ala–F_2Pmp–Ala–Gly–Gly–Ser–Ala–NH_2 [149]. The biological

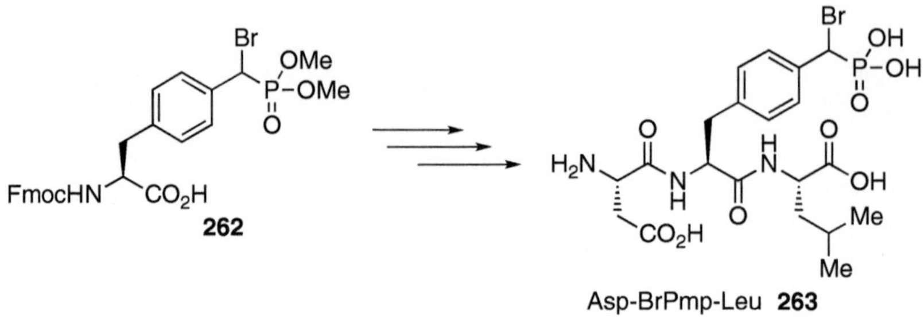

Scheme 12.28 Fmoc-L-BrPmp(OMe_2)-OH **262** as a building block for the tripeptide **263**.

activity (IC_{50} values) of both peptides toward PTP1B (phosphatase involved in diabetes and obesity) was shown.

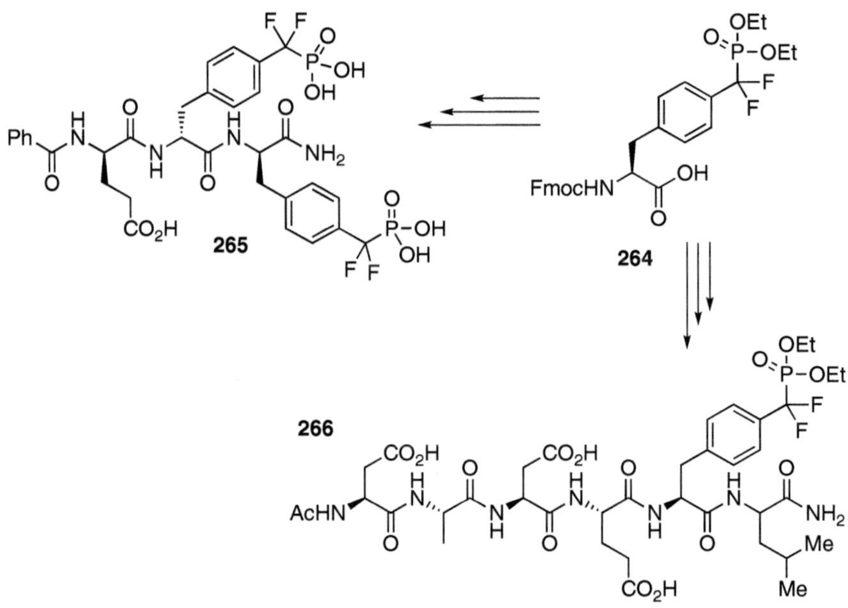

Scheme 12.29 *N*-Fmoc-F_2Pmp-OH **264** as a building block for peptides **265** and **266**. Source: Asante-Appiah et al. [148] / American Chemical Society.

Hiratake's group prepared a keto analog of fluorinated aspartic acid **267** as a potential inhibitor of γ-glutamylcysteine synthetase (γ-GCS) and glutamine synthetase (GS) [150]. They showed that deprotection of **267** induced the formation of imine **269** under the reaction conditions, and the yield of the product was 71% (from **267**) (Scheme 12.30).

Among non-classical examples of aminophosphonic acids, Brynaert-Marchand and coworkers developed the synthesis of valuable α-epoxy-δ-aminophosphonate **273** (Scheme 12.31) [151]. The synthetic route started from a cyclic diol **270**, which after the protection of C(4)-OH with *tert*-butyldimethylsilyl group and mesylation

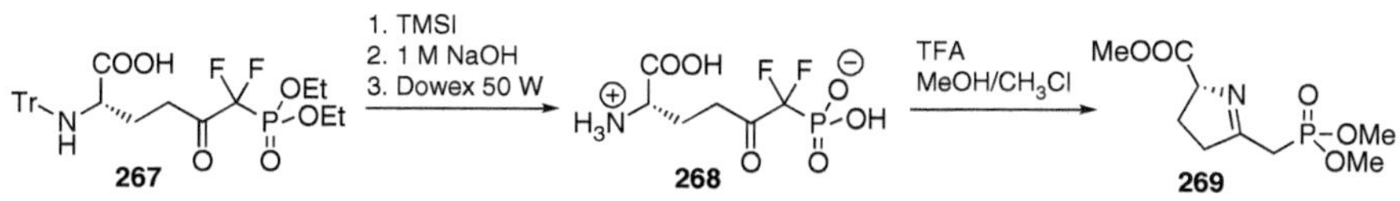

Scheme 12.30 Formation of imine **269** starting from **267**.

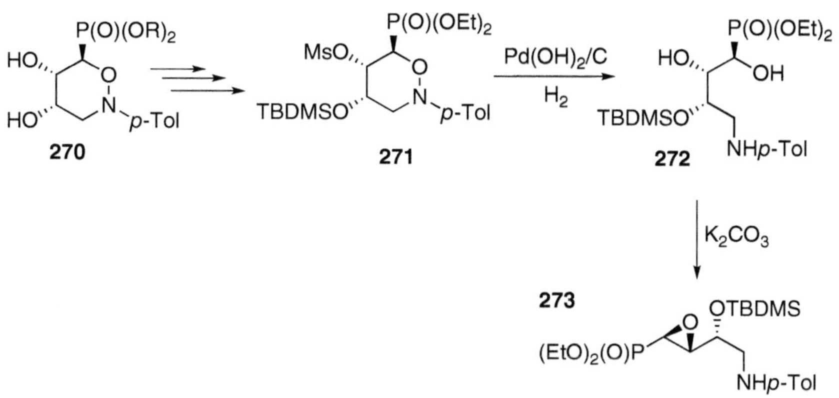

Scheme 12.31 The synthesis of α-epoxy-δ-aminophosphonate **273**. Source: Monbaliu et al. [151] / American Chemical Society.

of C(3)-OH afforded a cyclic phosphonate **271**. Ring-opening via catalytic hydrogenation resulted in functionalized δ-amino-phosphonate **272**, and intramolecular epoxidation induced by K_2CO_3 led to **273**.

Davis et al. presented another example of synthesizing less explored aminophosphonic acid **277** using 3-keto-phosphonoproline **274** as a building block (Scheme 12.32) [152].

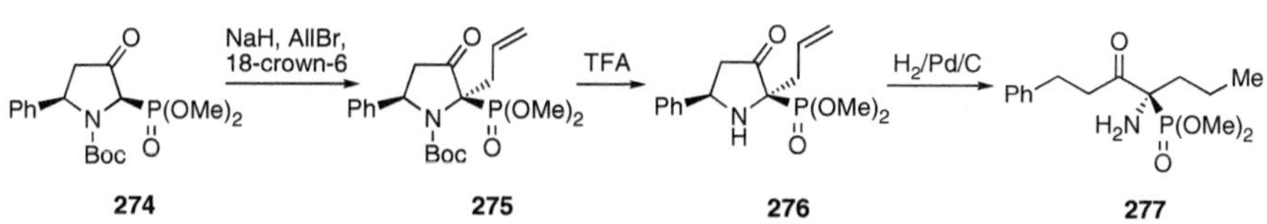

Scheme 12.32 The synthesis of amino-β-ketophosphonate (*R*)-**274** starting from **277**. Source: Davis et al. [152] / American Chemical Society.

Cyclic aminophosphonic acid (2*R*,5*R*)-**274** was alkylated by allyl bromide in the presence of 18-crown-6 to form (2*R*,5*R*)-**275** possessing a stereogenic quaternary carbon atom in 35% yield. Deprotection of the Boc group in **275** by trifluoroacetic acid (TFA) afforded pyrrolidine phosphonate (2*R*,5*R*)-**276** in 76% yield as a single diastereoisomer. A subsequent hydrogenation of **276** afforded the acyclic amino-β-ketophosphonate (*R*)-**277** in a 95% yield.

Luxen and coworkers elaborated the synthesis of two natural peptides: rhizocticin A and plumbemycin A, by condensing (*S*,*Z*)-2-amino-5-phosphono-3-pentenoic acid (APPA **278**) with the appropriate amino acids (Scheme 12.33) [153].

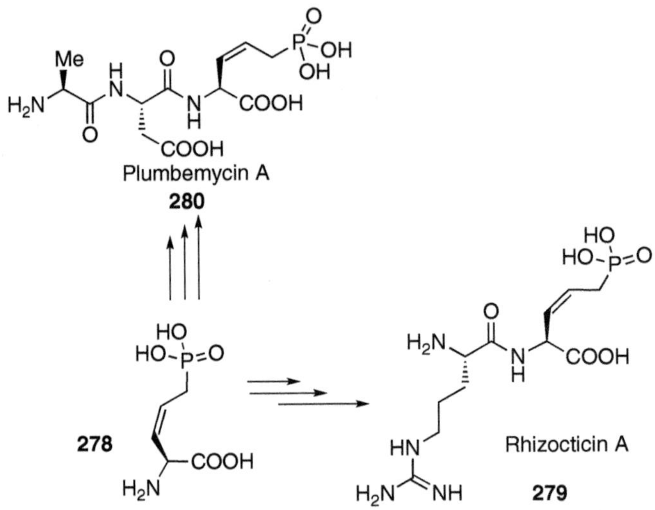

Scheme 12.33 The synthesis of two natural peptides: Rhizocticin A and Plumbemycin A.Source: Gahungu et al. [153] / Elsevier.

(*S*,*Z*)-APPA **278** is known as a threonine synthase inhibitor interfering with threonine biosynthesis in yeasts and bacteria. Compounds **279** and **280**, obtained by the authors, were evaluated as potential agents limiting the growth of a set of different bacteria and fungi.

The examples discussed above show that chiral aminophosphonic acids play a vital role in synthesizing compounds possessing biological activity. In most cases, chirality in aminophosphonic acids used to synthesize target molecules is placed at the α carbon atom. However, compounds possessing a chirality center at a different carbon atom were also discussed.

12.5 Stereogenic Organophosphorus Compounds with P—N and/or P—O Bonds

Compounds possessing phosphorus–nitrogen and/or phosphorus–oxygen bonds are frequently found in scientific literature due to their easy synthesis from readily available substrates. Regarding the potential arrangement of different substituents around the P(III) phosphorus atom, the different types of organophosphorus compounds formed are impressive (Figure 12.9).

Frequently used term "aminophosphines" is confused with those recently mentioned in the literature whose structures do not contain a direct P(III)—N bond, but still the phosphine moiety [P(III)—C bond system] contains one or more

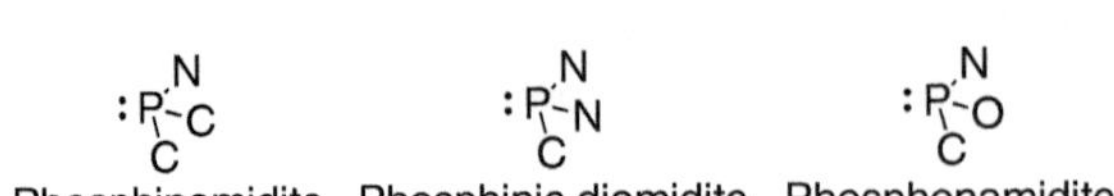

Figure 12.9 Organophosphorus compound with P—N and/or P—O bonds.

amine residues at the other end of the molecule. However, reports have addressed this confusion by giving names such as α and β or 2-aminophosphines in which the direct P(III)—N bond is absent, and only P—C—N and P—C—C—N bonds are present, respectively [154–157].

The synthesis of these compounds is mainly accomplished by a reaction between the phosphorus electrophile and nitrogen or oxygen nucleophile. When a chiral nucleophile is used, the formed organophosphorus compound may possess the newly formed chirality center at the phosphorus atom. For example, chiral cyclic triamides or diamidochloridites derived from *C*-chiral diamines were synthesized by Alexakis and coworkers (Scheme 12.34) [158].

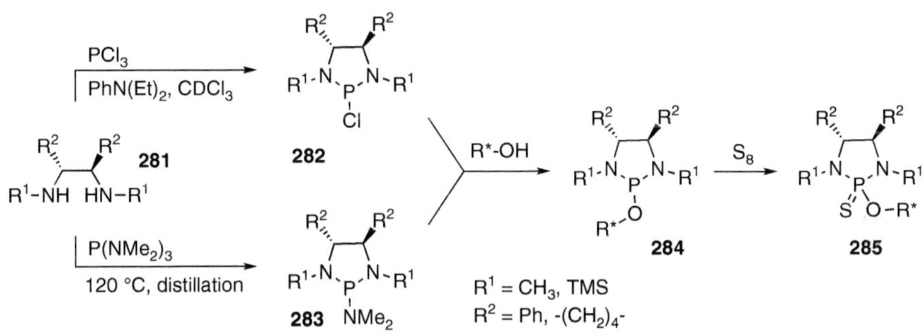

Scheme 12.34 Synthesis chiral cyclic triamides or diamidochloridites.

The major task here was the development of derivatization agents for enantiomer differentiation of carbinols using the NMR technique. *P*-Chiral phosphonous diamides or triamides (either **282** or **283**) were submitted to a reaction with a series of racemic carbinols and sulfur, which resulted in the formation of **285**. The formed products were analyzed using the NMR technique to determine the signal differentiation of the carbinolic proton.

Linclau and coworkers used a chiral phosphorous acid monoester diamide as a substrate in the Michaelis–Arbuzov reaction (Scheme 12.35) [159].

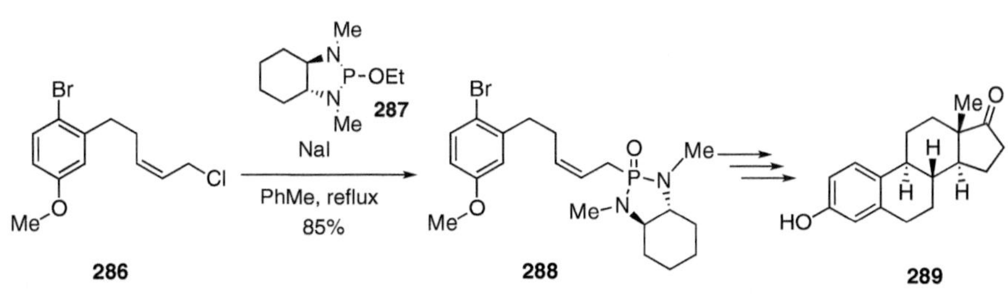

Scheme 12.35 Total synthesis of estrone **289**. Source: Foucher et al. [159] / American Chemical Society.

Compound **287** was used to increase the diastereoselectivity of the consecutive conjugate addition-alkylation of **288** in the synthesis of steroid **289**. Further experiments showed that triethyl phosphite maintained the same diastereoselection level.

Seltzer and coworkers designed and prepared a set of enantiopure phosphino-phosphonites with chirality located in a carbon scaffold of the phosphonite fragment (Scheme 12.36) [160].

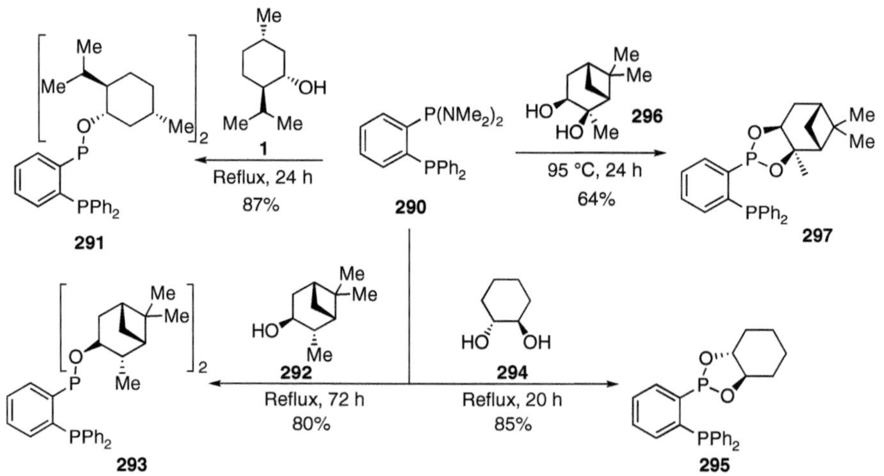

Scheme 12.36 Synthesis of enantiopure phosphino-phosphonites. Source: Kottsieper et al. [160] / Elsevier.

The substrate **290**, possessing two phosphorus functionalities (phosphino and phosphonito) of different reactivity, has been submitted to a reaction with a set of alcohols and diols affording the appropriate phosphonites via P–N/P–O nucleophilic substitution. In the case of α-pinene derived diol **296**, an additional chirality center is formed at the phosphorus atom due to the non-symmetric nature of diol.

Efficient methods for synthesizing *P*-stereogenic phosphinous acid amides are based on the reaction of chiral nucleophiles with racemic chlorophosphines. The reaction of chlorophosphine **298** with α-methylbenzylamine proceeds with chirality transfer from the chiral amine to the phosphorus atom (Scheme 12.37) [161].

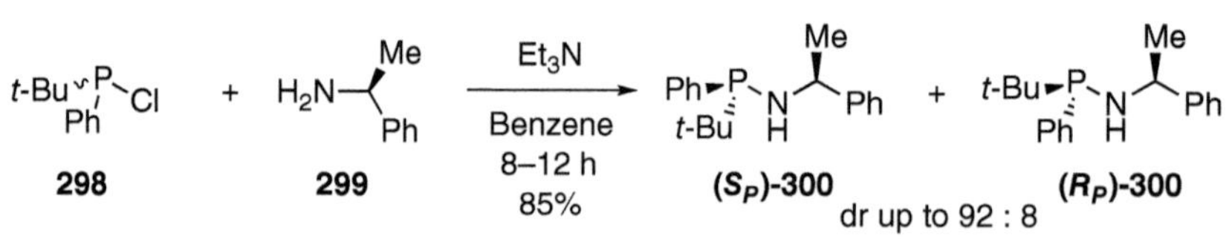

Scheme 12.37 Stereoselective formation of aminophosphines. Source: Kolodiazhnyi et al. [161] / Taylor & Francis.

The influence of various factors on the stereochemistry of the reaction of **298** with α-methylbenzylamine in the presence of the organic bases allowed some conclusions to be drawn about the mechanism of the reaction: (i) the stereoselectivity of a reaction depends on the type of solvent used; (ii) the stereoselectivity of the reaction depends on the type of organic base; (iii) an increase in the amount of chlorophosphine in the reaction mixture increases the stereoselectivity of the reaction, while an increase in the concentration of α-methylbenzylamine decreases the stereoselectivity; (iv) lowering the temperature decreases the stereoselectivity of the reaction; (v) the reaction of chlorophosphine **298** with (*S*)-α-methylbenzylamine afforded phosphinites with R_P configuration, while (*R*)-α-methylbenzylamine afforded the product with S_P configuration [162, 163].

Data indicate a significant influence of tertiary organic base on the nucleophilic substitution process at the trivalent phosphorus atom. The influence of temperature on stereoselectivity indicates that the reaction is not kinetically controlled [164, 165]. Furthermore, the influence of various factors on the stereochemical outcome of the reaction indicate a competition between the formation of a pentacoordinate intermediate and its deprotonation. Hence, the mechanism of nucleophilic substitution at trivalent phosphorus atom involves Berry pseudorotation and ligand exchange at the pentacoordinated phosphorus, leading to a thermodynamically more stable diastereoisomer.

P-Stereogenic phosphinous acid amides are a valuable starting material for synthesizing enantiomerically pure compounds (Scheme 12.38) [166]. Treatment of (R_P)-**301** with BH_3–THF complex allowed the formation of a stable crystalline borane analog (R_P)-**302**. The BH_3 group in (R_P)-**302** can be readily removed by diethylamine, which leads to the recovery of (R_P)-**301**. This protocol has been frequently used to protect sensitive P(III) compounds. Quaternization of (R_P)-**301** with MeI is a stereospecific process that led to the formation of phosphonium

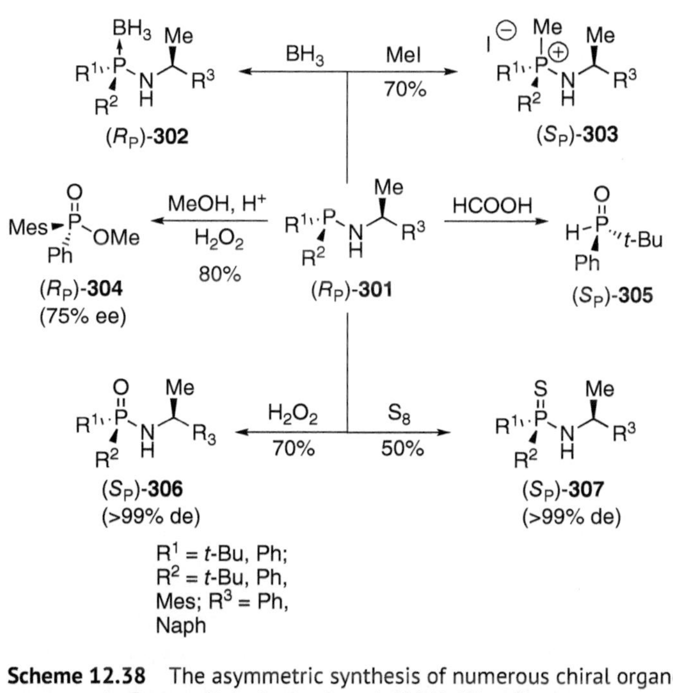

Scheme 12.38 The asymmetric synthesis of numerous chiral organophosphorus compounds. Source: Kolodiazhnyi et al. [166] / Elsevier.

salt (S_P)-**303**, preserving absolute configuration at the phosphorus atom [166]. Substitution of an amino group with a methoxyl group by refluxing in methanol in the presence of sulfuric acid allowed the formation of the appropriate phosphinic acid ester (R_P)-**304**. Moreover, acidolysis of compound (R_P)-**301** yielded the enantiomerically enriched SPO (S_P)-**305**. Oxidation of (R_P)-**301** with hydrogen peroxide was fully stereospecific and led to the appropriate amide (S_P)-**306**. Some phosphinous acid amides were submitted to a reaction with elemental sulfur, providing the corresponding thioanalogue (S_P)-**307** in a highly stereospecific manner [163–167].

Modifications of *P*-stereogenic amides can also occur at a side chain bonded to the phosphorus atom (Scheme 12.39) [167].

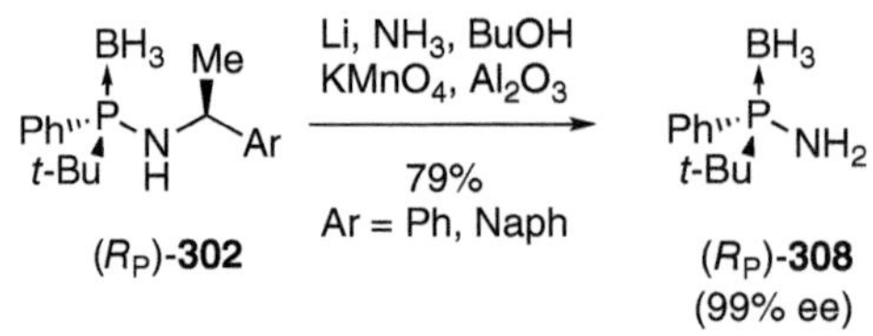

Scheme 12.39 Synthesis of primary aminophosphines. Source: Revés et al. [167] / John Wiley & Sons.

Hydrogenolysis of (R_P)-**302** by a solvated electron solution (lithium dissolved in liquid ammonia) afforded the desired product after cleavage of the benzyl fragment in amide moiety. The phenyl group attached to the phosphorus atom underwent a simultaneous Birch-type reduction. However, oxidation of the formed diene with $KMnO_4$ on alumina afforded the appropriate PAB amide (R_P)-**308**. Reductive cleavage of naphthylethyl fragment in (R_P)-**302** occurs prior to the undesired Birch-type reduction of the phenyl group. This behavior, which can be attributed to the difference in reduction potentials of the naphthalene and benzene rings, enabled the efficient synthesis of primary and secondary PAB amides (R_P)-**308**.

Chiral oxazolidinones have also been used to prepare several *P*-stereogenic organophosphorus compounds **311** with good yields and high enantioselectivities (Scheme 12.40) [168].

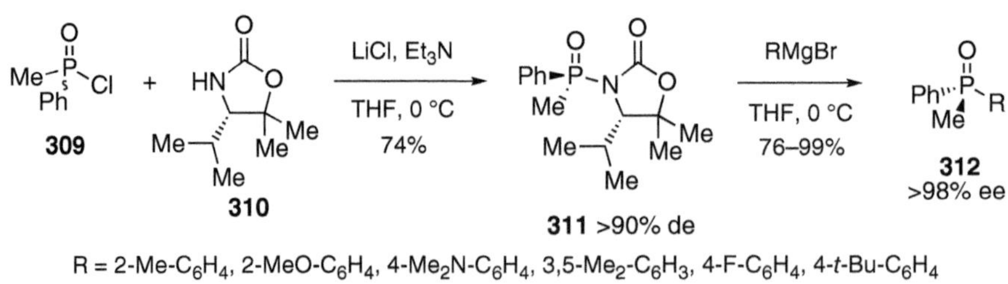

Scheme 12.40 Diastereoselective formation of *N*-phosphinoyl oxazolidinones. Source: Adams et al. [168] / American Chemical Society.

In the presence of a suitable base and a lithium salt, oxazolidinone **310** underwent a reaction with racemic phosphinic acid chloride **309** to form *N*-phosphinoyl oxazolidinone **311** in a good yield and high diastereoselectivity. The major diastereomer could be separated by column chromatography or crystallization. The diastereopure **311** was converted into various *P*-stereogenic phosphine oxides **312** through organometallic reagents.

Aminoalcohols are frequently used for chirality induction at the phosphorus atom among all known chiral compounds. These compounds are simple derivatives of naturally occurring substances derived from amino acids. Massive exploration of this research area was done in 1980–2000, and thus it is not mentioned here. However, a few new records have been reported since then.

A strategy involving chiral amine alcohols has several advantages over the methods above, using amine, diamine, or diol. The selection of naturally occurring chiral amine alcohols is much wider than both diols and diamines. Most of these compounds exist as single enantiomers, so there is no need to resolve them. Amino and hydroxyl functions vary in reactivity. Thus, a chemoselective substitution is possible, preventing double substitution with the same electrophile, which would cause a loss of chirality at the phosphorus center.

At the beginning of the 1990s, Jugé and coworkers developed a method to synthesize single *P*-epimers of oxazaphospholidine-boranes with (−)-ephedrine as a chirality-inducing agent [169]. This method is still in use and was successfully applied many times to synthesize *P*-stereogenic organophosphorus compounds in a nonracemic form. The same authors published several papers describing different *P*-chiral scaffolds built with both antipodes of ephedrine. Some of the constructed lattices were used directly in catalysis, and others were submitted to post-transformations (Scheme 12.41) [170, 171].

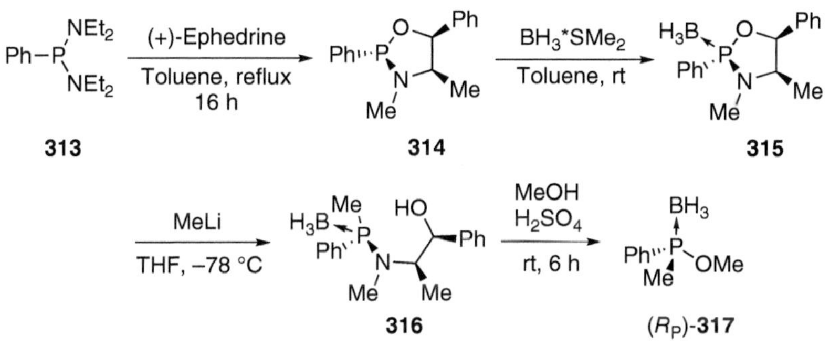

Scheme 12.41 Ephedrine-assisted formation of chiral phosphorous stereocenters.

A complete synthetic procedure for the transformation of chiral oxazaphospholidine-boranes to chiral phosphines and their derivatives was described by Jugé. However, using all intermediates as ligands for asymmetric catalysis is also possible. Boranes **315** and **316** were used as ligands in conjugate addition to 2-cyclohexenone catalyzed by copper(I) complexes.

Compound **315** is a convenient precursor for *P*-chiral molecules with two phosphorus atoms and *P*-chiral aminophosphane-phosphinites **319** (Scheme 12.42) [172–176] The synthesis of symmetrical compounds may consist of alcoholysis of

316, leading to a PAB ester (R_P)-**317**. Oxidative coupling of α-carbanion derived from (R_P)-**317** provides the appropriate dimeric product (R_P,R_P)-**320**. The synthesis of **319** proceeds via a nucleophilic opening of the oxazaphospholidine ring with an organometallic reagent, and the formed intermediate **318** is then subjected to a reaction with P(III) electrophile.

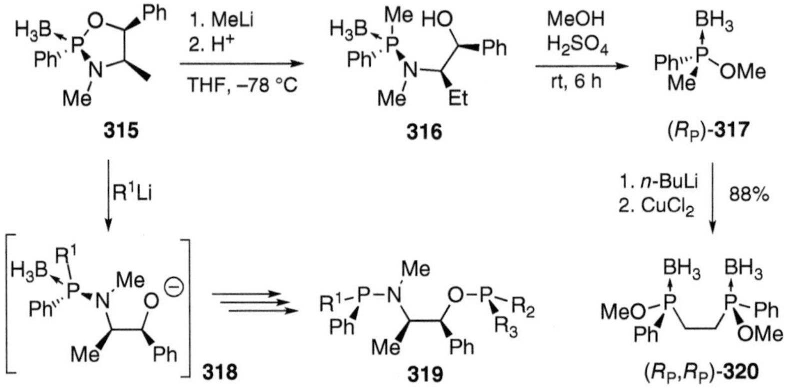

Scheme 12.42 Pseudoephedrine derived *bis*-P-chiral scaffolds.

After BH_3 removal, target compounds were applied to several asymmetric catalytic processes like asymmetric hydrogenation or catalytic reductive coupling of alkynes and aldehydes.

Other amino alcohols like pseudoephedrine [177] and (*1S,2R*)-*cis*-1-amino-2-indanol [178, 179] were also used as chiral auxiliaries in preparation of the corresponding oxazaphospholidines with similar efficiency. Oxazaphospholidine, built from *S*-(+)-prolinol, was used by Buono and coworkers to synthesize both enantiomers of **305** (Scheme 12.43) [180].

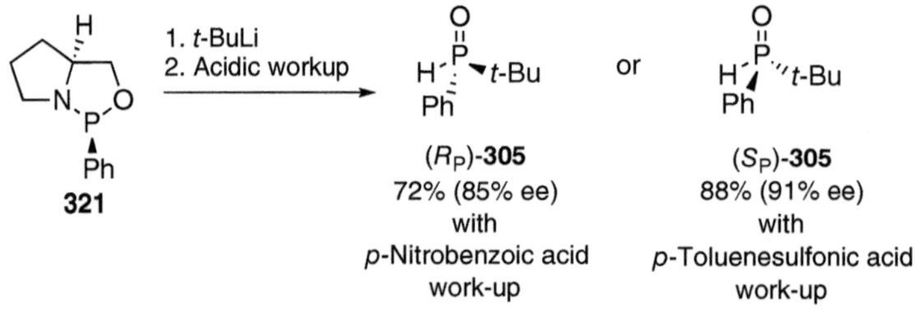

Scheme 12.43 Acid-dependent formation of secondary phosphine oxide antipodes. Source: Leyris et al. [180] / Elsevier.

Functionalization of the nitrogen atom in amino alcohols with the tosyl group was reported to facilitate the ring-opening process described by Han and coworkers (Scheme 12.44) [181].

Oxazaphospholidine formation from **322** and phosphorus electrophile was followed by hydrolysis where P—N and not P—O bond was cleaved in a highly

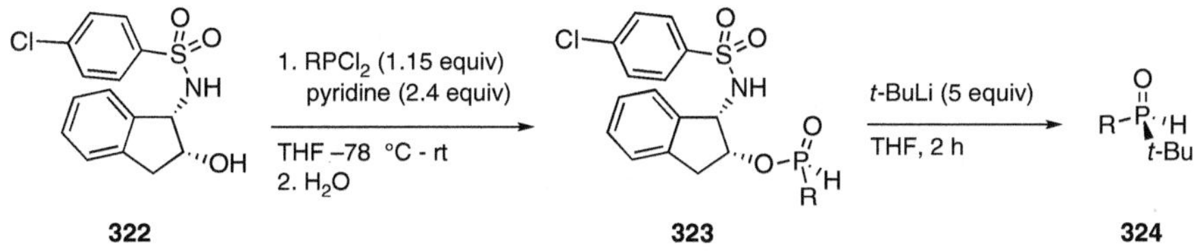

Scheme 12.44 1-Amino-2-indanol assisted synthesis of chiral *sec*-Phosphine oxides. Source: Han et al. [181] / American Chemical Society.

stereoselective manner, leading to the corresponding P-stereogenic phosphinate **323**. This intermediate served as a substrate in nucleophilic P—O bond displacement with organometallic reagent leading to the target SPO **324** in an enantiomerically enriched form.

Oxazaphospholidine fragment was also successfully used as a directing group in deprotonation-*ortho*-lithation of ferrocene (Scheme 12.45) [182].

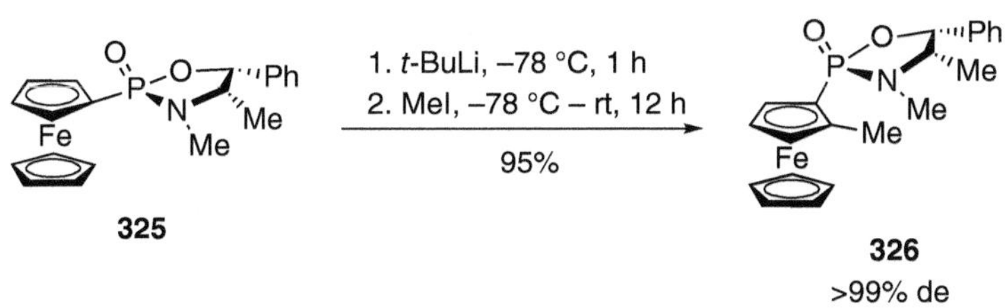

Scheme 12.45 Diastereoselective *ortho*-lithiation of ferrocene **325**. Source: Vinci et al. [182] / American Chemical Society.

In the presence of a strong base, **325** underwent *ortho*-deprotonation of ferrocene, affording the oxazaphospholidine-stabilized carbanion. In a subsequent step, adding an electrophile provided the disubstituted ferrocene **326** as a product. The deprotonation step created a new chirality element in the molecule – a planar chirality. The chiral oxazaphospholidine fragment in starting ferrocene directs not only the formation of carbanion but also the stereospecific creation of a new form of chirality.

12.6 Summary

The examples discussed in this chapter represent the most recent developments in preparing chiral organophosphorus compounds and their applications in stereoselective synthesis. This includes compounds possessing a chirality center located either in the carbon skeleton or the phosphorus atom. Irrespectively of the nature of chirality, chiral organophosphorus compounds are the source of strong development in many branches of organic synthesis, serving as substrates for the synthesis of biologically active compounds, pharmaceuticals, or ligands for asymmetric catalysis.

References

1 Knowles, W.S. (2004). *Asymmetric Catalysis on Industrial Scale*. Wiley-Blackwell.

2 Knowles, W.S., Sabacky, M.J., and Vineyard, B.D. (1972). Catalytic asymmetric hydrogenation. *J. Chem. Soc., Chem. Commun.* 10–11. https://doi.org/10.1039/C39720000010.

3 Munro, H.D. and Horner, L. (1970). Organophosphorus compounds-LXVI: the influence of some solvents on the racemization of (+)-methyl-*n*-propylphenylphosphine. *Tetrahedron* 26: 4621–4655. https://doi.org/10.1016/S0040-4020(01)93108-4.

4 Ye, X., Peng, L., Bao, X. et al. (2021). Recent developments in highly efficient construction of *P*-stereogenic centers. *Green Synth. Catal.* 2: 6–18. https://doi.org/10.1016/j.gresc.2020.12.002.

5 Dutartre, M., Bayardon, J., and Jugé, S. (2016). Applications and stereoselective syntheses of *P*-chirogenic phosphorus compounds. *Chem. Soc. Rev.* 45: 5771–5794. https://doi.org/10.1039/C6CS00031B.

6 Harveya, J.S. and Gouverneur, V. (2010). Catalytic enantioselective synthesis of *P*-stereogenic compounds. *Chem. Commun.* 46: 7477–7485. https://doi.org/10.1039/C0CC01939A.

7 Börner, A. (ed.) (2008). *Phosphorus Ligands in Asymmetric Catalysis*. Weinheim: Wiley-VCH.

8 Grabulosa, A., Granell, J., and Mulle, G. (2007). Preparation of optically pure *P*-stereogenic trivalent phosphorus compounds. *Coord. Chem. Rev.* 251: 25–90. https://doi.org/10.1016/j.ccr.2006.05.009.

9 Woźniak, L.A. and Okruszek, A. (2003). The stereospecific synthesis of *P*-chiral biophosphates and their analogues by the Stec reaction. *Chem. Soc. Rev.* 32: 158–169. https://doi.org/10.1039/B207207F.

10 Mislow, K., Donohue, J., Mandel, N. et al. (1971). Retention stereochemistry in a Grignard displacement reaction at chiral phosphorus. Absolute configuration of *O*-menthyl *S*-methyl phenylphosphonothiolate. *J. Am. Chem. Soc.* 93: 3792–3793. https://doi.org/10.1021/ja00744a053.

11 Farnham, W.B., Murray, R.K. Jr.,, and Mislow, K.J. (1970). Stereospecific alkylation of menthyl phenylphosphinate. *J. Am. Chem. Soc.* 92: 5809–5810. https://doi.org/10.1021/ja00722a083.

12 Lewis, R.A. and Mislow, K. (1969). Direct configurational correlation of trialkyl- and triarylphosphine oxides. *J. Am. Chem. Soc.* 91: 7009–7012. https://doi.org/10.1021/ja01053a020.

13 Korpiun, O., Lewis, R.A., Chickos, J., and at al. (1968). Synthesis and absolute configuration of optically active phosphine oxides and phosphinates. *J. Am. Chem. Soc.* 90: 4842–4846. https://doi.org/10.1021/ja01020a017.

14 Korpiun, O. and Mislow, K. (1967). New route to the preparation and configurational correlation of optically active phosphine oxides. *J. Am. Chem. Soc.* 89: 4784–4786. https://doi.org/10.1021/ja00994a037.

15 Emmick, T.L. and Letsinger, R.L. (1968). Unsymmetrical secondary phosphine oxides. Synthetic, isotopic exchange, and stereochemical studies. *J. Am. Chem. Soc.* 90: 3459–3465. https://doi.org/10.1021/ja01015a030.

16 Reiff, L.P. and Aaron, H.S. (1970). Stereospecific synthesis and reactions of optically active isopropyl methylphosphinate. *J. Am. Chem. Soc.* 92: 5275–5276. https://doi.org/10.1021/ja00720a077.

17 Nudelman, A. and Cram, D.J. (1971). Stereochemical reaction cycle with chiral phosphorus. *J. Am. Chem. Soc.* 36: 335–337. https://doi.org/10.1021/jo00801a020.

18 Gbubele, J.D. and Olszewski, T. (2021). Asymmetric synthesis of organophosphorus compounds using H–P reagents derived from chiral alcohols. *Org. Biomol. Chem.* 19: 2823–2846. https://doi.org/10.1039/d1ob00124h.

19 Lemouzy, S., Giordano, L., Herault, D. et al. (2020). Introducing chirality at phosphorus atoms: an update on the recent synthetic strategies for the preparation of optically pure *P*-stereogenic molecules. *Eur. J. Org. Chem.* 2020: 3351–3366. https://doi.org/10.1002/ejoc.202000406.

20 Chen, T. and Han, L.-B. (2015). Optically active *H*-phosphinates and their stereospecific transformations into optically active *P*-stereogenic organophosphoryl compounds. *Synlett* 26: 1153–1163. https://doi.org/10.1055/s-0034-1379996.

21 Wang, W.M., Liu, L.-J., Zhao, C.-Q. et al. (2015). Diastereoselective hydrolysis of asymmetric P–Cl species and synthesis of optically pure (R_P)-(−)-menthyl *H*-phenylphosphinate. *Eur. J. Org. Chem.* 2015: 2342–2345. https://doi.org/10.1002/ejoc.201500126.

22 Gatineau, D., Giordano, L., and Buono, G. (2011). Bulky, optically active *P*-stereogenic phosphine–boranes from pure *H*-menthylphosphinates. *J. Am. Chem. Soc.* 133: 10728–10731. https://doi.org/10.1021/ja2034816.

23 (a) Berger, O. and Montchamp, J.-L. (2016). General synthesis of *P*-stereogenic compounds: the menthyl phosphinate approach. *Org. Biomol. Chem.* 14: 7552–7562. https://doi.org/10.1039/C6OB01413E. (b) Berger, O. and Montchamp, J.-L. (2013). A general strategy for the synthesis of *P*-stereogenic compounds. *Angew. Chem. Int. Ed.* 52: 11377–11380. https://doi.org/10.1002/anie.201306628.

24 Berger, O., Gavara, L., and Montchamp, J.-L. (2012). Chemistry of the versatile (hydroxymethyl)phosphinyl P(O)CH_2OH functional group. *Org. Lett.* 14 (13): 3404–3407. https://doi.org/10.1021/ol3013793.

25 Benschop, H.P. and Van der Berg, G.R. (1970). Stereospecific inclusion in cycloamyloses: partial resolution of isopropyl methylphosphinate and related compounds. *J. Chem. Soc. D* 21: 1431–1432. https://doi.org/10.1039/C29700001431.

26 Sun, Y.-M., Xin, N., Xu, Z.-Y. et al. (2014). Addition of optically pure *H*-phosphinate to ketones: selectivity, stereochemistry and mechanism. *Org. Biomol. Chem.* 12: 9457–9465. https://doi.org/10.1039/C4OB01574F.

27 Chen, T., Zhou, Y., Guo, C. et al. (2013). Selective addition of P(O)–H bonds to alkynes catalyzed by transition metals immobilized on polystyrene-bound triphenylphosphine. *Chem. Lett.* 42 (9): 1065–1067. https://doi.org/10.1246/cl.130374.

28 Han, L.-B. and Zhao, C.-Q. (2005). Stereospecific addition of H–P bond to alkenes: a simple method for the preparation of (R_P)-phenylphosphinates. *J. Org. Chem.* 70: 10121–10123. https://doi.org/10.1021/jo051582b.

29 Zhou, Y., Yang, J., Chen, T. et al. (2014). Stereospecific aerobic oxidative dehydrocoupling of P(O)–H bonds with amines catalyzed by copper. *Bull. Chem. Soc. Jpn.* 87: 400–406. https://doi.org/10.1246/bcsj.20130310.

30 Xiong, B., Zhou, Y., Zhao, C. et al. (2013). Systematic study for the stereochemistry of the Atherton–Todd reaction. *Tetrahedron* 69 (54): 9373–9380. https://doi.org/10.1016/j.tet.2013.09.001.

31 Wang, G., Shen, R., Xu, Q. et al. (2010). Stereospecific coupling of *H*-phosphinates and secondary phosphine oxides with amines and alcohols: a general method for the preparation of optically active organophosphorus acid derivatives. *J. Org. Chem.* 75 (11): 3890–3892. https://doi.org/10.1021/jo100473s.

32 Yang, M., Sun, Y.-M., Hou, Q.G. et al. (2012). (R_P)-5-Methyl-2-(propan-2-yl) cyclo-hexyl phen-yl{phen-yl[(1-phenyl-eth-yl)amino]-meth-yl}phosphinate. *Acta Crystallogr., Sec. E: Cryst. Commun.* 68: 2328–2328. https://doi.org/10.1107/S1600536812028243.

33 Wang, W.-M., Liu, L.-J., Yao, L. et al. (2016). Stereospecific preparations of *P*-stereogenic phosphonothioates and phosphonoselenoates. *J. Org. Chem.* 81: 6843–6847. https://doi.org/10.1021/acs.joc.6b01192.

34 Rassukana, Y.V., Stanko, O.V., and Onys'ko, P.P. (2019). Enantiomeric *O,O*-dimenthyl α-iminotrifluoroethylphosphonates: novel chiral building blocks in asymmetric synthesis of α-trifluoromethylated α-aminophosphonic acid derivatives. *J. Fluorine Chem.* 219: 123–128. https://doi.org/10.1016/j.jfluchem.2019.01.007.

35 Kolodiazhnyi, O.I., Grishkun, E.V., Sheiko, S. et al. (1998). Chiral symmetric phosphoric acid esters as sources of optically active organophosphorus compounds. *Tetrahedron: Asymmetry* 9: 1645–1649. https://doi.org/10.1016/S0957-4166(98)00166-9.

36 Nesterov, V.V. and Kolodiazhnyi, O.I. (2005). Enantioselective synthesis of α-hydroxyalkylphosphonates. *Russ. J. Gen. Chem.* 75: 1161–1162. https://doi.org/10.1007/s11176-005-0386-8.

37 Wang, J.-P., Nie, S.-Z., Zhou, Z.-Y. et al. (2016). Preparation of optically pure tertiary phosphine oxides via the addition of *P*-stereogenic secondary phosphine oxide to activated alkenes. *J. Org. Chem.* 81: 7644–7653. https://doi.org/10.1021/acs.joc.6b01371.

38 Yan, B.-X., Zheng, H.-X., Ye, J.-J. et al. (2020). Stereoselective thermal addition of R_P and/or S_P-menthyl-deriving secondary phosphine oxides to active alkynes, the mechanism, stereochemistry and generation of *P,C*-chirogenic bis-phosphine derivatives. *Asian J. Org. Chem.* 9: 566–570. https://doi.org/10.1002/ajoc.202000079.

39 Zhang, H., Sun, Y.-M., Zhao, Y. et al. (2015). One-pot process that efficiently generates single stereoisomers of 1,3-bisphosphinylpropanes having five chiral centers. *Org. Lett.* 17: 142–145. https://doi.org/10.1021/ol503371r.

40 Xie, P., Guo, L., Xu, L. et al. (2016). Asymmetric P–C bond formation: diastereoselective synthesis of adjacent *P,C*-stereogenic allylic phosphorus compounds. *Chem. Asian J.* 11: 1353–1356. https://doi.org/10.1002/asia.201600108.

41 Yang, M., Xu, H., Zhou, Z.-Y. et al. (2016). Asymmetric induction in the addition of enantiomerically pure *H*-phosphinate to chiral aldimines: diastereoselective generation of α-amino phosphinates with *P,C*-stereogenic centers. *Tetrahedron: Asymmetry* 27: 815–822. https://doi.org/10.1016/j.tetasy.2016.06.022.

42 Zhou, Z.-Y., Zhang, H., Yao, L. et al. (2016). Double asymmetric induction during the addition of (R_P)-menthyl phenyl phosphine oxide to chiral aldimines. *Chirality* 28: 132–135. https://doi.org/10.1002/chir.22549.

43 Qiu, M.-R., Zheng, H.-X., Ye, J.-J. et al. (2020). Stereoselective preparation of P, axial-stereogenic allenyl bisphosphine oxides via chirality-transfer. *Org. Biomol. Chem.* 18: 3017–3021. https://doi.org/10.1039/D0OB00390E.

44 Zhang, H., Sun, Y.-M., Yao, L. et al. (2014). Stereogenic phosphorus-induced diastereoselective formation of chiral carbon during nucleophilic addition of chiral H–P species to aldehydes or ketones. *Chem. Asian J.* 9: 1329–1333. https://doi.org/10.1002/asia.201301650.

45 Pietrusiewicz, K.M., Koprowski, M., Drzazga, Z. et al. (2020). Efficient oxidative resolution of 1-phenylphosphol-2-ene and Diels–Alder synthesis of enantiopure bicyclic and tricyclic *P*-stereogenic C–P heterocycle. *Symmetry* 12: 346. https://doi.org/10.3390/sym12030346.

46 Dziuba, K., Lubańska, M., and Pietrusiewicz, K.M. (2020). Enantiodivergent synthesis of both PAMPO enantiomers using *l*-menthyl chloroacetate and stereomutation at P in classical quaternisation reactions. *Synthesis* 52: 909–916. https://doi.org/10.1055/s-0039-1691531.

47 Zagidullin, A.A., Ganushevich, Y.S., Miluykov, V.A. et al. (2020). Synthesis and asymmetric [4+2] cycloaddition reaction of 3,4,5-triphenyl-1-((1*R*,2*S*,5*R*)-menthyl)oxymethyl-1,2-diphosphole. *J. Organomet. Chem.* 914: 121218. https://doi.org/10.1016/j.jorganchem.2020.121218.

48 Ye, J.-J., Yan, B.-X., Wang, J.-P. et al. (2020). The construction of three C–P bonds of P-stereogenic tertiary phosphines containing (*l*)-menthyl. *Organ. Chem. Front.* 7: 2063–2068. https://doi.org/10.1039/D0QO00453G.

49 Imamoto, T., Oshiki, T., Onozawa, T. et al. (1990). Synthesis and reactions of phosphine-boranes. Synthesis of new bidentate ligands with homochiral phosphine centers via optically pure phosphine-boranes. *J. Am. Chem. Soc.* 112: 5244–5252. https://doi.org/10.1021/ja00169a036.

50 Imamoto, T., Kusumoto, T., Suzuki, N. et al. (1985). Phosphine oxides and lithium aluminum hydride-sodium borohydride-cerium(III) chloride: synthesis and reactions of phosphine-boranes. *J. Am. Chem. Soc.* 107: 5301–5303. https://doi.org/10.1021/ja00304a061.

51 Xu, Q., Zhao, C.-Q., and Han, L.-B. (2008). Stereospecific nucleophilic substitution of optically pure H-phosphinates: a general way for the preparation of chiral P-stereogenic phosphine oxides. *J. Am. Chem. Soc.* 130: 12648–12655. https://doi.org/10.1021/ja804412k.

52 Moradeda, D., Gatineau, D., Martin, D. et al. (2008). A simple route to chiralphosphinous acid–boranes. *Chem. Commun.* 3031–3033. https://doi.org/10.1039/B802817F.

53 Włodarczyk, A., Kozioł, A.E., and Stankevič, M. (2018). L-menthol-assisted synthesis of *P*-stereogenic phosphinous acid amides and phosphine-boranes. *Eur. J. Org. Chem.* 13: 1589–1600. https://doi.org/10.1002/ejoc.201800065.

54 Reetz, M.T. and Guo, H. (2005). Mixtures of monodentate P-ligands as a means to control the diastereoselectivity in Rh-catalyzed hydrogenation of chiral alkenes. *Beilstein J. Org. Chem.* 3: 1–3. https://doi.org/10.1186/1860-5397-1-3.

55 Gardiner, J.M., Crewe, P.D., Smith, G.E. et al. (2003). Synthesis, stereostructure, and conformations of novel bi- and trifunctional (+)-isomenthone derivatives. *Org. Lett.* 5: 467–470. https://doi.org/10.1021/ol027434e.

56 Tanaka, M. and Ogata, I.A. (1975). A novel route to menthyldiphenylphosphine. *Bull. Chem. Soc. Jpn.* 48: 1094–1094. https://doi.org/10.1246/bcsj.48.1094.

57 Neuffer, J. and Richter, W.J. (1986). Optisch aktive phosphine durch asymmetrische substitution prochiraler, homochiral substituierter phosphonate. *J. Organomet. Chem.* 301: 289–297. https://doi.org/10.1016/0022-328X(86)80033-X.

58 Malkov, A.V., Bella, M., Stara, I.G. et al. (2001). Modular pyridine-type *P,N*-ligands derived from monoterpenes: application in asymmetric heck addition. *Tetrahedron Lett.* 42: 3045–3048. https://doi.org/10.1016/S0040-4039(01)00369-0.

59 Hiroi, K. and Watanabe, K. (2001). Palladium-catalyzed asymmetric Diels–Alder reactions with novel chiral imino-phosphine ligands. *Tetrahedron: Asymmetry* 12: 3067–3070. https://doi.org/10.1016/S0957-4166(01)00546-8.

60 Carmona, D., Vega, C., Lahoz, F.J. et al. (1999). Imino–rhodium(III) and –ruthenium(II) compounds with Stereogenic metal centers. *Organometallics* 18: 3364–3369. https://doi.org/10.1021/om9810448.

61 Benetsky, E.B., Zheglov, S.V., Grishina, T.B. et al. (2007). Various P^*-chiral phosphite-type ligands: their synthesis, stereochemistry and use in pd-catalysed allylation. *Tetrahedron Lett.* 48: 8326–8330. https://doi.org/10.1016/j.tetlet.2007.09.125.

62 Bunlaksananusorn, T., Polborn, K., and Knochel, P. (2003). New *P,N* ligands for asymmetric Ir-catalyzed reactions. *Angew. Chem. Int. Ed.* 42: 3941–3943. https://doi.org/10.1002/anie.200351936.

63 Kmieciak, A. and Krzemiński, M.P. (2019). Chiral terpene auxiliaries V: synthesis of new chiral γ-hydroxyphosphine oxides derived from α-pinene. *Beilstein J. Org. Chem.* 15: 2493–2499. https://doi.org/10.3762/bjoc.15.242.

64 Kmieciak, A. and Krzemiński, M.P. (2017). Chiral terpene auxiliaries IV: new monoterpene PHOX ligands and their application in the catalytic asymmetric transfer hydrogenation of ketones. *Tetrahedron: Asymmetry* 28: 467–472. https://doi.org/10.1016/j.tetasy.2017.02.003.

65 Zhang, X. (2002). Ortho substituted chiral phosphines and phosphintes and their use in asymmetric catalytic reactions. US Patent 6, 653,485, filed 16 November 2001.

66 Widhalm, M. and Mereiter, K. (2003). Metallation reactions on 2,2′-bis (diphenylphosphinoyl)-1,1′-binaphthyl [$BINAP(O)_2$]. *Bull. Chem. Soc. Jpn.* 76: 1233–1244. https://doi.org/10.1246/bcsj.76.1233.

67 Kant, M., Bischoff, S., Siefken, R. et al. (2001). Synthesis and characterization of 4- and 4,4′-phosphorylated2,2′-bis(diphenylphosphanyl)-1,1′-binaphthyls. *Eur. J. Org. Chem.* 3: 477–481. https://doi.org/10.1002/1099-0690(200102)2001:3<477::AID-EJOC477>3.0.CO;2-3.

68 Dayoub, W., Favre-Reguillon, A., Berthod, M. et al. (2012). An improved and safer synthesis of (*R*)- and (*S*)-4,4′-diaminomethyl-BINAP. *Eur. J. Org. Chem.* 2012 (16): 3074–3078. https://doi.org/10.1002/ejoc.201200219.

69 Hu, A., Ngo, H.L., and Lin, W. (2004). Remarkable 4,4′-substituent effects on Binap: highly enantioselective Ru catalysts for asymmetric hydrogenation of β-aryl ketoesters and their immobilization in room-temperature ionic liquids. *Angew. Chem. Int. Ed.* 43: 2501–2504. https://doi.org/10.1002/anie.200353415.

70 Shu, Q.-D., Kong, S.-N., Wei, Y.-Z. et al. (2016). Two-dimensional coordination polymers based on functionalized chiral binaphthyl-derived ligands with phosphine oxide groups and zinc(II) and cadmium(II) ions. *Polyhedron* 116: 96–102. https://doi.org/10.1016/j.poly.2016.08.008.

71 Wahyu Lestari, W., Zaake-Hertling, H., Lönnecke, P. et al. (2013). 4,4′- and 5,5′-functionalized (*S*)- and (*R*)-2,2′-bis(diphenylphosphanyl)-1, 1′-binaphthyl oxide derivatives. *Z. Anorg. Allg. Chem.* 639: 2589–2596. http://dx.doi.org/10.1002/zaac.201300278.

72 Huang, Y.-Y., Yang, X., Feng, Y. et al. (2014). Chiral Ru/Ir bimetallic dendronized polymer catalysts constructed through sequential metal coordination and applied in asymmetric hydrogenation of quinaldine. *J. Mol. Catal. A: Chem.* 393: 150–155. https://doi.org/10.1016/j.molcata.2014.06.013.

73 Berthod, M., Saluzzo, C., Mignani, G. et al. (2004). 4,4′ and 5,5′-DiamBINAP as a hydrosoluble chiral ligand: syntheses and use in Ru(II) asymmetric biphasic catalytic hydrogenation. *Tetrahedron: Asymmetry* 15 (4): 639–645. https://doi.org/10.1016/j.tetasy.2003.12.033.

74 Wang, T., Lyu, Y., Xiong, K. et al. (2017). Chiral BINAP-based hierarchical porous polymers as platforms for efficient heterogeneous asymmetric catalysis. *Chin. J. Catal.* 38: 890–897. https://doi.org/10.1016/S1872-2067(17)62826-2.

75 Berthod, M., Mignani, G., and Lemaire, M. (2004). New perfluoroalkylated BINAP usable as a ligand in homogeneous and supercritical carbon dioxide asymmetric hydrogenation. *Tetrahedron: Asymmetry* 15 (7): 1121–1126. https://doi.org/10.1016/j.tetasy.2004.02.004.

76 Köckritz, A., Bischoff, S., Kant, M. et al. (2001). Asymmetric hydroformylation and hydrogenation catalyzed by chiral rhodium and ruthenium complexes of phosphorylated 2,2′-bis(diphenyl-phosphino)-1,1′-binaphthyls. *J. Mol. Catal. A: Chem.* 174: 119–126. https://doi.org/10.1016/S1381-1169(01)00193-5.

77 Hu, A., Ngo, H.L., and Lin, W. (2003). Chiral porous hybrid solids for practical heterogeneous asymmetric hydrogenation of aromatic ketones. *J. Am. Chem. Soc.* 125: 11490–11491. https://doi.org/10.1021/ja0348344.

78 Falkowski, J.M., Sawano, T., Zhang, T. et al. (2014). Privileged phosphine-based metal–organic frameworks for broad-scope asymmetric catalysis. *J. Am. Chem. Soc.* 136: 5213–5216. https://doi.org/10.1021/ja500090y.

79 Kesanli, B. and Lin, W. (2004). Mesoporous silica anchored Ru catalysts for highly enantioselective hydrogenation of β-ketoesters. *Chem. Commun.* (20): 2284–2285. https://doi.org/10.1039/B406697A.

80 Ding, Y., Wang, T., Lyu, Y. (2017). The preparation and application of two kinds of chiral ligands, chiral porous organic polymer. China patent CN106854220, filed 09 December 2015.

81 Wang, X., Li, J., Lu, S. et al. (2015). Efficient enantioselective hydrogenation of quinolines catalyzed by conjugated microporous polymers with embedded chiral BINAP ligand. *Chin. J. Catal.* 36: 1170–1174. https://doi.org/10.1016/s1872-2067(15)60903-2.

82 Okano, T., Kumobayashi, H., Akutagawa, S. et al. (1987). 5,5′-Diamino or 5,5′-diacetamido-2,2′-bis(diphenylphosphino)-1,1′-binaphthyls. US patent 4, 705,895, filed 21 February 1986.

83 Lemaire, M., Saluzzo, C., Berthod, M. (2004). New 4,4′-disubstituted 2,2′-diphosphino-1,1′-binaphthyl derivatives, useful as ligands in transition metal complex catalysts for asymmetric reactions such as hydrogenation. FR patent 2, 849,037A1, filed 18 December 2002.

84 Shimada, T., Suda, M., Nagano, T. et al. (2005). Facile preparation of a new BINAP-based building block, 5,5′-DiiodoBINAP, and its synthetic application. *J. Organ. Chem.* 70: 10178–10181. https://doi.org/10.1021/jo0517186.

85 Seki, T., McEleney, K., Crudden, C. M. (2012). Enantioselective catalysis with a chiral, phosphane-containing PMO material. *Chem. Commun.* 48(51): 6369–6371. https://doi.org/10.1039/c2cc31247f.

86 Kumobayashi, H. (1992). Complexe soluble dans l'eau de binaphtylphosphine, substitué par desulfonates de métaux alcalins, avec métaux de transition et son utilisation dans un procédé d'hydrogénation enantiosélective. European patent EP 0, 544,455B1, filed 19 November 1992.

87 Jin, X., Kong, F., Yang, Z. et al. (2013). Synthesis of BINAP ligands with imidazole tags for highly enantioselective Ru-catalyzed asymmetric hydrogenation of *β*-keto esters in ionic liquid systems. *J. Mol. Catal. A: Chem.* 374, 375: 22–26. https://doi.org/10.1016/j.molcata.2013.03.015.

88 Fan, Q., Ren, C., Yeung, C. et al. (1999). Highly effective soluble polymer-supported catalysts for asymmetric hydrogenation. *J. Am. Chem. Soc.* 121 (32): 7407–7408. https://doi.org/10.1021/ja9837497.

89 Bergens, S. H. (2015) Catalyst systems for use continuous flow reactors and methods of manufacture and use thereof. WO 2, 015,013829 A1, filed 02 August 2013.

90 Sun, Q., Meng, X., Liu, X. et al. (2012). Mesoporous cross-linked polymer copolymerized with chiral BINAP ligand coordinated to a ruthenium species as an efficient heterogeneous catalyst for asymmetric hydrogenation. *Chem. Commun.* 48 (85): 10505–10507. https://doi.org/10.1039/c2cc35192g.

91 McDonald, A. R., Müller, C., Vogt, D. et al. (2008). BINAP-Ru and -Rh catalysts covalently immobilised on silica and their repeated application in asymmetric hydrogenation. *Green Chem.* 10(4): 424–432. https://doi.org/10.1039/b714189k.

92 Wu, P., Deng, G., Fan, Q. et al. (2002). Assembling behavior of BINAP derivative. *Chem. Lett.* 31 (7): 706–707. https://doi.org/10.1246/cl.2002.706.

93 Wang, Z.-J., Deng, G.-J., Li, Y. et al. (2007). Enantioselective hydrogenation of quinolines catalyzed by Ir(BINAP)-cored dendrimers: dramatic enhancement of catalytic activity. *Org. Lett.* 9 (7): 1243–1246. https://doi.org/10.1021/ol0631410.

94 He, Y.-M., Feng, Y., and Fan, Q.-H. (2014). Asymmetric hydrogenation in the core of dendrimers. *Acc. Chem. Res.* 47 (10): 2894–2906. https://pubs.acs.org/doi/10.1021/ar500146e.

95 Huang, Y.-Y., He, Y.-M., Zhou, H.-F. et al. (2006). Thermomorphic system with non-fluorous phase-tagged Ru(BINAP) catalyst: facile liquid/solid catalyst separation and application in asymmetric hydrogenation. *J. Organ. Chem.* 71 (7): 2874–2877. https://doi.org/10.1021/jo052092m.

96 Fan, Q.-H., Liu, G.-H., Deng, G.-J. et al. (2001). New soluble bifunctional polymeric chiral ligands for enantioselectively catalytic reactions. *Tetrahedron Lett.* 42(51): 9047–9050. doi:https://doi.org/10.1016/s0040-4039(01)02004-4.

97 Wang, T., Wang, W., Lyu, Y. et al. (2017). Porous rh/BINAP polymers as efficient heterogeneous catalysts for asymmetric hydroformylation of styrene: enhanced enantioselectivity realized by flexible chiral nanopockets. *Chin. J. Catal.* 38 (4): 691–698. https://doi.org/10.1016/s1872-2067(17)62790-6.

98 Maegawa, Y., Nagano, T., Yabuno, T. et al. (2007). Preparation of functionalized aryl(diallyl)ethoxysilanes and their palladium-catalyzed coupling reactions giving sol–gel precursors. *Tetrahedron* 63 (46): 11467–11474. https://doi.org/10.1016/j.tet.2007.08.011.

99 Shimada T., Takenaka N., Goshima G. et al. (2010). Polymer having bis(diphenylphosphino)binaphthyl groups. US Patent US2, 010,002,280,0A1, filed, 12 March 2007.

100 Zhang, Z., Qian, H., Longmire, J. et al. (2000). Synthesis of chiral Bisphosphines with tunable bite angles and their applications in asymmetric hydrogenation of β-ketoesters. *J. Organ. Chem.* 65 (19): 6223–6226. https://doi.org/10.1021/jo000462v.

101 Qiu, L., Kwong, F.Y., Wu, J. et al. (2006). A new class of versatile chiral-bridged atropisomeric diphosphine ligands: remarkably efficient ligand syntheses and their applications in highly enantioselective hydrogenation reactions. *J. Am. Chem. Soc.* 128 (17): 5955–5965. https://doi.org/10.1021/ja0602694.

102 Fernández-Pérez, H., Mon, I., Frontera, A. et al. (2015). Enantiopure bis-phosphine ligands with appended crown ether groups as regulation sites for rh-mediated hydrogenations. *Tetrahedron* 71 (26, 27): 4490–4494. https://doi.org/10.1016/j.tet.2015.02.073.

103 Kean, Z.S., Akbulatov, S., Tian, Y. et al. (2014). Photomechanical actuation of ligand geometry in enantioselective catalysis. *Angew. Chem. Int. Ed.* 53 (52): 14508–14511. https://doi.org/10.1002/anie.201407494.

104 Wang, X.-B. and Zhou, Y.-G. (2008). Synthesis of tunable bisphosphine ligands and their application in asymmetric hydrogenation of quinolines. *J. Organ. Chem.* 73 (14): 5640–5642. https://doi.org/10.1021/jo800779r.

105 Deng, G.-J., Li, G.-R., Zhu, L.-Y. et al. (2006). Dendritic BIPHEP: synthesis and application in asymmetric hydrogenation of β-ketoesters. *J. Mol. Catal. A: Chem.* 244 (1, 2): 118–123. https://doi.org/10.1016/j.molcata.2005.08.036.

106 Graff, J., Łastawiecka, E., Guénée, L. et al. (2015). Asymmetric bromine-lithium exchange: application toward the synthesis of new biaryl-diphosphine ligands. *Adv. Synth. Catal.* 357 (13): 2833–2839. https://doi.org/10.1002/adsc.201500323.

107 Tang, W., Wang, W., Chi, Y. et al. (2003). A bisphosphepine ligand with stereogenic phosphorus centers for the practical synthesis of β-aryl-β-amino acids by asymmetric hydrogenation. *Angew. Chem. Int. Ed.* 42 (30): 3509–3511. https://doi.org/10.1002/anie.200351465.

108 Kasák, P., Mereiter, K., and Widhalm, M. (2005). Chiral α-branched mono phosphine auxiliaries, reversal of sense of asymmetric induction upon substitution. *Tetrahedron: Asymmetry* 16 (20): 3416–3426. https://doi.org/10.1016/j.tetasy.2005.09.008.

109 Kasák, P., Arion, V.B., and Widhalm, M. (2006). A chiral phosphepine–olefin rhodium complex as an efficient catalyst for the asymmetric conjugate addition. *Tetrahedron: Asymmetry* 17 (22): 3084–3090. https://doi.org/10.1016/j.tetasy.2006.11.022.

110 Nareddy, P., Mantilli, L., Guénée, L. et al. (2012). Atropoisomeric (*P,N*) ligands for the highly enantioselective pd-catalyzed intramolecular asymmetric α-arylation of α-branched aldehydes. *Angew. Chem. Int. Ed.* 51 (16): 3826–3831. https://doi.org/10.1002/anie.201108061.

111 Humbert, N., Larionov, E., Mantilli, L. et al. (2013). Highly modular C_1-symmetric chiral (*P,N*) ligands with a stereolabile P center: experimental and theoretical studies. *Chem. Eur. J.* 20 (3): 745–751. https://doi.org/10.1002/chem.201303146.

112 Franzoni, I., Guénée, L., and Mazet, C. (2014). Chiral monodentate phosphine ligands for the enantioselective α- and γ-arylation of aldehydes. *Tetrahedron* 70 (27, 28): 4181–4190. https://doi.org/10.1016/j.tet.2014.02.079.

113 Fujii, K., Ito, S., and Mikami, K. (2019). Synthetic methodologies for perfluoroaryl substituted (diaryl)methylphosphonates, -phosphinates via S_NAr reaction. *J. Organ. Chem.* https://doi.org/10.1002/chem.201303146 84 (19): 12281–12291.

114 Ziegler, D.T., Riesgo, L., Ikeda, T. et al. (2014). Biphenyl-derived Phosphepines as chiral nucleophilic catalysts: enantioselective [4+1] annulations to form functionalized cyclopentenes. *Angew. Chem. Int. Ed.* 53 (48): 13183–13187. https://doi.org/10.1002/anie.201405854.

115 Kukhar, V.P. and Romanenko, V.D. (2009). Chemistry of aminophosphonic acids and phosphonopeptides. In: *Aminoacids, Peptides and Proteins in Organic Chemistry*, vol. 2 (ed. A.B. Hudges), 189–260. Wiley-VCH.

116 Arribat, M., Cavelier, F., and Rémond, E. (2020). Phosphorus-containing amino acids with a P–C bond in the side chain or a P–O, P–S or P–N bond: from synthesis to applications. *RSC Adv.* 10: 6678–6724. https://doi.org/10.1039/C9RA10917J.

117 Mucha, A., Kafarski, P., and Berlicki, Ł. (2011). Remarkable potential of the α-aminophosphonate/phosphinate structural motif in medicinal chemistry. *J. Med. Chem.* 54: 5955–5980. https://doi.org/10.1021/jm200587f.

118 Cao, H.-Q., Li, J.-K., Zhang, F.-G. et al. (2021). Asymmetric synthesis of chiral amino carboxylic-phosphonic acid derivatives. *Adv. Synth. Catal.* 363: 688–729. https://doi.org/10.1002/adsc.202001345.

119 Kafarski, P. and Lejczak, B. (2001). Aminophosphonic acids of potential medical importance. *Curr. Med. Chem. Anticancer Agents* 1: 301–312. https://doi.org/10.2174/1568011013354543.

120 Kafarski, P. (2020). Phosphonopeptides containing free phosphonic groups: recent advances. *RSC Adv.* 10: 25898–25910. 10.1039/D0RA04655H.

121 Ordonez, M., Rojas-Cabrera, H., and Cativiela, C. (2009). An overview of stereoselective synthesis of α-aminophosphonic acids and derivatives. *Tetrahedron* 65: 17–49. https://doi.org/10.1016/j.tet.2008.09.083.

122 Carraro, M., Sartorel, A., Scorrano, G. et al. (2008). Chiral Strandberg-type molybdates $[(RPO_3)_2Mo_5O_{15}]^{2-}$ as molecular gelators: self-assembled fibrillar nanostructures with enhanced optical activity. *Angew. Chem. Int. Ed.* 47: 7275–7279. https://doi.org/10.1002/ange.200801629.

123 Carraro, M., Modugno, G., Sartorel, A. et al. (2009). Optically active polyoxotungstates bearing chiral organophosphonate substituents. *Eur. J. Inorg. Chem.* 2009: 5164–5174. https://doi.org/10.1002/ejic.200900728.

124 Modugno, G., Fabbretti, E., Vedove, A.D. et al. (2018). Tracking fluorescent polyoxometalates within cells. *Eur. J. Inorg. Chem.* 2018: 4955–4961. https://doi.org/10.1002/ejic.201800802.

125 Bougioukou, D.J., Mukherjee, S., and van der Donk, W.A. (2013). Revisiting the biosynthesis of dehydrophos reveals a tRNA-dependent pathway. *Proc. Natl. Acad. Sci. U.S.A.* 110: 10952–10957. https://doi.org/10.1073/pnas.1303568110.

126 Ulrich, E.C., Bougioukou D.J., van der Donk, W.A. (2018). Investigation of amide bond formation during dehydrophos biosynthesis. *ACS Chem. Biol.* 13: 537–541. https://doi.org/10.1021/acschembio.7b00949.

127 Mukherjee, S., Bougioukou, D.J., Ting, C.P. et al. (2019). Use of the dehydrophos biosynthetic enzymes to prepare antimicrobial analogs of alaphosphin. *Org. Biomol. Chem.* 17: 822–829. https://doi.org/10.1039/C8OB02860E.

128 Bovill, R., Evans, P.G., Howse, G.L., and Osborn, H.M.I. (2016). Synthesis and biological analysis of novel glycoside derivatives of L-AEP, as targeted antibacterial agents. *Bioorg. Med. Chem. Lett.* 26: 3774–3779. https://doi.org/10.1016/j.bmcl.2016.05.052.

129 Viveros-Ceballos, J.L., Ordónez, M., Sayago, F. et al. (2016). First synthesis of (*R*)- and (*S*)-1,2,3,4-tetrahydroisoquinoline-3-phosphonic acid (Tic^P) using a Pictet–Spengler reaction. *Eur. J. Org. Chem.* 2016: 2711–2719. https://doi.org/10.1002/ejoc.201600313.

130 Feng, Y., Park, J., Li, S.-G. et al. (2019). Chirality-driven mode of binding of α-aminophosphonic acid-based allosteric inhibitors of the human farnesyl pyrophosphate synthase (hFPPS). *J. Med. Chem.* 62: 9691–9702. https://doi.org/10.1021/acs.jmedchem.9b01104.

131 Yang, S., Froeyen, M., Lescrinier, E. et al. (2011). 3-Phosphono-L-alanine as pyrophosphate mimic for DNA synthesis using HIV-1 reverse transcriptase. *Org. Biomol. Chem.* 9: 111–119. 10.1039/C0OB00554A.

132 Watanabe, B., Tabuchi, Y., Wada, K. et al. (2017). Synthesis and evaluation of the inhibitory activity of the four stereoisomers of the potent and selective human γ-glutamyl transpeptidase inhibitor GGsTop. *Bioorg. Med. Chem. Lett.* 27: 4920–4924. https://doi.org/10.1016/j.bmcl.2017.09.017.

133 Cortes-Clerget, M., Gager, O., Monteil, M. et al. (2016). Novel easily recyclable bifunctional phosphonic acid carrying tripeptides for the stereoselective Michael addition of aldehydes with nitroalkenes. *Adv. Synth. Catal.* 358: 34–40. https://doi.org/10.1002/adsc.201500794.

134 Cortes-Clerget, M., Jover, J., Dussart, J. et al. (2017). Bifunctional tripeptide with a phosphonic acid as a bronsted acid for Michael addition: mechanistic insights. *Chem. Eur. J.* 23: 6654–6662. https://doi.org/10.1002/chem.201700604.

135 Rigger, L., Schmidt, R.L., Holman, K.M. et al. (2013). The synthesis of methylated, phosphorylated, and phosphonated 3′-aminoacyl-tRNASec mimics. *Chem. Eur. J.* 19: 15872–15878. doi: 10.1002/chem.201302188.

136 Caroff, E., Hubler, F., Meyer, E. et al. (2015). 4-((*R*)-2-{[6-((*S*)-3-Methoxypyrrolidin-1-yl)-2-phenylpyrimidine-4-carbonyl]amino}-3-phosphonopropionyl) piperazine-1-carboxylic acid butyl ester (ACT-246475) and its prodrug (ACT-281959), a novel $P2Y_{12}$ receptor antagonist with a wider therapeutic window in the rat than clopidogrel. *J. Med. Chem.* 58: 9133–9153. https://doi.org/10.1021/acs.jmedchem.5b00933.

137 Panigrahi, K., Eggen, M.J., Maeng, J.-H. et al. (2009). The α,α-difluorinated phosphonate L-pSer-analogue: an accessible chemical tool for studying kinase-dependent signal transduction. *Chem. Biol.* 16: 928–936. https://doi.org/10.1016/j.chembiol.2009.08.008.

138 Berkowitz, D.B., Shen, Q., and Maeng, J.-H. (1994). Synthesis of the (α,α-difluoroalkyl) phosphonate analogue of phosphoserine. *Tetrahedron Lett.* 35: 6445–6448. https://doi.org/10.1016/S0040-4039(00)78242-6.

139 Arrendale, A. Kim, K., Choi, J. Y. et al. (2012). Synthesis of a phosphoserine mimetic prodrug with potent 14-3-3 protein inhibitory activity. *Chem. Biol.* 19: 764–771. https://doi.org/10.1016/j.chembiol.2012.05.011.

140 Zhang, W., Sha, W., Pajkert, R. et al. (2017). β-Amino-γ,γ-difluoro-ω-phosphonoglutamic acid derivatives: an unexplored, multifaceted structural type of tailor-made α-amino acids. *Eur. J. Org. Chem.* 2017: 3451–3456. https://doi.org/10.1002/ejoc.201700570.

141 Otaka, A., Mitsuyama, E., Watanabe, H. et al. (2000). Development of new methodology for the synthesis of functionalized α-fluorophosphonates and its practical application to the preparation of phosphopeptide mimetics. *Chem. Commun.* 1081–1082. https://doi.org/10.1039/B002188L.

142 Foss, F.W., Snyder, A.H., Davis, M.D. et al. (2007). Synthesis and biological evaluation of γ-aminophosphonates as potent, subtype-selective sphingosine 1-phosphate receptor agonists and antagonists. *Bioorg. Med. Chem.* 15: 663–677. https://doi.org/10.1016/j.bmc.2006.10.060.

143 Hymel, D. and Burke, T.R. (2017). Phosphatase-stable phosphoamino acid mimetics that enhance binding affinities with the polo-box domain of polo-like kinase 1. *ChemMedChem* 12: 202–206. https://doi.org/10.1002/cmdc.201600574.

144 Duan, H.-Z., Chen, H.-X., Yu, Q. et al. (2019). Stereoselective synthesis of a phosphonate pThr mimetic via palladium-catalyzed γ-C(sp^3)–H activation for peptide preparation. *Org. Biomol. Chem.* 17: 2099–2102. 10.1039/C8OB02999G.

145 Chen, H.-X., Kang, J., Chang, R. et al. (2018). Synthesis of α,α-difluorinated phosphonate pSer/pThr mimetics via rhodium-catalyzed asymmetric hydrogenation of β-difluorophosphonomethyl α-(acylamino)acrylates. *Org. Lett.* 20: 3278–3281. https://doi.org/10.1021/acs.orglett.8b01151.

146 Li, P., Zhang, M., Peach, M.L. et al. (2003). Concise and enantioselective synthesis of Fmoc-Pmp(Bu^t)$_2$-OH and design of potent Pmp-containing Grb2-SH2 domain antagonists. *Org. Lett.* 5: 3095–3098. https://doi.org/10.1021/ol035078+.

147 Tulsi, N.S., Downey, A.M., and Cairo, C.W. (2010). A protected L-bromophosphonomethylphenylalanine amino acid derivative (BrPmp) for synthesis of irreversible protein tyrosine phosphatase inhibitors. *Bioorg. Med. Chem.* 18: 8679–8686. https://doi.org/10.1016/j.bmc.2010.09.040.

148 Asante-Appiah, E., Patel, S., Dufresne, C. et al. (2002). The structure of PTP-1B in complex with a peptide inhibitor reveals an alternative binding mode for bisphosphonates. *Biochemistry* 41: 9043–9051. https://doi.org/10.1021/bi0259554.

149 Meyer, C. and Köhn, M. Efficient scaled-up synthesis of *N*-α-Fmoc-4-phosphono (difluoromethyl)-L-phenylalanine and its incorporation into peptides. *Synthesis* 2011, 20: 3255–3260. https://doi.org/10.1055/s-0030-1260206.

150 Inoue, M., Hiratake, J., and Sakata, K. (1999). Synthesis and characterization of intermediate and transition-state analogue inhibitors of γ-glutamyl peptide ligases. *Biosci. Biotechnol., Biochem.* 63: 2248–2251. https://doi.org/10.1271/bbb.63.2248.

151 Monbaliu, J.-C., Tinant, B., and Marchand-Brynaert, J. (2010). [4+2]cycloadditions of 1-phosphono-1,3-butadienes with nitroso heterodienophiles: a versatile synthetic route for polyfunctionalized aminophosphonic derivatives. *J. Org. Chem.* 75: 5478–5486. https://doi.org/10.1021/jo100230r.

152 Davis, F.A., Xu, H., Wu, Y., and Zhang, J. (2006). Asymmetric synthesis of polyfunctionalized pyrrolidines from sulfinimine-derived pyrrolidine 2-phosphonates. Synthesis of pyrrolidine 225C. *Org. Lett.* 8: 2273–2276. https://doi.org/10.1021/ol060521c.

153 Gahungu, M., Arguelles-Arias, A., Fickers, P. et al. (2013). Synthesis and biological evaluation of potential threonine synthase inhibitors: Rhizocticin a and Plumbemycin a. *Bioorg. Med. Chem.* 21: 4958–4967. https://doi.org/10.1016/j.bmc.2013.06.064.

154 Bar-Nir Ben-Aroya, B. and Portnoy, M. (2002). Preparation of α-aminophosphines on solid support: model studies and parallel synthesis. *Tetrahedron* 58 (25): 5147–5158. https://doi.org/10.1016/S0040-4020(02)00471-4.

155 Abdur-Rashid, K., Guo, R., Lough, A.J. et al. (2005). Synthesis of ruthenium hydride complexes containing β-aminophosphine ligands derived from amino acids and their use in the H_2-hydrogenation of ketones and imines. *Adv. Synth. Catal.* 347 (4): 571–579. https://doi.org/10.1002/adsc.200404274.

156 Fu, X., Jiang, Z., and Tan, C.-H. Bicyclic guanidine-catalyzed enantioselective phospha-Michael reaction: synthesis of chiral β-aminophosphine oxides and β-aminophosphines. *Chem. Commun.* 50 (47): 5058–5060. https://doi.org/10.1039/B713151H.

157 Roennholm, P., Soedergren, M., and Hilmersson, G. (2007). Improved and efficient synthesis of chiral *N,P*-ligands via cyclic sulfamidates for asymmetric addition of butyllithium to benzaldehyde. *Org. Lett.* 9 (19): 3781–3783. https://doi.org/10.1021/ol701504c.

158 Chauvin, A.-S., Bernardinelli, G., and Alexakis, A. (2004). Determination of the absolute configuration of chiral aryl–alkyl carbinols using organophosphorus diamine derivatizing agents by ^{31}P NMR spectroscopy. *Tetrahedron: Asymmetry* 15 (12): 1857–1879. https://doi.org/10.1016/j.tetasy.2004.04.031.

159 Foucher, V., Guizzardi, B., Groen, M.B. et al. (2010). Versatile D→BCD steroid construction strategy, illustrated by the enantioselective total synthesis of estrone. *Org. Lett.* 12 (4): 680–683. https://doi.org/10.1021/ol902638w.

160 Kottsieper, K.W., Kühner, U., and Stelzer, O. (2001). Synthesis of enantiopure C_1 symmetric diphosphines and phosphino-phosphonites with ortho-phenylene backbones. *Tetrahedron: Asymmetry* 12 (8): 1159–1169. https://doi.org/10.1016/S0957-4166(01)00175-6.

161 Kolodiazhnyi, O.I. and Kolodiazhna, A.O. (2017). Stereochemistry of nucleophilic substitution at trivalent phosphorus. *Phosphorus, Sulfur Silicon Relat. Elem.* 192 (6): 621–633. https://doi.org/10.1080/10426507.2017.1284842.

162 Kolodiazhnyi, O.I., Andrushko, N.V., and Gryshkun, E.B. (2004). Stereoselective reactions of optically active derivatives of α-methylbenzylaminophosphine. *Russ. J. Gen. Chem.* 74 (4): 515–522.

163 Gryshkun, E.V., Andrushko, N.V., and Kolodiazhnyi, O.I. (2004). Stereoselective reactions of chiral amines with racemic chlorophosphines. *Phosphorus, Sulfur Silicon Relat. Elem.* 179 (6): 1027–1046. https://doi.org/10.1080/10426500490459650.

164 Mikołajczyk, M. (1980). Optically active trivalent phosphorus acid esters: synthesis, chirality at phosphorus and some transformations. *Pure Appl. Chem.* 52 (4): 959–972. https://doi.org/10.1351/pac198052040959.

165 Izumi, A. and Tai, S. (1977). *Stereo-Differentiating Reactions: The Nature of Asymmetric Reactions* (trans. Rittai kubetsu hannō). New York/Tokyo: Kodansha Ltd/Academic Press.

166 Kolodiazhnyi, O.I., Gryshkun, E.V., Andrushko, N.V. et al. (2003). Asymmetric synthesis of chiral *N*-(1-methylbenzyl)aminophosphines. *Tetrahedron: Asymmetry* 14 (2): 181–183. https://doi.org/10.1016/s0957-4166(02)00750-4.

167 Revés, M., Ferrer, C., León, T. et al. (2010). Primary and secondary aminophosphines as novel P-stereogenic building blocks for ligand synthesis. *Angew. Chem. Int. Ed.* 49(49): 9452–9455. https://doi.org/10.1002/anie.201004041.

168 Adams, H., Collins, R.C., Jones, S. et al. (2011). Enantioselective preparation of *P*-chiral phosphine oxides. *Org. Lett.* 13 (24): 6576–6579. https://doi.org/10.1021/ol202916j.

169 Juge, S., Stephan, M., Laffitte, J.A. et al. (1990). Efficient asymmetric synthesis of optically pure tertiary mono and diphosphine ligands. *Tetrahedron Lett.* 31 (44): 6357–6360. https://doi.org/10.1016/s0040-4039(00)97063-1.

170 Darcel, C., Kaloun, E.-B., Merdés, R. et al. (2001). Direct use of chiral or achiral organophosphorus boranes as pro-ligands for transition metal catalyzed reactions. *J. Organomet. Chem.* 624: 333–343. https://doi.org/10.1016/S0022-328X(01)00666-0.

171 Jugé, S., Stephan, M., Merdès, R. et al. (1993). Stereochemistry of the P–C bond formation in an oxazaphospholidine borane complex. *J. Chem. Soc., Chem. Commun.* (6): 531–533. https://doi.org/10.1039/c39930000531.

172 Darcel, C., Moulin, D., Henry, J.-C. et al. (2007). Modular P-chirogenic aminophosphane-phosphinite ligands for Rh-catalyzed asymmetric hydrogenation: a new model for prediction of enantioselectivity. *Eur. J. Org. Chem.* 2007 (13): 2078–2090. https://doi.org/10.1002/ejoc.200600966.

173 Chaux, F., Frynas, S., Laureano, H. et al. (2010). Enantiodivergent synthesis of *P*-chirogenic phosphines. *C.R. Chim.* 13 (8, 9): 1213–1226. https://doi.org/10.1016/j.crci.2010.06.001.

174 Den Heeten, R., Swennenhuis, B.H.G., van Leeuwen, P.W.N.M. et al. (2008). Parallel synthesis and screening of polymer-supported phosphorus-stereogenic aminophosphane-phosphite and -phosphinite ligands. *Angew. Chem.* 120 (35): 6704–6707. https://doi.org/10.1002/ange.200801689.

175 Colby, E.A. and Jamison, T.F. (2003). *P*-chiral, monodentate Ferrocenyl phosphines, novel ligands for asymmetric catalysis. *J. Organ. Chem.* 68 (1): 156–166. https://doi.org/10.1021/jo0264123.

176 Jaillet, A., Darcel, C., Bayardon, J. et al. (2020). Design of *P*-chirogenic aminophosphine–phosphinite ligands at both phosphorus centers: origin of enantioselectivities in pd-catalyzed allylic reactions. *J. Organ. Chem.* 85 (22): 14391–14410. https://doi.org/10.1021/acs.joc.0c00536.

177 Rippert, A.J., Linden, A., and Hansen, H.-J. (2000). Formation of diastereoisomerically pure oxazaphospholes and their reaction to chiral phosphane-borane adducts. *Helv. Chim. Acta* 83 (2): 311–321. https://doi.org/10.1002/(sici)1522-2675(20000216)83:2<311::aid-hlca311>3.0.co;2-i.

178 León, T., Riera, A., and Verdaguer, X. (2011). Stereoselective synthesis of *P*-stereogenic aminophosphines: ring opening of bulky oxazaphospholidines. *J. Am. Chem. Soc.* 133 (15): 5740–5743. https://doi.org/10.1021/ja200988c.

179 Zijlstra, H., León, T., de Cózar, A. et al. (2013). Stereodivergent S_N2@P reactions of borane oxazaphospholidines: experimental and theoretical studies. *J. Am. Chem. Soc.* 135 (11): 4483–4491. https://doi.org/10.1021/ja400208t.

180 Leyris, A., Nuel, D., Giordano, L. et al. (2005). Enantioselective synthesis of both enantiomers of *tert*-butylphenylphosphine oxide from (*S*)-prolinol. *Tetrahedron Lett.* 46 (50): 8677–8680. https://doi.org/10.1016/j.tetlet.2005.10.058.

181 Han, Z.S., Wu, H., Xu, Y. et al. (2017). General and stereoselective method for the synthesis of sterically congested and structurally diverse *P*-stereogenic secondary phosphine oxides. *Org. Lett.* 19 (7): 1796–1799. https://doi.org/10.1021/acs.orglett.7b00568.

182 Vinci, D., Mateus, N., Wu, X. et al. (2006). Oxazaphospholidine-oxide as an efficient *ortho*-directing group for the diastereoselective deprotonation of ferrocene. *Org. Lett.* 8 (2): 215–218. https://doi.org/10.1021/ol0523704.

13

Organosulfur Compounds as Chiral Building Blocks

Maria A.M. Capozzi[1] *and Cosimo Cardellicchio*[2]

[1] *Università degli Studi di Bari, Dipartimento di Chimica, Via Orabona 4, Bari 70125, Italy*
[2] *Università degli Studi di Bari, CNR ICCOM, c/o Dipartimento di Chimica, Via Orabona 4, Bari 70125, Italy*

13.1 State of the Art

"Organosulfur compounds in asymmetric synthesis" is an excellent 2008 Wiley book [1] that covers the main results within this chemistry. Two chapters of this book are relevant for the reader of the present work [2, 3]. This book has been accompanied by highly reputed recent review articles published in journals, among which three are mentioned [4–6]. The present work aims to update the Wiley book with the most recent results. However, the present work is not exhaustive but selective due to the editorial limitations. In this selection, meaningful reviews were preferred to many original papers. Within these constraints, it was decided to write a guide to extricate the vast literature on this topic by focusing on the most successful sulfur chiral building blocks and briefly analyzing their merits. To this end, a short introduction to past work is needed to understand the recent progress fully.

13.2 Introduction

Enantiopure sulfinyl compounds have been employed widely and successfully as valuable chiral building blocks in many synthetic strategies, leading to natural and bioactive compounds [3]. This success hinges on the high configuration stability of the stereogenic sulfur center and the high stereoinduction exerted by the sulfinyl moiety.

The chiral starting materials have been currently prepared by many different approaches [2]. Among them, the two most successful synthetic strategies have stood out: (i) the enantiospecific displacement with organometallic reagents (e.g. a Grignard reagent) of suitable leaving groups on an *S*-resolved sulfinyl compound, and (ii) the enantioselective oxidation of sulfides.

A survey of the most relevant organosulfur chiral building blocks is reported.

Chiral Building Blocks in Asymmetric Synthesis: Synthesis and Applications, First Edition.
Edited by Elżbieta Wojaczyńska and Jacek Wojaczyński.

13.3 The Tradition

13.3.1 Menthyl (R_S)- or (S_S)-*p*-toluenesulfinate (Andersen's Reagent)

Menthyl *p*-toluenesulfinate is the most employed chiral building block, and its success continues until today. This compound is the history of asymmetric synthesis and the chiral building blocks in general.

It was first synthesized at the beginning of the twentieth century, and it was re-discovered in the 1962–1965 period by Andersen and Mislow [7]. It was synthesized by reacting natural (1*R*, 2*S*, 5*R*)-menthol with *p*-toluenesulfinyl chloride, yielding an almost 1 : 1 mixture of the (R_S)- and (S_S)-epimers at the stereogenic sulfur center. The less soluble (S_S)-epimer was obtained after some crystallization steps (Scheme 13.1).

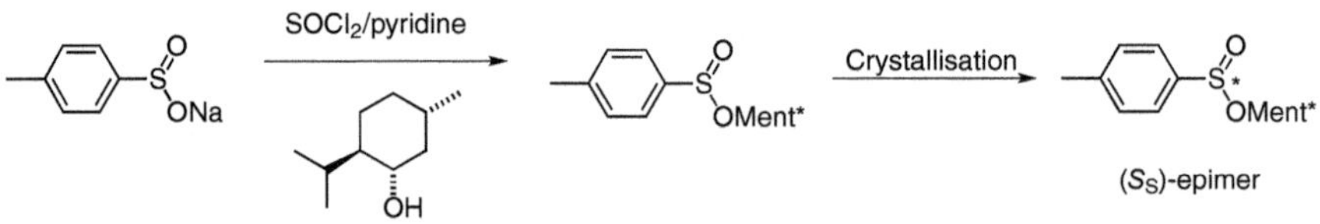

Scheme 13.1 Classical synthesis of menthyl (S_S)-*p*-toluenesulfinate.

In an alternative synthesis [7], Sharpless reported that the almost equimolar mixture of the (R_S)- and (S_S)-epimers of menthyl *p*-toluenesulfinate could be obtained by reacting natural menthol with *p*-toluenesulfonyl chloride in the presence of phosphite as a reducing agent (Scheme 13.2).

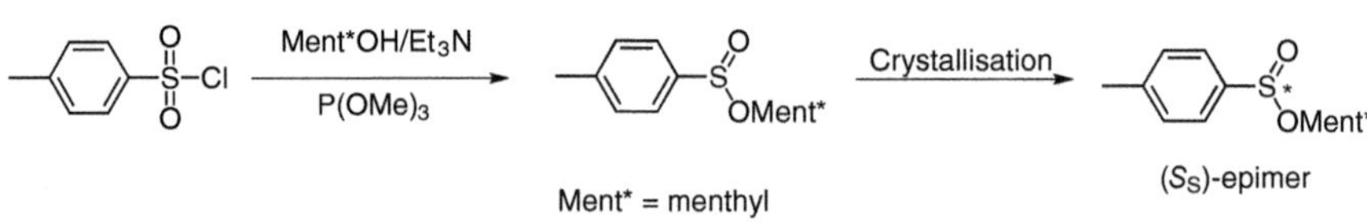

Scheme 13.2 Alternative synthesis of menthyl (S_S)-*p*-toluenesulfinate.

In Andersen's chemistry, menthyl (S_S)-*p*-toluenesulfinate reacts with alkyl or aryl Grignard reagents, yielding alkyl or aryl *p*-tolyl sulfoxides. From the stereochemical point of view, this reaction is a nucleophilic displacement at the stereogenic sulfur center, and it occurs with a clean and complete inversion at the sulfur atom (Scheme 13.3).

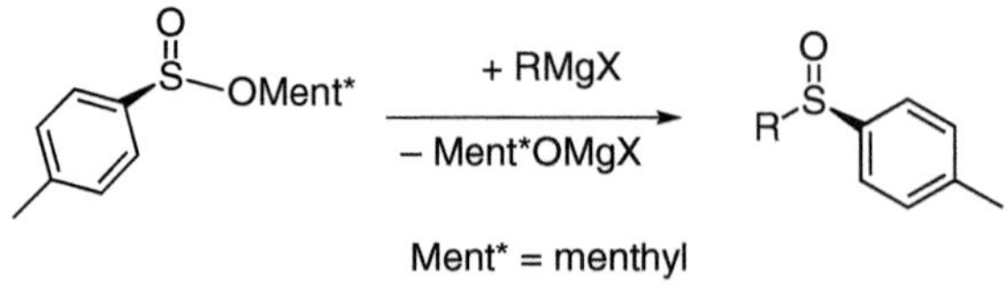

Scheme 13.3 Synthesis of alkyl or aryl *p*-tolyl sulfoxides by using menthyl (*S*)-*p*-toluenesulfinate.

Nowadays, alkyl and aryl *p*-tolyl sulfoxides can be easily obtained with Andersen's chemistry. A valuable modification in the production of menthyl (S_S)-*p*-toluenesulfinate was reported by Solladiè [7]. The more soluble (R_S)-epimer, which was discarded in older works, could be epimerized in acidic media, yielding a fresh 1 : 1 mixture of (R_S)- and (S_S)-epimers. This mixture can be recrystallized, yielding a further batch of the (S_S)-Andersen's reagent. This procedure can be repeated to recover additional amounts of the desired stereoisomer.

A similar approach yields the (R_S)-epimer if the synthetic scheme starts from the unnatural (1*S*, 2*R*, 5*S*)-menthol.

The Solladiè improvement of the preparation of this building block transformed the menthyl (R_S)- and (S_S)-*p*-toluenesulfinates into inexpensive and commercially available sulfur chiral building blocks. As a result, many sulfinyl substrates have been proposed as an alternative to Andersen's reagent [2], but the success of this molecule has remained unchanged over the decades.

Nucleophiles softer than the Grignard reagents (such as enolates or other lithiated species) react with the Andersen's reagent with the same stereochemical outcome, yielding many other sulfinyl compounds. New sulfinyl compounds bearing useful substituents were obtained and largely used to synthesize chiral complex molecular architectures [3].

Three main synthetic themes from past works were chosen to show these highly successful building blocks [8]. Representative examples of these synthetic strategies are collected in Scheme 13.4. Route (1): The synthesis of β-keto-sulfoxides; the highly diastereoselective transformation of these compounds into β-sulfinyl-alcohols. Next, route (2): The synthesis of β-sulfinyl unsaturated systems, cyclic or acyclic; the highly diastereoselective addition of a suitable nucleophile to the activated double bond. Finally, route (3): The highly diastereoselective Diels–Alder reaction involving β-sulfinyl unsaturated systems.

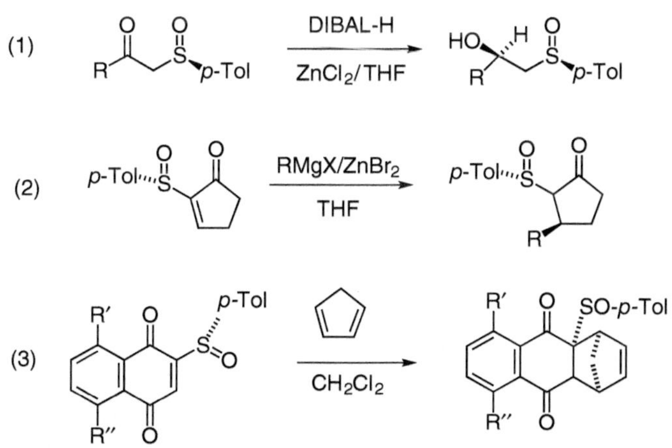

Scheme 13.4 Highly successful synthetic strategies from past works involving chiral sulfoxides.

The reason for the success of the sulfinyl intermediates deriving from Andersen's reagent hinges on the high asymmetric induction obtained in the formation of a new stereogenic center and the facility of removing the inducer of chirality at the end of the synthesis when its function is not needed anymore [3, 8].

In the last four decades, many chiral sulfinyl intermediates have been synthesized with different methods. However, in the years covered by the present work, several papers related to the employment of all the other chiral sulfinyl building blocks cannot reach the number of papers that have continued to use Andersen's reagent as the starting material.

As first recent examples, methyl *p*-tolyl sulfoxide, obtained from menthyl *p*-toluenesulfinate, was lithiated, and the following nucleophilic species was reacted with suitable electrophiles. For example, the approach of lithiating the *p*-tolyl methyl sulfoxide as the first step of a more complex process was brought in the synthesis of pavettamine [9] depicted in Scheme 13.5. The synthetic idea resembles route 1 in Scheme 13.4, i.e. the high asymmetric induction obtained in the DIBAL-H reduction of a β-ketosulfoxide.

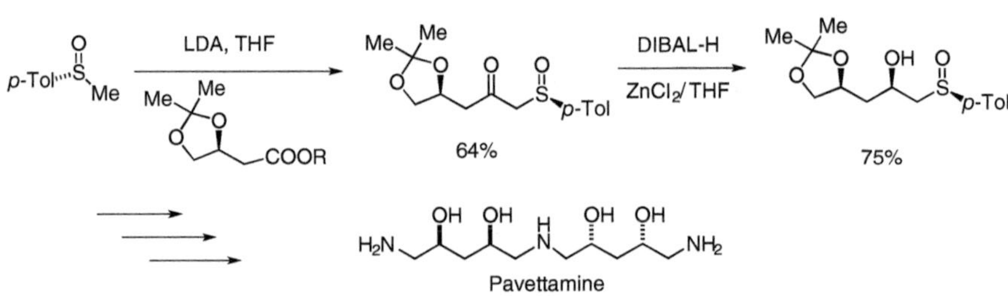

Scheme 13.5 First steps involving chiral sulfinyl compounds in the synthesis of pavettamine.

The same synthetic project of Scheme 13.5 that is: (i) metalation of methyl *p*-tolyl sulfoxide, (ii) reaction with an ester, and (iii) reduction with DIBAL-H was repeated in the synthesis of a fragment of amphidinol 3 [10].

A similar approach, with only slight differences, was followed in Scheme 13.6 to obtain a precursor of octalactins [11].

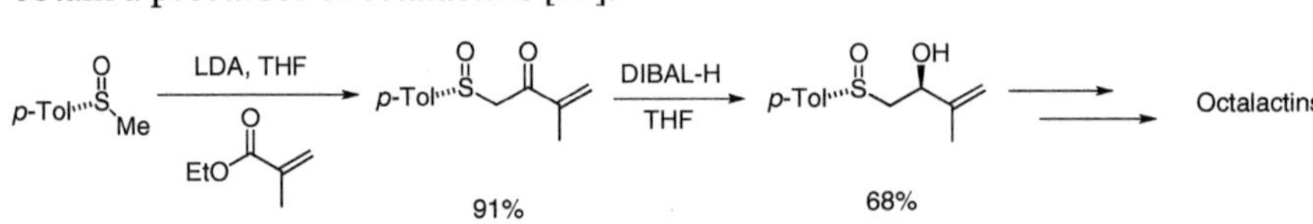

Scheme 13.6 Relevant chiral intermediates in the synthesis of octalactins.

In classical organomagnesium reagents, Fernandez de la Pradilla [12] reported that the protected form of 3-butyn-1-ol reacted with ethylmagnesium bromide, yielding the corresponding alkynylmagnesium reagent. This organomagnesium reagent was treated with menthyl (*S*)-*p*-toluenesulfinate (Scheme 13.7). The resulting sulfinyl compound was transformed into allylic dihydropirans, which were used to synthesize a fragment of *ent*-Dysiherbaine or deoxymalayamicin A [12], respectively.

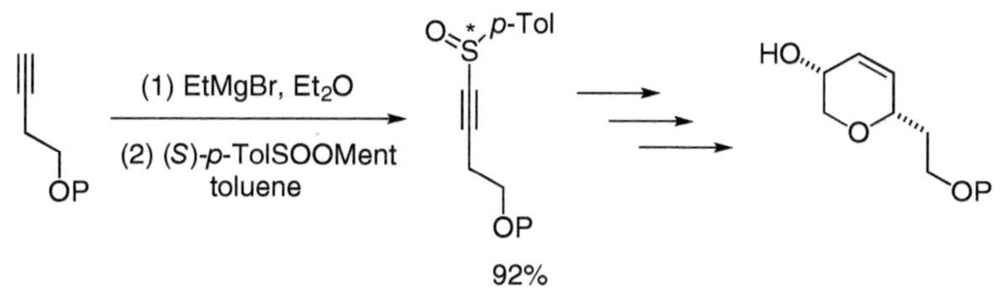

Scheme 13.7 Synthesis of chiral dihydropirans.

The chemistry of Scheme 13.7 was also applied to the total synthesis of ethyl deoxymonate B [13].

Menthyl *p*-toluenesulfinate was used to synthesize a sulfinylamine, a direct precursor of (+)-crispine A [14] (Scheme 13.8).

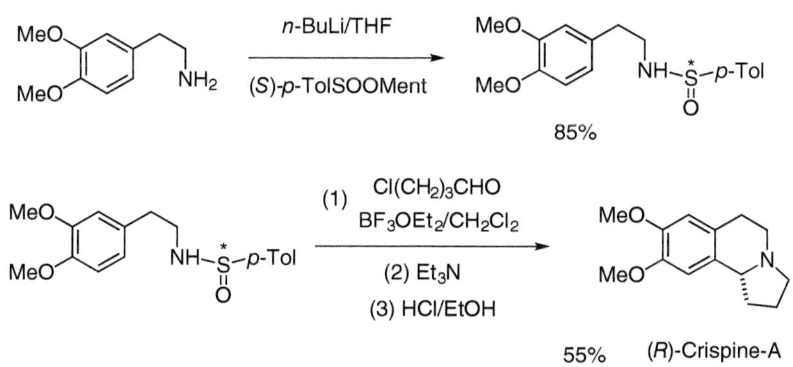

Scheme 13.8 Synthesis of (*R*)-crispine A.

The *ortho*-lithiation of activated aryl fragments provided innovative nucleophilic species that can react with Andersen's reagent to yield new useful building blocks for more complex molecular architectures. Many examples of this chemistry have been reported in the years covered by the present work.

As a representative synthetic scheme, Garcia Ruano's group reacted 1-(3,4-dimethoxyphenyl)-*N*,*N*-dimethylmethanamine with *n*-buthyllithium and treated the resulting species with menthyl (*S*)-*p*-toluenesulfinate to give the crucial (*S*)-1-[3,4-dimethoxy-2-(*p*-tolylsulfinyl)phenyl]-*N*,*N*-dimethylmethanamine (Scheme 13.9) [15].

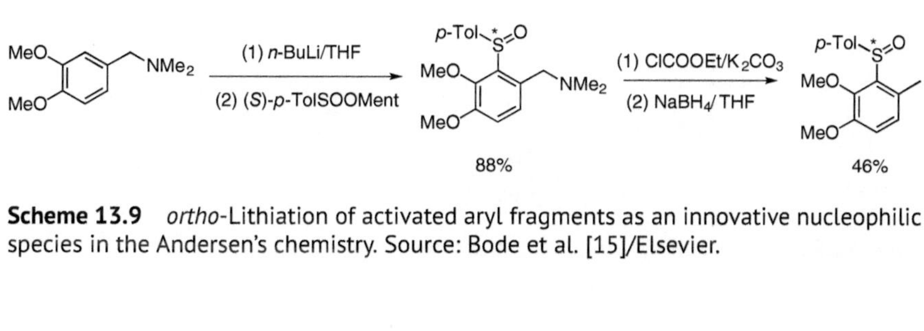

Scheme 13.9 *ortho*-Lithiation of activated aryl fragments as an innovative nucleophilic species in the Andersen's chemistry. Source: Bode et al. [15]/Elsevier.

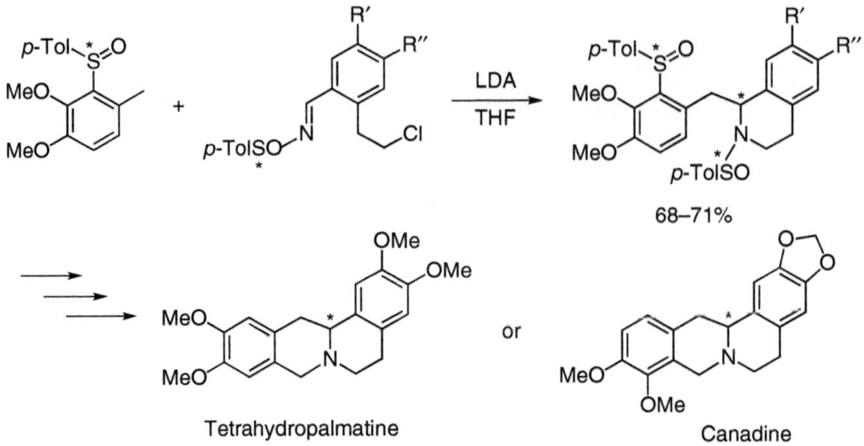

Scheme 13.10 Chiral sulfinyl intermediates in the synthesis of (*S*)-tetrahydropalmatine or (*S*)-canadine.

The (*S*)-1-[3,4-dimethoxy-2-(*p*-tolylsulfinyl)phenyl]-*N*,*N*-dimethylmethanamine represented in Scheme 13.9 was treated with a sulfinylimine (see Section 13.3.2), with the formation of the precursor of (*S*)-tetrahydropalmatine or (*S*)-canadine (Scheme 13.10).

The same research group extended the chemistry shown in Scheme 13.10 to yield the direct precursor of (*S*)-(−)-xylopinine (Scheme 13.11) [16].

The *ortho*-lithiation of activated aryl fragments was also used to synthesize sulfinyl quinonoid structures (Scheme 13.12 that were investigated as inhibitors of CDC25 phosphatases) [17].

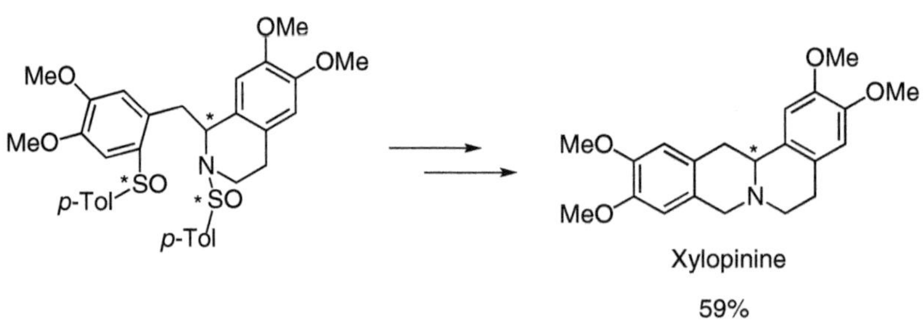

Scheme 13.11 Chiral sulfinyl intermediates in the synthesis of xylopinine. Source: Mastranzo et al. [16]/American Chemical Society.

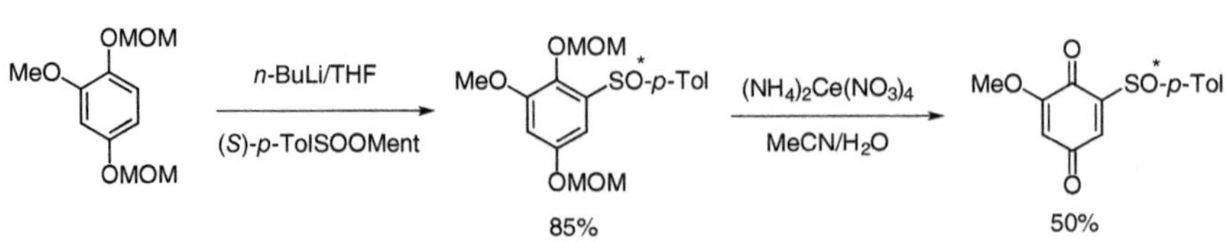

Scheme 13.12 Chiral sulfinyl quinonoid structures.

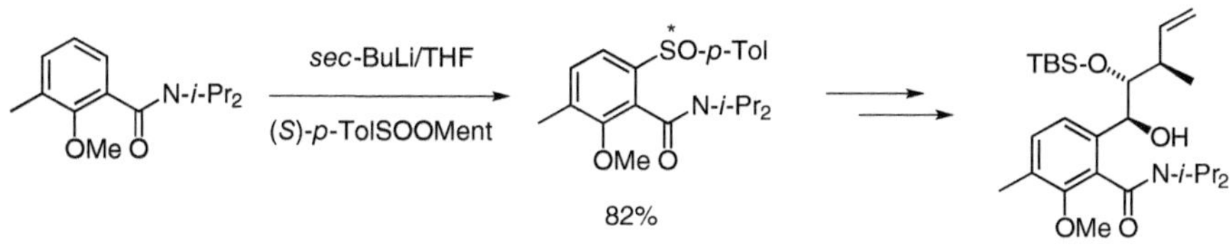

Scheme 13.13 Sulfinyl-based synthesis of the core of ajudazol. Source: Essig and Menche [18]/American Chemical Society.

In Menche's group, the reaction of menthyl *p*-toluenesulfinate with an *ortho*-lithiated aryl species was used to obtain the isochromanone core of ajudazol (Scheme 13.13) [18], and later to accomplish the total synthesis of this potent respiratory chain inhibitor [19].

This valuable synthetic scheme was repeated with a different aryl starting material to obtain a chiral precursor of atropisomeric building blocks (Scheme 13.14) that eventually originated isoplagiochin D [20, 21].

6-Bromo-5-iodobenzo[*d*][1,3]dioxole was reacted with *iso*-propylmagnesium chloride. The corresponding organomagnesium reagent was reacted with menthyl *p*-toluenesulfinate to yield an intermediate in the synthesis of steganone (Scheme 13.15) [22].

N-(2-bromophenyl)pivalamide was lithiated with *n*-butyllithium and then reacted with menthyl *p*-toluenesulfinate (Scheme 13.16), yielding the valuable 2-(*p*-tolylsulfinyl)aniline that could be used in the synthesis of bioactive cyclopropanes [23].

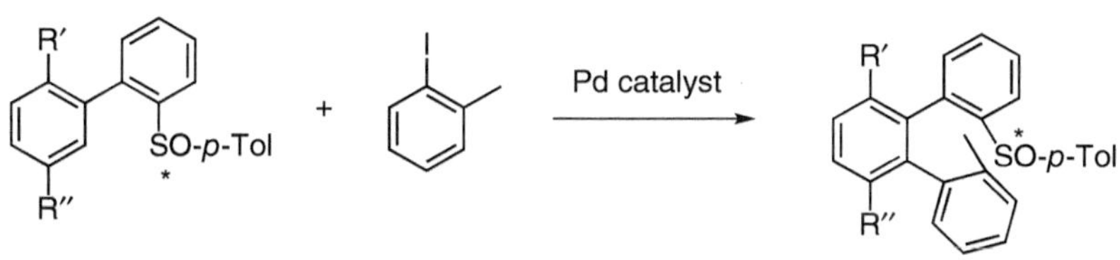

Scheme 13.14 Chiral sulfoxides in the synthesis of atropisomeric structures.

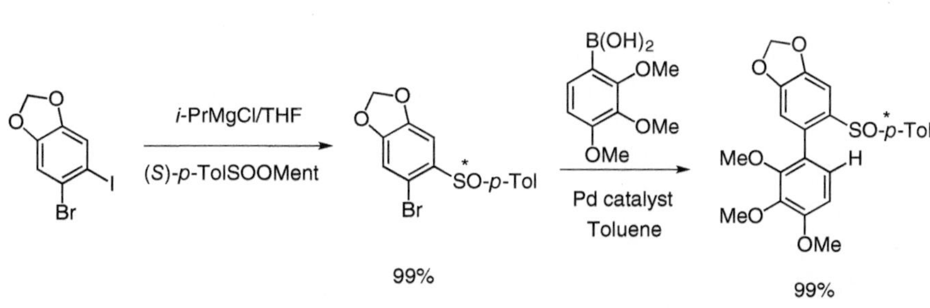

Scheme 13.15 A sulfinyl route for the synthesis of steganone. Source: Dherbassy et al. [22]/Elsevier.

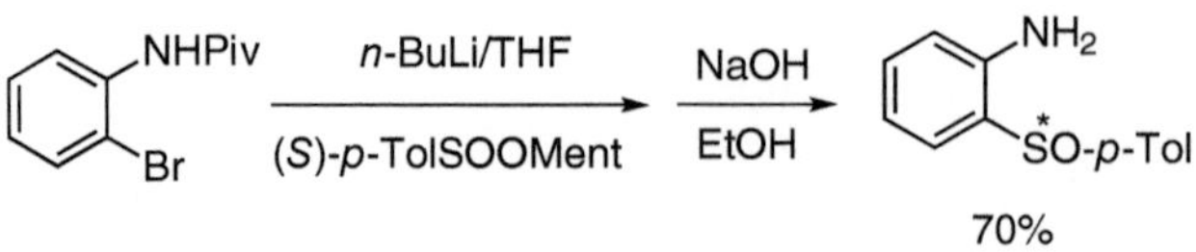

Scheme 13.16 Synthesis of chiral 2-(*p*-tolylsulfinyl)aniline.

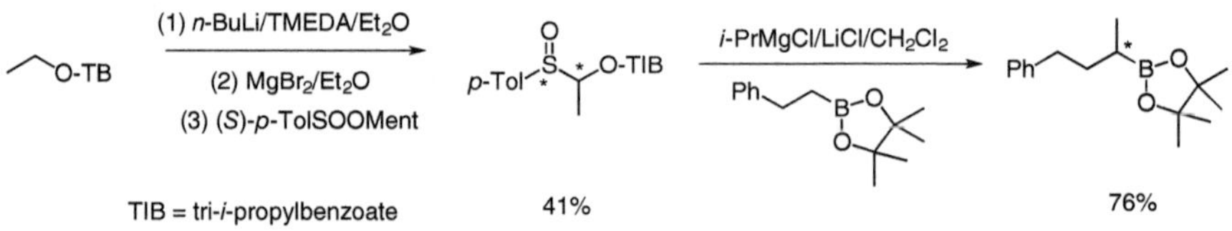

Scheme 13.17 Homologation of boronic esters with chiral sulfinyl compounds.

Other potentially useful sulfinyl building blocks were synthesized from menthyl *p*-toluenesulfinate, such as chiral sulfinyl bearing ferrocenes [24] or furans [25].

In a different research, sulfinyltriisopropylbenzoates, obtained with the Andersen procedure starting from menthyl *p*-toluenesulfinate, were employed as precursors to lithium and magnesium carbenoids (Scheme 13.17) for the stereoselective iterative homologation of boronic esters [26]. Magnesium carbenoids had a better stereochemical profile compared with lithium analogs.

Enantiopure sulfoxides have been used as chiral ligands in asymmetric synthesis. It is worth mentioning that a sulfoxide-oxazoline ligand of the palladium was successfully used in a branch-selective allylic C–H carboxylation of terminal alkynes [27], shown in Scheme 13.18. In other cases, this sulfinyl ligand was also obtained starting from the menthyl *p*-toluenesulfinate.

In summary, the extensive employment of the Andersen's reagent in asymmetric synthesis is expected. However, there is a clear distinction between a strategy that aims to synthesize a specific sulfoxide (for example, a blockbuster drug, such as (*S*)-omeprazol) and the employment of a chiral sulfinyl building block that has to be effective in the formation of a new stereogenic center.

In the first case, it is appropriate if hundreds of literature reports explore every possibility to obtain better yields or better stereochemical profiles. On the other hand, a sulfinyl chirality inductor is often effective independently on a group bound to it. Thus, it is hard to justify substituting the *p*-tolyl group of the menthyl *p*-toluenesulfinate with a different one, proposing a new chirality inductor. The

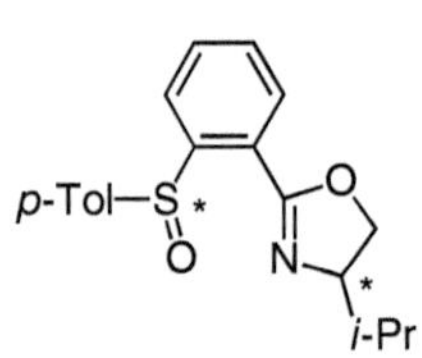

Scheme 13.18 A sulfoxide-oxazoline ligand.

Andersen's reagent is cheap and easily available. It is "natural" to be examined as the first choice in planning a new asymmetric synthesis.

13.3.2 (*R*)-*p*-Toluenesulfinamide (Davis' Reagent)

(*R*)-*p*-Toluenesulfinamide was introduced in the asymmetric synthesis by Davis [28]. Like many other sulfinyl chiral building blocks, it was synthesized starting from the menthyl (*S*)-*p*-toluenesulfinate (Scheme 13.19) upon reaction with lithium hexamethyldisilazane (LiHMDS) [29]. Even though it was obtained from the Andersen's reagent, in this work, it is treated in a separate chapter because the presence of the amino moiety close to the sulfinyl group transforms its reactivity and opens the route to the synthesis of many natural and bioactive compounds, in which the nitrogen atom is often present.

A typical development is the reaction of the (*R*)-*p*-toluenesulfinamide with aldehydes to yield sulfinylimines (Scheme 13.19, right). In this strategy, it must be stressed that the sulfinyl moiety also activates the C=N bond of the sulfinylimines.

This valuable synthetic scheme has had many applications in synthesizing bioactive compounds by Davis' group and many other research groups. In recent years, this idea was applied to the synthesis of (*S*)-(+)-cocaine [30]. The first synthetic steps involve the preparation of a multi-carbonyl sulfinyl intermediate by using the induction of the sulfinyl group in the addition of the enolate, as shown in Scheme 13.20.

A synthetic strategy analogous to Scheme 13.20 was used in the first steps of precursors of stemofoline alkaloids [31] and the synthesis of (−)-euphococcinine and (−)-adaline [32].

Another strategy to introduce a carbonyl moiety in the framework of the molecule is the addition of the enolate of a Weinreb's amine to a Davis's *N*-sulfinyl imine (Scheme 13.21) [33].

Davis's chemistry was also used in the total synthesis of the (−)-lirinine, an aporphine alkaloid (Scheme 13.22) [34].

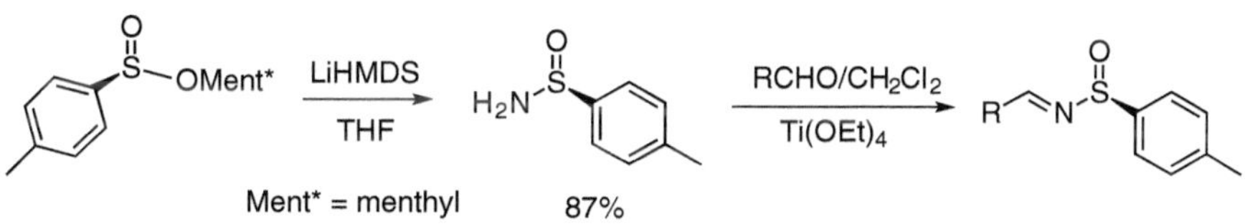

Scheme 13.19 Synthesis of *p*-toluenesulfinamide and its conversion into the corresponding imines.

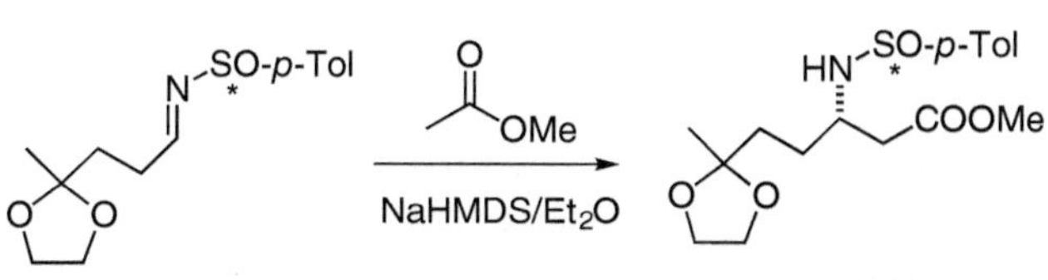

Scheme 13.20 First synthetic steps in the synthesis of (*S*)-(+)-cocaine.

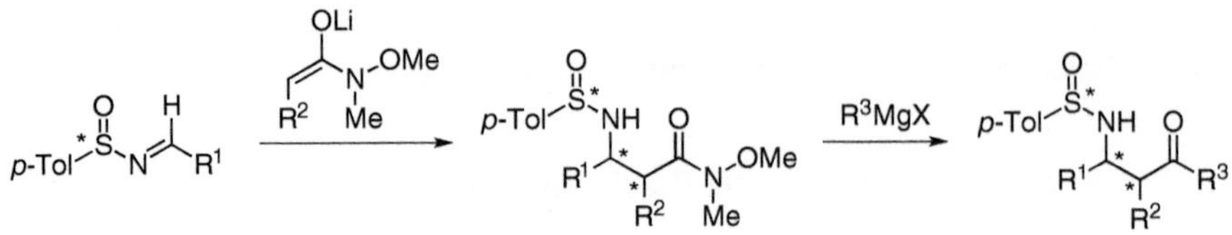

Scheme 13.21 Application of the enolate of a Weinreb's amine in the synthesis of sulfinyl bearing carbonyl compounds. Source: Davis and Xu [33]/American Chemical Society.

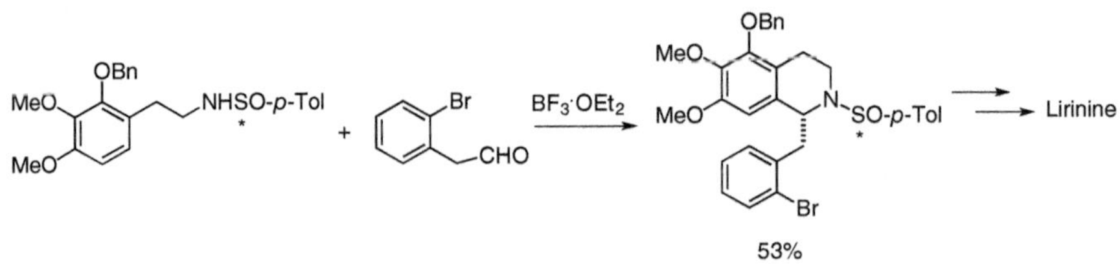

Scheme 13.22 An early step in the synthesis of lirinine. Source: Hellal et al. [34]/Elsevier.

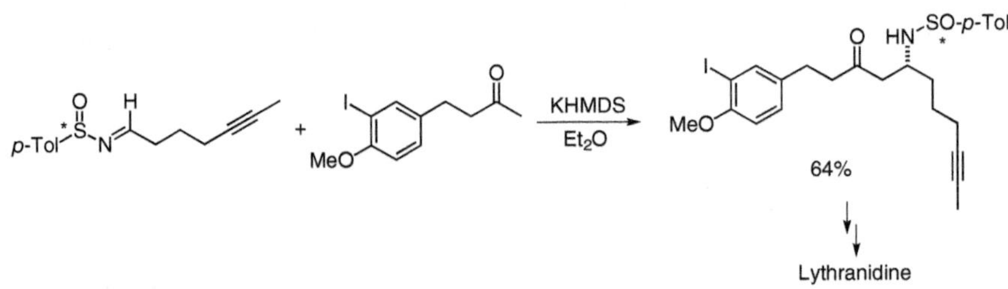

Scheme 13.23 An early step in the synthesis of lythranidine. Source: Gebauer and Furstner [35]/American Chemical Society.

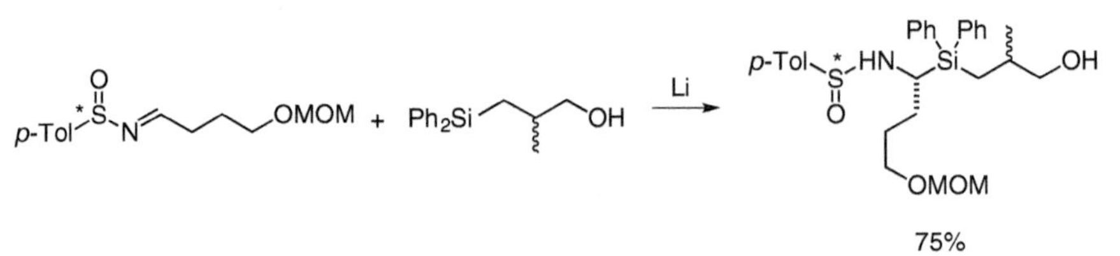

Scheme 13.24 An early step in the synthesis of serine protease inhibitors.

A synthesis of the (–)-lythranidine alkaloid used the same starting material (Scheme 13.23) [35].

A different chemistry involving silicon compounds and a Davis' imine was used to synthesize a precursor of inhibitors of the serine protease coagulation cascade (Scheme 13.24) [36].

Eventually, the Davis' reagent was applied in synthesizing Ibrexafungerp, a β-1,3-glucan synthesis inhibitor [37]. As far as the sulfinyl chemistry is used in constructing this complex molecular building, the sulfinylimine of Scheme 13.25 was transformed into an aziridine by reacting it with trimethyl sulfoxonium chloride in the presence of *n*-buthyllithium.

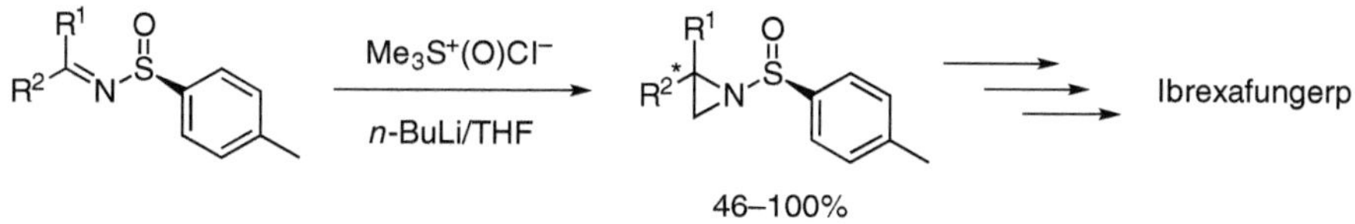

Scheme 13.25 An early step in the synthesis of ibrexafungerp.

13.3.3 (1*S*, 2*R*, 5*S*)-Menthyl (*S*)-*p*-Bromophenylsulfinate

Menthyl (*S*)-*p*-bromophenylsulfinate was reported by Capozzi et al. [38] in their synthetic and mechanistic investigation on the chemistry of carbanionic leaving groups [39]. Enantiopure dialkyl sulfoxide can be obtained starting from this substrate upon two displacement reactions with alkyl Grignard reagents: the first being the conventional displacement of the menthoxy group, as in the Andersen–Mislow procedure, and the second is the less known displacement of the *p*-bromophenyl group as a carbanionic leaving group. This substrate was synthesized starting from the corresponding *p*-bromobenzenesulfonyl chloride in the presence of phosphite as a reducing agent. The (S_S)-epimer was obtained with a Solladiè's cycle (crystallization of the (S_S)-epimer, racemization of the mother liquor, and further recovery of new batches of the (S_S)-epimer).

This interesting intermediate was used differently from the original proposal of the authors because the bromine atom on the aryl group can be considered the attachment point of new developments (such as cross-coupling reactions). In contrast, the menthyl sulfinate moiety maintains the conventional Andersen's chemistry (Scheme 13.26).

For example, Gallina and coworkers [40] transformed the menthoxy moiety of this molecule into the corresponding sulfinamide, as in Davis' chemistry. On the other hand, the bromine atom was the attachment point of a cross-coupling reaction with 4-methoxyphenylboronic acid to yield selective inhibitors of MMP8 at the end of the synthesis.

Bolm et al. transformed the menthyl *p*-bromophenylsulfinate into a sulfoxyimine, whereas the bromine atom was aminated according to the Buchwald–Hartwig procedure [41].

In the years covered by the present work, *para*-substituted Andersen's reagents were used to synthesize sulfinyl bearing ligands for ruthenium [42] and palladium [43], such as those depicted in Scheme 13.27, employed in asymmetric allylic alkylation.

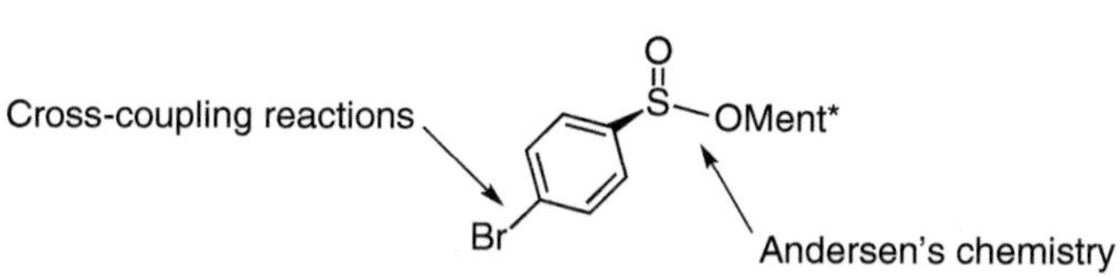

Scheme 13.26 A different menthyl arylsulfinate.

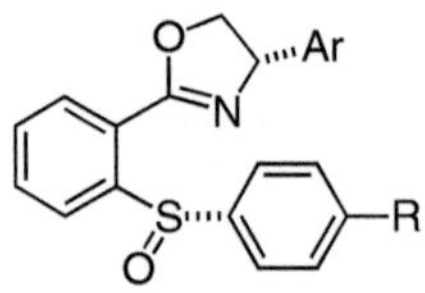

R = *tert*-Bu, Aryl

Scheme 13.27 Chiral sulfinyl ligands.

13.3.4 (*R*)-*tert*-Buthyl *tert*-Butanethiosulfinate (Ellman's Reagent)

In their important contribution to asymmetric synthesis, Ellman's group introduced *tert*-butyl *tert*-butanethiosulfinate [44], which is obtained by enantioselective vanadium-catalyzed mono-oxidation of di-*tert*-butyl sulfide (Scheme 13.28). The asymmetric oxidation of the disulfide was improved through the years by using different chiral ligands [45].

The Ellman's reagent reacts with organometallic reagents with the formation of the corresponding *tert*-butyl sulfoxide as a consequence of the enantiospecific nucleophilic substitution of the *tert*-butylthio moiety (Scheme 13.29, left).

This reagent is an innovative sulfinyl chiral building block obtained with enantioselective oxidation instead of separating a couple of diastereoisomers.

Moreover, this valuable chiral building block was transformed into the *tert*-butanesulfinamide (Scheme 13.29, right), as it occurs for the Davis's chemistry [45]. Many applications in the synthesis of valuable synthetic targets were reported for this last compound, and an excellent review is available [46].

However, special care must be undertaken in every step in which the Ellman's chemistry is applied to avoid the formation of foul-smelling *tert*-buthanethiol.

In the years covered by this work, Ellman's reagent was treated with the lithiated trimethoxytoluene (Scheme 13.30) in the first step of the enantioselective total synthesis of Yangambin [47].

A bromo-benzannulated spiroketal was lithiated and then treated with Ellman's reagent (Scheme 13.31) in the asymmetric synthesis of spiroketals, the spirocyclic core of the rubromicin antibiotics [48].

Using Ellman's reagent, Liao's group synthesized a series of *tert*-butylsulfinyl phosphine that were successfully employed in an asymmetric rhodium-catalyzed

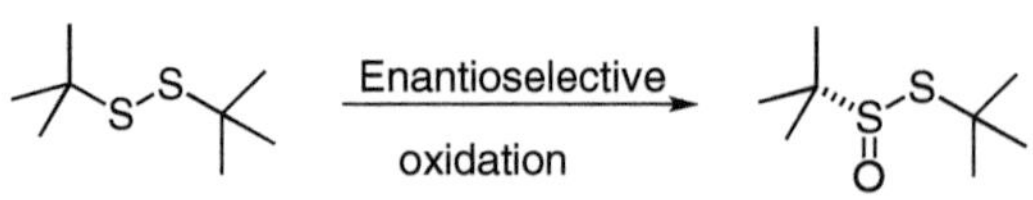

Scheme 13.28 Synthesis of (*S*)-*tert*-butyl *tert*-butanethiosulfinate.

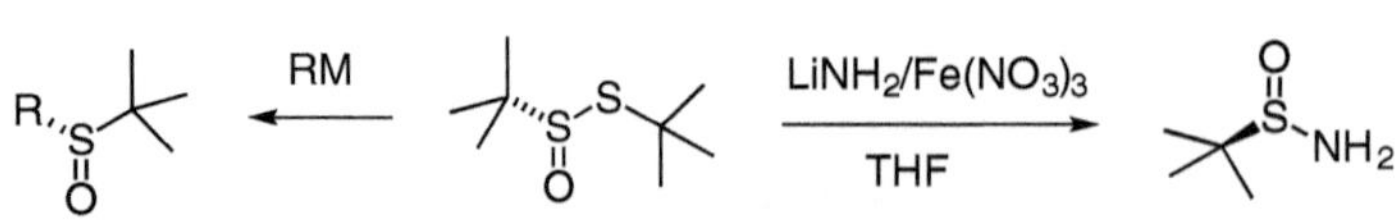

Scheme 13.29 Basic reactivity of the Ellman's reagent.

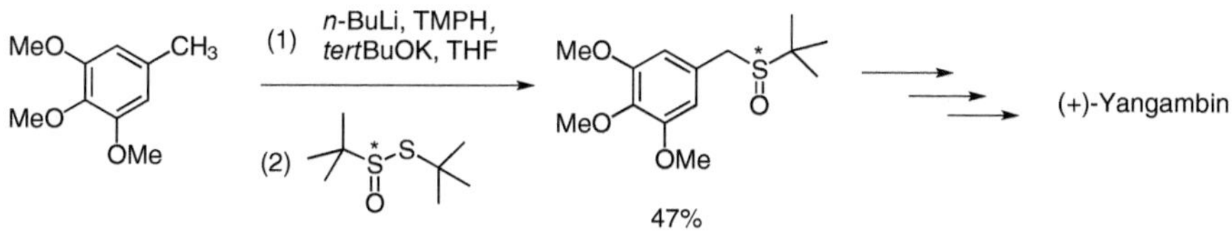

Scheme 13.30 Synthesis of a precursor of (+)-yangambin with the aid of the Ellman's reagent.

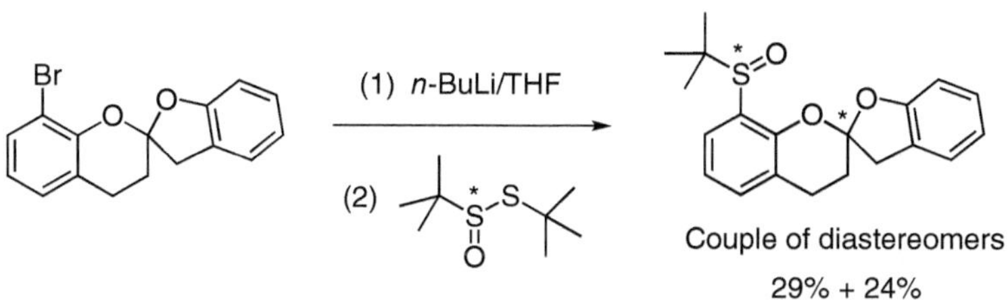

Scheme 13.31 Spiroketal intermediates with the aid of the Ellman's reagent.

asymmetric arylation [49], in an asymmetric copper-catalyzed borylstannation [50], in a copper-catalyzed asymmetric 1,6 boration [51], and in a cooperative-Cu/Pd catalyzed 1,4-arylboration [52].

Moreover, *tert*-butanesulfinamide can be used to synthesize a chiral sulfoximine, a new and potential pharmaceutical agent [53], and of a precursor of almorexant, an antagonist of the human orexin receptors [54]. One early synthetic step is depicted in Scheme 13.32.

tert-Butanesulfinamide was also reacted with a fluorinated diol in the early synthetic steps of a precursor of a potent inhibitor of *Pseudomonas aeruginosa* deacetylase (Scheme 13.33) [55].

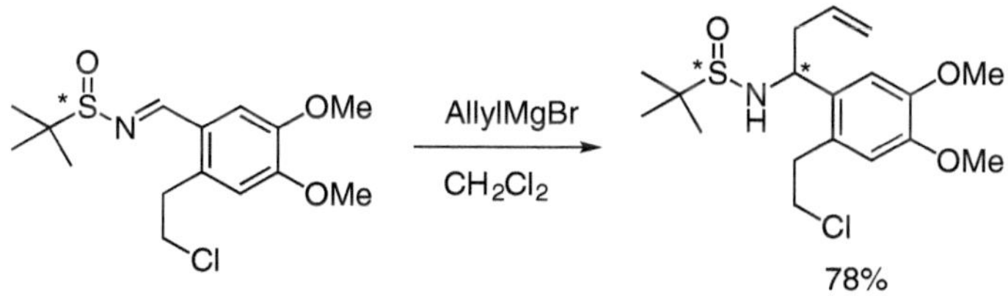

Scheme 13.32 An early step in the synthesis of Almorexant.

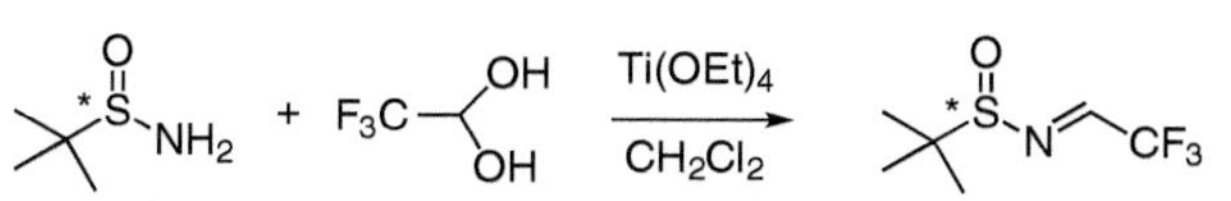

Scheme 13.33 An early step in the synthesis of an inhibitor of *Pseudomonas aeruginosa* deacetylase.

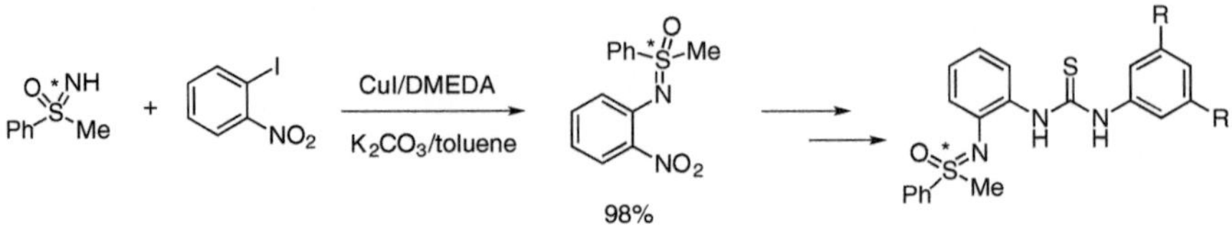

Scheme 13.34 Chiral methyl phenyl sulfoximine as a precursor of thioureas.

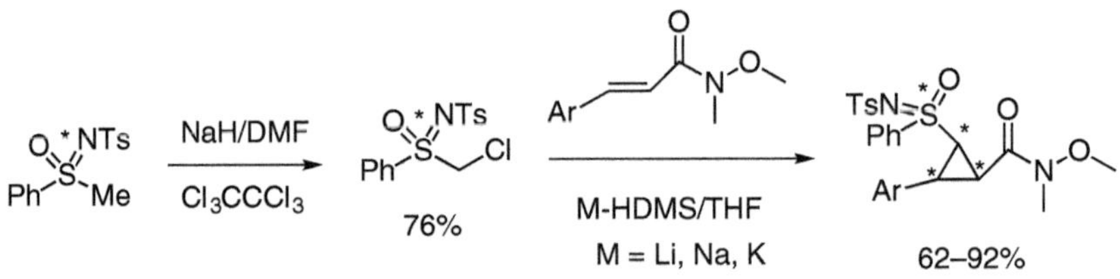

Scheme 13.35 Chiral sulfoximines as a precursor of chiral cyclopropanes.

13.3.5 Sulfoximines

A special interest was attached to these innovative bioactive compounds [56]. However, their applications have been limited since toxic or dangerous reagents were required to synthesize them. Only recently, new and mild reaction conditions [56] allowed to obtain sulfoximines safely, and an easy procedure was made [57]. This new availability will provide new reports of these interesting compounds in the coming years.

In this work, a sulfoximine as a chiral building block has the advantage of already having the sulfinyl-nitrogen bond, as it occurs in the Davis' or Ellman's chemistry.

The cross-coupling reaction with a chiral sulfoximine is a common starting point of many synthetic schemes, which occurred in the work of Bolm's group [58]. The chiral (*S*)-methyl phenyl sulfoximine was treated with 2-iodo-nitrobenzene in the presence of a copper catalyst to obtain an intermediate in the synthesis of chiral thioureas (Scheme 13.34).

(*R*)-*N*-Tosyl methyl phenyl sulfoximine was chlorinated, and the resulting chloromethyl species was reacted with an unsaturated Weinreb amide (Scheme 13.35) to yield cyclopropanes, valuable potential bioactive compounds [59].

13.4 Ideas for the Future

13.4.1 DAG-Chemistry

Since 1992, the Alcudia-Fernandez group reported the synthesis of alkyl (diacetoneglucose)sulfinate as an alternative to the Andersen's menthyl *p*-toluenesulfinate, in which the diacetoneglucose moiety substituted the original menthol. It is possible to obtain both the (R_S)- and (S_S)-stereoisomers by tuning the reaction conditions of the synthesis of this intermediate.

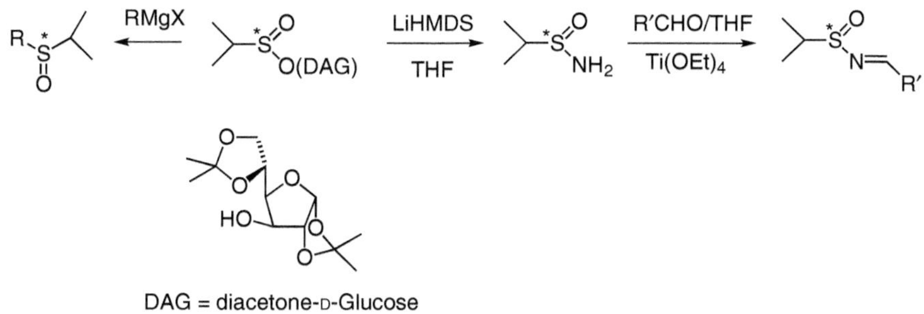

Scheme 13.36 Diacetone-D-glucose sulfinates and their employment in asymmetric synthesis.

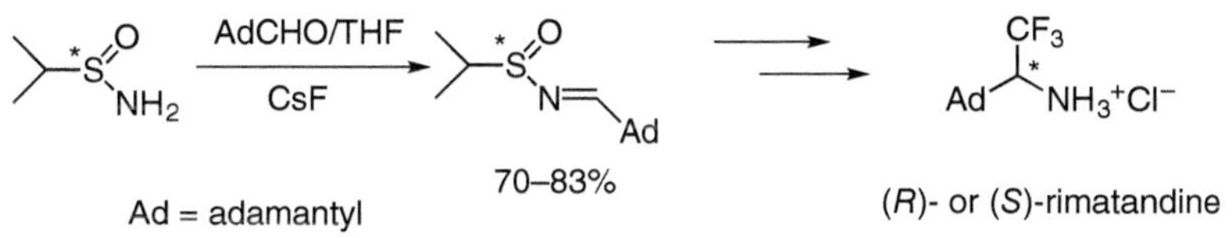

Scheme 13.37 DAG-sulfinates as precursor of rimatandine.

According to Andersen's chemistry, alkyl sulfoxides were synthesized upon substituting the diacetoneglucose moiety (Scheme 13.36, left) by Grignard reagents [60]. These substrates were also transformed to the corresponding sulfinylimine according to Davis' or Ellman's chemistry (Scheme 13.36, right). Among the various alkyl (DAG)-sulfinate, diacetoneglucose iso-propylsulfinate was claimed to have superior possibilities in synthesizing sterically hindered molecules [61]. As an example of this claim, fluorinated (*R*)- and (*S*)-rimatandine [61] were obtained with this chiral building block (Scheme 13.37).

The fact that chromatography was used to separate the epimeric (R_S)- and (S_S)-stereoisomers of alkyl (DAG)-sulfinate does not seem to encourage the employment of these chiral molecules beyond the proposing group.

13.4.2 Aryl Benzyl Sulfoxides

Some chiral sulfinyl compounds bearing two different leaving groups have been reported [2]. These building blocks yield sulfoxides with suitable reagents upon two sequential enantiospecific substitution reactions. The present work shows that these intermediates have not received particular attention outside the research groups that proposed them.

However, attention should be focused on this work due to the large family of enantiopure aryl benzyl sulfoxides that have been synthesized over the years with straightforward enantioselective oxidation. (*R*)- or (*S*)-benzyl *p*-bromophenyl sulfoxide were obtained as precursors of dialkyl sulfoxides upon two enantiospecific displacement reactions with alkyl Grignard reagents [39, 62].

It must be emphasized that the enantiopure benzyl *p*-bromophenyl sulfoxide was obtained by a favorable enantioselective titanium-catalyzed oxidation, using

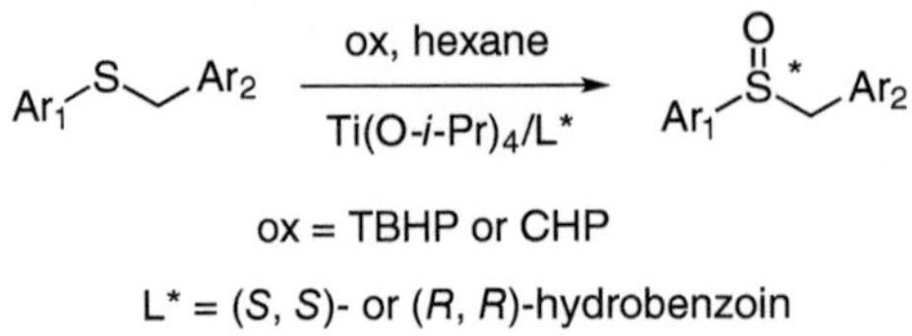

Scheme 13.38 Enantioselective oxidation of a large family of aryl benzyl sulfides by using a chiral titanium catalyst.

hydroperoxides as oxidants and the easily available (*S*, *S*)- or (*R*, *R*)-hydrobenzoin as the chirality inducer (Scheme 13.38).

Merits of this oxidation are: (i) titanium is cheap and a low-toxicity metal, (ii) a low-toxicity solvent is used, (iii) the reaction is performed at room temperature, and (iv) the crude reaction mixture can be purified without resorting to chromatography.

This oxidation was investigated from a mechanistic, synthetic, and computational point of view [63]. Further results were also summed up in the last paper of this series [64]. The final result is that this oxidation system tolerates many different substituents on the aryl groups. Thus, more than 50 enantiopure aryl benzyl sulfoxides have been easily synthesized through the years.

It is strange that this large family of structurally related chiral building blocks, in which one can choose the substrate more appropriate for the project, has had no application in synthesizing more complex molecular structures.

13.5 Conclusions

The application of chiral sulfinyl compound in asymmetric synthesis is the history of this science. Sometimes, it is indicated as an old-fashioned technology. However, the present work shows unequivocally that the application of chiral sulfinyl compounds is still a sophisticated, elegant, and successful synthetic strategy, probably the first option that should be considered when a new complex molecular building should be obtained.

References

1 Toru, T. and Bolm, C. (ed.) (2008). *Organosulfur Chemistry in Asymmetric Synthesis*. Weinheim: Wiley.

2 Kagan, H.B. (2008). Asymmetric synthesis of chiral sulfoxides. In: *Organosulfur Chemistry in Asymmetric Synthesis* (ed. T. Toru and C. Bolm), 1–29. Weinheim: Wiley.

3 Garcia Ruano, J.L., Alemán, J., Cid, M.B. et al. (2008). Asymmetric transformations mediated by sulfinyl groups. In: *Organosulfur Chemistry in Asymmetric Synthesis* (ed. T. Toru and C. Bolm), 55–159. Weinheim: Wiley.

4 Wojaczyńska, E. and Wojaczyński, J. (2020). Modern stereoselective synthesis of chiral sulfinyl compounds. *Chem. Rev.* 120: 4578–4611. https://doi.org/10.1021/acs.chemrev.0c00002.

5 Han, J., Soloshonok, V.A., Klika, K.D. et al. (2018). Chiral sulfoxides: advances in asymmetric synthesis and problems with the accurate determination of the stereochemical outcome. *Chem. Soc. Rev.* 47: 1307–1350. https://doi.org/10.1039/c6cs00703a.

6 Wojaczyńska, E. and Wojaczyński, J. (2010). Enantioselective synthesis of sulfoxides 2000–2009. *Chem. Rev.* 110: 4303–4356. https://doi.org/10.1021/cr900147h.

7 Oertling, H., Reckziegel, A., Surburg, H., Bertram, H.-J. (2007). Applications of menthol in synthetic chemistry. *Chem. Rev.* 107: 2136–2164. https://doi.org/10.1021/cr068409f

8 Carreño, M.C. (1995). Applications of sulfoxides to asymmetric synthesis of biologically active compounds. *Chem. Rev.* 95: 1717–1760. https://doi.org/10.1021/cr00038a002.

9 Bode, M.L., Gates, P.J., Gebretnsae, S.Y., Vleggaar, R. (2010). Structure elucidation and stereoselective total synthesis of pavettamine, the causal agent of gousiekte. *Tetrahedron* 66:2026–2036. https://doi.org/10.1016/j.tet.2010.01.043

10 Rival, N., Hazelard, D., Hanquet, G., Kreuzer, T., Bensoussan, C., Reymond, S., Cossy, J., Colobert, F. (2012). Diastereoselective synthesis of the C17–C30 fragment of amphidinol 3. *Org. Biomol. Chem.* 10:9418–9428. https://doi.org/10.1039/c2ob26641e

11 Bauder, C. (2015). A non-aldol preparation of enantiopure propionate-derived motifs with the assistance of chiral sulfoxides: application to a convergent synthesis of the lactone core of octalactins. *Eur. J. Org. Chem.*: 5402–5413. https://doi.org/10.1002/ejoc.201500527

12 Fernandez de la Pradilla, R., Lwoff, N., del Águila, M.A. et al. (2008). [2,3]-Sigmatropic rearrangements of 3-sulfinyl dihydropyrans: application to the syntheses of the cores of *ent*-dysiherbaine and deoxymalayamicin A. *J. Org. Chem.* 73: 8929–8941. https://doi.org/10.1021/jo8015709.

13 Fernandez de la Pradilla, R., Lwoff, N., and Viso, A. (2009). Stereoselective functionalization of dihydropyran-3-ols: application to the synthesis of enantiopure ethyl deoxymonate B. *Eur. J. Org. Chem.* 2312–2322. https://doi.org/10.1002/ejoc.200801133.

14 Sanchez-Obregon, R., Ortiz, B., Mastranzo, V.M., Yuste, F., Garcia Ruano, J.L., (2013). Short stereoselective synthesis of (+)-crispine A via an N-sulfinyl Pictet–Spengler reaction. *Tetrahedron Lett.* 54:1893–1896. https://doi.org/10.1016/j.tetlet.2013.01.121

15 Mastranzo, V.M., Olivares Romero, J.L., Yuste, F., Ortiz, B., Sanchez-Obregon, R., Garcia Ruano, J.L., (2012). Asymmetric synthesis of (*S*)-(−)-tetrahydropalmatine and (*S*)-(−)-canadine via a sulfinyl-directed Pictet-Spengler cyclization. *Tetrahedron* 68: 1266–1271. https://doi.org/10.1016/j.tet.2011.11.046

16 Mastranzo, V.M., Yuste, F., Ortiz, B., Sanchez-Obregon, R., Toscano, R.A., Garcia Ruano, J.L. (2011). Asymmetric synthesis of (*S*)-(−)-xylopinine. Use of the sulfinyl group as an *ipso* director in aromatic S_E. *J. Org. Chem.* 76:5036–5041. https://doi.org/10.1021/jo2007237

17 Evain-Bana, E., Schiavo, L., Bour, C., Lanfranchi, D.A., Berardozzi, S., Ghirga, F., Bagrel, D., Botta, B., Hanquet, G., Mori, M. (2017). Synthesis, biological evaluation and molecular modeling studies on novel quinonoid inhibitors of CDC25 phosphatases. *J. Enzyme Inhib. Med. Chem.* 32:113–118. https://doi.org/10.1080/14756366.2016.1238364

18 Essig, S. and Menche, D. (2016). Stereoselective synthesis of isochromanones by an asymmetric ortho-lithiation strategy: synthetic access to the isochromanone core of the ajudazols. *J. Org. Chem.* 81: 1943–1966. https://doi.org/10.1021/acs.joc.5b02781.

19 Wollnitzke, P., Essig, S., Gölz, J.P., von Schwarzemberg, K., Menche, D. (2020) Total synthesis of ajudazol A by a modular oxazole diversification strategy. *Org. Lett.* 22:6344–6348. https://doi.org/10.1021/acs.orglett.0c02188

20 Dherbassy, Q., Djukic, J.-P., Wencel-Delord, J., Colobert, F. (2018). Two stereoinduction events in one C–H activation step: a route towards terphenyl ligands with two atropisomeric axes. *Angew. Chem. Int. Ed.* 57:4668–4672. https://doi.org/10.1002/anie.201801130

21 Meidlinger, D., Marx, L., Bordeianu, C., Choppin, S., Colobert, F., Speicher, A. (2018). Access to the enantiopure axially chiral cyclophane isoplagiochin D through atropo-diastereoselective Heck coupling. *Angew. Chem. Int. Ed.* 57:9160–9164. https://doi.org/10.1002/anie.201803677

22 Dherbassy, Q., Wencel-Delord, J., and Colobert, F. (2016). Asymmetric C–H activation as a modern strategy towards expedient synthesis of steganone. *Tetrahedron* 72: 5238–5245. https://doi.org/10.1016/j.tet.2016.03.060.

23 Jerhaoui, S., Chahdoura, F., Rose, C. et al. (2016). Enantiopure sulfinyl aniline as a removable and recyclable chiral auxiliary for asymmetric $C(sp^3)$–H bond activation. *Chem. A Eur. J.* 22: 17397–17406. https://doi.org/10.1002/chem.201603507.

24 Utepova, I.A., Chupakin, O.N., Serebrennikova, P.O., Musikhina, A.A., Charushin, V.N. (2014). Two approaches in the synthesis of planar chiral azinylferrocenes. *J. Org. Chem.* 79:8659–8667. https://doi.org/10.1021/jo5014299

25 Montaña, A.M., Grima, P.M., Batalla, C., Kociok-Köhn, G. (2014). Synthesis of C3-substituted enantiopure 2-(*p*-tolylsulfinyl)-furans: the sulfoxide group as a chiral inductor for furan dienes as precursors of a wide variety of chiral intermediates. *Tetrahedron Asymmetry* 25:677–689. https://doi.org/10.1016/j.tetasy.2014.03.013

26 Casoni, G., Kucukdisli, M., Fordham, J.M. et al. (2017). α-Sulfinyl benzoates as precursors to Li and Mg carbenoids for the stereoselective iterative homologation of boronic esters. *J. Am. Chem. Soc.* 139: 11877–11886. https://doi.org/10.1021/jacs.7b05457.

27 Kondo, H., Yu, F., Yamaguchi, J., Liu, G., Itami, K. (2014). Branch-selective allylic C–H carboxylation of terminal alkenes by Pd/sox catalyst. *Org. Lett.* 16: 4212–4215. https://doi.org/10.1021/ol5019135

28 Davis, F.A., Zhang, Y., Andemichael, Y. et al. (1999). Improved synthesis of enantiopure sulfinimines (thiooxime *S*-oxides) from *p*-toluenesulfinamide and aldehydes and ketones. *J. Org. Chem.* 64: 1403–1406. https://doi.org/10.1021/jo9820622.

29 Fanelli, D.L., Szewczyk, J.M., Zhang, Y., Reddy, G.V., Burns, D.M., Davis, F.A. (2000). Sulfinimines (thiooximine *S*-oxides): asymmetric synthesis of methyl (*R*)-(+)-β-phenylalanate from (*S*)-(+)-*N*-(benzylidene)-*p*-toluenesulfinamide. *Org. Synth.* 77:50. https://doi.org/10.15227/orgsyn.077.0050

30 Davis, F.A., Theddu, N., and Edupuganti, R. (2010). Asymmetric total synthesis of (*S*)-(+)-cocaine and the first synthesis of cocaine C-1 analogs from *N*-sulfinyl β-amino ester ketals. *Org. Lett.* 12: 4118–4121. https://doi.org/10.1021/ol1017118.

31 Shanahan, C.S., Fuller, N.O., Ludolph, B., and Martin, S.F. (2011). Toward a total synthesis of the stemofoline alkaloids: advancement of a 1,3-dipolar cycloaddition strategy. *Tetrahedron Lett.* 52: 4076–4079. https://doi.org/10.1016/j.tetlet.2011.05.121.

32 Davis, F.A. and Edupuganti, R. (2010). Asymmetric synthesis of substituted homotropinones from *N*-sulfinyl β-amino ketone ketals. (−)-euphococcinine and (−)-adaline. *Org. Lett.* 12: 848–851. https://doi.org/10.1021/ol902910w.

33 Davis, F.A. and Xu, P. (2011). Asymmetric synthesis of *anti*-α-substituted β-amino ketones from sulfinimines. *J. Org. Chem.* 76: 3329–3337. https://doi.org/10.1021/jo2002352.

34 Hellal, M., Singh, S., and Cuny, G.D. (2012). Monoligated Pd(0)-catalyzed intramolecular *ortho*- and *para*-arylation of phenols for the synthesis of aporphine alkaloids. Synthesis of (−)-lirinine. *Tetrahedron* 68: 1674–1681. https://doi.org/10.1016/j.tet.2011.12.022.

35 Gebauer, K., Fürstner, A. (2014). Total synthesis of the biphenyl alkaloid (−)-lythranidine. *Angew. Chem. Int. Ed.* 53:6393–6396. https://doi.org/10.1002/anie.201402550

36 Duong, H.Q. and Sieburth, S.M.N. (2018). Asymmetric synthesis of silanediol inhibitors for the serine protease coagulation cascade enzyme FXIa. *J. Org. Chem.* 83: 5398–5409. https://doi.org/10.1021/acs.joc.8b00116.

37 Apgar, J.M., Wilkening, R.R., Parker, D.L. Jr., et al. (2021). Ibrexafungerp: an orally active β-1,3-glucan synthesis inhibitor. *Bioorg. Med. Chem. Lett.* 32: 127661. https://doi.org/10.1016/j.bmcl.2020.127661.

38 Capozzi, M.A.M., Cardellicchio, C., Naso, F., Spina, G., Tortorella, P. (2001). Highly stereoselective route to dialkyl sulfoxides based upon the sequential displacement of oxygen and carbon leaving groups by Grignard reagents on sulfinyl compounds. *J. Org. Chem.* 66:5933–5936. https://doi.org/10.1021/jo010334m

39 Capozzi, M.A.M., Cardellicchio, C., and Naso, F. (2004). Enantioselective routes to sulfoxides based upon the use of carbanionic leaving groups. *Eur. J. Org. Chem.* 1855–1863. https://doi.org/10.1002/ejoc.200300497.

40 Pochetti, G., Gavuzzo, E., Campestre, C., Agamennone, M., Tortorella, P., Consalvi, V., Gallina, C., Hiller, O., Tschesche, H., Tucker, P.A., Mazza. F. (2006). Structural insight into the stereoselective inhibition of MMP-8 by enantiomeric sulfonamide phosphonate. *J. Med. Chem.* 49: 923–931. https://doi.org/10.1021/jm050787

41 Cho, G.Y., Okamura, H., and Bolm, C. (2005). Synthesis and palladium catalyzed coupling reactions of enantiopure *p*-bromophenyl methyl sulfoximine. *J. Org. Chem.* 70: 2346–2349. https://doi.org/10.1021/jo047940c.

42 Trost, B.M., Rao, M., and Dieskau, A.P. (2013). A chiral sulfoxide-ligated ruthenium complex for asymmetric catalysis: enantio- and regioselective allylic substitution. *J. Am. Chem. Soc.* 135: 18697–18704. https://doi.org/10.1021/ja411310w.

43 Liu, W., Ali, S.Z., Ammann, S.E., White, M.C. (2018). Asymmetric allylic C–H alkylation via palladium(II)/*cis*-ArSOX catalysis. *J. Am. Chem. Soc.* 140:10658–10662. https://doi.org/10.1021/jacs.8b05668

44 Cogan, D.A., Liu, G., Kim, K. et al. (1998). Catalytic asymmetric oxidation of *tert*-butyl disulfide. Synthesis of *tert*-butanesulfinamides, *tert*-butyl sulfoxides, and *tert*-butanesulfinimines. *J. Am. Chem. Soc.* 120: 8011–8019. https://doi.org/10.1021/ja9809206.

45 Weix, D.J., Ellman, J.A. (2005). (R_S)-(+)-2-Methyl-2-propanesulfinamide [*tert*-butanesulfinamide]. *Org. Synth.* 82: 157–165. https://doi.org/10.15227/orgsyn.082.0157.

46 Robak, M.T., Herbage, M.A., Ellman, J.A. (2010). Synthesis and application of *tert*-butanesulfinamide. *Chem. Rev.* 110: 3600–3740. https://doi.org/10.1021/cr900382t

47 Syed, M.K., Murray, C., and Casey, M. (2014). Stereoselective synthesis of lignans of three structural types from a common intermediate, enantioselective synthesis of (+)-yangambin. *Eur. J. Org. Chem.* 5549–5556. https://doi.org/10.1002/ejoc.201402584.

48 Aitken, H.R.M., Furkert, D.P., Hubert, J.G. et al. (2013). Enantioselective access to benzannulated spiroketals using a chiral sulfoxide auxiliary. *Org. Biomol. Chem.* 11: 5147–5155. https://doi.org/10.1039/C3OB41065J.

49 Wang, J., Wang, M., Cao, P., Jiang, L., Chen, G., Liao, J. (2014). Rhodium-catalyzed asymmetric arylation of β,γ-unsaturated α-ketoamides for the construction of nonracemic γ,γ-diarylcarbonyl compounds. *Angew. Chem. Int. Ed.* 53:6673–6677. https://doi.org/10.1002/anie.201403325.

50 Jia, T., Cao, P., Wang, D. et al. (2015). Copper-catalyzed asymmetric three-component borylstannation: enantioselective formation of C–Sn bond. *Chem. A Eur. J.* 21: 4918–4922. https://doi.org/10.1002/chem.201500060.

51 Lou, Y., Cao, P., Jia, T. et al. (2015). Copper-catalyzed enantioselective 1,6-boration of *para*-quinone methides and efficient transformation of *gem*-diarylmethine boronates to triarylmethanes. *Angew. Chem. Int. Ed.* 54: 12134–12138. https://doi.org/10.1002/anie.201505926.

52 Liao, Y, Yin, X., Wang, X., Yu, W., Fang, D., Hu, L., Wang, M., Liao, J. (2020). Enantioselective synthesis of multisubstituted allenes by cooperative Cu/Pd-catalyzed 1,4-arylboration of 1,3-enynes. *Angew. Chem. Int. Ed.* 59: 1176–1180. https://doi.org/10.1002/anie.201912703.

53 Aota, Y., Kano, T., and Maruoka, K. (2019). Asymmetric synthesis of chiral sulfoximines through the *S*-alkylation of sulfinamides. *Angew. Chem. Int. Ed.* 58: 17661–17665. https://doi.org/10.1002/anie.201911021.

54 Senkar Reddy, N.S. and Subba Reddy, B.V. (2014). Stereoselective total synthesis of almorexant. *Tetrahedron Lett.* 55: 3157–3159. https://doi.org/10.1016/j.tetlet.2014.03.130.

55 Piizzi, G., Parker, D.T., Peng, Y., Dobler, M., Patnaik, A., Wattanasin, S., Liu, E., Lenoir, F., Nunez, J., Kerrigan, J., McKenney, D., Osborne, C., Yu, D., Lanieri, L., Bojkovic, J., Dzink-Fox, J., Lilly, M.-D., Sprague, E.R., Lu, Y., Wang, H., Ranjitkar, S., Xie, L., Wang, B., Glick, M., Hamann, L.G., Tommasi, R., Yang, X., Dean, C.R. (2017). Design, synthesis, and properties of a potent inhibitor of *Pseudomonas aeruginosa* deacetylase LpxC. *J. Med. Chem.* 60:5002–5014. https://doi.org/10.1021/acs.jmedchem.7b00377

56 Han, Y., Xing, K., Zhang, J. et al. (2021). Application of sulfoximines in medicinal chemistry from 2013 to 2020. *Eur. J. Med. Chem.* 209: 112885. https://doi.org/10.1016/j.ejmech.2020.112885.

57 Zenzola, M., Doran, R., Degennaro, L., Luisi, R., Bull, J.A. (2016). Transfer of electrophilic NH using convenient sources of ammonia: direct synthesis of NH sulfoximines from sulfoxides. *Angew. Chem. Int. Ed.* 128:1–6. https://doi.org/10.1002/anie.201602320.

58 Frings, M., Thomé, I., Bolm, C. (2012). Synthesis of chiral sulfoximine-based thioureas and their application in asymmetric organocatalysis. *Beilstein J. Org. Chem.* 8: 1443–1451. https://doi.org/10.3762/bjoc.8.164

59 Shen, X., Liu, Q., Zhang, W., Hu, J. (2016) Stereoselective synthesis of (sulfonimidoyl)cyclopropanes with (*R*)-PhSO(NTs)CH_2Cl and α,β-unsaturated Weinreb amides: tuning the of selectivity between C–Cl and C–S bond cleavage. *Eur. J. Org. Chem.* 906–909. https://doi.org/10.1002/ejoc.201501611.

60 Fernández, I., Khiar, N. (2003). Recent developments in the synthesis and utilization of chiral sulfoxides. *Chem. Rev.* 103: 3651–3706. https://doi.org/10.1021/cr990372u

61 Moreno, N., Recio, R., Valdivia, V., Khiar, N., Fernandez, I. (2019). *N*-Isopropylsulfinylimines vs. *N*-*tert*-butylsulfinylimines in the stereoselective synthesis of sterically hindered amines: an improved synthesis of enantiopure (*R*)- and (*S*)-rimantadine and the trifluoromethylated analogues. *Org. Biomol. Chem.* 17: 9854–9858. https://doi.org/10.1039/c9ob02241d

62 Capozzi, M.A.M., Cardellicchio, C., Naso, F., and Rosito, V. (2002). A straightforward enantioselective route to dialkyl sulfoxides based upon two carbon-for-carbon substitution reactions on the sulfinyl groups. *J. Org. Chem.* 67: 7289–7294. https://doi.org/10.1021/jo025974j.

63 Naso, F., Capozzi, M.A.M., Bottoni, A., Calvaresi, M., Bertolasi, V., Capitelli, F., Cardellicchio, C. (2009). A combined theoretical and experimental investigation on the enantioselective oxidation of aryl benzyl sulfides in the presence of a chiral titanium catalyst. *Chem. A Eur. J.* 15: 13417–13426. https://doi.org/10.1002/chem.200902110

64 Capozzi, M.A.M., Frascaro, V., Pescitelli, G., Cardellicchio, C. (2019). New insights into the titanium-mediated enantioselective oxidation of fluorinated aryl benzyl sulfides and aryl phenacyl sulfides. *Tetrahedron* 75: 2406–2412. https://doi.org/10.1016/j.tet.2019.03.016

14

Organoselenium Compounds as Chiral Building Blocks

Luana Bagnoli and Claudio Santi

University of Perugia, Group of Catalysis, Synthesis and Organic Green Chemistry, Department of Pharmaceutical Sciences, Via del Liceo, 1, 06123 Perugia, Italy

14.1 Introduction

The richness of organoselenium compounds, their structural and electronic diversity, and their specific chemical behavior make them versatile building blocks, that play a significant role in modern organic synthesis. Organoselenium reagents are commonly used in nucleophilic, electrophilic, and radical reactions, generally combining high chemo-, regio-, and stereoselectivity under mild experimental conditions. Moreover, selenium functionalities can be easily reduced, eliminated, or transformed, introducing new functional groups into an organic scaffold. The main aspects of organoselenium chemistry have been recently described in a series of books and review articles [1–3]. This chapter will consider some general aspects of the chemistry of selenium compounds related to asymmetric synthesis. The possibilities of using chiral diselenides as building blocks in an electrophilic selenylation reaction and catalysts in asymmetric bromolactonization and the asymmetric alkylation of aldehydes mediated by dialkyl zinc reagents were also largely collected and reported [4–8].

These aspects will be discussed shortly in this chapter, paying more attention to the electrophilic asymmetric selenofunctionalization representing the history and pioneer approach of organoselenium chemistry to asymmetric synthesis. This chapter will concentrate on using vinyl selenones as a versatile building block to develop various and powerful methods of enantioselective synthesis because it is an aspect scarcely covered by previous review articles.

14.2 Asymmetric Selenofuctionalization Reactions Promoted by Electrophilic Selenium Reagents

In the last 30 years, optically pure diselenides have been successfully used as chiral building blocks in asymmetric electrophilic addition and cyclofunctionalization

Chiral Building Blocks in Asymmetric Synthesis: Synthesis and Applications, First Edition.
Edited by Elżbieta Wojaczyńska and Jacek Wojaczyński.

reactions starting from variously decorated and functionalized olefins [9–13]. After the first example reported by Tomoda et al. [14], several aromatic [9, 10, 15–21], and aliphatic diselenides [22–31] become accessible thanks to the development of new efficient procedures for the selenylation of aromatic rings functionalized (usually in *ortho* position) with chiral moieties, or into aliphatic scaffold derived from functionalized terpenes.

All these compounds were used as precursors of electrophilic reagents that can be easily obtained through the Se–Se oxidative cleavage. The methoxyselenenylation of styrene was the reaction generally used to compare the stereoselectivity of different selenenylating reagents and/or the effects of different reaction conditions (Scheme 14.1).

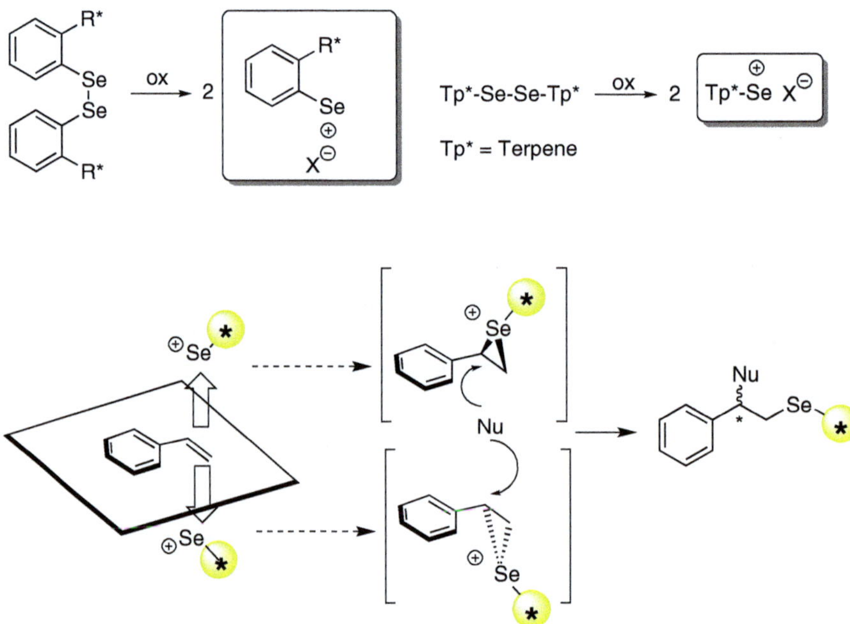

Scheme 14.1 Chiral diselenides in asymmetric synthesis.

Electrophilic addition reaction passing from the intermediate formation of a seleniranium ion is known as a highly stereoselective process. When the selenenylating reagent is chiral, adding a C=C bond leads to forming a couple of diastereoisomeric seleniranium intermediates. They will be formed depending on the energy of the corresponding transition states, and a specific selectivity in favor of one of the two pathways could be observed.

The presence of nonbonding interaction between the positively charged selenium atom and a neighbor heteroatom has been demonstrated to be crucial for the transfer of chirality starting from aromatic diselenides, and several authors investigated it for a large number of organochalcogen systems using a combination of techniques, such as X-ray crystallography, NMR, and DFT calculations. Furthermore, it was also

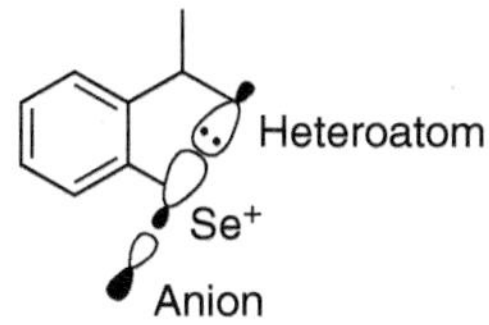

Figure 14.1 Selenium-heteroatom nonbonding interaction. Source: Based on Mukherjee et al. [32] and Tiecco et al. [33].

reported that the nature of the anion and the most apparent temperature considerably affect diastereoselectivity of the overall process [32, 33] (Figure 14.1).

Considering diselenides, it is possible to generalize that the enantioselectivity produced by the aliphatic substituents at the selenium atom in electrophilic selenofunctionalization reactions is usually lower in respect to those obtained using aromatic ones, and also the nature of the substrate and nucleophile plays a role in the stereoselection. Some examples referring- to the methoxyselenylation of styrene are shown in Scheme 14.2, evidencing the ability of a sulfur atom to coordinate the electrophilic selenium atom during the addition. It was supposed that a stronger nonbonding interaction forces the chiral moiety close to the reactive center, bringing to more rigid transition states, increasing the difference in their stability and stereoselectivity [34–38].

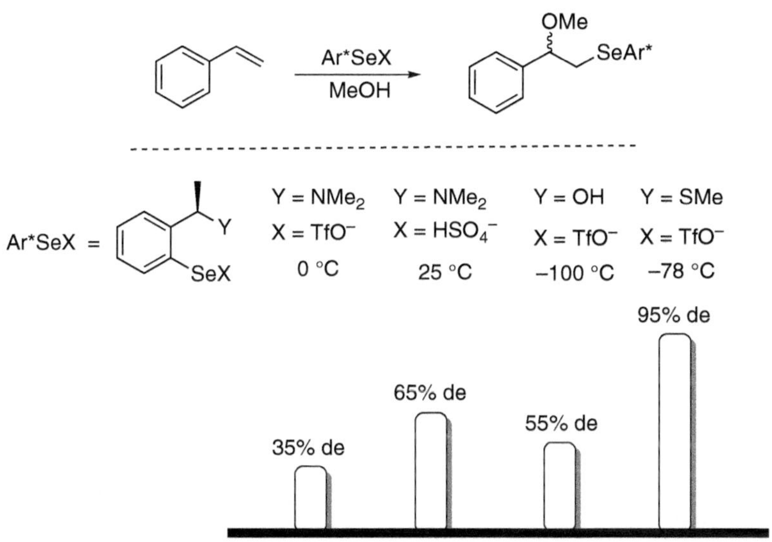

Scheme 14.2 Selected examples of asymmetric methoxyselenylation of styrene.

Similarly, when the nucleophile is embedded in the substrate at a distance suitable to attack the seleniranium ion intermediate, the selenylation of the double bond leads to an electrophilic cyclization reaction, which has been employed in the synthesis of heterocycles. Selected examples of selenocyclization reactions promoted by chiral diselenides are reported in Scheme 14.3 [9, 10, 30, 39–49]. The variety of heterocycles reported indicates the robustness and potency of the synthetic method in the construction of molecules of pharmaceutical interest. Furthermore, in some cases, the possibility of separating the formed diastereoisomers by chromatography results in enantiomerically pure compounds.

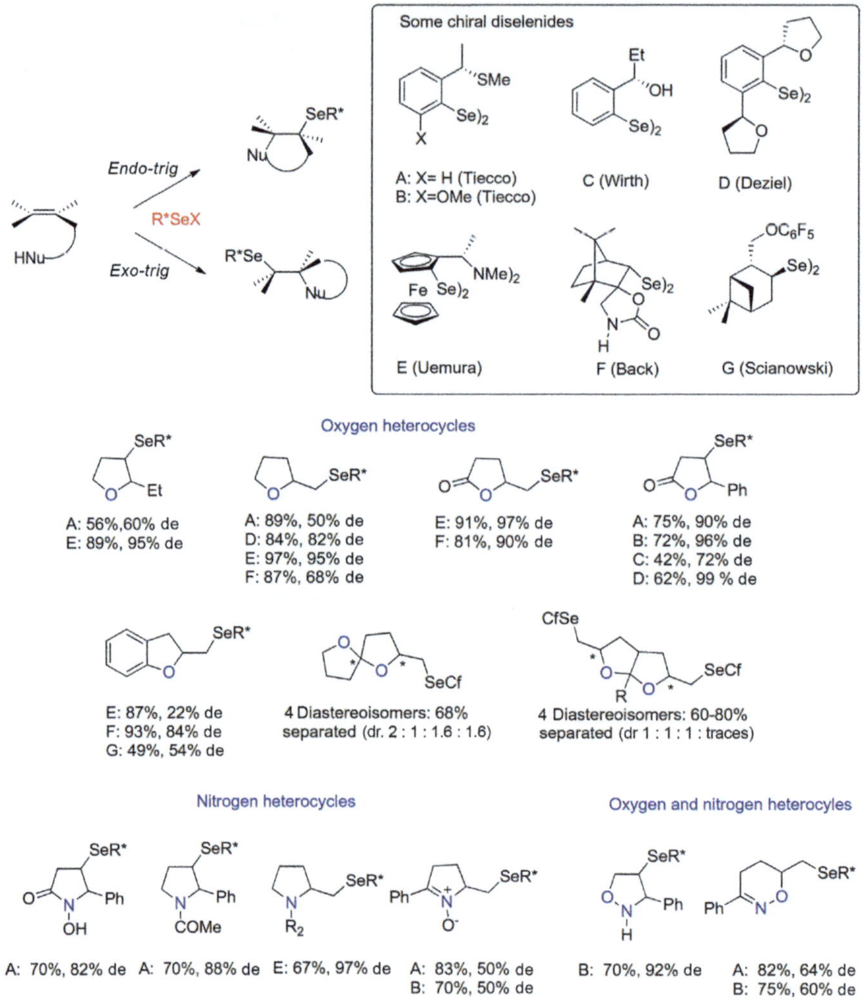

Scheme 14.3 Chiral selenium-mediated cyclization. Source: Based on Refs. [9, 10, 30, 39–49].

Optically active heterocyclic compounds can also be synthesized using a different approach in which achiral phenylselenium reagents and enantiomerically pure substrates are employed. Different mono- and byciclic oxygen heterocycles were synthesized from commercial enantiopure compounds through simple conversions promoted by organoselenium reagents. Enantiopure tetrahydrofurans were obtained through 5-*exo-trig* selenocyclization reactions promoted by the selenium electrophiles starting from enantiopure alkenols, synthesized by the opening reaction of enantiopure epoxides, and a subsequent substitution of the PhSe group by an allyl group [50]. Starting from Garder aldehyde, enantiomerically pure hydroxysubstituted tetrahydrofuran derivatives were obtained through a 5-*endo-trig* selenocyclizations of olefinic diols promoted by *N*-phenylselenophthalimide (N-PSP). The further substitution of the PhSe group in the C-4 position with an allyl group leads to the formation of allyl derivatives, which are alkenes containing a nucleophilic OH group in an appropriate position to provide a second electrophilic selenium cyclization. Enantiopure perhydrofuro[3,4-*b*]pyrans and perhydrofuro[3,4-*b*]furans were obtained starting with the compounds containing the OH and the allyl groups *trans* or *cis* to each other [51]. Enantiomerically pure 1,4-dioxanes were obtained by a short synthetic sequence with the selenium-promoted addition of enantiopure diols to β,γ-unsaturated esters and nitriles as the key step. After chromatographic separation, the enantiomerically pure diastereomeric selenoethers were transformed into the corresponding allylic ethers by deselenylation processes and finally into dioxanes by an intramolecular conjugate addition promoted by sodium hydride [52]. Selected examples are reported in Scheme 14.4.

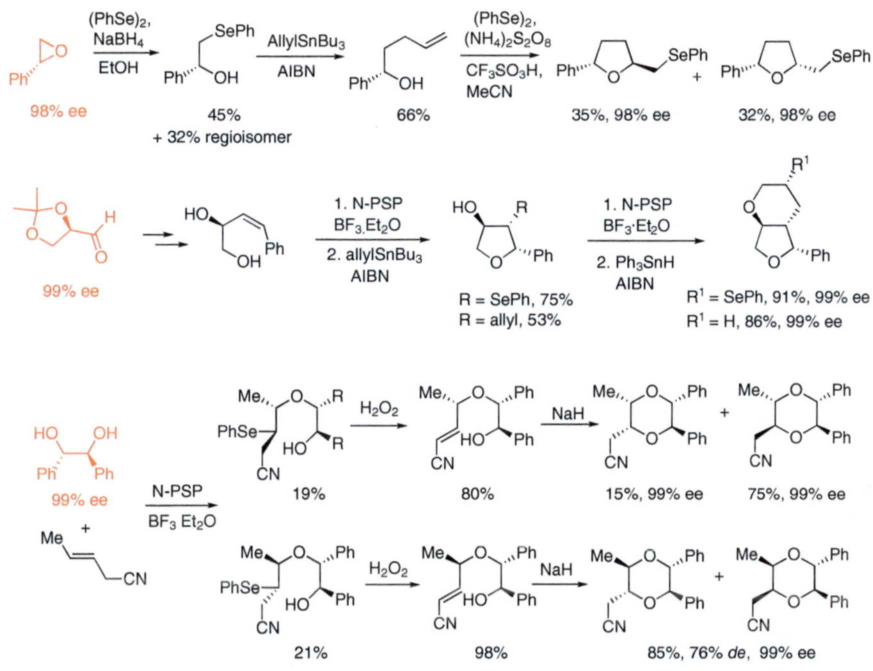

Scheme 14.4 Reagent-controlled asymmetric selenocyclizations for the synthesis of mono- and byciclic oxygen heterocycles.

Food and Drug Administration (FDA) databases indicate the structural importance of nitrogen-based heterocycles in drug design [53, 54]. Starting from commercially available enantiomerically pure amino alcohols and using simple conversion promoted by organoselenium reagents, different enantiomerically pure *cis*- and *trans*-2,5-disubstituted pyrrolidines were prepared through a *5-exo-trig* selenocyclization of *N*-Boc protected δ-alkenyl amines [55]. The radical replacement of the phenylphenyl group with an allyl group by treatment with allyl tributyltin and AIBN provides allylated pyrrolidines, containing a double bond and internal nitrogen, and a cyclization reaction results in formation of bicyclic derivatives. Hexahydro-1*H*-pyrrolizines and octahydroindolizine were isolated after chromatography in the enantiomerically pure form [56]. These azabicyclic alkaloids are widely represented in nature and exhibit interesting biological activities. Hexahydro-1*H*-pyrrolizines possess properties as histaminic H1 and H3 antagonists and, together with octahydroindolizines, represent a new class of non-opiate antinociceptive agents. A selected example is reported in Scheme 14.5.

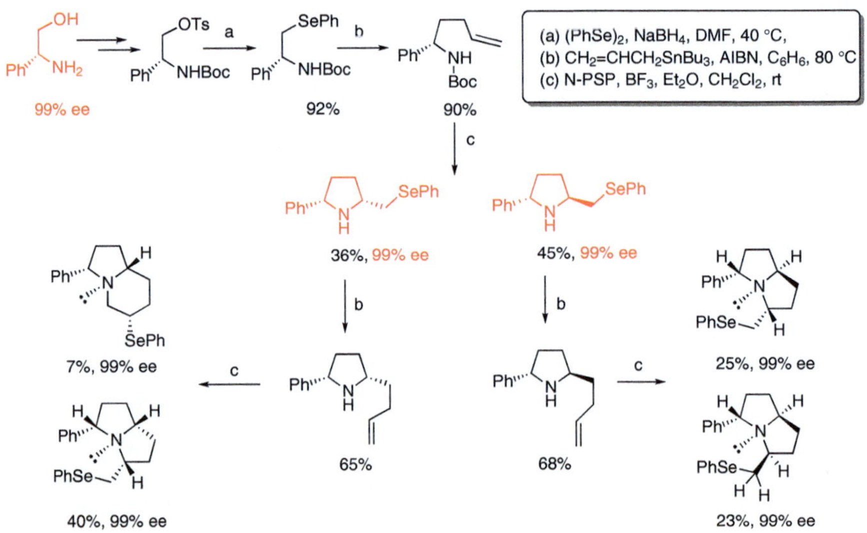

Scheme 14.5 Selenium promoted synthesis of enantiopure pyrrolidines, octahydroindolizines and hexahydro-1*H*-pyrrolizines. DMF, *N*,*N*-dimethylformamide.

The selenium moiety can be easily removed by radical reduction with triphenyltin hydride and AIBN, or it can be converted into a selenone, which undergoes substitution reactions with different nucleophiles giving pyrrolidines containing azido, methylthio, cyano, and iodo group in good yields and without loss of enantiomeric purity (Scheme 14.6) [55].

The ability of the selenonyl group to act as a good leaving group was used in the construction of enantiomerically pure substituted azetidines. The key step is a ring closure reaction, which occurs by a stereospecific intramolecular nucleophilic substitution of the selenonyl group by the nitrogen atom of arylsulfonamides (Scheme 14.7) [57].

Another reagent-controlled asymmetric selenocyclization was reported for carbohydrates. Structural analogs of the trisaccharide repeating unit from *Streptococcus pneumoniae* 19F, which is responsible for respiratory tract infections and meningitis, were prepared. The strategy involves a cross-metathesis reaction between two distinct sugar-olefins followed by the intramolecular selenocyclization of the obtained heterodimers as a key step (Scheme 14.8) [58]. The cyclized products were

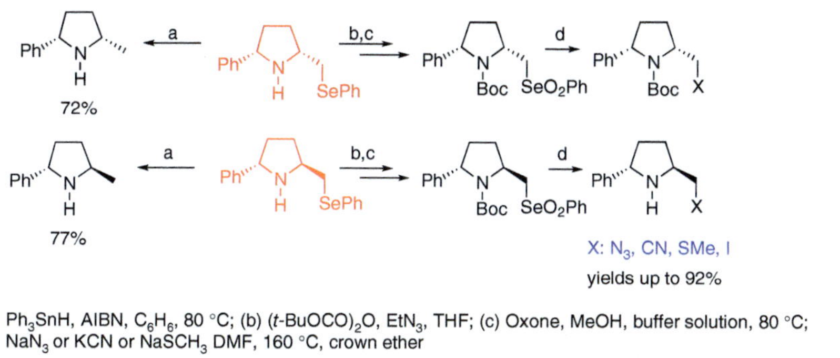

Scheme 14.6 Possible conversions of selenium functionality. Source: Tiecco et al. [55]/Elsevier.

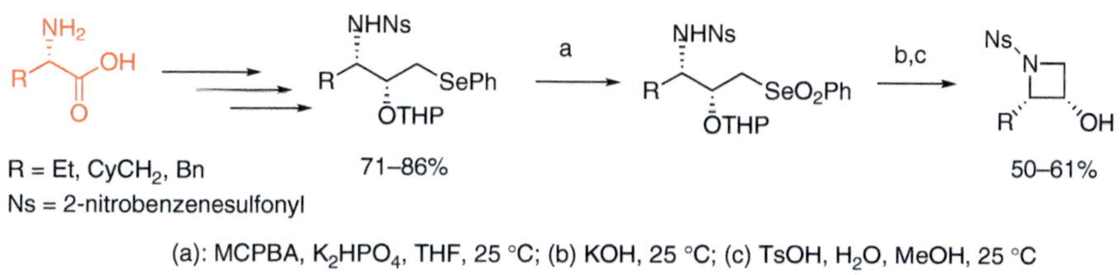

Scheme 14.7 Asymmetric synthesis of substituted azetidines. THF, tetrahydrofuran. Source: Tiecco et al. [57]/Royal Society of Chemistry.

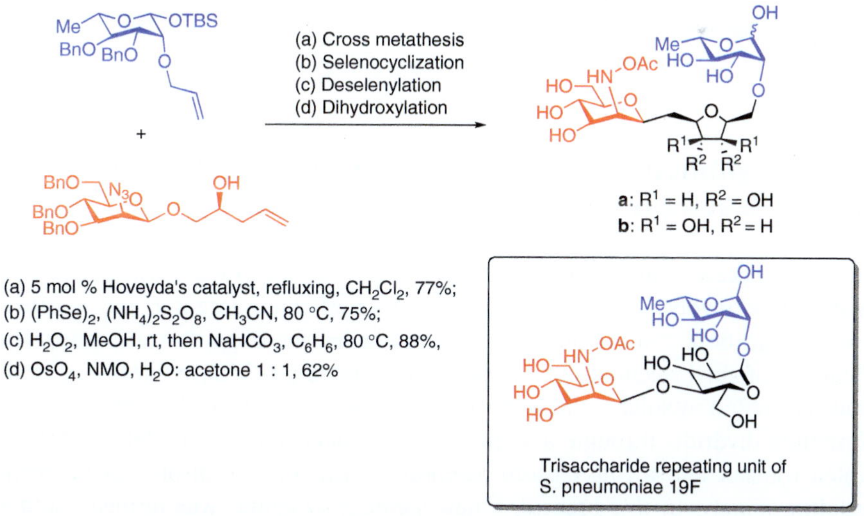

Scheme 14.8 Synthesis of bioactive oligosaccharides. Source: Ronchi et al. [58]/American Chemical Society.

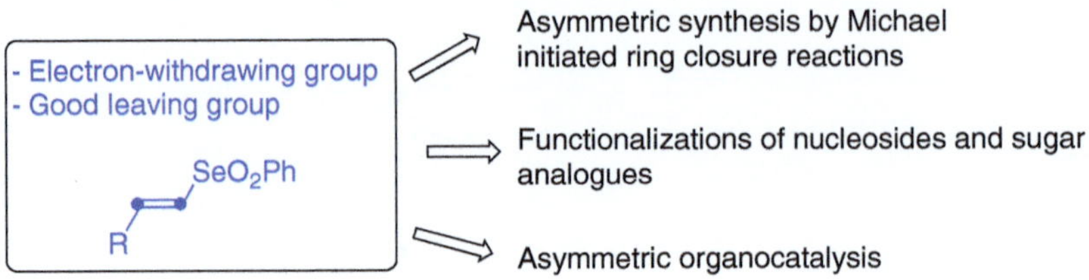

Figure 14.2 Vinyl selenones as building blocks in asymmetric synthesis.

easily subjected to the elimination reaction by oxidation of selenium species with hydrogen peroxide and finally hydroxylated via osmilation.

14.3 Vinyl Selenones as Important Building Blocks in Asymmetric Processes

Vinyl selenones represent a class of compounds containing a hexavalent, tetra-coordinated selenium atom. Although, in certain aspects, the structure and reactivity of these compounds are similar to the corresponding vinyl sulfones, some peculiar features of the chemistry of selenium favor their use. The phenylselenonyl moiety is a strong electron-withdrawing group that activates the C—C double bond to nucleophilic addition. However, the weak C—Se bond donates to phenylseleno moiety, a better leaving group, for further substitution or elimination reactions. Recent applications of vinyl selenones in domino or sequential *one-pot* processes for the asymmetric synthesis of carbo- and heterocyclic compounds using chiral auxiliarity or substrates and for the manipulation of sugars analogs and nucleosides as well as organocatalyzed enantioselective construction of all-carbon quaternary stereocenters will be reported in this section (Figure 14.2).

14.4 Asymmetric Synthesis by Michael-Initiated Ring Closure Reactions from Vinyl Selenones

Vinyl selenones are particularly suited for Michael-Initiated Ring Closure Reactions (MIRC) that involve a conjugate addition to a Michael acceptor and a subsequent cyclization. Small rings like cyclopropanes were synthesized using the chemistry of vinyl selenones and malonates containing (−)-bornyl or (−)-menthyl groups as chiral auxiliaries (Scheme 14.9). The reaction proceeds in the presence of sodium hydride through a deprotonation/Michael addition, followed by a proton transfer and a ring closure reaction by nucleophilic displacement of the selenium moiety. In this process, a new stereogenic center was formed. Even if these reactions occurred with low diastereoselectivities, the two diastereoisomers were easily separated by chromatography, giving access to enantiomerically pure compounds. Some products were converted into α-cyclopropane-α-amino acids

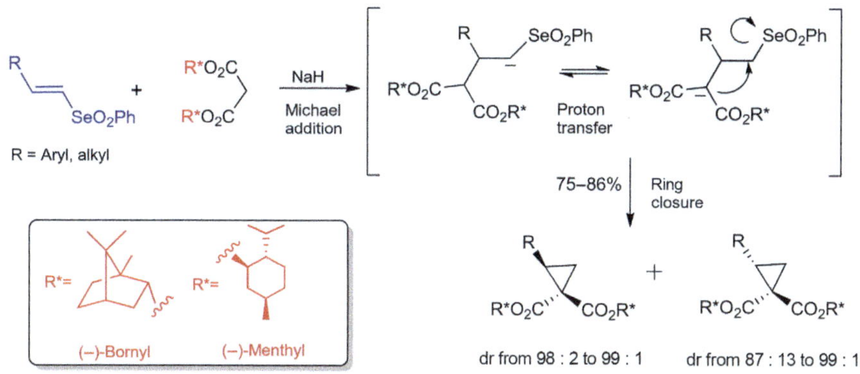

Scheme 14.9 Asymmetric cyclopropanations of chiral methylene compounds with vinyl selenones.

(ACCs), which have interesting applications in pharmacology and bioorganic chemistry [59].

Asymmetric domino processes were developed to access six-membered heterocycles using vinyl selenones as Michael acceptors and commercially available chiral 1,2-diols, *N*-protected amino alcohols, amino thiols, or diamines as *bis*-nucleophiles in the presence of a base (Scheme 14.10) [60]. The initial formation of carbanions and the subsequent proton transfer produces the negative charge displacement on the other heteroatoms, which operates intramolecular substitution of the selenonyl group. Enantiomerically pure 1,4-dioxanes, morpholines, thiomorpholine, and piperazines obtained are privileged scaffolds present in several pharmacologically active compounds like some piperazines in phase II clinical trials.

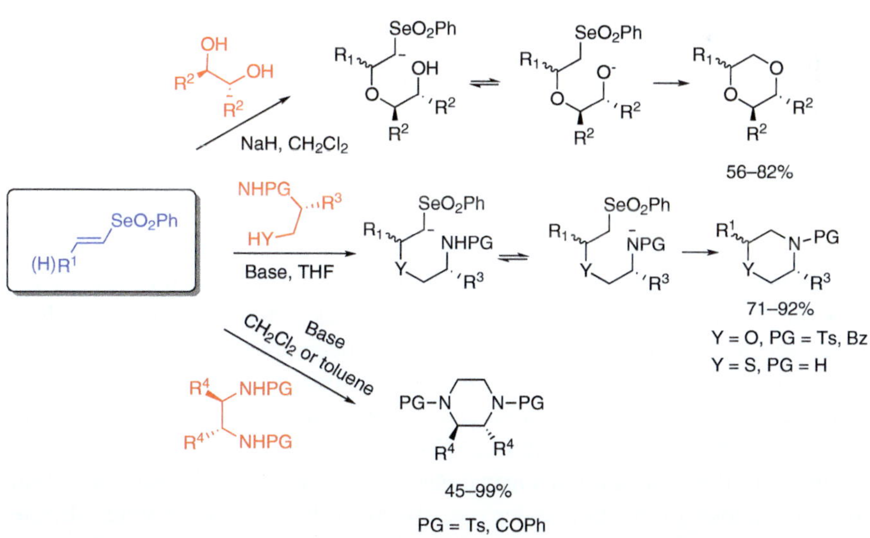

Scheme 14.10 Synthesis of chiral 1,4-dioxanes, morpholines, thiomorpholine, and piperazines. PG, protecting group. Source: Bagnoli et al. [60]/John Wiley & Sons.

Similar *one-pot* conjugate addition/cyclization was developed to synthesize chiral pyrazino indoles and pyrroles using enantiopure indole and pyrrole carboxamides as Michael donors and vinyl selenone as Michael acceptors (Scheme 14.11) [61].

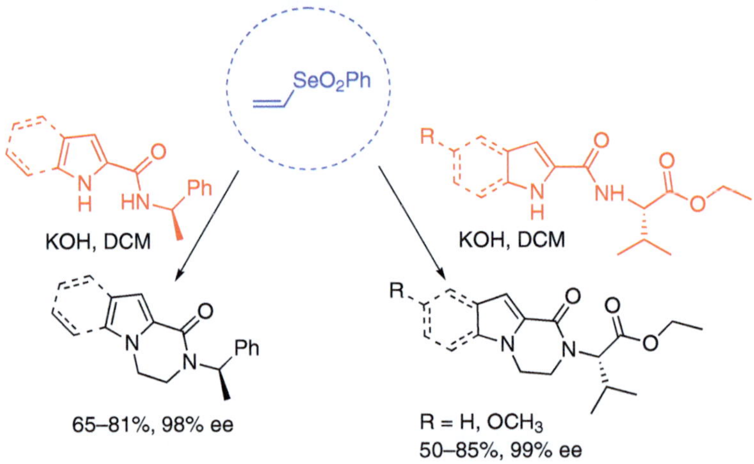

Scheme 14.11 Synthesis of enantiopure pyrazino-indoles and pyrroles. DCM, dichloromethane. Source: Palomba et al. [61]/Elsevier.

Indole and pyrrole carboxamides were prepared through coupling reactions employing enantiopure amines or amino esters (ee > 98%). The introduction of an amino acid portion on the amidic function is tolerated without racemization in the presence of potassium hydroxide. These heterocycle-fused derivatives represent privileged scaffolds and are common in drugs and natural products.

14.5 Functionalization of Vinyl Selenones of Carbohydrates and Nucleotides

The combination of chirality of biomolecules, like carbohydrates and nucleotides, with the high reactivity of vinyl selenones generate complex molecular scaffolds with defined stereochemistry. Chattopadhyaya and coworkers [62–65] reported some simultaneous functionalizations of 2′ and 3′ carbons of β-D-nucleosides as a part of the drug discovery program. In these processes, vinyl selenones act as chiral Michael acceptors with various achiral sulfur, nitrogen, oxygen, and carbon nucleophiles. Uridine derivatives obtained are potential inhibitors of HIV-specific reverse transcriptase. Selected examples of functionalization of 2′,3′ene-2′-phenylselenonyl nucleoside are reported in Scheme 14.12 [63].

In line with the utilization of sugar molecules as "chiral pool" reagents, Pathak designed D-xilose or D-fructose-derived alkenyl selenones to explore Michael initiated functionalization of sugars with small carbocycles and heterocycles rings. Methyl-α-D-2-selenonyl pent-2-enofuranoside was employed with different nucleophiles in diversity-oriented synthesis (DOS) to generate bicyclic cyclopropanated

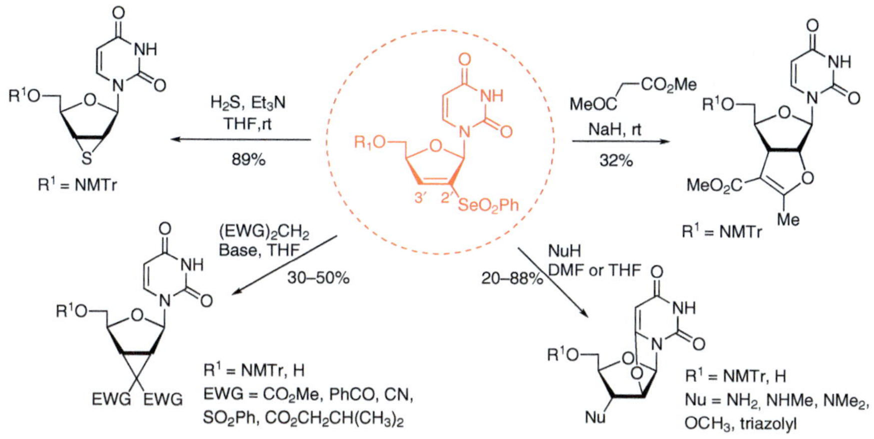

Scheme 14.12 Functionalization of 2′,3′-ene-2′-phenylselenonyl nucleosides through selenone-based transformation. Source: Tong et al. [63]/Elsevier.

carbohydrate, azasugar, dihydrofuran- and dihydroisoxazole- substituted furanosides, and isonucleosides. Different planar heterocycles, such as pyrrole, imidazole, 1,2,4-triazole, 1-*H*-tetrazole, and the nucleobase thymine, reacted with the vinyl selenone as Michael acceptor, followed by elimination of phenyl seleninic acid ($PhSeO_2H$) and concomitant elimination of methanol to afford furan-linked heterocycles (Scheme 14.13) [66].

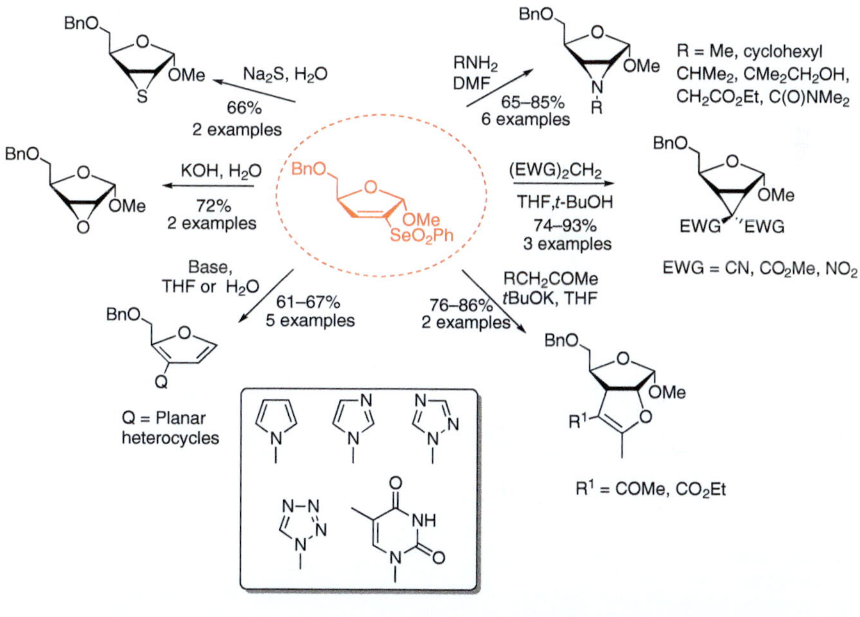

Scheme 14.13 A reactive selenosugar for the synthesis of enantiomerically pure heterocycles and carbocycles. Source: Bhaumik et al. [66]/American Chemical Society.

Similar domino reactions have been developed by constructing of vinyl selenones on the furanose and pyranose skeletons of D-fructose [67]. This practical and powerful route provides a new platform for fructochemistry (Scheme 14.14).

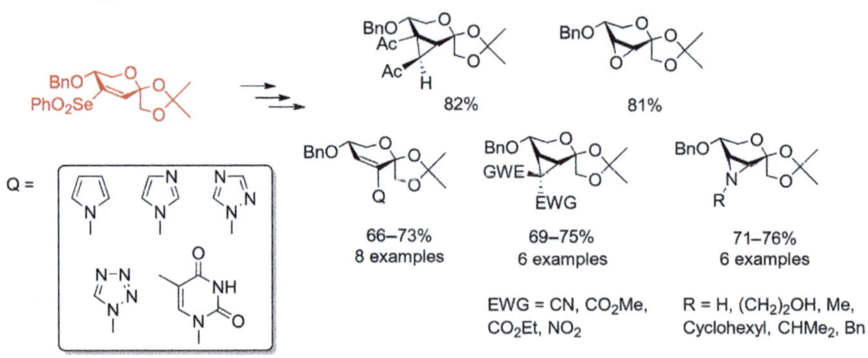

Scheme 14.14 Vinyl selenones derived from D-xylose and D-fructose in diverse orientated synthesis.

Moreover, Pathak and coworkers reported the conversion of vinyl selenone-modified carbohydrates derived from D-xylose and D-glucose into enantiomerically pure 1,4,5-trisubstituted 1,2,3-triazoles by organic azides [68]. A regioselective 1,3-dipolar cycloaddition initiated a cascade process with organic azides, followed by the elimination of benzenseleninic acid. For the opening of the sugar ring due to the cleavage of the acetal bond, this approach affords densely functionalized triazoles, which are susceptible to further modifications because of the presence of an aldehyde group and a secondary hydroxyl group (Scheme 14.15).

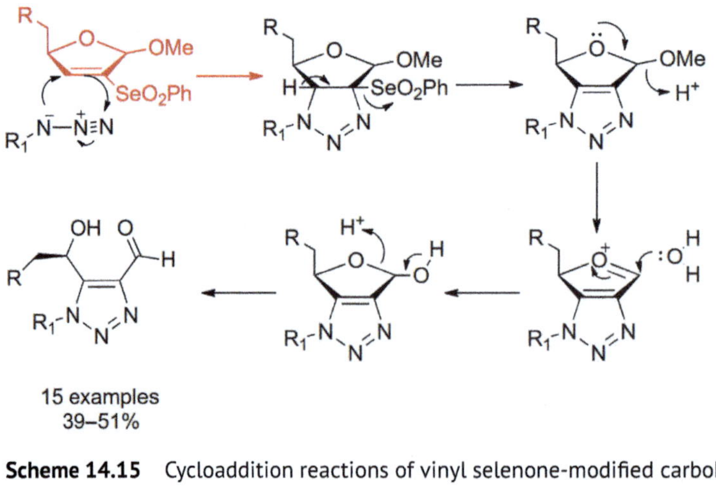

Scheme 14.15 Cycloaddition reactions of vinyl selenone-modified carbohydrates.

14.6 Asymmetric Organocatalytic Transformations Starting with Vinyl Selenones

In asymmetric organocatalytic transformations, the conversion of prochiral or racemic substrates into highly enantioenriched products is realized by the effect of a substoichiometric amount of chiral organocatalysts. These compounds are small bifunctional molecules bearing a hydrogen bond donor group, which activates the electrophile through hydrogen bonding. A basic site, usually a tertiary amine, generates the nucleophile by deprotonation. Simultaneously, the chiral catalyst activates both the Michael donor and acceptor, allowing the addition to occur with an excellent level of stereocontrol [69]. These processes are becoming an increasingly important segment of organic chemistry, offering several advantages over metal-based and bioorganic methods. Moreover, common advantages include metal-free conditions, inertness toward moisture and oxygen, availability, low cost, and low toxicity. These features are particularly useful in the production of pharmaceutical intermediates. In addition, privileged chiral organocatalysts like ureas, thioureas, C6′-hydroxyl cinchona derivatives, or squaramides are employed in Michael-type reactions between vinyl selenones and different pro-nucleophiles.

In 2009, Marini and coworkers reported the first organocatalytic preparation of densely functionalized compounds containing all-carbon quaternary stereocenters. Nowadays, the enantioselective formation of these stereocenters is one of the most exciting areas in asymmetric catalysis. Using racemic α-substituted cyanoacetates and unsubstituted phenyl vinyl selenone in the presence of a thioureidic catalyst, highly functionalized addition products were isolated in high yields (90–97%) with good to excellent enantioselectivity (76–90% ee). These Michael adducts were smoothly converted into synthetically useful polyfunctional compounds by taking advantage of the excellent leaving group ability of the selenone group. The formation of the all-carbon quaternary stereocenter bearing halogen, azido, or a terminal double bond is realized without loss of enantiomeric purity (Scheme 14.16, route 1) [70]. A sequential *one-pot* organocatalyzed enantioselective conjugate addition/cyclopropanation cascade was observed when less reactive β-substituted vinyl selenones were employed. Enantioenriched Michael adducts cyclized through an intramolecular alkylation induced by a de-ethoxycarbonylation. Two different conditions were used: Krapcho-type protocol with LiCl in HMPA or EtONa in EtOH. The sequential *one-pot* strategy generates cyclopropanes as a single *Z* isomer in good yields and enantioselectivities (Scheme 14.16, route 2) [71].

Similar asymmetric conjugate addition was reported by Zhu using a C6′-OH quinine-derived catalyst bearing an *O*-*n*-butyl group (C6′OH-Q-O*n*Bu) [72]. In all cases, the α-aryl-α-(2′-phenylselenonylethyl)-α-isocyanoacetates were obtained in good to excellent yields with an enantiomeric excess ≥74%. As reported in the selected example shown in Scheme 14.17, the resulting Michael adduct was subsequently converted into α,α-disubstituted α-amino acid and in pharmaceutically relevant heterocycles such as oxindole and pyrrolidinone. A possible transition state consistent with the observed enantioselectivity was also shown in the same scheme.

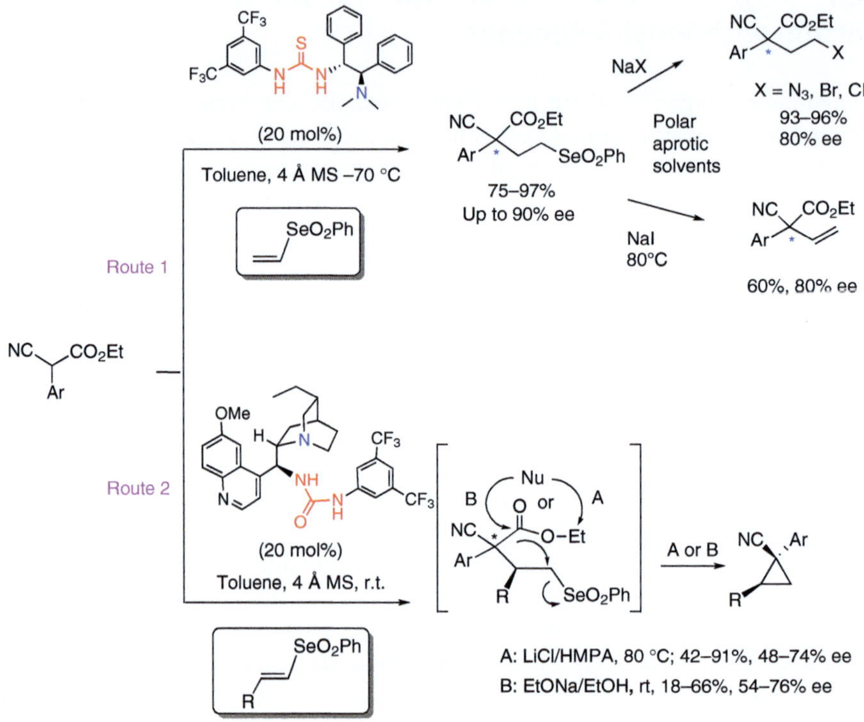

Scheme 14.16 Organocatalyzed reactions for the preparation of densely functionalized compounds and highly substituted cyclopropanes. MS, molecular sieves. Source: Marini et al. [70, 71].

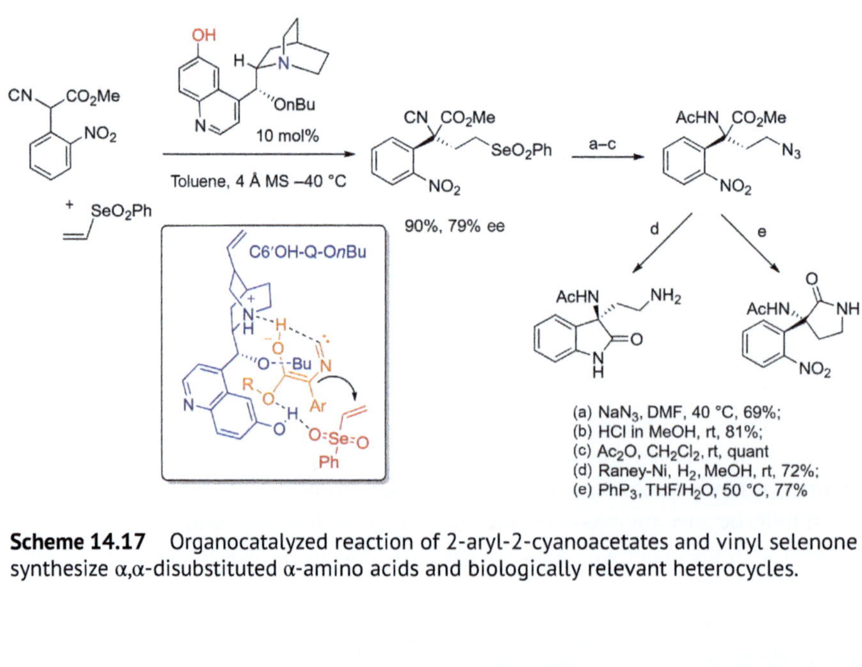

Scheme 14.17 Organocatalyzed reaction of 2-aryl-2-cyanoacetates and vinyl selenone to synthesize α,α-disubstituted α-amino acids and biologically relevant heterocycles.

This methodology was applied in the enantioselective total synthesis of both (+)- and (−)-Trigonoliimine A, an alkaloid compound that has attracted the attention of the synthetic community for its fascinating molecular structure, containing an unusual hexacyclic skeleton and a quaternary carbon center [72]. The key step is represented by the enantioselective Michael addition of isocyanoacetates to an unsubstituted vinyl selenone to generate an adduct that can readily be converted into a functionalized amino acid and after further transformations into the Trigonoliimine A (Scheme 14.18).

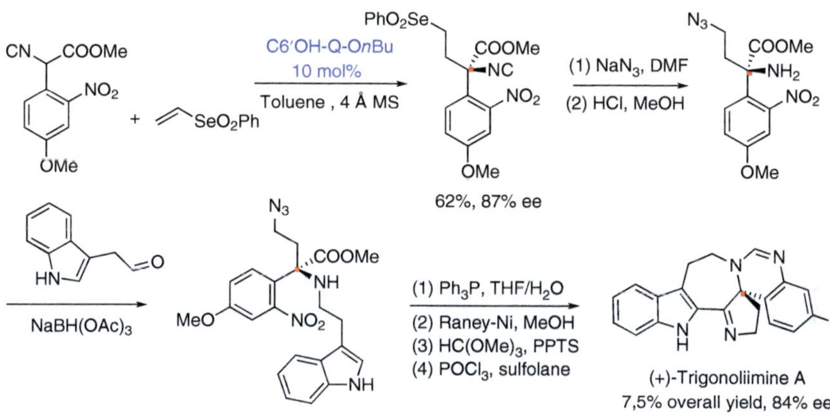

Scheme 14.18 Total synthesis of (+)-Trigonoliimine A.

Zhu and coworkers reported another interesting application in the field of natural product chemistry in the total synthesis of (+)-Hinckdentine A. This marine alkaloid was isolated from the bryozoan *Hincksinoflustra denticulate*. A key step of the synthesis is a catalytic enantioselective Michael addition of an α-aryl α-isocyanatate to phenyl vinyl selenone [73] in the presence of a quinidine-derived bifunctional catalyst (Scheme 14.19).

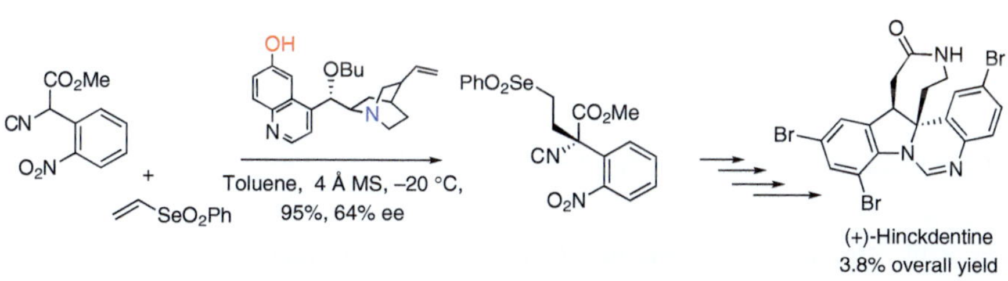

Scheme 14.19 A key step for the total synthesis of Hinckdentine A.

Using the same catalyst starting from α-substituted α-isocyanates with phenyl vinyl selenones, the 4,4-disubstituted 1,3-oxazinan-2-one was generated with an excellent enantiocontrol and in an excellent yield. This heterocyclic ring is a privileged scaffold found in bioactive and natural compounds displaying antibacterial, anti-inflammatory, anti-diabetes, and anti-HIV activities [74] (Scheme 14.20).

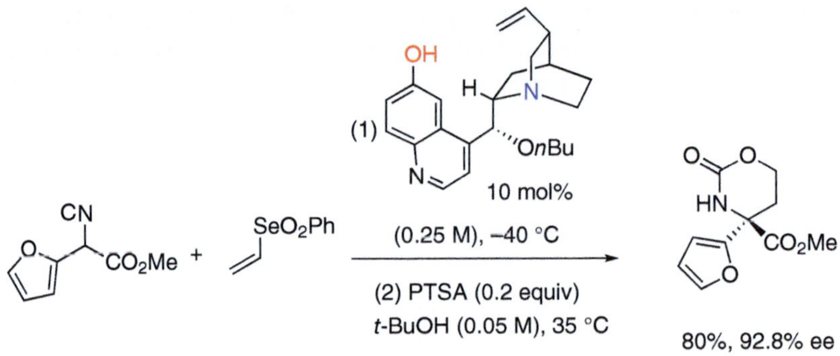

Scheme 14.20 One-pot enantioselective synthesis of 1,3-oxazinan-2-one. PTSA, *p*-toluenesulfonic acid. Source: Buyck et al. [74]/American Chemical Society.

In Scheme 14.21, the phenyl selenonyl group has a triple role in this reaction: it acts consecutively as an activator for the Michael addition, a leaving group in the intramolecular cyclization, and a latent oxidant to generate 1,3-oxazinan-2-one with the elimination of benzeneselenenic acid.

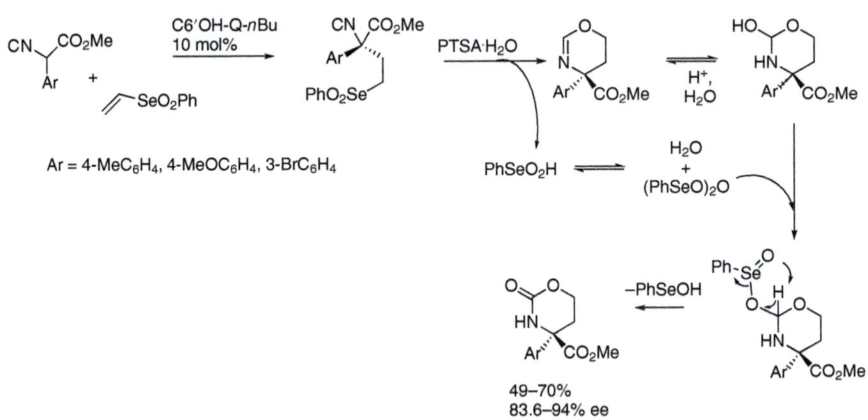

Scheme 14.21 Triple role of phenylselenonyl group.

In 2016, Zhu and coworkers developed a Cinchona alkaloid-catalyzed Michael addition between methyl α-alkyl-α-nitroacetate and phenyl vinyl selenone. The resulting adducts were obtained in good yields and enantioselectivity. Subsequently, they were transformed into α,α-disubstituted cyclic and acyclic quaternary α-amino acids taking advantage of the chemical versatility of nitro and phenylselenonyl functionalities [75] (Scheme 14.22).

Enantioselective vinylogous addition of deconjugated substituted butenolides to vinyl selenone was developed by Simlandy and Mukherjee employing a bifunctional

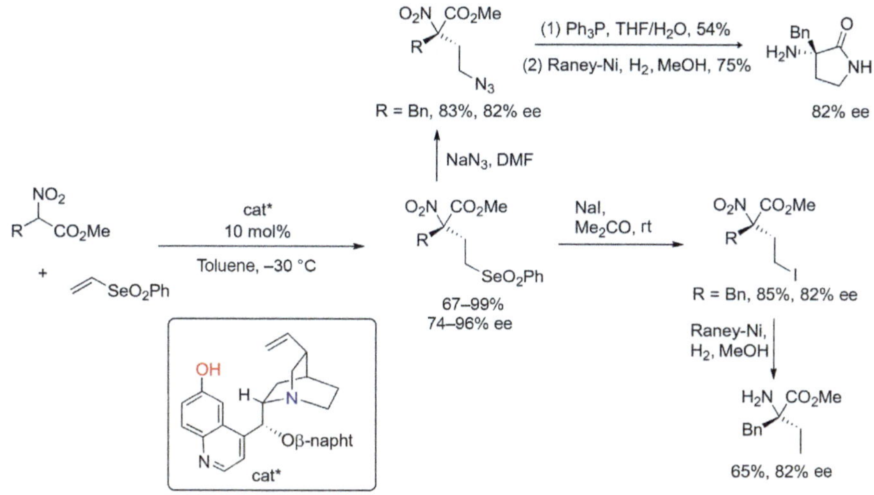

Scheme 14.22 Enantioselective synthesis of quaternary α-amino acids employing the versatility of phenylselenonyl group. Source: Clemenceau et al. [75]/John Wiley & Sons.

tertiary amino-thiourea derivative [76]. Densely functionalized γ,γ-disubstituted butenolides containing a quaternary stereogenic center were obtained in good yields and enantioselectivity. As reported in Scheme 14.23, the phenylselenonyl group of the resulting Michael adducts can be converted into an azide derivative and finally into a triazole by Cu-catalyzed click-reaction with phenylacetylene or into γ-iodoethyl butenolides using sodium iodide as the nucleophile.

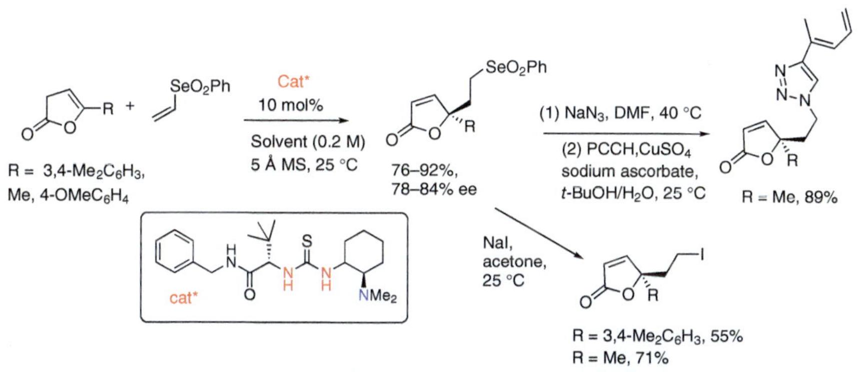

Scheme 14.23 Enantioselective vinylogous Michael addition to vinyl selenones in the synthesis of densely functionalized butenolides. PhCCH, phenylacetylene.

Asymmetric organocatalytic processes were also employed to construct polycyclic compounds such as spirolactones [77] and polycyclic pyrrolidines [78], which are present in natural products and biologically relevant compounds. The key step in both processes is the conjugate addition of cyclic *tert*-butyl-β-ketoesters to vinyl phenyl selenone catalyzed by 6′-OH 9-*O*-(-9′-phenanthryl)ether quinine derivative.

Spirolactones were obtained in good yields and excellent enantioselectivity in the presence of silica gel through a *one-pot* Michael addition/cyclization reaction, exploiting the excellent leaving ability of the selenone group and the easy removal of *tert*-butyl residue (Scheme 14.24, route 1). When a pseudoenantiomeric catalyst with the opposite configurations at C8 and C9 was employed, spirolactones with comparable yields and enantiomeric excess were obtained with the opposite enantioselectivity. Alternatively, the Michael adducts were transformed into highly enantioenriched dihydroindeno[1,2-*b*]pyrroles through a three-step sequential *one-pot* process, excluding purification steps (Scheme 14.24, route 2). At first, the Michael adducts were transformed into alkyl azides by an intermolecular nucleophilic substitution of the selenone with sodium azide. Subsequently, a Staudinger/intramolecular aza-Wittig sequence gave dihydroindeno[1,2-*b*]pyrroles with excellent enantioselectivity. Finally, cyclic β-aminoesters of indeno[1,2-*b*]pyrrolidines were obtained by reduction with sodium borohydride.

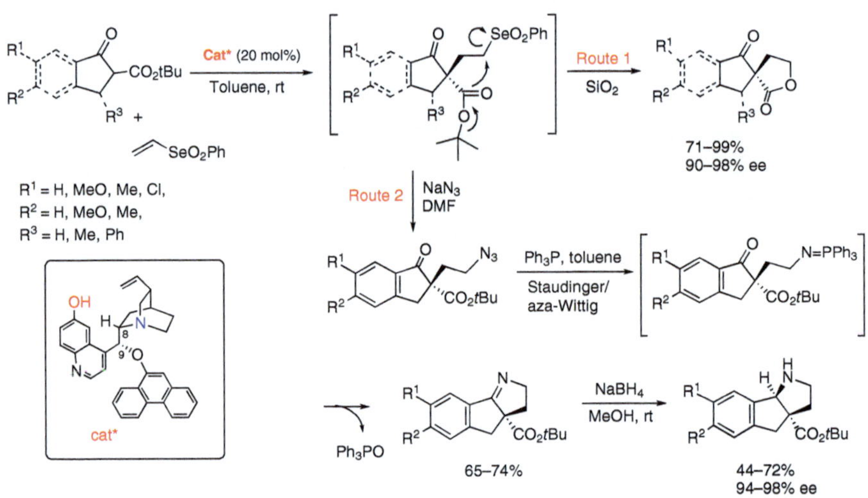

Scheme 14.24 Organocatalyzed enantioselective Michael addition/cyclization cascade to synthesize of spirolactones and cyclic β-aminoesters indeno[1,2-*b*] pyrrolidines.

Chen and coworkers reported other interesting organocatalytic processes for synthesizing polycyclic compounds. Variously substituted 2-oxindoles were added to phenyl vinyl selenone in the presence of a *Cinchona* alkaloid-based thiourea in a pyridine-based ionic liquid [79]. The Michael adducts were obtained with high yields (80–91%) and excellent enantioselectivities (85–95% ee). As shown in Scheme 14.25, the enantioenriched adduct can be converted into optically active pyrroloindoline by a nucleophilic substitution of phenylselenonyl group with sodium azide, followed by reduction and cyclization. This heterocyclic scaffold is present in (−)-physostigmine, showing wide biological activities, and has been clinically used as medicine.

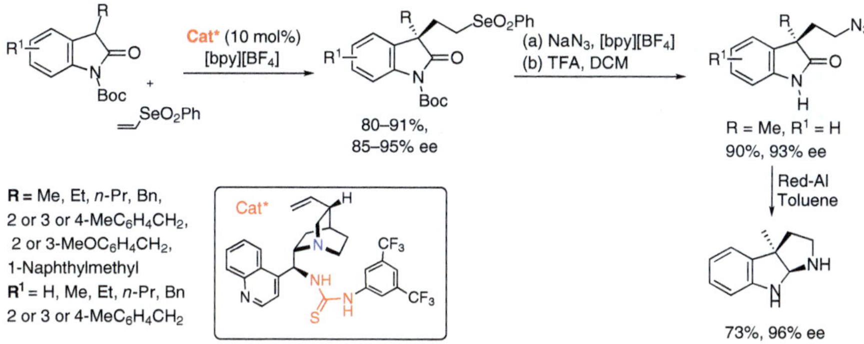

Scheme 14.25 Organocatalyzed Michael additions of oxindoles to vinyl selenone in ionic liquid and application to synthesize pyrroloindoline. TFA, Trifluoroacetic acid.

14.7 Conclusion

The results discussed in this chapter demonstrate that organic selenium reagents can be successfully employed in asymmetric selenofunctionalization reactions inducted by chiral selenium reagents or by chiral substrates. Moreover, the reactivity of vinyl selenones has also been explored in various Michael-initiated multiple bond-forming reactions and organocatalyzed processes. The ease of handling, the mild reaction conditions, and the various possible transformations make the selenium reagents interesting building blocks with useful applications in constructing heterocycles, natural products, and densely functionalized compounds.

References

1 Wirth, T. (2011). *Organoselenium Chemistry: Synthesis and Reactions*. Weinheim: Wiley-VCH Verlag GmbH & Co.

2 Santi, C. (2014). *Organoselenium Chemistry: Between Synthesis and Biochemistry*. Sharjah: Bentham Science Publishers.

3 Lenardão, E.J., Santi, C., and Sancineto, L. (2018). *New Frontiers in Organoselenium Compounds*. Cham: Springer International Publishing AG.

4 Makhal, P.N., Nandi, A., and Rao Kaki, V. (2021). Insights into the recent synthetic advances of organoselenium compounds. *ChemistrySelect* 6 (4): 663–679. https://doi.org/10.1002/slct.202004029.

5 Amol, D.S. and Mamoru, K. (2021). Organocatalysed synthesis of selenium containing scaffolds. *Curr. Organocat.* 8 (1): 5–26. https://doi.org/10.2174/2213337207999200414143605.

6 Lihao, L., Xiaodan, Z. (2021). Modern organoselenium catalysis: opportunities and challenges *Synlett* 32 (13): 1262–1268. DOI: https://doi.org/10.1055/a-1506-5532.

7 Santi, C. and Tidei, C. (2013). Electrophilic Se/Te reagents: reactivity and their contribution to "Green Chemistry". In: *The Chemistry of Organic Selenium and Tellurium Compounds* (ed. Z. Rappoport), 569–655. Chichester: Wiley.

8 Santi, C. and Tidei, C. (2014). Addition reaction for the formation of C–S and C–Se bond. In: *Comprehensive Organic Synthesis*, 2e (ed. P. Knochel), 605–637. Elsevier https://doi.org/10.1016/B978-0-08-097742-3.00724-2.

9 Wirth, T. (1999). Chiral selenium compounds in organic synthesis. *Tetrahedron* 55: 1–28. https://doi.org/10.1016/S0040-4020(98)00946-6.

10 Wirth, T. (2000). Organoselenium chemistry in stereoselective reactions. *Angew. Chem. Int. Ed.* 39: 3740–3749. https://doi.org/10.1002/1521-3773(20001103)39:21%3C3740::aid-anie3740%3E3.0.co;2-n.

11 Browne, D.M. and Wirth, T. (2006). New developments with chiral electrophilic selenium reagents. *Curr. Org. Chem.* 10: 1893–1903. https://doi.org/10.2174/138527206778521213.

12 Freudendahl, D.M., Shahzad, S.A., and Wirth, T. (2009). Recent advances in organoselenium chemistry. *Eur. J. Org. Chem.* 11: 1649–1664. https://doi.org/10.1002/ejoc.200801171.

13 Freudendahl, D.M., Santoro, S., Shahzad, S.A. et al. (2009). Green chemistry with selenium reagents: development of efficient catalytic reactions. *Angew. Chem. Int. Ed.* 48: 8409–8411. https://doi.org/10.1002/anie.200903893.

14 Tomoda, S., Fujita, K., and Iwaoka, M. (1992). Asymmetric oxyselenenylation of olefins using optically active selenobinaphthyls and *d*-menthol as a nucleophile. *Chem. Lett.* 1123–1124. https://doi.org/10.1246/cl.1992.1123.

15 Déziel, R., Goulet, S., Grenier, L. et al. (1993). Asymmetric selenomethoxylation of olefins involving a chiral C2 symmetrical electrophilic organoselenium reagent. *J. Org. Chem.* 58: 3619–3621. https://doi.org/10.1021/jo00066a010.

16 Déziel, R. and Malenfant, E. (1995). Asymmetric ring closure reactions mediated by a chiral C2 symmetrical organoselenium reagent. *J. Org. Chem.* 60: 4660–4662. https://doi.org/10.1021/jo00119a055.

17 Déziel, R., Malenfant, E., and Bélanger, G. (1996). Practical synthesis of (*R,R*)- and (*S,S*)-bis[2,6-bis(1-ethoxyethyl)phenyl] diselenide. *J. Org. Chem.* 61: 1875–1876. https://doi.org/10.1021/jo951919h.

18 Fukuzawa, S., Takahashi, K., Kato, H., and Yamazaki, H. (1997). Asymmetric methoxyselenenylation of alkenes with chiral ferrocenylselenium reagents. *J. Org. Chem.* 62: 7711–7716. https://doi.org/10.1021/jo970982z.

19 Fujita, K., Iwaoka, M., and Tomoda, S. (1994). Synthesis of diaryl diselenides having chiral pyrrolidine rings with C_2 symmetry. Their application to the asymmetric methoxyselenenylation of *trans*-β-methylstyrenes. *Chem. Lett.* 923–926. https://doi.org/10.1246/cl.1994.923.

20 Fujita, K., Murata, K., Iwaoka, M., and Tomoda, S. (1995). Asymmetric methoxyselenenylation of olefins using an optically active diaryl diselenide derived from D-mannitol. *Tetrahedron Lett.* 36: 5219–5222. https://doi.org/10.1016/0040-4039(95)00976-J.

21 Fujita, K., Murata, K., Iwaoka, M., and Tomoda, S. (1997). Design of optically active selenium reagents having a chiral *tertiary* amino group and their application to asymmetric inter- and intramolecular oxyselenenylations. *Tetrahedron* 53: 2029–2048. https://doi.org/10.1016/S0040-4020(96)01166-0.

22 Back, T.G. and Dyck, B.P. (1996). Asymmetric cyclization of unsaturated alcohols and carboxylic acids with camphor-based selenium electrophiles. *J. Chem. Soc., Chem. Commun.* 2567–2568. https://doi.org/10.1039/CC9960002567.

23 Back, T.G., Dyck, B.P., and Nan, S. (1999). Asymmetric electrophilic methoxyselenylations and cyclizations with 3-camphorseleno derivatives. *Tetrahedron* 55: 3191–3208. https://doi.org/10.1016/S0040-4020(98)01133-8.

24 Back, T.G., Moussa, Z., and Parvez, M. (2002). Asymmetric methoxyselenenylations and cyclizations with 3-camphorseleno electrophiles containing oxime substituents at C-2. Formation of an unusual oxaselenazole from an oxime-substituted selenenyl bromide. *J. Org. Chem.* 67: 499–509. http://dx.doi.org/10.1021/jo016061c.

25 Ścianowski, J., Rafiński, Z., and Wojtczak, A. (2006). Syntheses and reactions of new optically active terpene dialkyl diselenides. *Eur. J. Org. Chem.* 14: 3216–3225. https://doi.org/10.1002/ejoc.200600044.

26 Rafiński, Z., Ścianowski, J., and Wojtczak, A. (2008). Asymmetric selenocyclization with the use of dialkyl monoterpene diselenides. *Tetrahedron: Asymmetry* 19: 223–230. https://doi.org/10.1016/j.tetasy.2007.11.032.

27 Rafiński, Z. and Ścianowski, J. (2008). Synthesis and reactions of enantiomerically pure dialkyl diselenides from the *p*-menthane group. *Tetrahedron: Asymmetry* 19 (10): 1237–1244. https://doi.org/10.1016/j.tetasy.2008.04.027.

28 Ścianowski, J. and Wełniak, M. (2009). Syntheses of the optically active terpene hydroxyphenylselenides. *Phosphorus, Sulfur Silicon Relat. Elem.* 184: 1440–1447. https://doi.org/10.1080/10426500902947682.

29 Rafiński, Z., Ścianowski, J., and Wojtczak, A. (2009). Synthesis and reactions of the optically active dialkyl diselenides from the pinane group. *Lett. Org. Chem.* 6: 321–328. https://doi.org/10.2174/157017809788489846.

30 Ścianowski, J., Rafiński, Z., Szuniewicz, A., and Wojtczak, A. (2009). New chiral selenium electrophiles derived from functionalized terpenes. *Tetrahedron* 65: 10162–10174. https://doi.org/10.1016/j.tet.2009.10.005.

31 Ścianowski, J., Rafiński, Z., Wojtczak, A., and Burczyński, K. (2009). Syntheses and reactions of terpene β-hydroxyselenides and β-hydroxydiselenides. *Tetrahedron: Asymmetry* 20: 2871–2879. https://doi.org/10.1016/j.tetasy.2009.12.001.

32 Mukherjee, A.J., Zade, S.S., Singh, H.B., and Sunoj, R.B. (2010). Organoselenium chemistry: role of intramolecular interactions. *Chem. Rev.* 110: 4357–4416. https://doi.org/10.1021/cr900352j.

33 Tiecco, M., Testaferri, L., Santi, C. et al. (2006). Intramolecular non bonding interactions between selenium and sulfur-spectroscopic evidence and importance in asymmetric synthesis. *Eur. J. Org. Chem.* 21: 4867–4873. http://dx.doi.org/10.1002/ejoc.200600517.

34 Santi, C., Fragale, G., and Wirth, T. (1988). Synthesis of a new chiral nitrogen containing diselenide as a precursor for selenium electrophiles. *Tetrahedron: Asymmetry* 9: 3625–3628. https://doi.org/10.1016/S0957-4166(98)00380-2.

35 Wirth, T. and Fragale, G. (1997). Asymmetric addition reactions with optimized selenium electrophiles. *Eur. J. Org. Chem.* 3: 1894–1902. https://doi.org/10.1002/chem.19970031123.

36 Fragale, G., Neuburger, M., and Wirth, T. (1998). New and efficient selenium reagents for stereoselective selenenylation reactions. *Chem. Commun.* 1867–1868. https://doi.org/10.1039/A804264K.

37 Tiecco, M., Testaferri, L., Bagnoli, L. et al. (2000). Efficient asymmetric selenomethoxylation and selenohydroxylation of alkenes with a new sulfur containing chiral diselenide. *Tetrahedron Lett.* 41: 3241–3245. https://doi.org/10.1016/S0040-4039(00)00358-0.

38 Tiecco, M., Testaferri, L., Santi, C. et al. (2002). Preparation of a new chiral non-racemic sulfur-containing diselenide and applications in asymmetric synthesis. *Chem. Eur. J.* 8: 1118–1124. https://doi.org/10.1002/1521-3765(20020301)8:5%3C1118::AID-CHEM1118%3E3.0.CO;2-2.

39 Tiecco, M., Testaferri, L., Bagnoli, L. et al. (2006). Enantioselective synthesis of heterocyclic compounds mediated by organoselenium reagents. *Arkivoc* 186–206. https://doi.org/10.3998/ark.5550190.0007.715.

40 Tiecco, M., Testaferri, L., Bagnoli, L. et al. (2006). Organoselenium mediated asymmetric cyclizations. Synthesis of enantiomerically pure 1,6-dioxaspiro[4.4]nonanes. *Tetrahedron: Asymmetry* 17: 2768–2774. https://doi.org/10.1016/j.tetasy.2006.10.021.

41 Tiecco, M., Testaferri, L., Bagnoli, L. et al. (2005). Synthesis of enantiomerically pure perhydrofuro[2,3-*b*]furans. *Tetrahedron: Asymmetry* 16: 2429–2435. https://doi.org/10.1016/j.tetasy.2005.06.016.

42 Tiecco, M., Testaferri, L., Bagnoli, L. et al. (2001). Efficient asymmetric selenocyclizations of alkenyl oximes into cyclic nitrones and 1,2-oxazines promoted by sulfur containing diselenides. *Tetrahedron: Asymmetry* 12: 3297–3304. https://doi.org/10.1016/S0957-4166(02)00013-7.

43 Back, T.G., Dyck, B.P., and Nan, S. (1999). Asymmetric electrophilic methoxyselenenylations and cyclizations with 3-camphorseleno derivatives. *Tetrahedron* 55: 3191–3208. https://doi.org/10.1016/S0040-4020(98)01133-8.

44 Tiecco, M., Testaferri, L., Marini, F. et al. (2001). A sulfur-containing diselenide as an efficient chiral reagent in asymmetric selenocyclization reactions. *Tetrahedron: Asymmetry* 12: 1493–1502. https://doi.org/10.1016/S0957-4166(01)00248-8.

45 Nishibayashi, Y., Srivastava, S.K., Takada, H. et al. (1995). Highly selective asymmetric intramolecular selenocyclisation. *Chem. Commun.* 2321–2322. https://doi.org/10.1039/C39950002321.

46 Déziel, R., Malefant, E., Thibault, C. et al. (1997). 2,6-bis[(2*S*)-tetrahydrofuran-2-yl]phenyl diselenide: an effective reagent for asymmetric electrophilic addition reactions to olefins. *Tetrahedron Lett.* 38: 4753–4756. https://doi.org/10.1016/S0040-4039(97)01030-7.

47 Takada, H., Nishibayashi, Y., and Uemura, S. (1999). The use of chiral diferrocenyl diselenides for highly selective asymmetric intramolecular

selenocyclisation. *J. Chem. Soc., Perkin Trans. 1* 1511–1516. https://doi.org/10.1039/A901087D.

48 Tiecco, M., Testaferri, L., Marini, F. et al. (2005). Asymmetric syntheses promoted by organoselenium reagents. *Phosphorus, Sulfur Silicon Relat. Elem.* 180: 729–740. https://doi.org/10.1080/10426500590907462.

49 Tiecco, M., Testaferri, L., Santi, C. et al. (2000). New nitrogen containing chiral diselenides: synthesis and asymmetric addition reactions to olefins. *Tetrahedron: Asymmetry* 11: 4645–4650. https://doi.org/10.1016/S0957-4166(00)00469-9.

50 Tiecco, M., Testaferri, L., Bagnoli, L. et al. (2004). Synthesis of enantiomerically pure substituted tetrahydrofurans from epoxides and phenylselenium reagents. *Tetrahedron: Asymmetry* 15: 405–412. https://doi.org/10.1016/j.tetasy.2003.10.042.

51 Tiecco, M., Testaferri, L., Bagnoli, L. et al. (2004). Synthesis of enantiomerically pure perhydrofuro [3,4-*b*] pyrans and perhydrofuro [3,4-*b*] furans. *Tetrahedron: Asymmetry* 15: 1949–1955. https://doi.org/10.1016/j.tetasy.2004.05.009.

52 Tiecco, M., Testaferri, L., Marini, F. et al. (2003). Synthesis of enantiomerically pure 1,4-dioxanes from alkenes promoted by organoselenium reagents. *Tetrahedron: Asymmetry* 14: 1095–1102. https://doi.org/10.1016/S0957-4166(03)00124-1.

53 Vitaku, E., Smith, D.T., and Njardarson, J.T. (2014). Analysis of the structural diversity, substitution patterns, and frequency of nitrogen heterocycles among US FDA approved pharmaceuticals: miniperspective. *J. Med. Chem.* 57: 10257–10274. https://doi.org/10.1021/jm501100b.

54 Cabrele, C. and Reiser, O. (2016). The modern face of synthetic heterocyclic chemistry. *J. Org. Chem.* 81: 10109–10125. http://dx.doi.org/10.1021/acs.joc.6b02034.

55 Tiecco, M., Testaferri, L., Bagnoli, L. et al. (2007). Selenium promoted synthesis of enantiopure pyrrolidines starting from chiral aminoalcohols. *Tetrahedron: Asymmetry* 18: 2758–2767. https://doi.org/10.1016/j.tetasy.2007.11.003.

56 Tiecco, M., Testaferri, L., Bagnoli, L., and Scarponi, C. (2008). Selenium-promoted synthesis of enantiopure octahydroindolizines, hexahydro-1*H*-pyrrolizines and hexahydro-3*H*-pyrrolizin-3-ones. *Tetrahedron: Asymmetry* 19: 2411–2416. https://doi.org/10.1016/j.tetasy.2008.10.006.

57 Tiecco, M., Testaferri, L., Temperini, A. et al. (2007). Stereocontrolled synthesis of substituted *N*-arenesulfonyl azetidines from γ-(phenylseleno)alkyl arylsulfonamides. *Org. Biomol. Chem.* 5: 3510–3519. https://doi.org/10.1039/B712861D.

58 Ronchi, P., Scarponi, C., Salvi, M. et al. (2013). Synthesis of a structural analogue of the repeating unit from *Streptococcus pneumoniae* 19F capsular polysaccharide based on the cross-metathesis–Selenocyclization reaction sequence. *J. Org. Chem.* 78: 5172–5183. https://doi.org/10.1021/jo4001146.

59 Bagnoli, L., Scarponi, C., Testaferri, L., and Tiecco, M. (2009). Preparation of both enantiomers of cyclopropane derivatives from the reaction of vinyl selenones with di-(−)-bornyl malonate. *Tetrahedron: Asymmetry* 20: 1506–1514. https://doi.org/10.1016/j.tetasy.2009.05.019.

60 Bagnoli, L., Scarponi, C., Rossi, M.G. et al. (2011). Synthesis of enantiopure 1,4-dioxanes, morpholines, and piperazines from the reaction of chiral 1,2-diols, amino alcohols, and diamines with vinyl selenones. *Chem. Eur. J.* 17: 993–999. https://doi.org/10.1002/chem.201002593.

61 Palomba, M., Sancineto, L., Marini, F. et al. (2018). A domino approach to pyrazino-indoles and pyrroles using vinyl selenones. *Tetrahedron* 74: 7156–7163. https://doi.org/10.1016/j.tet.2018.10.044.

62 Wu, J.-C. and Chattopadhyaya, J. (1989). Michael addition reactions of αβ-ene-3′-phenylselenone of uridine. New synthesis of 2′,3′-dideoxy-ribo-aziridino-, 2′,3′-dideoxy-2′, 3′-ribo-cyclopropyl- and 2,2′-O-anhydro-3′-deoxy-3′-amino uridine derivatives. *Tetrahedron* 45: 4507–4522. https://doi.org/10.1016/S0040-4020(01)89086-4.

63 Tong, W., Xi, Z., Gioeli, C., and Chattopadhyaya, J. (1991). Synthesis of new 2′, 3′-modified uridine derivatives from 2′,3′-ene-2′-phenylselenonyl uridine by Michael addition reactions. *Tetrahedron* 47: 3431–3450. https://doi.org/10.1016/S0040-4020(01)86406-1.

64 Wu, J.-C. and Chattopadhyaya, J. (1990). A new stereospecific synthesis of [3.1.0]bicyclic cyclopropano analog of 2′3′-dideoxyuridine. *Tetrahedron* 46: 2587–2592. https://doi.org/10.1016/S0040-4020(01)82038-X.

65 Tong, W., Wu, J.-C., Sandström, A., and Chattopadhyaya, J. (1990). Synthesis of new 2′,3′-dideoxy-2′,3′-α-fused-heterocyclic uridines, and some 2′,3′-ene-2′-substituted uridines from easily accessible 2′,3′-ene-3′ phenylselenonyl uridine. *Tetrahedron* 46: 3037–3060. https://doi.org/10.1016/S0040-4020(01)88395-2.

66 Bhaumik, A. and Pathak, T. (2015). Methyl-α-D-2-selenonyl pent-2-enofuranoside: a reactive selenosugar for the diversity-oriented synthesis of enantiomerically pure heterocycles, carbocycles, and isonucleosides. *J. Org. Chem.* 80: 11057–11064. https://doi.org/10.1021/acs.joc.5b01192.

67 Bhaumik, A., Das, A., and Pathak, T. (2016). Vinyl selenones derived from *d*-fructose: a new platform for fructochemistry. *Asian J. Org. Chem.* 5: 1048–1062. https://doi.org/10.1002/ajoc.201600183.

68 Bhaumik, A., Samanta, S., and Pathak, T. (2014). Enantiopure 1,4,5-trisubstituted 1,2,3-triazoles from carbohydrates: applications of organoselenium chemistry. *J. Org. Chem.* 79: 6895–6904. https://doi.org/10.1021/jo5009564.

69 Dalko, P.I. (2013). *Comprehensive Enantioselective Organocatalysis: Catalysts, Reactions, and Applications*. Weinheim: Wiley-VCH Verlag GmbH & Co. KGaA.

70 Marini, F., Sternativo, S., Del Verme, F. et al. (2009). Enantioselective organocatalytic Michael addition of α-substituted cyanoacetates to α, β-unsaturated selenones. *Adv. Synth. Catal.* 351: 103–106. https://doi.org/10.1002/adsc.200800592.

71 Marini, F., Sternativo, S., Del Verme, F. et al. (2009). A new stereoselective synthesis of cyclopropanes containing quaternary stereocentres via organocatalytic Michael addition to vinyl selenones. *Adv. Synth. Catal.* 351: 1801–1806. https://doi.org/10.1002/adsc.200900222.

72 Buyck, T., Wang, Q., and Zhu, J. (2013). Catalytic enantioselective Michael addition of α-aryl-α-isocyanoacetates to vinyl selenone: synthesis of α,α-disubstituted α-amino acids and (+)- and (−)-trigonoliimine A. *Angew. Chem. Int. Ed.* 52: 12714–12718. https://doi.org/10.1002/anie.201306663.

73 Torres-Ochoa, R.O., Buyck, T., Wang, Q., and Zhu, J. (2018). Heteroannulation of arynes with α-amino imides: synthesis of 2,2-disubstituted indolin-3-ones and application to the enantioselective total synthesis of (+)-Hinckdentine a. *Angew. Chem. Int. Ed.* 57: 5679–5683. https://doi.org/10.1002/anie.201800746.

74 Buyck, T., Wang, Q., and Zhu, J. (2014). Triple role of phenylselenonyl group enabled a one-pot synthesis of 1,3-oxazinan-2-ones from α-isocyanoacetates, phenyl vinyl selenones, and water. *J. Am. Chem. Soc.* 136: 11524–11528. https://doi.org/10.1021/ja506031h.

75 Clemenceau, A., Wang, Q., and Zhu, J. (2016). Enantioselective synthesis of quaternary α-amino acids enabled by the versatility of the phenylselenonyl group. *Chem. Eur. J.* 22: 18368–18372. https://doi.org/10.1002/chem.201604781.

76 Simlandy, A.K. and Mukherjee, S. (2016). Catalytic asymmetric formal γ-allylation of deconjugated butenolides. *Org. Biomol. Chem.* 14: 5659–5664. https://doi.org/10.1039/C5OB02362A.

77 Sternativo, S., Calandriello, A., Costantino, F. et al. (2011). A highly enantioselective one-pot synthesis of spirolactones by an organocatalyzed Michael addition/cyclization sequence. *Angew. Chem. Int. Ed.* 50: 9382–9385. https://doi.org/10.1002/anie.201104819.

78 Sternativo, S., Walczak, O., Battistelli, B. et al. (2012). Organocatalytic Michael addition of indanone carboxylates to vinyl selenone for the asymmetric synthesis of polycyclic pyrrolidines. *Tetrahedron* 68: 10536–10541. https://doi.org/10.1016/j.tet.2012.08.077.

79 Zhang, T., Cheng, L., Hameed, S. et al. (2011). Highly enantioselective Michael addition of 2-oxindoles to vinyl selenone in RTILs catalyzed by a Cinchona alkaloid-based thiourea. *Chem. Commun.* 47: 6644–6646. https://doi.org/10.1039/C1CC10880H.

15

Allenes as Chiral Building Blocks in Asymmetric Synthesis

Rafał Loska and Alicja Wasilewska-Rosa

Institute of Organic Chemistry, Polish Academy of Sciences, Kasprzaka 44/52, 01-224 Warsaw, Poland

15.1 Introduction

The ability to synthesize complex organic compounds in an enantioselective manner is crucial in achieving the desired biological, pharmacological, and physicochemical properties. It is also used to determine the actual absolute configuration of natural products. Allenes are excellent but still quite underexplored, chiral building blocks. They can undergo many reactions typical of unsaturated compounds [1], still carrying the stereochemical information from the starting materials to the products. This is possible due to the inherent axial chirality of allenes substituted with two different groups at each end of the system of cumulated double bonds. Exploiting allene axial chirality in enantioselective reactions enabled several innovative transformations that dramatically increased the efficiency of the synthetic strategies toward complex target products. This review aims to highlight examples of chiral, enantioenriched allenes acting as the source of chirality in the enantioselective synthesis of natural products, bioactive compounds, or other valuable substances, particularly those dating from the last 20 years, with emphasis on the reaction mechanisms responsible for the successful chirality transfer (CT; Scheme 15.1). In the past, a few reviews dealing with various aspects of these phenomena appeared in the literature [2–6].

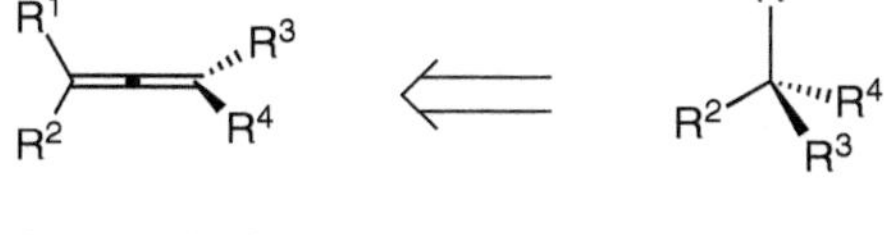

Scheme 15.1 Axial chirality of allenes as the source of central chirality of products in asymmetric synthesis.

Considering that chiral allenes are almost completely unavailable as natural products, it is evident that advances in their application in asymmetric synthesis were possible thanks to efficient methods of their preparation [7], especially in their enantioenriched form. Synthesis with chiral allene building blocks always relies

Chiral Building Blocks in Asymmetric Synthesis: Synthesis and Applications, First Edition.
Edited by Elżbieta Wojaczyńska and Jacek Wojaczyński.

upon successful translating point chirality of allene precursors into chirality of the final products. For many years, the most reliable and popular method of accessing allenes in a stereodefined form were S_N2' reactions of chiral derivatives of propargyl alcohols [8–11], but the repertoire of the available efficient methods has increased significantly in recent years [12–18]. Their comprehensive review is outside the range of this chapter. However, they will be mentioned in the discussion of axial CT examples to explain the origin of chiral information in each case.

15.2 Nucleophilic Addition and Substitution

Allenes bearing electron-accepting substituents, particularly 2,3-allenoic esters, are susceptible to Michael addition. In its iron-catalyzed version, this reaction has been reported to provide regioselectively β,γ-unsaturated esters, although with no stereoselectivity [19]. A more elaborate example of a Michael-type reaction of 2 equiv of 2,3-allenoate with organozinc compounds to give cyclohexenones was developed by Ma's group (Scheme 15.2) [20]. Importantly, the reaction of highly enantioenriched allenoates proceeded with complete CT to the final cyclic products. It has been rationalized by assuming that the reaction begins with a Michael addition of Et_2Zn to 2,3-allenoate with the formation of α-zincated, optically active alkenoate. Subsequent nucleophilic attack of its γ carbon on another molecule of 2,3-allenoate gives a more complex zincated alkenoate, which undergoes rotation around the C(α)—C(β) single bond and cyclization via nucleophilic 1,2-addition to the ester group. Control experiments ruled out the possibility of the racemic zinc 1,3-dienolate reacting with allenoate.

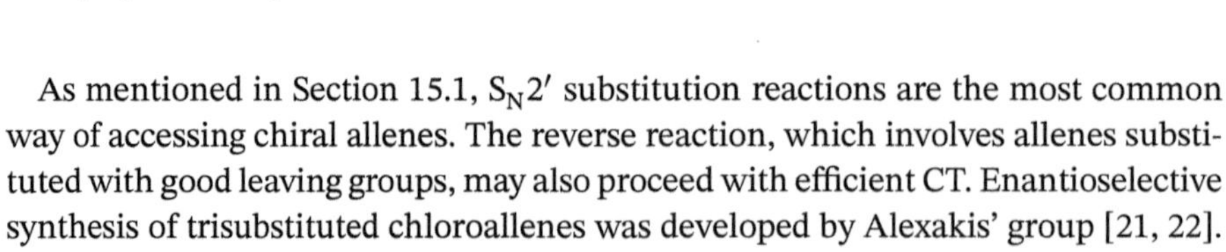

Scheme 15.2 Enantiospecific, dialkylzinc-initiated cyclization of 2,3-allenoates. Source: Lu et al. [20]/John Wiley & Sons.

As mentioned in Section 15.1, S_N2' substitution reactions are the most common way of accessing chiral allenes. The reverse reaction, which involves allenes substituted with good leaving groups, may also proceed with efficient CT. Enantioselective synthesis of trisubstituted chloroallenes was developed by Alexakis' group [21, 22]. S_N2' substitution of one chlorine atom in 1,1-dichloro-2-alkynes with alkyl-, aryl-, or silylmagnesium reagents in the presence of catalytic amounts of CuBr and chiral aminophosphine ligand gave the expected products with excellent levels of enantioselectivity. Such chiral chloroallenes were valuable building blocks for further functionalization to obtain more complex allenes, for example, through Suzuki and Sonogashira couplings. Importantly, they were also amenable for enantiospecific

Cu-catalyzed nucleophilic substitution of the remaining chlorine atom with the formation of either trisubstituted allenes or terminal alkynes with a quaternary carbon stereogenic center. Interestingly, the product depended entirely on the solvent used (tetrahydrofuran (THF) or dichloromethane (DCM)).

Allene chirality has also been exploited for directing the stereochemical course of nucleophilic addition to the adjacent carbonyl group [23]. The addition of nucleophiles to the *s-trans* conformer of allenyl aldehydes occurs preferentially from the side occupied by the smaller terminal allene substituent. Marshall observed excellent stereoselectivity in additions of Grignard reagents to nonracemic allenals bearing a terminal TBDPS or ^{t}Bu substituent or in L-Selectride reductions of similar allenyl methyl ketones. The model substrates were prepared by [2,3] Wittig rearrangement of chiral propargyltin substrates, followed by Swern oxidation.

15.3 Allenylzinc and Indium Reagents, Allenylsilanes and Allenylstannanes

Reactions of allylmetal reagents with carbonyl compounds are among the most frequently applied transformations in stereoselective synthesis [24]. Analogous allenylmetal reagents, which have the additional advantage of possible axial chirality, were introduced by Marshall as a powerful tool for the synthesis of polyketide natural products [25, 26]. Allenylzinc reagents can be generated *in situ* from chiral propargyl mesylates in the Pd-catalyzed reaction with diethylzinc (Scheme 15.3).

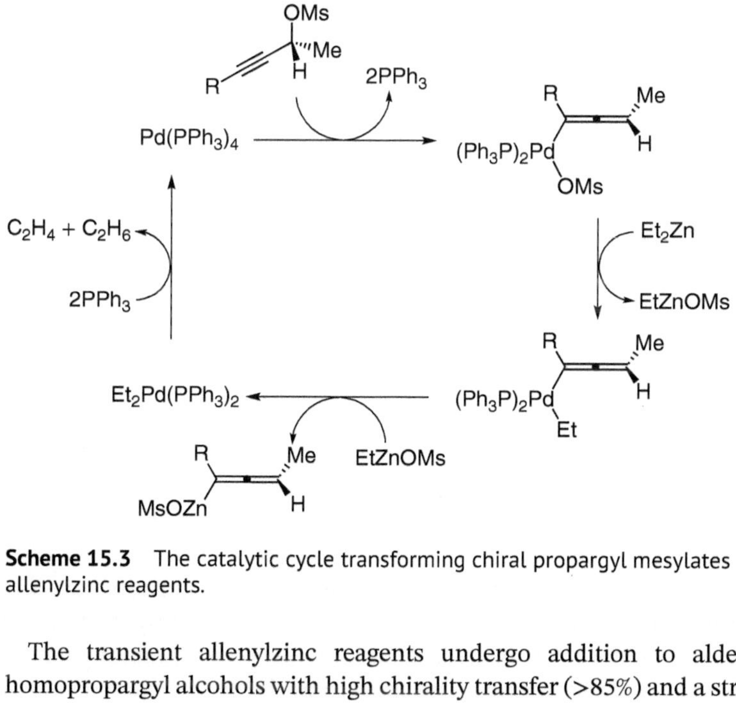

Scheme 15.3 The catalytic cycle transforming chiral propargyl mesylates into chiral allenylzinc reagents.

The transient allenylzinc reagents undergo addition to aldehydes to give homopropargyl alcohols with high chirality transfer (>85%) and a strong preference for the *anti*-configured products. The most useful variant of this reaction is the addition of chiral aldehydes, which usually proceeds with very high diastereo- and enantioselectivity. Importantly, stereoselectivity is strongly reagent-controlled. The configuration of the two stereocenters of the formed stereotriad is determined by the chirality of the starting mesylate and the intermediate allenylzinc. This may be explained by cyclic transition states in which the terminal substituent of the allene moiety is eclipsed with the aldehyde hydrogen (Scheme 15.4).

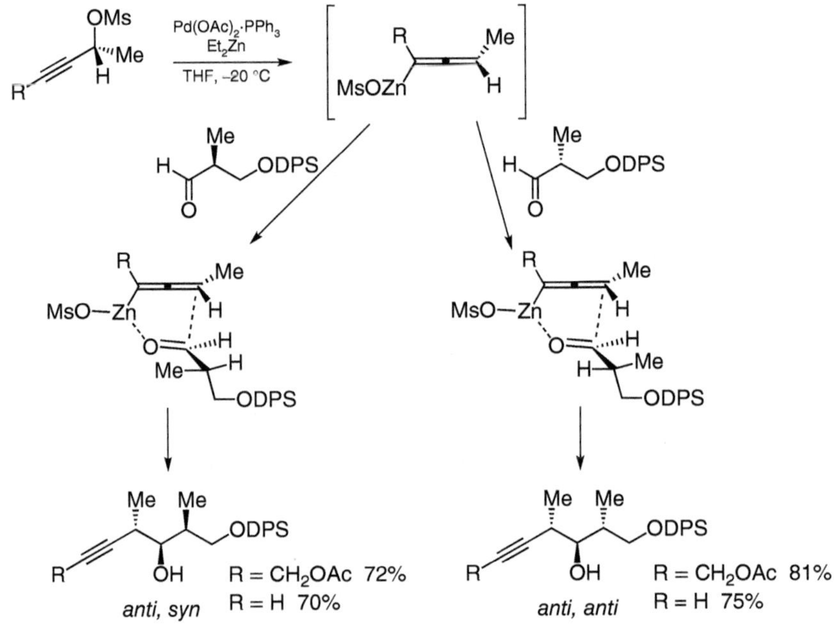

Scheme 15.4 Stereospecific reactions of chiral allenylzinc reagents with enantiomers of α-chiral aldehyde. DPS = dimethylphenylsilyl.

Similar reactions of *N*-Boc-protected (*R*)-alaninal with (*R*) or (*S*)-configured propargyl mesylates, transformed *in situ* into allenylzinc or allenylindium reagents, gave predominantly *anti,anti* or *anti,syn* stereotriads, respectively, with diastereomeric ratio usually exceeding 95 : 5 [27]. On the other hand, racemic mesylates led to equimolar mixtures of both diastereoisomers. This methodology was further exploited in synthesizing the C(20)–C(26) fragment of superstolide A.

The complementary methodology that allows for the preparation of *syn,syn* stereotriads is based on allenylstannanes or allenylsilanes [28]. Due to their lower reactivity, their reaction with aldehydes requires catalysis with Lewis acids such as $BF_3 \cdot OEt_2$ or $TiCl_4$. The actual stereochemical outcome of the addition to chiral aldehydes bearing oxygen substituents depends on the nature of the protecting group: an aldehyde bearing a bulky DPS group mainly or exclusively gave the expected *syn,syn* adducts, whereas, with the smaller *O*-benzyloxy group, the configuration of the central hydroxylated stereocenter depended primarily on the configuration of the aldehyde substrate. This is probably due to the intermediacy of the Ti-chelated transition states in the reaction of the benzylated substrate, rather than open Felkin–Ahn transition structures typically postulated for the silylated substrate.

A striking demonstration of the versatility of the above methodology in the synthesis of polyketides was the total synthesis of the natural cytotoxic product (+)-membrenone C to determine its absolute configuration [29]. The synthetic strategy relied upon two consecutive reactions with allenylmetal reagents that installed four stereogenic centers out of five centers present in the target molecule (Scheme 15.5). The power of the allenylmetal methodology was further highlighted by the synthesis of the enantiomeric (−)-membrenone C starting from the same *S*-aldehyde and enantiomers of both allenylmetal reagents but applied in reverse order.

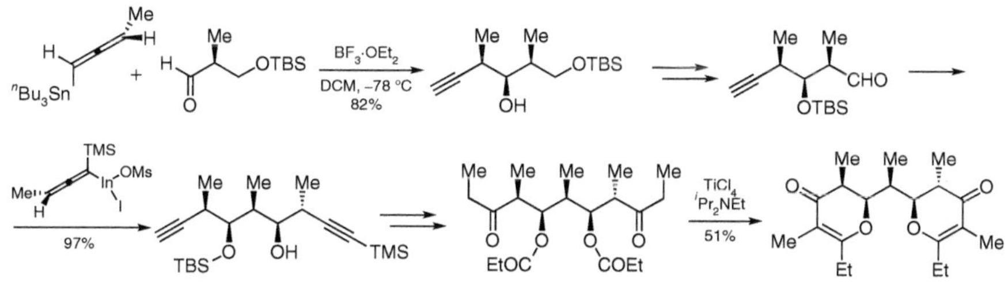

Scheme 15.5 Allenylmetal-based approach to the enantioselective synthesis of (+)-membrenone C.

More recently, chiral allenylsilane was successfully applied by Mulzer in synthesizing the C(13)–C(18) fragment of the antibiotic branimycin [30]. (*R*)-3-Methoxy-2-triisopropylsililoxypropanal was propargylated with enantiopure 2-trimethylsilyl-2,3-pentadiene in good yield and excellent diastereoselectivity (20 : 1) to provide the desired stereotriad in *syn,syn* configuration, apparently from an antiperiplanar, anti-Felkin–Ahn transition state.

In the following years, allenylsilane chemistry was pursued by Panek's group, who disclosed the synthesis of chiral allenylsilanes by Johnson–Claisen rearrangement [31], stereoselective addition of allylsilanes to *N*-sulfonylimines [32], a cascade reaction of allenylsilane addition to azidoaldehydes, followed by thermal azide–alkyne cycloaddition [33], or finally, a Ferrier allylic rearrangement to control the configuration of the stereocenter directly attached to the dihydropyran rings of the target C-glycosides [34]. Further, this reaction was found to be highly diastereospecific, with facial selectivity of chiral allenylsilane overriding the effect of the chiral glycal ring. The reaction scope also included furanose-derived dihydrofurans.

Another type of allenyl analog of allylsilanes is silylmethyl-substituted allenes. The first example of moderately enantiospecific electrophilic substitution in such a system was reported by Hiemstra [35]. Ogasawara's group developed an enantioselective Tsuji–Trost reaction of 3-bromo-5-trimethylsilyl-1,3-pentadienes with carbon and nitrogen nucleophiles, catalyzed with palladium complexes with chiral phosphines [36]. The products, chiral enantioenriched trimethylsilylmethylallenes, were subjected to reactions with carbocationic electrophiles generated from acetals or aldehydes in the presence of $TiCl_4$, forming alkoxyalkyl-substituted 1,3-dienes with good levels of chirality transfer in some cases. The stereoselectivity of this S_E' reaction was explained by minimization of sterical interactions of the approaching electrophile with substituents attached to the double bond of allene, which does not participate in the reaction. Later, the stereoselectivity of this methodology was improved using bulkier silyl substituents [37].

15.4 Epoxidation, Aziridination, and Silacyclopropanation of Allenes

An interesting feature of allene reactivity is facile oxidation to dioxides, also termed spirodiepoxides, upon treatment with oxidants such as dimethyldioxirane (DMDO) or MCPBA [38]. The reaction begins with the more substituted double bond of allene. Oxidation of the second double bond of the intermediate vinyl ether is usually faster than the first. Formation of spirodiepoxides from unsymmetrically substituted allenes tends to be highly diastereoselective, with the first oxygen delivered at the face of the double bond, which is less hindered by the substituents of the other double bond. The second oxidation occurs at the less hindered side of the already formed oxirane ring. Opening spirodiepoxide rings leads mainly to α-hydroxyketones and follows typical rules of epoxide ring-opening: nucleophiles attack the least substituted carbon of the spirodiepoxide moiety, and acid-catalyzed reactions proceed through the most stabilized carbocations.

The above considerations imply that transforming chiral allenes into spirodiepoxides may be an efficient way of translating allene axial chirality into central chirality of hydroxyketone derivatives, which is indeed the case. One of the early examples comes from the work of Marshall, who investigated the diastereoselective addition of nucleophiles to axially chiral allenyl aldehydes (α,β,γ-diunsaturated butanals) [23]. With aldehydes bearing a bulky substituent at the terminal allene carbon (^{t}Bu or better TBDPS), high facial selectivity of addition of Grignard reagents was observed, explainable by the addition of a nucleophile to the less hindered side of the *s-trans* conformer of the substrate. Concerning spiroepoxides, some adducts were further oxidized with MCPBA with a complete facial selectivity to give unstable intermediates, which underwent a spontaneous transformation to single diastereomers of hydroxyenones, presumably by the eliminative, acetate-assisted opening of two epoxide rings (Scheme 15.6). This methodology was later employed in the stereoselective synthesis of carbohydrate precursors [39].

Some imaginative synthetic strategies toward complex natural products based upon spirodiepoxide chemistry have been disclosed in literature in the following years. Williams studied the regioselective opening of the spirodiepoxide ring system with nitrogen nucleophiles, including a cascade reaction that involved further formation of another epoxide ring in an intramolecular S_N2 reaction [40]. S_N2' substitution of propargylic mesylate (95% ee) transformed it into chiral allene,

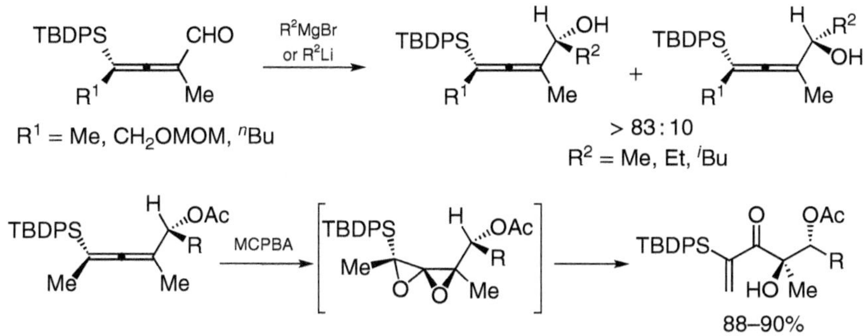

Scheme 15.6 Face-selective 1,2-addition of Grignard reagents to chiral allenals and synthesis of enantiopure α-hydroxyketone via spirodiepoxide.

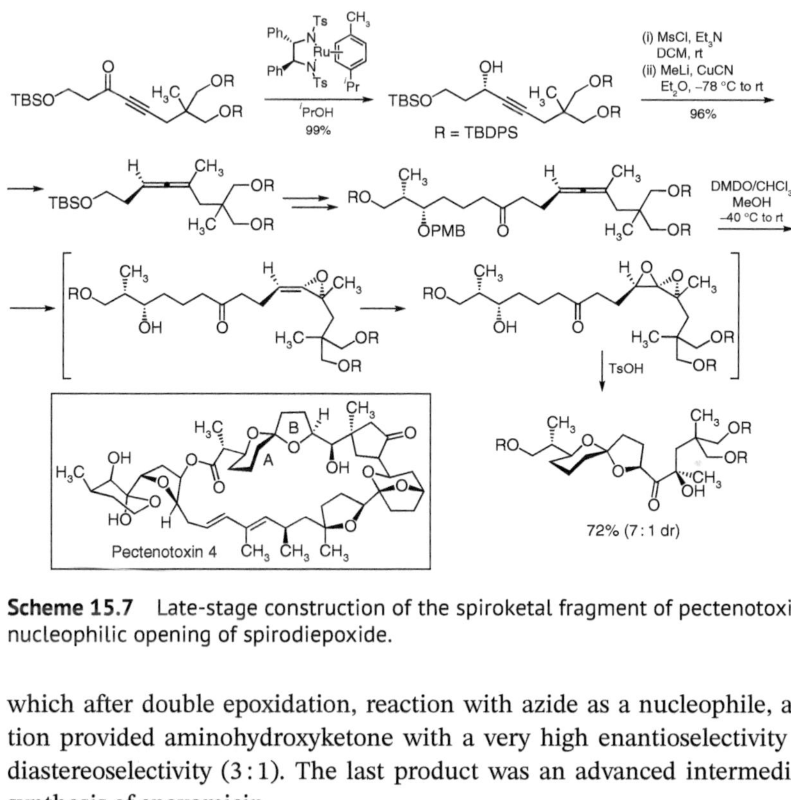

Scheme 15.7 Late-stage construction of the spiroketal fragment of pectenotoxin 4 by nucleophilic opening of spirodiepoxide.

which after double epoxidation, reaction with azide as a nucleophile, and reduction provided aminohydroxyketone with a very high enantioselectivity and good diastereoselectivity (3 : 1). The last product was an advanced intermediate in the synthesis of epoxomicin.

An intramolecular opening of a spirodiepoxide with an oxygen nucleophile is the key step of the Williams total synthesis of the C(1)–C(15) fragment of a macrolide natural product pectenotoxin 4 [41]. In the synthetic sequence (Scheme 15.7), chirality was introduced during enantioselective (>95 : 5 er) reduction of alkynone to propargylic alcohol in the presence of a chiral ruthenium catalyst. Typical S_N2' chemistry and chain elongation provided the key allene intermediate with appropriately positioned oxygen functional groups. DMDO oxidation cleaved the

PMB protecting group and at the same time oxidized the allene to spirodiepoxide. In the presence of TsOH, the formation of a cyclic hemiacetal occurred, followed by the opening of the spirodiepoxide rings initiated by the attack of the hemiacetal hydroxyl, with the formation of the target spirodiketal fragment of pectenotoxin 4 with 7 : 1 diastereoselectivity. The key to the success of this transformation was the highly face-selective epoxidation of allene (20 : 1 for the first, more substituted double bond, and 7 : 1 for the second).

The same group disclosed an alternative, even more, selective approach to a larger fragment of pectenotoxin 4, involving spirodiepoxide opening with a hydroxyl originating from initial epoxide opening a few years later [42]. Stereoselective pyrane ring closure by a hydroxyl nucleophile attack on spirodiepoxide was also the critical step in synthesizing psymberin [43].

The synthetic potential of spirodiepoxides was expanded considerably with the development of regio- and stereoselective ring-opening with organocuprates [44]. As mentioned earlier, oxidation of tri-substituted chiral allenes usually proceeds with excellent facial selectivity for the first epoxidation and with moderate for the second. The subsequent regioselective attack of RCu(CN)Li reagents on the less substituted carbon resulted in hydroxyketones with two stereocenters at α and α' positions. This method was applied to concisely synthesize a stereotetrad present in erythromycin. Other important contributions to enantioselective synthesis exploiting spirodiepoxide chemistry that came from the Williams research group was the investigation of the influence of an additional stereogenic center on the stereoselectivity of double epoxidation [45], of directing effects during epoxidation of allenylsilanes and opening of the corresponding spirodiepoxide intermediates [46] and stereospecific rearrangement of spirodiepoxides into oxetan-3-ones [47].

The robustness of the spirodiepoxide methodology is particularly evident in the Williams synthetic approach to erythrynolides [48]. The aim of the initial model studies on a macrocycle containing only two identical stereotriads of dihydroerythronolide A was to introduce these two key fragments via stereoselective oxidation of bis(allenyl)macrocycle, followed by the regioselective opening of the intermediate bis(spiroepoxide) (Scheme 15.8). Two diastereomeric macrolactones were prepared, one containing two allenyl moieties of the same sense of axial chirality and the other with mutually enantiomeric allene units. Oxidation of both allene fragments in the former diastereomer, followed by the regioselective opening of both spirodiepoxides with methyl cuprate, produced a macrocycle with four noncontiguous stereocenters with correct relative configuration in 22% yield, which is quite impressive considering the number of actual reaction steps involved in this transformation. The other diastereomeric bis(allene) behaved similarly, and both allenic macrocycles reacted even more efficiently with thiobenzamide as nucleophile instead of organocopper reagent. A significant difference in susceptibility of the two allene moieties allowed for the step-wise introduction of two different nucleophiles.

These studies were subsequently extended to bis(allene) macrolactone precursors decorated with additional functional groups and stereocenters present in erythromycin [49]. During epoxidation to a spirodiepoxide, dihydroxylation, and electrophilic bromination, this intermediate gave various erythromycin analogs in

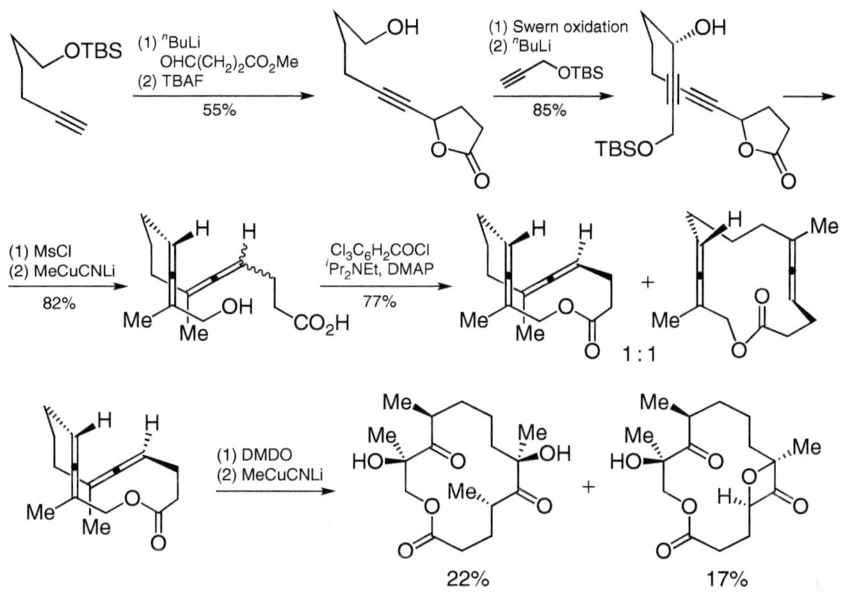

Scheme 15.8 Bis(spirodiepoxide) strategy for the erythrynolide macrolactone synthesis.

high yields with high stereoselectivity controlled by the axial chirality of the allene fragments.

More recently, the chemistry of allene oxides has been explored by Guo, Zhou, and coworkers, who examined the oxidation of allenic phosphine oxides with MTO [50]. This reaction proceeded as monoepoxidation of the more electron-rich double bond, probably followed by oxirane opening assisted by the phosphorous-bound oxygen. Hydrolysis of the resulting five-membered heterocycle led to the formation of hydroxyketones. Such a reaction course explained the very high stereospecificity of this transformation (Scheme 15.9).

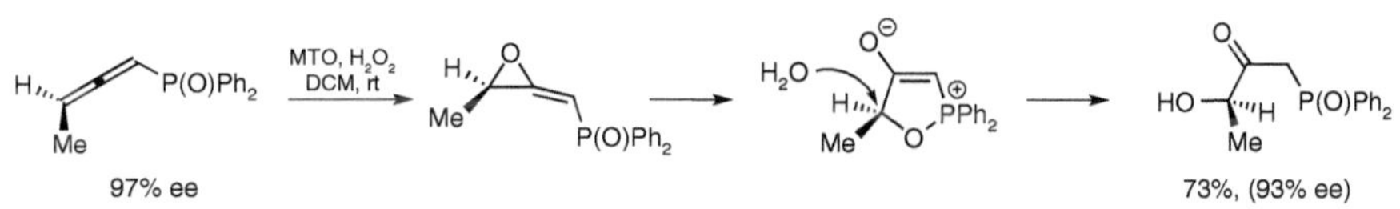

Scheme 15.9 The synthesis of chiral phosphine oxides by allene epoxidation.

Stereospecific aziridination of allenes with nitrenes allows for the preparation of compounds containing nitrogen-substituted stereocenters. Stoll and Blakey developed intramolecular addition of metallonitrenes to allenes, leading to 2-amidoallyl cation intermediates (Scheme 15.10) [51]. The dipolar nature of the latter species could be demonstrated in [3+2] cycloaddition with benzaldehyde or [3+3] with nitrone. In the absence of any dipolarophile, 2-amidoallyl cation rearranged to strained cyclopropylimine that could be trapped with Grignard reagents that added selectively at the less hindered, convex face of the bicyclic system. The whole process is highly diastereoselective, and with a chiral allenic substrate, an efficient transfer of chirality could be obtained.

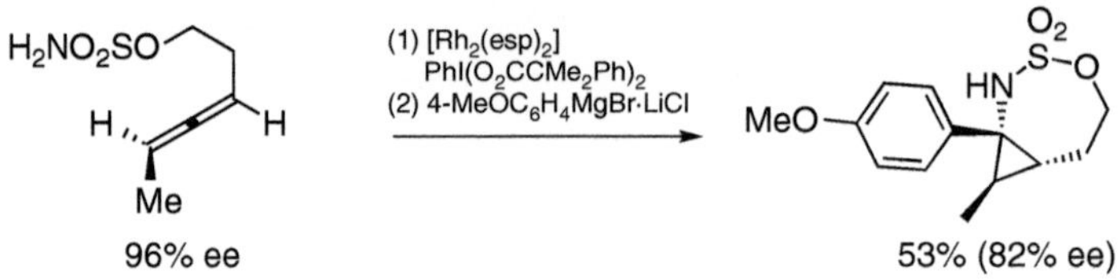

Scheme 15.10 CT in Rh-catalyzed nitrene addition to allene. Source: Stoll et al. [51]/American Chemical Society.

A similar methodology was developed by Schomaker's group, who investigated an intramolecular Rh-catalyzed aziridination of chiral allenyl carbamates with the formation of bicyclic methyleneaziridines (Scheme 15.11) [52]. Further aziridination of the remaining double bond led to spirodiaziridines with high CT from the starting allene.

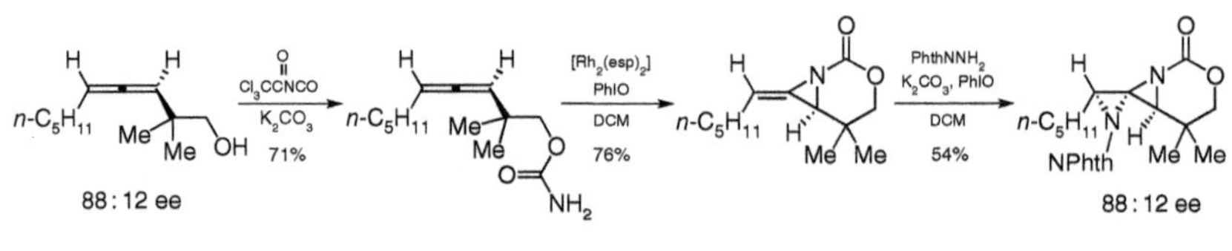

Scheme 15.11 Stereospecific aziridination of chiral allene. Source: Rigoli et al. [52]/American Chemical Society.

Even more complex products in terms of the number of new stereocenters formed could be obtained by cyclization of allenic sulfamates, followed by the nucleophilic opening of the aziridine ring, face-selective addition of an electrophile to the exocyclic double bond, and finally, the addition of another nucleophile to the cyclic *N*-sulfonyliminium ion (Scheme 15.12) [53].

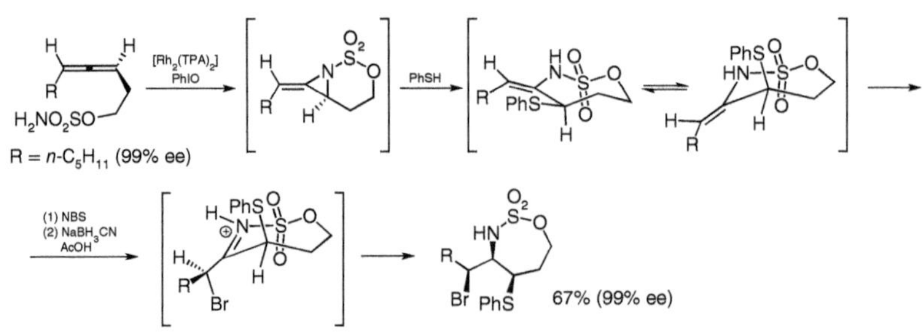

Scheme 15.12 Cyclization of sulfamate-allenes to bicyclic aziridines for the synthesis of cyclic stereotriads. Source: Adams et al. [53]/American Chemical Society.

Similar to the above processes is the reaction of silylene transfer to allenes, including chiral ones [54]. Treatment of allenes with saturated silacyclopropane in the presence of silver catalysts transformed them into moderately stable alkylidenesilacyclopropanes, which were subjected to insertion of carbonyl compounds, either thermal or catalyzed by CuI, to give alkylideneoxasilacyclopentanes in high yields and excellent regio- and stereoselectivity (Scheme 15.13). Moreover, diastereoselective epoxidation of the exocyclic double bond and base-mediated opening of both rings provided a stereotetrad as a single isomer in high yield.

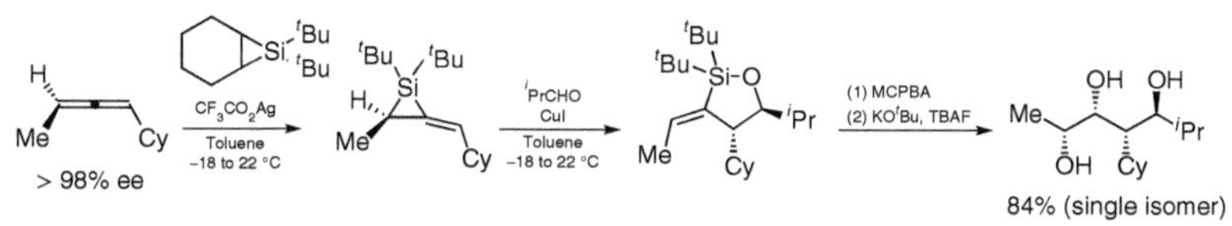

Scheme 15.13 Silylene transfer to chiral allene and ring expansion in the reaction with aldehydes.

15.5 Nazarov Cyclization of Allenyl Vinyl Ketones

Enantiospecific Nazarov cyclization of allenyl vinyl ketones was developed by Tius' group [55–57]. In this reaction, the transfer of the axial chirality of allene to the stereogenic center in the cyclopentanone ring is enabled by controlling the sense of the conrotatory ring closure (Scheme 15.14). The favored direction of conrotation is when the bulky substituent at the terminal carbon of the allene moiety moves "away" from the formed carbon–carbon bond, producing a *Z* exocyclic double bond and determining the configuration of the ring sp^3 carbon. Assuming that the carbocationic intermediates of the Nazarov cyclization do not undergo rearrangements and that the reaction occurs through a concerted fashion exclusively, unidirectional conrotation should result in complete chirality transfer.

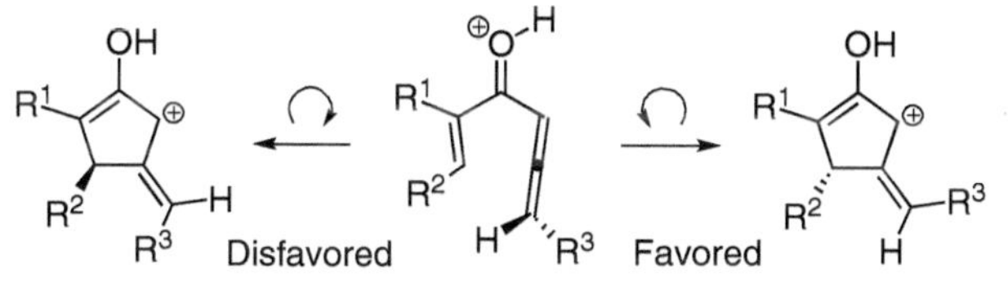

Scheme 15.14 The concept of torquoselectivity controlled by axial chirality of allene.

This concept was successfully demonstrated with chiral methoxymethyl-substituted allenes, as shown in Scheme 15.15. ^{n}BuLi-promoted isomerization of methoxymethyl propargyl ethers resulted in methoxymethylallenes that were again deprotonated and converted into carboxylic acids upon treatment with CO_2 and then into morpholine amides. Each racemic allene was then resolved by chiral HPLC. The addition of vinyllithium reagent set the stage for the key Nazarov cyclization reaction, which occurred spontaneously upon acidic workup with

NaH_2PO_4 to give *Z* cyclopentenones with good levels of CT for the ^{n}Bu-substituted allene and very high for the more sterically hindered ^{t}Bu substrate. According to expectation, small amounts of *E* isomer of cyclopentenone obtained from ^{n}Bu-substituted allene had the opposite configuration at the ring stereogenic center.

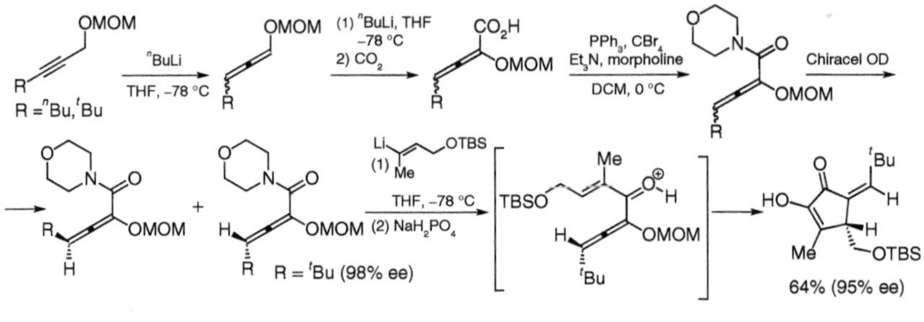

Scheme 15.15 The synthesis of chiral methoxymethyl allenes and their enantiospecific Nazarov cyclization.

Control of the chiral sense of conrotation in the Nazarov cyclization of allenyl vinyl ketones has also been achieved by using an allene lacking axial chirality but substituted with a chiral auxiliary [58]. The sense of conrotation of chiral racemic allenyl vinyl ketones in a Nazarov reaction interrupted by [4+2] cycloaddition was investigated more recently by Boyd, Burnell, and coworkers [59].

A different stereochemical outcome has been reported for cyclizing systems similar to allenyl vinyl ketones but containing an oxygen-substituted stereocenter between the two unsaturated moieties and a lithium allenolate. In such systems, the sense of conrotation and configuration of the final cyclopentenone is dictated by the point chirality of the sp^3 stereocenter and not by axial chirality of the allene fragment [60, 61].

15.6 Nucleophilic Cyclization and Addition Promoted by Electrophiles

Electrophilic activation of allenes by protonation, halogenation, or reversible formation of π-complexes with silver and gold, followed by the addition of nucleophiles, is perhaps the most extensively studied aspect of the allene chemistry. Particularly well-explored are *endo* cyclizations with the formation of unsaturated five- or sometimes six-membered heterocyclic rings. Enantioselectivity in such reactions can be achieved using metal complexes and chiral ligands [62]. However, several reactions with excellent CT levels, particularly gold-catalyzed ones [63], have also been reported.

Stereospecific *endo*-cyclization of α-allenols to dihydrofurans has been studied since the 1990s [64]. Silver-mediated stereospecific cycloisomerization was found to occur preferentially with tertiary hydroxy groups, followed by secondary and primary, due to steric effect, with electronic effects (formation of more stabilized carbocation) playing a lesser role [65]. In one of the studies of Krause's group, allenic hydroxyesters, obtained by oxidation of titanium anions of 3,4-dienoates with DMDO, underwent cyclization upon treatment with gaseous HCl in chloroform in high yields and with a complete transfer of allene axial chirality into the C(5) stereogenic center [66]. A method more suitable for cyclizing acid-sensitive α-hydroxyallenes (2-hydroxy-3,4-dienoates) to 2,5-dihydrofurans in the presence of $AuCl_3$ in good or excellent yields and a complete transfer of chirality was later reported by the same group [67]. A similar study on silver-mediated cyclization of 2-hydroxy-2-trifluoromethyl-3,4-dienoates has also been performed [68].

Allene axial chirality was the controlling factor of the stereochemistry of the final product of the total synthesis of an unnatural but RNA-translatable aminoacid (+)-furanomycin (Scheme 15.16) [69]. Chelation-controlled addition of optically pure lithium acetylide to Garner aldehyde in the presence of $ZnBr_2$ provided alcohol with 9 : 1 diastereoselectivity. Lithium aluminium hydride (LAH) reduction of the propargyl ether moiety directed by the free hydroxyl group formed a stereodefined allene, which in the next step underwent silver-promoted cyclization to 2,5-*trans*-disubstituted dihydrofuran. The synthesis was accomplished by a few deprotection and redox steps.

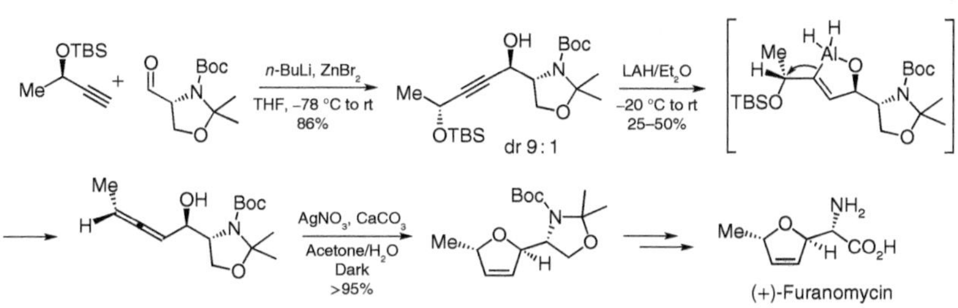

Scheme 15.16 Application of silver-promoted allenol cyclization in synthesizing a biologically active dihydrofuran derivative. Source: VanBrunt and Standaert [69]/American Chemical Society.

The synthetic utility of Au-catalyzed cycloisomerization of allenols to dihydrofurans was demonstrated in a concise synthesis of natural products, (−)-isocyclocapitelline and (−)-isochrysotricine [70, 71]. Starting from ethylenebenzyloxy derivative of methyl propiolate, reduction, Horner-Wadsworth-Emmons (HWE) reaction, reduction of ester to alcohol, oxidation to ketone, and Grignard addition provided enyne alcohol (Scheme 15.17). This racemic intermediate was then kinetically resolved using Katsuki–Sharpless epoxidation to give enyne with an excellent optical purity (98% ee). Another Katsuki–Sharpless reaction, but with opposite enantiomer of DET, provided (*R*,*R*,*R*)-epoxyalcohol. After transforming the terminal hydroxy group into a tertiary alcohol, treatment with MeMgCl in the presence of CuCN and triphenylphosphite ligand gave the key chiral allenic diol. With only 0.05 mol% of $AuCl_3$ catalyst, this allene was then cyclized to the expected dihydrofuran with 96% de. Further functional group transformations and construction of the carbazole ring in a Pictet–Spengler reaction with tryptamine accomplished the synthesis, confirming the absolute configuration of natural substances.

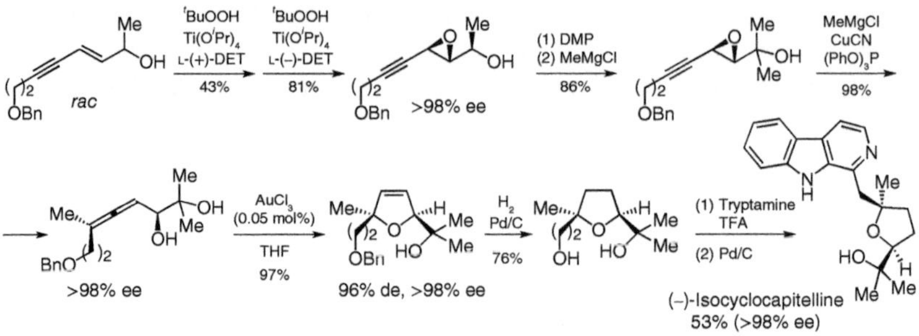

Scheme 15.17 Regio- and diastereospecific cyclization of allenediol in the synthesis of (−)-isocyclocapitelline.

Chemoenzymatic separation of racemic bis(hydroxymethyl)allene provided its butanoic ester with a high optical purity [72]. $AgNO_3/SiO_2$-catalyzed cycloisomerization led to piperonyl-substituted dihydrofuran of the same optical purity, which was then transformed into natural products hyperione A and its epimer.

Enantiomerically pure β-halobutenolides were obtained in CuX_2-mediated cyclization of 2,3-allenoic acids that were first prepared in their optically pure form by crystallizing the respective salts with a natural amine L-(−)-cinchonidine [73]. Direct treatment of the salt of *S*-allene (without the prior release of free carboxylic acid) with $CuBr_2$ or $CuCl_2$ in acetone-water mixture gave β-bromo- and β-chlorobutenolide, respectively, in excellent yields and with very high ee values. The procedure using both enantiomers of α-methylbenzylamine was similarly efficient, and it provided an access to both enantiomers of the target butenolides. An efficient CT was also observed in synthesizing chiral butenolides in a Pd-catalyzed cross-coupling and cyclization of 2,3-allenoic acids [74]. Salts of 2-alkyl-2,3-allenoic acids bearing a phenyl or *n*-heptyl group at C(3) with L-(−)-cinchonidine or both enantiomers of α-methylbenzylamine were reacted with aryl iodides in the presence of $Pd_2(dba)_3CHCl_3$, PPh_3, and TBAB at low temperature (5 °C). High ee of the product was also observed in the reaction of the salt of chiral allene and diisopropylethylamine (DIPEA), which proved that the origin of stereoselectivity was indeed the axial chirality of allene and not the presence of the chiral amine component. This process probably involved an initial oxidative addition of palladium to ArI, coordination to the 3,4-double bond of allene, cyclization, and reductive elimination. Such a reaction pathway was crucial for the observed stereoselectivity, as similar conditions promoting the formation of Pd-allyl intermediate led to less efficient chirality transfer.

Nelson's group developed an efficient preparation method of chiral β-lactones bearing an alkynyl substituent at β carbon based upon acyl halide–aldehyde cyclocondensation (AAC) in the presence of a chiral Al catalyst [75, 76]. These compounds were subsequently used as excellent precursors of chiral allenes functionalized with a methylenecarboxy group in a highly regioselective and stereospecific S_N2' reaction with alkylmagnesium reagents, catalyzed by CuCN·2LiBr or CuBr·dimethyl sulphide (DMS) [77]. Nucleophiles such as ester zinc enolates or magnesium salt of MeCN could also be used. The driving force for this process is the opening of the strained β-lactone ring. The usefulness of this methodology was demonstrated in a concise, asymmetric synthesis of a natural product, (−)-malyngolide (Scheme 15.18). AAC reaction of propionyl bromide and 4-benzyloxybutynal furnished alkynyllactone in high yield (85%) and stereoselectivity (*cis:trans* 91 : 1, 94% ee), which was then opened with *n*-nonyl Grignard reagent to give the key allene intermediate. Activation of the more substituted double bond with Ag(I) resulted in a 6-*endo-trig* formation of the lactone ring. Hydrogenation of the double bond with concomitant deprotection of benzyl group provided the target product with excellent stereoselectivity (94% ee, 100% de), proving an efficient chirality transfer from the allene intermediate.

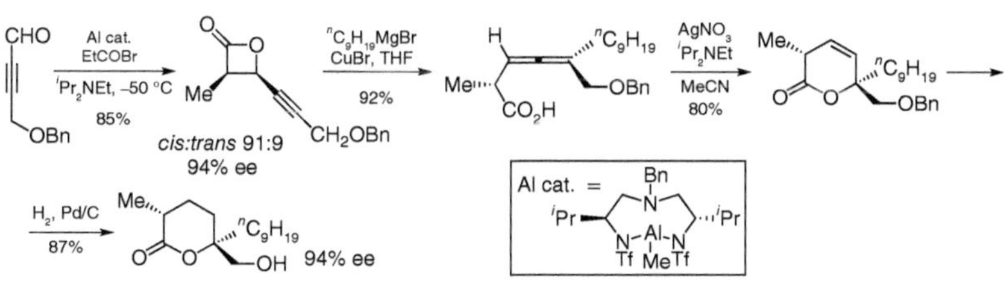

Scheme 15.18 The synthesis of (−)-malyngolide.

Sulfinylallenes (1,2-allenic sulfoxides) react smoothly with iodine and water to give products of regioselective iodohydroxylation of the terminal (more nucleophilic) double bond (Scheme 15.19). Inversion of the sulfur chirality center accompanying this reaction and its complete *E* selectivity prompted Ma and coworkers to propose the mechanism involving the participation of the sulfinyl oxygen and formation of a cyclic intermediate, followed by hydrolysis [78]. This type of mechanism implies retention of allene chirality in the reaction course. This was demonstrated by iodohydroxylation of chiral 1,2-allenic sulfoxides, bromohydroxylation with Br_2 or $CuBr_2$, and chlorohydroxylation with $CuCl_2·2H_2O$. Vinyl halogen and sulfinyl functions could be employed in Pd and Ni-catalyzed cross-coupling reactions, thus enabling a general and fully stereoselective synthesis of allyl alcohols.

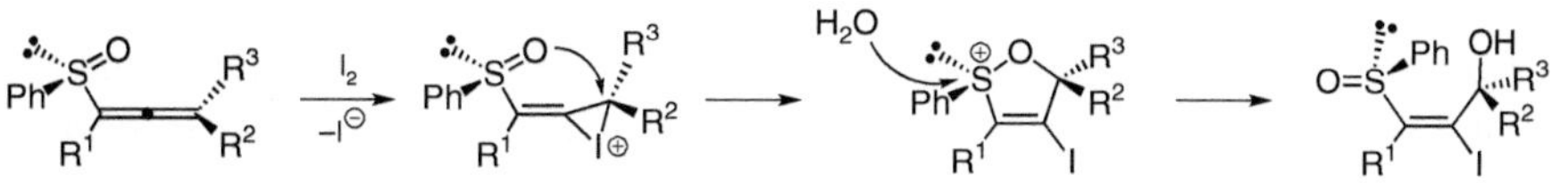

Scheme 15.19 Formation of an iodonium cation from allene and its sulfoxide-assisted opening.

In later years, Ma's group studied the stereoselective opening of iodonium ions during iodolactonization of longer chain allenols [79, 80].

Endo-cyclization of allenes with nitrogen nucleophiles as a means of pyrroline synthesis was reported already in 1980 [81]. Silver-mediated cyclization of 3-allenylazetidinones to carbapenems proceeded in a highly stereospecific manner. Two epimeric propargyl alcohols were converted into stereodefined allenes by adding methylcopper reagents. After cyclization, the configuration of the new stereogenic center was, in each case, consistent with an *anti* attack of the ring nitrogen atom on the cyclic silver cation. The reaction of the more sterically hindered substrate in the presence of excess $AgBF_4$ led to the diastereoisomer configuration inconsistent with the above mechanism, probably due to silver-mediated isomerization of either the substrate or the product.

Silver- or palladium-catalyzed 5-*endo-trig* cyclization of aminomethylene allenes is a useful method of preparing trisubstituted 3,4-pyrrolines [82]. Enantioenriched allene (83% ee), prepared from chiral (99% ee) propargyl mesylate by S_N2' substitution with aminomethylene copper reagent, underwent cyclization to enantioenriched pyrroline with a moderate chirality transfer (70% ee of the final product).

Concerning other nucleophiles, enantiospecific cyclization to dihydrothiophenes has been disclosed by Krause's group [83]. Intramolecular hydroarylation of allenes with electron-rich aromatic rings in the presence of a dinuclear, achiral Au catalyst has been developed by Ma's group [84]. Initially, cyclization of racemic benzyl allenes to 1,4-dihydroarenes was explored, resulting in an efficient method that exploited the Ph_3PAuCl/AgOTf catalytic system. However, the same catalyst failed to secure high levels of chirality transfer when a chiral substrate was employed. The solution came with Au complexes with a bulkier phosphine ligand and a dinuclear gold complex [(dppm)Au_2Cl_2]. Such complexes exhibit higher selectivity for binding to the double bond further from the aromatic ring in the *anti* mode and the subsequent electrophilic attack on the ring, minimizing the side reaction of racemization via allylgold complexes (Scheme 15.20).

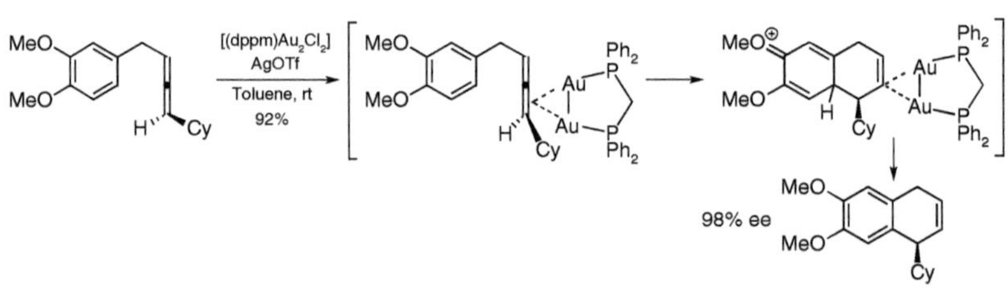

Scheme 15.20 Intramolecular hydroarylation of chiral benzylallenes.

Hydroarylation of allenes directly connected to an aromatic ring provides indenes. Egi and coworkers designed a silver-catalyzed reaction of ynamides with tertiary propargyl alcohols in which allenyl acetamides were formed in good yield and very efficient transfer of point chirality of alcohol to axial chirality of allene [85]. In the presence of Au and In catalysts, isomerization of these allenes to 1,3-dienes predominated, but $PtCl_2$ smoothly transformed them into chiral indenones with a quaternary carbon center in the five-membered ring, again with very high levels of retention of chiral information.

Nucleophilic *exo* cyclization of allenes is usually observed when the nucleophilic center is connected with an allene with a longer tether. However, this is not always the determining factor. Interesting observations on the influence of the reaction conditions on the exact type of products obtained in the metal-catalyzed cyclization of α-aminoallenes were reported by Ibuka and coworkers [86]. Cyclization of two diastereomeric aminomethylallenes of opposite axial chirality and identical configuration of the amino-substituted stereocenter, performed in the presence of silver salts in acetone or palladium catalysts in DMF, led exclusively to the two expected diastereomeric pyrrolines. On the other hand, Pd-catalyzed reactions conducted in dioxane exhibited a completely different chemoselectivity, producing 2-vinylaziridines. *cis/trans* Selectivity of aziridine formation still depended largely on the configuration of the allene moiety, but it was not as high as for five-membered ring formation.

Early studies on silver-mediated, stereospecific *exo* cyclization of an allenyl amine to a six-membered ring were performed by Lathbury and Gallagher [87]. Coordination of silver cation to the 5,6 double bond of *N*-benzylocta-5,6-dienylamine (80% ee) provided a 2-vinylpiperidine derivative in 86% yield. Reductive deprotection and catalytic hydrogenation of the double bond transformed this product into (−)-coniine containing *R* and *S* isomers in an 8 : 1 ratio, confirming an efficient chirality transfer upon cyclization. Ha and Cha used a similar approach to the stereoselective construction of nitrogen heterocycles in a much more advanced synthetic task: total synthesis of alkaloids clavepictines A and B (Scheme 15.21) [88]. Chemo- and stereoselectivity of the nucleophilic attack of the piperidine nitrogen on the silver-activated allene moiety (introduced by the Johnson–Claisen rearrangement of chiral propargyl alcohol) was directed by the allene chirality and the configuration of substituents already present in the existing piperidine ring. In addition to forming the bicyclic system, the same transformation also introduced the double bond present in the target product.

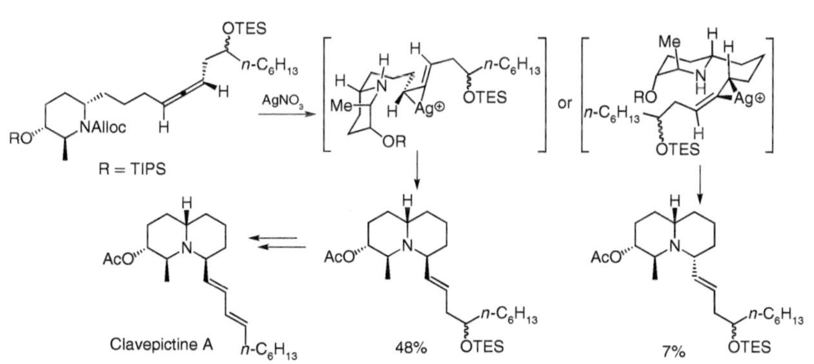

Scheme 15.21 Silver-promoted *exo* cyclization of allenylamine.

In 2006, Widenhoefer found that the catalytic system composed of equimolar amounts of Au[P(^{t}Bu)$_2$(*o*-biphenyl)]Cl and AgOTf, originally designed for *exo*-hydroamination of *N*-alkenyl carbamates, was very efficient in promoting an intramolecular, highly chemo- and regioselective *exo* hydroamination of allenyl carbamates or allenylindoles with the formation of vinylpyrrolidines and vinylpiperidines or vinyltetrahydrocarbazoles, respectively [89]. A similar *exo*-selective cyclization of 4,5-dien-1-ols required exchanging AgOTf for AgOTs. Otherwise, mixtures of products resulting from *exo* and *endo* attacks on the double bond proximal to the hydroxyl group were obtained. Selective formation of *E* alkenes during hydroamination of terminally substituted allenes inspired the authors to examine enantioenriched substrates. Indeed, high levels of CT were obtained in all three reaction types. The observed product configuration could be explained by an outer sphere mechanism in which coordination of Au to either side of the allene double bond gives two complexes that quickly equilibrate. The addition of a nucleophile then occurs in the *cis* complex, *anti* to the terminal alkyl substituent. Protodeauration of the formed alkenylgold complex gives the final chiral *E* alkene. In the case of intramolecular hydroalkoxylation, the formation of both *E* and *Z* products (each of them with a high enantioselectivity) has been attributed to the small size of the hydroxyl nucleophile. The same group observed a similar relationship between axial stereochemistry of allene and *E*/*Z* configuration of the double bond in the products of an intramolecular hydroalkoxylation catalyzed by chiral, dinuclear gold complexes. However, in this case, the absolute configuration of the sp^3 center in the heterocyclic ring was governed entirely by the catalyst [90].

Synthesis of chiral trisubstituted allenes containing a hydroxy group at the end of a methylene tether via copper-catalyzed S_N2' substitution of propargylic phosphates with alkylboranes enabled enantiospecific synthesis of 2-vinyltetrahydrofurans and 2-vinyltetrahydropyrans with a tetrasubstituted stereogenic center in the ring (Scheme 15.22) [91]. Initially, moderate levels of chirality transfer were obtained with the [AuP(^{t}Bu)$_2$(*o*-biphenyl)]OTs used previously by Widenhoefer. This could be explained based on control experiments by faster racemization of the starting allene with the catalyst. The situation could be ameliorated by employing gold complexes with bulky phosphines, particularly ^{t}Bu$_3$P and moderately coordinating ligands, such as tosylate. In the presence of ^{t}Bu$_3$PAuCl/AgOTs, the reaction proceeded with a complete *exo* and *E* selectivity to form the target oxygen heterocycles with a very high CT. The product of the expected chirality could be obtained even from allenols containing an additional stereocenter (although the "mismatched" case exhibited a slightly lower CT) or a disubstituted allene.

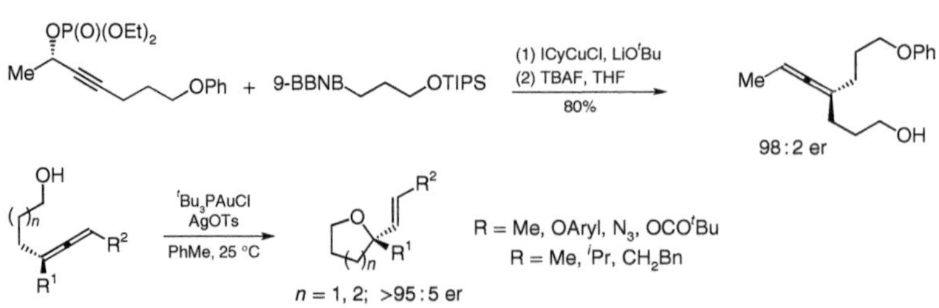

Scheme 15.22 Enantiosepecific cyclization of allenols to vinyltetrahydrofuranes and pyranes. ICy = 1,3-dicyclohexylimidazolylidene. Source: Cox et al. [91]/John Wiley & Sons.

After a slight modification of the reaction conditions (exchange of tosyl ligand for 4-methoxycarbonylbenzoate), an efficient cyclization with phenolic hydroxy groups was also achieved.

An intramolecular Au-catalyzed *exo* hydroarylation of allene with a pyrrole ring played a pivotal role in Nelson's synthesis of (−)-rhazinilam (Scheme 15.23) [92]. A chiral trimethylenepyrrole-substituted allene was first obtained in the alkaloid-catalyzed AAC reaction between EtCOCl and 2-pentynal, followed by S_N2' ring opening with the appropriate pyrrolyl Grignard reagent. The key to an efficient chirality transfer was the coordinating ester substituent in *anti* relation to the incoming pyrrole nucleophile. A high yield and enantiomeric excess of the target tetrahydroindolizine were achieved with $Ph_3PAuOTf$ or $AuCl_3/AgOTf$. In contrast, palladium catalysts led to a significant racemization due to their relatively low Lewis acidity, which confirmed that stereoselectivity of the cyclization step was not associated with the additional Me-substituted stereocenter. The synthesis of (−)-rhazinilam was accomplished by a few more steps that included acylation of the pyrrole ring to attenuate its nucleophilicity, oxidative cleavage of the ester side chain and installation of a chain of appropriate length in the HWE reaction, catalytic hydrogenation, iodination of the pyrrole ring, Suzuki cross-coupling with an NHBoc-substituted phenylboronic acid and finally, amide bond formation that established the appropriate atropisomeric configuration around the aryl–aryl bond.

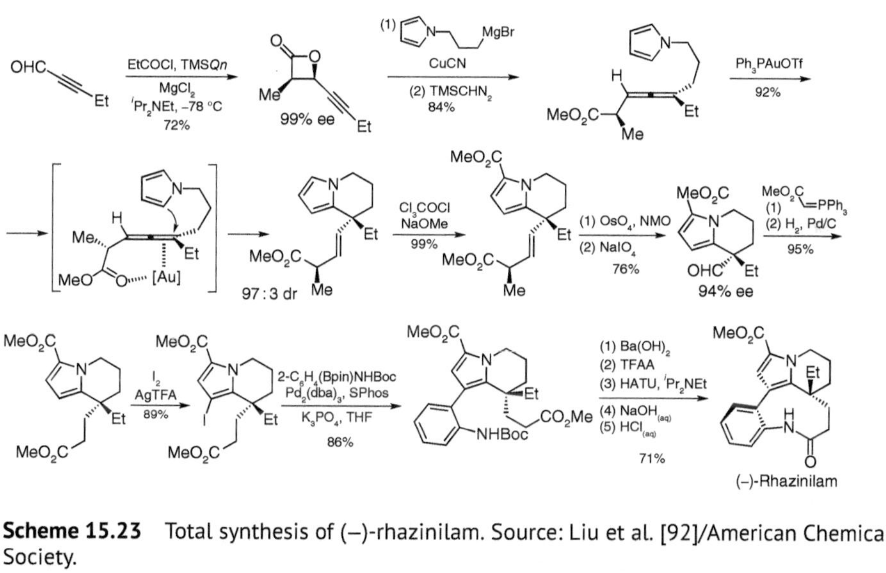

Scheme 15.23 Total synthesis of (−)-rhazinilam. Source: Liu et al. [92]/American Chemical Society.

Late transition metal-catalyzed hydroamination of multiple carbon–carbon bonds is one of the most important amine synthesis methods [93]. Intermolecular allene hydroamination, performed in a stereospecific manner to access chiral allylamines, is a process much more challenging than its intramolecular version, particularly due to the fast racemization of allenes observed in the presence of Au catalysts [94]. Nevertheless, a successful reaction of this type was disclosed by Nishina and Yamamoto. They reported a room temperature Au(III)-catalyzed reaction of anilines with mono- and disubstituted allenes, including chiral ones, which provided allylamines in good yields and with a high degree of chirality transfer (Scheme 15.24) [95]. The observed enantioselectivity could be explained by the initial formation of the gold–amine complex and its *syn* addition to the less hindered side of the alkyl-substituted double bond of allene. Nucleophilic attack of amine on Au-activated allene would probably result in the opposite configuration of the product, and Au-allyl intermediates would lead to racemization.

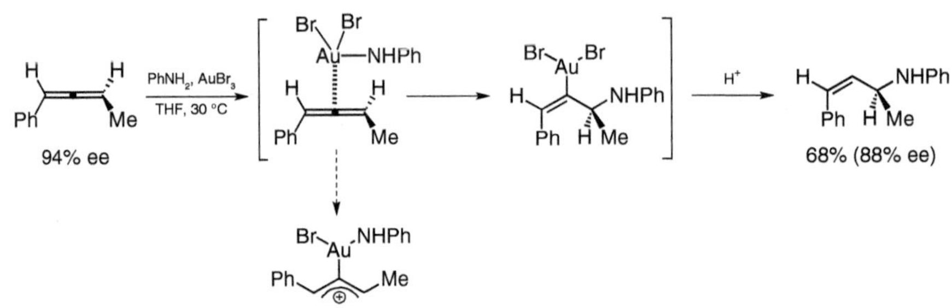

Scheme 15.24 CT in intermolecular Au-catalyzed hydroamination of allenes. Source: Nishina and Yamamoto [95]/John Wiley & Sons.

An intermolecular hydroamination and hydroalkoxylation of allenes in the presence of the $Ph_3PAuCl/AgOTf$ catalytic system appear to proceed according to different mechanisms. However, hydroamination begins with amine-gold formation and further complexation of allene. In the case of hydroalkoxylation, lower nucleophilicity of alcohol results in the initial formation of a complex between gold and allene, leading to its fast racemization [96]. This effect was largely attenuated when the NHC gold complex was employed in the reaction of 1,3-disubstituted allene as fairly efficient CT was observed by Widenhoefer in the hydroalkoxylation of (*S*)-pent-2,3-dienyl benzoate (97% ee) with BnOH [97]. Performing the reaction with BnOH as a nucleophile in toluene at rt, they obtained the expected allylic ether with 64% or 79% ee, depending on the reaction time and concentration.

A general method for highly regio- and enantiospecific, intermolecular allene alkoxylation was designed by the Lee group [98]. Performing the reactions in DMF at 0 °C, they succeeded in reducing the activity of an NHC gold complex enough to prevent racemization of the starting chiral allenes completely while maintaining its catalytic activity in hydroalkoxylation. Interestingly, in an earlier work, similar reaction conditions allowed regioselective hydroalkoxylation of the more substituted double bond of allenes and avoided isomerization of the kinetic tertiary allylic ethers into more stable primary ones [99].

A detailed mechanistic study of gold-catalyzed allene hydroamination was reported by Toste's group [100]. According to their conclusions, the rate-determining step of the reaction is the reversible formation of a π complex between allene and mononuclear Ph_3PAu^+ species. Then, a transition state is reached. The gold ion is bound to the central atom of allene, and the allene molecule becomes partially planar while retaining its chiral information and minimizing the nascent 1,3-allylic interactions. The interaction of this species in the course of a two-step, no intermediate-type process with a nucleophile leads to another transition state in which the carbon–nitrogen bond is formed. At low nucleophile concentration, the first TS collapses to a planar vinylgold complex, providing the route for fast Au-catalyzed allene racemization.

Carbopalladation of chiral 1-alkyl-3-arylallene with aryl iodides, followed by the addition of a malonate nucleophile, was reported to proceed in a highly enantiospecific manner [101]. The stereochemical structure of the intermediate allylpalladium complexes was explained by initial carbopalladation of the phenyl substituted double bond *anti* with respect to the alkyl substituent, followed by isomerization to a complex, minimizing the 1,3-allylic strain and nucleophilic addition from the side opposite to palladium.

Early attempts at intermolecular enantioselective hydroarylation of allenes with indoles with 3-allylindoles formation and exploiting substrate axial chirality were reported by Che and coworkers [102]. Reactions of chiral 1,3-diphenylallene (98% ee) with *N*-methylindole in various Au or Au/Ag catalytic systems provided the expected 3-allylindole derivative in excellent yields but without any enantioselectivity. The authors attributed this disappointing result to the racemization of allene with gold complexes. Consequently, they turned their attention to processes catalyzed by gold in the presence of chiral phosphine ligands with a considerable success.

An efficient chirality transfer in allene hydroarylation was later achieved by Lee's group, who envisioned that the key to avoiding Au-mediated racemization via a planar vinylgold complex was a careful design of the allene substrate, placing an alkyl group at one end of the allene system and an inductively electron-withdrawing substituent, such as CH_2NR_2, CH_2OR, CH_2CN, at the other (Scheme 15.25) [103]. Such a substrate design was also beneficial for the regioselectivity of the reaction. The reaction scope included several allene and indole derivatives (with 1,3-dimethylindole reacting at C(2)), as well as pyrroles, 1,3,5-trimethoxy- and 1,3-dimethoxybenzene. Importantly, an efficient chirality transfer was achieved in nearly all cases (especially with more nucleophilic arenes). A high stereoselectivity and formation of (*R,E*)-configured products could be explained by a preferential attack *anti* to the coordinated gold catalyst and *anti* to the CH_2-bound heteroatom substituent. Control experiments confirmed that allene quickly racemized in the absence of a nucleophile. On the other hand, optical purity erosion was much

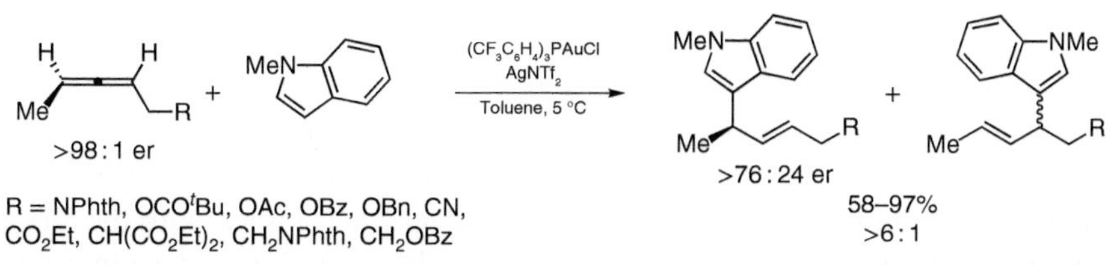

Scheme 15.25 Asymmetric intermolecular hydroarylation of allenes with *N*-methylindole. Source: Sutherland et al. [103]/John Wiley & Sons.

slower under the standard reaction conditions. A reaction with a nucleophile was much faster than racemization, which could be slowed down by the reversible binding of Au catalyst to the indole substrate.

15.7 Cycloisomerization and Other Reactions Catalyzed by Transition Metals

Sato and coworkers investigated the reaction of diisopropoxy(η-2-propene)titanium with allenol carbonates. Coordination of titanium to the proximal double bond of allene resulted in a nucleophilic 5-*exo-trig* attack on the carbonate carbonyl and intramolecular nucleophilic acyl substitution. Chiral, enantiomerically enriched substrates formed β,γ-unsaturated esters with nearly the same optical purity [104]. The same research group has observed the complete retention of chiral information of allene-ynes protected with TMS on the alkyne terminus upon their titanium-mediated cyclization to vinylcyclopentanones [105]. The reaction course involves the attack of the titanacycle on the internal double bond of the allene moiety, *anti* to the silyl group (Scheme 15.26). Removal of titanium upon protonation gives chiral *Z*-alkene. Insertion of carbon monoxide into the intermediate complex resulted in the formation of 5,5-bicyclic products as single diastereomers of optical purity reflecting that of the starting allenylsilane. An analogous cyclization of substrates with a shorter tether, 1,2-dien-6-ynes, for stereoelectronic reasons proceeded via an attack on the silicon-substituted allenic double bond with a formation of a titanacyclopentene intermediate which reacted further with carbonyl compounds with very high stereoselectivity.

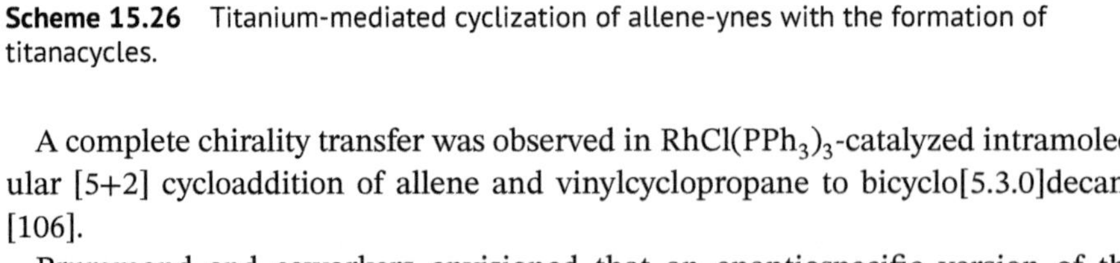

Scheme 15.26 Titanium-mediated cyclization of allene-ynes with the formation of titanacycles.

A complete chirality transfer was observed in $RhCl(PPh_3)_3$-catalyzed intramolecular [5+2] cycloaddition of allene and vinylcyclopropane to bicyclo[5.3.0]decane [106].

Brummond and coworkers envisioned that an enantiospecific version of the allenic Pauson–Khand reaction for the synthesis of nonracemic α-alkylidene cyclopentenones could be achieved if the reaction proceeded predominantly via the less sterically hindered conformer of the metal alkyne complex, with the allene substituent pointing away from the triple bond, at the same time, determining the *E*/*Z* configuration of the exocyclic double bond in the product [107]. Each of the geometrical isomers of the product should exhibit high enantiopurity, assuming that the metal catalyst does not cause their isomerization. In the initial experiments, zirconium catalyst gave the most promising results in five vs. six-membered ring formation regioselectivity and facial selectivity of the metal toward the two sides of the allene moiety, as evidenced by the exclusive formation of *E*-alkylidene cyclopentenone. Cyclization of enantioenriched (95% ee) alkynyl allene in the presence of zirconocene complex, ^{n}BuLi and carbon monoxide gave chiral alkylidene cyclopentanone exclusively as *E* isomer, but with lower ee (85%). Experiments involving protonation of the intermediate zirconacycles demonstrated that the facial selectivity of cyclization was incomplete (10 : 1). However, the minor product with *Z* double bond and a "wrong" configuration on the stereogenic center readily isomerized to *E* during acidic workup. A better facial selectivity was obtained with molybdenum hexacarbonyl-mediated cyclizations and larger substituents at the allene terminus. Particularly, DPS-allene-yne (95% ee) could be transformed into a mixture of *E* (95% ee) and *Z*-alkylidene (63% ee) in an 80% overall yield.

Later, the same group disclosed Rh(I)-catalyzed enantioselective allenic Pauson–Khand reaction, this time involving the distal double bond of the allene-ynes and forming a new stereogenic center α to the carbonyl group (Scheme 15.27) [108]. The required chiral starting allenes were prepared to employ the Johnson–Claisen rearrangement of the respective chiral propargylic alcohols that were in turn enantioenriched by enzymatic kinetic resolution. After a brief optimization, 5,7-bicyclic products of cyclocarbonylation were obtained in high yields. Substrates containing trisubstituted allene moiety reacted with a complete chirality transfer. Enantiopurity considerably lower than expected was observed in products formed from 1,3-disubstituted allenes, particularly those connected to the alkyne part with an oxygen tether. Measurements of the enantiomeric excess of products obtained after only partial conversion indicated that the enantiopurity erosion occurred with the substrate rather than the product.

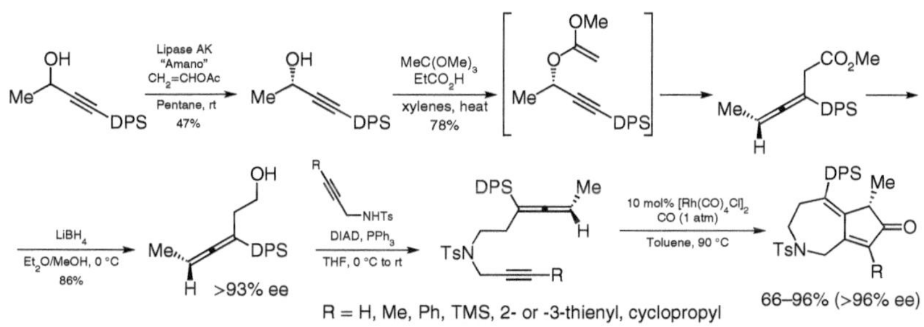

Scheme 15.27 Allenic Pauson–Khand reaction leading to 5,7-bicyclic systems. Source: Grillet and Brummond [108]/American Chemical Society.

More recently, an excellent enantioselectivity of transformation of allene-ynes with disubstituted allene fragment into 6,5-bicyclic cyclopentenones in the presence of [RhCl(CO)$_2$]/AgSbF$_6$ was reported by Ma's group [109]. The stereochemical course of this reaction can be explained by the formation of a less sterically crowded rhodacycle (Scheme 15.28).

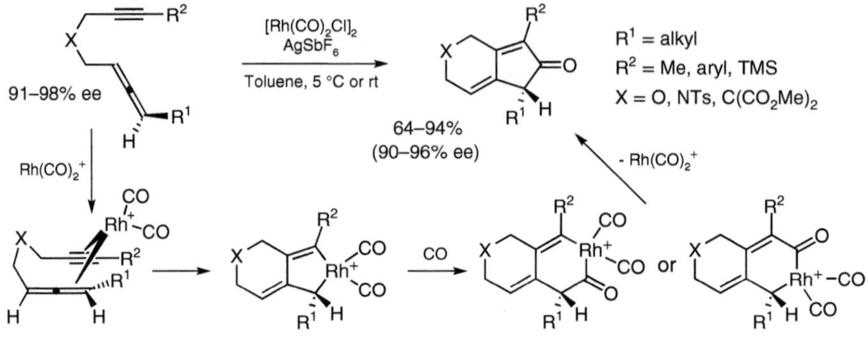

Scheme 15.28 Allenic Pauson–Khand reaction catalyzed by a cationic rhodium complex.

An efficient transfer of axial chirality has also been achieved in a C–H activation reaction, Rh(III)-catalyzed hydroarylation of allenes with *N*-methoxybenzamides [110]. The presence of the hydroxymethyl group in the allene substrate caused lactamization of the primary C–H activation product. This mild (room temperature or below) reaction proceeded via a π-complex of allene and metallacycle formed by amide-directed C–H activation of the aromatic ring. A preference for the complex exhibiting a lower steric congestion between the bulkier substituent at the allene terminus, and *N*-methoxyl of the amide group can explain the high stereospecificity of this reaction (Scheme 15.29).

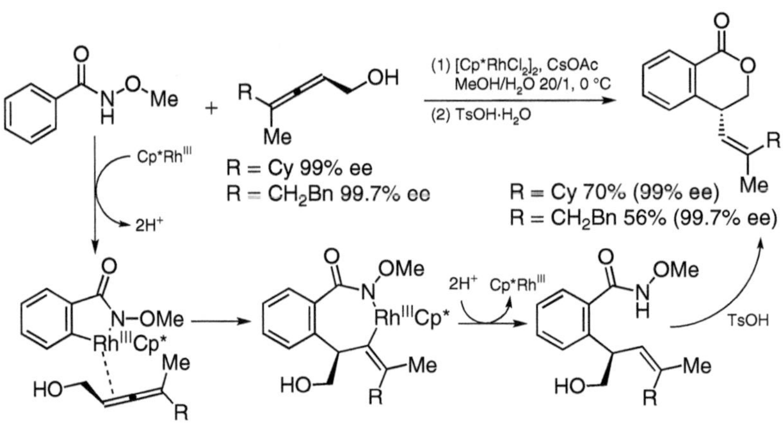

Scheme 15.29 CT in C–H activation of methoxybenzamides followed by allenol incorporation and lactonization.

Jamison developed the Ni-catalyzed reaction between allenes and aldehydes, which involved an electrophilic attack on the central allene atom, leading to allylic alcohol derivatives (Scheme 15.30) [111, 112]. A combination of Ni(cod)$_2$ and NHC carbene ligand catalyzed a three-component reaction between 1,3-disubstituted allenes, aldehydes, and organosilane with a complete *Z*-selectivity of the remaining double bond and, importantly, with very high CT. In unsymmetrical allenes, addition across the more substituted double bond was exclusively observed. The reaction with Et_3SiD established the exact position of incorporating silane hydrogen in the product molecule and enabled the authors to formulate the mechanism explaining the unusual selectivity of the process. According to this proposal, the reaction is initiated by coordinating Ni(0) to the less hindered double bond of allene, *anti* to the larger substituent. Coordination of aldehyde on the face opposite to the methyl substituent and addition of the Ni complex across the double carbon–oxygen bond gives a cyclic complex, reduction of which provides an allylnickel complex, reducing the steric interactions between the terminal substituents. A reductive elimination delivers the hydrogen atom with the less sterically congested alkene formation.

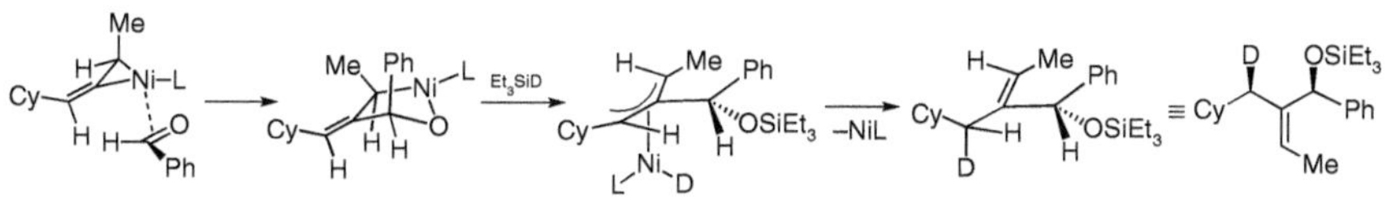

Scheme 15.30 Origin of enantiospecificity in Ni-catalyzed reaction of allenes, aldehydes, and triethylsilane.

Ni(0)-catalyzed alkylative cyclization of enone and chiral alkene was envisioned as a possible strategy of synthesis of the kainoid natural product, (−)-domoic acid [113]. Starting from a chiral oxazolidinone *N*-propargyl derivative, a mixture of two diastereomeric vinylallenes was prepared, differing in the configuration of allene moieties. Disappointingly, one of these isomers underwent a spontaneous intramolecular Diels–Alder reaction between the vinylallene and enone double bond present in the remote part of the molecule. At the same time, the other one met the same fate during the attempted nickel-catalyzed cyclization. Notably, the adduct was formed in a highly stereospecific manner in both cases. The second generation of substrates lacking a pendant vinyl group was then prepared. Their independent cyclization proceeded satisfactorily in each case, producing the expected vinylcyclopentanone precursor of (−)-domoic acid and its diastereomer.

A stereoselective synthesis of β,γ-unsaturated nitriles via nickel(0)-catalyzed hydrocyanation of chiral allenes with acetone cyanohydrin was described [114]. The highest levels of chirality transfer were observed with the ^{t}Bu-substituted substrate. Based upon control experiments, ee erosion in some cases was mainly due to isomerization of the starting allenes without the participation of the cyanide source. The reaction mechanism probably involved generating a cyano hydrido Ni(II) complex followed by hydronickelation of the allene moiety, with the hydrogen atom forming a bond to the allene central sp carbon. Among the four possible modes of such addition leading to allylnickel complexes, two modes involving hydronickelation of the double bond conjugated with the aryl ring account for the formation of the major product.

15.8 Cycloaddition of Allenes

Allene–alkene [2+2] cycloaddition, especially in an intramolecular setting, is a powerful carbon–carbon bond formation [115]. Photoinduced [2+2] cycloaddition of axially chiral allenes can involve an efficient axial-to-point chirality transfer, as demonstrated already in the 1990s by Carreira, who developed a highly enantioselective, intramolecular, and photoinduced [2+2] cycloaddition of allenes with enones and enoates (Scheme 15.31) [116]. The concept was inspired by earlier results concerning alkene-enone photocycloaddition obtained by Corey and Becker. They observed that cyclization of biradical intermediates could be faster than rotation about single bonds. In order to examine the possibility of efficient axial chirality transfer in allene photocycloaddition, the requisite chiral allene substrates were prepared from alkynones by an enantioselective reduction with (*S*)-Alpine Borane, followed by stereospecific mesylate S_N2' substitution. Additionally, a diastereoselective allene synthesis from glyceraldehyde diethyl ketal was performed to examine the absolute sense of stereoselection of photocycloaddition reactions. The obtained chiral allenols were subjected to a condensation with 1,3-dicarbonyl compounds and the Mitsunobu coupling with 4-hydroxycoumarine. In the photocycloaddition reactions of these reactants, involving proximal double bonds of allenes, excellent levels of CT were observed for cyclohexenone and coumarin-derived substrates and somewhat lower for cyclopentenone derivative. This methodology was later modified by replacing the terminal ^{t}Bu as a stereocontrolling group with TMS, easily removed from the photocycloaddition products [117].

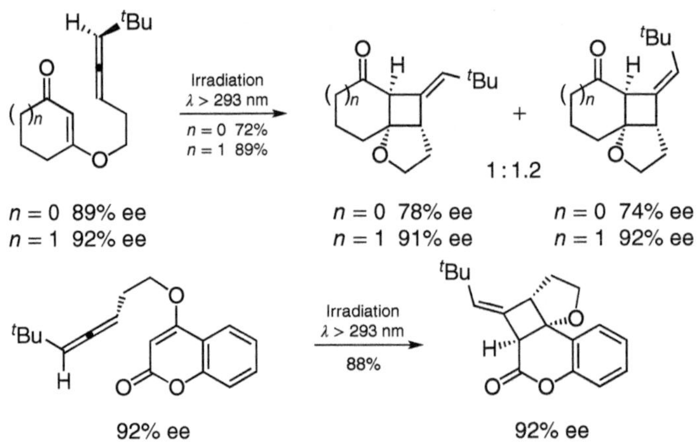

Scheme 15.31 Photoinduced intramolecular [2+2] cycloaddition of allenes and enones. Source: Carreira et al. [116]/American Chemical Society.

In the 1980s, Okamura's group investigated enantioselective, sigmatropic reactions controlled by axial chirality of vinylallenes, including intramolecular Diels–Alder reaction [118, 119]. Examination of molecular models suggested that cyclization of homoallyl vinylallene to hydrindane bicyclic system should be particularly facial and amenable for the stereoselective synthesis of tricyclic terpene (+)-sterpurene (Scheme 15.32). An enantiomerically pure propargyl alcohol, obtained by resolution of diastereomeric carbamate derivatives, was first coupled with iodocyclobutene. Treatment with phenylsulfenyl chloride and

[2,3]-sigmatropic rearrangement provided allenic sulfoxide, spontaneous [4+2] cycloaddition formed the 4,6,5-tricyclic skeleton of the natural product. The synthesis was accomplished by Ni-catalyzed cross-coupling with MeMgBr and partial reduction of a diene to provide (+)-sterpurene of optical rotation identical with the natural sample.

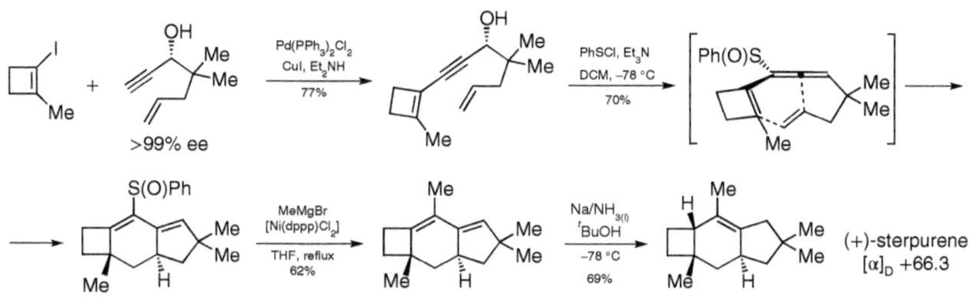

Scheme 15.32 Allene–alkene Diels–Alder route to (+)-sterpurene.

Diels–Alder cycloaddition between 3-trimethylsililoxyfuran and methyl 5-hydroxypenta-2,3-dienoate with hydroxy function protected with PMB was the basis of the stereoselective synthesis of the precursor of the A ring of vitamin D [120]. This reaction, which engaged the ester-conjugated double bond of allene exclusively, proceeded with complete *endo* and *E,Z* selectivity, establishing the correct stereochemistry of the C(1) stereocenter (vitamin D numbering). Further functional group manipulations included stereoselective ketone reduction, probably directed by the C(1) hydroxyl, to establish the A ring's C(3) stereocenter.

Complete CT has also been observed in an intramolecular Diels–Alder reaction with furan ring, providing a complex oxatricyclic skeleton as a single diastereomer [121].

Trost prepared a series of highly enantioenriched (ee above 85%) allenes by Pd-catalyzed dynamic kinetic asymmetric alkylation of racemic allenylmethyl acetates with malonates in the presence of a *trans*-1,2-diaminocyclohexane-derived phosphine ligand [122]. The authors demonstrated the utility of their method by performing an intramolecular Diels–Alder reaction of sorbyl-substituted allenes catalyzed by a cationic Rh(I) complex (Scheme 15.33). As a result, 5,6-bicyclic products were obtained as single diastereoisomers and with complete chirality transfer.

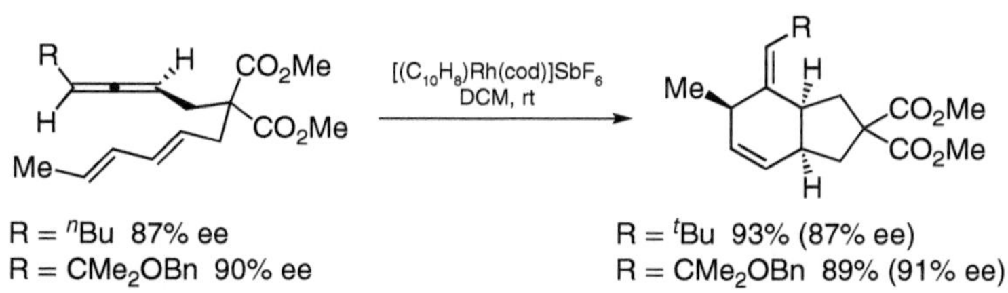

Scheme 15.33 Intramolecular allene–diene Diels–Alder cycloaddition.

Another example of nearly complete chirality transfer in Diels–Alder cycloaddition of allenes with cyclopentadiene is the work of Huang, Tan, and coworkers [123].

In this case, chiral allenes were obtained by isomerization of 3-alkynoates with a chiral derivative of guanidine. Mixtures of alkyne and chiral allenoate were then subjected to the reaction with diene, providing mainly *endo* cycloadducts. Alkynes were recovered unreacted from the reaction mixtures. The same group also examined highly stereospecific transformations of the obtained allenoates into butenolides upon the action of NIS, pyridinium tribromide, or Au(I) salt.

Excellent facial selectivity has also been observed in 1,3-dipolar cycloaddition of carbonyl ylides with allenic esters [124]. The product resulting from addition to the less hindered face of the double bond conjugated with the ester group was observed exclusively, with a preference for *endo* addition. However, this study utilized racemic allenes as substrates.

An innovative approach to the preparation of enantioenriched allenes was designed by Bach [15]. Photocatalytic deracemization of racemic, pyrrolidinone-derived allenes was performed upon irradiation with visible light and in a specially tailored photocatalyst, capable of hydrogen bond formation with amide groups. Some of the highly enantioenriched allenes (>86% ee) were then engaged in enantiospecific Diels–Alder reaction with cyclopentadiene, displaying a preference for *exo* products (Scheme 15.34) [125]. The same facial selectivity governed

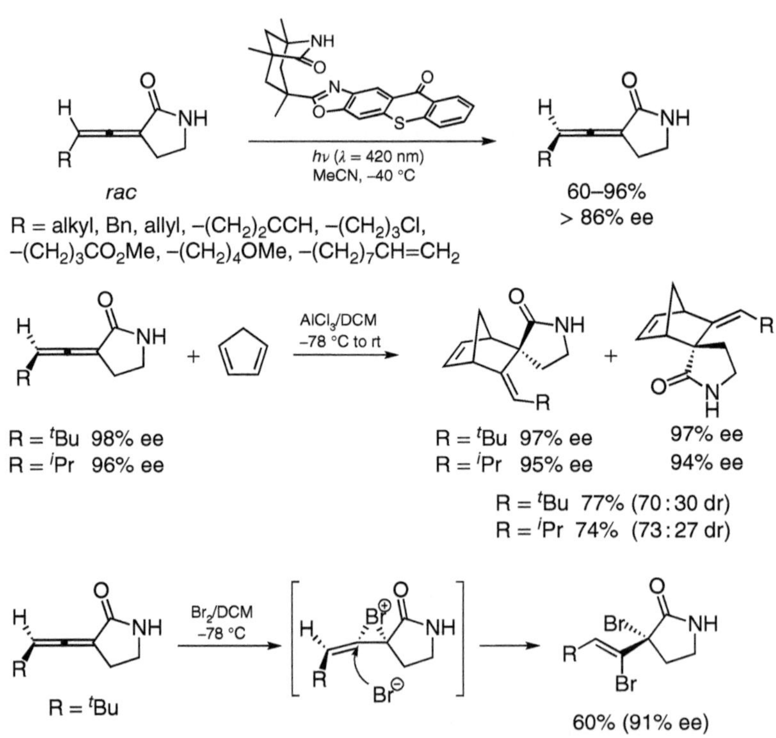

Scheme 15.34 Photocatalytic deracemization of allene and its enantiospecific cycloaddition with cyclopentadiene and bromination. Source: Plaza et al. [125]/John Wiley & Sons.

bromination, which formed a bromonium cation from the double bond adjacent to the amide carbonyl.

A few more efficient and diastereoselective synthetic methods based upon allene cycloadditions have been designed. However, they either use racemic allenes in reactions with substrates bearing the source of chirality [126], or chiral information embedded in the allene substrate is lost in the course of metal complex formation [127–129].

Acknowledgments

We are grateful to the National Science Centre, Poland (Grant 2017/25/B/ST5/00214 "Stereoselective synthesis of allenes") for financial support.

References

1 Alcaide, B. and Almendros, P. (2014). *Chem. Soc. Rev.* 43: 2886–2887.
2 Hoffmann-Röder, A. and Krause, N. (2002). *Angew. Chem. Int. Ed.* 41: 2933–2935.
3 Campolo, D., Gastaldi, S., Roussel, C. et al. (2013). *Chem. Soc. Rev.* 42: 8434–8466.
4 Neff, R.K. and Frantz, D.E. (2015). *Tetrahedron* 71: 7–18.
5 Alonso, J.M., Quirós, M.T., and Muñoz, M.P. (2016). *Org. Chem. Front.* 3: 1186–1204.
6 Lassaletta, J.M. (ed.) (2019). *Atropoisomerism and Axial Chirality*. World Scientific.
7 Brummond, K.M. and DeForrest, J.E. (2007). *Synthesis* 795–818.
8 Krause, N. and Hoffmann-Röder, A. (2004). *Tetrahedron* 60: 11671–11694.
9 Ogasawara, M. (2009). *Tetrahedron: Asymmetry* 20: 259–271.
10 Wang, H., Luo, H., Zhang, Z.-M. et al. (2020). *J. Am. Chem. Soc.* 142: 9763–9771.
11 Zhu, W.-R., Su, Q., Diao, H.-J. et al. (2020). *Org. Lett.* 22: 6873–6878.
12 Kolakowski, R.V., Manpadi, M., Zhang, Y. et al. (2009). *J. Am. Chem. Soc.* 131: 12910–12911.
13 Lo, V.K.-Y., Zhou, C.-Y., Wong, M.-K., and Che, C.-M. (2010). *Chem. Commun.* 46: 213–215.
14 Dai, J., Duan, X., Zhou, J. et al. (2018). *Chin. J. Chem.* 36: 387–391.
15 Hölzl-Hobmeier, A., Bauer, A., Vieira Silva, A. et al. (2018). *Nature* 564: 240–243.
16 Zheng, W.-F., Zhang, W., Huang, C. et al. (2019). *Nat. Catal.* 2: 997–1005.
17 Huang, X. and Ma, S. (2019). *Acc. Chem. Res.* 52: 1301–1312.
18 Gao, M., Gao, Q., Hao, X. et al. (2020). *Org. Lett.* 22: 1139–1143.
19 Lu, Z., Chai, G., and Ma, S. (2007). *J. Am. Chem. Soc.* 129: 14546–14547.
20 Lu, Z., Chai, G., and Ma, S. (2008). *Angew. Chem. Int. Ed.* 47: 6045–6048.

21 Li, H., Müller, D., Guénée, L., and Alexakis, A. (2012). *Org. Lett.* 14: 5880–5883.

22 Li, H., Grassi, D., Guénée, L. et al. (2014). *Chem. Eur. J.* 20: 16694–16706.

23 Marshall, J.A. and Tang, Y. (1993). *J. Org. Chem.* 58: 3233–3234.

24 Yus, M., González-Gómez, J.C., and Foubelo, F. (2011). *Chem. Rev.* 111: 7774–7854.

25 Marshall, J.A. (2000). *Chem. Rev.* 100: 3163–3185.

26 Yu, S. and Ma, S. (2012). *Angew. Chem. Int. Ed.* 51: 3074–3112.

27 Marshall, J.A. and Mulhearn, J.J. (2005). *Org. Lett.* 7: 1593–1596.

28 Marshall, J.A. and Maxson, K. (2000). *J. Org. Chem.* 65: 630–633.

29 Marshall, J.A. and Ellis, K.C. (2003). *Org. Lett.* 5: 1729–1732.

30 Felzmann, W., Castagnolo, D., Rosenbeiger, D., and Mulzer, J. (2007). *J. Org. Chem.* 72: 2182–2186.

31 Brawn, R.A. and Panek, J.S. (2007). *Org. Lett.* 9: 2689–2692.

32 Brawn, R.A. and Panek, J.S. (2009). *Org. Lett.* 11: 4362–4365.

33 Brawn, R.A., Welzel, M., Lowe, J.T., and Panek, J.S. (2010). *Org. Lett.* 12: 336–339.

34 Brawn, R.A. and Panek, J.S. (2010). *Org. Lett.* 12: 4624–4627.

35 Mentink, G., van Maarseveen, J.H., and Hiemstra, H. (2002). *Org. Lett.* 4: 3497–3500.

36 Ogasawara, M., Ueyama, K., Nagano, T. et al. (2003). *Org. Lett.* 5: 217–219.

37 Ogasawara, M., Ge, Y., Okada, A., and Takahashi, T. (2012). *Eur. J. Org. Chem.* 1656–1663.

38 Crandall, J.K., Batal, D.J., Sebesta, D.P., and Lin, F. (1991). *J. Org. Chem.* 56: 1153–1166.

39 Marshall, J.A. and Tang, Y. (1994). *J. Org. Chem.* 59: 1457–1464.

40 Katukojvala, S., Barlett, K.N., Lotesta, S.D., and Williams, L.J. (2004). *J. Am. Chem. Soc.* 126: 15348–15349.

41 Lotesta, S.D., Hou, Y., and Williams, L.J. (2007). *Org. Lett.* 9: 869–872.

42 Joyasawal, S., Lotesta, S.D., Akhmedov, N.G., and Williams, L.J. (2010). *Org. Lett.* 12: 988–991.

43 Shangguan, N., Kiren, S., and Williams, L.J. (2007). *Org. Lett.* 9: 1093–1096.

44 Ghosh, P., Lotesta, S.D., and Williams, L.J. (2007). *J. Am. Chem. Soc.* 129: 2438–2439.

45 Zhang, Y., Cusick, J.R., Ghosh, P. et al. (2009). *J. Org. Chem.* 74: 7707–7714.

46 Ghosh, P., Cusick, J.R., Inghrim, J., and Williams, L.J. (2009). *Org. Lett.* 11: 4672–4675.

47 Sharma, R. and Williams, L.J. (2013). *Org. Lett.* 15: 2202–2205.

48 Ghosh, P., Zhang, Y., Emge, T.J., and Williams, L.J. (2009). *Org. Lett.* 11: 4402–4405.

49 Liu, K., Kim, H., Ghosh, P. et al. (2011). *J. Am. Chem. Soc.* 133: 14968–14971.

50 Hou, J., Chen, Y., Ma, D. et al. (2015). *Chem. Commun.* 51: 7439–7442.

51 Stoll, A.H. and Blakey, S.B. (2010). *J. Am. Chem. Soc.* 132: 2108–2109.

52 Rigoli, J.W., Boralsky, L.A., Hershberger, J.C. et al. (2012). *J. Org. Chem.* 77: 2446–2455.

53 Adams, C.S., Boralsky, L.A., Guzei, I.A., and Schomaker, J.M. (2012). *J. Am. Chem. Soc.* 134: 10807–10810.
54 Buchner, K.M., Clark, T.B., Loy, J.M.N. et al. (2009). *Org. Lett.* 11: 2173–2175.
55 Hu, H., Smith, D., Cramer, R.E., and Tius, M.A. (1999). *J. Am. Chem. Soc.* 121: 9895–9896.
56 Harrington, P.E., Murai, T., Chu, C., and Tius, M.A. (2002). *J. Am. Chem. Soc.* 124: 10091–10100.
57 Tius, M.A. (2014). *Chem. Soc. Rev.* 43: 2979–3002.
58 Bee, C. and Tius, M.A. (2003). *Org. Lett.* 5: 1681–1684.
59 Morgan, T.D.R., LeBlanc, L.M., Ardagh, G.H. et al. (2015). *J. Org. Chem.* 80: 1042–1051.
60 Schultz-Fademrecht, C., Tius, M.A., Grimme, S. et al. (2002). *Angew. Chem. Int. Ed.* 41: 1532–1535.
61 Nieto Faza, O., Silva López, C., Álvarez, R., and de Lera, Á.R. (2005). *Chem. Commun.* 4285–4287.
62 Widenhoefer, R.A. (2008). *Chem. Eur. J.* 14: 5382–5391.
63 Krause, N. and Winter, C. (2011). *Chem. Rev.* 111: 1994–2009.
64 Marshall, J.A. and Pinney, K.G. (1993). *J. Org. Chem.* 58: 7180–7184.
65 Aurrecoechea, J.M. and Solay, M. (1998). *Tetrahedron* 54: 3851–3856.
66 Krause, N., Laux, M., and Hoffmann-Röder, A. (2000). *Tetrahedron Lett.* 41: 9613–9616.
67 Hoffmann-Röder, A. and Krause, N. (2001). *Org. Lett.* 3: 2537–2538.
68 Golubev, A.S., Sergeeva, N.N., Hennig, L. et al. (2003). *Tetrahedron* 59: 1389–1394.
69 VanBrunt, M.P. and Standaert, R.F. (2000). *Org. Lett.* 2: 705–708.
70 Volz, F. and Krause, N. (2007). *Org. Biomol. Chem.* 5: 1519–1521.
71 Volz, F., Wadman, S.H., Hoffmann-Röder, A., and Krause, N. (2009). *Tetrahedron* 65: 1902–1910.
72 Sapu, C.M. and Deska, J. (2013). *Org. Biomol. Chem.* 11: 1376–1382.
73 Ma, S. and Wu, S. (2001). *Chem. Commun.* 441–442.
74 Ma, S. and Shi, Z. (2002). *Chem. Commun.* 540–541.
75 Nelson, S.G., Peelen, T.J., and Wan, Z. (1999). *J. Am. Chem. Soc.* 121: 9742–9743.
76 Nelson, S.G. and Wan, Z. (2000). *Org. Lett.* 2: 1883–1886.
77 Wan, Z. and Nelson, S.G. (2000). *J. Am. Chem. Soc.* 122: 10470–10471.
78 Ma, S., Ren, H., and Wei, Q. (2003). *J. Am. Chem. Soc.* 125: 4817–4830.
79 Jiang, X., Fu, C., and Ma, S. (2008). *Chem. Eur. J.* 14: 9656–9664.
80 Lü, B., Jiang, X., Fu, C., and Ma, S. (2009). *J. Org. Chem.* 74: 438–441.
81 Prasad, J.S. and Liebeskind, L.S. (1988). *Tetrahedron Lett.* 29: 4253–4256.
82 Dieter, R.K. and Yu, H. (2001). *Org. Lett.* 3: 3855–3858.
83 Morita, N. and Krause, N. (2006). *Angew. Chem. Int. Ed.* 45: 1897–1899.
84 Qiu, Y., Zhou, J., Li, J. et al. (2015). *Chem. Eur. J.* 21: 15939–15943.
85 Egi, M., Shimizu, K., Kamiya, M. et al. (2015). *Chem. Commun.* 51: 380–383.
86 Ohno, H., Toda, A., Miwa, Y. et al. (1999). *J. Org. Chem.* 64: 2992–2993.
87 Lathbury, D. and Gallagher, T. (1986). *J. Chem. Soc., Chem. Commun.* 114–115.

88 Ha, J.D. and Cha, J.K. (1999). *J. Am. Chem. Soc.* 121: 10012–10020.

89 Zhang, Z., Liu, C., Kinder, R.E. et al. (2006). *J. Am. Chem. Soc.* 128: 9066–9073.

90 Zhang, Z. and Widenhoefer, R.A. (2007). *Angew. Chem. Int. Ed.* 46: 283–285.

91 Cox, N., Uehling, M.R., Haelsig, K.T., and Lalic, G. (2013). *Angew. Chem. Int. Ed.* 52: 4878–4882.

92 Liu, Z., Wasmuth, A.S., and Nelson, S.G. (2006). *J. Am. Chem. Soc.* 128: 10352–10353.

93 Huang, L., Arndt, M., Gooßen, K. et al. (2015). *Chem. Rev.* 115: 2596–2697.

94 Butler, K.L., Tragni, M., and Widenhoefer, R.A. (2012). *Angew. Chem. Int. Ed.* 51: 5175–5178.

95 Nishina, N. and Yamamoto, Y. (2006). *Angew. Chem. Int. Ed.* 45: 3314–3317.

96 Nishina, N. and Yamamoto, Y. (2008). *Tetrahedron Lett.* 49: 4908–4911.

97 Zhang, Z. and Widenhoefer, R.A. (2008). *Org. Lett.* 10: 2079–2081.

98 Webster, S., Sutherland, D.R., and Lee, A.-L. (2016). *Chem. Eur. J.* 22: 18593–18600.

99 Hadfield, M.S. and Lee, A.-L. (2010). *Org. Lett.* 12: 484–487.

100 Wang, Z.J., Benitez, D., Tkatchouk, E. et al. (2010). *J. Am. Chem. Soc.* 132: 13064–13071.

101 Kato, F. and Hiroi, K. (2004). *Chem. Pharm. Bull.* 52: 95–103.

102 Wang, M.-Z., Zhou, C.-Y., Guo, Z. et al. (2011). *Chem. Asian J.* 6: 812–824.

103 Sutherland, D.R., Kinsman, L., Angiolini, S.M. et al. (2018). *Chem. Eur. J.* 24: 7002–7009.

104 Yoshida, Y., Okamoto, S., and Sato, F. (1996). *J. Org. Chem.* 61: 7826–7831.

105 Urabe, H., Takeda, T., Hideura, D., and Sato, F. (1997). *J. Am. Chem. Soc.* 119: 11295–11305.

106 Wender, P.A., Glorius, F., Husfeld, C.O. et al. (1999). *J. Am. Chem. Soc.* 121: 5348–5349.

107 Brummond, K.M., Kerekes, A.D., and Wan, H. (2002). *J. Org. Chem.* 67: 5156–5163.

108 Grillet, F. and Brummond, K.M. (2013). *J. Org. Chem.* 78: 3737–3754.

109 Han, Y., Zhao, Y., and Ma, S. (2019). *Chem. Eur. J.* 25: 9529–9533.

110 Zeng, R., Fu, C., and Ma, S. (2012). *J. Am. Chem. Soc.* 134: 9597–9600.

111 Ng, S.-S. and Jamison, T.F. (2005). *J. Am. Chem. Soc.* 127: 7320–7321.

112 Ng, S.-S. and Jamison, T.F. (2005). *Tetrahedron* 61: 11405–11417.

113 ElDouhaibi, A.S., Kassab, R.M., Song, M., and Montgomery, J. (2011). *Chem. Eur. J.* 17: 6326–6329.

114 Amako, Y., Arai, S., and Nishida, A. (2017). *Org. Biomol. Chem.* 15: 1612–1617.

115 Padwa, A., Lipka, H., Watterson, S.H., and Murphree, S.S. (2003). *J. Org. Chem.* 68: 6238–6250.

116 Carreira, E.M., Hastings, C.A., Shepard, M.S. et al. (1994). *J. Am. Chem. Soc.* 116: 6622–6630.

117 Shepard, M.S. and Carreira, E.M. (1997). *J. Am. Chem. Soc.* 119: 2597–2605.

118 Okamura, W.H., Peter, R., and Reischl, W. (1985). *J. Am. Chem. Soc.* 107: 1034–1041.

119 Gibbs, R.A., Bartels, K., Lee, R.W.K., and Okamura, W.H. (1989). *J. Am. Chem. Soc.* 111: 3717–3725.

120 De Schrijver, J. and De Clercq, P.J. (1993). *Tetrahedron Lett.* 34: 4369–4372.

121 Jung, M.E. and Min, S.-J. (2005). *J. Am. Chem. Soc.* 127: 10834–10835.

122 Trost, B.M., Fandrick, D.R., and Dinh, D.C. (2005). *J. Am. Chem. Soc.* 127: 14186–14187.

123 Liu, H., Leow, D., Huang, K.-W., and Tan, C.-H. (2009). *J. Am. Chem. Soc.* 131: 7212–7213.

124 Rout, L. and Harned, A.M. (2009). *Chem. Eur. J.* 15: 12926–12928.

125 Plaza, M., Jandl, C., and Bach, T. (2020). *Angew. Chem. Int. Ed.* 59: 12785–12788.

126 Schlessinger, R.H. and Bergstrom, C.P. (1995). *J. Org. Chem.* 60: 16–17.

127 Gung, B.W., Craft, D.T., Bailey, L.N., and Kirschbaum, K. (2010). *Chem. Eur. J.* 16: 639–644.

128 Mei, L.-Y., Wei, Y., Tang, X.-Y., and Shi, M. (2015). *J. Am. Chem. Soc.* 137: 8131–8137.

129 Fang, R., Feng, Z., Kirillov, A.M., and Yang, L. (2020). *Organometallics* 39: 1782–1789.

16

The Synthesis and Application of BINOL Derivatives as Effective Building Blocks for Catalysts Employed in Enantioselective Synthesis

Janusz Jurczak, Patryk Niedbała, and Agata Tyszka-Gumkowska

Institute of Organic Chemistry, Polish Academy of Sciences, Kasprzaka 44/52, Warsaw 01-224, Poland

16.1 Introduction

Since 1973, when Cram and coworkers [1] for the first time applied and reported BINOL-based crown ethers in enantioselective recognition of chiral ammonium salts, the application of this platform has allowed for an incredible development both in the chiral recognition [2–4] and, more importantly, in the asymmetric catalysis [5–8]. For over three decades, BINOL molecule is one of the most widely used ligands in catalytic asymmetric reactions, both in the broadly understood organocatalysis [9] and in the transition-metal catalysis [10]. In recent years, many scientists have used this chiral molecule to create a new group of catalysts that has been shown to be suitable for effective asymmetric induction. Herein, we would like to present how versatile and useful is [1,1′-binaphthalene]-2,2′-diol.

Asymmetric organic synthesis is nowadays one of the most rapidly developing branches of modern chemistry, continually broadening our ability to create functionalized chiral molecules with high application potential [11–13]. Stereoselective methods for forming carbon–carbon and carbon–heteroatom bonds, using asymmetric catalysis, have found wide applications, not only in laboratory practice but also in industry [14, 15]. This finds reflection in the significant growth in the number of research publications dedicated to stereoselective synthesis and in the continual need for important, useful, and biologically active synthetic compounds. In parallel, the constant demand for obtaining chemical compounds characterized by high enantiomeric purity, as well as the increasing sense of ecological awareness, has been propelling the development of what is called "green chemistry." These trends come together in organocatalysis, which requires the use of catalysts that are resistant to external conditions (water, oxygen) and above all do not bring in additional impurities in the form of heavy metals. Organocatalysis makes it possible to effectively generate biologically active substances – especially pharmaceutical compounds, which require high standards of analytical purity, and often also enantiomeric purity. The importance of organocatalysis in modern chemistry is undeniably demonstrated by the selection of year 2021 Nobel Prize winners. The Royal Swedish Academy of Sciences has decided to award in chemistry Benjamin List and David MacMillan "for

Chiral Building Blocks in Asymmetric Synthesis: Synthesis and Applications, First Edition.
Edited by Elżbieta Wojaczyńska and Jacek Wojaczyński.

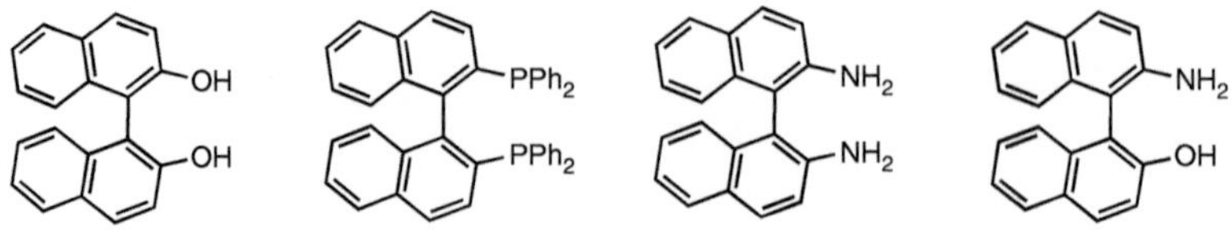

Figure 16.1 BINOL and its derivatives.

the development of asymmetric organocatalysis." Organocatalysis is regarded as a new precision tool in molecular design that directly affects pharmaceutical research and the development of "green chemistry."

Undisputedly, one of the most popular and widely used platforms in asymmetric catalysis is BINOL ([1,1′-binaphthalene]-2,2′-diol). It is characterized by the presence of axial chirality, resulting from the inhibition of rotation. Unlike the very popular platform of *Cinchona* alkaloids, BINOL is not a naturally occurring compound. Its structure was first described in 1926 by Pummerer and coworkers [16], while in enantiomerically pure form it was synthesized in 1971 by Viterbo and coworkers [17]. Since then, many derivatives of BINOL have also been prepared, including bis(diphenylphosphino)-1,1′-binaphthyl (BINAP) [18], 2,2′-bis(diphenylphosphinoamino)-1,1′-binaphthyl (BINAM) [19], and 2-amino-2′-hydroxy-1,1′-binaphthyl (NOBIN) [19] (Figure 16.1).

BINOL platform was employed in organic synthesis for the first time in the mid-1970s. Cram used BINOL-derived polycyclic crown ethers as chiral receptors [1]. A few years later, Noyori and coworkers [18] in turn presented the use of BINAP in the asymmetric hydrogenation of olefins. Since then, BINOL, as well as the synthesis of its derivatives, has experienced very intense development. The aforementioned Cram group [20] developed the synthesis of chiral ligands and their ability to bind alkali metals, ammonium, and alkylammonium cations. In parallel, Noyori et al. [21] utilized BINOL in the reduction of carbonyl compounds. In the following review, we have chosen to present the application of BINOL derivatives in three main streams related to asymmetric catalysis. The first one concerns the use of simple substituted BINOL derivatives with free hydroxyl groups. The second utilizes phase-transfer catalysis (PTC), using onium salts as charged catalysts capable of accelerating reactions carried out in heterogeneous systems. On the other hand, the third essential trend is the application of a chiral phosphoric acid derived from BINOL.

16.2 BINOL Derivatives with Free Hydroxyl Groups

In addition to the extraordinary value of BINOL-based ligands in the transition metal [22, 23], the unprotected hydroxyl groups of this chiral scaffold can serve themselves as a versatile catalytic center. Among the asymmetric organocatalytic methods that involve non-covalent interactions with substrates, diol-based catalytic systems represent a versatile chiral tool in various types of transformations such as conjugated addition, alkylation, or multicomponent reactions [24, 25]. In this part, we aim to give the reader an overview of the recent achievements in the application of certain BINOL-based organocatalysts in the synthesis of interesting chiral building blocks with organoboron compounds as a starting material.

16.2.1 Allylboration Reactions

One of the most fundamental reactions in this context is the allylboration of carbonyl compounds. In 2006, the group of Schaus [26] reported the first organocatalytic variants of enantioselective allylboration of ketones **1**, leading to an important class of chiral building blocks **3** (Scheme 16.1). Among many tested chiral diols, BINOL derivatives ***C1*** showed the best catalytic activity. A proposed catalyst was obtained in a multistep reaction from enantiomerically pure (*S*)-BINOL, involving methylation of hydroxyl groups, bromination, and a subsequent demethylation.

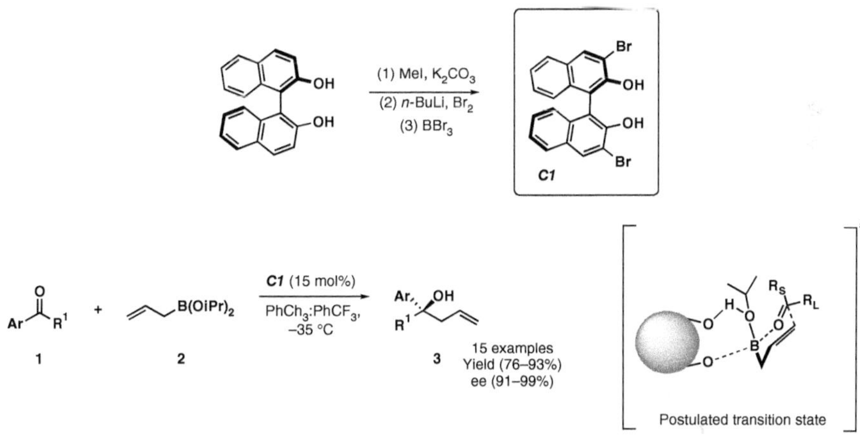

Scheme 16.1 Pioneering application of organocatalytic allylboration of ketones by Schaus. Source: Lou et al. [26] / American Chemical Society.

The mechanism of this transformation was based on the transesterification reaction occurring between the isopropoxy group of **2** with BINOL ligand ***C1***, resulting in highly diastereo- and enantioselective formation of product **3**, according to Zimmerman–Traxler model shown in Scheme 16.1. Excellent results were obtained with a variety of electron-rich and electron-deficient ketones. After this successful attempt, a similar approach has also been found very valuable in the allylboration of acyl aldimines [27] and multicomponent condensation [28].

Lately, Szabo and coworkers [29] expanded the application of this approach by challenging asymmetric propargyl- and allylboration of hydrazonoesters **4** leading to α-amino acid derivatives. Thus, this study presents for the first time *de novo* formation of stereocenters in amino acid products **7** and **8** under metal-free conditions with BINOL ***C2*** as inexpensive catalyst (Scheme 16.2).

Based on the previous studies regarding propargylation of ketones [30], esterification of boronic acid leads to a more Lewis acidic product and accelerates the reaction

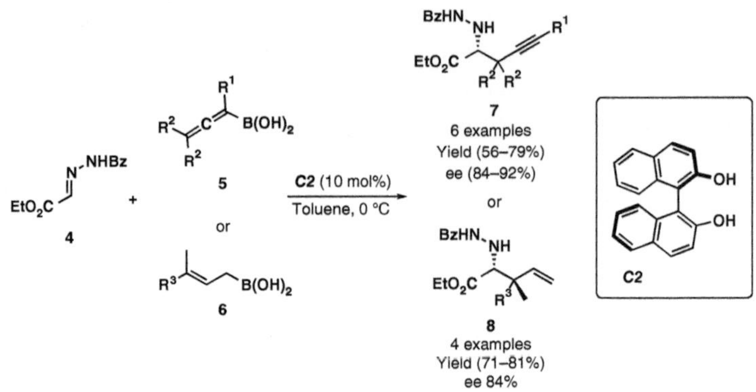

Scheme 16.2 Stereoselective propargyl- and allylboration of hydrazonoesters by Szabo. Source: Jonker et al. [29] / Royal Society of Chemistry.

rate. Moreover, the explanation of stereoselectivity was also supported by DFT calculations.

16.2.2 Conjugated Addition Reactions

According to a similar mode of substrate activation, May and coworkers [31] reported the functionalization of heterocycle-appended enones **9** via BINOL catalyst ***C3*** in conjugated addition (Scheme 16.3). Authors presented also a three-step synthesis of the catalyst, reaching 71% overall yield. Promoted by earlier successful examples of such transformations with boronate esters and acids [32], in a recent report they were able to optimize reaction conditions also for more stable trifluoroborates **10**. Such efforts made this protocol more suitable from the practical point of view, but it also allowed for an impressive extension of the substrate scope and easy utilization of this method in natural products synthesis.

An efficient system for the enantioselective preparation of triarylmethanes **14** via organocatalytic 1,6-addition of aryloboronic acids **13** to *para*-quinone methides **12** in the presence of BINOL catalyst ***C4*** has been developed by Weng and Lu [33] (Scheme 16.4). This interesting structural motif, frequently found in natural products and functional materials [34], represents a challenging target for the developing of suitable catalytic systems, owing to three similar aryl groups in the stereocenter.

In this case, the *ortho*-hydroxy substituent in the substrate **12** as a directing group realizes the remote stereocontrol of the reaction and is essential for this 1,6-addition. Authors highlighted a broad scope of substrates including both electron-rich and electron-deficient arylboronic acids and the possibility to easily scale up the protocol to emphasize the robustness of the presented method.

16.2.3 Other Examples

Multicomponent processes, such as Petasis reaction discovered in 1993 [35], have increased considerably as sustainable synthetic methods in medical chemistry

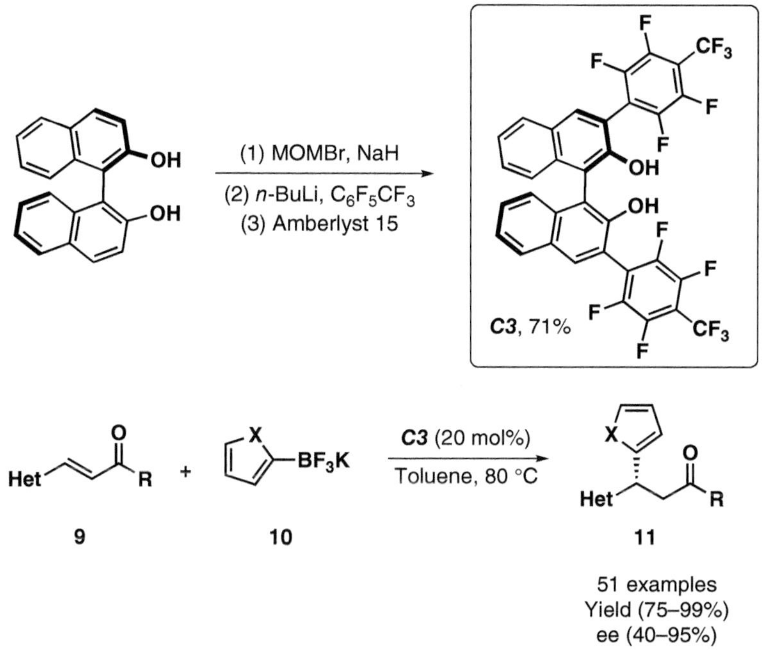

Scheme 16.3 Application of BINOL in asymmetric conjugated addition of trifluoroborates to enones by May. Source: Shin et al. [31] / John Wiley & Sons.

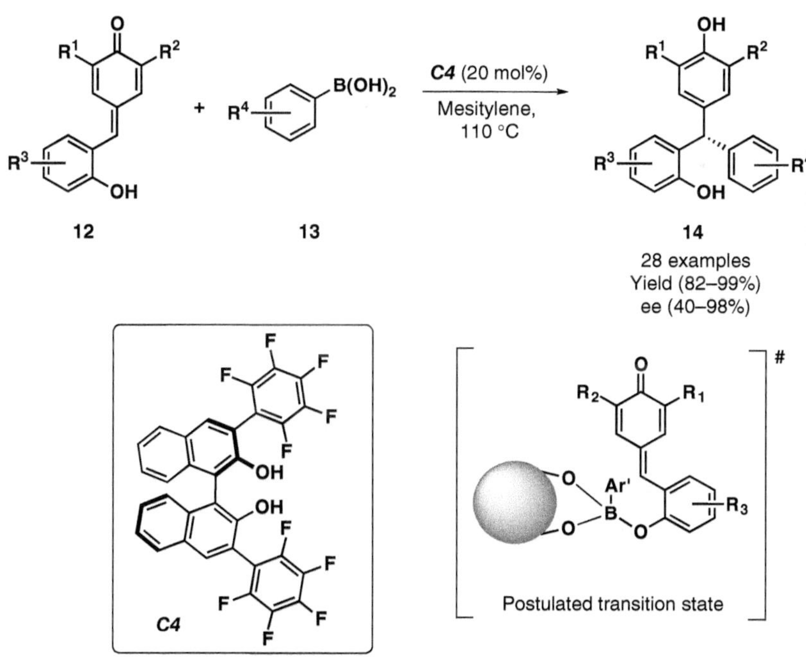

Scheme 16.4 Organocatalytic 1,6-addition of arylboronic acids to para-quinone methides by Weng and Lu. Source: Huang et al. [33] / John Wiley & Sons.

over the last 20 years. Therefore, asymmetric versions of these transformations are highly supported by the pharmaceutical industry, since the number of drugs contain advanced amine moieties [36, 37]. Thus, the most successful attempts have been focused on the applications of the BINOL chiral scaffold as catalyst or chiral auxiliary.

The procedure reported recently by Marques and Burke [38] represents a valuable example of the stereoselective Petasis reaction between isatin-based boronic esters **15**, *ortho*-hydroxy aldehydes **16**, and secondary amines **17** (Scheme 16.5).

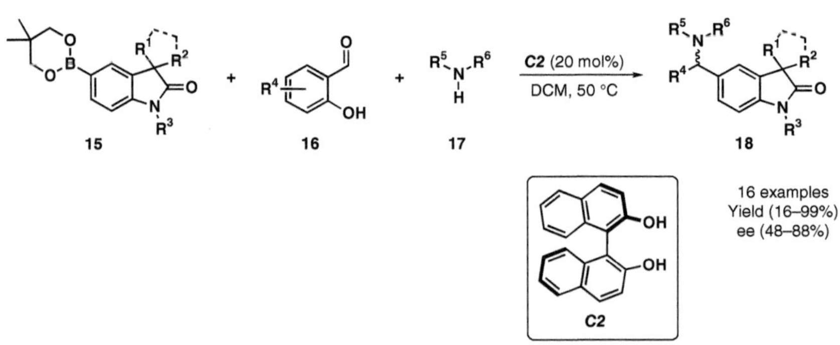

Scheme 16.5 Stereoselective Petasis reaction by Marques and Burke. DCM, dichloromethane. Source: Marques et al. [38] / John Wiley & Sons.

A variety of organocatalysts were tested during optimization of the reaction conditions; nevertheless, the best results were obtained for the simple and relatively inexpensive BINOL catalyst ***C2***. The reaction is characterized by excellent yields and enantiomeric excess values for a broad scope of the substrates, even in the gram-scale protocol.

Synthesis of chiral boronic acids has been considered as an immense challenge in organic synthesis [39]. Novel concept of the Szabo group [40] is based on the similar transesterification reaction discussed above but coupled with stereoselective homologation of the chain with diazoalkanes (Scheme 16.6).

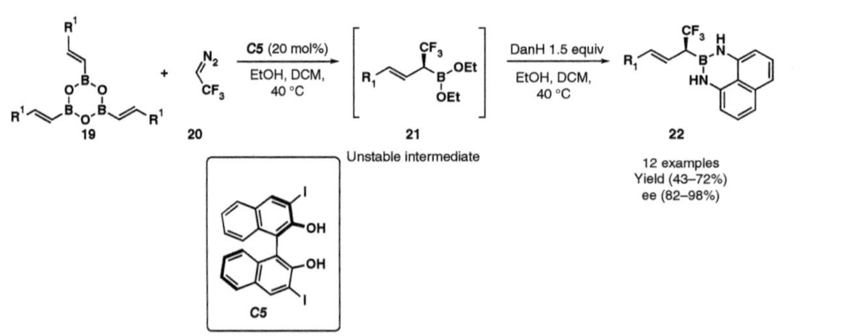

Scheme 16.6 Organocatalytic synthesis of trifluoromethyl allylboronic acids by Szabo. DCM, dichloromethane. Source: Jonker et al. [40] / American Chemical Society.

Alkenylboroxine **19** in the presence of catalytic amount of BINOL ***C5*** and EtOH can readily react with diazo compound **20** to form oxygen-sensitive intermediate **21** that can be easily protected by diaminonaphthalene (DanH). Purified final compounds were characterized as strongly reactive with various electrophiles, which only extended the concept of further application of the method developed.

16.3 Onium Salts as Charged Catalysts in PTC

In recent years, with the growing demand for green and sustainable chemistry [41, 42], PTC, characterized by exceptionally mild reaction protocols and the absence of complicated equipment, is undoubtedly a prevailing methodology for the development of asymmetric synthesis [43]. Considering the convenient possibility for carrying out large-scale preparations with a small amount of chemical waste [44], PTC has found wide application in both academic and industrial laboratories [45]. Since the early studies on highly enantioselective alkylation promoted by phase-transfer catalysts, namely, quintenary ammonium salts [46], utilization of BINOL as a chiral scaffold still represents one of the most promising general concepts in this area [47].

Pioneering studies reported by Cram and Sogah [48] were focused on non-covalent interactions between catalysts and substrates. Authors presented this concept on a model Michael addition of methyl vinyl ketone **24** to β-ketoester **23** (Scheme 16.7). Facial selectivity toward the formed Michael adduct proceeds by the formation of diastereomeric host/guest complex with BINOL-based macrocyclic catalyst ***C6***. Owing to this specific interaction, it was possible to control the stereoselectivity of the reaction in an unprecedented manner.

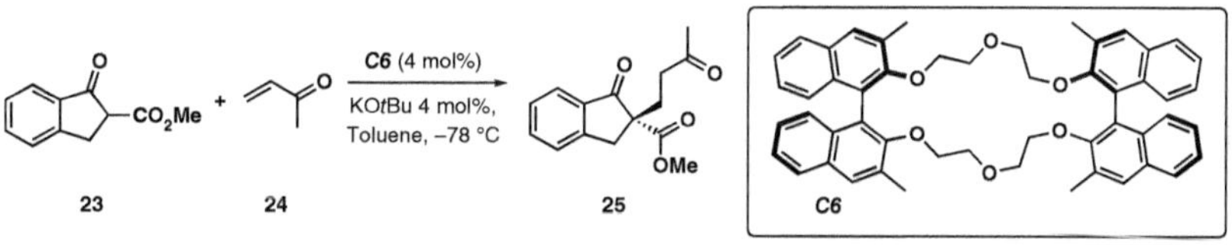

Scheme 16.7 Asymmetric Michael addition catalyzed by BINOL-based crown ethers by Cram. Source: Cram et al. [48] / Royal Society of Chemistry.

This discovery allowed for significant progress in the field of asymmetric synthesis, as in later years even more efforts were devoted to the use of BINOL in the construction of catalysts. Undoubtedly, it was also associated with the relatively easy functionalization of BINOL core and the availability of its both enantiomers, which is a general problem with chiral platforms originated from natural sources.

More recently, in the field of PTC, noteworthy advancements have been made with Maruoka's catalysts in the form of *N*-spiroammonium salts (Scheme 16.8) [49].

The catalysts ***C7–C9*** were obtained in a six-step sequence from commercially available (*S*)-binaphthol. In principle, during reactions carried out under PTC conditions, the phase-transfer catalyst as a lipophilic salt forms a diastereomeric

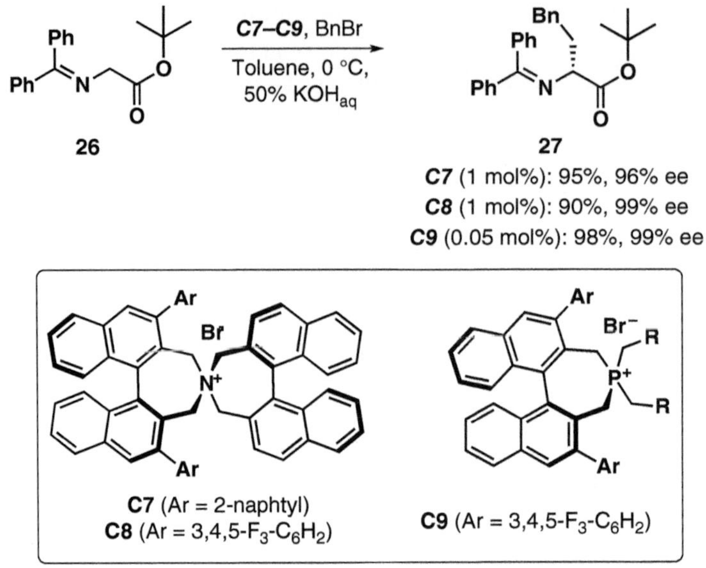

Scheme 16.8 Enantioselective benzylation of glycine derivative by Maruoka's catalysis. Source: Ooi et al. [49] / American Chemical Society.

ion pair with carbanion in the interfacial region causing an enantioselective course of the reaction [50]. In the case of Maruoka's system, the very small catalyst loads (0.05–1 mol%) are worth underlining, ensuring high yields and enantiomeric excesses of products in alkylation of prochiral protected glycine derivative – one of the benchmark reactions in the PTC area. Similar results were obtained in many other types of chemical reactions promoted by similar BINOL derivatives [51].

This high catalytic activity significantly differentiates these catalysts from the conventionally used systems based on *Cinchona* alkaloids that are usually added to the reaction in amounts of 5–10 mol%. However, the disadvantage of the Maruoka's BINOL systems is a multistage synthesis and high molar mass of the catalysts.

Modified hybrid phase-transfer catalysts of this type, possessing hydrogen bond donors in the structure, usually ensure even better control over the selectivity in a demanding asymmetric reaction. Such a modification allows for simultaneous activation of the nucleophilic and the electrophilic reaction partner. For instance, this approach was successfully used in the enantioselective sulfenylation and chlorination of β-ketoester **28** with the amide-based catalyst ***C10*** (Scheme 16.9) [52]. In comparison with the analogous catalyst with a free hydroxyl group ***C11***, the amide derivatives exhibit significant improvement of the reaction efficiency due to better defined and stronger hydrogen bond formed. This type of catalyst was prepared in a multistep synthetic procedure, starting from BINOL, via cyano and amine derivative, and a subsequent quaternization. Even 0.1 mol% of the amide catalyst ensured high yield (98%) and excellent enantiomeric excess (94% ee) in the model sulfenylation reaction.

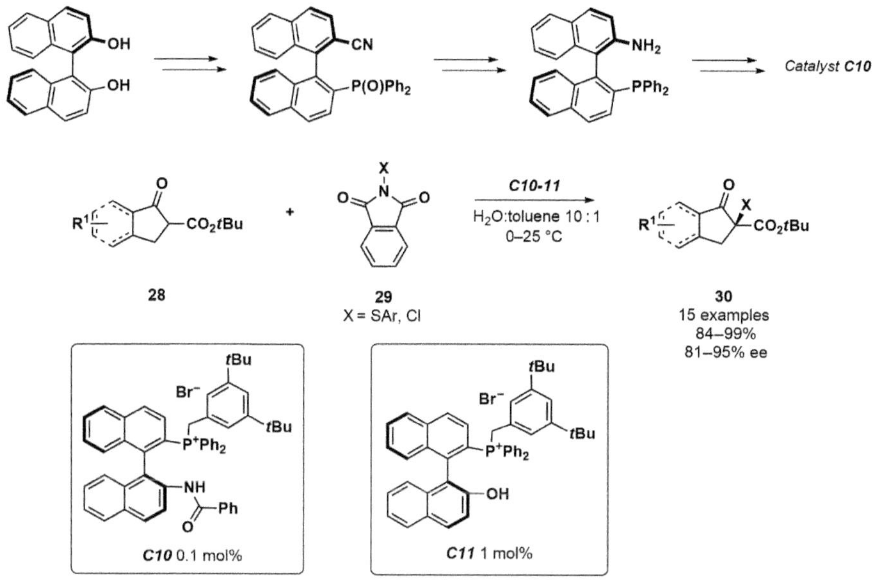

Scheme 16.9 Maruoka's hybrid phase-transfer catalysis. Source: Shirakawa et al. [52] / American Chemical Society.

The urea derivatives of BINOL-based phosphonium salts ***C12*** were also tested as effective catalysts for the nucleophilic aromatic substitution reaction, leading to the 3,3′-diaryloxyindole derivatives **33** (Scheme 16.10) [53]. High yields and enantiomeric excesses in the synthesis of the library of biologically interesting compounds were obtained with 5 mol% of catalyst ***C12***. It was expected by the authors that the urea moiety interacts with the nitroarene **32** via two hydrogen bonds, thus giving a well-organized transition state that provides high stereocontrol of the reaction course.

It was also presented that the nitro group in products **33** can be readily transformed into other functionalities. Furthermore, it was the first example of an organocatalytic method for synthesizing chiral triarylmethanes possessing an all-carbon quaternary center after systems based on metal complexes [54].

The Gouverneur's group [55], while working on the asymmetric nucleophilic fluorination of β-bromosulfides **34**, decided to combine two essential strategies of modern asymmetric synthesis, namely, the interactions through hydrogen-bond and solid–liquid PTC (Scheme 16.11). In the presented example, excellent results were obtained using CsF as a safe inorganic fluorine source, which represents an interesting and very useful expansion of the fluorination methodology [56].

Non-covalent interactions between di-urea catalyst ***C13*** (obtained from commercially available BINAM and appropriate isocyanate) and fluoride from its cesium salt, enabling highly efficient and enantioselective opening of episulfonium ion formed *in situ*. Importantly, this catalytic manifold allowed for the first time for

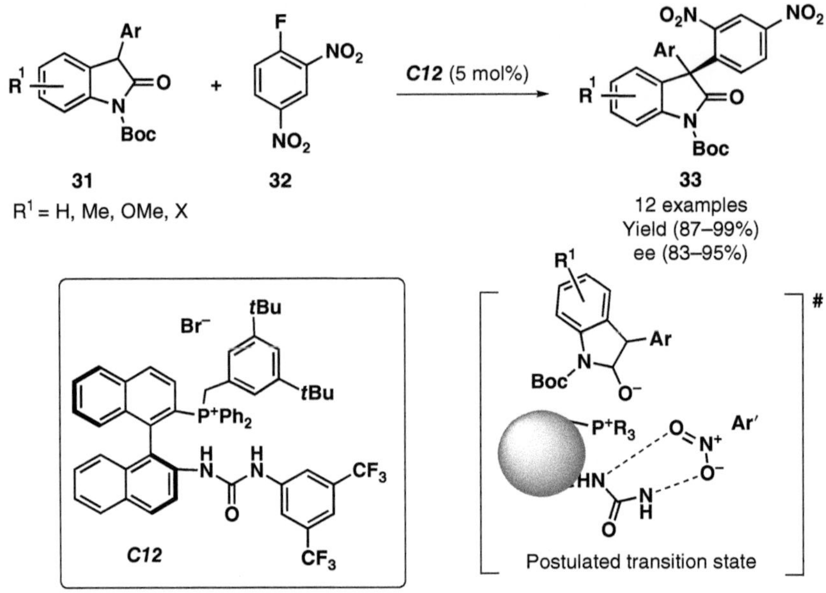

Scheme 16.10 Nucleophilic aromatic substitution catalyzed by BINOL-based phosphonium salts by Maruoka. Source: Shirakawa et al. [53] / John Wiley & Sons.

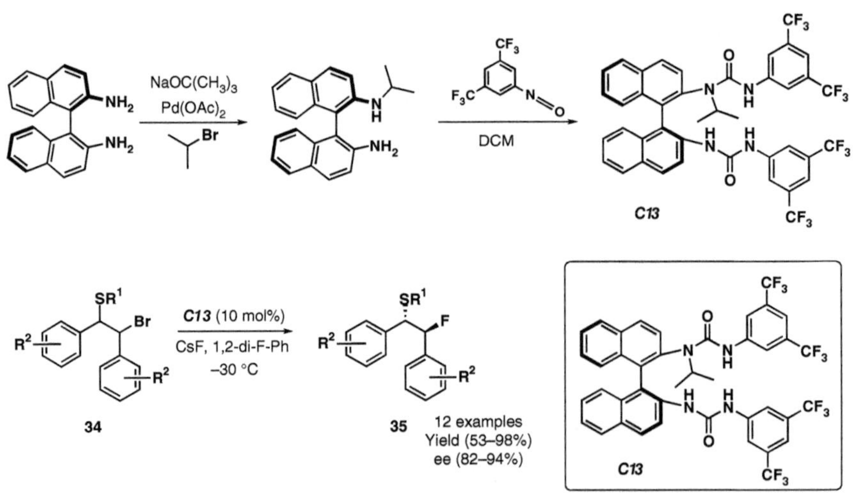

Scheme 16.11 Asymmetric nucleophilic fluorination under hydrogen bonding PTC by Gouverneur. DCM, dichloromethane. Source: Pupo et al. [55] / American Association for the Advancement of Science.

the utilization of anionic nucleophiles insoluble in organic solvents. Moreover, investigation of the reaction mechanism was supported by DFT calculations.

Recently, enantioselective syntheses of β- and γ-fluoroamines were also explored by the same group, leading to the replacement of expensive CsF with more accessible and safe KF [57, 58].

16.4 Chiral Phosphoric Acids Derived from BINOL Platform

Binaphthyl phosphoric acids (BPAs), characterized by the presence of axial chirality, are very often used as chiral Brønsted acids, capable of catalyzing a range of organic reactions, with the formation of enantiomerically enriched products of scientific, industrial, and biological significance. Such compounds are often referred to as bifunctional catalysts. In addition to the conformational rigidity, they have an acidic proton in their structure with a basic phosphoryl group in its immediate vicinity. The unique properties of this type of catalysts are related to the formation of a seven-membered ring, which prevents a free rotation around the P—O bond and thus establishes the Brønsted acid conformation. This structural property is not reflected in analogous carboxylic and sulfonic acids. The first application of such bifunctional catalysts was demonstrated in 2004 by the Akiyama [59] and Terada [60] groups in the formation of new C—C bonds. Since then, a very large number of applications of this type of catalysts have been reported in the literature, capable of accelerating a variety of reactions, including Michael reaction, Friedel–Crafts reaction, Diels–Alder reaction, 1,3-dipolar cycloadditions, multicomponent reactions (Mannich, Strecker, Biginelli), Nazarov cyclizations, etc.

16.4.1 Michael Reaction

This is one of the mildest methods of generation of a new C—C bond in an asymmetric approach demonstrated in 2009 by Akiyama [61]. Presented phosphoric acid derivative ***C14*** was obtained in a three-step synthesis with overall yield 35%. The presented reaction of β-ketoesters **36** and methyl vinyl ketone **37** in the presence of 10 mol% of catalyst ***C14*** resulted in obtaining indanone derivatives with very good yields and high ee (Scheme 16.12).

In this reaction, the plausible mechanism shows that both substrates create the complex with the catalyst that promotes a chiral environment. In 2013, the Dixon group [62] proposed a Michael reaction/iminium ion cyclization of urea derivative of tryptamines **39** with enones **40**, using BPA catalyst ***C15*** (Scheme 16.13).

The authors showed a broad scope of products (22 examples) that were obtained in good yields (up to 78%) and enantioselectivities (up to 96% ee). The utility of BPA catalysts in asymmetric reactions has also been tested in other Michael reaction variations, for example, phospha-Michael-type addition reaction. In 2020, Terada and coworkers [63] presented the addition of diarylphosphine oxides **42** with alkenyl benzimidazoles **43** in the presence of catalyst ***C16*** well known in this group (Scheme 16.14). This catalyst containing large 9-anthryl substituent was obtained in a six-step synthesis and very high overall yield reaching 55%.

The authors obtained products **44** (17 examples) possessing phosphorus and benzimidazoles in their structures. These products were formed in high yields and excellent enantiomeric excess (up to 99% ee). Moreover, the subsequent reduction of the phosphine oxide moiety resulted in the chiral phosphine, which is a potential

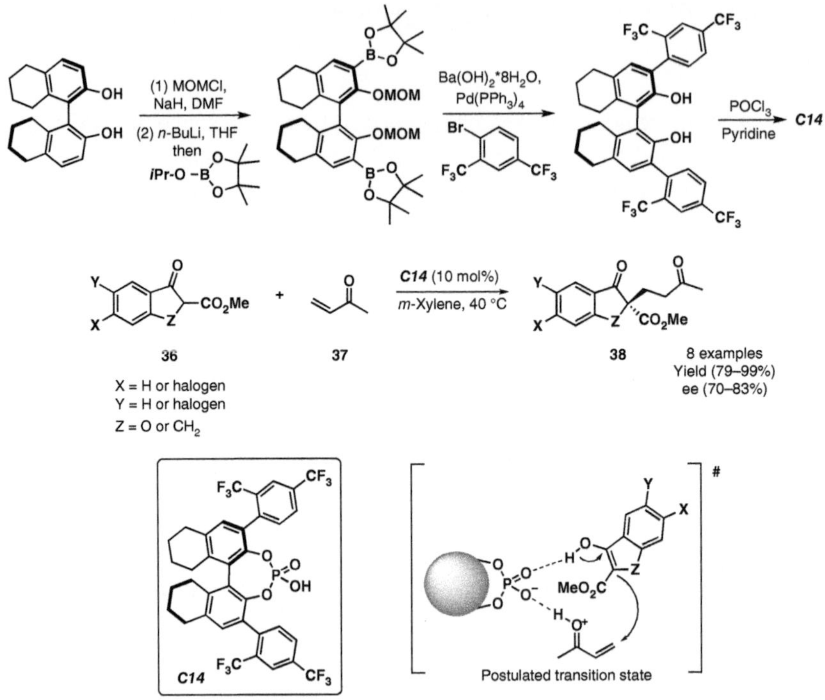

Scheme 16.12 Michael reaction of indanone derivatives by Akiyama. Source: Akiyama et al. [61] / John Wiley & Sons.

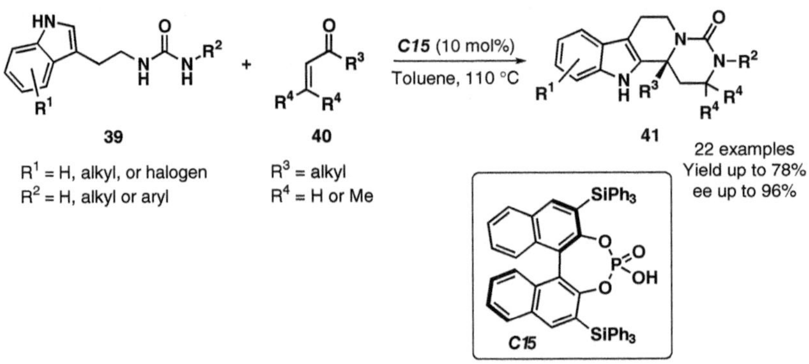

Scheme 16.13 Michael reaction/iminium ion cyclization by Dixon. Source: Aillaud et al. [62] / American Chemical Society.

benzimidazole-based chiral P,N-ligand **45** for metal catalysts, without any loss of enantiomeric excess.

16.4.2 Friedel–Crafts Reaction

The Friedel–Crafts reaction is a common method to introduce substitution to aromatic systems by forming new C—C bonds. The interest of this reaction in the

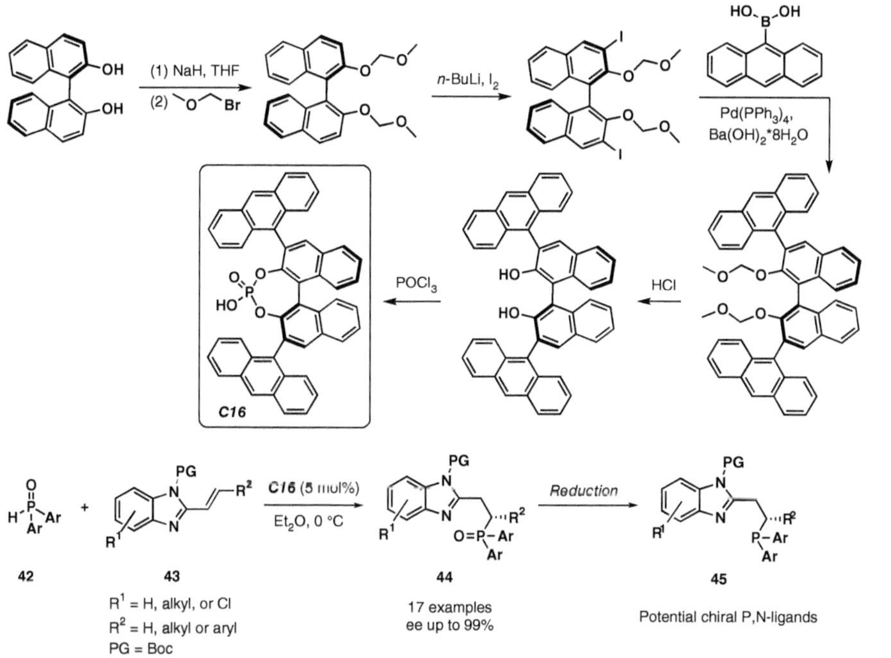

Scheme 16.14 Phospha-Michael-type addition by Terada. Source: Hou et al. [63] / American Chemical Society.

asymmetric variant using BPA catalyst started in 2004 when Terada and coworkers [64] reported an aza-Friedel–Crafts reaction of *N*-Boc imines **47** and methoxyfuran **46**, using 2 mol% of catalyst ***C17*** (Scheme 16.15).

Reported furan-2-ylamine derivatives **48** were obtained in a highly enantioselective fashion, up to 97% ee. Moreover, the authors presented the synthetic utility of this transformation, derivatizing the furan ring to γ-butenolide **49**, a common building block in the synthesis of various natural products.

In 2010, Bolm and coworkers [65] reported a highly enantioselective Friedel–Crafts reaction catalyzed by BPA. In their work, *N*-Boc-protected ethyl trifluoropyruvate imine **50** was reacted with a variety of indole derivatives **51** and activated by 6 mol% of catalyst ***C18*** (Scheme 16.16).

The authors obtained a number of quaternary α-amino acids in excellent yields (up to 99%) and high enantioselectivities (up to 96% ee). The subsequent deprotection of the Boc-derivative was accomplished, giving the amino acid ester in excellent yield while retaining the stereochemical information.

16.4.3 Diels–Alder Reaction

The Diels–Alder reaction is one of the most useful reactions in organic chemistry, enabling the formation of new six-membered rings. It is therefore not surprising that this reaction has become the subject of many reviews, both in the classical variant and in other variants of this reaction. The first ever aza-Diels–Alder reaction

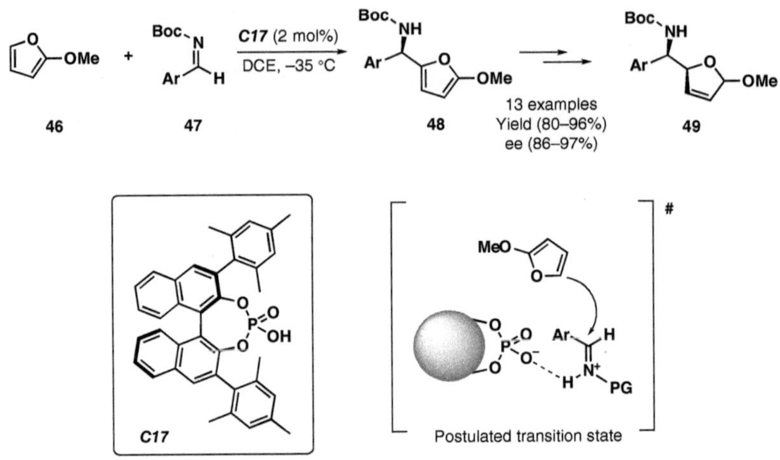

Scheme 16.15 Aza-Friedel–Crafts reaction by Terada. DCE, dichloroethane. Source: Uraguchi et al. [64] / American Chemical Society.

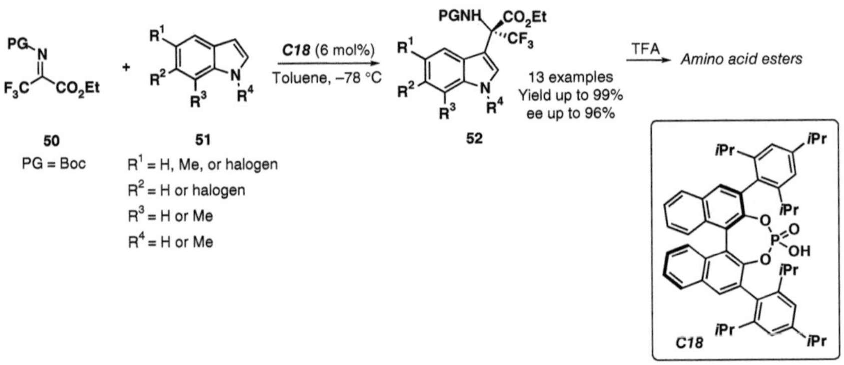

Scheme 16.16 Aza-Friedel–Crafts reaction by Bolm. TFA, trifluoroacetic acid. Source: Husmann et al. [65] / American Chemical Society.

under asymmetric conditions was presented in 2006 by Akiyama and coworkers [66] (Scheme 16.17).

In their work, authors presented the reaction between 2-hydroxyphenyl imines **53** and Brassard's diene **54**. The enantioenriched lactams **55** (16 examples) were obtained in high yields and enantioselectivities – up to 98% ee. These compounds may be precursors of piperidine, which derivatives are of great interest for the pharmaceutical industry, as building blocks of alkaloids, peptides, etc. Presented aza-Diels–Alder reaction was carried out with 3 mol% of catalyst ***C19*** in the form of pyridinium salt, showing a good tolerance to functional groups in imine. Moreover, it was proved that the presence of hydroxyl group attached to aromatic imine **53** helps in obtaining high enantioselectivities. This is most likely due to the additional hydrogen bonding that stabilizes the resulting complex with the catalyst.

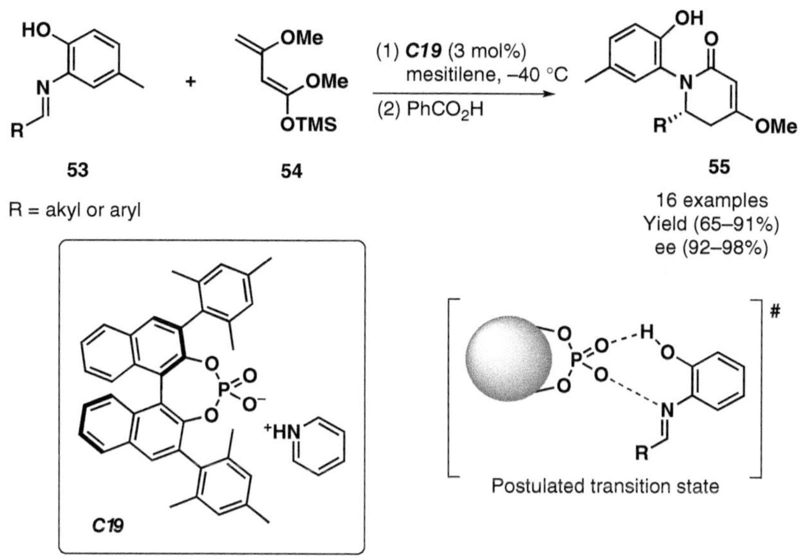

Scheme 16.17 Aza-Diels–Alder reaction by Akyiama. Source: Itoh et al. [66] / John Wiley & Sons.

Not only imine substrates can be activated by chiral BPA. There are also some reports indicating that carbonyl compounds may also be activated in this manner. One of the first studies on the asymmetric variant of the oxo-Diels–Alder reaction was that published in 2009 by Terada and coworkers [67]. Authors presented the reaction between glyoxylate **56** and a number of dienes **57**, catalyzed by ***C20*** (Scheme 16.18).

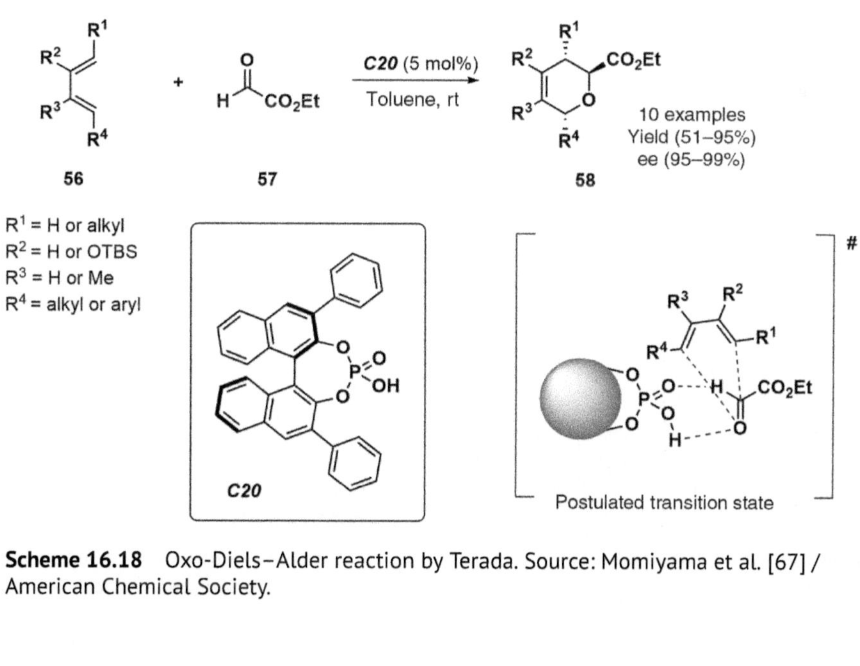

Scheme 16.18 Oxo-Diels–Alder reaction by Terada. Source: Momiyama et al. [67] / American Chemical Society.

The enantioenriched products in a form of substituted dihydropyrans **58** were obtained in very good yields (up to 95%) and excellent enantioselectivities (95–99% ee). The mechanism is similar to the former one, presented for the aza-Diels–Alder reaction. The aldehyde proton is involved in the interaction with the basic part of the catalyst. At the same time, there is an additional interaction between the acidic proton of the catalyst and the oxygen atom of the aldehyde.

16.4.4 1,3-Dipolar Cycloaddition

BPA are capable of catalyzing reactions in which they can form interactions with the substrate molecule stabilizing both positive and negative charges. When the substrate is a molecule having two charges separated from each other, and especially in the 1,3-relationship, the formation of a complex with the catalyst is very efficient, which in turn is very likely to lead to a product with high enantioselectivity. These types of interactions were examined in 2008 by Gong and coworkers [68]. The authors presented the new dimeric catalyst ***C21*** possessing two BINOL platforms. They showed the reaction using aldehydes **59**, amines **60**, and electron-deficient alkenes **61** (Scheme 16.19).

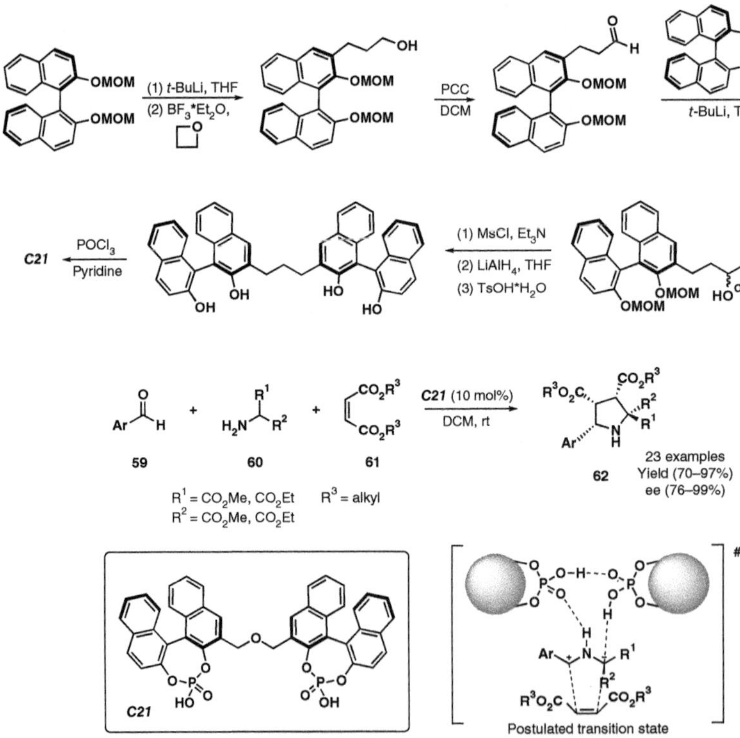

Scheme 16.19 1,3-Dipolar cycloaddition by Gong. DCM, dichloromethane; PCC, pyridinium chlorochromate; THF, tetrahydrofuran. Source: Chen et al. [68] / American Chemical Society.

In the first step, the imine intermediate was obtained through the condensation of **59** and **60**. Then in the presence of chiral diphosphoric catalyst ***C21***, the azomethine ylide was generated, which in turn reacted with alkene **61** in an enantioenriched way. The large scope of products of this reaction – multisubstituted pyrrolidines **62** – was obtained in very good yields and good to excellent enantioselectivities (up to 99% ee).

Another interesting example of asymmetric 1,3-dipolar cycloaddition was presented in 2016 by Wang and coworkers [69]. The authors described the synthesis of biologically important 3,3′-pyrrolidonyl spirooxindole scaffolds (Scheme 16.20).

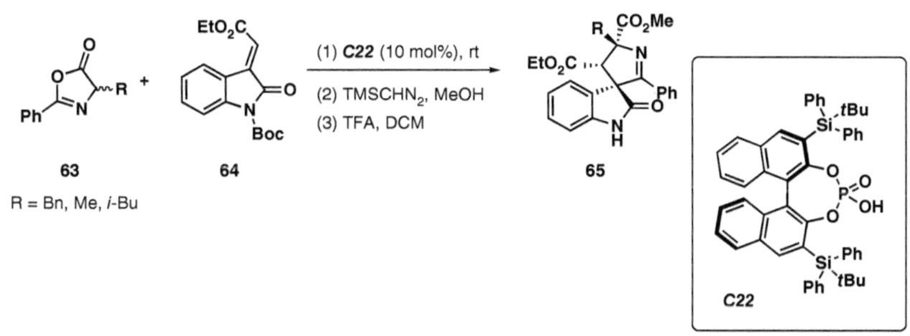

Scheme 16.20 1,3-Dipolar cycloaddition of azlactones and methyleneindolinones by Wang. DCM, dichloromethane; TFA, trifluoroacetic acid. Source: Zhang et al. [69] / Royal Society of Chemistry.

The presented reaction of azlactones **63** as *N*-protonated 1,3-dipoles and methyleneindolinones **64** occurred in the presence of 10 mol% of catalyst ***C22***. The authors obtained 3′-pyrrolidonyl spirooxindole with high yields of up to 93% and, what is even more important, excellent enantioselectivities (up to 99% ee).

16.4.5 Multicomponent Reactions

Multicomponent reactions are a group of desirable reactions, as determined by the sequential processes possible in a one-pot system. Because several reaction components are simultaneously present in the system, the design of a multicomponent reaction carried out under enantiodifferentiating conditions presents additional challenges. The first examples of asymmetric Mannich reaction catalyzed by BPA were reported in 2004 by Akyiama [59] and Terada [60] groups.

Akiyama's group presented an enantioselective Mannich-type reaction between imines **66** and ketene silyl acetals **67**, in the presence of ***C23*** catalyst (10 mol%), obtaining β-aminoesters **68** with an enantiomeric excess reaching 96% ee (Scheme 16.21) [50].

The authors showed 15 examples, pointing out the importance of the hydroxyl group in the imine structure. The acidic proton of the catalyst activates the imine while an additional non-covalent interaction with the Lewis basic site of the catalyst is possible. This results in the formation of a stronger complex, giving high enantiomeric excesses.

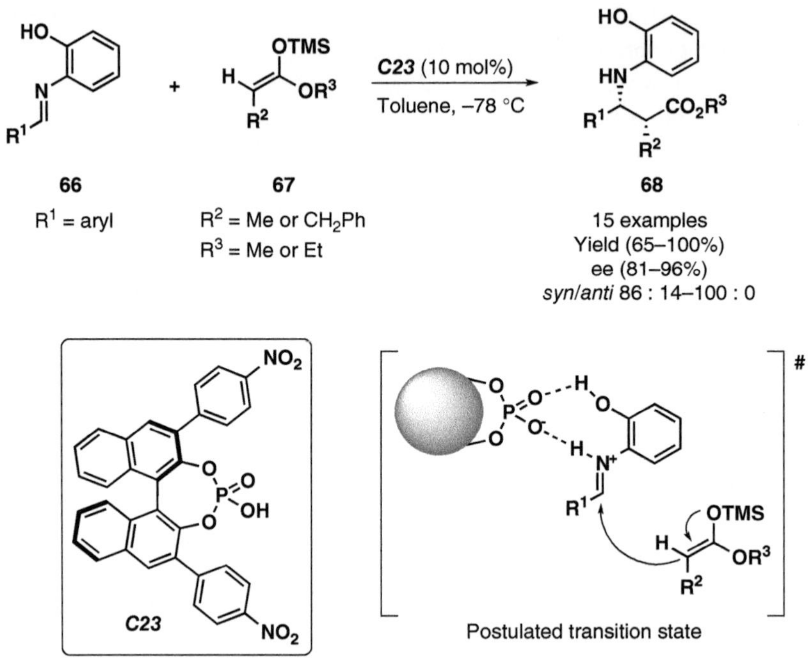

Scheme 16.21 Mannich-type reaction presented by Akiyama. Source: Akiyama et al. [59].

Almost simultaneously, Terada and coworker [60] demonstrated the use of a geometrically similar catalyst ***C24*** in a standard Mannich reaction, using *N*-Boc-protected aryl imines **69** and acetylacetone **70** (Scheme 16.22) [51].

The authors presented six examples, with enantiomeric excesses reaching 98% ee. Furthermore, it was noted that substitution of the catalyst at the 3,3′-position with large substituents improves the reaction yield and the enantiomeric excess. In the following year, the same authors [70] presented the development of their new research. They demonstrated the alkylation of α-diazoesters **72** with *N*-acylimines **73** in the presence of 2 mol% of the catalyst ***C16***. The authors obtained the corresponding products **74** in very good yields and enantiomeric excess up to 97% ee (Scheme 16.23).

The authors proposed a mechanism in which imine activation occurs, followed by the attack of the diazo compound. The alternative mechanism proposed by Peng and coworkers [71] few years later assumes the catalytic activation of both the imine and the azo species.

This multicomponent process is one of the most fundamental reactions in organic chemistry, leading to versatile intermediates. Another very useful multicomponent process, which leads to dihydropyrimidinones (an important building block of pharmacological compounds), namely, the Biginelli reaction, was presented in 2008 by Gong and coworkers [72] (Scheme 16.24).

Using 10 mol% of partially hydrogenated catalyst ***C25*** allowed to carry out the reactions between unsaturated aldehydes **75**, amines **76**, and β-keto esters

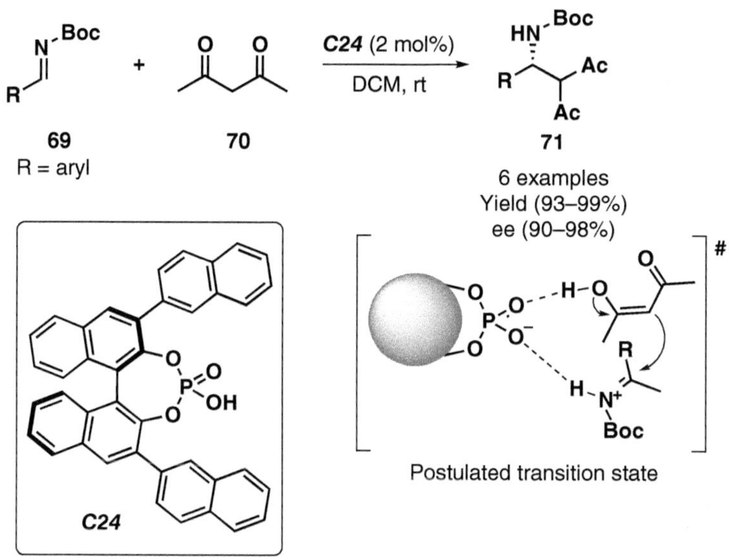

Scheme 16.22 Mannich reaction of acetylacetone by Terada. DCM, dichloromethane. Source: [60] / John Wiley & Sons.

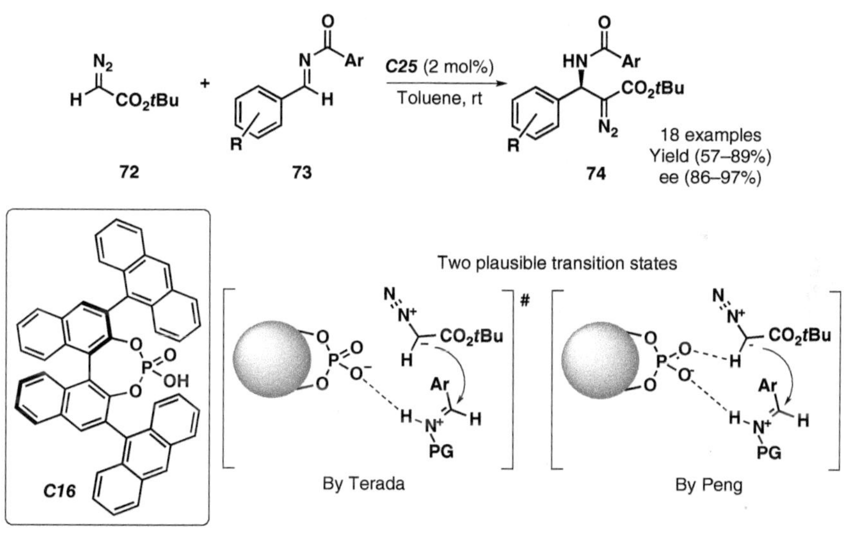

Scheme 16.23 Alkylation of α-diazoesters with *N*-acylimines by Terada and Peng [70, 71].

77 with average to very good yields and good to excellent enantioselectivities (up to 98% ee). The authors obtained a broad scope of condensed products **78**. The mechanism of the catalyzed reaction involves activation of the generated imine by chiral phosphoric acid and formation of an additional hydrogen bond between the enol formed in the system and by the Lewis basic site on the catalyst.

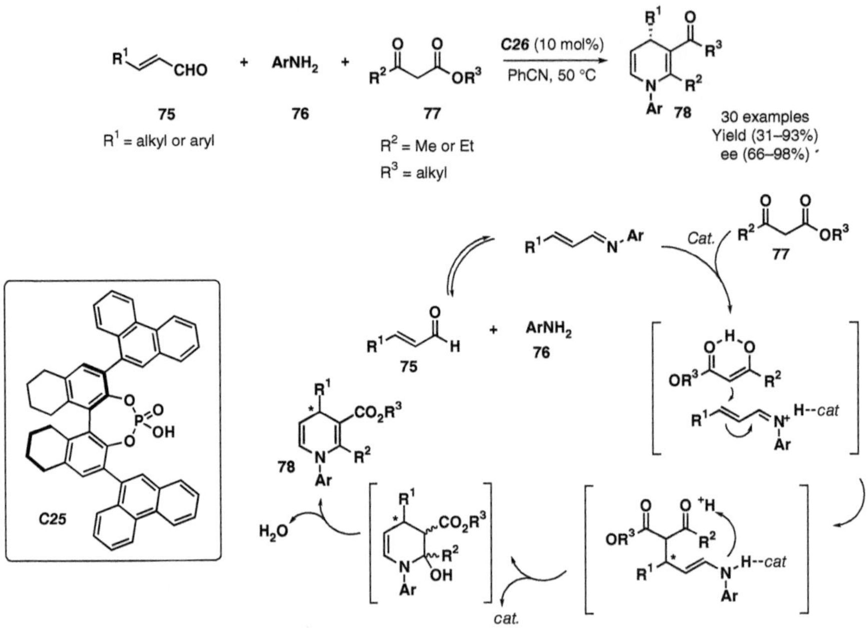

Scheme 16.24 The Biginelli reaction by Gong. Source: Jiang et al. [72] / John Wiley & Sons.

One of the most useful reactions providing natural and unnatural α-amino acid derivatives is the Strecker reaction. It involves the addition of hydrogen cyanide or its equivalent to the generated in situ imine. For the first time, a BPA catalyst was used in this type of reaction in 2006 by Rueping et al. [73] (Scheme 16.25).

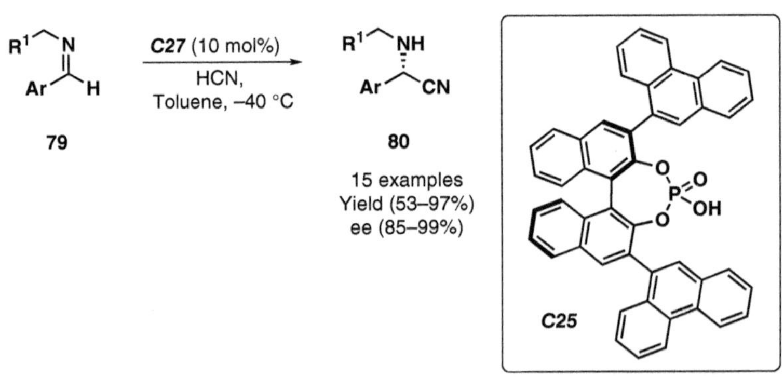

Scheme 16.25 Strecker reaction by Rueping. HCN, hydrogen cyanide. Source: Rueping et al. [73] / John Wiley & Sons.

Strecker reaction products were obtained from various benzyl protected imines **79** and HCN in the presence of 10 mol% of catalyst ***C25***. The authors obtained products **80** with excellent enantiomeric excess – up to 99% ee. Moreover, they observed that in this case the use of a suitable solvent was crucial. The aromatic solvents toluene and benzene proved to be the most advantageous. The conversion of the obtained products to amino acids occurred without a decrease in enantiopurity. In the following year, the same authors [74] extended the applicability of the reactions to ketimine substrates. The reaction still proceeded with very good yields, but a slight decrease in enantioselectivity was observed.

16.4.6 Other Examples

In 2006, MacMillan and coworkers [75] used chiral phosphoric acids in the reductive amination reaction, one of the most important methods in the synthesis of amines. The simple combination of aromatic amine **81** and carbonyl compounds (aromatic ketones **82**), in the presence of catalyst ***C15*** and Hantzsch ester, a hydride donor, resulted in the synthesis of a number of secondary amines **83** in good yields and excellent enantiomeric excesses (Scheme 16.26).

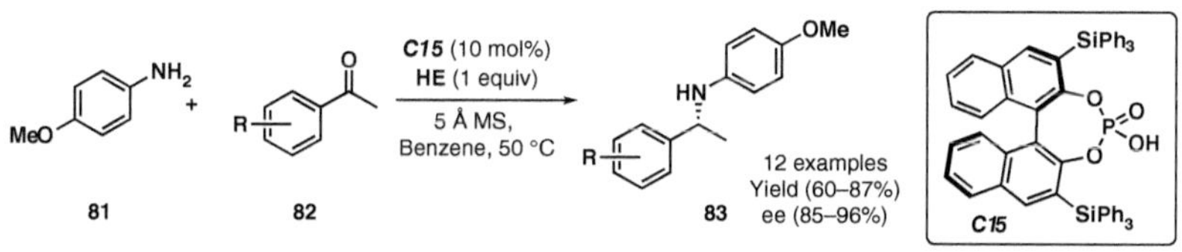

Scheme 16.26 Reductive amination by MacMillan. MS, molecular sieves. Source: Storer et al. [75] / American Chemical Society.

The authors checked also this type of reaction, employing alkyl ketones; however, they obtained slightly lower enantioselectivities. It was suggested that some additional non-covalent interactions with aromatic rings of the catalyst occurred. In 2010, List and coworkers [76] improved the functionality of this transformation, optimizing the reaction conditions. They managed to use α-aminotoluene as a substrate. The authors showed an effective method to obtain a number of enantioenriched primary amines.

In 2006, on the other hand, List and coworkers [77] presented an example for asymmetric Pictet–Spengler reaction (Scheme 16.27). This transformation enables the synthesis of tetrahydro-β-carbolines and tetrahydroisoquinolines – building blocks of many naturally occurring compounds with useful biological activity.

The authors used a range of tryptamine derivatives **84**. In the presence of aldehyde (1 equiv) and 20 mol% of catalyst ***C18***, they received the corresponding alkaloids **85** with moderate to excellent yields and up to 96% ee. The plausible mechanism consists the bifunctional activation with coordination to the protonated imine and the NH group of the indole ring.

Despite the fact that nearly 50 years have passed since Cram first reported the use of BINOL as a synthetic platform in organic chemistry, the topics of using this building block as a fragment of chiral receptors and/or catalysts are of enduring interest to scientists. These compounds are widely used in academic and industrial research, which is their essence in basic research. The importance of this topic can

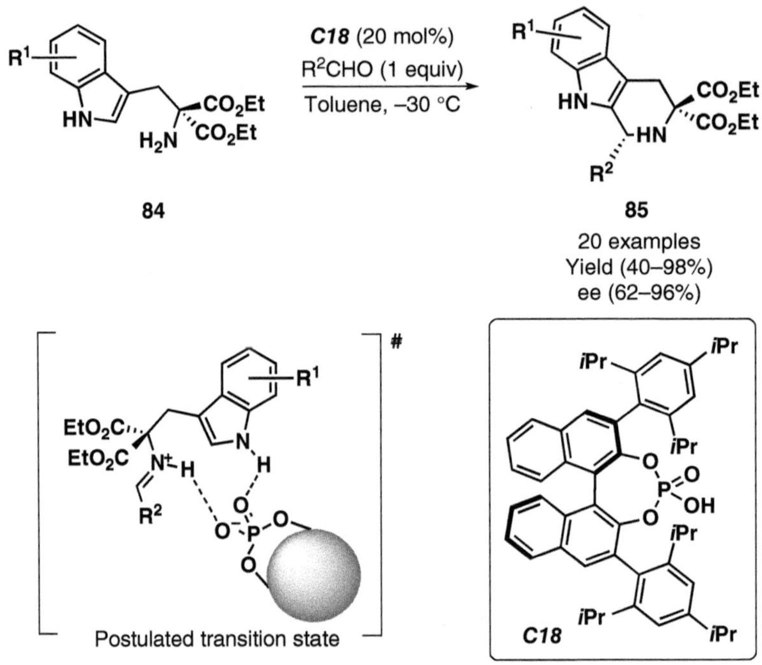

Scheme 16.27 Pictet–Spengler reaction by List. Source: Seayed et al. [77] / American Chemical Society.

also be seen in the number of publications on the use of BINOL as a versatile chiral synthetic platform. The structure of the transition state depends on the subtle interactions between the substrate and the catalyst. This plays a key role in the use of organocatalysts in asymmetric synthesis. The research problem presented here, namely, the search for efficient catalysts in asymmetric reactions, has undoubtedly influenced the complexity and structural diversity of BINOL-based catalysts. The development of this field of organic chemistry and catalysis largely depends on the ability to find correlations between the structure of chiral catalysts and their activity and the mechanism of the reaction taking place. Rational and efficient design of such catalysts comes from the ability to analyze structurally diverse classes of catalysts and examples of catalyzed reaction types in supramolecular catalysis systems. The analysis of electrophile and/or nucleophile association processes by supramolecular catalyst molecules is a key aspect for a better understanding of reaction processes. This phenomenon also affects the possibility of searching for new transformations conducted under conditions of asymmetric catalysis.

Acknowledgments

We are grateful to the National Science Centre, Poland (Grant 2017/25/B/ST5/00214 "Stereoselective synthesis of allenes") for financial support.

References

1 Kyba, E.P., Siegel, M.G., Sousa, L.R. et al. (1973). Chiral, hinged, and functionalized multiheteromacrocycles. *J. Am. Chem. Soc.* 95 (8): 2691–2692.

2 Pu, L. (2012). Enantioselective fluorescent sensors: a tale of BINOL. *Acc. Chem. Res.* 45 (2): 150–163.

3 Yu, S. and Pu, L. (2015). Recent progress on using BINOLs in enantioselective molecular recognition. *Tetrahedron* 71 (5): 745–772.

4 Yu, F., Chen, Y., Jiang, H. et al. (2020). Recent advances of BINOL-based sensors for enantioselective fluorescence recognition. *Analyst* 145 (21): 6769–6812.

5 Brunel, J.M. (2005). BINOL: a versatile chiral reagent. *Chem. Rev.* 105 (3): 857–898.

6 Shibasaki, M. and Matsunaga, S. (2006). Design and application of linked-BINOL chiral ligands in bifunctional asymmetric catalysis. *Chem. Soc. Rev.* 35 (3): 269–279.

7 Zamfir, A., Schenker, S., Freund, M. et al. (2010). Chiral BINOL-derived phosphoric acids: privileged Brønsted acid organocatalysts for C–C bond formation reactions. *Org. Biomol. Chem.* 8 (23): 5262–5276.

8 Parmar, D., Sugiono, E., Raja, S. et al. (2014). Complete field guide to asymmetric BINOL-phosphate derived Brønsted acid and metal catalysis: history and classification by mode of activation; Brønsted acidity, hydrogen bonding, ion pairing, and metal phosphates. *Chem. Rev.* 114 (18): 9047–9153.

9 Rueping, M., Sugiono, E., Steck, A. et al. (2010). Synthesis and application of polymer-supported chiral Brønsted acid organocatalysts. *Adv. Synth. Catal.* 352 (2, 3): 281–287.

10 Castelló, L.M., Hornillos, V., Vila, C. et al. (2015). Pd-catalyzed cross-coupling of aryllithium reagents with 2-alkoxy-substituted aryl chlorides: mild and efficient synthesis of 3,3′-diaryl BINOLs. *Org. Lett.* 17 (1): 62–65.

11 Pellissier, H. (2010). *Recent Developments in Asymmetric Organocatalysis.* Cambridge: RSC Publishing Thomas Graham House.

12 Dalko, P.I. (2013). *Comprehensive Enantioselective Organocatalysis.* Weinheim: Wiley-VCH.

13 Alemán, J. and Cabrera, S. (2013). Applications of asymmetric organocatalysis in medicinal chemistry. *Chem. Soc. Rev.* 42 (2): 774–793.

14 Busacca, C.A., Fandrick, D.R., Song, J.J. et al. (2011). The growing impact of catalysis in the pharmaceutical industry. *Adv. Synth. Catal.* 353 (11, 12): 1825–1864.

15 Blaser, H.U. and Schmidt, E. (2004). *Asymmetric Catalysis on Industrial Scale: Challenges, Approaches and Solutions.* Weinheim: Wiley-VCH.

16 Pummerer, R., Prell, E., and Rieche, A. (1926). Preparation of binaphthylene dioxide. *Chem. Ber.* 59 (8): 2159–2161.

17 Jacques, J., Fouquey, C., and Viterbo, R. (1971). Enantiomeric cyclic binaphthyl phosphoric acids as resolving agents. *Tetrahedron Lett.* 12 (48): 4617–4620.

18 Miyashita, A., Yasuda, A., Takaya, H. et al. (1980). Synthesis of 2,2'-bis(diphenylphosphino)-1,1′-binaphthyl (BINAP), an atropisomeric chiral bis(triaryl)phosphine, and its use in the rhodium(I)-catalyzed asymmetric hydrogenation of α-(acylamino)acrylic acids. *J. Am. Chem. Soc.* 102 (27): 7932–7934.

19 Smrčina, M., Lorenc, M., Hanuš, V. et al. (1991). A facile synthesis of 2-amino-2′-hydroxy-1,1′-binaphthyl and 2,2′-diamino-1,1′-binaphthyl by oxidative coupling using copper(II) chloride. *Synlett* 1991 (4): 231–232.

20 Artz, S.P., DeGrandpre, M.P., and Cram, D.J. (1985). Host-guest complexation. 33. Search for new chiral hosts. *J. Org. Chem.* 50 (9): 1486–1496.

21 Noyori, R., Tomino, I., Tanimoto, Y. et al. (1984). Asymmetric synthesis *via* axially dissymmetric molecules. 6. Rational designing of efficient chiral reducing agents. Highly enantioselective reduction of aromatic ketones by binaphthol-modified lithium aluminum hydride reagents. *J. Am. Chem. Soc.* 106 (22): 6709–6716.

22 Christoffers, J., Koripelly, G., Rosiak, A. et al. (2007). Recent advances in metal-catalyzed asymmetric conjugate additions. *Synthesis* 2007 (9): 1279–1300.

23 Panetti, G.B., Robinson, J.R., Schelter, E.J. et al. (2021). Expanding the rare-earth metal BINOLate catalytic multitool beyond enantioselective organic synthesis. *Acc. Chem. Res.* 54 (11): 2637–2648.

24 Nguyen, T.N., Chen, P.-A., Sethakarn, K. et al. Chiral diol-based organocatalysts in enantioselective reactions. *Molecules* 23 (9): 2317–2353.

25 Nájera, C., Sasano, J.M., and Saá, J.M. (2009). Bifunctional binols: chiral 3,3′-bis(aminomethyl)-1,1′-bi-2-naphthols (binolams) in asymmetric catalysis. *Eur. J. Org. Chem.* 2009 (15): 2385–2400.

26 Lou, S., Moquist, P.N., and Schaus, S.E. (2006). Asymmetric allylboration of ketones catalyzed by chiral diols. *J. Am. Chem. Soc.* 128 (39): 12660–12661.

27 Lou, S., Moquist, P.N., and Schaus, S.E. (2007). Asymmetric allylboration of acyl imines catalyzed by chiral diols. *J. Am. Chem. Soc.* 129 (49): 15398–15404.

28 Barbaro, K.S., Luan, Y., Ramella, D. et al. (2015). Enantioselective multicomponent condensation reactions of phenols, aldehydes, and boronates catalyzed by chiral biphenols. *Org. Lett.* 17 (23): 5812–5815.

29 Jonker, S.J.T., Diner, C., Schulz, G. et al. (2018). Catalytic asymmetric propargyl- and allylboration of hydrazonoesters: a metal-free approach to sterically encumbered chiral α-amino acid derivatives. *Chem. Commun.* 54 (91): 12852–12855.

30 Alam, R., Vollgraff, T., Eriksson, L. et al. (2015). Synthesis of adjacent quaternary stereocenters by catalytic asymmetric allylboration. *J. Am. Chem. Soc.* 137 (35): 11262–11265.

31 Shih, J.-L., Nguyen, T.S., and May, J.A. (2015). Organocatalyzed asymmetric conjugate addition of heteroaryl and aryl trifluoroborates: a synthetic strategy for discoipyrrole D. *Angew. Chem. Int. Ed.* 54 (34): 9931–9935.

32 Le, P.Q., Nguyen, T.S., and May, J.A. (2012). A general method for the enantioselective synthesis of α-chiral heterocycles. *Org. Lett.* 14 (23): 6104–6107.

33 Huang, G.-B., Huang, W.-H., Guo, J. et al. (2019). Enantioselective synthesis of triarylmethanes *via* organocatalytic 1,6-addition of arylboronic acids to *Para*-quinone methides. *Adv. Synth. Catal.* 361 (6): 1241–1246.

34 Mondal, S. and Panda, G. (2014). Synthetic methodologies of achiral diarylmethanols, diaryl and triarylmethanes (TRAMs) and medicinal properties of diaryl and triarylmethanes-an overview. *RSC Adv.* 4 (54): 28317–28358.

35 Petasis, N.A. and Akritopoulou, I. (1993). The boronic acid Mannich reaction: a new method for the synthesis of geometrically pure allylamines. *Tetrahedron Lett.* 34 (4): 583–586.

36 Wu, P. and Nielsen, T.E. (2018). Petasis three-component reactions for the synthesis of diverse heterocyclic scaffolds. *Drug Discovery Today* 29: 27–33.

37 Wu, P., Givskov, M., and Nielsen, T.E. (2019). Reactivity and synthetic applications of multicomponent Petasis reactions. *Chem. Rev.* 119 (20): 11245–11290.

38 Marques, C.S., McArdle, P., Erxleben, A. et al. (2020). Accessing new 5-α-(3,3-disubstituted oxindole)-benzylamine derivatives from isatin: stereoselective organocatalytic three component Petasis reaction. *Eur. J. Org. Chem.* 2020 (24): 3622–3634.

39 Diner, C. and Szabo, K.J. (2017). Recent advances in the preparation and application of allylboron species in organic synthesis. *J. Am. Chem. Soc.* 139 (1): 2–14.

40 Jonker, S.J.T., Jayarajan, R., Kireilis, T. et al. (2020). Organocatalytic synthesis of α-trifluoromethyl allylboronic acids by enantioselective 1,2-borotropic migration. *J. Am. Chem. Soc.* 142 (51): 21254–21259.

41 Wende, R.C. and Schreiner, P.R. (2012). Evolution of asymmetric organocatalysis: multi- and retrocatalysis. *Green Chem.* 14 (7): 1821–1849.

42 Krištofíková, D., Modrocká, V., Mečiarová, M. et al. (2020). Green asymmetric organocatalysis. *ChemSusChem* 13 (11): 2828–2858.

43 Shirakawa, S. and Maruoka, K. (2013). Recent developments in asymmetric phase-transfer reactions. *Angew. Chem. Int. Ed.* 52 (16): 4312–4348.

44 Tan, J. and Yasuda, N. (2015). Contemporary asymmetric phase transfer catalysis: large-scale industrial applications. *Org. Process Res. Dev.* 19 (11): 1731–1746.

45 Starks, C.M., Liotta, C.L., and Halpern, M. (1994). *Phase-Transfer Catalysis: Fundamentals, Applications and Industrial Perspectives*. New York: Chapman & Hall.

46 O'Donnell, M.J., Benett, W.D., and Wu, S. (1989). The stereoselective synthesis of α-amino acids by phase-transfer catalysis. *J. Am. Chem. Soc.* 111 (6): 2353–2355.

47 O'Donnell, M. (2004). The enantioselective synthesis of α-amino acids by phase-transfer catalysis with achiral Schiff base esters. *Acc. Chem. Res.* 37 (8): 506–517.

48 Cram, D.J. and Sogah, G.D.Y. (1981). Chiral crown complexes catalyze Michael addition reactions to give adducts in high optical yields. *J. Chem. Soc., Chem. Commun.* 13: 625–628.

49 Ooi, T., Kameda, M., and Maruoka, K. (1999). Molecular design of a C_2-symmetric chiral phase-transfer catalyst for practical asymmetric synthesis of α-amino acids. *J. Am. Chem. Soc.* 121 (27): 6519–6520.

50 Mąkosza, M. and Fedoryński, M. (2012). Phase transfer catalysis – basic principles, mechanism and specific features. *Curr. Catal.* 1 (2): 79–87.

51 Kano, T. and Maruoka, K. (2015). Binaphthyl-derived cyclic amines and their salts as asymmetric organocatalysts, Chapter 17. In: *Sustainable Catalysis:*

Without Metals or Other Endangered Elements, Part 2 (ed. M. North), 135–163. York: Thomas Graham House.

52 Shirakawa, S., Tokashi, T., Kasai, A. et al. (2013). Design of chiral bifunctional quaternary phosphonium bromide catalysts possessing an amide moiety. *Org. Lett.* 15 (13): 3350–3353.

53 Shirakawa, S., Koga, K., Tokuda, T. et al. (2014). Catalytic asymmetric synthesis of 3,3′-diaryloxindoles as triarylmethanes with a chiral all-carbon quaternary center: phase-transfer-catalyzed S_NAr reaction. *Angew. Chem. Int. Ed.* 53 (24): 6220–6223.

54 Nishimura, T., Noishiki, A., Tsui, G.C. et al. (2012). Asymmetric synthesis of (triaryl)methylamines by rhodium-catalyzed addition of arylboroxines to cyclic N-sulfonyl ketimines. *J. Am. Chem. Soc.* 134 (11): 5056–5059.

55 Pupo, G., Ibba, F., Ascough, D.M.H. et al. (2018). Asymmetric nucleophilic fluorination under hydrogen bonding phase-transfer catalysis. *Science* 360 (6389): 638–642.

56 Yang, X., Wu, T., Phipps, R.J. et al. (2015). Advances in catalytic enantioselective fluorination, mono-, di-, and trifluoromethylation, and trifluoromethylthiolation reactions. *Chem. Rev.* 115 (2): 826–870.

57 Pupo, G., Vicini, A.C., Ascough, D.M.H. et al. (2019). Hydrogen bonding phase-transfer catalysis with potassium fluoride: enantioselective synthesis of β-fluoroamines. *J. Am. Chem. Soc.* 141 (7): 2878–2883.

58 Roagna, G., Ascough, D.M.H., Ibba, F. et al. (2020). Hydrogen bonding phase-transfer catalysis with ionic reactants: enantioselective synthesis of γ-fluoroamines. *J. Am. Chem. Soc.* 142 (33): 14045–14051.

59 Akiyama, T., Itoh, J., Yokota, K. et al. (2004). Enantioselective Mannich-type reaction catalyzed by a chiral Brønsted acid. *Angew. Chem. Int. Ed.* 43 (12): 1566–1568.

60 Uraguchi, D. and Terada, M. (2004). Chiral Brønsted acid-catalyzed direct Mannich reactions *via* electrophilic activation. *J. Am. Chem. Soc.* 126 (17): 5356–5357.

61 Akiyama, T., Katoh, T., and Mori, K. (2009). Enantioselective Robinson-type annulation reaction catalyzed by chiral phosphoric acids. *Angew. Chem. Int. Ed.* 48 (23): 4226–4228.

62 Aillaud, I., Barber, D.M., Thompson, A.L. et al. (2013). Enantioselective Michael addition/iminium ion cyclization cascades of tryptamine-derived ureas. *Org. Lett.* 15 (12): 2946–2949.

63 Hou, L., Kikuchi, J., Ye, H. et al. (2020). Chiral phosphoric acid-catalyzed enantioselective phospha-Michael-type addition reaction of diarylphosphine oxides with alkenyl benzimidazoles. *J. Org. Chem.* 85 (22): 14802–14809.

64 Uraguchi, D., Sorimachi, K., and Terada, M. (2004). Organocatalytic asymmetric aza-Friedel–Crafts alkylation of furan. *J. Am. Chem. Soc.* 126 (38): 11804–11805.

65 Husmann, R., Sugiono, E., Mersmann, S. et al. Enantioselective organocatalytic synthesis of quaternary α-amino acids bearing a CF_3 moiety. *Org. Lett.* 13 (5): 1044–1047.

66 Itoh, J., Fuchibe, K., and Akiyama, T. (2006). Chiral Brønsted acid catalyzed enantioselective aza-Diels–Alder reaction of Brassard's diene with imines. *Angew. Chem. Int. Ed.* 45 (29): 4796–4798.

67 Momiyama, N., Tabuse, H., and Terada, M. (2009). Chiral phosphoric acid-governed anti-diastereoselective and enantioselective hetero-Diels–Alder reaction of glyoxylate. *J. Am. Chem. Soc.* 131 (36): 12882–12883.

68 Chen, X.-H., Zhang, W.-Q., and Gong, L.-Z. (2008). Asymmetric organocatalytic three-component 1,3-dipolar cycloaddition: control of stereochemistry *via* a chiral Brønsted acid activated dipole. *J. Am. Chem. Soc.* 130 (17): 5652–5653.

69 Zhang, Z., Sun, W., Zhu, G. et al. (2016). Chiral phosphoric acid catalyzed enantioselective 1,3-dipolar cycloaddition reaction of azlactones. *Chem. Commun.* 52 (7): 1377–1380.

70 Uraguchi, D., Sorimachi, K., and Terada, M. (2005). Organocatalytic asymmetric direct alkylation of α-diazoester *via* C–H bond cleavage. *J. Am. Chem. Soc.* 127 (26): 9360–9361.

71 Zhang, H., Wen, X., Gan, L. et al. (2012). Highly efficient asymmetric Mannich reaction of dialkyl α-diazomethylphosphonates with *N*-carbamoyl imines catalyzed by chiral Brønsted acids. *Org. Lett.* 14 (8): 2126–2129.

72 Jiang, J., Yu, J., Sun, X.-X. et al. (2008). Organocatalytic asymmetric three-component cyclization of cinnamaldehydes and primary amines with 1,3-dicarbonyl compounds: straightforward access to enantiomerically enriched dihydropyridines. *Angew. Chem. Int. Ed.* 47 (13): 2458–2462.

73 Rueping, M., Sugiono, E., and Azap, C. (2006). A highly enantioselective Brønsted acid catalyst for the Strecker reaction. *Angew. Chem. Int. Ed.* 45 (16): 2617–2619.

74 Rueping, M., Sugiono, E., and Moreth, S.A. (2007). Metal-free, enantioselective Strecker reactions catalyzed by chiral BINOL and TADDOL catalysts. *Adv. Synth. Catal.* 349 (4, 5): 759–764.

75 Storer, R.I., Carrera, D.E., Ni, Y. et al. (2006). Enantioselective organocatalytic reductive amination. *J. Am. Chem. Soc.* 128 (1): 84–86.

76 Wakchaure, V.N., Nicoletti, M., Ratjen, L. et al. (2010). Towards a practical Brønsted acid catalyzed and Hantzsch ester mediated asymmetric reductive amination of ketones with benzylamine. *Synlett* 2010 (18): 2708–2710.

77 Seayad, J., Seayad, A.M., and List, B. (2006). Catalytic asymmetric Pictet–Spengler reaction. *J. Am. Chem. Soc.* 128 (4): 1086–1087.

17

Chiral Acenes: Synthesis and Applications

Andrea Nitti, Giovanni Preda, and Dario Pasini

University of Pavia, Department of Chemistry and INSTM Research Unit, Via Taramelli 12, 27100 Pavia, Italy

17.1 Introduction

Acenes are a class of organic compounds and polycyclic aromatic hydrocarbons made up of linearly fused benzene rings. Linear acenes are rather unique, since they consist of only one Clar sextet, regardless of the number of annulated rings. Naphthalene is the smallest acene, followed by anthracene, tetracene, pentacene, and so on. They have been the subject of extensive studies for over a century because of their application as optoelectronic organic materials, fluorescent bioimaging probes, and bioactive compounds [1, 2]. They exhibit excellent physical properties due to their extended π-conjugation, small HOMO–LUMO gaps, long absorption and emission wavelengths, and effective π–π interactions for imparting mechanical strengths. Pentacene, for example, has been long studied as semiconductor and has been incorporated into organic field-effect transistors, reaching high charge carrier mobilities [3].

The term "acene" is intuitively associated to a flat, rigid, and achiral molecule. This is frequently not the case, because the crystal structures of most acenes and their derivatives show some deviations from planarity [4]. Once the rings are more than two, the multiplicity of geometrical pathways and the potential deviation from the rectilinear arrangement for fusing further benzene or heteroaromatic rings can give rise to a series of fascinating structures, including graphenes and nanographenes. Indeed, the bottom-up approach to the synthesis of molecular nanographenes, nanotubes, and nanobelts is a very hot topic of current chemical research, with high relevance to the fields of supramolecular chemistry and organic materials [5, 6].

Helical chirality differs from point chirality: it is a property of chiral systems that do not contain stereocenters, which are characterized by four nonequivalent points at the vertices of a tetrahedron [7]. The presence of the helical stereogenic unit warrants for the chemical structure to be not superimposable on its mirror image. This type of chirality is also known as axial chirality because of the presence of a stereogenic axis instead of a center. In molecules that are not inherently helically chiral, helicity can be induced. DNA, for example, can be folded into a helical conformation by specific directional non-covalent interactions. In rigid molecules, helical

Chiral Building Blocks in Asymmetric Synthesis: Synthesis and Applications, First Edition.
Edited by Elżbieta Wojaczyńska and Jacek Wojaczyński.

conformations can arise if unfavorable steric interactions, or strain, are present in their non-helical conformations, which is the driving force toward formation of the energetically favored helical conformations. This second type of helical chirality is commonly found in chiral acenes; the induced deformation is typically spread over a large number of bonds, which makes the core relatively flexible. Previous reviews on helically chiral strained polyaromatic compounds have been published [8–12].

In this chapter, we give an overview of the most important classes of chiral acenes, namely, helicenes, twistacenes, and nanobelts, focusing on recent developments on their synthesis and applications.

17.2 Chiral Carbo[*n*]helicenes

According to "Golden Book," the IUPAC Compendium of Chemical Terminology, carbo[*n*]helicenes are *ortho*-fused polycyclic aromatic compounds in which all rings (minimum five) are angularly annulated to give helically-shaped chiral molecules [13]. Carbohelicenes are composed only of benzene rings in the backbone having general molecular structures shown in Figure 17.1, differently from heterohelicenes that contain at least one heteroatom in the framework. Figure 17.1 illustrates a beginning of homologous series starting from the progenitor with five-membered fused rings to end with homologue having seven-membered fused rings by easy adding of a benzene rings.

According to IUPAC, only structures where *n* is at least 5 are considered helicenes. To simplify the IUPAC nomenclature, in 1956, Newman and Lednicer first introduced the name of hexahelicene ([6]helicene) as a substitution of phenantro[3,4-*c*]phenantrene. Today, the systematic naming accepted by IUPAC for this class of compounds is based on the number of rings in square brackets [*n*] followed by a word *helicene*. Thus, for the pentahelicene, we write [5]helicene (see Figure 17.1). In this chapter, unsubstituted carbo[*n*]helicene will be indicated

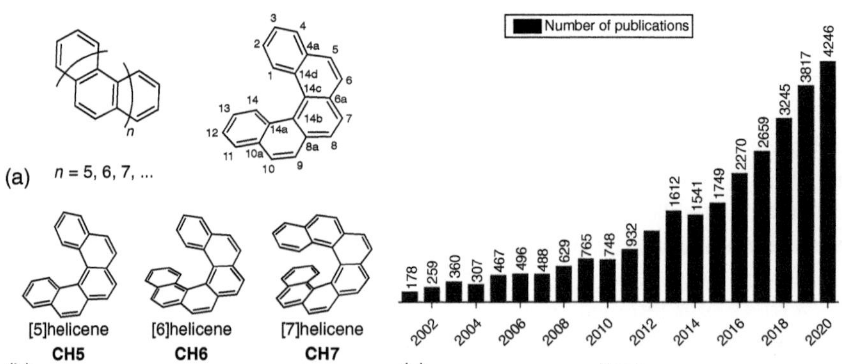

Figure 17.1 (a) Molecular structures of carbo[*n*]helicenes and helicene numbering of carbon atoms. (b) Homologue series of **CH*n***. (c) Number of publications per year using "helicene" as topic. Source: From Web of Science®.

with abbreviation **CH*n***, where *n* indicates the number of fused-aromatic rings. For instance, [5]helicene becomes **CH5**, [8]helicene becomes **CH8**, [14]helicene becomes **CH14**, and so on.

Carbohelicenes belong to a class of fascinating, helically chiral molecules, with a rich history in chemistry since the very beginning of the twentieth century. In the 1950s, the study by Newman et al. reported the synthesis and resolution of hexahelicene [14]. During the following 30 years, Wynberg [15], Martin [16], Laarhoven and Prinsen [17], and Katz and coworkers [18] carried out pioneering studies of the synthesis, spectral properties, and structures of helicenes. In the 1990s, the Diels–Alder synthetic approach, overcoming Martin's photochemical methods, brought about an important breakthrough in the preparation of helicenes on a large scale [19]. Since the late 1990s, the development of new strategies for the enantioselective synthesis of helicenes and the discovery of new fields of applications have led to an impressive increase of the publication numbers per year (see Figure 17.1). Helicene chemistry is being considered as an expanding and modern field, leading to several applications in supramolecular chemistry, in nanoscience, in chemical biology, and in polymers and materials science [20–33]. In particular, a recent book edited by Chen and Shen can be viewed as a comprehensive synthetic compendium of helicenes chemistry [34].

To better describe the helicene chemistry, we divided this chapter into four sections. In the first one, we describe structural properties (topological aspect, conjugation, and chirality). In the second part, we illustrate several synthetic approaches, followed by optical resolution. The third part is devoted to modern asymmetric synthesis reported in the literature. Finally, in the last section, we discuss the principal applications in materials science.

17.2.1 Structure and Properties of Helicenes

[*n*]Helicenes (**CH*n***) have a twisted structure of fused rings that give them a propeller-like configuration, so it is easy to state that the defining property of a helicene is its helical structure. The tilt of ring planes coming from the attenuation of the steric hindrance between the two terminal rings makes energetically favorable to wrap around a C_2-symmetrical axis as shown in Figure 17.2.

17.2.1.1 Topological Description

From topological point of view, the helical structure of [*n*]helicenes can be defined by followed descriptors: the *helical pitch*, the distance between two ends; *the C—C bond length of both inner and outer helix*; the *torsional angle*, a dihedral angle between the four adjacent inner carbon atoms (a, b, c, d); and the *interplanar angle*, defined as the angle between the two terminal benzene rings. According to Figure 17.2, helicenes present three helical pathways: one is formed by atoms in the inner rim (labeled a, b, c, d), the next is composed by center atoms (labeled α, β, γ, δ), and the last one is composed by outer atoms (labeled A, B, C, D).

The helical pitch r_p is a superior parameter for globally understanding the features of a helical structure. Figure 17.2 shows the trend of two pitch lengths calculated

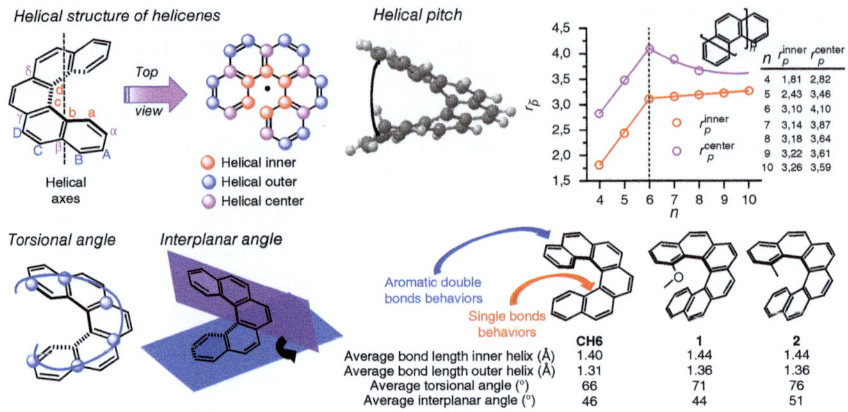

Figure 17.2 Helicene topological descriptors.

by using the inner carbon atoms (r_p^{inner}) and the mean center of each benzene ring (r_p^{center}) against the rings number (n). Both pitch lengths progressively increase until the overlap of aromatic rings begins ($n \leq 6$) and after that show distinctly different dependencies indicating a strong impact of ring overlap on the helix structure. The linear and exponential extrapolation of the inner and center pitches to infinite n led to the ultimate value of $r_p^{\text{inner}} = r_p^{\text{center}} = 3.5$ Å for [∞]helicene.

The distortion of benzene rings in the helicene structure causes a loss of planarity with consequent alteration of the lengths of C—C bonds. For instance, the lengths of the bonds in the inner helix (between carbon atoms a, b, c, d) are increased, while the lengths of the bonds in the outer helix (connecting between carbon atoms A, B, C, D) are shorter. *As a result of torsional strain, the bond lengths in the skeleton are different, with different C—C bonds having the features of a single bond or a double bond.* In comparison with the bond length of benzene (1.39 Å), in the [6]helicene, the average bond length of the C—C bonds in the outer helix is shortened to 1.36 Å, while the average length of the C—C bonds in the inner helix is lengthened 1.43 Å, which is close to length of a single bond (1.48 Å).

The torsional angle is greatly influenced by the steric hindrance of the substituent presents. For example, for [5]helicenes of Figure 17.2, the total torsional angle trend is **2** > **1** > **CH6** as a consequence of the steric hindrance of methyl groups, twisting the helical structure more and enlarging the interplanar angle between the terminal rings [35, 36]. The interplanar angle is affected by the steric hindrance of substituents and the length of helical skeleton. Generally, it increases from the frameworks bearing four benzene rings to that with six benzene rings and decreases as the helical structure is lengthened further [17].

17.2.1.2 Aromaticity and Optoelectronic Properties

Compared with the corresponding linear acenes, helicenes show a greater aromatic distortion. However, little loss of the local aromaticity has been found, showing that the σ–π interactions (inner helix vs. outer helix) do not significantly interfere with the π-electron delocalization and the two terminal rings of helicenes are the most aromatic [37]. In other word, *the local aromaticity does not decrease so much because the π-electrons could be delocalized* via *the twisted structure*. Calculated aromatic index on optimized molecular geometry is the best evidence that corroborates this sentence. Typical theoretical descriptors are the harmonic oscillator model of aromaticity (HOMA), and the nucleus-independent chemical shifts (NICS) at the ring center [38].

As the number of the benzene rings increases, the wavelength of the absorption maximum does not change so much. This indicates that the carbohelicenes have an HOMO–LUMO gap (E_{gap}), which could not be reduced by fusing benzene rings to the helical skeletons (Figure 17.3), introducing pyrene moieties into the helicene structure. Helicene moieties embedded into large polycyclic aromatic systems were recently synthetized, and their structure are reported in Figure 17.3. Nakakuki and Matsuda reported the synthesis and the properties of compound **3**, which represents a primary substructure of the helical graphenes [40]. The π-extended helicoid structure possesses an intramolecular π–π interaction, determined by X-ray crystallography. Corannulene–[*n*]helicenes, such as compound **4**, were synthetized, and their properties studied by Scott and coworkers [41]. The UV–Vis absorption spectrum presents the absorption maximum at 418 nm, while the emission is peaked at 447 nm with a photoluminescence quantum yield (PLQY) of 6%. Evans and Martín extended the conjugation of the terminal rings with a nanographene [42]. The folded nanographene **5** is composed of two hexa-*peri*-hexabenzocoronene (HBC) layers fused to a [10]helicene, with an helical pitch of 3.6 Å. The rigidity of the helicene linker forces the layers to adopt a nearly aligned conformation, rarely observed in few-layer graphene.

By combining the advantages of nanographenes and helicenes, the authors have constructed a bilayer system of 30-fused and perfectly conjugated benzene rings. Compound **5** shows an absorption maximum at 380 nm while the emission is peaked at 540 nm recording a Stokes shift of 160 nm. Campaña and coworkers reported the synthesis and characterization of [7]helicene fused with HBC (compound **6**). Remarkably, nonlinear photophysical analysis demonstrated a two-photon absorption cross section of 870 GM at 800 nm and a perfect overlap between linear, nonlinear, and chiral emissions [39]. Liu and Feng synthesized helical nanographenes **7** and **8** containing an azulene unit [43]. The embedded azulene units in the helical structure have a high twisting degree because of the contiguous steric repulsion at the helical inner rim. These helical nanographenes possess narrow energy gaps (1.88 eV for **7**; 2.03 eV for **8**) and manifest a global aromatic structure while the inner azulene unit exhibits weak antiaromatic character.

The optical properties can be more profoundly modified using a pull–push approach by incorporation of donor and acceptor substituents. The modification of optoelectronic properties by constructing pull–push structures is more practical, because the electron-donating or electron-withdrawing groups can be more effective in changing the desired characteristics. Furthermore, their installation on the helicene scaffold is relatively easy, such as in the cases of structures **9–10** [44]. Very recently Ravat and coworkers synthetized a novel class of helical diimide molecules **11–13** shown in Figure 17.3, in which two imide units are connected via

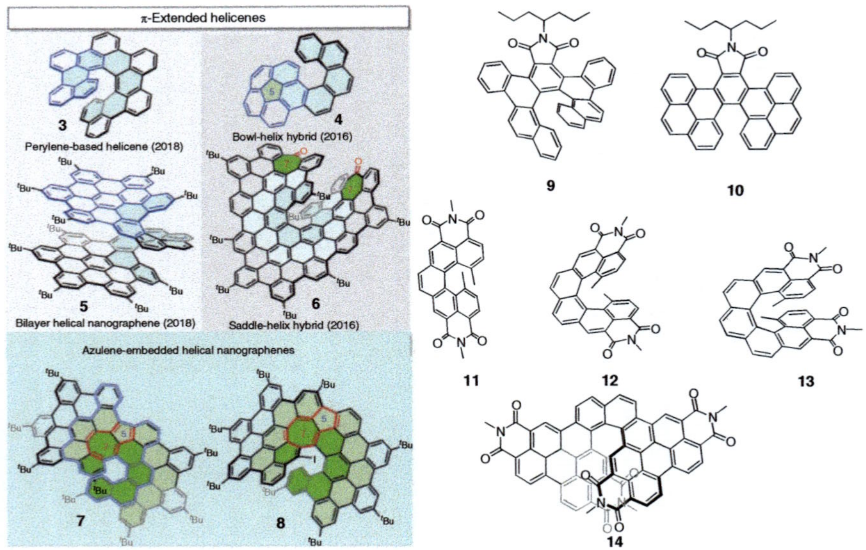

Figure 17.3 Helicene scaffold embedded into large polycyclic aromatic systems and helicenes with push–pull architecture. Source: Adapted with permission from Cruz et al. [39] (copyright 2016 American Chemical Society).

[5]-, [6]-, and [7]helicene skeletons [45]. All three compounds showed a significant bathochromic shift in their absorption (400, 425, and 450 nm, respectively) and emission spectra (480, 482 and 500 nm, respectively) compared with those of corresponding [*n*]helicenes. The PLQY of **11** and **13** were 0.22 and 0.12, which are more than five times higher than those of **CH5** and **CH7**. Nuckolls and coworkers synthetized a [7]helicene **14** in which the terminal benzene rings are fused in a perylene diimide acceptor structures [46]. Impressive redshift of absorption and emission was attributed to the push–pull effect and to the extension of conjugation.

17.2.1.3 Chirality of Helicene

The twisted structure of [*n*]helicenes confers them a chirality even though they do not possess any stereocenters. According to the Cahn, Ingold, and Prelog rules, the helicity descriptors are *M* to the left-handed helix and *P* to the right-handed one (Figure 17.4). Helicenes have very high specific optical rotation, which, according to a general empirical relationship, can be related with the *P*/*M* absolute configuration. For instance, (*P*)-helicenes are dextrorotatory (with positive $[\alpha]_\lambda^T$), while (*M*)-helicenes are levorotatory (with negative $[\alpha]_\lambda^T$).

The electronic circular dichroism (ECD) spectra could be used to determine the *P*/*M* absolute configuration [47]. Typical ECD spectra of (+)-[*n*]helicenes are shown in Figure 17.4, where strong negative and positive Cotton effects (CEs) are clearly visible at around 250 and 320 nm. These apparently coupled CEs were commonly observed in all carbo[*n*]helicenes, and the signs of these CEs can be readily used as a tool for determining the absolute configuration of helicenes. Because of the C_2-symmetric structure, the main transitions of [*n*]helicenes can be categorized into the *A* (C_2-axis polarized) and *B* (perpendicular to the *A*-axis) symmetry groups. According to the symmetry groups, the strong negative CEs at around 250 nm are assigned as $1B_a$ transitions, while the positive CEs at around 320 nm are assigned

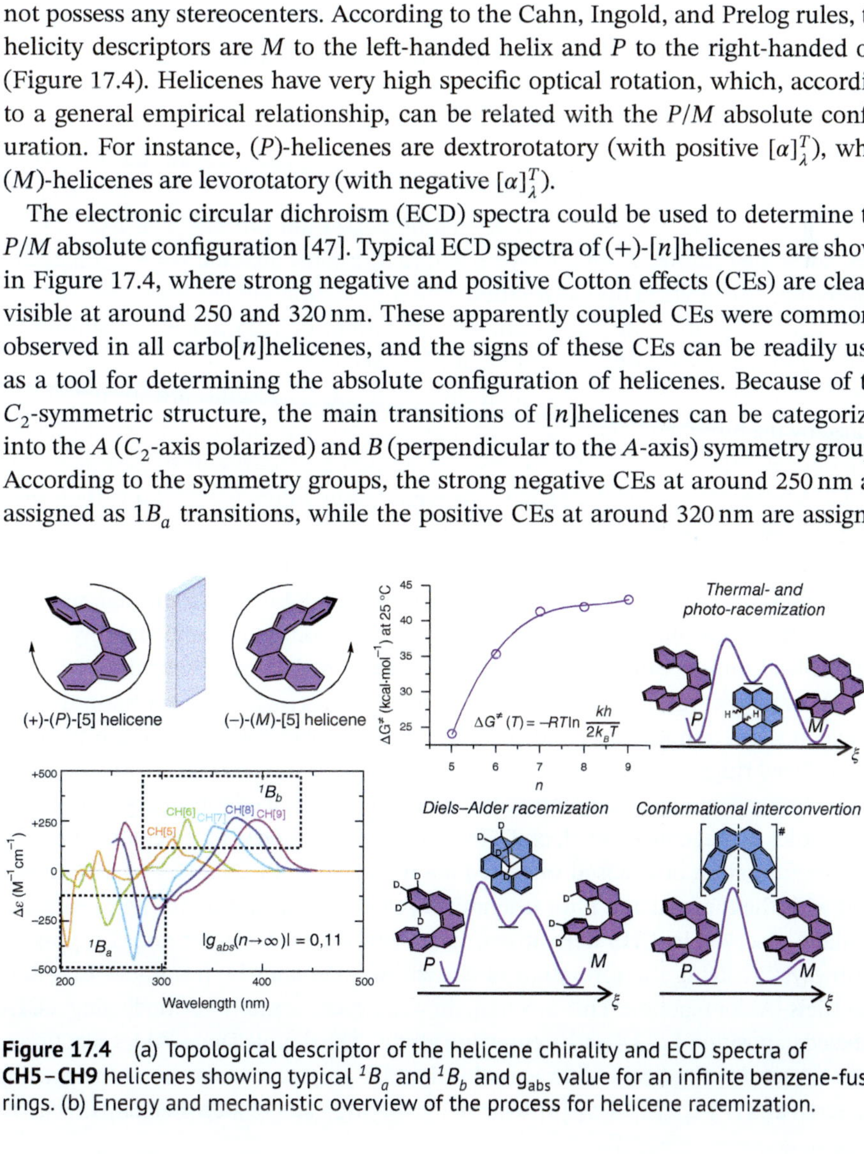

Figure 17.4 (a) Topological descriptor of the helicene chirality and ECD spectra of **CH5–CH9** helicenes showing typical ^1B_a and ^1B_b and g_{abs} value for an infinite benzene-fused rings. (b) Energy and mechanistic overview of the process for helicene racemization.

as ^1B_b transitions. The ^1B_b transition is mostly from HOMO or HOMO−1 to LUMO or LUMO+1. The relevant orbitals associated with the ^1B_b transitions (HOMO, HOMO−1, LUMO, and LUMO+1) are the typically of the π and π^* types. For the ^1B_a transitions, the orbitals of much lower and higher energy levels were found to be contributing in a complex manner.

Direct comparison of the circular dichroism (CD) spectra in Figure 17.4 of **CH5–CH9** reveals some general trends for the first strong CE at the ^1B_b transition. Thus, from excitation energies point of view of ^1B_a and ^1B_b transitions, the helicene series suffered the effect of delocalization to show systematic bathochromic shifts of the bands, whereas the CD intensity does not change systematically. The g factor linearly correlates with n, but where the helix forms a single turn and both ends start to overlap ($n = 6$), the tendency shows a discontinuous point. An extrapolation of the straight line to infinite n affords an ultimate g factor as large as 0.11 for [∞]helicene.

Although helicenes are composed by aromatic rings, the helical skeleton is not as rigid as it may seem. In fact, helicenes could be completely converted from M- to P-helicity, and vice versa. Their stereodynamics can be quantified in terms of the Gibbs activation energy ($\Delta G^{\neq}(T)$) of enantiomerization process. The $\Delta G^{\neq}(T)$ values can be determined from the racemization rate constants that were estimated by observing the change in optical rotation with time at different temperatures. According to $\Delta G^{\neq}(T)$ value, an [n]helicene can or cannot be resolved into enantiomers under ambient conditions. Therefore, a high $\Delta G^{\neq}(T)$ is necessary to access chiral functional materials based on [n]helicenes. The $\Delta G^{\neq}(T)$ values are strongly influenced by the number of benzene-fused rings. As shown in the plot of Figure 17.4, for [n]helicenes ($n = 4$–7), the $\Delta G^{\neq}$ values at 25 °C increase sharply, until the upper limit of 43.1 kcal mol^{-1} observed for **CH9**. The enantiomerization barrier for [n]helicenes ($n \geq 10$) has not yet been determined experimentally, but the theoretical calculations predict a sharp increase [48]. The unsubstituted **CH4** has not been resolved into enantiomers owing to its poor configurational stability at room temperature. **CH5** is the borderline case, which displays partial configurational stability. It can be resolved into enantiomers, but an enantioenriched sample racemizes within a couple of days at room temperature [49]. [n]Helicenes ($n \geq 6$) are configurationally stable at room temperature, and $\Delta G^{\neq}(T)$ increases with the increasing number of ortho-fused rings.

The first mechanistic insight into the interconversion of enantiomers comes from the studies of Stegemeyer et al. on **CH5**. The authors proposed a mechanism involving the formation of a closed intermediate in the position C1 and C14 and consistent with the thermal and photochemical cyclodehydrogenation in the presence of oxidant observed in **CH5** derivatives. Later, Martin et al. hypothesized a possible pathways involving the formation of closed intermediates by intramolecular double Diels–Alder reaction. This mechanism was experimentally verified using a **CH6** labeled in position 1,2,3,4 with deuterium atoms. The double Diels–Alder reaction of (M)-1,2,3,4-d_4-[6]helicene after equilibration did give a 1 : 1 mixture of the starting material and (P)-1,2,13,14-d_4-[6]helicene, involving a single-step mechanism via the conformational transformation in the transition state. The involvement of the conformational pathway also justifies the low energy barrier for smaller helicenes as the steric demand increases with the added number of benzene rings.

17.2.2 Racemic Synthesis and Optical Resolution Methods

The racemic synthesis followed by the optical resolution is a usual practice to obtain optically active helicenes because enantioselective syntheses do not often guarantee high enantiomeric excess.

17.2.2.1 Metal-Free Photocyclization Reactions

The photocyclization of stilbene moieties followed by dehydrogenation, firstly introduced by Martin et al. [50], is one the most important methods for the synthesis of many helicene homologues (**CH5**–**CH14**) and derivatives, because the stilbene precursors are easy to prepare (Figure 17.5).

First reports of photocyclization conducted on substrate **16** exhibited low yields as consequence of a low regioselectivity. In fact, photocyclization can take place at either the 2- or the 3-position of the precursor, leading to a mixture of isomers difficult to separate, as well as parasite reactions between the hydroiodic acid formed in the reaction mixture with the double bond of stilbenes. Such limitations were overcome by Katz and coworkers, who developed an efficient bromine-directed

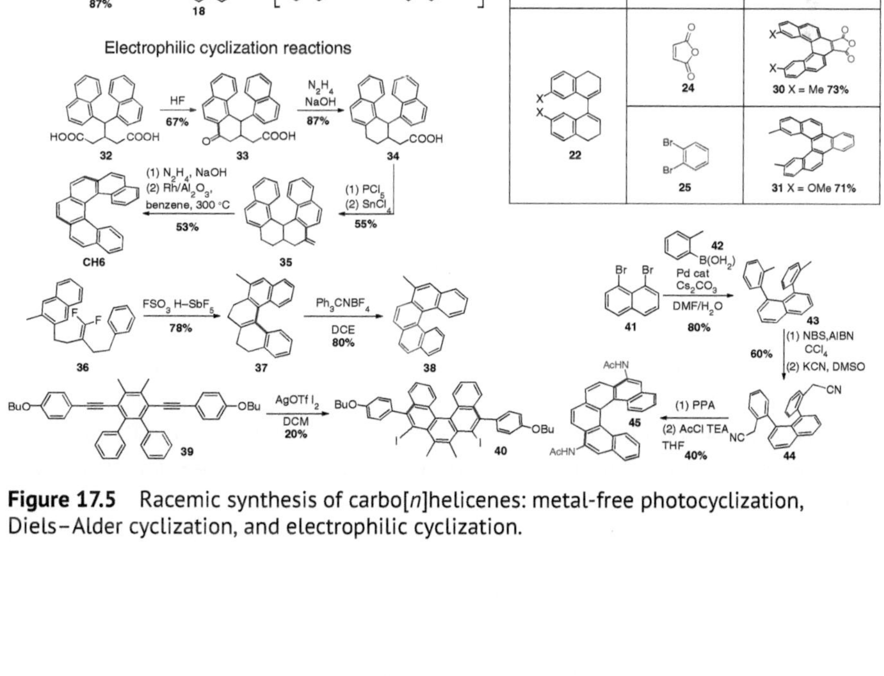

Figure 17.5 Racemic synthesis of carbo[*n*]helicenes: metal-free photocyclization, Diels–Alder cyclization, and electrophilic cyclization.

photocyclization method. It is nowadays considered the standard procedure for the photocyclization of stilbenoid precursors [51]. This protocol makes use of a brominated precursor **17**, which ensures a regioselective cyclization of the 1- to 2-positions, with an excess of propylene oxide in combination with a stoichiometric amount of iodine in an inert atmosphere. Koning and coworkers described another photocyclization reaction starting from precursor **19** for the preparation of **CH5** in good yield [52]. The reaction involves the deprotonation of the methyl group followed by transformation into an enolate, which undergoes photocyclization through isomerization. Photocyclization is difficult to employ in a large-scale preparation and lacks tolerance to some functional groups, such as amino and nitro groups.

17.2.2.2 Diels–Alder Cycloaddition

Diels–Alder synthetic approaches are the breakthrough in a large-scale preparation of helicenes. First report that uses this synthetic approach by Katz and coworkers [53] leads to [5]helicene **26** and **27** in gram scale quantities by a reaction between *p*-divinylbenzene **20** and *p*-benzoquinone **23**.

According to Diels–Alder theory, the presence of strongly electron-donating substituents on the diene and electron-withdrawing groups on the dienophile increases the efficiency of Diels–Alder reactions. Therefore, electron-rich dienes are the best choice for the synthesis of helicenes in combination with electron-poor dienophiles such as *p*-benzoquinone **23**, maleic anhydride **24**, and benzyne precursor **25**. Electron-donating substituents have been introduced in different sites of dienes, and a variety of helicenes (**26**–**31**) have been synthesized by Diels–Alder reactions (see Figure 17.5) [19, 54–57].

17.2.2.3 Friedel–Crafts-Type Cyclization

Friedel–Crafts-type reactions (acylation, alkylation) followed by an aromatization step are other strategies explored for the synthesis of [*n*]helicenes, limited only to symmetric helicenes (see Figure 17.5). Newman and coworkers [58] proposed a double acylation as a key step for the synthesis of **CH6**. Not only classical Friedel–Crafts acylations are employed to construction of helicenes, but also Ichikawa and coworkers described an efficient electrophilic cyclization of **36** followed by an aromatization, using the "magic acid" to generate a carbocation species [59]. Another example of a functionalized [5]helicene derivative is compound **40** prepared by a directed electrophilic cyclization of conjugated ethynyl phenylene **39** with iodonium species [60]. The presence of two *ortho*-methyl groups in **39** is important to facilitate the domino cyclization. They not only direct the reaction to form a helical structure but also enhance the nucleophilicity of the aromatic moieties. A functionalized [6]helicene **45** with acetamido groups was synthesized from electrophilic cyclization with polyphosphoric acid (PPA) of acetonitrile moieties [61]. Precursor **44** was obtained in three synthetic steps that involve a Pd-catalyzed Suzuki–Miyaura cross-coupling on 1,8-dibromonaphthalene **43** followed by a radical bromination of the methyl groups and cyanation.

17.2.2.4 Metal-Catalyzed Reactions

Metal-catalyzed synthetic approaches for the synthesis of helicenes were widely explored in past 20 years. The best studied metal-catalyzed reactions for the synthesis of helicenes are cross-coupling reactions (direct arylation, Stille, and Suzuki protocols), [2+2+2] cycloisomerization of alkynes, and ring-closing olefin metathesis.

A double intramolecular Pd-catalyzed C–H arylation (DHA reactions) was employed in the synthesis of [5]-helicenes **47a–c** (Figure 17.6) [63]. Two methoxy groups direct the cyclization reactions toward the helical framework. This approach has a good tolerance for the presence of electron-withdrawing substituents.

The annulation reaction was developed using stannane and boronates. Scott and Xue reported the use of a double Stille coupling to prepare helicene **51** in good yield using 2,20-dibromo-1,10-binaphthyl **48** (Figure 17.6) [64], while Shimizu and coworkers described a double Suzuki–Miyaura coupling to achieve [5]helicene **53** between bis(pinacolatoboryl)alkenes **52** and binaphthyl compound **48** (Figure 17.6) [62]. Although these three methods are convenient and the precursors can be easily obtained, the products are difficult to further functionalize, and the synthesis is limited to [5]- and [6]helicenes.

Pd-catalyzed intermolecular [2+2+2] cyclotrimerization between alkyne and arynes or between arynes alone was first reported by the Pérez and Guitián groups [65]. Starting from precursors **54a–c**, the authors generated in mild conditions aryne species able to co-trimerize in the presence of $Pd_2(dba)_3$, giving several polycyclic aromatic compounds **55–64** containing multiple helicene structures (Figure 17.6). The regioselectivity of the cyclotrimerization of arynes depends on the structure. For instance, the homotrimerization of arynes of **54b** and **54c** led to the regioselective formation of compounds **60** and **64**, respectively, while those of aryne coming from **54a** gave a regioisomeric mixture of compounds **55** and **56**. Using two equivalents of alkynes, in combination with one equivalent of **54c**, the authors achieved the regioselective synthesis of tetrasubstituted [4]helicene **63** in high yields.

To avoid the formation of mixture of regioisomers, alkynes can be implemented in one substrate to make an intramolecular [2+2+2] cyclization. The first reports on intramolecular [2+2+2] cyclizations by Starý and Stará groups are shown in Figure 17.7. The authors synthesized functionalized tetrahydrohelicenes **67a–c** using two complementary routes starting from easily prepared precursors **66a–c**. *Route a* made use of $CpCo(CO)_2$ complex in combination with PPh_3, decane as solvent at 130 °C under irradiation with a Xe-lamp. *Route b* was performed with $Ni(cod)_2$ complex, PPh_3, in THF at room temperature. Compounds **67a–c** were aromatized with 2,3-dichloro-5,6-dicyano-1,4-benzoquinone (DDQ) or Ph_3CBF_4 affording the parent helicenes **CH5–CH7** [66]. Intramolecular [2+2+2] cyclization was an efficient approach to synthetize helicene **CH11** [67]. This methodology is efficient (100% atom economy, good to excellent yields, fast reaction) but also functional to rapid construction of nonsymmetric and multifunctionalized (tetrahydro)helicenes.

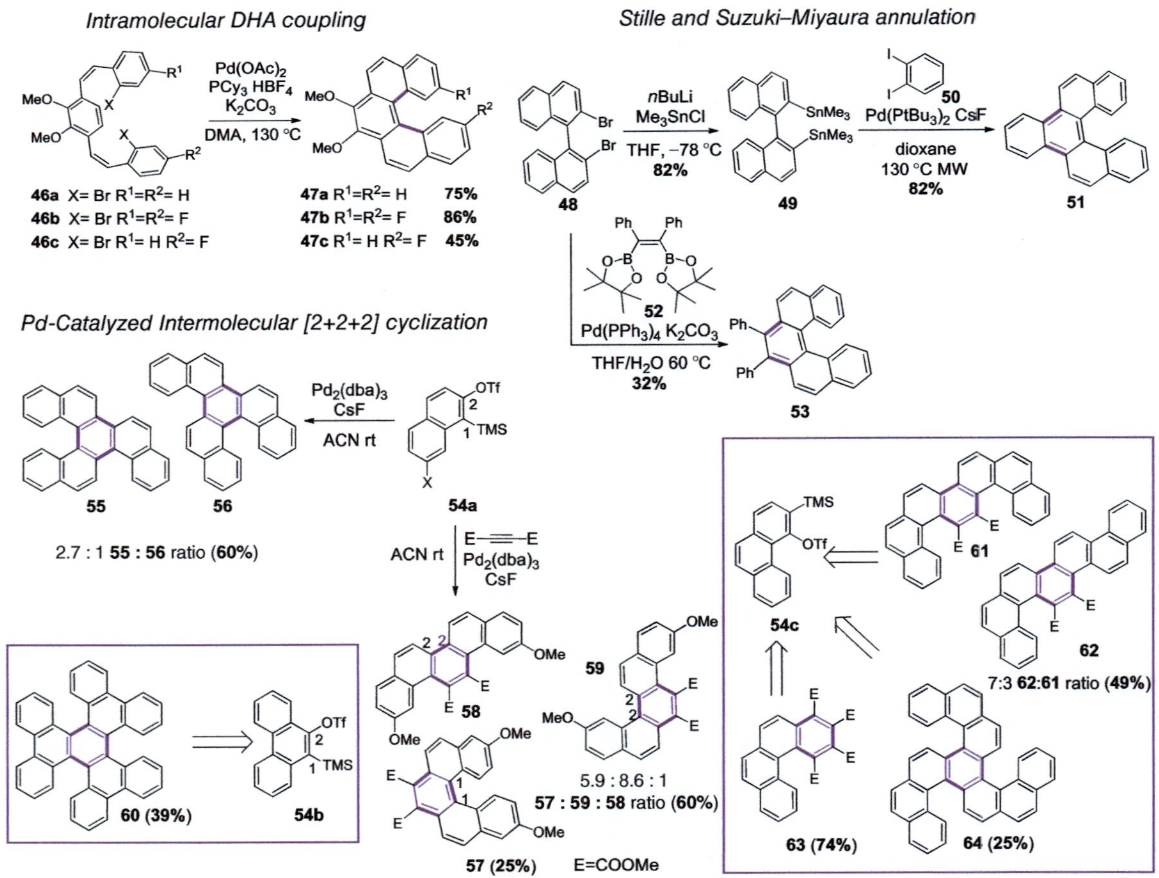

Figure 17.6 Racemic synthesis of carbo[*n*]helicenes through Pd-catalyzed reactions: intramolecular DHA reactions, Stille and Suzuki–Miyaura annulations, and intermolecular [2+2+2] cyclization. Source: Shimizu et al. [62]/American Chemical Society.

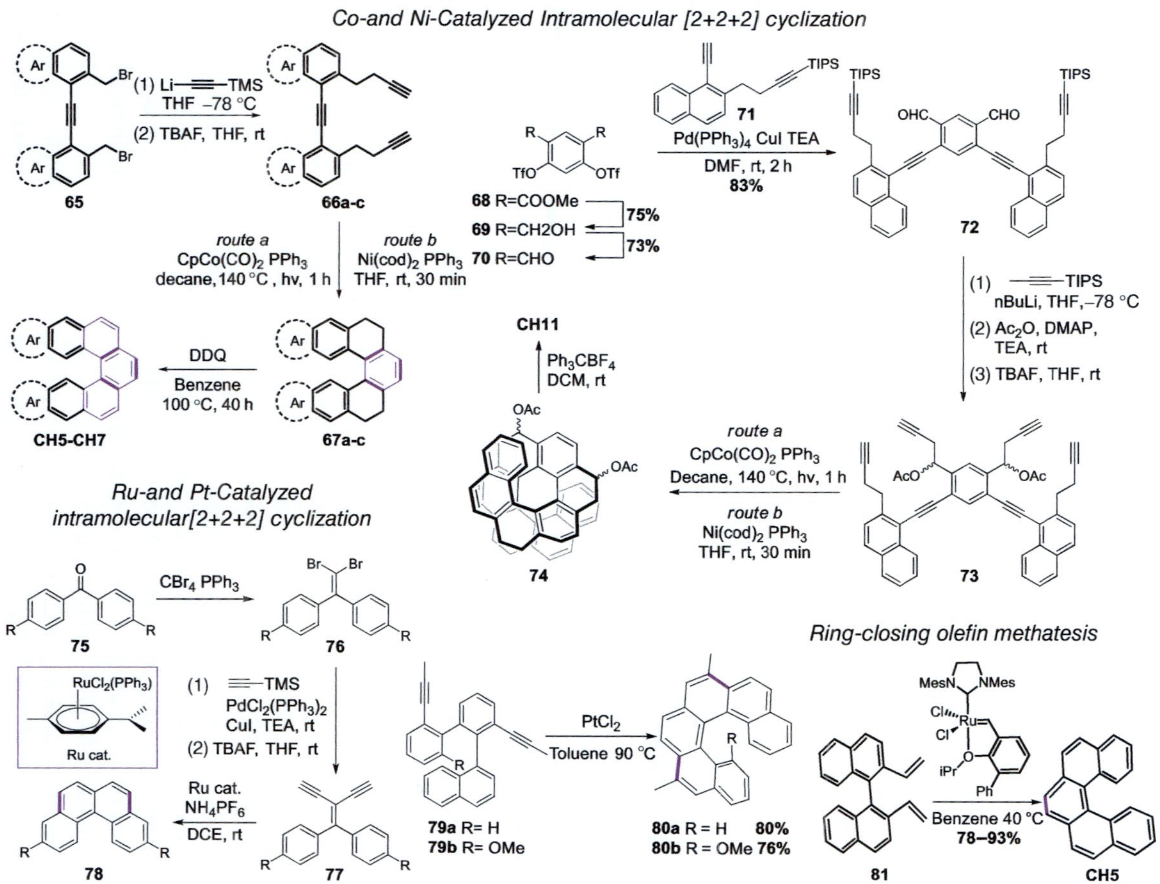

Figure 17.7 Intramolecular Co-, Ni-, Ru-, and Pt-catalyzed [2+2+2] cyclization.

A Ru-catalyzed double cyclization on 1,1-diaryl-2,2-diethynylethylenes **77** was disclosed by Scott and Donovan, and can be applied for the synthesis of longer helicenes [68]. Precursor **77** was easily obtained from aryl ketones **75**. Compound **77** was isomerized to form **78** in the presence of a Ru(II) catalyst (Figure 17.7).

Fürstner's group reported the synthesis of **80a-b** in good yields via double Pt-catalyzed [2+2+2] cyclization of biphenylyl-naphthalene precursor **79a-b** synthesized by a convergent and modular assembly method [69]. These easily prepared building blocks allow the introduction of substituents at the most sterically hindered position of the helical backbone.

The ring-closing olefin metathesis (RCM) approach was explored initially by Collins and co-workers [70]. The authors described the synthesis of several helicenes in mild conditions (40 °C) in 78–93% yield after purification. From readily modified 1,10-binaphthyl frameworks, this method, with good tolerance and high efficiency, was found to be a facile way to prepare substituted [5]-, [6]-, and [7]helicenes. Recently the Matsuda and Hirose groups reported the synthesis of π-expanded helicene (**CH13**), that is, the helically twisted analogue of kekulene, using a sixfold RCM reaction as a key step [71]. Figure 17.7 showed the representative synthesis of helicene **CH5** starting from compound **81**.

17.2.2.5 Optical Resolution

To investigate and utilize optically active helicenes, efficient ways for racemate resolutions are of great significance since they are still the most practical way to obtain pure enantiomers. In fact, helicenes obtained from racemic synthesis were resolved by recrystallization as chiral CT complex, direct HPLC resolution, chiral auxiliary strategy, and kinetic resolution with enzymes.

Charge-transfer (CT) complexation is a non-covalent bond interaction observed between an electron-rich π-compound (donor) and an electron-deficient π-compound (acceptor). When a chiral donor and an acceptor are employed, a difference in affinity in CT complexation appears between the stereoisomers, and separation can be achieved after precipitation of CT complexes. Recrystallization in the presence of a chiral reagent was the principal method employed for optical resolution because it does not suffer from limitations of scale (Figure 17.8). Both (+)- and (−)-TAPA **82** have been widely used to separate helicene enantiomers, because they form charge-transfer complexes in which (*S*)-(+)-TAPA exhibits a stronger interaction with (*M*)-(−)-helicenes and vice versa [72]. Other enantiopure compounds were used for optical resolution of helicenes: alkaloids quinine **83** [73], cinchonidine **84** [74], and brucine **85** [75].

Direct resolution by HPLC is a widely used resolution technique. The first study dates back to 1976 by Mikeš and coworkers [76] that was optimized by Kawazura's group modifying a silica gel column with 11% (S)-(+)-TAPA [77], Since then, a variety of commercial and modified chiral stationary phases have been utilized to separate the enantiomers of different helicenes.

Different *chiral auxiliaries* for optical resolution of helicenes by HPLC or column chromatography have been reported (Figure 17.8). Chiral acid was used for optical

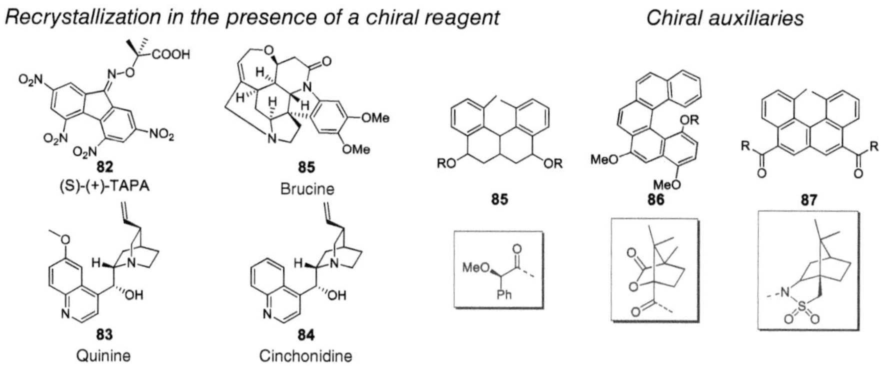

Figure 17.8 Optical resolution strategies. (a) chiral agents used for recrystallization, (b) chiral auxiliaries used for optical resolution of [4]helicenes.

resolution of helicenes containing hydroxyl groups such as (*R*)-**85** [78]. Katz and coworkers utilized (*S*)-camphanoyl chloride to separate **86** [79].

Enzymatic resolutions were also developed as method for optical resolution of helicenes. Liu and Katz reported an enzymatic hydrolysis on hemiketal precursors which afforded the optically active helicene in 62% ee [53]. The key step, kinetic resolution, was employed with bovine pancreas enzyme (Novozyme 435).

17.2.3 Asymmetric Syntheses

Asymmetric synthesis (both catalytic and stoichiometric) is an alternative to racemic synthesis followed by optical resolution. Preparation of enantiopure helicenes on a multigram scale is a challenging task. The first enantioselective synthesis of helicenes was reported by Kagan et al. and later reexamined by Fuchter and Kuimova. The authors reported that circularly polarized light can induce an enantioselective photocyclization of stilbene precursors to give enantioenriched helicenes [80]. For instance, the cyclization of precursor **88** is reported in Figure 17.9. Helicenes have also been prepared from optically pure substrates bearing centers or axes of chirality in the main scaffold. For instance, Gingras et al. described the synthesis of bromohelicene (*M*)-**93** starting from the biaryl precursor **90**, which could be resolved on a large scale. The hydroxy moieties of (−)-**90** were converted to the tosylates and then to methyl groups via Negishi cross-coupling to yield compound (−)-**91**. Finally, enantiopure (*M*)-**93** was obtained with an ee > 97% via benzylic coupling in the presence of *t*-BuOK starting from (−)-**92** (Figure 17.9) [82].

A widely used approach to asymmetric synthesis makes use of chiral auxiliaries. The chiral auxiliaries are incorporated into synthetic routes to control the absolute configuration of the products generated during the reaction. After the diastereoselective transformations, the auxiliary is removed under conditions that do not cause racemization of the desired products. Carreño and Urbano groups developed a method using diastereoselective Diels–Alder reactions using enantiopure dienophile (+)-**95** to achieve the (*P*)-**96** in good yields with high enantioselectivity (ee > 98%) [83].

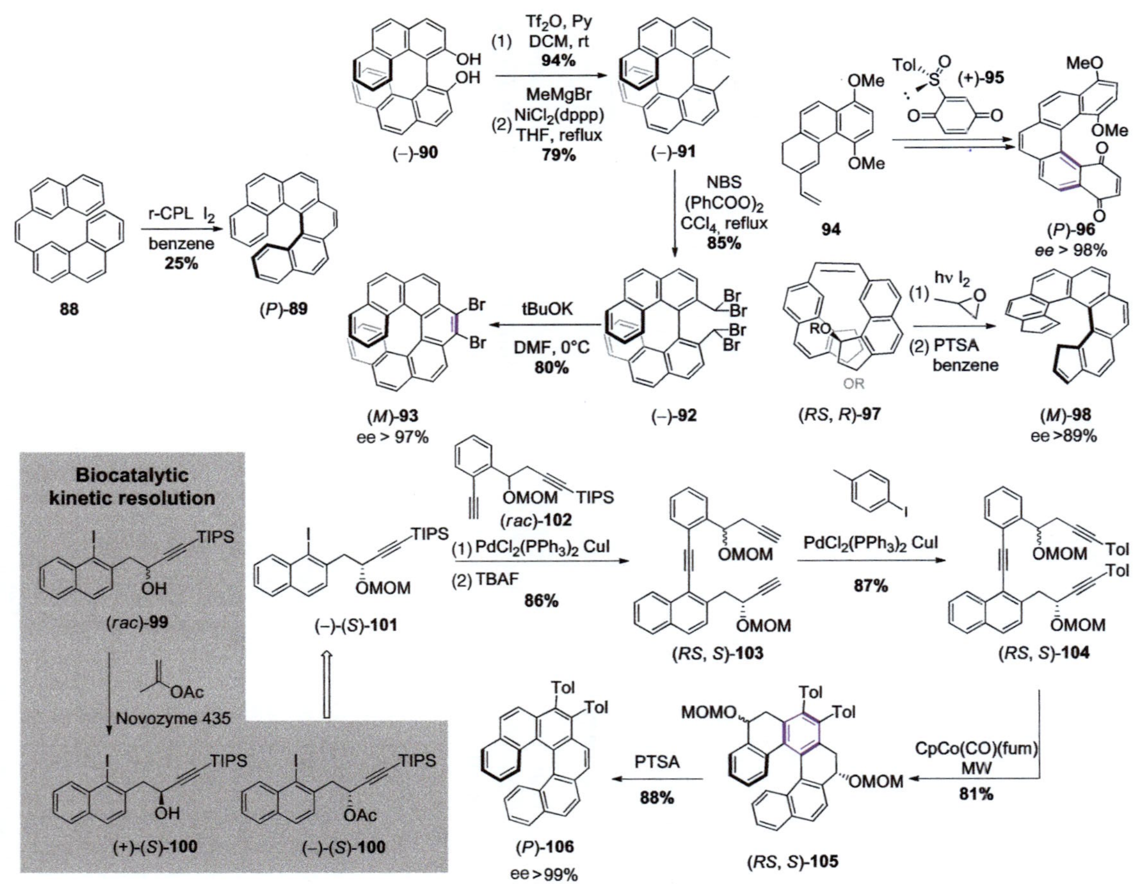

Figure 17.9 Selected examples of asymmetric synthesis from optically pure substrates and with chiral auxiliary approach. Source: Šámal et al. [81]/Royal Society of Chemistry.

Katz and Sudhakar introduced a large silane group on one terminal ring of **97** that directed the photocyclization by chiral induction. After that, (*M*)-**98**, having two cyclopentadiene as terminal aromatic rings, was obtained in good yield and enantioselectivity after treatment with *p*-toluenesulfonic acid (PTSA) [84].

Stará and Starý group developed a general diastereoselective route for the preparation of enantiopure [5]-, [6]-, and [7]helicenes [81]. The authors proposed a tandem [2+2+2] cycloisomerization on precursor (*RS*, *S*)-**104** as the key step; the starting substrate has one of the two stereocenters with a defined chirality, which is able to efficiently transfer the chiral information on the helix in construction, leading to the diastereomerically pure tetrahydrohelicene derivative (*RS*, *S*)-**105**. Finally, the chiral auxiliary is removed in the final aromatization step, leading to the (*P*)-**106** with ee > 99%. Importantly, the key building blocks such as (+)-(*R*)- or (−)-(*S*)-**101** can be synthetized optically pure through enantioselective biocatalysis on a multigram scale (grey box in Figure 17.9).

When the synthesis of chiral precursor is difficult, asymmetric catalytic strategies are preferred. These methods make use of chiral metal centers, obtained using chiral ligands, as chiral inductors in the final product. The catalyst load is reduced when compared with the stoichiometric amount of chiral precursor required in the case of the chiral auxiliaries.

A plethora of efficient enantioselective catalytic [2+2+2] cycloisomerizations employing transition metal complexes (Co, Ni, Rh, Ru, Pd, Au, Pt) to form manifold helicene scaffolds has been described in the literature. Chiral Co^{I} complexes were employed to synthetize nonracemic tetrahydrohelicenes, but the chirality transfer is only moderate (25% ee) [85]. The use of Ni^{0} catalysis in combination with chiral phosphorus or *N*-heterocyclic carbene (NHC) ligands led to much better enantiodiscrimination such as a **CH6** with ee > 99% [86]. Alternatively, Rh^{I} and Ir^{I} catalysts with chiral bidentate diphosphines were developed. For instance, using a chiral Rh^{I} complex, enantiomeric excess of 93% was obtained [87]. To become more practical, enantioselective [2+2+2] cycloisomerization of π-electron systems requires the development of more efficient and easily accessible chiral ligands (preferably for Ni^{0} or Rh^{I}) providing a wide range of helicenes in high ee's. Metathesis has also been used by Collins and Grandbois to achieve the enantioselective synthesis of a [7]helicene in high ee [88].

17.2.4 Application of Chiral Carbo[*n*]helicenes

The development of protocols for obtaining helicenes in a gram scale as well as with high optical purity has opened the way for their use in a plethora of applications ranging from catalysis to electronics and from sensing to molecular switches.

17.2.4.1 Helicenes in Catalysis

Helicene derivatives have been applied as chiral ligands in catalytic asymmetric reactions. First report by Reetz group in 1997 described the synthesis of compound **107** and its use in enantioselective Rh^{+}-catalyzed hydrogenation of itaconic ester **110** to afford product **111** in 57% yield and 39% ee (Figure 17.10) [89]. Later, using

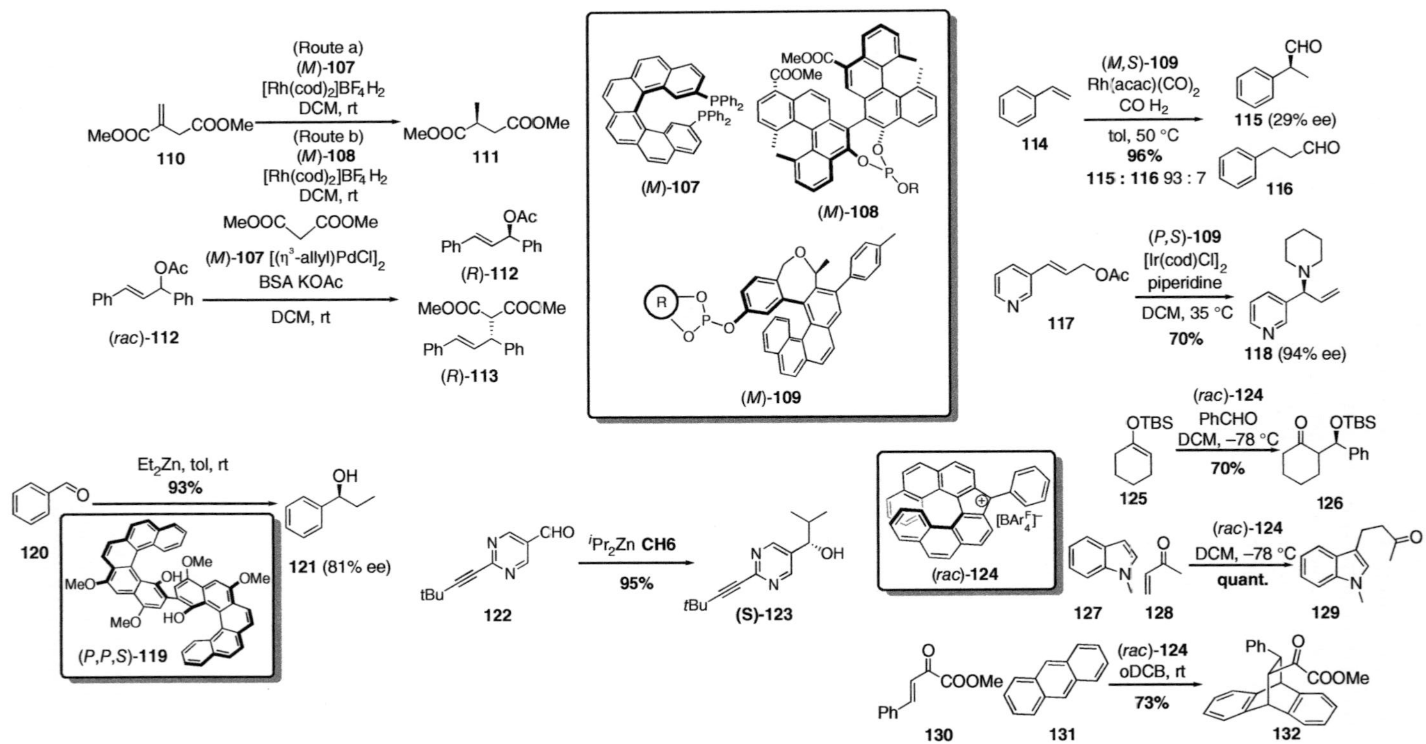

Figure 17.10 Helicenes in catalysis.

helicene-based analogue of BINOL **108**, Nakano and Yamaguchi reported the same transformation with a quantitative yield and 96% ee [90]. Bis-phosphine **107** was efficiently used for the kinetic resolution of **112** in Pd^{2+}-mediated substitution with dimethyl malonate by Reetz and Sostmann [91]. The kinetic resolution of compound **112** was excellent in terms of yield and ee. According to this study, **107** acts as a monodentate ligand, because the distance between the two phosphorus atoms is mismatched with coordination of the Pd^{2+} cation in chelating manner. Stará and coworkers reported the synthesis of helicene phosphates and their application in Rh-catalyzed hydroformylation and Ir-catalyzed allylic amination [92]. The asymmetric allylic amination in the presence of (*P*,*S*)-**109** as a catalyst was achieved with up to 94% ee and excellent regioselectivity (Figure 17.10). A bis-helicene **119** in which two **CH5** subunits established an axial chiral analogue to those of BINOL and had two hydroxyl groups able to coordinate a Zn^{2+} cation was prepared by Katz and coworkers [93]. This bidentate ligand was used in the addition of Et_2Zn reagent to aldehydes, leading (in a model reaction) to the corresponding alcohol in 93% yield and 81% ee.

The use of helicene in catalysis is not limited to helicene-based chiral ligands for transition metals. Gross and Oestreich reported a trityl cation embedded into a **CH7**-like backbone as Lewis acid organocatalyst **124** that was able to promote Mukaiyama aldol condensation and Diels–Alder and Friedel–Crafts reactions with indole **127** [94]. The nonfunctionalized **CH5** and **CH6** were utilized by Soai and coworkers as chiral inducers for the enantioselective addition of $^{i}Pr_2Zn$ reagent to pyrimidine-based carbaldehyde **122** [95]. As shown in Figure 17.10, the (*P*)-**CH6** can transfer its chiral information in the transition state of transformation to yield secondary alcohol (*S*)-**123** with excellent enantioselectivity. Authors explained the high ee values and the absolute configurations of **123** by a mechanism in which **CH6** induced asymmetry in the initially formed zinc alkoxide of **123**, followed by an autocatalytic asymmetric amplification by the zinc alkoxide of **123**. The absolute configuration of **123** was controlled by the *P* or *M* helicity of helicenes.

17.2.4.2 Helicenes in Organic Electronics

Due to the good thermal and chemical stability and strong and easy tunable fluorescence or phosphorescence, helicenes have found a wide use in organic light-emitting diodes (OLEDs), organic field-effect transistors (OFETs), and organic photovoltaics (OPVs).

Sooksimuang and coworkers reported an OLED with the configuration using helicene **133** as the active layer (Figure 17.11). Helicene **133** exhibits a good thermal stability and no glass transition temperature that ensures high lifetime of devices and strong greenish-blue emission with a PLQY of 27% [96]. Moorthy and coworkers developed helicene **134**, a **CH5** functionalized with fluorescent phenylanthryl moieties, as blue-emissive materials in non-doped OLED devices [97]. Authors showed that **134** displayed good fluorescence properties (PLQY of 27%) and thermal stabilities in analogy with observations of Sooksimuang group. In devices, a dual

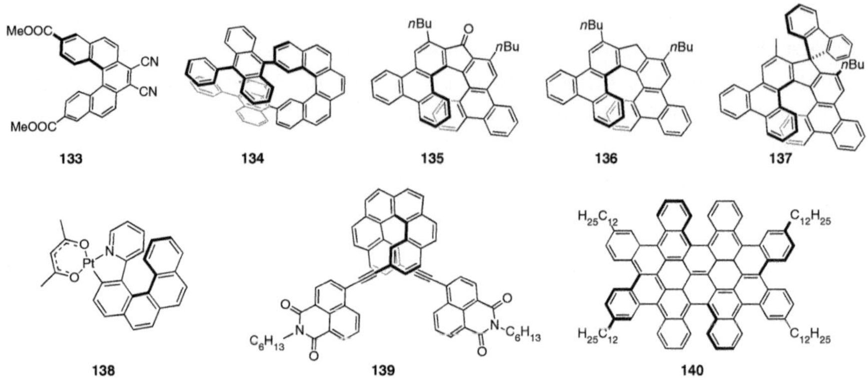

Figure 17.11 Helicenes used for organic electronics.

use of **134** serving as both excellent deep-blue emissive material and host material was described.

Due to their helical chirality and rigid conjugated structures, helicenes can be applied as chiral luminophores. In fact, suitably functionalized optically pure helicenes show circularly polarized luminescence (CPL) phenomenon that can be used for construction of CP-OLED devices. For instance, Tanaka and coworkers applied an asymmetric Rh-catalyzed [2+2+2] cycloaddition for the preparation of the optically active helicenes **135–137** with a **CH7** scaffold bearing both triphenylene and fluorene cores [87]. All compounds were obtained in several synthetic steps in 63–71% yield and 91–93% ee. Helicene **135** showed large redshifts of the absorption and emission maxima as compared with those of **136** and **137**, while **136** and **137** presented higher PLQY (32.0% and 29.6%, respectively) than **135** (2%). All helicenes shown high g_{lum} value of −0.032. CP-phosporescent-OLEDs (CP-PhOLEDs) were fabricated using helicene scaffold as chiral moiety. The main advantage of CP-PhOLED devices is their intrinsic ability to emit light from triplet excited states, achieving therefore theoretically higher efficiencies than CP-OLED. Fuchter and coworkers developed a platinahelicene complex **138** to construct a CP-PhOLED that achieved both a display level brightness and a high g_{EL} factor of 0.38 [98].

OPVs and OFETs were another field of application for helicenes. For instance, preparation of helicenes end-capped with naphthalimide and their application as non-fullerene acceptors in OPV devices was investigated by Crassous and coworkers [99]. Impact of the enantiopurity on device performance was investigated through the synthesis of racemic and enantiomerically pure **139**. Authors have recorded a very strong increase of the device performance (PCE), from 0.4% to 2.0%, switching from the racemic to the enantiopure **139** when embedded with poly(3-hexylthiophene) (P3HT) as donor material.

Nuckolls and coworkers prepared highly contorted graphene-like helicene **140**, bearing four [4]helicene and two [5]helicene units, and examined the electronic properties in OFET device [100]. Compound **140** was found to be a hole transporting semiconductor with the mobility of 0.002 $cm^2\ V^{-1}s^{-1}$.

17.2.4.3 Helicenes in Biochemistry

The applications of helicenes in biochemistry include their interactions with biomolecules and cell imaging. It was found that the helicene amine shows the chiral recognition toward DNA/nucleosides [101]. Studies on the selective binding of helicenes to Z-DNA and B-DNA suggested that the amino groups on helicene skeletons are essential for the interaction with Z-DNA [102]. Moreover, the investigation on the interactions between spermine-functionalized optically pure [5]helicene and B- and Z-DNA indicated that (*P*)-isomer showed a strong affinity to B-DNA, while (*M*)-isomer displayed a preference for Z-DNA [103].

Because of the π-conjugated helical structures, helicenes can be also applied to cell imaging. A series of tetrahydrohelicene nanoparticles for fluorescence cell imaging shows an interesting structure-dependent distribution, which provides a new perspective for the design of fluorescence probes [104–106].

17.3 Twistacenes

Twisted acenes, called twistacenes, were first introduced by Pascal [4]. While the parent acenes are planar, they are twisted out of planarity upon substitution, and the resulting reduction of symmetry can yield helical chirality. They tend to be more stable and soluble compared with their parent acenes, and twisting does not drastically inhibit π-conjugation [107–111].

Most twistacenes are fluctional (Figure 17.12, top). The twisting introduces *M* and *P* helicities, and the molecules in fact exist as racemic mixtures, since they readily racemize at room temperature. The fast equilibrium hinders efforts to obtain enantiomerically pure twistacenes. The first example in which the chirality in twistacenes has been addressed and compounds resolved into their two enantiomers was reported by Gidron and coworkers [112]. They synthesized and characterized a series of twistacenes **107** having an anthracene backbone diagonally tethered by alkyl bridges of differing lengths, from 3 to 6 carbon atoms (Figure 17.12, top). The bridge induces a twist of various angles, and the authors were able to systematically monitor the effect of twisting on electronic and optical properties. The absorption is bathochromically shifted with increasing twist, while fluorescence quantum efficiency drops dramatically (from 0.3 to 0.07). The authors were able to isolate the tethered twistacenes in enantiomerically pure form using chiral HPLC. No racemization was observed even upon prolonged heating. CD spectroscopy (Figure 17.12, top right) showed mirror imaged spectra and a strong anisotropy factor.

In a second recent example, Li et al. synthesized a decatwistacene by bottom-up synthesis using palladium-catalyzed Suzuki cross-coupling and C–H activation. The twisting is induced by steric hindrance between imide groups and neighboring annulated benzene rings. Decatwistacene **108** (Figure 17.12, bottom) was obtained as a purple-black solid, and it has a high solubility in common organic solvents, including dichloromethane, toluene, and tetrahydrofuran; this solubility is directly related to the twisting deformation.

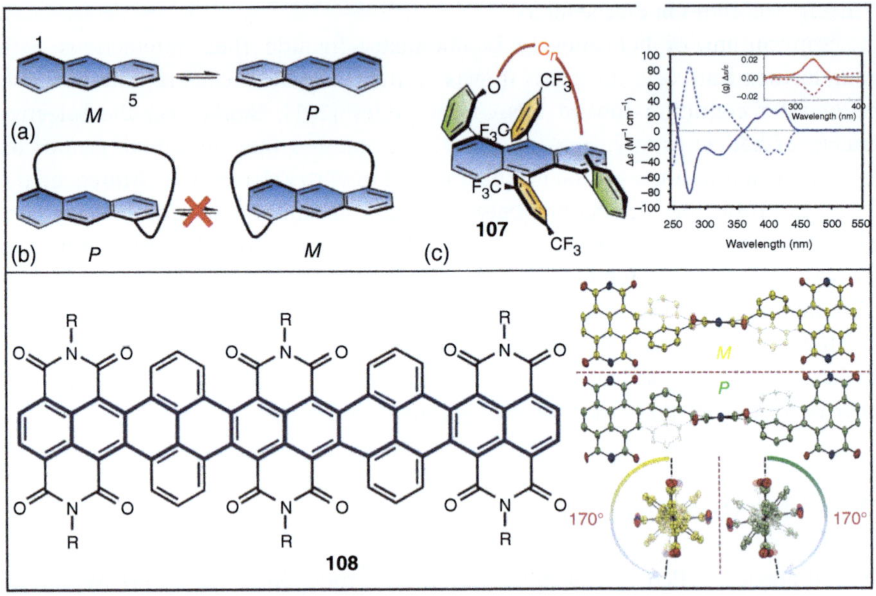

Figure 17.12 Top: Chiral twistacenes. Source: Adapted with permission from Bedi et al. [112] (copyright 2018 American Chemical Society). Bottom: Decatwistacene and its crystal structure [113].

Single-crystal X-ray analyses revealed that decatwistacene, which is the longest twistacene reported, exhibits an astonishing overall end-to-end torsion angle of about 170°, the largest torsion angle reported. Twistacene **108** has an enhanced stability with respect to light and oxygen. Owing to its large twisting deformation and the LUMO levels caused by the introduction of imide groups, the compound bears a great potential for the construction of chiral graphene nanoribbons with good stability and processability [113].

17.4 Chiral Nanobelts

Even before the discovery of fullerene and carbon nanotubes, the structures of aromatic hydrocarbon nanobelts (double-stranded and belt-shaped molecules consisting of fully unsaturated benzenoid rings) have been proposed and discussed in the chemical literature. Heilbronner was the first to propose cycloacenes as molecules for theoretical investigations. After him, Vogtle proposed an armchair belt as a synthetic target, which became known as the Vogtle belt [114]. A few years later, [12]collarenes (potential precursors to cycloacenes) were synthesized, but the actual synthesis and isolation of carbon nanobelts remained an elusive goal, despite numerous efforts. After a long period of time (2008), Jasti et al. reported the first synthesis of a cycloparaphenylene (a carbon nanoring) revitalizing interest in approaches to achieve the synthesis of curved and conjugated benzenoid rings [115].

In 2017, Itami reported the sensational synthesis of the first fully unsaturated armchair carbon nanobelt (other carbon nanobelts were reported by the same group using the same strategy the following year), while, in 2019, Miao reported the first synthesis of a chiral carbon nanobelt representing a horizontal sidewall segment of a chiral (18,12)carbon nanotube [116, 117]. In 2020, Wang observed for the first time the formation of a cycloacene by MALDI mass spectrum, and in 2021, finally Chi and Itami independently reported the synthesis of the first zigzag hydrocarbon nanobelts through an iterative sequence of diastereoselective Diels–Alder reactions followed by reductive aromatization [118–120].

The synthesis of the only chiral nanobelt **116** (Figure 17.13) reported to date starts from **109** with the nucleophilic addition of 2-bromo-6-lithionaphthalene to 2,5-di(benzyloxy)-1,4-benzoquinone to yield the *syn* diol, which was subsequently methylated to give compound **110**. Palladium-mediated coupling of **110** with bis(pinacolate)diboron provided the pinacol boronate derivative **111**. Subsequent macrocyclization was achieved by Suzuki coupling of **111** with 2,5-diaryl-1,4-dibromobenzene **112** and after reductive aromatization the trimeric macrocycle **113** was obtained. The success of this macrocyclization relies on the directing angle (74°) of the derivative **111**. After further chemical elaborations, the polyarylated macrocycle **115** was obtained by another Suzuki coupling with (3,4-dopropoxyphenyl)boronic acid (83%). After some tests, the best conditions for the Scholl reaction of **115** were obtained by treatment with DDQ/CH_3SO_3H/Sc(OTf)$_3$ in toluene refluxing to afford nanobelt **116** with a yield of 66%. This yield corresponds to a 96.6% yield for the formation of each individual C—C bond. The success of the Scholl reaction in synthesizing the chiral nanobelt can be attributed to the presence of the strongly electron donating propyloxy groups, which lower the oxidation potential and act as directing groups. The molecular structure of chiral nanobelt **116** was determined by NMR spectra and high-resolution mass spectroscopy. The nanobelt **116** showed strong blue luminescence (464 and 492 nm, quantum yield 46%) upon UV irradiation in CH_2Cl_2, as shown together with the UV–vis spectrum (Figure 17.13). The nanobelt solutions were stable when kept in dark for 14 days, while by exposure to ambient light they showed significant changes in the UV–vis spectrum. The resolution of the structure could not be obtained by X-ray diffraction, and the structure was observed by scanning tunneling microscopy (STM).

Many of the properties of single-walled carbon nanotubes (SWCNTs), such as electrical conductivity and band gap, are strongly dependent on the chiral indices (*n*,*m*), which determine the diameter and chirality of SWCNTs. The chirality-controlled synthesis of SWCNTs is a formidable challenge in nanocarbon science. Furthermore, although important advances have been recently achieved in chirality-controlled synthesis of SWCNTs, such as the selective growth of (12,6)SWCNTs with abundance greater than 92% using solid alloy catalysts, organic synthesis has been proposed as among the most promising bottom-up approaches to achieve such a goal. Nanocarbon fragments (including chiral hydrocarbon nanobelts) can potentially be used as templates for the molecular-controlled synthesis of SWCNTs with potential repercussions in many areas of science [121]. In

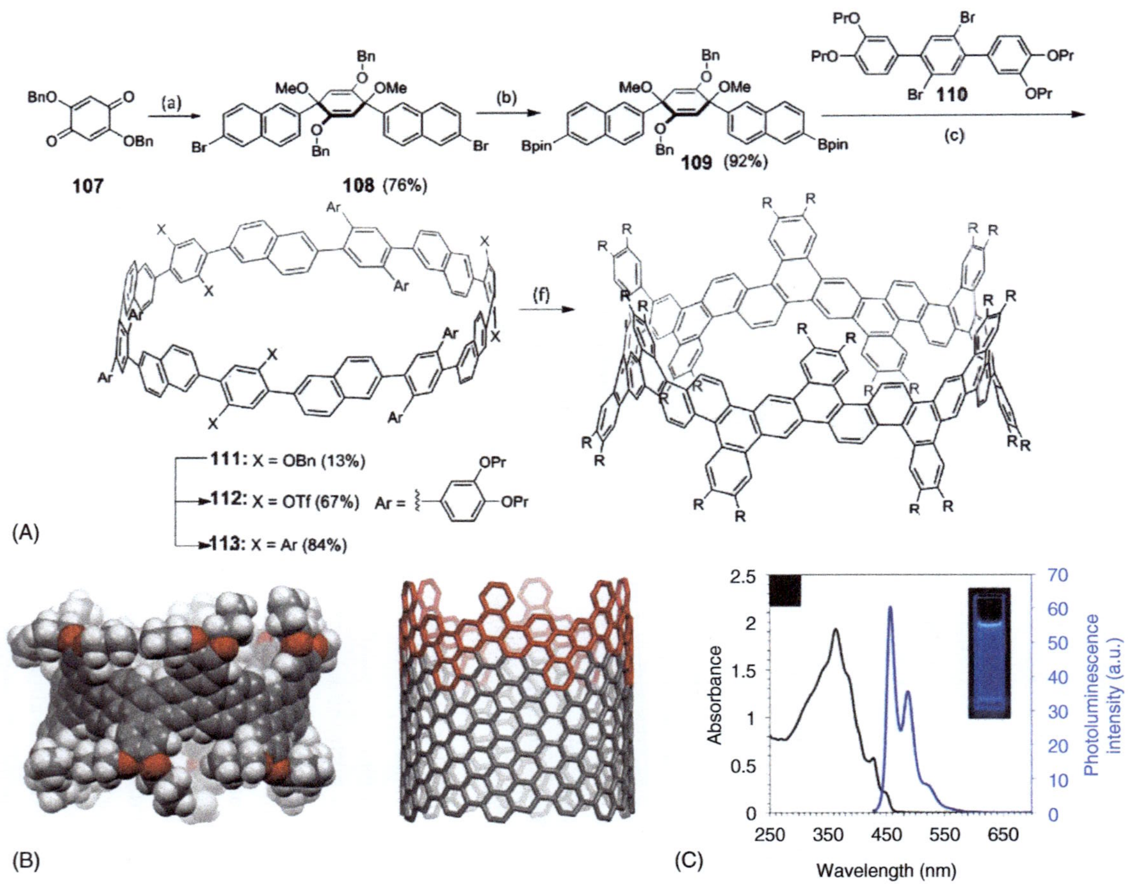

Figure 17.13 (A) synthesis of **116**. Reagents and conditions: (a) (i) 2,6-dibromonaphthalene, BuLi, THF, −84 °C, and (ii) iodomethane, sodium hydride, THF, room temperature (RT). (b) Bis(pinacolato)diboron, $Pd_2(dba)_3$, $PCy_3{\cdot}HBF_4$, KOAc, dioxane, reflux. (c) (i) $Pd_2(dba)_3$, SPhos, K_2CO_3, Bu_4NBr, toluene/H_2O, 80 °C, and (ii) H_2SnCl_4, THF, RT. (d) (i) BCl_3, 1,2,4,5-tetramethylbenzene, CH_2Cl_2, 84 °C, and (ii) pyridine, Tf_2O, CH_2Cl_2, RT. (e) (3,4-Dipropoxyphenyl)boronic acid, $Pd(PPh_3)_4$, $K_3PO_4{\cdot}1.5\ H_2O$, dioxane, reflux. (f) DDQ, $Sc(OTf)_3$, toluene, reflux. (B) Polycyclic backbones of CNBs **116** (shown in red) mapped on the models of (18,12)CNTs and UV–vis and PL spectra of **116**. Source: Adapted with permission from Shi et al. [120].

addition, the synthesis of chiral cycloacenes will serve to achieve molecular-defined nanocarbons with nonambiguous spectroscopic properties (instead of compound mixtures obtained by the classical methods currently employed, see chemical vapor deposition) going to refine the current knowledge on aromaticity, electron delocalization, and molecular strain. They will find application in physics and material science, and also their unusual host–guest properties will be extensively studied.

17.5 Concluding Remarks

Chiral acenes represent a relatively recent addition to the pool of unnatural, asymmetrical building blocks. In this chapter, we have shown that the possibilities to introduce chirality into acenes can be very diverse, giving rise to very different classes of compounds (helicenes, twistacenes, carbon nanobelts). All these classes possess enormous potential for a variety of applications. We have briefly discussed some of those, focusing on the use of helicenes in catalysis, in organic electronics, and in biochemistry.

References

1 Anthony, J.E. (2008). The larger acenes: versatile organic semiconductors. *Angew. Chem. Int. Ed.* 47: 452–483.

2 Ito, H., Ozaki, K., and Itami, K. (2017). Annulative π-extension (APEX): rapid access to fused arenes, heteroarenes, and nanographenes. *Angew. Chem. Int. Ed.* 56: 11144–11164.

3 Wang, C., Dong, H., Hu, W. et al. (2006). Semiconducting π-conjugated systems in field-effect transistors: a material Odyssey of organic electronics. *Chem. Rev.* 112: 2208–2267.

4 Pascal, R.A. (2006). Twisted acenes. *Chem. Rev.* 106: 4809–4819.

5 Preda, G., Nitti, A., and Pasini, D. (2020). Chiral triptycenes in supramolecular and materials chemistry. *ChemistryOpen* 9: 719–727.

6 Pasini, D. and Ricci, M. (2007). Macrocycles as precursors for organic nanotubes. *Curr. Org. Synth.* 4: 59–80.

7 Mislow, K. (1999). Molecular chirality. In: *Topics in Stereochemistry*, vol. 22 (ed. S.E. Denmark) 1–82. Hoboken: Wiley.

8 Gingras, M. (2013). One hundred years of helicene chemistry. Part 1: non-stereoselective syntheses of carbohelicenes. *Chem. Soc. Rev.* 42: 968–1006.

9 Gingras, M., Felix, G., and Peresutti, R. (2013). One hundred years of helicene chemistry. Part 2: stereoselective syntheses and chiral separations of carbohelicenes. *Chem. Soc. Rev.* 42: 1007–1050.

10 Rickhaus, M., Mayor, M., and Jurıcek, M. (2017). Chirality in curved polyaromatic systems. *Chem. Soc. Rev.* 46: 1643–1660.

11 Rickhaus, M., Mayor, M., and Jurıcek, M. (2016). Strain-induced helical chirality in polyaromatic systems. *Chem. Soc. Rev.* 45: 1542–1556.

12 Schmuck, C. (2003). Molecules with helical structure: how to build a molecular spiral staircase. *Angew. Chem. Int. Ed.* 42: 2448–2452.

13 McNaught, A.D. and Wilkinson, A. (1997). *Compendium of Chemical Terminology IUPAC Recommendations Gold Book*. Blackwell Science. Inc. https://goldbook.iupac.org.

14 Newman, M.S., Lutz, W.B., and Lednicer, D. (1955). A new reagent for resolution by complex formation; the resolution of phenanthro-[3,4-*c*] phenanthrene [5]. *J. Am. Chem. Soc.* 77: 3420–3421.

15 Wynberg, H. (1971). Some observations on the chemical, photochemical, and spectral properties of thiophenes. *Acc. Chem. Res.* 4: 65–73.

16 Martin, R.H. (1974). The helicenes. *Angew. Chem. Int. Ed.* 13: 649–660.

17 Laarhoven, W.H. and Prinsen, W.J.C. (1984). Carbohelicenes and heterohelicenes. *Top. Curr. Chem.* 125: 63–130.

18 Liberko, C.A., Miller, L.L., Katz, T.J., and Liu, L.B. (1993). The electronic structure of helicene-bisquinone anion radicals. *J. Am. Chem. Soc.* 115: 2478–2483.

19 Paruch, K., Katz, T.J., Incarvito, C. et al. (2000). First Friedel–Crafts diacylation of a phenanthrene as the basis for an efficient synthesis of nonracemic [7]helicenes. *J. Organomet. Chem.* 65: 7602–7608.

20 Katz, T.J. (2000). Syntheses of functionalized and aggregating helical conjugated molecules. *Angew. Chem. Int. Ed.* 39: 1921–1923.

21 Urbano, A. (2003). Recent developments in the synthesis of helicene-like molecules. *Angew. Chem. Int. Ed.* 42: 3986–3989.

22 Hoffmann, N. (2008). Photochemical reactions as key steps in organic synthesis. *Chem. Rev.* 108: 1052–1103.

23 Jørgensen, K.B. (2010). Photochemical oxidative cyclisation of stilbenes and stilbenoids – the Mallory-reaction. *Molecules* 15: 4334–4358.

24 Dou, G.-L. and Shi, D.-Q. (2011). Advances in the synthesis of helicenes. *Chin. J. Org. Chem.* 31: 1989–1996.

25 Shen, Y. and Chen, C.-F. (2012). Helicenes: synthesis and applications. *Chem. Rev.* 112: 1463–1535.

26 OuYang, J. and Crassous, J. (2018). Chiral multifunctional molecules based on organometallic helicenes: recent advances. *Coord. Chem. Rev.* 376: 533–547.

27 Lin, W.-B., Fang, L., Chen, C.-F. et al. (2018). Recent progress on multidimensional construction of helicenes. *Chin. Chem. Lett.* 29: 40–46.

28 Fernández-García, J.M., Martín, N. et al. (2019). Chiral molecular carbon nanostructures. *Acc. Chem. Res.* 52: 1565–1574.

29 Zhao, W.-L., Chen, C.-F. et al. (2019). Advances in helicene derivatives with circularly polarized luminescence. *Chem. Commun.* 55: 13793–13803.

30 Stará, I.G. and Starý, I. (2020). Helically chiral aromatics: the synthesis of helicenes by [2+2+2] cycloisomerization of π-electron systems. *Acc. Chem. Res.* 53: 144–158.

31 Jakubec, M. and Storch, J. (2020). Recent advances in functionalizations of helicene backbone. *J. Organomet. Chem.* 85: 13415–13428.

32 Ravat, P. (2021). Carbo[*n*]helicenes restricted to enantiomerize: an insight into the design process of configurationally stable functional chiral PAHs. *Chem. Eur. J.* 27: 3957–3967.

33 Mori, T. (2021). Chiroptical properties of symmetric double, triple, and multiple helicenes. *Chem. Rev.* 121: 2373–2412.

34 Chen, C.-F. and Shen, Y. (2018). *Helicene Chemistry*. Springer.

35 Kuroda, R. (1982). Crystal and molecular structure of [5]helicene: crystal packing modes. *J. Chem. Soc., Perkin Trans. 2* 7: 789–794.

36 Yamamoto, K., Usui, K. et al. (2013). Synthesis of [5]helicenes with a substituent exclusively on the interior side of the helix by metal-catalyzed cycloisomerization. *Org. Lett.* 15: 1806–1809.

37 Schulman, J.M. and Disch, R.L. (1999). Aromatic character of [*n*]helicenes and [*n*]phenacenes. *J. Phys. Chem. A* 103: 6669–6672.

38 Chen, Z., von Ragué Schleyer, P. et al. (2005). Nucleus-independent chemical shifts (NICS) as an aromaticity criterion. *Chem. Rev.* 105: 3842–3888.

39 Cruz, C.M., Campaña, A.G. et al. (2018). Undeca-benzo[7]superhelicene: a helical nanographene ribbon as a circularly polarized luminescence emitter. *Angew. Chem. Int. Ed.* 57: 14782–14786.

40 Nakakuki, Y., Matsuda, K. et al. (2018). Hexa-peri-hexabenzo[7]helicene: homogeneously π-extended helicene as a primary substructure of helically twisted chiral graphenes. *J. Am. Chem. Soc.* 140: 4317–4326.

41 Fujikawa, T., Itami, K., Scott, L.T. et al. (2016). Corannulene–helicene hybrids: chiral π-systems comprising both bowl and helical motifs. *Org. Lett.* 18: 3992–3995.

42 Evans, P.J., Martín, N. et al. (2018). Synthesis of a helical bilayer nanographene. *Angew. Chem. Int. Ed.* 57: 6774–6779.

43 Ma, J., Feng, X. et al. (2020). Helical nanographenes containing an azulene unit: synthesis, crystal structures, and properties. *Angew. Chem. Int. Ed.* 59: 5637–5642.

44 Selected examples: (a)Dougherty, K.J., Pascal, R.A. et al. (2015). Helical mesobenzanthrones: a class of highly luminescent helicenes. *Tetrahedron* 71: 1694–1699. (b) Bock, H., Durola, F. et al. (2014). Helicenes from diarylmaleimides. *Org. Lett.* 16: 1546–1549. (c) Li, Y.-Y., Chen, C.-F. et al. (2014). Dihydroindeno[2,1-*c*]fluorene-based imide dyes: synthesis, structures, photophysical and electrochemical properties. *J. Organomet. Chem.* 79: 2139–2147.

45 Saal, F., Ravat, P. et al. (2020). [*n*]Helicene diimides ($n = 5$, 6, and 7): through-bond versus through-space conjugation. *J. Am. Chem. Soc.* 142: 21298–21303.

46 (a) Schuster, N.J., Nuckolls, C. et al. (2020). The structural origins of intense circular dichroism in a waggling helicene nanoribbon. *J. Am. Chem. Soc.* 142: 7066–7074. (b) Xiao, X., Nuckolls, C. et al. (2021). Chirality amplified: long, discrete helicene nanoribbons. *J. Am. Chem. Soc.* 143: 983–991.

47 Nakai, Y., Mori, T., and Inoue, Y. (2012). Theoretical and experimental studies on circular dichroism of carbo[*n*]helicenes. *J. Phys. Chem. A* 116: 7372–7385.

48 Barroso, J., Cabellos, J.L., Pan, S. et al. (2018). Revisiting the racemization mechanism of helicenes. *Chem. Commun.* 54: 188–191.

49 Goedicke, C. and Stegemeyer, H. (1970). Resolution and racemization of pentahelicene. *Tetrahedron Lett.* 11: 937–940.

50 Flammang, M., Nasielsk, J., and Martin, R.H. (1967). Synthesis of heptahelicene (1) benzo [*c*] phenanthro [4,3-g]phenanthrene. *Tetrahedron Lett.* 8: 743–744.

51 Liu, L.B. and Katz, T.J. (1991). Bromine auxiliaries in photosyntheses of [5]helicenes. *Tetrahedron Lett.* 32: 6831–6834.

52 Pathak, R., de Koning, C.B. et al. (2004). The synthesis of angularly fused polyaromatic compounds by using a light-assisted, base-mediated cyclization reaction. *Org. Biomol. Chem.* 2: 3504–3508.

53 Liu, L.B. and Katz, T.J. (1990). Simple preparation of a helical quinone. *Tetrahedron Lett.* 31: 3983–3986.

54 Willmore, N.D., Hoic, D.A., and Katz, T.J. (1994). Diels–Alder reactions of.Alpha.-substituted styrenes with p-benzoquinone. *J. Organomet. Chem.* 59: 1889–1891.

55 Fox, J.M., Goldberg, N.R., and Katz, T.J. (1998). Efficient synthesis of functionalized [7]helicenes. *J. Organomet. Chem.* 63: 7456–7462.

56 Carreño, M.C., Urbano, A. et al. (1999). Enantioselective approach to both enantiomers of helical bisquinones. *J. Organomet. Chem.* 64: 1387–1390.

57 Nuckolls, C., Katz, T.J., Castellanos, L. et al. (1999). Synthesis and aggregation of a conjugated helical molecule. *J. Am. Chem. Soc.* 121: 79–88.

58 Newman, M.S. and Wolf, M. (1952). A new synthesis of benzo(*c*)phenanthrene: 1,12-dimethylbenzo(*c*)phenanthrene. *J. Am. Chem. Soc.* 74: 3225–3228.

59 Ichikawa, J., Yokota, M., Kudo, T., and Umezaki, S. (2008). Efficient helicene synthesis: Friedel-Crafts-type cyclization of 1,1-difluoro-1-alkenes. *Angew. Chem. Int. Ed.* 47: 4870–4873.

60 Goldfinger, M.B., Crawford, K.B., and Swager, T.M. (1997). Directed electrophilic cyclizations: efficient methodology for the synthesis of fused polycyclic aromatics. *J. Am. Chem. Soc.* 119: 4578–4593.

61 Pieters, G., Gaucher, A., Prim, D., and Marrot, J. (2009). First expeditious synthesis of 6,11-diamino-[6]carbohelicenes. *Chem. Commun.* 4827–4828.

62 Shimizu, M., Hiyama, T. et al. (2008). Palladium-catalyzed annulation of *vic*-bis(pinacolatoboryl)alkenes and -phenanthrenes with 2,2′-dibromobiaryls: facile synthesis of functionalized phenanthrenes and dibenzo[g,*p*]chrysenes. *Angew. Chem. Int. Ed.* 47: 8096–8099.

63 Kamikawa, K., Matsuzaka, H. et al. (2007). Synthesis of helicenes utilizing palladium-catalyzed double C–H arylation reaction. *J. Organomet. Chem.* 72: 7406–7408.

64 Xue, X. and Scott, L.T. (2007). Thermal cyclodehydrogenations to form 6-membered rings: cyclizations of [5]helicenes. *Org. Lett.* 9: 3937–3940.

65 Peña, D. and Castedo, L. (1999). Synthesis of hexabenzotriphenylene and other strained polycyclic aromatic hydrocarbons by palladium-catalyzed cyclotrimerization of arynes. *Org. Lett.* 1: 1555–1557.

66 Stará, I.G., Starý, I., Šaman, D. et al. (1999). Transition metal catalysed synthesis of tetrahydro derivatives of [5]-, [6]- and [7]helicene. *Tetrahedron Lett.* 40: 1993–1996.

67 Sehnal, P., Stará, I.G., Starý, I. et al. (2009). An organometallic route to long helicenes. *PNAS* 106: 13169–13174.

68 Donovan, P.M. and Scott, L.T. (2004). Elaboration of diaryl ketones into naphthalenes fused on two or four sides: a naphthoannulation procedure. *J. Am. Chem. Soc.* 126: 3108–3112.

69 Mamane, V., Hannen, P., and Fürstner, A. (2004). Synthesis of phenanthrenes and polycyclic heteroarenes by transition-metal catalyzed cycloisomerization reactions. *Chem. Eur. J.* 10: 4556–4575.

70 Collins, S.K., Côté, J. et al. (2006). Preparation of helicenes through olefin metathesis. *Angew. Chem. Int. Ed.* 45: 2923–2926.

71 Nakakuki, Y., Hirose, T., and Matsuda, K. (2018). Synthesis of a helical analogue of kekulene: a flexible π-expanded helicene with large helical diameter acting as a soft molecular spring. *J. Am. Chem. Soc.* 140: 15461–15469.

72 Nakagawa, H., Kawazura, H. et al. (1982). Correlation between the stability and structure of diastereomeric charge-transfer complexes in solution. (*P*)-[7]thiaheterohelicene and (*S*)- and (*R*)-[[(tetranitrofluorenylidene)amino]oxy]propionic acid pairs. *J. Phys. Chem.* 86: 2311–2314.

73 Okubo, H., Yamaguchi, M., and Kabuto, C. (1998). Macrocyclic amides consisting of helical chiral 1,12-dimethylbenzo[*c*]phenanthrene-5,8-dicarboxylate. *J. Organomet. Chem.* 63: 9500–9509.

74 Newman, M.S. and Wise, R.M. (1956). The synthesis and resolution of 1,12-dimethylbenzo[*c*]phenanthrene-5-acetic acid. *J. Am. Chem. Soc.* 78: 450–454.

75 Gao, J.P., Wang, Z.Y. et al. (1999). Synthesis, resolution and racemization study of helically twisted *o*-terphenyls. *Chem. Commun.* 1281–1282.

76 Mikeš, F., Boshart, G., and Gil-Av, E. (1976). Resolution of optical isomers by high-performance liquid chromatography, using coated and bonded chiral charge-transfer complexing agents as stationary phases. *J. Chromatogr.* 122: 205–221.

77 Nakagawa, H., Kawazura, H. et al. (1981). Optical resolution of heterohelicenes by high performance liquid chromatography. *Bull. Chem. Soc. Jpn.* 54: 1903–1904.

78 Cheung, J.L., Sternhell, S. et al. (1997). Synthesis of a chiral helical molecular template based on trans-5,6,6a,7,8,12b-hexahydro-1,12-dimethylbenzo[*c*]phenanthrene-5,8-dione. *J. Organomet. Chem.* 62: 62–66.

79 Nuckolls, C., Katz, T.J., and Castellanos, L. (1996). Aggregation of conjugated helical molecules. *J. Am. Chem. Soc.* 118: 3767–3768.

80 Richardson, R.D., Fuchter, M.J. et al. (2015). Dual wavelength asymmetric photochemical synthesis with circularly polarized light. *Chem. Sci.* 6: 3853–3862.

81 Šámal, M., Stará, I.G., Starý, I. et al. (2015). An ultimate stereocontrol in asymmetric synthesis of optically pure fully aromatic helicenes. *J. Am. Chem. Soc.* 137: 8469–8474.

82 Terrasson, V., Gingras, M. et al. (2014). Benzylic-type couplings provide an important asymmetric entry to functionalized, non-racemic helicenes. *RSC Adv.* 4: 32412–32414.

83 Hoyo, A.M., Urbano, A., and Carreño, M.C. (2016). Enantioselective synthesis of four stereoisomers of sulfinyl ferrocenyl quinones with central, planar, and helical chirality. *Org. Lett.* 18: 20–23.

84 Sudhakar, A. and Katz, T.J. (1986). Directive effect of bromine on stilbene photocyclizations. An improved synthesis of [7]helicene. *Tetrahedron Lett.* 27: 2231–2234.

85 Heller, B., Starý, I., Stará, I.G. et al. (2013). Chiral Cobalt^{I} and Nickel^{0} complexes in the synthesis of nonracemic helicenes through the enantioselective [2+2+2] cyclotrimerisation of alkynes. *J. Organomet. Chem.* 723: 98–102.

86 Jančařk, A., Stará, I.G., Starý, I. et al. (2013). Rapid access to dibenzohelicenes and their functionalized derivatives. *Angew. Chem. Int. Ed.* 52: 9970–9975.

87 Sawada, Y., Tanaka, K. et al. (2012). Rhodium-catalyzed enantioselective synthesis, crystal structures, and photophysical properties of helically chiral 1,1′-bitriphenylenes. *J. Am. Chem. Soc.* 134: 4080–4083.

88 Grandbois, A. and Collins, S.K. (2008). Enantioselective synthesis of [7]helicene: dramatic effects of olefin additives and aromatic solvents in asymmetric olefin metathesis. *Chem. Eur. J.* 14: 9323–9329.

89 Reetz, M.T., Beuttenmuller, E.W., and Goddard, R. (1997). First enantioselective catalysis using a helical diphosphane. *Tetrahedron Lett.* 38 (18): 3211–3214.

90 Nakano, D. and Yamaguchi, M. (2003). Enantioselective hydrogenation of itaconate using rhodium bihelicenol phosphite complex. Matched/mismatched phenomena between helical and axial chirality. *Tetrahedron Lett.* 44 (27): 4969–4971.

91 Reetz, M.T. and Sostmann, S. (2000). Kinetic resolution in Pd-catalyzed allylic substitution using the helical PHelix ligand. *J. Organomet. Chem.* 603 (1): 105–109.

92 Krausová, Z., Stará, I.G., Starý, I. et al. (2011). Helicene-based phosphite ligands in asymmetric transition-metal catalysis: exploring Rh-catalyzed hydroformylation and Ir-catalyzed allylic amination. *Eur. J. Org. Chem.* 3849–3857.

93 Dreher, S.D., Katz, T.J., Rheingold, A.L. et al. (2000). Application of the Russig–Laatsch reaction to synthesize a bis[5]helicene chiral pocket for asymmetric catalysis. *J. Organomet. Chem.* 65 (3): 815–822.

94 Gross, B.M. and Oestreich, M. (2021). The trityl cation embedded into a [7]helicene-like backbone: preparation and application as a Lewis acid catalyst. *Synthesis* 53 (14): 2512–2516.

95 Sato, I., Soai, K. et al. (2001). Asymmetric induction by helical hydrocarbons: [6]- and [5]helicenes. *Angew. Chem. Int. Ed.* 40 (6): 1096.

96 Sahasithiwat, S., Sooksimuang, T. et al. (2010). 3,12-Dimethoxy-7,8-dicyano-[5]helicene as a novel emissive material for organic light-emitting diode. *Synth. Met.* 160 (11–12): 1148–1152.

97 Jhulki, S., Moorthy, J.N. et al. (2017). Carbo[5]helicene versus planar phenanthrene as a scaffold for organic materials in OLEDs: the electroluminescence of anthracene-functionalized emissive materials. *New J. Chem.* 41: 14730–14737.

98 Brandt, J.R., Fuchter, M.J. et al. (2016). Circularly polarized phosphorescent electroluminescence with a high dissymmetry factor from PHOLEDs based on a platinahelicene. *J. Am. Chem. Soc.* 138 (31): 9743–9746.

99 Josse, P., Crassous, J. et al. (2017). Enantiopure versus racemic naphthalimide end-capped helicenic non-fullerene electron acceptors: impact on organic photovoltaics performance. *Chem. Eur. J.* 23 (26): 6277–6281.

100 Xiao, S., Nuckolls, C. et al. (2013). Supersized contorted aromatics. *Chem. Sci.* 4 (5): 2018–2023.

101 Nakagawa, H., Yamada, K.I. et al. (2003). Study of chiral recognition of bilayered phosphatidylcholine vesicles using a helicene probe: characteristic function of cholesterol. *Chirality* 15 (8): 703–708.

102 Honzawa, S., Kumagai, I. et al. (2002). Chiral recognition in the binding of helicenediamine to double strand DNA: interactions between low molecular weight helical compounds and a helical polymer. *Bioorg. Med. Chem.* 10 (10): 3213–3218.

103 Tsuji, G., Kawakami, K., and Sasaki, S. (2013). Enantioselective binding of chiral 1,14-dimethyl[5] helicene–spermine ligands with B- and Z-DNA. *Bioorg. Med. Chem.* 21 (19): 6063–6068.

104 Latterini, L., Caronna, T. et al. (2011). Fluorescence properties of aza-helicenium derivatives for cell imaging. *J. Photochem. Photobiol., A* 222 (2–3): 307–313.

105 Li, M., Feng, L.-H., Chen, C.-F. et al. (2014). Tetrahydro[5]helicene-based nanoparticles for structure-dependent cell fluorescent imaging. *Adv. Funct. Mater.* 24 (28): 4405–4412.

106 Cauteruccio, S. and Chiellini, F. (2015). A nanostructured PLGA system for cell delivery of a tetrathiahelicene as a model for helical DNA intercalators. *ChemPlusChem* 80 (3): 490–493.

107 Xiao, J.C., Wudl, F., Zhang, Q.C. et al. (2012). Synthesis and structure characterization of a stable nonatwistacene. *Angew. Chem. Int. Ed.* 51: 6094–6098.

108 Yano, Y., Itami, K. et al. (2016). Helically twisted tetracene: synthesis, crystal structure, and photophysical properties of hexabenzo[*a,c,fg,j,l,op*]tetracene. *Synlett* 27: 2081–2084.

109 Suzuki, S., Itami, K., and Yamaguchi, J. (2017). Synthesis of octaaryl naphthalenes and anthracenes with different substituents. *Angew. Chem. Int. Ed.* 56: 15010–15013.

110 Clevenger, R.G., Kilway, K.V. et al. (2018). Synthesis and structure of a longitudinally twisted hexacene. *Chem. Eur. J.* 24: 3113–3116.

111 Norton, J.E. and Houk, K.N. (2005). Electronic structures and properties of twisted polyacenes. *J. Am. Chem. Soc.* 127: 4162–4163.

112 Bedi, A., Shimon, L.J.W., and Gidron, O. (2018). Helically locked tethered twistacenes. *J. Am. Chem. Soc.* 140: 8086–8090.

113 Fan, W., Wang, Z. et al. (2017). A decatwistacene with an overall 170° torsion. *Angew. Chem. Int. Ed.* 56: 15373–15377.

114 Vogtle, F. (1983). Cyclophanes II. Concluding remarks. *Top. Curr. Chem.* 115: 157–159.

115 Jasti, R., Bertozzi, C.R. et al. (2008). Synthesis, characterization, and theory of [9]-, [12]-, and [18]cycloparaphenylene: carbon nanohoop structures. *J. Am. Chem. Soc.* 130: 17646–17647.

116 Povie, G., Itami, K. et al. (2017). Synthesis of a carbon nanobelt. *Science* 356: 172–175.

117 Gui, K.Y., Miao, Q. et al. (2019). Synthesis of armchair and chiral carbon nanobelts. *Chem* 5: 838–847.

118 Shi, T.H., Guo, Q.H., Tong, S., and Wang, M.X. (2020). Toward the synthesis of a highly strained hydrocarbon belt. *J. Am. Chem. Soc.* 142: 4576–4580.

119 Cheung, K.Y., Watanabe, K., Segawa, Y., and Itami, K. (2021). Synthesis of a zigzag carbon nanobelt. *Nat. Chem.* 13: 255–259.

120 Shi, T.H., Tong, S., and Wang, M.X. (2020). Construction of hydrocarbon nanobelts. *Angew. Chem. Int. Ed.* 59: 7700–7705.

121 Yang, F., Wang, X., Zhang, D. et al. (2014). Chirality-specific growth of single-walled carbon nanotubes on solid alloy catalysts. *Nature* 510: 522–524.

18

2-Aza-21-Carbaporphyrin in Construction of Chiral Supramolecular Assemblies

Sebastian Koniarz and Piotr J. Chmielewski

University of Wrocław, Department of Chemistry, F. Joliot-Curie 14, Wrocław 50383, Poland

18.1 Preface

Among numerous organic chemistry processes that appear to be straightforward and well understood, but in fact they are not, is a stochastic synthetic porphyrinogenesis. The formation of a symmetric macrocycle with, invariantly, four pyrroles linked by methylene and, after oxidation, methine bridges is by no means obvious under the condition of acid-catalyzed condensation of aryl aldehydes and pyrrole (Figure 18.1). A careful examination of the reaction condition in relation with the macrocycle yield allowed optimization of the process outcome and minimization of the amounts of other, mostly linear, shorter or longer oligomers concurrently formed [1–6]. At present, the highly symmetric *meso*-tetraaryl porphyrin is like an "all-purple Rubik's Cube" to the synthetic chemists because its preparation is no longer challenging. Conversely, optimization of syntheses of non-symmetric tetrapyrroles, i.e. porphyrin systems of an effective point group symmetry lower than D_{2h} (or D_{4h} for planar metal complexes, dianions, dications, and dynamic systems, Figure 18.1), such as those comprising different substituents at either *meso*- or pyrrole β-carbons, is still in progress. Such lower symmetry porphyrins need to be assembled from predefined building blocks and/or separated from the complex reaction mixtures due to a reversible character of the annulation under the acidic condition resulting in a scrambling (C—C bond acidolysis!) [7–22]. Alternatively, the symmetry of the macrocycle may be disturbed by post-annulated modification of the porphyrin perimeter or interior in various substitution or addition reactions.

Significantly, a one-point substitution taking place at a β-carbon changes a highly-symmetric macrocycle into a prochiral system of point group symmetry C_s, i.e. with only one symmetry plane defined by the mean plane of the porphyrin. Figure 18.2 summarizes approaches toward chirality of the systems consisting of the regular porphyrin [23–46].

☆ This work was supported by the Polish National Science Center (2018/29/B/ST5/00280).

Chiral Building Blocks in Asymmetric Synthesis: Synthesis and Applications, First Edition.
Edited by Elżbieta Wojaczyńska and Jacek Wojaczyński.

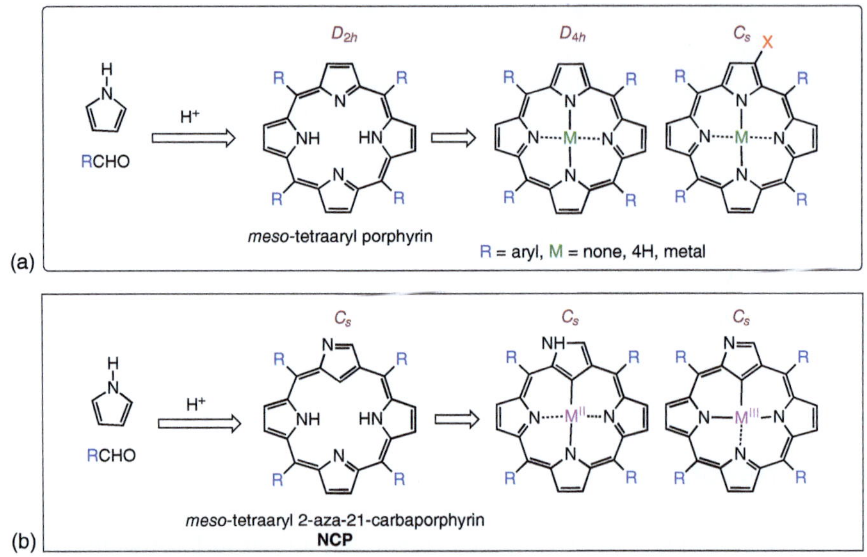

Figure 18.1 *meso*-Tetraaryl porphyrins (a) and *meso*-tetraaryl-2-aza-21-carbaporphyrin (b, **NCP**) and their complexes.

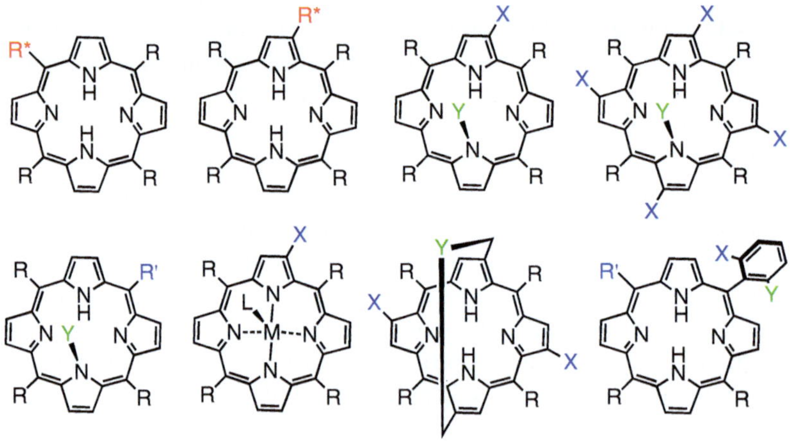

Figure 18.2 Chiral porphyrins, their complexes, and dimers.

Almost 90 years ago, the symmetric tetraarylporphyrin thus emerged, and over that period, the aromatic tetrapyrroles have become ubiquitous in organic, inorganic, and supramolecular chemistry with the stochastic condensation as a main source of these macrocycles. Suddenly, in the mid-1990s, it came to the porphyrin chemists that there are more types of macrocyclic systems formed upon acid-catalyzed condensation of the pyrrole and aryl aldehyde than merely porphyrin. Over the time, it appeared that by changing its conditions, the reaction may deliver macrocycles of expanded or shrank ring and porphyrin isomers. In fact, the formation of the macrocycles isomeric to a "regular" porphyrin had been assumed already by Rothemund who discovered the method of porphyrin synthesis back in the 1930s, but none ever actually established the formation of other than tetrapyrrole system in the reaction mixture. It all started in 1994 when two groups, one operating in Poland and the other in Japan, independently and serendipitously discovered at the same time another porphyrin-like system that was isomeric with tetraarylporphyrin but of C_s symmetry (Figure 18.1) [47, 48].

meso-Tetraaryl-2-aza-21-carbaporphyrin, termed initially also *inverted porphyrin*, and finally known under the well-established trivial name of *N-confused porphyrin* (**NCP**), was the first porphyrin isomer and for some time, the only one in which the structure of [18]porphyrin(1.1.1.1) was retained. In the beginning, it was considered merely a structural curiosity, mostly because until the end of the twentieth century it was obtained with very low yields not exceeding 5%, with rather poor perspectives for an improvement [49]. Moreover, the **NCP** separation out of the "Rothemund's soup" consisting mostly of a tar and the regular porphyrin was tedious and time consuming. All those problems have been overcame by application of the proper acidic catalyst and optimization of the reaction conditions including temperature, concentration, and the reaction time [49, 50]. At the moment, **NCP** is as readily obtainable as the regular porphyrin in the gram-scale syntheses and can be used in various types of systems as a replacement for the regular porphyrin. Thus, the regular tetraarylporphyrin and **NCP** have common origin and (sometimes) common fate [51–53] giving rise to identical degradation products. Obviously, different coordination cores (NNNN donor set for the regular and CNNN donor set for the confused porphyrin) and ability of **NCP** to act as a trianionic ligand make its coordination chemistry particularly rich and versatile [54].

The most attractive feature of this aromatic porphyrinoid from a point of view of structural modification is its ability to react in various reaction types that do not occur for regular porphyrins. These processes involve both macrocycle's interior and perimeter with a special contribution of the *confused* pyrrole where most of the reactions take place for the porphyrin free base and its metal complexes (Figure 18.3). Some of these modifications result in chirality of the molecule by making porphyrin faces diastereotopic; the others give rise to a chirality center. Moreover, the nitrogen donor on the macrocyclic perimeter may be involved in several interactions, including H-bonding, metal ion coordination, and covalent bonding. Thus, interactions between the **NCP** units or between molecules of its metal complexes are not limited to the coordination core but may involve also the atoms located at the perimeter. That potentially gives rise to molecular or

Figure 18.3 Substitution and addition sites in modified **NCP**.

supramolecular structures involving more than one NCP. In this context, the low symmetry of this naturally prochiral porphyrinoids appears to be a drawback because connection of two such subunits may remove the symmetry plane of each of them resulting in a mixture of stereoisomers.

Thus, a relatively small structural alteration stemming from different ring closure (seven C_{meso}–C_α bonds and one C_{meso}–C_β bond in **NCP** instead of eight C_{meso}–C_α bonds in regular porphyrins) results in big differences in coordination, spectroscopic, and redox properties. The chiral systems, based on the intrinsic chirality of the **NCP** ring, are formed easily, and quite often but frequently the importance of that is not appreciated by the researcher dealing with this macrocycle. The chirality is a burden that may complicate composition of the product pool upon assembling of the chiral subunits. On the other hand, some enantiopure **NCP** derivatives and complexes can be obtained quite easily, and their synthesis is inexpensive. Thus, **NCP** can be considered as potential chiral seed in catalysis, including organocatalysis.

Over the 28 years of the NCP existence, there were many chiral monomeric and oligomeric systems reported, including purely organic free bases and metal complexes of **NCP**. This chapter is devoted to identification and discussion of the systems that contain N-confused porphyrin skeleton and are chiral, regardless if any attempt of enantiomer separation had been undertaken by the researches reporting them.

18.2 Monomers

18.2.1 Free Bases

The very first X-ray structure of N-confused porphyrin revealed chirality of this macrocycle in the solid state [48]. Unlike for regular porphyrin where the core comprises two hydrogen-bearing nitrogens, the **NCP** interior is more sterically crowded due to the presence of additional CH group. As a consequence of that, the macrocycle is distorted with the *confused* pyrrole tipped from the mean porphyrin plane (Figure 18.4). Although such a planar chirality has never been indicated in solution of **NCP**, in fact, it never has been excluded, either. ^{1}H NMR (nuclear magnetic resonance) that is indicative for the occurrence of diastereotopicity in rigid systems shows here equivalence of both porphyrin faces reflected by the equivalence of *ortho*- or *meta*-protons in each of the *meso*-phenyls. That suggests achiral

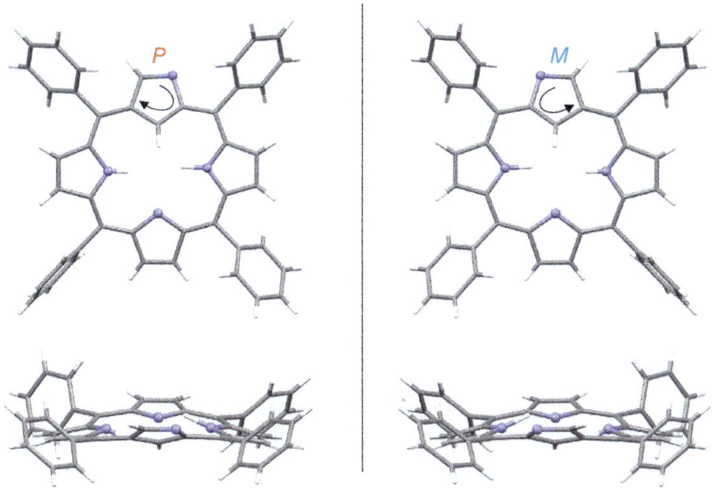

Figure 18.4 Two projections of enantiomers of *meso*-tetraphenyl-2-aza-21-carbaporphyrin in the solid state [48].

character of the system due to dynamics of the ring that changes conformation with a rate that is above the NMR timescale. It is believed that such a conformation change is possible in solution owing to a tautomery involving the external nitrogen.

The tautomer of **NCP** with 2-NH dominates in polar solvents like dimethylformamide (DMF) or pyridine [55], but it is likely present also in the solvents such as chloroform or dichloromethane, though in minute amount and under the fast exchange condition it cannot be observed as an individual species by ^{1}H NMR. The absence of one of the hydrogens in the porphyrin crevice of 2-NH**NCP** tautomer allows for the *confused* pyrrole conformation change with respect to the rest of the macrocyclic ring. All and all, the **NCP** free base appears to be configurationally unstable, and it is rather no chance to separate the enantiomers or observe any optical activity of it.

The case of configurational instability of **NCP** shows that the presence of a relatively big substituent inside the porphyrin crevice is needed to prevent fast movement of the *confused* pyrrole. The first **NCP** free base for which chirality has been established by chiroptical methods comprised alkyl groups attached to C21, i.e. internal carbon of the *confused* pyrrole (Figure 18.5) [56]. This macrocycle was obtained by demetalation of the enantiomers obtained by a chiral stationary phase high performance liquid chromatography (HPLC) of 21-methylated nickel(II) complex of **NCP** [57]. The absolute configuration assignment of these enantiomers was based on the simulation of circular dichroism (CD) spectra by time-dependent DFT (TD DFT) method. Apparently, methyl or benzyl substituents at C21 prevent the *confused* pyrrole conformation changes and stabilize configuration of the porphyrin. Thus, the system has a plane of chirality that can be observed also by a differentiation of some *ortho*- and *meta*-H signals of the *meso*-aryls in ^{1}H NMR.

The free base **NCP** derivatives bearing different substituents at C21 such as NO_2 or halogens and phosphine derivatives are strongly distorted and inherently chiral,

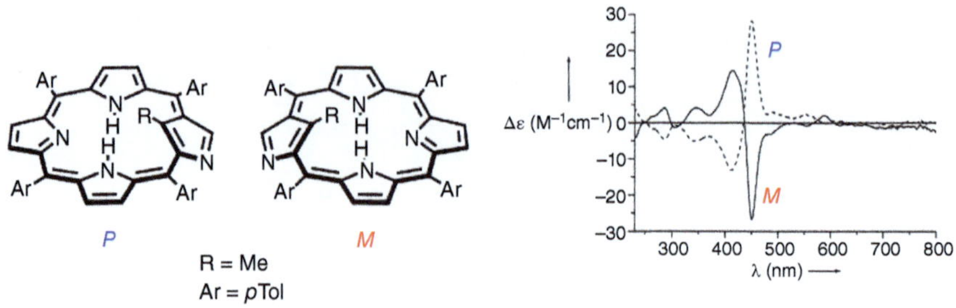

Figure 18.5 Enantiomers of 21-alkylated **NCP** and their CD spectra of dichloromethane solutions.

though they crystallize as racemates and any attempt to separate the enantiomers has never been undertaken (Figure 18.6) [58, 59, 62, 63]. Also, intrinsically chiral ligands with 21-(2-thiopyridyl) [60] or 21-OH [61] substituents were obtained by addition of 2-thiopyridyl to the nickel(II) or oxygenation of iron(II) or manganese complexes, followed by demetalation (Figure 18.6). Another chiral derivative comprising 21-carboxymethyl moiety was obtained serendipitously via reaction of nickel(II) complex with azo-radical [64].

Oxygenation of **NCP** in the presence of primary alcohols gives rise to tris(alkoxylated) derivatives 3-OR-21-$(OR)_2$**NCP**, bearing a ketal functionality on the internal carbon (Figure 18.7). Unlike many other reactions resulting in substitution at C21, this one does not involve a metal complex intermediate. Since the alkoxy groups are *trans* to each other, the molecule is of C_s symmetry and, thus, achiral. By reductive elimination of one of the substituent from C21, the faces of the macrocycle become diastereotopic, and the molecule gains chirality clearly indicated by differentiation of certain proton signals in the ^{1}H NMR. Separation of the enantiomers by a chiral stationary-phase HPLC allows observation of their electronic circular dichroism (ECD) spectra [65]. Interestingly, the exchange of the 21-alkoxy group (transetherification) is facile for the primary alcohols and proceeds with the configuration inversion. Application of enantiopure secondary alcohol (e.g. 1-phenylethanol) results in an induction of some diastereomeric excess related to unidirectional configurational change of the macrocycle.

The readily formed 21-Br**NCP** can be easily and reversibly transformed into a prochiral N-fused porphyrin [66] that have been subjected to further Stille or Sonagashira couplings and other nucleophilic substitution reactions [67, 68] giving rise to a plethora of *internally* modified **NCP**s, including 21-cyano, 21-phenylethynyl, or 21-trimethylsilylethynyl-**NCP**. All of them are chiral and, thus far, racemic (Figure 18.8). Among these systems, the most fascinating are internally bridged derivatives with ethene linkage between C21 and N22 [67, 69]. Interestingly, such an approach involving fusion step of the synthesis gives rise exclusively to C21, N22-bridging system (Figure 18.9). Conversely, the imine bridge formed from 21-amino-**NCP** and aryl aldehyde links C21 and either N22 or N24. The both diastereomers are chiral in the solid state and indicate distinct structural and spectroscopic properties. Interestingly, a unidirectional transition

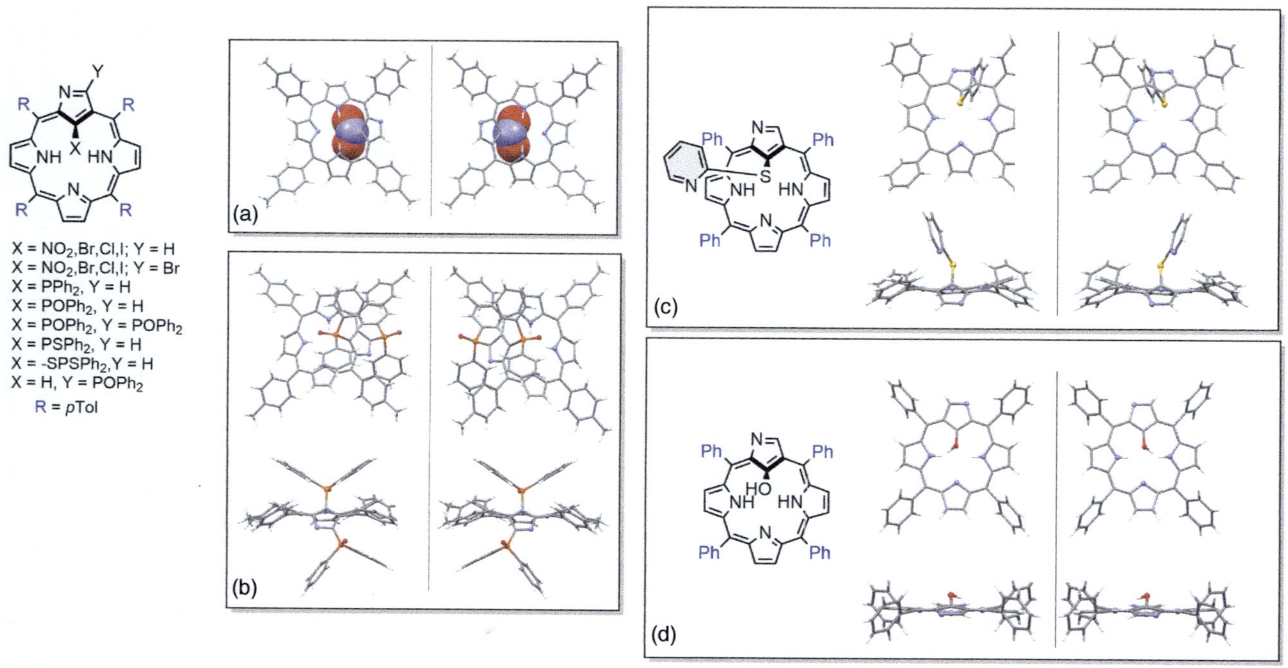

Figure 18.6 21-substituted **NCP** and solid-state structures of enantiomers of 21-NO_2-**NCP** (a) [58, 59] and two projections of 3,21-$(POPh_2)_2$-**NCP** (b) [58, 59], 21-(2′-S-py)**NCP** (c) [60], and 21-OH-**NCP** [61]. The nitro group in (a) is depicted with space filling representation.

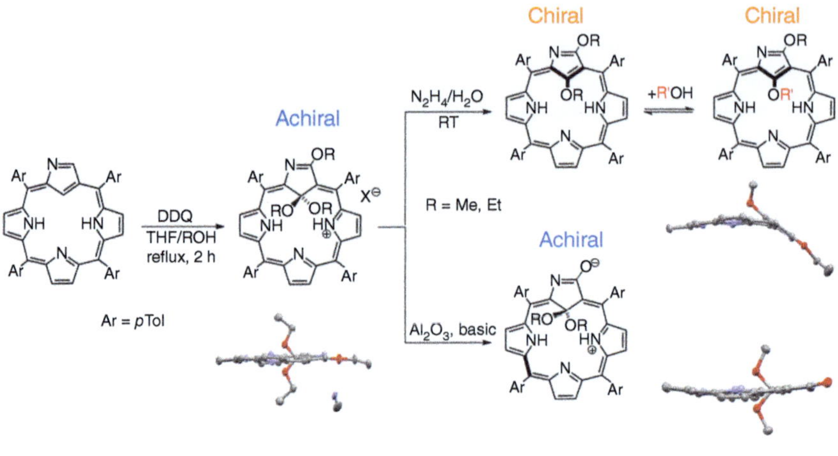

Figure 18.7 Formation and reactivity of 2,21-alkoxy **NCP** [65].

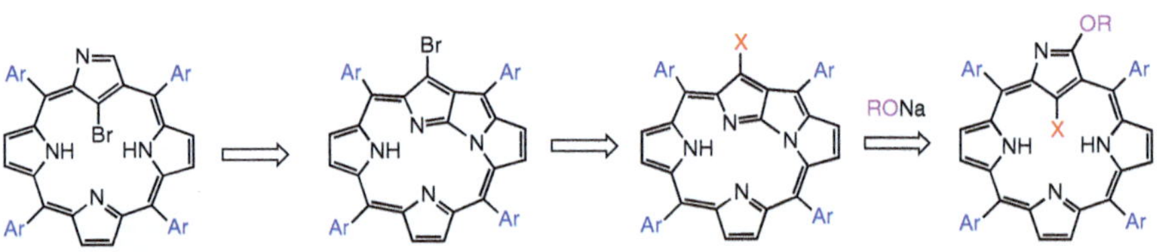

Figure 18.8 Application of N-fused porphyrin in the synthesis of 21-substituted NCP derivatives [67].

toward the more stable isomers was observed upon heating (Figure 18.9) [70]. The internally bridged **NCP** derivatives, obtained in a reaction of **NCP**Ag^{III} complex with dimethylamine comprise 21-dimethylamine group bound either to N22 or to N24. For these systems, which crystalize as racemic mixtures, the diastereomers are in equilibrium that is fast in the ^{1}H NMR timescale (Figure 18.9) [71].

A very simple method of the internal ring fusion is a direct and highly regioselective (90%) carbonylation of **NCP** with paraformaldehyde [72]. The chiral derivatives thus formed comprise a lactam functionality with carbonyl bridging C21 and N24 (Figure 18.10). The enantiomers of CO-**NCP** are easily separable by means of a chiral stationary-phase HPLC. The internally carbonylated *meso*-(3′,5′-methoxyphenyl)-**NCP** under acidic condition undergoes a fusion involving C3 and *ortho* -carbon of the adjacent aryl giving rise to a chiral system of five fused rings (Figure 18.10b). The absolute configurations of the enantiomers have been assigned by TD DFT simulations of the ECD spectra. CO-**NCP** interacts enantioselectively with camphorsulfonic acid and with chiral alcohols. Interestingly, the system racemizes under acidic conditions that is related with a pre-hydrolytic bonding of water molecule to the internal bridging carbon (Figure 18.10c).

Configurationally stable 21-arylazo-**NCP** has been synthesized in a multistep process starting from 21-NO_2-**NCP** [73]. An alternative one-step method involving a direct azotization of **NCP** with isoamyl nitrite under mild conditions appears

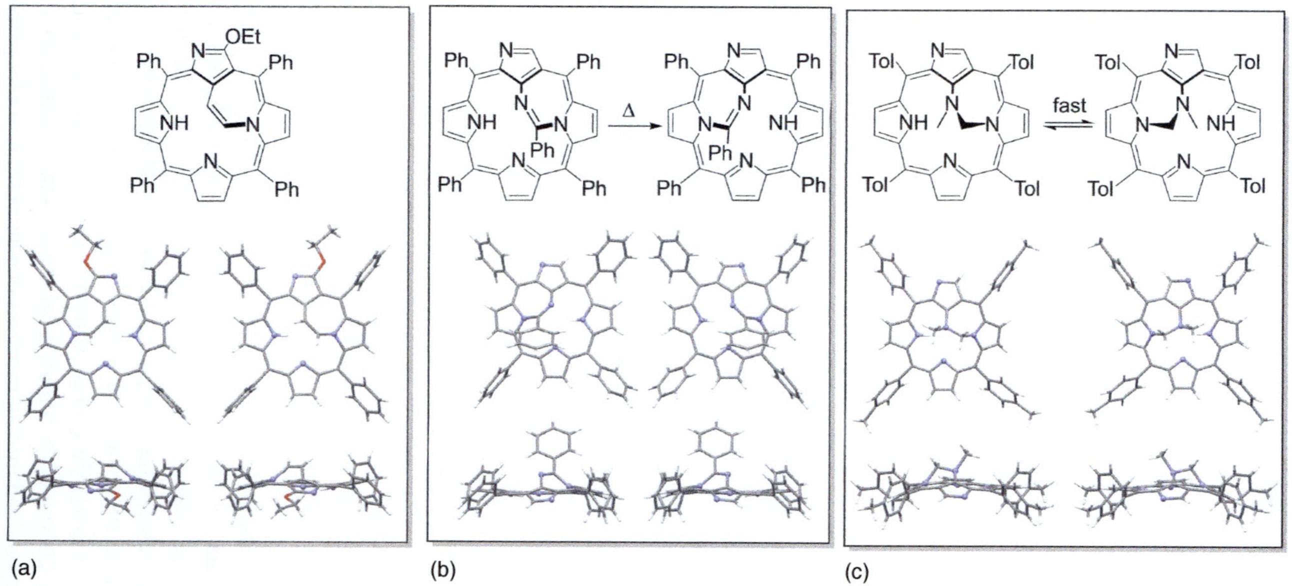

Figure 18.9 Stereoisomers of internally bridged **NCP** derivatives: with 21,22-ethene (a), 21-imino-22-phenylmethyl (b), and 21-methylamino-22(24)-methylene-bridge (c). Stick representations show enantiomers in the solid state; schematic structures present diastereomers. Source: Kashiwagi et al. [70]/American Chemical Society.

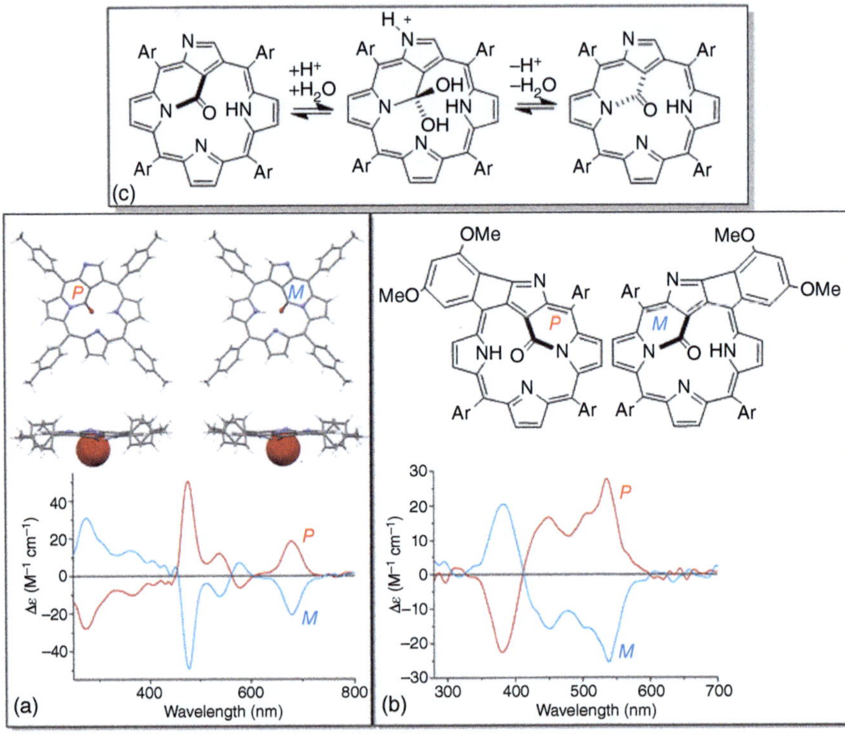

Figure 18.10 Solid-state structures of enantiomers (a), schematic structure of the externally fused derivative (b), and CD spectra (a and b) and racemization scheme of internally carbonylated **NCP**s (c) [72].

to be effective only for anthranilic acid [74]. For other anilines, 3,21-aminated **NCP** or 3-amino-21-imino-**NCP** derivatives are formed under such conditions, and isoamyl nitrite can be replaced by other oxidants, e.g. *p*-chloranil. The systems bearing arylazo-, arylamino-, or imino-functionality on C21 are configurationally stable that allows separation of the enantiomers and assignment of the absolute configuration based on comparison of experimental and TD DFT simulated CD spectra (Figure 18.11).

Virtually all modifications of **NCP** involve the *confused* pyrrole either at C21, C3, or N2 or combination of these sites [75–80]. Since **NCP** derivatives gain chirality by losing their effective planarity, it is unlikely that molecules become intrinsically chiral merely by attachment of substituent on the porphyrin perimeter. Diastereotopic faces of the macrocycle may be theoretically generated by the unsymmetrical (not necessarily chiral) substituent that is devoid of rotational freedom (vide infra). On the other hand, an *addition* taking place at any of pyrrole carbons or in the *meso*-position may result in desymmetrization by generation of a chirality center [81]. A very rare case of the structural modification of **NCP** that **does not** involve the *confused* pyrrole is represented with 1,3-dipolar cycloaddition of 2,6-dichlorobenzonitrile oxide to 2-methylated **NCP** [82]. The reaction affords a

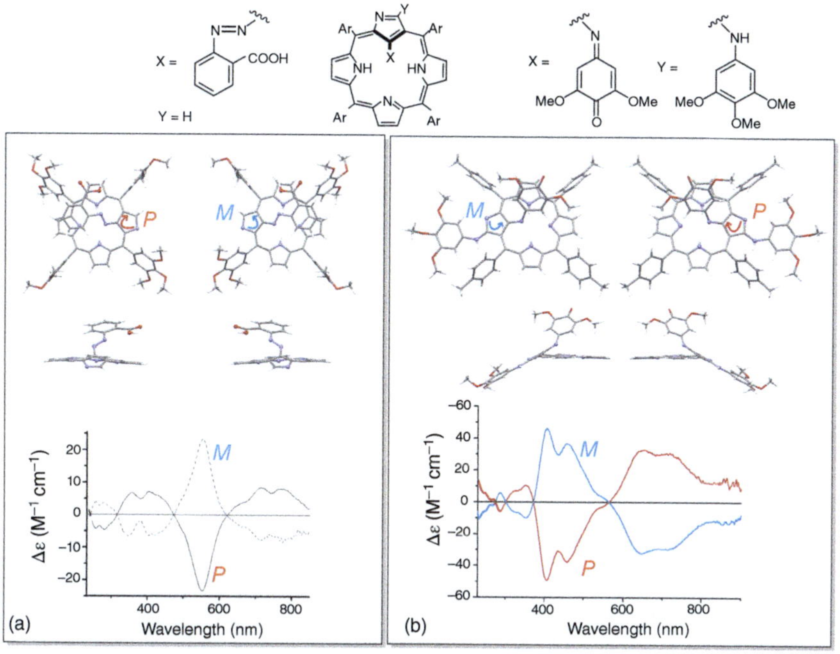

Figure 18.11 Enantiomers of 21-(*o*-carboxyphenyl)azo-**NCP** (a) and 3-amine-21-imine-**NCP** (b) and their CD spectra [74].

mixture of four regio- or stereoisomeric monoadducts of chlorin type (oxazolocarbachlorins, **OC**), i.e. consisting of a pyrrole ring that is hydrogenated at β-carbons with retained macrocyclic aromaticity (Figure 18.12). Interestingly, in the same reaction, the formation of a bis-adduct, i.e. bis(oxazolo)bacteriocarbachlorin (**OB**), is completely regio- and stereoselective, though the occurrence of eight isomeric bacteriochlorins is theoretically possible (Figure 18.12). Separation of chlorin stereoisomers has been made possible by a chiral stationary phase HPLC after their metalation with nickel(II). Several enantiomeric pairs have been separated, and their optical activity is established by ECD. Conversely, an attempt of the **OB** enantiomers separation fails, likely due to a similarity of the isomers and their *meso*-like structures, i.e. occurrence of two pairs of structurally similar stereocenters of opposite configurations (Figure 18.12).

The addition to the *meso*-position breaks macrocyclic conjugation but may result in a stereogenic center formation. The systems comprising two the same substituents at the *meso*-C [83, 84], although containing a *confused* pyrrole, are effectively C_s symmetric and, thus, achiral. Such systems, which are termed N-confused calix[4]phyrins, are in fact nonaromatic phlorins that are isomeric with aromatic chlorins. The chiral N-confused[4]calixphyrins are formed upon addition/cyclization reaction involving **NCP** and acid chlorides, *in situ* generated nitrilimines, or dialkyl acetylene diesters (Figure 18.13) [85, 86]. The exocyclic rings link external nitrogen of the *confused* pyrrole (N2) and the adjacent *meso*-carbon (C20). The lactam-comprising products of the **NCP** reaction with acyl chlorides

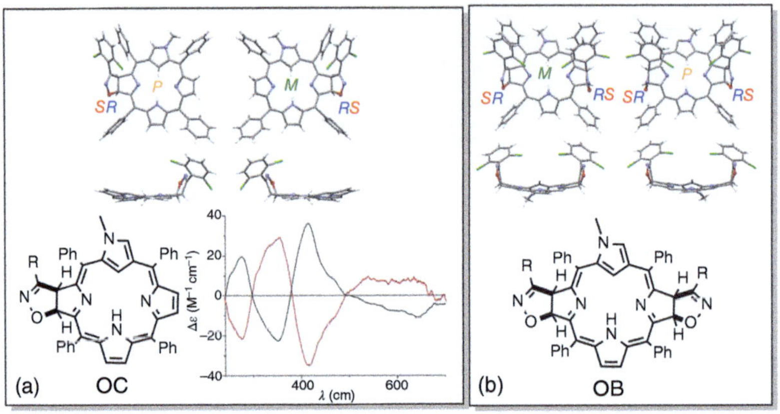

Figure 18.12 Enantiomers and CD spectra of 2-methyl-7,8-(2′-(2″,6″-dichlorophenyl) oxazolo-21-carbachlorin **OC** (a) and solid-state enantiomers of 2-methyl-7,8,17,18-bis(2′-(2″,6″-dichlorophenyl)oxazolo)-21-bacteriocarbachlorin **OB** (b) [82].

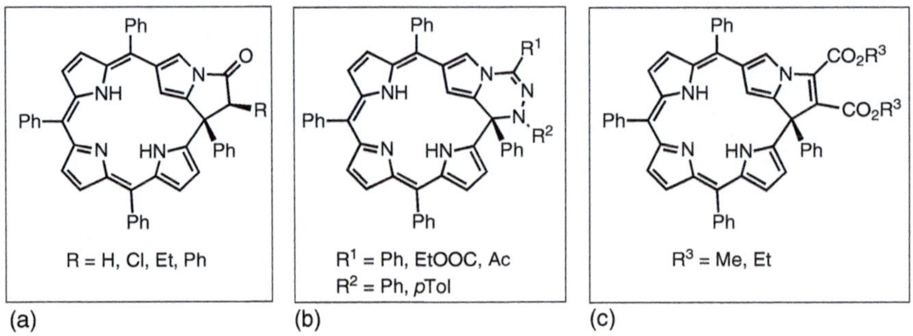

Figure 18.13 Structures of chiral phlorins obtained by 1,3-cycloadditions to **NCP** involving N2 and C20: product of addition of acyl chlorides (a), iminenitriles (b), and dialkyl acetylenedicarboxylates (c) [85, 86].

can theoretically form diastereomers of various configurations at the stereocenters. Despite that, only one NMR-defined stereoisomer with Ph and R in the *cis*-orientation has been observed for all derivatives. All these systems are configurationally stable, but their chiroptical properties remain unexplored.

The aforementioned N-confused[4]calixphyrins cannot be transformed to aromatic macrocycles without C—C or C—N bond fission, involving *meso*-carbon. On the other hand, a reaction of **NCP** with organic phosphites gives rise to phlorins that undergo dehydrogenation resulting in the appropriate 3-substituted **NCP** derivatives. The first stage of this substitution process involves addition of $PO(OR)_2$ moiety formed from the phosphite due to the presence of acetic acid as catalyst. The H-shift from C3 to the adjacent *meso*-carbon C5 is responsible for the stereogenic center formation at this carbon and for the lack of macrocyclic conjugation (Figure 18.14) [87]. The system is thus nonaromatic, but it is stable

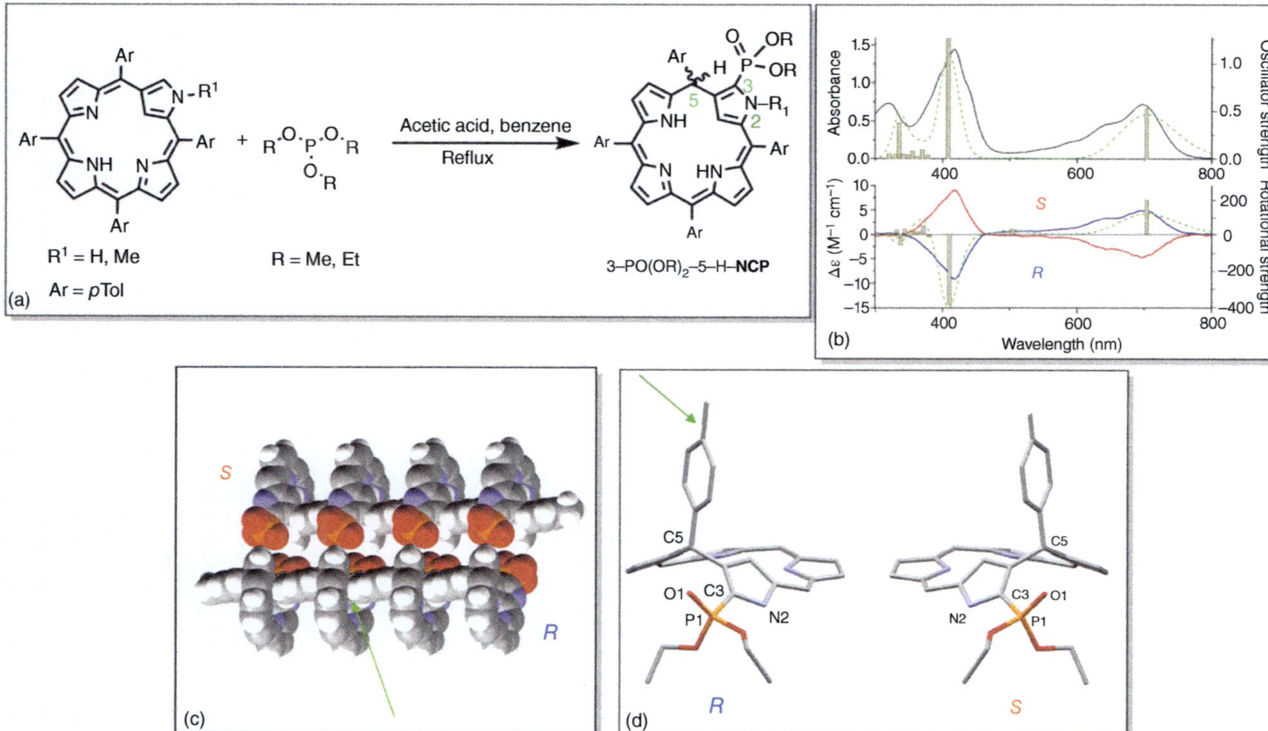

Figure 18.14 Synthesis of 3-P(O)(OEt)$_2$–5-H-**NCP** (a), optical absorption (black trace) and ECD spectra of separated enantiomers (blue and red traces) along with TD DFT simulations for DFT-calculated model of the enantiomer *R* (b), fragment of the spacefill representation of packing diagram of the racemate with all ethyl substituents and all *meso*-phenyls, except that at C5 removed for clarity (c), and enantiomers from the crystal structure of the racemate with all aryls but C5-Tol and all hydrogens removed (d). The green arrows in (c) and (d) indicate position of the methyl group contacting with other molecule in the solid state. Source: Liu et al. [87]/American Chemical Society.

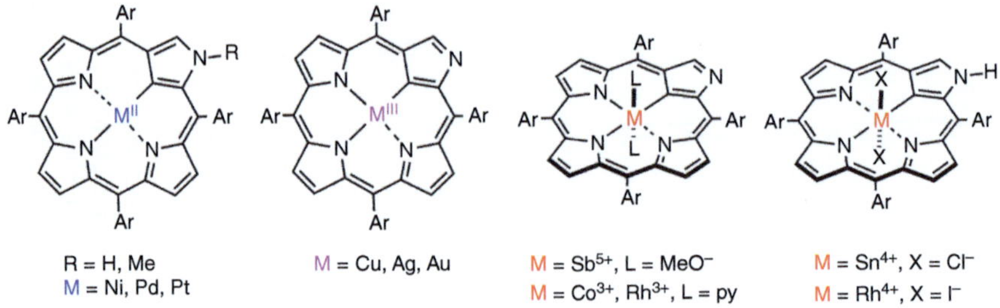

Figure 18.15 Examples of square planar and tetragonal bipyramid complexes of **NCP**.

toward autoxidation, and its enantiomers can be separated by means of chiral stationary phase HPLC. Interestingly, the phlorin 3-P(O)(OEt)$_2$-5-H-**NCP** crystallizes as racemate constituted by homochiral rods alternating in the crystal lattice. These stereoselective assemblies are formed due to interactions of methyl protons of *para*-tolyl substituent at C5 with localized π-electron density of the phlorin interior (Figure 18.14). The absolute configurations for the separated enantiomers have been assigned by TD DFT simulation of ECD spectra.

18.2.2 Metal Complexes

Coordination properties of unaltered **NCP** are by far different from those of the regular porphyrins, mostly due to the distinct donor atom set consisting of three nitrogens. Among them two are of amine type, i.e. bear removable protons, and the other is of imine type and pyridine-like donation mode. Obviously, the presence of CH inside the macrocyclic crevice makes the biggest difference. Generally, coordination mode of C21 depends on deprotonation of this site upon metal insertion. The other obvious distinction is related with an imine-like donor site at the macrocycle's perimeter. The nitrogen of the *confused* pyrrole may accept proton, thus changing the overall negative charge of **NCP** from −3 to −2 that may result in equal stabilization of divalent and trivalent metal ions in the square-pyramidal neutral complexes. Systems of this type are achiral due to the effectively planar structure of the metalated macrocycle as represented by d^8 metal complexes, such as those of nickel(II) [47, 88], palladium(II) [89], platinum(II) [90, 91], copper(III) [92], silver(III) [93], or gold(III) [94]. Similarly, six-coordinated complexes with metal in a square-bipyramidal geometry of the coordination sphere like those of manganese(III) [95], cobalt(III) [96], rhodium(III) [97], antimony(V) [98], or tin(IV) [99] are prochiral (Figure 18.15). The ligand not always employs its full capacity of the cation neutralization even when high oxidation state of the metal is expected to force the porphyrin to act as a triply negative ligand [100]. On the other hand, some derivatives that are not capable to neutralize trications form stable positively charged metal(III) complexes [81, 101]. Thus, coordination chemistry of **NCP** is rich and not always predictable.

There are several types of complexes where C21 does not form any σ-bond with metal ion. In the 12-group metal complexes, **NCP** acts as monoanionic [102, 103] or dianionic [102, 104, 105] ligand without deprotonation of the inner carbon of the *confused* pyrrole that only weakly interacts side-on with the metal ion, likely involving a delocalized π-electron system. The metal environment here can be described as heavily distorted square pyramidal or, disregarding the weak M–C(21)H bond, as distorted tetrahedral. Regardless of the description, these molecules are chiral, lacking any symmetry element (Figure 18.16). Many other metal complexes use **NCP** as tricoordinating ligand [106–110], though for some iron(II) complexes of **NCP**, an agostic interaction of 21-CH with metal ion has been established. Complexes of this type occasionally involve the external nitrogen in coordination to a metal center of other molecule giving rise to homo- or heterodimers (vide infra) [104, 110, 111]. Monomeric complexes of this type are exclusively formed for 2-N-substituted NCP [112, 113]. For some metals, such as manganese, iron, or cobalt, the coordination mode of **NCP** depends on oxidation state [108, 114] or type of the apical ligand [96, 115] that is connected with deprotonation of the inner carbon of the *confused* pyrrole. For these systems, pentacoordination of the central ion and, thus, inequivalence of the macrocycle's faces retain chirality of the complex even after deprotonation of C21.

Some of these complexes are sensitive to oxygenation at either C3 or C21 that finally leads to destruction of the porphyrin ring. Most of them are labile and do not survive chromatographic separation. That is probably the reason for a unique character of zinc and cadmium complexes of 3-P(O)(OR)-**NCP** for which enantiomers can be separated by a chiral stationary-phase HPLC [87]. Interestingly, in the crystal lattice of racemate, discrete heterochiral dimers are formed as a result of two-point N–H…O=P(OEt) hydrogen bonding (Figure 18.17).

Flexibility of the **NCP** as a ligand is noticeable also for the second and third row of transition metals. Unaltered **NCP** coordinates two bis(carbonyl)rhodium(I) [116] units as a monoanion using external nitrogen of the *confused* pyrrole and two internal nitrogens of the regular pyrroles (Figure 18.18). Such a system can be easily converted to the aforementioned achiral rhodium(III) complexes (Figure 18.15) where macrocyclic ligand is either di- or trianionic [97]. On the other hand, a reaction of **NCP** with iridium(I) bis(carbonyl)(*p*-toluidyne) chloride gives rise to a fascinating helically chiral bismetallic complex. All nitrogen atoms are involved here in coordination to Ir^{I}, though the *confused* pyrrole is inverted in such a way that 21-CH is pointing outward the macrocyclic cavity and, obviously, the porphyrin ring is drastically distorted, forming a short helix (Figure 18.18) [117].

An interesting bismetallic cationic complex is formed upon addition of pentamethylcyclopentadienyl ruthenium(II) ($[RuCp^*]^+$) to **NCP**Ag^{III} (Figure 18.19a) [119]. Introduction of a voluminous group onto one of the *meso* carbons clearly arrests rotation around *meso–ipso* bond, thus leading to differentiation of the porphyrin faces and chirality of the molecule. Even more complicated chiral complexes consisting of lanthanide (erbium(III) or yterbium(III)) coordinated to all inner donors of the **NCP** in the sitting-atop mode with cyclopentadienylatocobalt(I) decorating a system of three $OP(OEt)_2$ ligands have been reported by Wong and coworkers (Figure 18.19) [118].

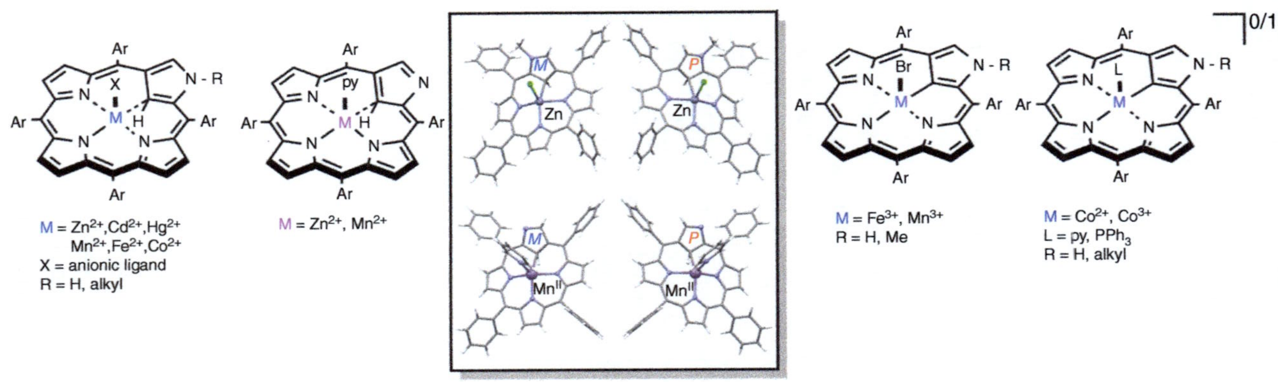

Figure 18.16 Schematic structures of chiral complexes of manganese, iron, cobalt, and the group 12 elements with **NCP** and examples of the solid-state enantiomers of zinc and manganese(II) complexes.

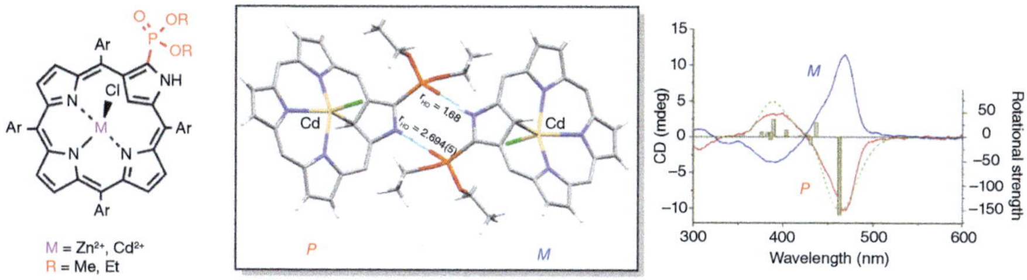

Figure 18.17 Schematic structure of zinc and cadmium complexes of 3-P(O)(OR)-**NCP**, heterochiral dimer of 3-P(O)(OR)-**NCP**-CdCl observed in the solid state of racemic crystal, and ECD spectra of separated enantiomers of this complex (blue and red traces) along with histogram of TD DFT calculated electronic transitions for a model of enantiomer *P* (yellow sticks) and Gauss-shaped convoluted calculated spectrum (dashed green trace).

The examples indicate that some metal complexes of **NCP** are intrinsically chiral without any modifications of the ligand directed toward desymmetrization. On the other hand, as it has been mentioned above, among the **NCP** derivatives, there are many ligands that are already fully asymmetric. Moreover, some prochiral complexes are easily convertible to chiral systems in the processes triggered by the presence of a metal ion. The first and the best example of such a reaction is methylation of the internal carbon of the *confused* pyrrole C21 coordinated to nickel(II) giving rise to a stereogenic center localized at C21 [57, 120]. The reactivity of coordinated carbanionic C21 toward haloalkanes [88, 101, 121, 122] is not limited to nickel(II) complexes but has been observed also for copper(II) [103], palladium(II), and platinum(II) [123]. In the 21-alkylated complexes, C21 can act as a sp^3-hybridized σ-bonded carbanion or as a neutral side-on η^1-coordinated fragment of the planar *confused* pyrrole analogous to that observed for zinc or cadmium complexes of **NCP** (Figures 18.16 and 18.17). In paramagnetic dimethylated 2,21-Me_2**NCP**NiX (X = halogenide), the latter coordination mode is solely observed, while for 21-Me**NCP**Ni upon protonation of N2, C21 changes hybridization from sp^3 in neutral form to sp^2 and, concurrently, the spin state from diamagnetic to paramagnetic. The 21-alkylated derivatives of **NCP** are thus chiral, either by the presence of chirality center on the pyramidal σ-coordinating carbon or by the chirality plane caused by diastereotopicity of the complex or macrocycle faces (Figure 18.20). Enantiomers of 21-R**NCP**Ni can be easily separated by means of a chiral stationary phase HPLC [56]. There is no stereoselectivity in the reaction of **NCP**Ni with enantiopure chiral iodoalkyl, i.e. (S)-(+)-1-Iodo-2-methylbutane [122], although analysis of nuclear overhauser effect (NOE) interactions allowed determination of configurations at C21 in the HPLC-separated diastereomers and experimental assignment of CD spectra to the absolute configurations of the complexes [56]. For many other metal ions, chiral 21-alkyl- or 2,21-dialkyl**NCP** complexes have been obtained, including iron(II) [124], iron nitrosyl [125], cobalt [126–128], cobalt nitrosyl [129], and rhenium(I) tricarbonyl [130], though the enantiomers were separated only for certain nickel(II) species (Figure 18.20).

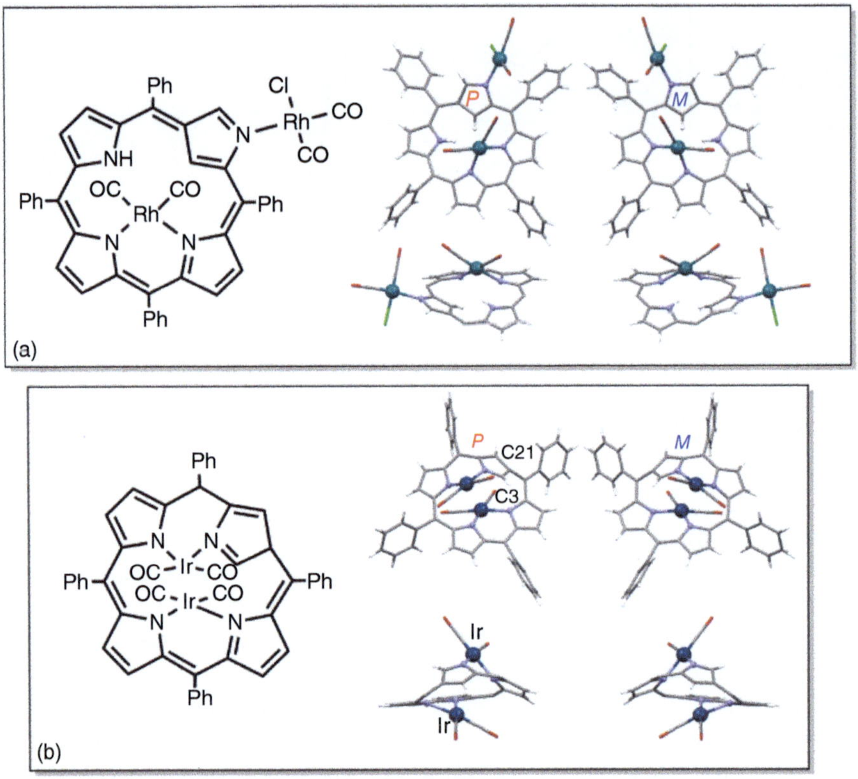

Figure 18.18 Schematic representation and solid-state enantiomers of the **NCP** complexes with bis(rhodium(I)) (a) and bis(iridium(I)) (b). Source: Toganoh et al. [117]/American Chemical Society.

Cyanation or oxygenation/hydroxylation of C21 in several metal complexes of **NCP** and its 2-alkylated derivatives also leads to chiral complexes [108, 126, 128, 131–133]. In the oxygenated complexes, the oxygen atom bound to the inner carbon coordinates to the central ion along with, or instead of C21. Coordination of a substituent introduced to the **NCP**'s interior has been established for 21-methylated iron nitrosyl **NCP** that undergoes oxidation with concurrent deprotonation of 21-Me. The methylene thus formed coordinates tightly to the metal center (1.974(2) Å) closing a ferracyclopropane ring (Figure 18.21) [125]. On the other hand, internally 22-N-methylated **NCP**, subjected to reaction with dirhenium decacarbonyl, gives rise to an internal carbene rhenium(I) tricarbonyl complex with a carbon bridge between C21 and N22 or N24 (Figure 18.21) [130, 134]. Interesting examples of chiral **NCP** complexes with appended apical ligand are those of already mentioned 21-(2′-thiopyridyl)**NCP** [60] that can be obtained in the reaction of **NCP**Ni with 2,2′-dithiodipyridine and then remetalated with other ions, e.g. cobalt(III). For the six-coordinated ruthenium(II) **NCP** complex with apical pyridine and carbonyl, a reaction with 2-mercaptopyridine gives rise to six-coordinated species with axial carbonyl and 21-appended *S*-pyridyl,

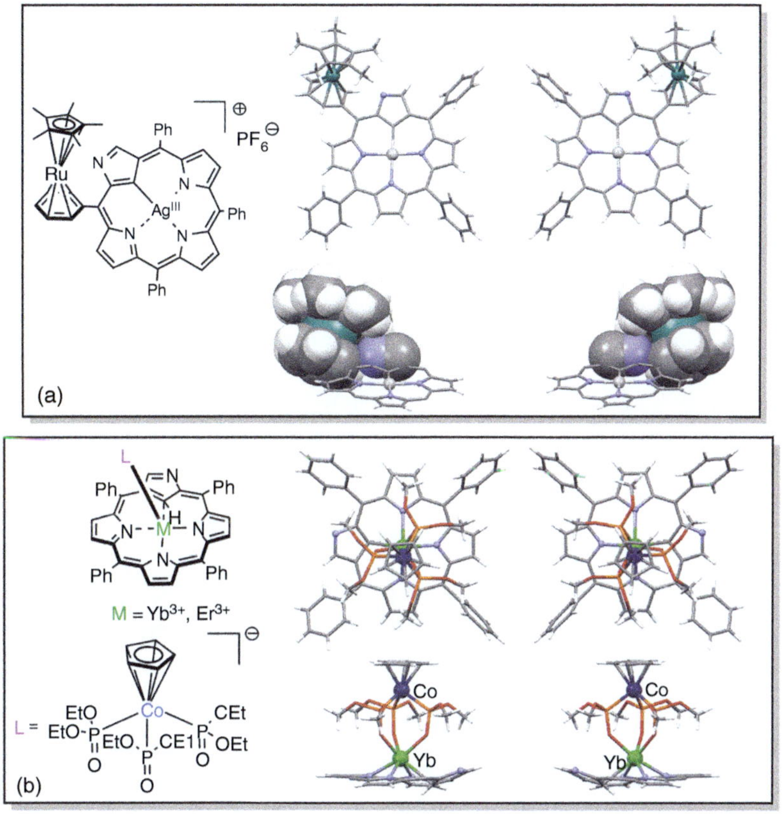

Figure 18.19 Schematic structure and solid-state enantiomers of **NCP**Ag(RuCp*)PF_6 (a, the anion is omitted) and lanthanide complexes of **NCP** (b). In the side views of the structure of [**NCP**Ag(RuCp*)]$^+$, all atoms of the N2–C3 fragment and those of 20-Ph-Ru-Cp* moiety are drawn with spacefill representation to illustrate the steric hindrance in this part of the molecule. Source: Zhu et al. [118]/American Chemical Society.

coordinating *trans* to it (Figure 18.21). The same ruthenium(II) complex reacting with 2-hydroxy- or 2-aminopyridine gives rise to 21-(1′-pyridine)-**NCP**RuCO, i.e. a system with pyridine covalently C–N bounded to the inner carbon of the *confused* pyrrole and with conveniently oriented ketone or imine group, respectively, coordinating in the axial position to the metal center *trans* to CO [135].

18.3 Dimers and Oligomers

The systems consisting of more than one **NCP** subunit can be chiral either owing to a rigid fashion of linkage or due to the predefined chirality of the subunits. Dimers consisting of prochiral subunits liked by a bridge of flexible nature are configurationally unstable. Thus, no effects related to their solid-state chirality can be observed in solution. In particular, a separation of the enantiomers is excluded,

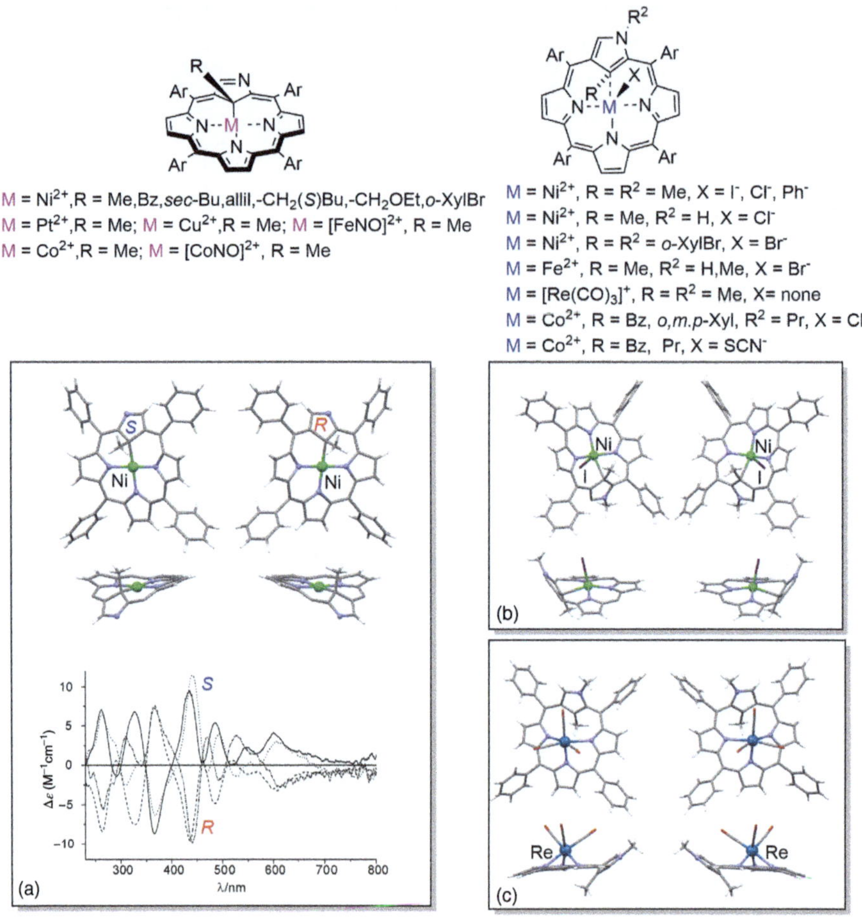

Figure 18.20 Schematic structures of 21- and 2,21-alkylated derivatives of the **NCP** transition metal complexes, solid-state enantiomers of 21-Me**NCP**Ni (a), 2,21-$(Me)_2$**NCP**Ni (b), and 2,21-$(Me)_2$**NCP**$Re(CO)_3$ (c). In the left panel, the CD spectra of the separated enantiomers of 21-Me**NCP**Ni and 21-Bz**NCP**Ni are presented. Source: Chmielewski et al. [56]/John Wiley & Sons.

and differentiation of porphyrin faces or diastereotopicity of methylene protons in ^{1}H NMR cannot be observed under fast exchange conditions. The subunits in an oligomer can be bound covalently or by coordination, and in both types of dimers the linkage may be bridgeless or may require a bridging organic group between macrocycles or a metal ion between donor atoms.

18.3.1 Coordinating Oligomers

The very special feature of **NCP** is the presence of the N-donor on the macrocyclic perimeter that can coordinate a metal ion. One of the first dimeric systems of **NCP** consisted of two zinc complexes linked by a relatively complicated system formed

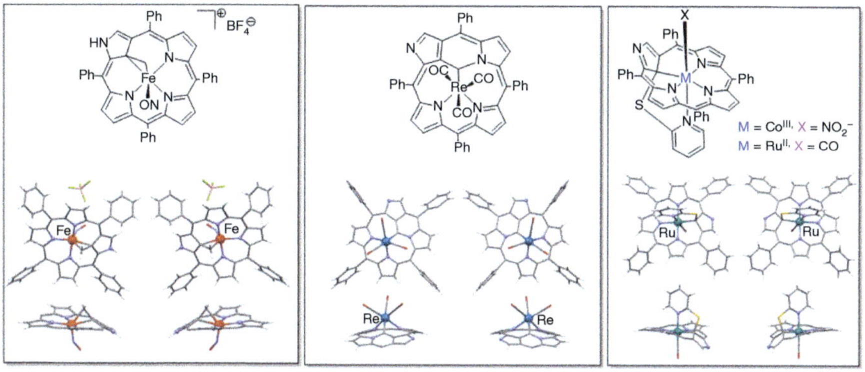

Figure 18.21 Schematic structures and solid-state enantiomers of selected **NCP** complexes with appended groups coordinated to the metal centers. Source: Ching et al, [125]/John Wiley & Sons.

by zinc acetate (Figure 18.22) [102]. The subunits in this complex are homochiral, and the formation of this dimer is diastereoselective. Similarly, the bridgeless coordinative dimers formed for the group 12 metals consist of the subunits of the same chirality [104] while for iron(II) [111], the dimeric complex of this type is achiral comprising heterochiral monomers (Figure 18.22). For manganese(II), both homo- and heterochiral dimers have been found in the solid state by X-ray diffraction analysis [110]. Apparently, these systems are labile, and for the metals of 12 group of the periodic table, the heteronuclear systems, i.e. containing different metal ions in the subunits, have been obtained by substitution. It seems that the lability prevents separation of stereoisomers of the dimers. For monoaryl **NCP** zinc complex, a homochiral self-assembling gives rise to a bridgeless trimer of C_3 symmetry (Figure 18.22) [136].

Another example of dimeric coordinatively linked system involving both porphyrin interior and external nitrogen has been found for palladium(II) that forms a chiral heterodimer, i.e. the metal ions coordinate differently to the porphyrin subunits making them distinct (Figure 18.23) [89]. In one subunit, the palladium(II) ion Pd1 is coordinated to N22a and N23a and to N2b and *ortho*-C of the adjacent *meso*-aryl (C20b) belonging to the other porphyrin subunit. In that other subunit, N23b and N24b are involved in coordination to the other metal ion Pd2 along with N2a and orthometalated aryl in the *meso* site C20a. Also a reaction of **NCP** with understoichiometric amount of platinum(II) chloride results in a heterodimer, though in this case only porphyrin peripheries are involved in coordination to one Pt^{II} ion [90]. One of the subunits coordinates with N2a and *ortho*-carbon of the aryl at C20, while the other – only with N2b. The square-planar environment of the metal is replenished with a chloride coordinating either *cis* or *trans* to the carbanion, though the former isomer is about 10 times more populated than the latter. Obviously, both palladium(II) and platinum(II) dimers are chiral and owing to their inertness could be subjected to enantiomer separation. Our preliminary approach to

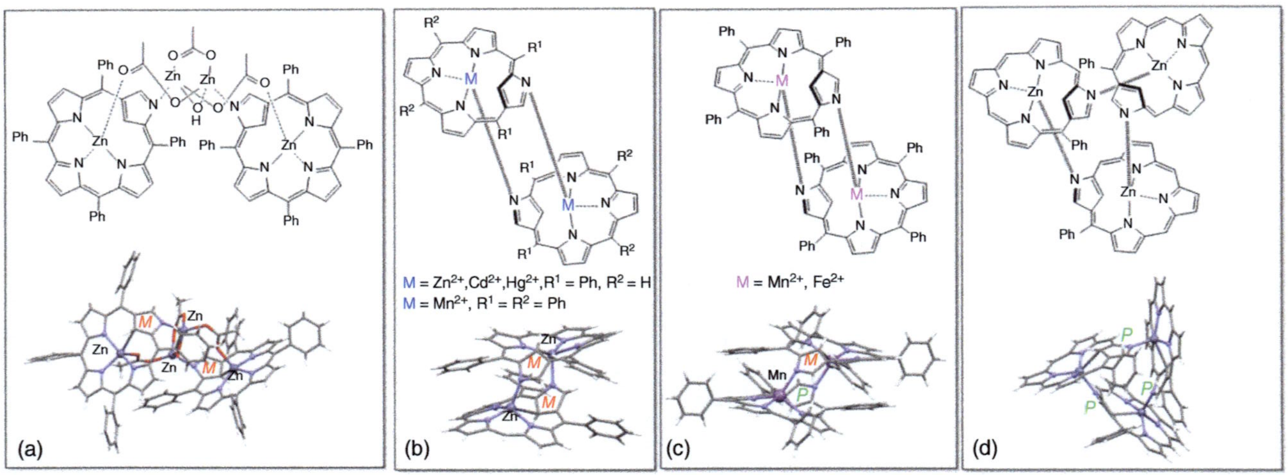

Figure 18.22 Schematic and selected solid-state structures of **NCP** dimers: bis(**NCP**Zn) bridged by $Zn_2(AcO)_3OH$ (a), homochiral bridgeless complexes of group 12 metals and manganese(II) (b), heterochiral bridgeless dimers of manganese(II) and iron(II) (c), and homochiral bridgeless trimer of zinc 5-phenyl-**NCP** (d).

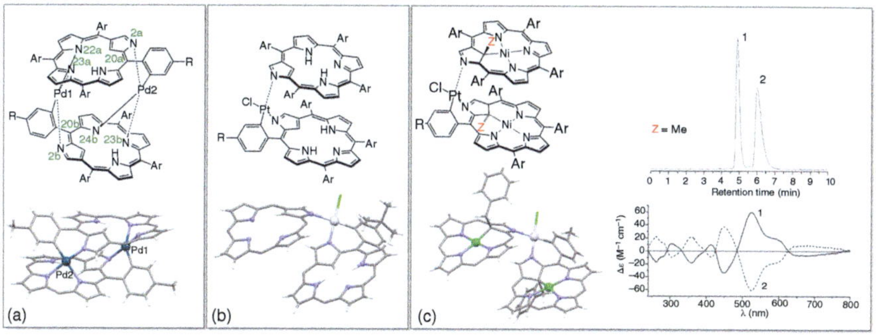

Figure 18.23 Schematic and selected solid-state structures of **NCP** dimers: bis(**NCP**Pd) (a), bis(**NCP**)PtCl (b), and bis(21-Z-**NCP**Ni)PtCl (c). In the crystal structures, all but orthomethalated *meso*-aryls are hidden for clarity. Panel (c) comprises a chiral stationary-phase HPLC profile for bis(21-Me-**NCP**Ni)PtCl and ECD spectra of the enantiomers [56].

separate bis(**NCP**Pd) appears to be successful, and these results will be published in due course. In the platinum(II)-bridged bis(**NCP**), the coordination cavities remain unoccupied, though already at the platinum : **NCP** ratio of 1 : 1, insertion rather than dimerization/bridging takes place. Conversely, for the nickel(II)-metalated 21-alkylated systems [57, 88, 121, 122], even application of an excess of the platinum(II) source (either $PtCl_2$ or $Pt(PhCN)_2Cl_2$) gives rise to a dimeric species of heterodimeric character just described for the "empty" platinum-bridged bis(**NCP**) [91]. The dimerization is diastereoselective with opposite configurations of C21 in the subunits and chloride exclusively at *cis*-position with respect to coordinating 20-*o*-carbon (Figure 18.23). Owing to a helical shape of the molecule and inertness of the platinum(II) ion, the enantiomers of bis(21-Z-**NCP**Ni)PtCl, where Z = Me or Bz, can be separated despite heterochiral character of the subunits (Figure 18.23) [56]. All and all, these three types of systems should not be regarded as dimers constituted by chiral subunits but rather as assemblies of their own stereogenic and chiroptical properties. Significantly, for helical bis(21-Z-**NCP**Ni)PtCl, the molar CD coefficients are about one order of magnitude higher than for monomeric 21-alkylated nickel(II) complexes (compare CD spectra in Figures 18.20 and 18.23).

Rhodium **NCP** dimerizes quite differently using two $Rh(CO)_2$ fragments to link porphyrin subunits by coordination of one rhodium to both external nitrogen atoms, while the other is involved in the $Rh_3(CO)_2$ cluster formation bridging macrocyclic interiors [137]. Such a gable-shaped complex doubly linked by coordination appears to be chiral, but no attempt has been undertaken to separate its stereoisomers (Figure 18.24). Also, μ-oxo-dimer of iron(III) 21-Me**NCP** crystallizes as a homochiral assembly [125]. However, the discrimination of homochirality vs. heterochirality is not definite in this case due to the ambiguity of N2 and C3 site assignments in the solid state. Such an assignment is necessary for the configuration determination of each of the subunits (Figure 18.24) and is often difficult due to a little difference of electron density on carbon and nitrogen and possible disorder at this part of the molecule.

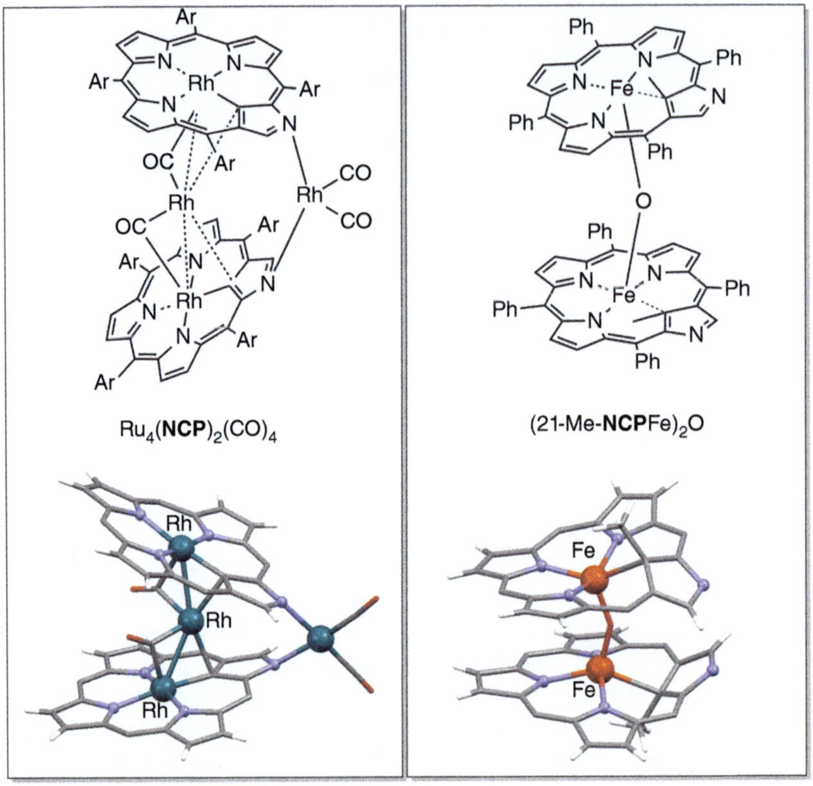

Figure 18.24 Tetrakis(rhodium) bis(**NCP**) tetracarbonyl (a) and μ-oxo-dimer of iron(III) 21-Me-**NCP** (b).

18.3.2 Covalently Linked Dimers

The very first example of the system consisting of two covalently-bridged **NCP** subunits was a heterodimer consisting of **NCP** nickel(II) complexes 2,21′-linked by methylene [121]. This system is formed very effectively and selectively in the reaction of **NCP**Ni with methylene bromide or iodide. As all other 21-substituted **NCP** derivatives, this heterodimer is chiral, and its enantiomers can be easily separated by a chiral stationary-phase HPLC [56]. Another chiral heterodimer 2-(21-*o*-xylene**NCP**Ni)-**NCP** comprises both nickel(II) and an "empty" porphyrin ligand [101]. On the other hand, the 2,2′-*o*-xylene-linked **NCP** dimers [81, 101] are flexible, and no chirality symptoms can be observed for them even at low temperatures. Metalation of such a dimeric ligand with zinc or cadmium gives rise to chiral systems that is a consequence of the chirality of their subunits. Apparently, such metalations are diastereoselective since only one form can be observed by ^{1}H NMR. Unfortunately, as it has been found for a monomers of the group 12 metals, the dimers are labile, and their stereoisomers cannot be separated by column chromatography due to their prompt decomposition. Conversely, the 21,21′-*o*-xylene(**NCP**Ni)$_2$ is sufficiently robust, and its stereoisomers can be easily

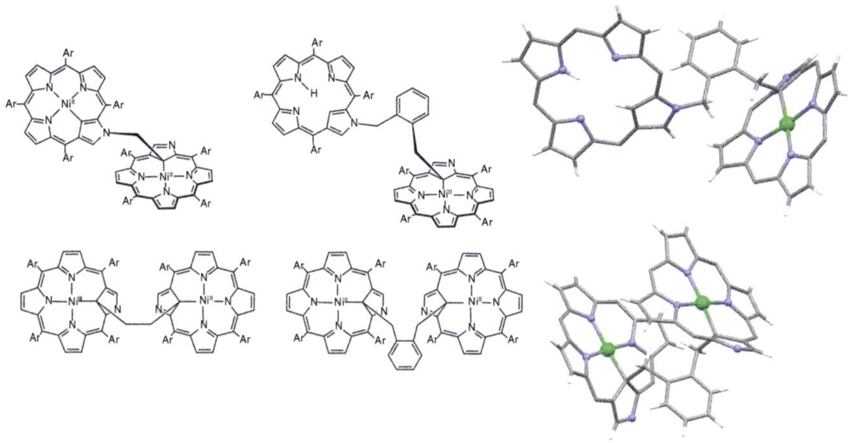

Figure 18.25 Schematic structures of several chiral covalently bridged bis(**NCP**) and solid-state structures of heterodimer 2-(21-*o*-xylene**NCP**Ni)-**NCP** and homodimer 21,21′-*o*-xylene-(**NCP**Ni)$_2$ with all *meso*-aryls removed for clarity.

separated [56]. Due to lack of the stereoselectivity of the alkylation at C21, there are two equally populated diastereomers observed by ^{1}H NMR that can be assigned to a racemic mixture (mixture of enantiomers consisting of homochiral subunits) and a *meso*-form (a heterochiral dimer of opposite configurations at C21 and C21′) [138]. Some of the chiral dimeric systems consisting of **NCP** linked by different bridges are presented in Figure 18.25.

The case of 21,21′-xylene-linked dimer is a good example to illustrate limitation of the X-ray diffraction analysis in the chirality recognition for the systems where determination of configuration relies on a differentiation between carbon and nitrogen as it is for the chiral 21-substituted **NCP** systems. For this dimer, the crystal lattice is constituted by both enantiomers and the diastereomeric *meso*-form, but the refined structure does not discriminate among them, mostly due to a disorder leading to a partial occupancy of N2 by C3 and vice versa, even in the crystals comprising pure enantiomer. Due to such ambiguity, even the presence of a heavy atom such as nickel(II) in the coordination core of the porphyrin does not allow an assignment of the absolute configuration by means of a Flack's analysis.

A readily obtainable bis(**NCP**) linked by dipyrrin moiety [139] consists of a functional bridge linking external carbons of the *confused* pyrrole of the subunits preserving internal and external coordination properties of the porphyrinoid. The dimer, though flexible in its nature, forms a chiral assembly of C_2 symmetry comprising two dimeric units of bis(silver(III)) complexes bridged by multimodal coordination of four silver(I) ions (Figure 18.26). The assembly consists of two homochiral helical units, and thus, the system is topologically chiral, though the enantiomers of the assembly cannot be separated by a chiral stationary phase HPLC due to a labile character of the silver(I) bridges. However, a flexible character of the dipyrrin-linked dimeric free base or bis(silver(III)) complex of this ligand allows induction of optical activity of these species. For example, the addition

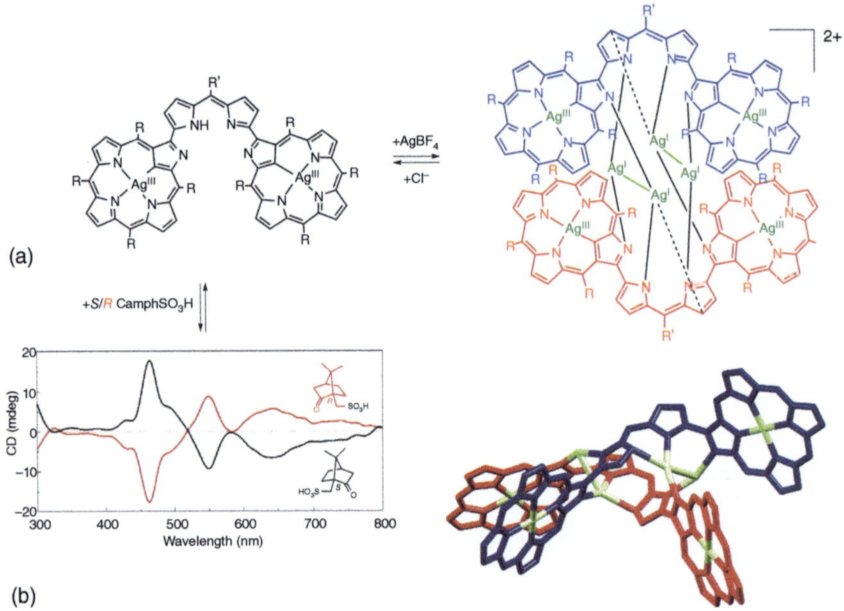

Figure 18.26 Schematic structures of silver(III) complex 2′,2″-bis(3-**NCP**Ag)-*meso*-aryl-dipyrrin, its dimeric assembly linked by silver(I) (a), CD spectra of the adducts with enantiopure camphorsulfonic acids, and solid-state structure of the dimeric assembly with all *meso*-substituents hidden for clarity (b).

of enantiopure camphorsulfonic acid to a dipyrrin-linked dimer consisting of **NCP**Ag^{III} in chloroform solution gives rise to a strong CD signal even at room temperature. Apparently, the external nitrogens of the **NCP** subunits and dipyrrin moiety interact with the acid molecule or camphorsulfonate anion and the helical dimer conforms unidirectional to that chiral seed (Figure 18.26).

The directly linked **NCP** dimers are known in two variants: involving either C21 [140] or C3 [141] sites of the subunits. The first of the bridgeless dimer type, i.e. 21,21′-linked species, has been obtained from the dimeric N-fused porphyrin by usual procedure of inversion under basic conditions as two stereoisomers (9 : 1) with OMe group at C3, while the 3,3′-linkage is formed under strongly acidic condition, starting from an unaltered **NCP**. Both dimers are presented in the Figure 18.27. In both species, the chirality relies on the axis along the unique bond linking the subunits, and thus, configurational stability of these systems depends on the restriction of the rotational freedom around this axis. It seems that enantiomers of 21,21′-(**NCP**)$_2$ should be configurationally stable since the rotation, which implies racemization, appears to be totally arrested by the presence of *meso*-aryls in the positions 5 and 20 of both subunits. Nevertheless, no attempt of the enantiomer separation has been reported to date. The 3,3′-linked dimer is configurationally unstable as is its bis(nickel(II)) complex. At this point, the dimer is similar to the directly β,β′-linked regular *meso*-aryl bis(porphyrin) that, by the way, has been discovered several years after 3,3′-(**NCP**)$_2$ and is also configurationally

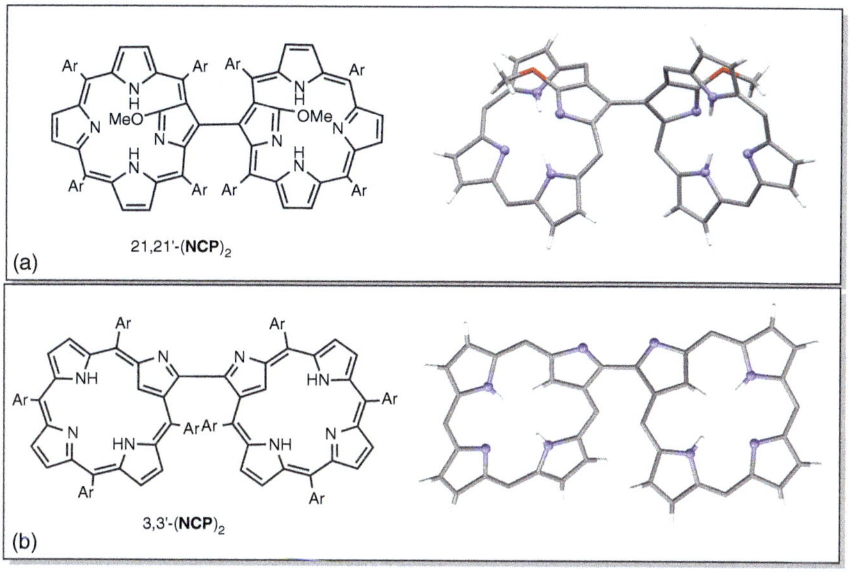

Figure 18.27 Schematic and the solid state structures of 21,21′-(**NCP**)$_2$ (a) and 3,3′-(**NCP**)$_2$ (b). In the solid-state structure representations, all *meso*-aryls are hidden for clarity.

unstable as a free base and in the form of some metal complexes [142, 143]. The subunits of the dimer retain the coordinating properties of the monomeric **NCP**. Thus, metalation and consecutive modifications known for the monomers and described above can be performed for 3,3′-(**NCP**)$_2$. The bis(silver(III)) complex is configurationally stable, and optical activity of its enantiomers has been recognized [144]. Also, alkylation of the bis(nickel(II)) complex with methyl iodide appears to be diastereoselective and hinder rotation around the 3,3′-bond, giving rise to a stable chiral system methylated at inner carbons of both *confused* pyrroles [145]. In fact, for the enantiomers of this system, the absolute configurations have been assigned by the single-crystal diffraction analysis of the enantiomorph obtained by HPLC separation of the enantiomers. That allowed to verify the assignments that have been made on the basis of TD DFT simulation of the CD spectra.

The 3,3′-(**NCP**)$_2$ free base is thus a flexible structure and its conformational variations reflect a mutual orientation of the subunits. There are two limiting conformations of 3,3′-(**NCP**)$_2$ that are related with a dihedral angle formed by the *confused* pyrroles: cisoid or synclinal and bent transoid or anticlinal (Figure 18.28) [146]. These conformations depend on the state of the subunits: for roughly planar system (free base, mono- and dication, silver(III), or nickel(II) complex), the cisoid conformation has been observed, while for the systems with the *confused* pyrroles strongly deviating from the mean plane of the rest of porphyrinoids (tetracation, zinc complexes), the anticlinal conformation is more stable. In both conformations the system is axially chiral. Its rotational flexibility allows induction of optical activity by application of enantiopure guest or ligand. In some cases, such an optical activity is related with generation of enantiomeric excess, but in other systems it

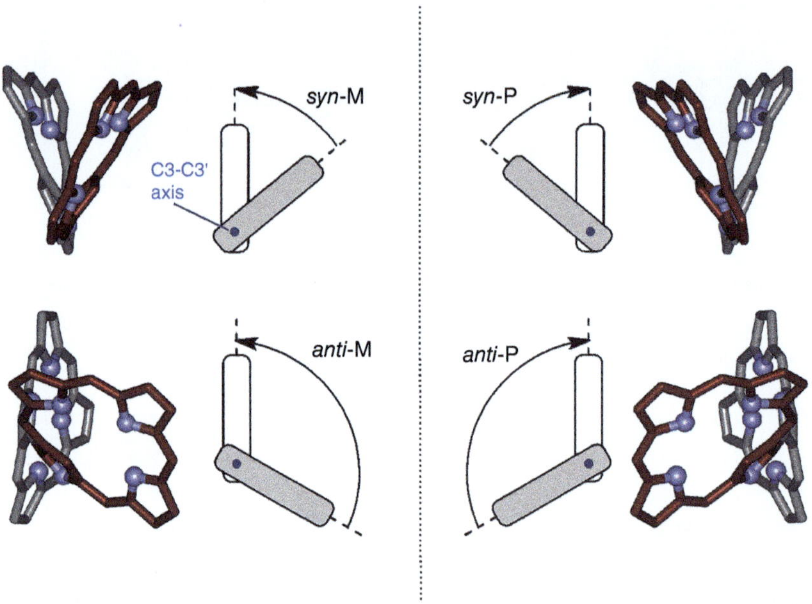

Figure 18.28 Definition of 3,3′-(**NCP**)$_2$ chirality for the synclinal (a) and anticlinal (b) orientations.

reflects small structural differences between the diastereomers formed. For example, exchange of the apical chloride in racemic 3,3′-(**NCP**ZnCl)$_2$ with enantiopure chiral carboxylate gives rise to the pure enantiomer of the complex, and no diastereomer formation is observed [147]. Similar unidirectional configuration change can be observed for a dication 3,3′-[(**NCP**H)$_2$]$^{2+}$ interacting with enantiopure carboxylates. The enantiomeric excess generated by a chiral guest or ligand remains to some extent upon metalation or transmetalation leading to configurationally stable bis(silver(III)) complexes [148] and can be recognized by ECD or HPLC. On the other hand, interaction of racemic bis(zinc) complex with enantiopure alcohols causes no diastereomeric excess, although quite strong circular dichroism signals are observed. Similarly, strong CD spectra are observed for racemic tetraprotonated dimer [3,3′-(**NCP**H$_2$)$_2$]Cl$_4$ upon the addition of enantiopure carboxylic acids or alcohols, but ^{1}H NMR indicates no, or very little, diastereomeric excess [144].

An interesting example of helical configurationally stable yet adaptable system is 3,3′-(**NCP**)$_2$Pt comprising platinum(II) or platinum(IV) ion coordinating to both external nitrogens of the subunits and orthometalating both adjacent aryls [149]. The enantiomers of 3,3′-(**NCP**)$_2$Pt^{II} can be easily separated by means of a chiral stationary phase HPLC and transformed into 3,3′-(**NCP**)Pt^{IV}MeI in the oxidative addition reaction that preserves original configuration and enantiomeric excess. Racemization for these systems is slow ($t_{1/2} \sim 30$ days), but it prevents the absolute configuration assignment by the crystallographic methods because racemic phase is always formed even when crystallization starts from enantiopure species. On the other hand, such a limited flexibility allows unidirectional configuration change

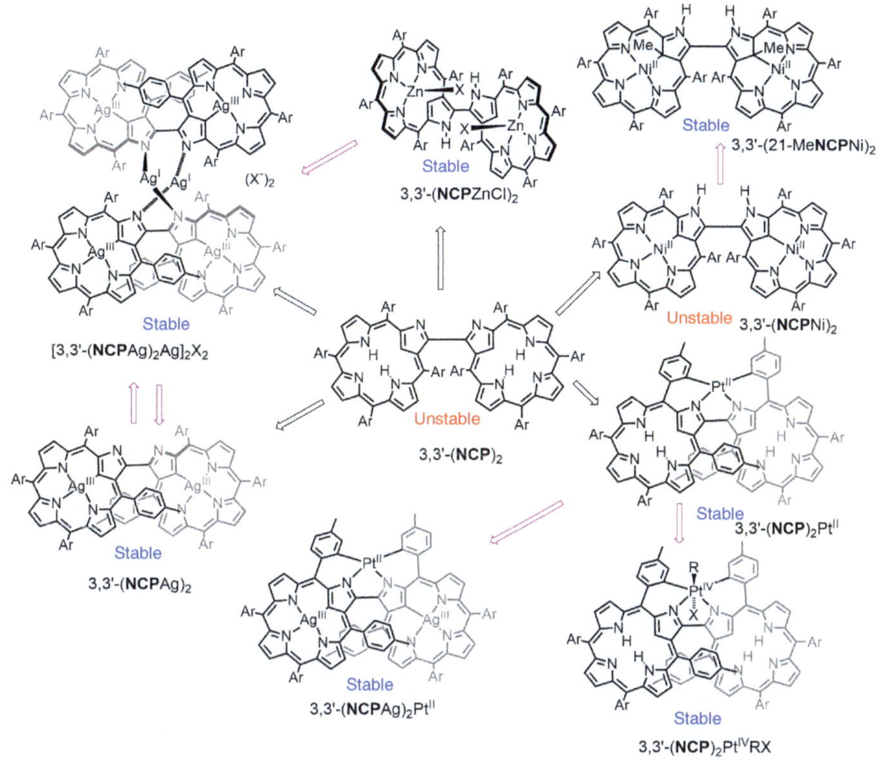

Figure 18.29 Metal complexes of 3,3′-(**NCP**)$_2$ with indication of their configurational stability.

stimulated by the presence of enantiopure monodentate anionic ligand attached to platinum(IV) and introduced by means of the oxidative addition reaction [149]. Conversely, metalation of macrocyclic crevice by silver(III) totally freezes configurational changes, and neither racemization nor induced enantiomerization can be observed for 3,3′-(**NCP**Ag)$_2$Pt^{II} even at elevated temperatures.

Figure 18.29 summarizes studies on the 3,3′-(**NCP**)$_2$ systems and indicates their configurational stability and relations among them.

18.4 Summary and Outlook

After 28 years from its discovery, **NCP** appears to be still undervalued as a chiral ligand, despite many indication of usability of other porphyrinoids in catalysis or supramolecular chemistry. Replacement of the regular porphyrin with its N-confused analogue may enrich a pool of the potential catalysts and create new possibilities of their application. Indeed, the coordination properties of **NCP** are different from those of regular porphyrin, and preparation of the enantiopure complexes may be challenging mostly due to their organometallic character related with involvement of C21 in coordination. On the other hand, some of these unique features, such as external coordination, external coordination with orthometalation, facile bridged or bridgeless oligomerization, stabilization of high oxidation states of the metals coordinated within the macrocyclic cavity, or distinct redox properties of some of the complexes, give opportunity of reshaping the role of this porphyrinoid in the reactions oriented toward stereoselectivity.

References

1 Rothemund, P. (1935). Formation of porphyrins from pyrrole and aldehydes. *J. Am. Chem. Soc.* 57: 2010–2011.

2 Rothemund, P.A. (1936). New porphyrin synthesis. The synthesis of porphin. *J. Am. Chem. Soc.* 58: 625–627.

3 Rothemund, P. and Menotti, A.R. (1941). Porphyrin studies. IV. The synthesis of *a,b,g,d*-tetraphenylporphine. *J. Am. Chem. Soc.* 63: 267–270.

4 Adler, A.D., Longo, F.R., and Shergalis, W. (1964). Mechanistic investigations of porphyrin syntheses. I. Preliminary studies on *ms*-tetraphenylporphin. *J. Am. Chem. Soc.* 86: 3145–3149.

5 Adler, A.D., Longo, F.R., Finarelli, J.D. et al. (1967). Simplified synthesis for *meso*-tetraphenylporphin. *J. Org. Chem.* 32: 476.

6 Adler, A.D. (1979). Some landmarks in the history of porphyrin related materials to 1950 AD. In: *Porphyrin Chemistry Advances* (ed. F.R. Longo), 1–7. Ann Arbor: Science Publishers, Inc.

7 Lindsey, J.S., Hsu, H.C., and Schreiman, I.C. (1986). Synthesis of tetraphenylporphyrins under very mild conditions. *Tetrahedron Lett.* 27: 4969–4970.

8 Lindsey, J.S., Schreiman, I.C., Hsu, H.C. et al. (1987). Rothemund and Adler–Longo reactions revisited: synthesis of tetraphenylporphyrins under equilibrium conditions. *J. Org. Chem.* 52: 827–836.

9 Wagner, R.W., Lawrence, D.S., and Lindsey, J.S. (1987). An improved synthesis of tetramesitylporphyrin. *Tetrahedron Lett.* 28: 3069–3070.

10 Lindsey, J.S. and Wagner, R.W. (1989). Investigation of the synthesis of *ortho*-substituted tetraphenylporphyrins. *J. Org. Chem.* 54: 828–836.

11 Lindsey, J.S. (1994). The synthesis of *meso*-substituted porphyrins. In: *Metalloporphyrins Catalyzed Oxidations* (ed. F. Montanari and L. Casella), 49–86. Dordrecht, The Netherlands: Kluwer Academic Publishers.

12 Lindsey, J.S., MacCrum, K.A., Tyhonas, J.S., and Chuang, Y.-Y. (1994). Investigation of a synthesis of *meso*-porphyrins employing high concentration conditions and an electron transport chain for aerobic oxidation. *J. Org. Chem.* 59: 579–587.

13 Lee, C.-H. and Lindsey, J.S. (1994). One-flask synthesis of *meso*-substituted dipyrromethanes and their application in the synthesis of *trans*-substituted porphyrin building blocks. *Tetrahedron* 50: 11427–11440.

14 Lee, C.-H., Li, F., Iwamoto, K. et al. (1995). Synthetic approaches to regioisomerically pure porphyrins bearing four different *meso*-substituents. *Tetrahedron* 51: 11645–11672.

15 Li, F., Yang, K., Tyhonas, J.S. et al. (1997). Beneficial effects of salts on an acid-catalyzed condensation leading to porphyrin formation. *Tetrahedron* 53: 12339–12360.

16 Ravikanth, M., Strachan, J.-P., Li, F., and Lindsey, J.S. (1998). *Trans*-substituted porphyrin building blocks bearing iodo and ethynyl groups for applications in bioorganic and materials chemistry. *Tetrahedron* 54: 7721–7734.

17 Littler, B.J., Ciringh, Y., and Lindsey, J.S. (1999). Investigation of conditions giving minimal scrambling in the synthesis of *trans*-porphyrins from dipyrromethanes and aldehydes. *J. Org. Chem.* 64: 2864–2872.

18 Geier, G.R. III,, Callinan, J.B., Rao, P.D., and Lindsey, J.S. (2001). A survey of acid catalysts in dipyrromethanecarbinol condensations leading to *meso*-substituted porphyrins. *J. Porphyrins Phthalocyanines* 5: 810–823.

19 Geier, G.R. III, and Lindsey, J.S. (2002). Effects of diverse acid catalysts on the reaction course in the two-step one-flask synthesis of *meso*-tetraphenylporphyrin. *J. Porphyrins Phthalocyanines* 6: 159–185.

20 Wallace, D.M. and Smith, K.M. (1990). Stepwise syntheses of unsymmetrical tetra-arylporphyrins. Adaptation of the macdonald dipyrrole self-condensation methodology. *Tetrahedron Lett.* 31: 7265–7268.

21 Wallace, D.M., Leung, S.H., Senge, M.O., and Smith, K.M. (1993). Rational tetraarylporphyrin syntheses: tetraarylporphyrins from the MacDonald route. *J. Org. Chem.* 58: 7245–7257.

22 Lindsey, J.S. (2000). Synthesis of *meso*-substituted porphyrins. In: *The Porphyrin Handbook* (ed. K.M. Kadish, K.M. Smith and R. Guilard), 45–118. San Diego: Academic Press.

23 Kubo, H., Aida, T., Inoue, S., and Okamoto, Y. (1988). Molecular asymmetry of an *N*-alkylporphyrin with enantiotopic faces. Resolution and spectroscopic characterizations of optical antipodes of *N*-methyletioporphyrin I. *J. Chem. Soc., Chem. Commun.* 1015–1017.

24 Konishi, K., Oda, K., Nishida, K. et al. (1992). Asymmetric epoxidation of olefins catalyzed by manganese complexes of chiral "strapped" porphyrins with diastereotopic faces. A novel strategy for stereochemical modeling of the active site of cytochrome P-450. *J. Am. Chem. Soc.* 114: 1313–1317.

25 Konishi, K., Takahata, Y., Aida, T., and Inoue, S. (1993). Chiral N-substituted porphyrins related to heme inactivation products. First crystallographic determination of absolute stereochemistry and correlation with circular dichroism. *J. Am. Chem. Soc.* 115: 1169–1170.

26 Aida, T. and Inoue, S. (2000). Metalloporphyrins as catalysts for precision macromolecular synthesis. In: *The Porphyrin Handbook* (ed. K.M. Kadish, K.M. Smith and R. Guilard), 133–156. San Diego: Academic Press.

27 Borovkov, V.V., Lintuluoto, J.M., and Inoue, Y. (2000). Supramolecular chirogenesis in bis(zincporphyrin): an absolute configuration probe highly sensitive to guest structure. *Org. Lett.* 2: 1565–1568.

28 Borovkov, V.V., Lintuluoto, J.M., and Inoue, Y. (2001). Supramolecular chirogenesis in zinc porphyrins: mechanism, role of guest structure, and application for the absolute configuration determination. *J. Am. Chem. Soc.* 123: 2979–2989.

29 Borovkov, V.V., Lintuluoto, J.M., and Inoue, Y. (2002). Stoichiometry-controlled supramolecular chirality induction and inversion in bisporphyrin systems. *Org. Lett.* 4: 169–171.

30 Borovkov, V.V., Hembury, G.A., and Inoue, Y. (2004). Origin, control, and application of supramolecular chirogenesis in bisporphyrin-based systems. *Acc. Chem. Res.* 37: 449–459.

31 Borovkov, V.V., Fujii, I., Muranaka, A. et al. (2004). Rationalization of supramolecular chirality in a bisporphyrin system. *Angew. Chem. Int. Ed.* 43: 5481–5485.

32 Borovkov, V.V., Muranaka, A., Hembury, G.A. et al. (2005). Chiral bis-chlorin: enantiomer resolution and absolute configuration determination. *Org. Lett.* 7: 1015–1018.

33 Borovkov, V.V., Hembury, G.A., and Inoue, Y. (2005). Supramolecular chirogenesis with bis-chlorin versus bis-porphyrin hosts: peculiarities of chirality induction and modulation of optical activity. *J. Org. Chem.* 70: 8743–8754.

34 Borovkov, V.V. and Inoue, Y. (2006). Supramolecular chiral recognition by bischlorins: a two-point interaction mode combined with the host's conformational modulation controlled by the guest's stereochemistry and bulkiness. *Org. Lett.* 8: 2337–2340.

35 Hodgson, M.C., Borovkov, V.V., Inoue, Y., and Arnold, D.P. (2006). A new type of chiral porphyrin: organopalladium porphyrins with chiral chelating diphosphine ligands. *J. Organomet. Chem.* 691: 2162–2170.

36 Bhyrappa, P., Borovkov, V.V., and Inoue, Y. (2007). Supramolecular chirogenesis in bis-porphyrins: interaction with chiral acids and application for the absolute configuration assignment. *Org. Lett.* 9: 433–435.

37 Hembury, G.A., Borovkov, V.V., and Inoue, Y. (2008). Chirality-sensing supramolecular systems. *Chem. Rev.* 108: 1–73.

38 Borovkov, V. and Inoue, Y. (2009). A versatile bisporphyrinoid motif for supramolecular chirogenesis. *Eur. J. Org. Chem.* 189–197.

39 Kawamata, M., Sugai, T., Minoura, M. et al. (2017). Nitrogen-bridged metallodiazaporphyrin dimers: synergistic effects of nitrogen bridges and *meso*-nitrogen atoms on structure and properties. *Chem. Asian J.* 12: 816–821.

40 Balaz, M., Holmes, A.E., Banedetti, M. et al. (2005). Synthesis and circular dichroism of tetraarylporphyrin-oligonucleotide conjugates. *J. Am. Chem. Soc.* 127: 4172–4173.

41 Ito, A., Konishi, K., and Aida, T. (1996). Free base of chiral N-substituted porphyrins as catalysts for asymmetric reaction. *Tetrahedron Lett.* 37: 2585–2588.

42 Aida, T. and Jiang, D.-L. (2000). Dendrimer porphyrins and metalloporphyrins: syntheses, structures and functions. In: *The Porphyrin Handbook* (ed. K.M. Kadish, K.M. Smith and R. Guilard), 369–384. San Diego: Academic Press.

43 Toyofuku, K., Alam, M.A., Tsuda, A. et al. (2007). Amplified chiral transformation through helical assembly. *Angew. Chem. Int. Ed.* 46: 6476–6480.

44 Barry, J., Campbell, L., Smith, D.W., and Kodadek, T.J. (1997). Synthesis and catalytic epoxidation activity of terpene-derived D4-symmetric metalloporphyrins. *Tetrahedron* 53: 7753–7776.

45 Labuta, J., Ishihara, S., Šikorský, T. et al. (2013). NMR spectroscopic detection of chirality and enantiopurity in referenced systems without formation of diastereomers. *Nat. Commun.* 4: 3188–3196.

46 Rose, E., Andrioletti, B., Zrig, S., and Quelquejeu-Etheve, M. (2005). Enantioselective epoxidation of olefins with chiral metalloporphyrin catalysts. *Chem. Soc. Rev.* 34: 573–583.

47 Chmielewski, P.J., Latos-Grażyński, L., Rachlewicz, K., and Głowiak, T. (1994). Tetra-*p*-tolylporphyrin with an inverted pyrrole ring: a novel isomer of porphyrin. *Angew. Chem. Int. Ed. Eng.* 33: 779–781.

48 Furuta, H., Asano, T., and Ogawa, T. (1994). "N-confused porphyrin": a new isomer of tetraphenylpoprhyin. *J. Am. Chem. Soc.* 116: 767–768.

49 Geier, G.R. III, and Lindsey, J.S. (1999). N-confused tetraphenylporphyrin and tetraphenylsapphyrin formation in one-flask syntheses of tetraphenylporphyrin. *J. Org. Chem.* 64: 1596–1603.

50 Geier, G.R. III,, Haynes, D.M., and Lindsey, J.S. (1999). An efficient one-flask synthesis of N-confused tetraphenylporphyrin. *Org. Lett.* 1: 1455–1458.

51 Wojaczyński, J., Popiel, M., Szterenberg, L., and Latos-Grażyński, L. (2011). Common origin, common fate: regular porphyrin and N-confused porphyrin yield an identical tetrapyrrolic degradation product. *J. Org. Chem.* 76: 9956–9961.

52 Wojaczyński, J. and Latos-Grażyński, L. (2010). Photooxidation of N-confused porphyr A route to N-confused biliverdin analogues. *Chem. Eur. J.* 16: 2679–2682.

53 Furuta, H., Maeda, H., and Osuka, A. (2002). Regioselective oxidative liberation of aryl-substituted tripyrrinone metal complexes from N-confused porphyrin. *Org. Lett.* 4: 181–184.

54 Chmielewski, P.J. and Latos-Grażyński, L. (2005). Core modified porphyrin – a macrocyclic platform for organometallic chemistry. *Coord. Chem. Rev.* 249: 2510–2533.

55 Furuta, H., Ishizuka, T., Osuka, A. et al. (2001). NH Tautomerism of N-confused porphyrin. *J. Am. Chem. Soc.* 123: 6207–6208.

56 Chmielewski, P.J., Szterenberg, L., and Siczek, M. (2011). Enantiomer resolution of intrinsically chiral C21-alkylated N-confused porphyrin complexes. *Chem. Eur. J.* 17: 1009–1020.

57 Chmielewski, P.J., Latos-Grażyński, L., and Głowiak, T. (1996). Reactions of nickel(II) 2-aza-21-carba-5,10,15,20-tetraphenylporphyrin with methyl iodide. The first structural characterization of a paramagnetic organometallic nickel(II) complex. *J. Am. Chem. Soc.* 118: 5690–5701.

58 Furuta, H., Ishizuka, T., Osuka, A., and Ogawa, T. (2000). "N-Fused Porphyrin": a new tetrapyrrolic porphyrinoid with a fused tri-pentacyclic ring. *J. Am. Chem. Soc.* 122: 5748–5757.

59 Grzegorzek, N., Latos-Grażyński, L., and Szterenberg, L. (2012). Regioselective phosphorylation and thiophosphorylation of N-confused porphyrin: a route to hybrid carbaporphyrinoids. *Org. Biomol. Chem.* 10: 8064–8075.

60 Miyazaki, T., Yamamot, T., Mashita, S. et al. (2018). N-confused porphyrin metal complexes with an axial pyridine directly tethered from an inner carbon: a bioinspired ligand as a versatile platform for catalysis. *Eur. J. Inorg. Chem.* 203–207.

61 Hung, C.-H., Wang, S.-L., Ko, J.-L. et al. (2004). Demetalation of the regioselective oxygenation product of an N-confused porphyrin complex. *Org. Lett.* 6: 1393–1396.

62 Furuta, H., Maeda, H., and Osuka, A. (2002). Confusion, inversion, and creation – a new spring from porphyrin chemistry. *Chem. Commun.* 1795–1804.

63 Ishikawa, Y., Yoshida, I., Akaiwa, K. et al. (1997). Nitration of N-confused porphyrin. *Chem. Lett.* 453–455.

64 Jiang, H.-W., Chen, Q.-Y., Xiao, J.-C., and Gu, Y.-C. (2009). The oxidation of Ni(II) N-confused porphyrins (NCPs) with azo radical initiators and an unexpected intramolecular nucleophilic substitution reaction via a proposed Ni(III) NCP intermediate. *Chem. Commun.* 3732–3734.

65 Li, X., Liu, B., Xu, X., and Chmielewski, P.J. (2012). DDQ-supported alkoxylation of 2.aza-21-carbaporphyrin and noncatalyzed transetherification of its 3,21-dialkoxy derivatives. *J. Org. Chem.* 77: 8206–8219.

66 Furuta, H., Ishizuka, T., Osuka, A., and Ogawa, T. (1999). "N-Fused Porphyrin" from N-confused porphyrin. *J. Am. Chem. Soc.* 121: 2945–2946.

67 Toganoh, M., Kimura, T., and Furuta, H. (2008). Endocyclic extension of porphyrin π-system by interior functionalization of N-confused porphyrins. *Chem. Eur. J.* 14: 10585–10594.

68 Toganoh, M. and Furuta, H. (2012). Blooming of confused porphyrinoids – fusion, expansion, contraction, and more confusion. *Chem. Commun.* 48: 937–954.

69 Toganoh, M., Kimura, T., and Furuta, H. (2008). Endocyclic extension of porphyrin p-system in etheno-bridged N-confused tetraphenylporphyrin. *Chem. Commun.* 102–104.

70 Kashiwagi, N., Akeda, T., Morimoto, T. et al. (2007). Synthesis and isomerization of imino-fused N-confused porphyrin. *Org. Lett.* 9: 1733–1736.

71 Grzegorzek, N., Pawlicki, M., and Latos-Grażyński, L. (2009). Regioselective amination of carbaporpholactone and N-confused porphyrin. *J. Org. Chem.* 74: 8547–8553.

72 Liu, B., Li, X., Maciołek, J. et al. (2014). Regioselective internal carbonylation of the 2-Aza-21-carbaporphyrin: access to configurationally stable chiral porphyrinoids. *J. Org. Chem.* 79: 3129–3139.

73 Toganoh, M., Hihara, T., Yonekura, K. et al. (2009). Synthesis and characterization of novel azo-embedded N-confused tetraphenylporphyrin. *J. Porphyrins Phthalocyanines* 13: 215–222.

74 Ren, D., Liu, B., Li, X. et al. (2019). Reactions of 2-aza-21-carbaporphyrin with aniline derivatives. *Org. Chem. Front.* 6: 908–918.

75 Chmielewski, P.J., Maciołek, J., and Szterenberg, L. (2009). Efficient regiospecific conjugated ring fusion in N-confused porphyrin. *Eur. J. Org. Chem.* 3930–3939.

76 Li, X., Liu, B., Yi, P. et al. (2011). Synthesis of N-confused porphyrin derivatives with a substituted 3-C position. *J. Org. Chem.* 76: 2345–2349.

77 Liu, B., Li, X., Zhang, J., and Chmielewski, P.J. (2013). L-proline catalyzed reaction of N-confused porphyrin and active methylene compounds. *Org. Biomol. Chem.* 11: 4831–4839.

78 Ren, D., Koniarz, S., Li, X., and Chmielewski, P.J. (2020). First imidazole-fused carbaporphyrinoid and its conversion to a N-heterocyclic carbene precursor. *Chem. Commun.* 56: 4856–4839.

79 Schmidt, I. and Chmielewski, P.J. (2001). An inverted porphyrin with a pendant pyrrole – identification of a tetraphenylsapphyrin isomer in the Rothemund synthesis. *Tetrahedron Lett.* 42: 1151–1154.

80 Schmidt, I. and Chmielewski, P.J. (2001). Application of 2-ethylpyrrole for a direct synthesis of 3-substituted inverted porphyrins. *Tetrahedron Lett.* 42: 6389–6392.

81 Chmielewski, P.J. (2005). Extension of N-confused porphyrin by an *o*-xylene fragment. *Org. Lett.* 7: 1789–1792.

82 Li, X., Liu, B., Yu, X. et al. (2012). 1,3-dipolar cycloaddition of 2,6-dichlorobenzonitrile oxide to 2-methyl-N-confused porphyrin. Regio- and stereoselective synthesis and structural characterization of 2-aza-21-carbabacteriochlorin and resolution of 2-aza-21-carbachlorin enantiomers. *J. Org. Chem.* 77: 2431–2440.

83 Depraetere, S., Smet, M., and Dehaen, W. (1999). N-confused calix[4]pyrroles. *Angew. Chem. Int. Ed.* 38: 3359–3361.

84 Furuta, H., Ishizuka, T., Osuka, A. et al. (2001). Metal complexes of an N-confused calix[4]phyrin derivative—the first X-ray structure of an organometallic compound of divalent copper. *Angew. Chem. Int. Ed.* 40: 2323–2325.

85 Li, X., Chmielewski, P.J., Xiang, J. et al. (2006). Synthesis and characterization of pyrrolidin-2-one fused N-confused calix[4]phyrins. *Org. Lett.* 8: 1137–1140.

86 Li, X., Chmielewski, P.J., Xiang, J. et al. (2006). Synthesis of N-confused phlorins via an addition/cyclization pathway. *J. Org. Chem.* 71: 9739–9742.

87 Liu, B., Li, X., Xu, X. et al. (2013). 3-(Dialkoxyphosphoryl)-N-confused phlorin and porphyrin. Synthesis, stereochemistry, and coordination properties. *J. Org. Chem.* 78: 1354–1364.

88 Schmidt, I., Chmielewski, P.J., and Ciunik, Z. (2002). Alkylation of the inverted porphyrin nickel(II) complex by dihalogenalkanes: formation of monomeric and dimeric derivatives. *J. Org. Chem.* 67: 8917–8927.

89 Furuta, H., Kubo, N., Maeda, H. et al. (2000). N-confused double-decker porphyrins. *Inorg. Chem.* 39: 5424–5425.

90 Furuta, H., Youfu, K., Maeda, H., and Osuka, A. (2003). Facile formation of N-confused porphyrin dimers by platinum(II) coordination to the outer-nitrogen atoms. *Angew. Chem. Int. Ed.* 42: 2186–2188.

91 Chmielewski, P.J. and Schmidt, I. (2004). Diastereoselective assembling of 21-C-alkylated nickel(II) complexes of inverted porphyrin on a platinum(II) template. *Inorg. Chem.* 43: 1885–1894.

92 Maeda, H., Ishikawa, Y., Matsueda, H. et al. (2003). Control of Cu(II) and Cu(III) states in N-confused porphyrin by protonation/deprotonation at the peripheral nitrogen. *J. Am. Chem. Soc.* 125: 11822–11823.

93 Furuta, H., Ogawa, T., Uwatoko, Y., and Araki, K. (1999). N-confused tetraphenylporphyrin-silver(III) complex. *Inorg. Chem.* 38: 2676–2682.

94 Toganoh, M., Niino, T., and Furuta, H. (2008). Luminescent Au(III) organometallic complex of N-confused tetraphenylporphyrin. *Chem. Commun.* 4070–4072.

95 Harvey, J.D. and Ziegler, C.J. (2003). Manganese N-confused porphyrin reactivity: CH bond activation and *meso* carbon reduction. *Chem. Commun.* 2890–2891.

96 Harvey, J.D. and Ziegler, C.J. (2004). Dianionic and trianionic macrocycles in cobalt N-confused porphyrin complexes. *Chem. Commun.* 1666–1667.

97 Niino, T., Toganoh, M., Andrioletti, B., and Furuta, H. (2006). Rhodium N-confused porphyrin-catalyzed alkene cyclopropanation. *Chem. Commun.* 4335–4337.

98 Ogawa, T., Furuta, H., Takahashi, M. et al. (2000). Synthesis and characterization of N-confused porphyrinatoantimony(V): toward a low energy gap molecular wire. *J. Organomet. Chem.* 611: 551–557.

99 Xie, Y., Morimoto, T., and Furuta, H. (2006). SnIV complexes of N-confused porphyrins and oxoporphyrins – unique fluorescence "Switch-On" halide receptor. *Angew. Chem. Int. Ed.* 45: 6907–6910.

100 Chmielewski, P.J. and Latos-Grażyński, L. (1997). EPR and ^{2}H NMR studies on the oxidation of nickel(II) 2-aza-21-carbatetraphenylporphyrin to form novel organometallic nickel(III) complexes. *Inorg. Chem.* 36: 840–845.

101 Chmielewski, P.J. (2009). Synthesis and characterization of transition metal complexes of dimeric N-confused porphyrin linked by an *o*-xylene fragment. *Inorg. Chem.* 48: 432–445.

102 Furuta, H., Ishizuka, T., and Osuka, A. (2002). Flexible inner and outer coordination of Zn(II) N-confused porphyrin complex. *J. Am. Chem. Soc.* 124: 5622–5623.

103 Chmielewski, P.J., Latos-Grażyński, L., and Schmidt, I. (2000). Copper(II) complexes of inverted porphyrin and its methylated derivatives. *Inorg. Chem.* 39: 5475–5482.

104 Furuta, H., Morimoto, T., and Osuka, A. (2004). Structures and ligand exchange of N-confused porphyrin dimer complexes with group 12 metals. *Inorg. Chem.* 43: 1618–1624.

105 Furuta, H., Ishizuka, T., and Osuka, A. (2003). Zinc complexes of N-confused calix[4]phyrin. *Inorg. Chem. Commun.* 6: 398–401.

106 Chen, W.-C. and Hung, C.-H. (2001). Synthesis and characterization of iron N-confused porphyrins: structural evidences of agostic interaction. *Inorg. Chem.* 40: 5070–5071.

107 Hung, C.-H., Chang, C.-S., Ching, W.-M., and Chuang, C.-H. (2006). Molecular assembling using axial phenolate on an iron N-confused porphyrin complex. *Chem. Commun.* 1866–1868.

108 Rachlewicz, K., Wang, S.-L., Ko, J.-L. et al. (2004). Oxidation and oxygenation of iron complexes of 2-aza-21-carbaporphyrin. *J. Am. Chem. Soc.* 126: 4420–4431.

109 Rachlewicz, K., Wang, S.-L., Peng, C.-H. et al. (2003). Remarkable paramagnetically shifted ^1H and ^2H NMR spectra of iron(II) complexes of 2-aza-21-carbaporphyrin – an evidence for agostic interaction. *Inorg. Chem.* 42: 7348–7350.

110 Harvey, J.D. and Ziegler, C.J. (2002). Dimeric and monomeric forms of manganese N-confused porphyrin. *Chem. Commun.* 1942–1943.

111 Hung, C.-H., Chen, W.-C., Lee, G.-H., and Peng, S.-M. (2002). Dimeric iron N-confused porphyrin complexes. *Chem. Commun.* 1516–1517.

112 Ge, Y., Cheng, G., Xu, N. et al. (2019). Zinc 2-N-methyl N-confused porphyrin: an efficient catalyst for the conversion of CO_2 into cyclic carbonates. *Catal. Sci. Technol.* 4255–4261.

113 Chen, Y.C., Tung, J.-Y., Liu, T.-K. et al. (2018). The h^1-H–C...Hg agostic interactions in the mercury complexes of N-confused porphyrin. *Dalton Trans.* 47: 14774–14784.

114 Bohle, D.S., Chen, W.-C., and Hung, C.-H. (2002). Metal oxidation promoted C–H activation in manganese complexes of N-confused porphyrin. *Inorg. Chem.* 41: 3334–3336.

115 Qu, W., Ding, T., Cetin, A. et al. (2006). Facile peripheral modification of N-confused porphyrin. *J. Org. Chem.* 71: 811–814.

116 Srinivasan, A., Furuta, H., and Osuka, A. (2001). The first bis-Rh(I) metal complex of N-confused porphyrin. *Chem. Commun.* 1666–1667.

117 Toganoh, M., Konagawa, J., and Furuta, H. (2006). Bis[iridium(I)] complex of inverted N-confused porphyrin. *Inorg. Chem.* 45: 3852–3854.

118 Zhu, X., Wong, W.-K., Lo, W.-K., and Wong, W.-Y. (2005). Synthesis and crystal structure of the first lanthanide complex of N-confused porphyrin with an η^2 agostic C–H interaction. *Chem. Commun.* 1022–1024.

119 Yamamoto, T., Mitsuno, K., Mori, S. et al. (2018). Two discrete RuCp* (Cp* = pentamethylcyclopentadienyl) binding modes of N-confused porphyrins: peripheral π complex and sitting atop ruthenocenophane complex by skeletal transformation. *Chem. Eur. J.* 24: 6742–6746.

120 Chmielewski, P.J. and Latos-Grażyński, L. (2000). Reactions of nickel(II) 2,21-dimethyl-2-aza-21-carbaporphyrin with phenyl grignard reagents, phenyllithium, and *n*-butyllithium. *Inorg. Chem.* 39: 5639–5647.

121 Schmidt, I. and Chmielewski, P.J. (2002). First example of a covalently bound dimeric inverted porphyrin. *Chem. Commun.* 92–93.

122 Schmidt, I. and Chmielewski, P.J. (2003). Nickel(II) complexes of 21-C-alkylated inverted porphyrins: synthesis, protonation, and redox properties. *Inorg. Chem.* 42: 5579–5593.

123 Won, D.-H., Toganoh, M., Uno, H., and Furuta, H. (2009). Pt(II) N-confused porphyrin: an expanded pyrrole that affords a stable π-anion. *Dalton Trans.* 6151–6158.

124 Rachlewicz, K., Gorzelańczyk, D., and Latos-Grażyński, L. (2006). Iron complexes of C- and N-methylated 2-aza-21-carbaporphyrin: NMR studies. *Inorg. Chem.* 45: 9742–9747.

125 Ching, W.-M. and Hung, C.-H. (2012). Interior aliphatic C–H bond activation on iron(II) N-confused porphyrin through synergistic nitric oxide binding and iron oxidation. *Chem. Commun.* 48: 4989–4991.

126 Hung, C.-H., Peng, C.-H., Shen, Y.-L. et al. (2008). Preparation and oxygenation of cobalt N-confused porphyrin nitrosyl complexes. *Eur. J. Inorg. Chem.* 1196–1199.

127 Lai, Y.Y., Chang, Y.-C., Chen, J.-H. et al. (2016). Axial zero-field splitting in mononuclear Co(II) 2-N-substituted N-confused porphyrin: Co(2-NC_3H_5-21-Y-$CH_2C_6H_4CH_3$-NCTPP)Cl (Y = o, m, p) and Co(2-NC_3H_5-21-$CH_2C_6H_5$-NCTPP)Cl. *Dalton Trans.* 45: 4854–4862.

128 Chen, Y.-C., Tung, J.-Y., Liu, T.-K. et al. (2019). The selectivity between inner C-cyanation and alkylation in cobalt(II) complexes of N-confused porphyrin with an axial NCS ligand. *Polyhedron* 159: 233–238.

129 Chuang, C.-H., Liaw, W.F., and Hung, C.-H. (2016). Conversion of nitric oxide into nitrous oxide as triggered by the polarization of coordinated NO by hydrogen bonding. *Angew. Chem. Int. Ed.* 55: 5190–5194.

130 Yamamoto, T., Toganoh, M., Mori, S. et al. (2012). Rhenium complexes of peripherally π-extended N-confused porphyrins. *Chem. Sci.* 3: 3241–3248.

131 Xiao, Z., Patrick, B.O., and Dolphin, D. (2003). Inner C-cyanide addition and nucleophilic addition to Ni(II) N-confused porphyrins. *Chem. Commun.* 1062–1063.

132 Xiao, Z., Patrick, B.O., and Dolphin, D. (2003). Ni(III) complex of an N-confused porphyrin inner C-oxide. *Inorg. Chem.* 42: 8125–8127.

133 Lin, W.-C., Hsaio, D.-Z., Chang, W.-P. et al. (2012). Molecular structures of Cu(II), Rh(III) complexes of 2-N substituted N-confused porphyrin inner C-oxide. *Polyhedron* 42: 241–248.

134 Toganoh, M., Hihara, T., and Furuta, H. (2010). N-heterocyclic carbene embedded in an N-confused porphyrin framework. *Inorg. Chem.* 49: 8182–8184.

135 Miyazaki, T., Fukuyama, K., Mashita, S. et al. (2019). Ruthenium N-confused porphyrins: selective reactivity for ambident 2-heteroatom-substituted pyridines serving as axial ligands. *ChemPlusChem* 84: 603–607.

136 Morimoto, T., Uno, H., and Furuta, H. (2007). Benzene ring trimer interactions modulate supramolecular structures. *Angew. Chem. Int. Ed.* 46: 3672–3675.

137 Toganoh, M., Niino, T., Maeda, H. et al. (2006). Unprecedented formation of a rhodium cluster triggered by rhodium-fastened N-confused gable porphyrin. *Inorg. Chem.* 45: 10428–10430.

138 Chmielewski, P.J. (2007). Synthesis, structure, and redox properties of N-confused bis(porphyrinatonickel(II)) linked by o-xylene. *Inorg. Chem.* 46: 1617–1626.

139 Wojaczyński, J., Maciołek, J., and Chmielewski, P.J. (2017). Dipyrrin–bis(N-confused porphyrin) conjugate: synthesis, synergetic ligation and chirality sensing. *Chem. Asian J.* 12: 643–647.

140 Ishizuka, T., Osuka, A., and Furuta, H. (2004). Inverted N-confused porphyrin dimer. *Angew. Chem. Int. Ed.* 43: 5077–5081.

141 Chmielewski, P.J. (2004). Synthesis and characterization of a directly linked N-confused porphyrin dimer. *Angew. Chem. Int. Ed.* 43: 5655–5658.

142 Bringmann, G., Rüdenauer, S., Götz, D.G.C. et al. (2006). Axially chiral directly β,β′-linked bisporphyrins: synthesis and stereostructure. *Org. Lett.* 8: 4743–4746.

143 Bringmann, G., Goetz, D.C.G., Gulder, T.A.M. et al. (2008). Axially chiral, β,β’-bisporphyrins: synthesis and configurational stability tuned by the central metals. *J. Am. Chem. Soc.* 130: 17812–17825.

144 Chmielewski, P.J., Siczek, M., and Stępień, M. (2015). Bis(N-confused porphyrin) as a semirigid receptor with a chirality memory: a two-way host enantiomerization through point-to-axial chirality transfer. *Chem. Eur. J.* 21: 2547–2559.

145 Chmielewski, P.J. and Maciołek, J. (2012). Diastereoselective methylation of bis(N-confused porphyrinatonickel(II)): access to configurationally stable chiral bis(porphyrinoid) and non-symmetric dimers. *Chem. Commun.* 48: 428–430.

146 Chmielewski, P.J., Siczek, M., and Szterenberg, L. (2011). Protonated N-confused porphyrin dimer: formation, structure, and guest binding. *Inorg. Chem.* 50: 6719–6736.

147 Siczek, M. and Chmielewski, P.J. (2007). Synthesis, characterization, and chirality of dimeric N-confused porphyrin-zinc complexes: toward the enantioselective synthesis of bis(porphyrinoid) system. *Angew. Chem. Int. Ed.* 46: 7432–7436.

148 Chmielewski, P.J. (2005). Synthesis and characterization of a cyclic bis-silver(I) assembly of four 2-aza-21-carbaporphyrinatosilver(III) subunits. *Angew. Chem. Int. Ed.* 44: 6417–6420.

149 Chmielewski, P.J., Durlej, B., Siczek, M., and Szterenberg, L. (2009). Helical bis(N-confused porphyrins) with subunits fused by double orthometalation with platinum: adaptability of an apparently rigid system. *Angew. Chem. Int. Ed.* 48: 8736–8739.

19

Catenane, Rotaxane, and Molecular Knot Chiral Building Blocks

Jean-Claude Chambron

University of Strasbourg, Institut de Chimie de Strasbourg, CNRS, UMR 7177, 1, rue Blaise Pascal, 67008 Strasbourg, France

19.1 Introduction

In chemistry, building blocks, from the most general viewpoint, are molecular fragments that can be found in molecular and macromolecular structures, of natural or synthetic origin, and that impart them with either structural properties at the molecular or supramolecular level (e.g. aromatic moiety or gelator, respectively) or physical properties resulting from the interaction of matter with light, electric, and magnetic fields. Functional groups are a type of building blocks of their own, as they confer chemical properties to the molecules. Chiral building blocks are a particular case of structural building blocks, in relation with symmetry properties. Examples of naturally occurring chiral building blocks are the aminoacids and the carbohydrates, which, among other families of compounds, such as the alkaloids, constitute the so-called chiral pool. Aminoacids and carbohydrates are building block components of natural polymers, often exhibiting helical chirality, the proteins for the former, and the nucleic acids for the latter. Typical chiral synthetic building blocks are those found in ligands employed in metal catalysis, e.g. chiral phosphines with central chirality (C_1-chiral at phosphorus or C_2-chiral at diphosphines) and axial chirality (2,2′-bis(diphenylphosphino)-1,1′-binaphthyl (BINAP) and related molecules), which played a central role in the development of asymmetric catalysis.

Chiral molecules lack symmetry planes and a center of inversion in all of their conformations, and their enantiomers cannot be converted into each other in normal conditions [1]. There are, however, chiral molecules, the enantiomers of which could not be exchanged into each other even if they were endowed with complete flexibility. These molecules are qualified as "topologically chiral," and typical examples can be found among catenanes and molecular knots [2, 3]. Since the pioneering works of Schill [4], the Nobel Prize winners Sauvage [5] and Stoddart [6], and their coworkers, who developed the first syntheses of these compounds, the field has expanded considerably. In particular, the use of dynamic coordination and covalent

Chiral Building Blocks in Asymmetric Synthesis: Synthesis and Applications, First Edition.
Edited by Elżbieta Wojaczyńska and Jacek Wojaczyński.

bonds represented recent breakthroughs, especially for the synthesis of higher-order catenanes and molecular knots. We shall only mention a couple of these molecules throughout this chapter. The interested reader can refer to the numerous comprehensive and excellent works that have been published recently [7–10]. In this essay, we shall show how catenanes, rotaxanes, and molecular knots can constitute chiral building blocks, that is, how they were used in the construction of more complex chiral molecular structures. It is important to note that the chirality and the particular stereoisomerism exhibited by the molecules discussed here result only from interlocking and threading followed by stoppering and knotting. The classification of all types of chiralities exhibited by catenanes and rotaxanes has been presented and discussed in extensive reviews recently [11, 12].

19.2 Elements of Molecular Topology

The modern representation of molecules uses graphs in which vertices stand for atoms and edges stand for bonds [3]. The molecular graph accounts for atom connectivity only, i.e. the *intrinsic* topology of the molecule, and must be completed by the knowledge of its geometrical properties, i.e. the distances between atoms and the angles between bonds. Distances between directly bonded atoms and angles between adjacent bonds can vary only within restricted limits (e.g. those allowed by vibrations), whereas when affecting atoms that are not directly interconnected or bonds that do not share any atom, the variations of the geometrical parameters may encompass much larger values, *while being finite*. These limits can be set by the mechanics of bond rotation, the fixed nature of the bond distances, and/or the steric hindrance. It is the variations within these limits that generate molecular conformations. Contrary to the actual molecular structure, when placed and handled in the 3D space (such an operation is called *embedding*), the molecular graph is endowed with complete flexibility and can take up an infinity of *presentations*. A graph is said to be planar when it is possible to find a presentation that can be projected in the plane without bond crossing.

There are several cases, however, for which such a molecular graph incompletely describes the molecular topology. These cases show up when the presentation irreducibly leaves at least one bond crossing whatever the deformations imparted and their magnitude [13–15]. Leaving aside the case of the rare molecules having *intrinsically non-planar* graphs, we shall focus on catenanes and molecular knots. Two examples are given in Figure 19.1. Figure 19.1a shows the chemical structure of the copper(I) complex $[Cu(\mathbf{1})]^+$ of [2]catenane **1**, a molecule made of two interlocking rings [16]. The molecular graph corresponding to a [2]catenane is a set of separated rings (Figure 19.2a). However, to have a complete description of the topology of the [2]catenane, it is necessary to *embed* its molecular graph into the 3D space. This operation produces a pair of interlocked rings, which represents the *extrinsic* topology of the [2]catenane (Figure 19.2b). Figure 19.1b shows the chemical structure of a synthetic molecular trefoil knot (**2**) [17]. Its intrinsic topology is called a circuit graph (Figure 19.2c). Again, the *extrinsic* topology of the trefoil knot

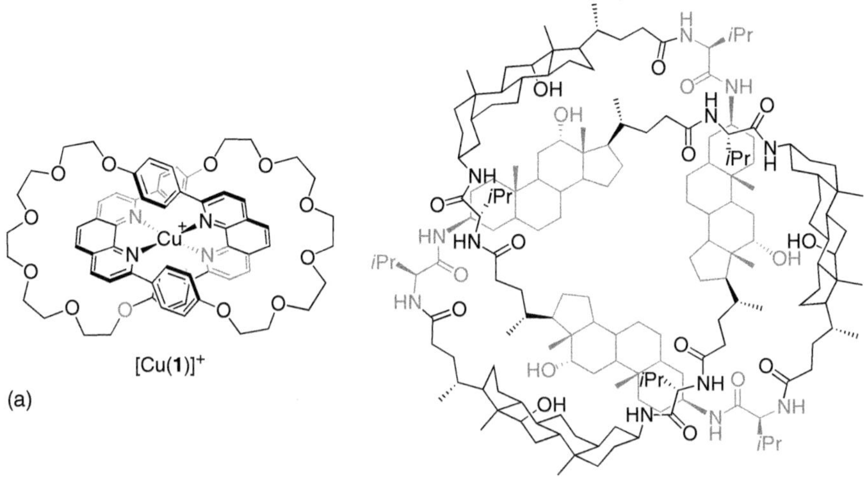

$[Cu(\mathbf{1})]^+$

(a)

(b) **2**

Figure 19.1 Chemical structures of (a) the copper(I) complex $[Cu(\mathbf{1})]^+$ of [2]catenane (**1**) and (b) the molecular trefoil knot **2** (Λ enantiomer).

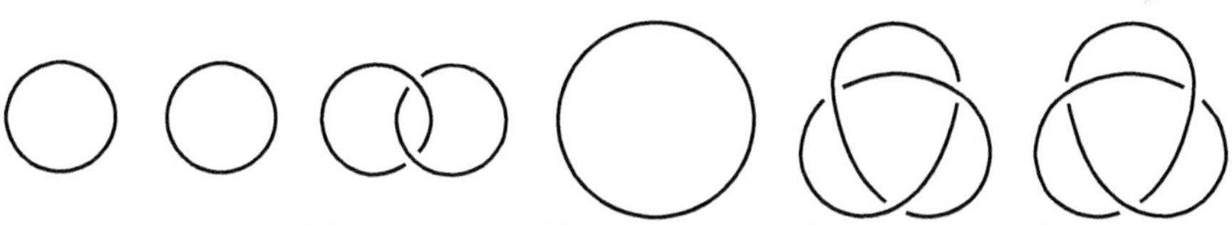

(a) (b) (c) (d) (e)

Figure 19.2 Schematic representations and embedded presentations of the molecular graphs of a [2]catenane and a molecular trefoil knot. (a) The graph of a [2]catenane is the set of its separated rings. (b) Embedded D_{2d} presentation of a [2]catenane. (c) The circuit graph is the graph of prime knots, including the trefoil knot. (d and e) Embedded D_3 presentations (also called diagrams) of the trefoil knot: Λ enantiomer (d) and Δ enantiomer (e).

will appear once its molecular graph is embedded in the 3D space. Figure 19.2d,e shows the highly D_3-symmetric presentations (in the case of a knot it is called a diagram) of the left-handed (Λ) and right-handed (Δ) enantiomers of the trefoil knot, respectively. For the cases illustrated here and analogous molecules, the constitutive chemical bonds represent impassable *frontiers*. In summary, the *extrinsic topology* of molecules with a circuit graph or a set of circuit graphs (which constitute their intrinsic topology) is completely described by embedding the molecular graphs into the 3D space. For example, each knot is a particular embedding of the circuit graph, the trefoil knot being the particular embedding the presentations that show a minimum of three crossings.

It is important to note that the foundations of molecular topology were established on molecules made from covalent bonds, the topology of which, under the conditions of observation (solvent, concentration, temperature) could be considered as an invariant, that of the associated model [3]. Half a century later, the situation has dramatically changed, as the most productive methods for making molecules exhibiting nontrivial topologies of increasing complexities (e.g. multiply interlocked rings, higher-order knots) rely on self-assembly using dynamic covalent and coordination bonds [18–25]. These processes, which operate at thermodynamic equilibrium, use reversible bonds, of either covalent or non-covalent nature. Examples of equilibria between structures of different topologies have been reported, as well as strategies to shift the equilibria to produce a molecule of a given topology at the expense of the others. In these conditions, the reversible chemical bonds no longer represent impassable frontiers, and the topology is no longer an invariant of the object molecule. Therefore, the topological model of a molecule and its effective topology will coincide only in given experimental conditions.

Rotaxanes are molecules that are basically composed of two types of components: rings and axles, the latter bearing terminal groups bulky enough to prevent the rings from unthreading from the axle. Figure 19.3 shows the example of the copper(I) complex [Cu(**3**)]$^+$ of the bis(porphyrin)-stoppered [2]rotaxane (**3**) [26]. In the 3D space, a [2]rotaxane has two equivalent embeddings, one in which the ring is threaded along the axle and another in which ring and dumbbell are separated. The analogy of the structural characteristics of rotaxanes and catenanes with those of objects of the macroscopic world that are part of mechanical devices and machines has led to the concept of the mechanical bond, which holds together the rings of a catenane and the rotaxane components [27]. Actually, the existence of rotaxanes relies on *finite* steric hindrance, which can be overcome by a temperature

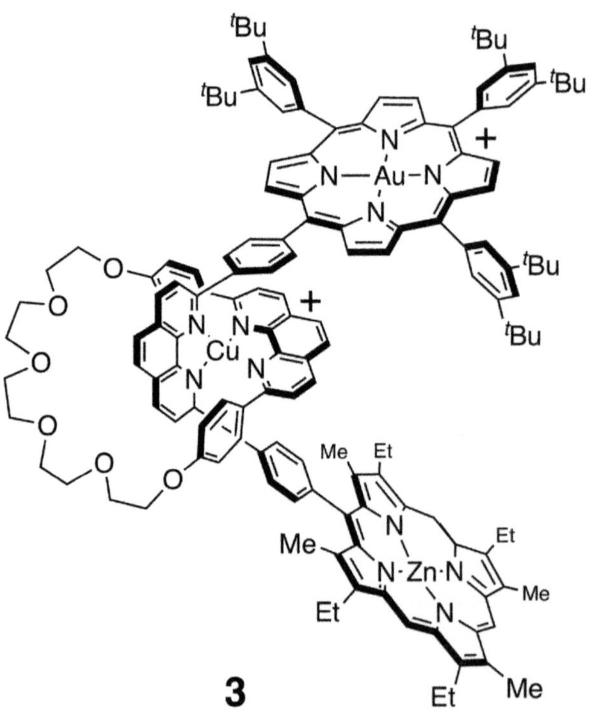

3

Figure 19.3 Chemical structure of porphyrin-stoppered [2]rotaxane **3**.

increase, for example. By contrast, rings of catenanes could be considered as held altogether by *infinite* steric hindrance (the impassable frontiers formed by the sequence of bonds). Therefore, from the chemical viewpoint and in conditions where the rings cannot pass over the stoppers, the rotaxanes can be considered as catenane analogues. For example, the projection of the 3D structure of the basic [2]rotaxane on a plane shows two crossings as the basic [2]catenane.

19.3 Topological Chirality and Chiral Catenanes

Catenanes and molecular knots are classified as topologically chiral if and only if their enantiomers cannot be interconverted by continuous deformations [3]. For example, to convert the left-handed trefoil knot into its right-handed enantiomer, it is clear that it would be necessary that the three pairs of intersecting bonds pass through each other. Topologically chiral knotted molecules owe their chirality to their embedding in the 3D space, without any other condition. Therefore, they are qualified as *intrinsically* topologically chiral. Catenanes, molecules made of interlocked rings, can be intrinsically or *conditionally* topologically chiral. For example, the most trivial embeddings of the simplest catenane, made of two interlocked rings, are those whose presentations show respectively two and four crossings. These embeddings are known as Hopf link and Solomon rings, respectively [7]. The two-crossing [2]catenane is clearly achiral (D_{2d} symmetry). However, if we impart an orientation to each of the rings, its symmetry is decreased to D_2, and it becomes topologically chiral, because the orientations of the rings cannot be reversed. Since it depends on ring orientation, the topological chirality of the two-crossing [2]catenane is *conditional* (Figure 19.4a, b). The four-crossing [2]catenane owes its chirality solely to its embedding; therefore, as the trefoil knot, it is intrinsically topologically chiral (Figure 19.4c, d). Since the early reports on topologically chiral [2]catenanes by Sauvage and coworkers [28–30], with few exceptions, most subsequent reports concerned the doubly interlocked [2]catenanes (Figure 19.4c, d), which were generally obtained in thermodynamically controlled reactions [31–35].

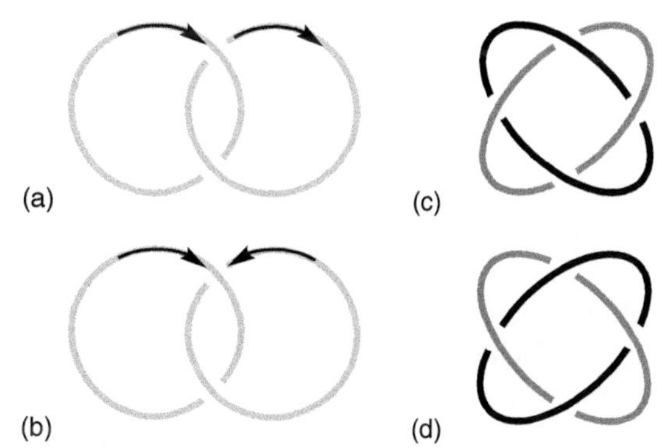

Figure 19.4 Embeddings of the enantiomers of topologically chiral (a and b) Hopf link with oriented rings and (c and d) Solomon rings. In the latter case, the rings have been made different for clarity.

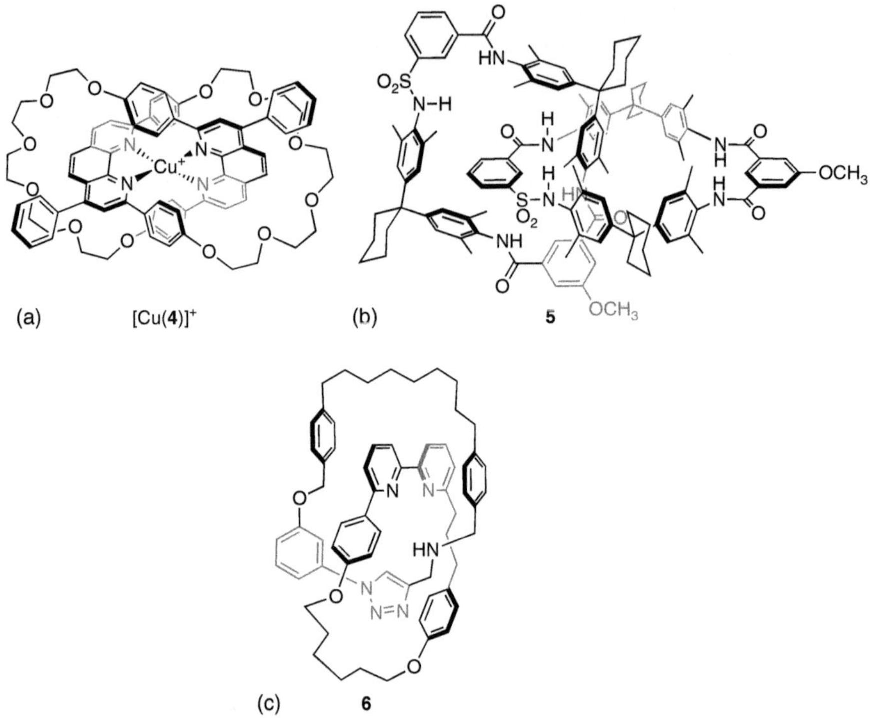

Figure 19.5 Chemical structures of conditionally topologically chiral [2]catenanes: (a) copper(I) complex [Cu(**4**)]$^+$, (b) amide-based [2]catenane **5**, and (c) [2]catenane **6** obtained by a stereoselective active template synthesis using a chiral auxiliary (one enantiomer shown in each case).

The very first topologically chiral [2]catenane was reported by Mitchell and Sauvage (Figure 19.5a) [28]. The catenane (**4**) was prepared following the copper(I) template methodology. The orientation of the 2,9-diphenyl-1,10-phenanthroline (dpp)-incorporating macrocyclic component was imparted by an extra single phenyl substituent, which was placed at the position 4 of the dpp chelate at the very beginning of the synthesis of the phenanthroline moiety, making the sequence (4,7) enantiomer of the sequence (7,4). Few years later, Vögtle and his group achieved successfully the preparation of a topologically chiral [2]catenane (**5**, Figure 19.5b) made from lactam macrocycles incorporating three amide bonds and a sulfonamide, the sequence of which conferred an orientation to the macrocycles [36]. Recently, Goldup and his group reported the stereoselective synthesis of the topologically chiral catenane **6** (Figure 19.5c) by an active metal template coupling performed in the presence of a chiral auxiliary [37]. It is interesting to compare the structures of the first (**4** and **5**) and the last (**6**) generation topologically chiral [2]catenanes. The orientation of the rings in the former is obtained by a minimal change of the overall structure of the molecule, which has D_2 average symmetry, whereas in the latter it is obtained from macrocycles of heterogeneous composition. The corresponding [2]catenanes are therefore asymmetric.

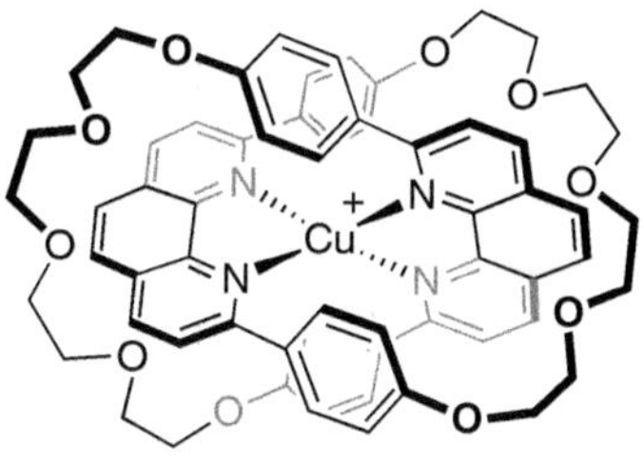

[Cu(**7**)]$^+$

Figure 19.6 The chiral copper(I) [2]catenane complex [Cu(**7**)]$^+$ (one enantiomer shown). Compared with [Cu(**1**)]$^+$ and [Cu(**4**)]$^+$, each constituent macrocycle of [Cu(**7**)]$^+$ contains one oxyethylene subunit less.

Actually, the very first [2]catenane identified as chiral (**7** of Figure 19.6) did not incorporate oriented rings, nor had a topologically chiral embedding [38]. However its chirality resulted from steric encumbering as a consequence of interlocking. **7** is the smallest member of the series of the [2]catenanes synthesized in the group of Sauvage and coworkers by the copper(I) template route, which requires that each macrocycle incorporates a chelate fragment derived from dpp as metal-recognition element [39]. The macrocycles of **7** contain 27 atoms in the shortest path along the ring vs. 30 for **1** and **4**. The chirality of **7** in solution was revealed by ^{1}H nuclear magnetic resonance (NMR) studies and was shown to arise from strain effects: each macrocycle was not large enough to accommodate both the Cu^+ cation and the dpp chelate of the other macrocycle coordinated to it, without deformations breaking the average D_{2d} symmetry of the complex into C_2. Another consequence of the strain induced by the threading of a [Cu(dpp)]$^+$ fragment of a macrocycle inside the other was the efficient sequestration of the metal cation, which could not be removed by an excess of cyanide. **1**, the rings of which contain 30 atoms along their internal circuit, is achiral. Therefore, the removal of an oxyethylene subunit from the 30 atom-membered rings of this [2]catenane makes a chiral catenane from an achiral one.

Taking advantage of the fact that the D_4-symmetric diagram of the Solomon rings of Figure 19.7a was equivalent to the D_2-symmetric diagram of Figure 19.7b, Sauvage and coworkers also reported the first preparation of a doubly interlocked [2]catenane [40] by using the helicate strategy that they had implemented for the synthesis of a molecular trefoil knot (see Section 19.5). The *m*-phenylene-bridged bis(phenanthroline) that they had designed for the high yielding synthesis of a molecular trefoil knot [41] was homologated to a tris(phenanthroline), which was subsequently functionalized with oligo(oxyethylene) chains bearing terminal alkene functions [42]. Dietrich-Buchecker and Sauvage showed that the functionalized tris(chelate) obtained could form a double-stranded helicate containing three Li^+ cations. Cyclization by alkene metathesis in Grubb's conditions, followed by exchange of Cu^+ for Li^+, chromatographic purification, hydrogenation of the double bonds, and demetalation of the cyclized helicate, afforded the doubly interlocked [2]catenane **8** in 30% yield.

Another example of a doubly interlocked [2]catenane obtained by using metal templation has been provided by the group of Leigh who used the D_4-symmetric presentation shown in Figure 19.7a [43]. Such a presentation could be reached

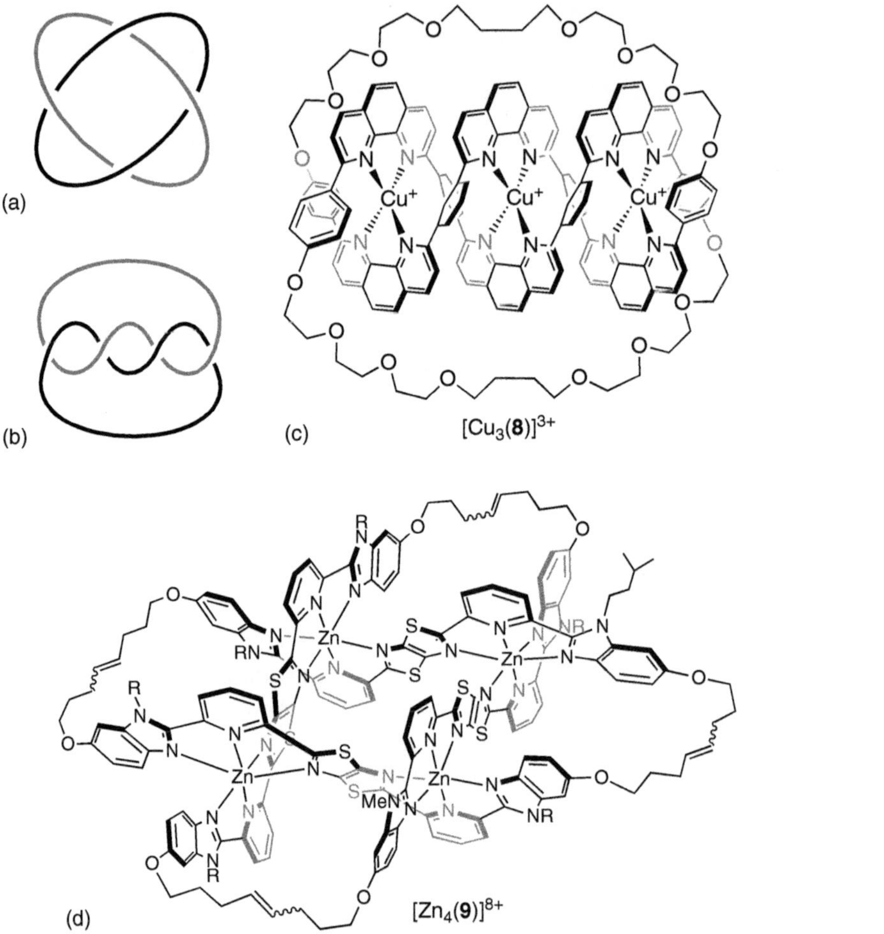

Figure 19.7 (a) D_4- and (b) D_2-symmetric presentations of the Solomon rings (left-handed enantiomers shown; the rings have been made different for clarity), (c) the trinuclear copper(I) complex $[Cu_3(\mathbf{8})]^{3+}$ of [2]catenane **8**, and (d) the tetranuclear zinc(II) complex $[Zn_4(\mathbf{9})]^{8+}$ of [2]catenane **9**.

through the self-assembly of a tetranuclear Zn^{2+} complex based on a highly preorganized bis-tridentate ligand into an interwoven molecular grid bearing terminal alkene chains. The tetranuclear zinc complex $[Zn_4(\mathbf{9})]^{8+}$ was obtained in 72% yield after ring-closing metathesis (Figure 19.7d). Demetalation by reaction with lithium sulfide afforded the doubly interlocked [2]catenane **9** in 52% yield. These two examples illustrate the importance of the presentation in the design of the synthesis strategy for topologically complex molecules [8, 44]. This will again be illustrated in the case of molecular knots (Section 19.5).

Since then, other doubly interlocked [2]catenanes have been obtained, for example, by dynamic covalent chemistry. Sanders and coworkers subjected to oxidation in mildly basic aqueous solutions a molecular strand incorporating two naphthalenediimide hydrophobic moieties separated by a hexamethylene chain and bearing cysteinyl thiol end groups [20]. They isolated a major product displaying the topology of a D_4-symmetric Solomon link and a minor product exhibiting the figure-of-eight knot topology (Figure 19.12c).

19.4 Chiral [2]Rotaxanes

The conditions stated for the topological chirality of a two-crossing [2]catenane can be directly transposed to a two-crossing [2]rotaxane, keeping in mind that the latter is not topologically chiral. The expression *mechanically chiral* has been recommended [27, 12] and used instead [45], but the preexisting concept of *inherent chirality* could work as well [46]. It was first defined for calix[4]arenes bearing different aryl substituents, the sequence of which confers an orientation to the macrocycle. The combination of the latter with the orientation of the calix[4]arene axis makes such a molecule chiral. Figure 19.8 compares schematically the inherent chirality of a [2]rotaxane made from oriented dumbbell and macrocyclic components and the topological chirality of a two-crossing [2]catenane made from oriented macrocyclic components.

The first chiral [2]rotaxanes ever synthesized incorporated α-cyclodextrin (α-CDX) as ring. α-CDX is a chiral, cone-shaped, oriented macrocycle. It has C_6-symmetry and is a case of cycloenantiomerism. As it provides a direction along the axle, if the latter is also oriented, diastereomers are generated, and the corresponding [2]rotaxanes show inherent diastereoisomerism. Ogino was the first in 1981 to report the formation of such [2]rotaxanes (**10**, Figure 19.9), which were made from α-CDX as ring, α,ω-diaminododecane as axle, and the chiral $[Co(en)_2Cl]^{2+}$ complex subunits as stoppers [47, 48]. The product was isolated as a mixture of diastereomers, which was partly resolved. With the stoppers being Δ- and Λ-chiral, if α-CDX is noted ">," there are four diastereoisomers: $\Delta > \Lambda$, $\Delta < \Lambda$, $\Delta > \Delta$, and $\Lambda > \Lambda$, of which two (**10a** = $\Lambda < \Lambda$ and **10b** = $\Lambda < \Delta$) are represented in Figure 19.9. **10b** is an inherently chiral [2]rotaxane in the sense defined in Figure 19.8b, as it has both an oriented macrocycle and an oriented dumbbell. Actually the synthesis, resolution, and study of the first inherently chiral [2]rotaxane was reported by Vögtle and coworkers, who used the same principle for making the [2]rotaxane **11** of Figure 19.10 chiral as the one devised for the topologically chiral [2]-catenane **5** [49]. Takata and coworkers reported the first attempt of synthesis of an inherently chiral rotaxane (**12**) based on the crown ether/ammonium interaction by asymmetric catalysis [50]. Hiratani and coworkers used an organic templation to make

(a) (b)

Figure 19.8 Schematical comparison of (a) a [2]catenane with oriented rings and (b) a [2]rotaxane having both ring and axle orientations (one enantiomer represented in each case).

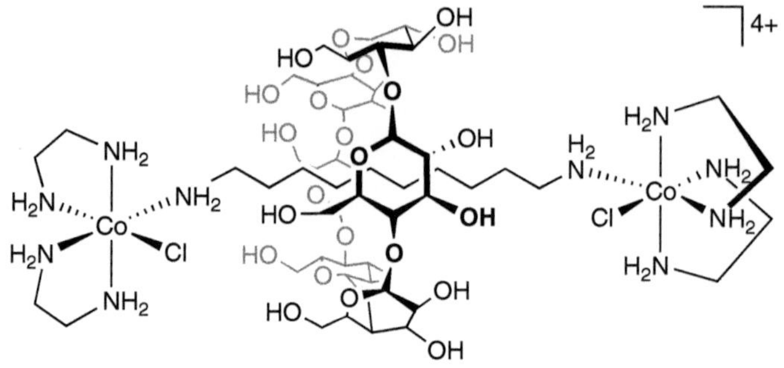

10a

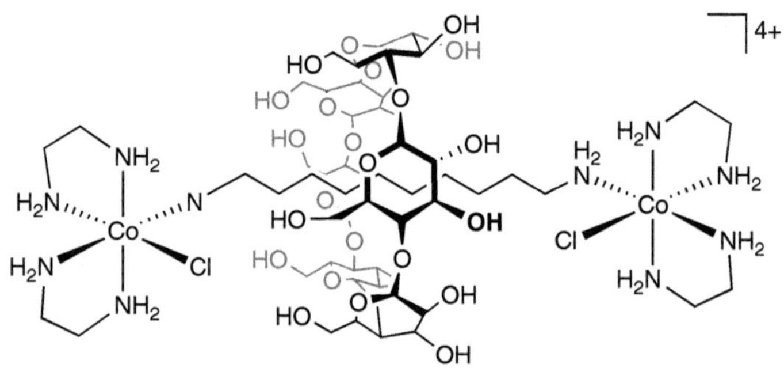

10b

Figure 19.9 Two isomers of the cyclodextrin-based [2]rotaxane **10**: **10a** corresponds to $\Lambda < \Lambda$ and **10b** corresponds to $\Lambda < \Delta$.

a chiral [2]rotaxane bearing two different stoppers (**13**) including a fluorescent anthracenyl stopper that was used for the detection of chiral substances [51, 52]. More recently, Goldup and coworker reported a very elegant asymmetric synthesis of inherently chiral [2]rotaxanes, such as **14** (Figure 19.10), applying the active template technique to an unsymmetrical 2,2′-bipyridine-incorporating macrocycle. The azide half-dumbbell precursor contained a D-glucose derivative, which, once the rotaxane formation had completed and the diastereomeric rotaxanes had been separated by conventional chromatography, was cleaved by ester/amide conversion, affording the rotaxane enantiomers [53].

Just as chirality may arise from asymmetric and dissymmetric conformations in simple molecules (e.g. biaryls, cyclophanes), which result from hindered rotations about single bonds, co-conformational chirality has been mainly observed in rotaxanes where it results from restricted translation of the macrocycle along the axle. Co-conformational chirality was first reported by Leigh and coworkers who designed an axle with a symmetry plane containing a *prochiral* atom substituted with an alcohol function for further derivatization (Figure 19.11) [54]. Once the [2]rotaxane was made, conversion of this function to a benzoyl ester afforded what could be called a *compartmental* [2]rotaxane in which the benzoyl ester group plays the role of a central stopper. The location of the macrocyclic component on

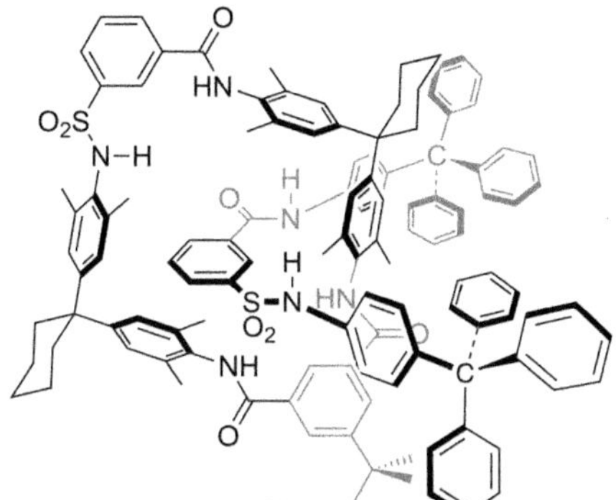

11

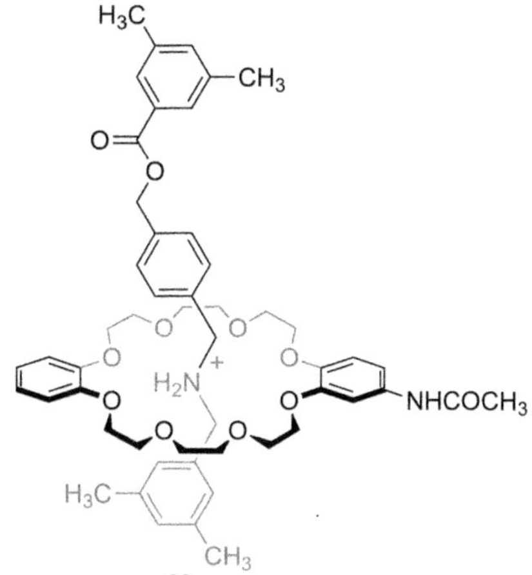

12

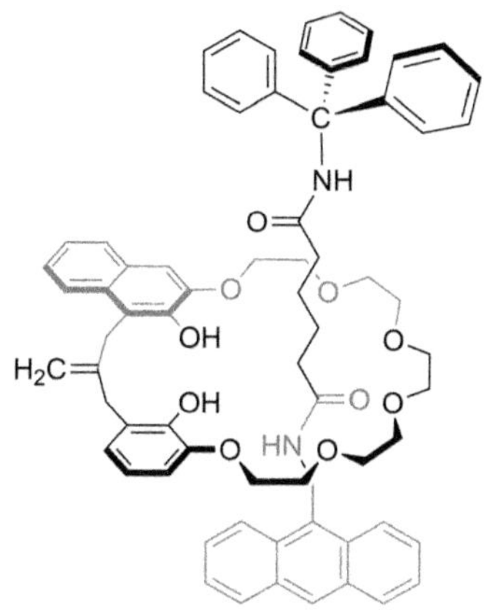

13

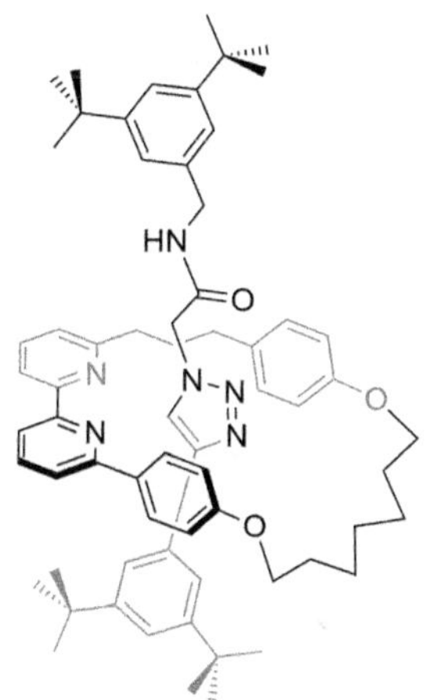

14

Figure 19.10 Inherently chiral [2]rotaxanes: **11** is based on the substitution of an amide function by a sulfonamide function in both macrocycle and axle, **12** has an unsymmetrical axle, **13** contains a dumbbell with two different stoppers, and **14** is made from a macrocycle and a dumbbell of heterogeneous compositions (one enantiomer is shown in each case).

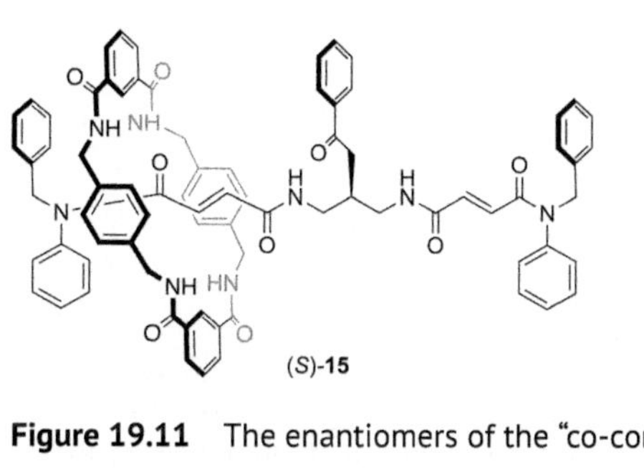

(*S*)-**15**

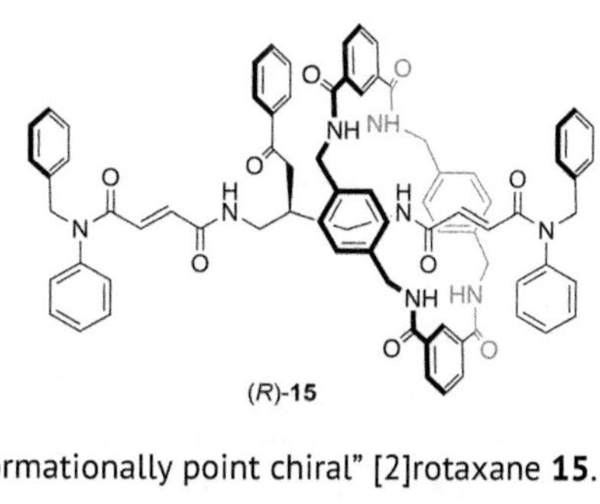

(*R*)-**15**

Figure 19.11 The enantiomers of the "co-conformationally point chiral" [2]rotaxane **15**.

either side of the axle makes the rotaxane **15** chiral, because the central atom of the axle is chirogenic. This kind of chirality in [2]rotaxanes has been qualified as "co-conformationally point chiral." The central carbon atom of the axle could be considered as having four different substituents, because the threaded macrocycle makes the halves of the axle different from each other.

19.5 Topologically Chiral Molecular Knots

The prime knots are classified according to the minimal number of crossings shown by their diagrams. Knots corresponding to different embeddings in 3D space but displaying the same number of crossings are distinguished by using subscript numbers. The first five prime knots are shown in Figure 19.12. Among these, ignoring the achiral unknot 0_1, the other knots are topologically chiral, except knot 4_1 (figure-of-eight knot), as it is possible to find a sequence of continuous deformations allowing the interconversion of its enantiomers [55].

Since the first report, by Sauvage, Dietrich-Buchecker, and coworkers, on the rational metal template synthesis of a molecule exhibiting the topology of a trefoil knot [56], other methods using passive and active metal templates have been developed for the synthesis of molecular trefoil knots. Several examples of these molecules were obtained using reactions under thermodynamic control using either dynamic covalent or coordination bonds [19, 57]. There are also examples of molecular knots that were discovered by serendipity, their formation being rationalized afterward. Higher-order knots were prepared by using circular helicates as templates. A particular family of knots is that of torus knots. By definition, a torus knot can be embedded on the surface of a torus, which means that its closed loop does not intersect itself [10]. All the torus knots, except the unknot, are topologically chiral. The linear, two-braid metallo-helicate approach to higher-order catenanes and knots turned out to be effective only for the syntheses of the four-crossing [2]catenane and the trefoil knot, both topologically chiral. Fortunately, its use as precursor, by Leigh et al., of a functionalized analogue of the circular pentanuclear Fe^{2+}-helicate initially reported by Lehn and coworkers [58] allowed these authors to synthesize a topologically chiral covalent pentafoil knot by metal complex templation and covalent connection of proximal extremities of

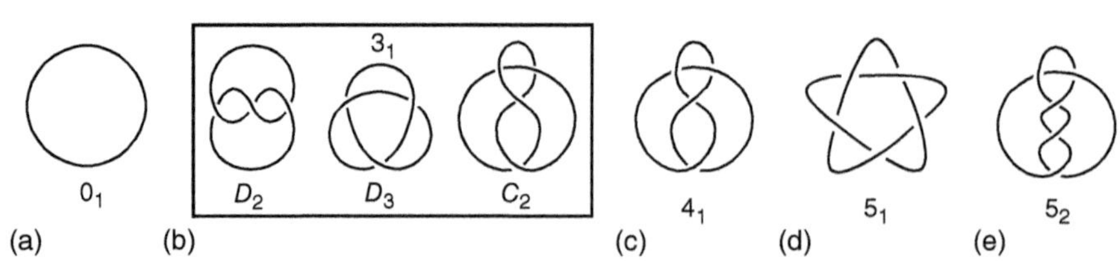

Figure 19.12 Classical diagrams of the first five prime knots including the unknot 0_1 (a). The right-handed trefoil knot (b) has been represented by three equivalent diagrams differing by their symmetry. Other knots represented are the figure-of-eight knot (c), the pentafoil knot (d), and the three-twist knot (e).

the tris-bipyridyl containing braids [59, 60]. Actually, a trefoil knot ligand backbone could also be obtained in the form of a two-braid circular trinuclear Zn^{2+}-helicate by imine bond formation between dibenzylamine-functionalized chelates and pyridine 2,6-carboxaldehyde [57]. Leigh and his group also used the circular helicate approach [61] and conducted the asymmetric synthesis of trefoil knots from trinuclear Co^{3+} and Zn^{2+} complexes, closing the knots by alkene metathesis [62]. The templation of circular helicates requires metal cations with higher coordination numbers than those used for making linear, two-braid metallo-helicates. In the latter case, four-coordinate metal cations of tetrahedral geometry (e.g. Cu^{+} and Li^{+}) are employed, whereas in the former cases, Fe^{2+}, Co^{3+}, and Zn^{2+} are six-coordinate metal cations (octahedral geometry). The highest-order molecular knot synthesized so far (noted 8_{19}) is a torus knot and was obtained by cyclization of a three-braid circular tetranuclear Fe^{2+}-helicate using alkene metathesis [63].

We shall focus on the topologically chiral trefoil knot, which is the simplest and the first knot to have been made in the form of a molecule. In their pioneering works, Sauvage and Dietrich-Buchecker succeeded in synthesizing a trefoil molecular knot based on the use of the D_2-symmetric diagram (Figure 19.12b) [56]. They took advantage of the fact that the former contains a double helix, a structure which could be obtained by homologation of the $[Cu(dpp)_2]^+$ template motif they had used earlier for the synthesis of [2]catenanes. The optimized thread, which gave the highest yields of molecular knots, was made by connecting the nine-positions of two 2-phenoxy-1,10-phenanthroline chelates with a *meta*-phenylene bridge [41, 56, 64]. The phenoxy substituents were used to anchor the precursors of the chains incorporating oligo(oxyethylene) atom sequences that closed the double helix into the trefoil knot **16** (represented as its copper(I) complex $[Cu_2(\mathbf{16})]^{2+}$ in Figure 19.13a). Noteworthy, the cyclization of the knotted molecule could be done sequentially, which turned out to be useful for the preparation of composite knots (see Section 19.6). The use of the olefin metathesis reaction to accomplish the cyclization of the double helix precursor turned out, as in many other instances, to improve dramatically the yields of product.

Application of a metal templation to direct the formation of a knot via the D_3-symmetric presentation was foreseen in the early 1970s by Sokolov [65] and first achieved by Hunter and coworkers through the synthesis of a tris-chelate molecular thread made by separating 5,5′-disubstituted bipyridines by rigid biphenylene hinges [66]. Upon coordination to a Zn^{2+} cation, the tris-bipyridine is spontaneously tied into the open precursor, which was subsequently locked into the trefoil knot **17** in complex $[Zn(\mathbf{17})]^{2+}$ (Figure 19.13b). A similar approach was taken by Leigh and coworkers, who designed a molecular thread incorporating three tridentate 2,9-pyridinedicarboxamide chelates separated from trioxyethylene-bridged naphthalene subunits by homochiral benzylic carbons. In this way, an open knot was obtained stereospecifically as a single enantiomer by reacting of the ligand strand with Eu^{3+}, a cation with high coordination numbers. It was then closed to the knot complex $[Eu(\mathbf{18})]^{3+}$ by alkene metathesis (Figure 19.13c) [67]. In Section 19.6, we shall see how such an open knot can be used for making composite knots. A really D_3-symmetric knot (complex $[Lu(\mathbf{19})]^{3+}$ of Figure 19.13d) had been made in the

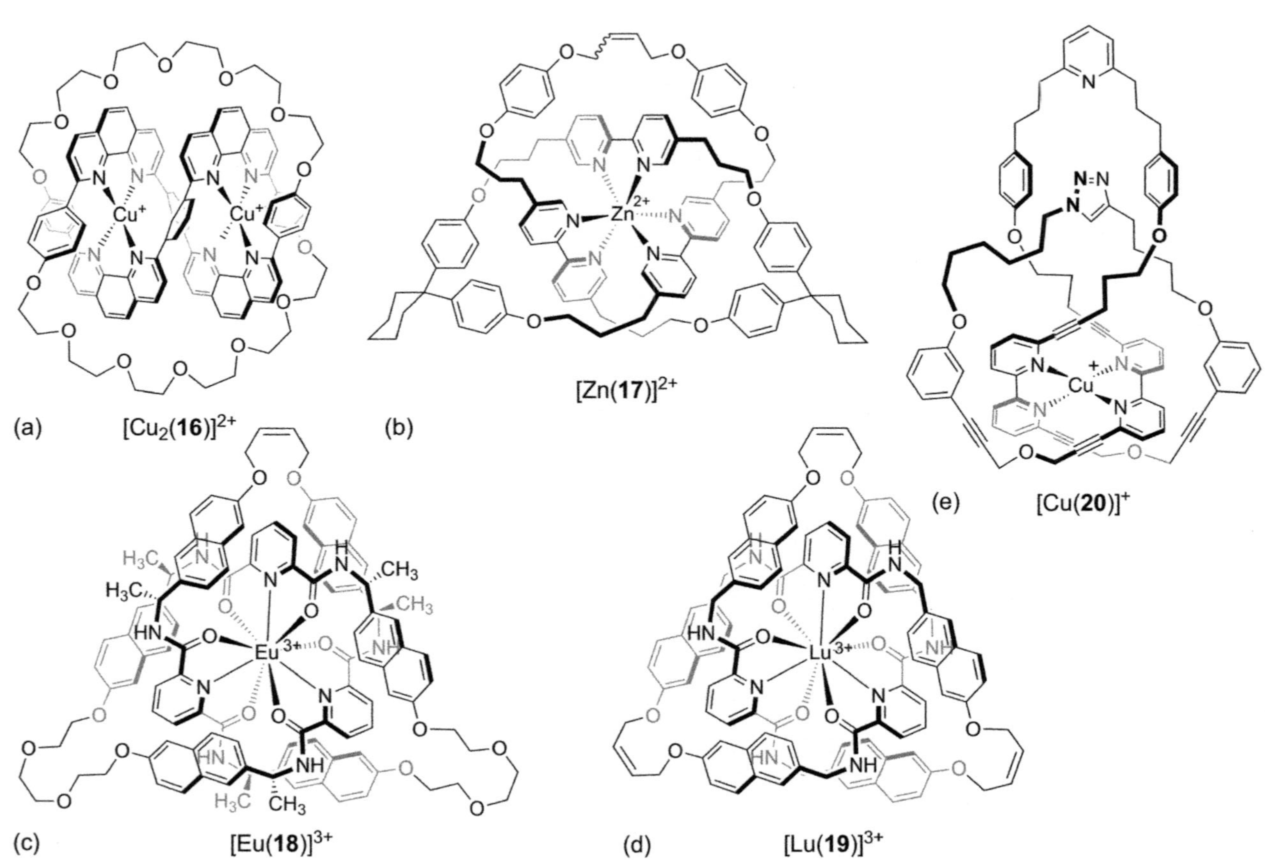

Figure 19.13 Chemical structures of complexes of the molecular trefoil knots obtained by metal templation (left-handed enantiomer shown in each case): (a) dinuclear copper(I) complex of knot **16** (D_2 diagram and symmetry), (b) zinc complex of knot **17** (D_3 diagram, but C_2 symmetry), (c) europium complex of knot **18** (D_3 diagram, but C_2 symmetry), (d) lutetium complex of knot **19** (D_3 diagram and symmetry), and (e) copper(I) complex of knot **19** (C_2 diagram, but asymmetric).

same group by triple closure of the spontaneous arrangement of three tridentate 2,9-pyridine-dicarboxamide chelates [68]. We shall not conclude the section devoted to metal templation without mentioning the rational use of the combination of an active-metal template and a passive metal template for the synthesis of a trefoil knot by Leigh and coworkers ([Cu(**20**)]$^+$ of Figure 19.13e). These authors designed a molecular strand incorporating a central pyridine ligand, a terminal alkyne and azide function, and a 6,6′-bipyridine (bipy) chelate between each function and the pyridine moiety [69]. The molecular strand is spontaneously wrapped around copper(I) by coordination of the bipy chelates. Upon binding to pyridine ligand subunit, the second Cu(I) cation was correctly positioned for the catalysis of the AAC reaction between the dangling alkyne and azide termini of the strand, which closed the trefoil knot through the loop developed between the central pyridine and the [Cu(bipy)$_2$]$^+$ complex subunit. This principle of knot formation uses the diagram C_2 shown in Figure 19.12b.

Vögtle and coworkers discovered that the reaction of the U-shaped diamine based on a central 2,6-benzenedicarboxamide flanked by rigid biphenylene hinges with 2,6-dichlorocarbonylpyridine produced the trefoil molecular knot **21** spontaneously in 20% yield [70]. Obtaining an X-ray crystal structure allowed the authors to detect a number of intramolecular hydrogen bonds that directed the knot formation and helped them to rationalize its formation (Figure 19.14) [71]. Feigel and coworkers, while attempting to cyclize an enantiomerically pure steroid trimer, actually obtained a trefoil knot (**2**), which formed as a single enantiomer by dimerization of the trimer (Figure 19.1b) [17]. Again, tying of the starting steroid-incorporating strand into a knot was driven by hydrogen bonding interactions. The last fully organic molecular knot we shall mention was also obtained in optically active form. Sanders and his coworkers designed a molecular strand incorporating three rigid naphthalenediimide subunits bridged by flexible homochiral propionate water-solubilizing groups and ended by cysteine-derived thiolates. The authors

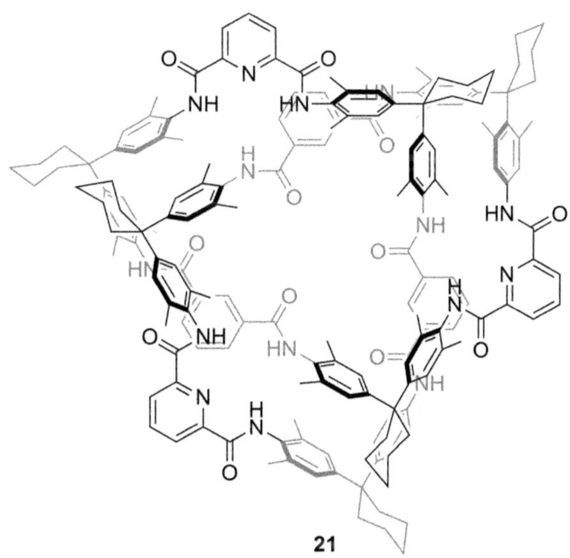

21

Figure 19.14 A molecular trefoil knot (**21**) obtained by hydrogen bond templation (see also Figure 19.1b).

showed that, upon air oxidation, concomitant trimerization and high yield tying of the strands into a homochiral knot occurred [20].

19.6 Catenanes, Rotaxanes, and Knots as Chiral Building Blocks

Just as classical diastereoisomerism results from the combination of two or more chirality elements (center, axis, and plane of chirality; helicity) in molecules with a planar graph, a particular form of diastereoisomerism will arise in molecules containing two or more catenane, rotaxane, and knot chiral building blocks. The incorporation of these nontrivial chiral building blocks in molecules can be done in either of two ways: (i) the building blocks are linked independently of each other in one molecule, and (ii) the building blocks are merged together so that parts of one building block belong to another. Figures 19.15–19.17 collect selected topologies that were obtained by connecting or merging chiral [2]catenanes, [2]rotaxanes, and knot building blocks. The corresponding real molecules are presented in the subsequent sections.

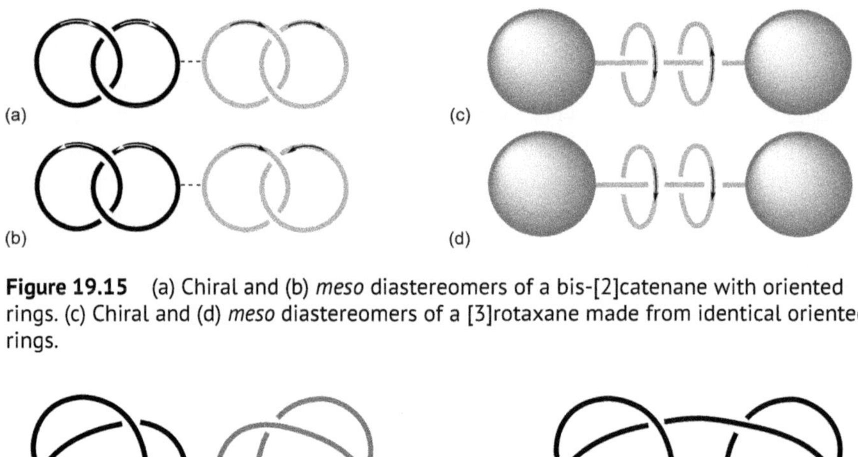

Figure 19.15 (a) Chiral and (b) *meso* diastereomers of a bis-[2]catenane with oriented rings. (c) Chiral and (d) *meso* diastereomers of a [3]rotaxane made from identical oriented rings.

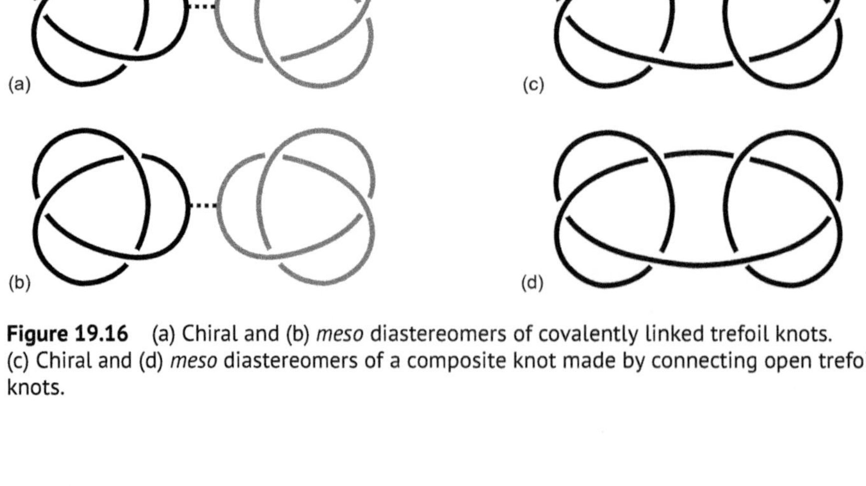

Figure 19.16 (a) Chiral and (b) *meso* diastereomers of covalently linked trefoil knots. (c) Chiral and (d) *meso* diastereomers of a composite knot made by connecting open trefoil knots.

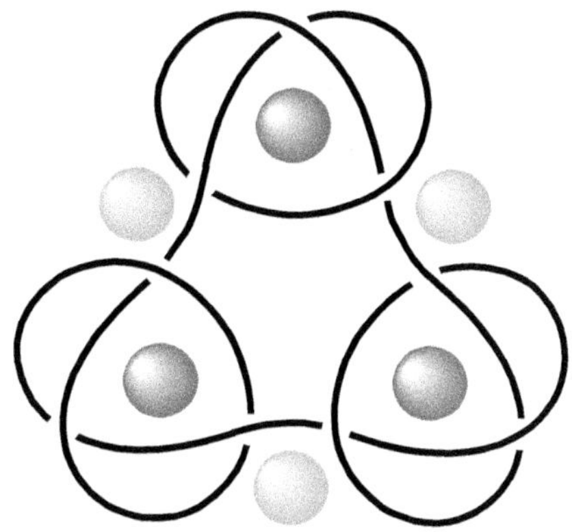

Figure 19.17 A composite topologically chiral knot made by merging three open trefoil knots of the same handedness using six template metal cations that are located in two different environments.

19.6.1 Catenanes

Connecting two topologically chiral [2]catenanes will produce a mixture of topological diastereomers, a pair of enantiomers, and a *meso* form (Figure 19.15a,b). Vögtle and coworkers synthesized an amide/sulfonamide [2]catenane, in which one of the sulfonamides was methylated, and obtained it as a mixture of enantiomers that could be separated by chiral high performance liquid chromatography (HPLC) [72]. As a result of the single methylation, the [2]catenane was formed selectively in a single co-conformation. Deprotonation of the sulfonamide group of the racemate followed by reaction with 1,2-bis(2-iodoethoxy)ethane allowed the authors to prepare compound **22** in which two independent topologically chiral [2]catenane building blocks were covalently bridged (Figure 19.18). They obtained a mixture of the *meso* form and the pair of enantiomers, from which the former could be separated by conventional chromatography. However, they were unable to resolve the pair of enantiomers, possibly because of the absence of unsubstituted sulfonamide functions, which seem essential for the resolution of this type of compound by chiral HPLC.

19.6.2 Rotaxanes

A [3]rotaxane containing a C_s-symmetric axle, but two oriented rings, will exist as three stereoisomers, a *meso* form in which the rings display the same orientation and a pair of enantiomers in which the rings display opposite orientations (Figure 19.15c,d). The principle of chirality of the chiral isomers of such a [3]rotaxane is similar to the principle of chirality of inherently chiral [2]rotaxanes: the second wheel is used for the desymmetrization of the axle. It is important to note that, in these conditions, the orientation of the second ring is not necessary; the [3]rotaxane would be chiral even if only one of the two rings was oriented. Vögtle and coworkers reported the synthesis and isolation of such [3]rotaxanes (**23**) constructed from axles incorporating either flexible or rigid spacers and triphenyl methane stoppers and from their classical amide-incorporating rings, the orientation of which is conferred by the replacement of one amide by a sulfonamide group (Figure 19.19a) [73]. The so-called bonnanes were obtained by covalent bridging of the macrocycles of [3]rotaxanes containing a central stilbene subunit [74].

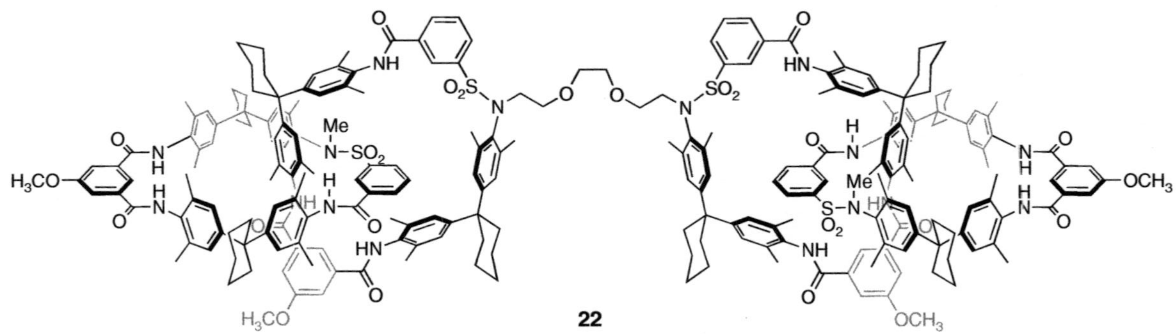

22

Figure 19.18 A chiral bis[2]catenane (**22**) made by covalently bridging two topologically chiral homochiral catenane building blocks.

(a)

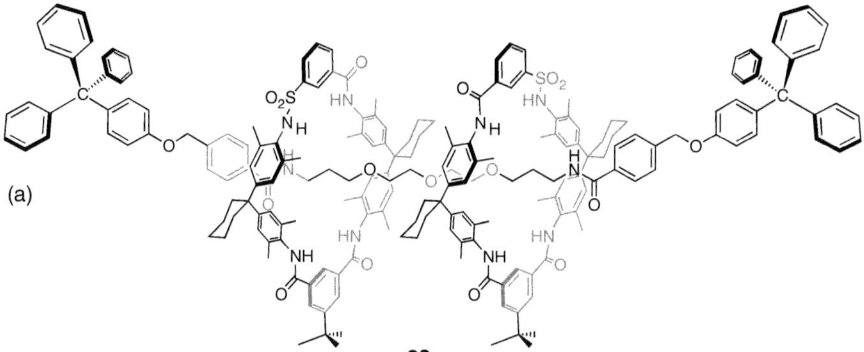

23

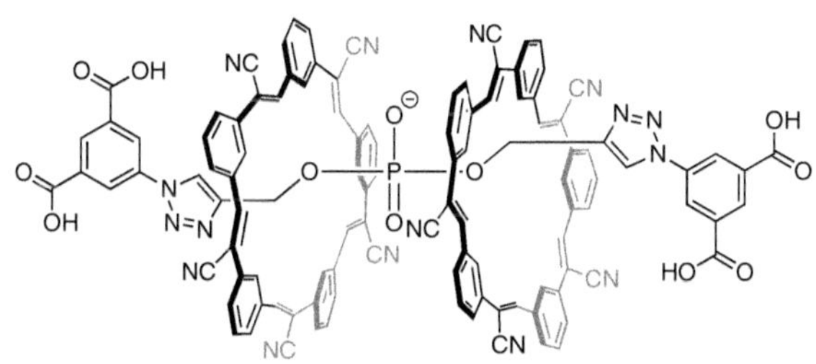

(b) 24

Figure 19.19 (a) A [3]rotaxane (**23**) incorporating two oriented macrocycles will exist in the form of a diastereomeric mixture of a pair of enantiomers, of which one is shown, and a *meso* form (not shown). (b) If, in addition to their orientation, the macrocycles have a direction, 10 possible stereoisomers are theoretically possible, but only three were obtained for **24**, the chiral pair of enantiomers, of which one is shown, and the *meso* form (not shown). Source: (a) [73] / John Wiley & Sons.

Flood and coworkers reported on the so-called cyanostar electrodeficient cyanostilbene-based macrocycle, the inward oriented protons of which (i.e. ethylenic and aromatic) bear a residual positive charge. These original, oriented macrocycles were shown to take up the shape of a bowl and to form complexes of 2 : 1 host/guest stoichiometry with weakly coordinating anions like BF_4^-, ClO_4^-, and PF_6^-, in which they sandwich the anion through their small rim. Cyanostar macrocycles were incorporated as rings in [3]rotaxanes made from of a very short axle based on a central dialkyl-substituted phosphate anion [75]. Due to the preferred mutual orientation of the macrocycles in their anion complexes, the [3]rotaxanes were formed diasteroselectively. Out of 10 possible stereoisomers, only three are formed, the achiral *meso* form, in which the macrocycles display the same orientation of the five double bonds, and the pair of enantiomers, in which the macrocycles display opposite orientations of the five double bonds, and the ratio was 8 : 2 (Figure 19.19b). The [3]rotaxane **24** was robust to slippage at room temperature even in the presence of a large excess of PF_6^- anions.

19.6.3 Molecular Knots

19.6.3.1 Covalent Linking of Trefoil Knot Building Blocks

The trefoil knot was used as a chiral building block in a number of discrete molecules, in two different ways, as outlined in the introduction of this section. The covalent connection of closed trefoil knots, shown schematically for two knots in Figure 19.16a,b, was explored by Vögtle and coworkers, who prepared a series of linear, branched, and cyclic knot conjugates [76, 77]. They first synthesized amide-based molecular knots incorporating outward protected phenol aromatic groups. After selective deprotection, they had at their disposal functionalized knots as chiral building blocks for the construction of larger molecules containing several knots. Compounds with various knot nuclearities were obtained. The simplest compound was a dumbbell, in which two knots are interconnected with a biphenyl spacer using sulfonic esters. If the knots have the same constitution, two diastereomers are obtained, the *meso* form, from knots of opposite chirality, and the pair of enantiomers from homochiral knots (**25**, Figure 19.20). If the knots differ by their constitution, a nonsymmetrical dumbbell-shaped molecule is obtained in the form of two pairs of enantiomers. The authors have drawn an interesting analogy between the trefoil knot and the center of chirality as chiral stereogenic units. A classical organic example of a molecule with a single center of chirality is glyceraldehyde. The knot-stoppered dumbbell, containing two knots of identical composition, is analogous to tartaric acid, the historical example of the occurrence of the diastereomeric *D–L* pair and the *meso* form; a knot-stoppered dumbbell containing two different knots is a topologically chiral analogue of erythrose and threose, which exist as two diastereomeric pairs of enantiomers. The authors used their knot-stoppered dumbbells for making [2]rotaxanes incorporating an oriented amide/sulfonamide macrocycle. They drew an analogy between the resulting "knotaxane," in which the chirality of the knot stoppers was combined with the orientation of the macrocycle, with trihydroxyglutaric acid [78]. Trihydroxyglutaric acid is the prototypical example of a molecule containing a pseudoasymmetric center. As a consequence, it exists as a *D–L* pair of enantiomers and two *meso* forms. Likewise, the *D–L* pair of enantiomers of the knotaxane differs by the chirality sense of the stoppers, which are homochiral, and the orientation of the macrocycle. However, the *meso* forms, the stoppers of which are heterochiral, differ only by the orientation of the macrocycle. Interconnection of two knot-stoppered dumbbell molecules produced a linear tetraknotane, the stereochemical properties of which were compared with those of tetrahydroxyadipic acid, which contains four asymmetric carbon centers and thus gives rise to 10 stereoisomers (i.e. four *D–L* pairs and two *meso* forms). The representative of the branched multi-knot containing system was a minimal dendritic compound (**26**, Figure 19.21) in which a central trefoil knot represented the core to which three peripheral trefoil knots were covalently bound. The branched tetraknotane was expected to exist in the form of four "*D–L*" pairs, that is, **Δ**ΔΔΔ/**Λ**ΛΛΛ, **Δ**ΔΔΛ/**Λ**ΛΛΔ, **Δ**ΔΛΛ/**Λ**ΛΔΔ, and **Δ**ΛΛΛ/**Λ**ΔΔΔ, the embolden letter representing the stereochemical descriptor of the knot located in the node of the tetraknotane.

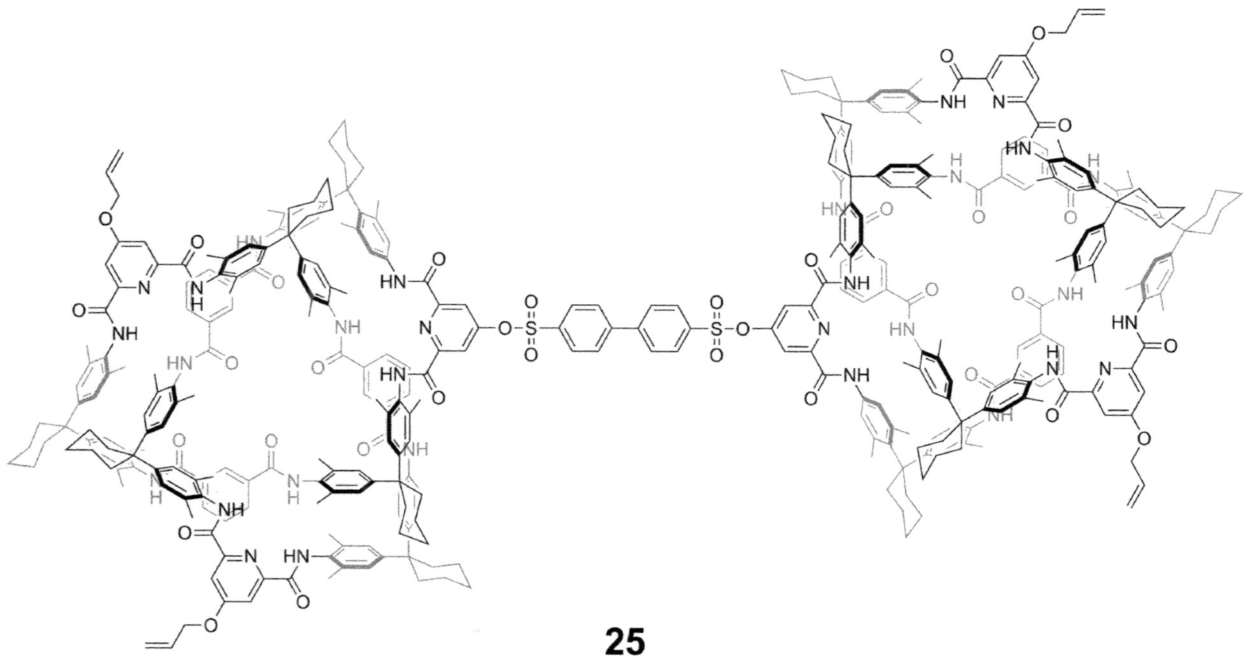

25

Figure 19.20 Covalent bridging of trefoil knots produces knot-stoppered dumbbell molecule. The isomer shown (**25**) is made from homochiral (Δ) trefoil knot subunits.

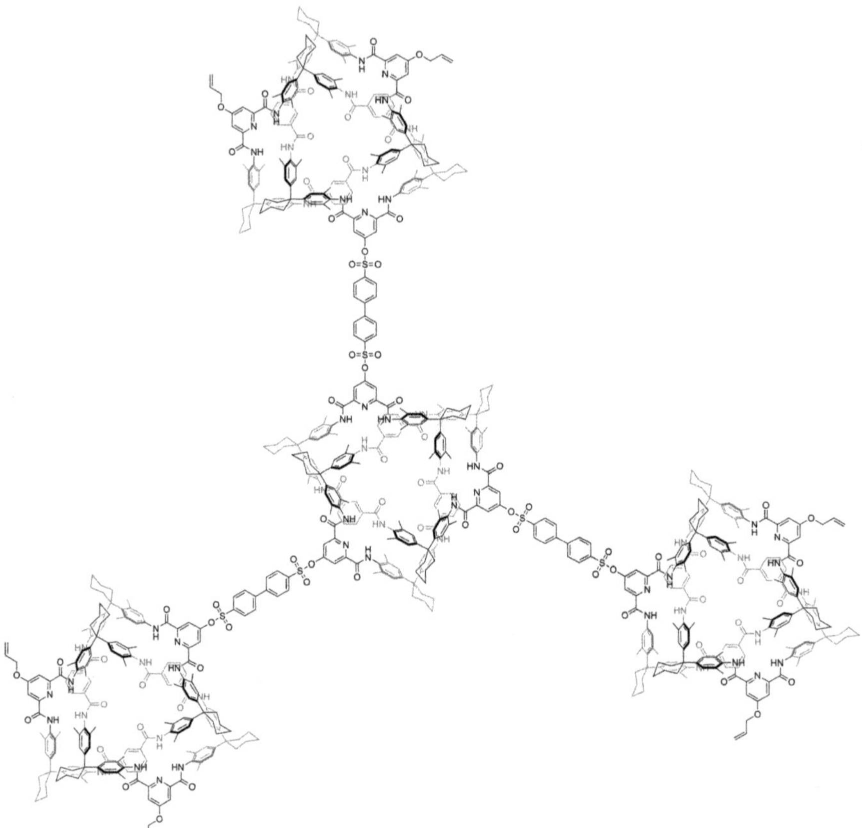

26

Figure 19.21 Chemical structure of the diastereomer of the branched tetraknotane **26** containing four homochiral (Δ) trefoil knot subunits.

Cyclic compounds, called "knotanophanes," and incorporating from two to four trefoil knots could be obtained in a one-pot reaction, with an overall yield of 65%, and separated by silica gel HPLC. The knotanophane dimer **27** (Figure 19.22) was isolated as a mixture of diastereomers (the two chiral forms, ΔΔ and ΛΛ, which were detected by chiral HPLC), and the *meso* form. The stereoisomers of the knotanophane trimer (**28** of Figure 19.23) and tetramer could not be detected individually, but a stereochemical analysis showed that the former should exist in the form of two diastereomers, each comprising two enantiomers, i.e. the ΔΔΔ/ΛΛΛ forms and the ΔΛΛ/ΛΔΔ forms, and that the latter should exist in the form of four diastereomers, two "*D–L*" pairs (ΔΔΔΔ/ΛΛΛΛ and ΔΔΔΛ/ΛΛΛΔ), and two *meso* forms (ΔΛΔΛ and ΔΔΛΛ).

19.6.3.2 Composite Knots

The so-called composite knots are obtained by combining prime knots. Sauvage and coworkers showed that homocoupling, in Glaser–Hay conditions, of a double helical dinuclear complex in which the *para* positions of the phenyl substituents of the

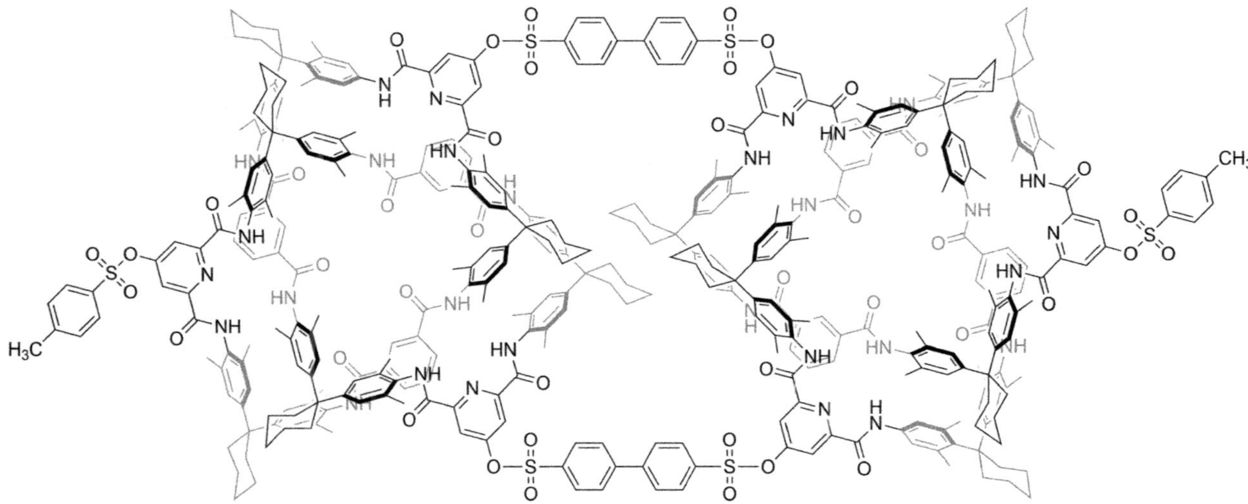

27

Figure 19.22 A covalent dimer of trefoil knots. The isomer shown (**27**) is chiral and made from homochiral (Δ) trefoil knot subunits.

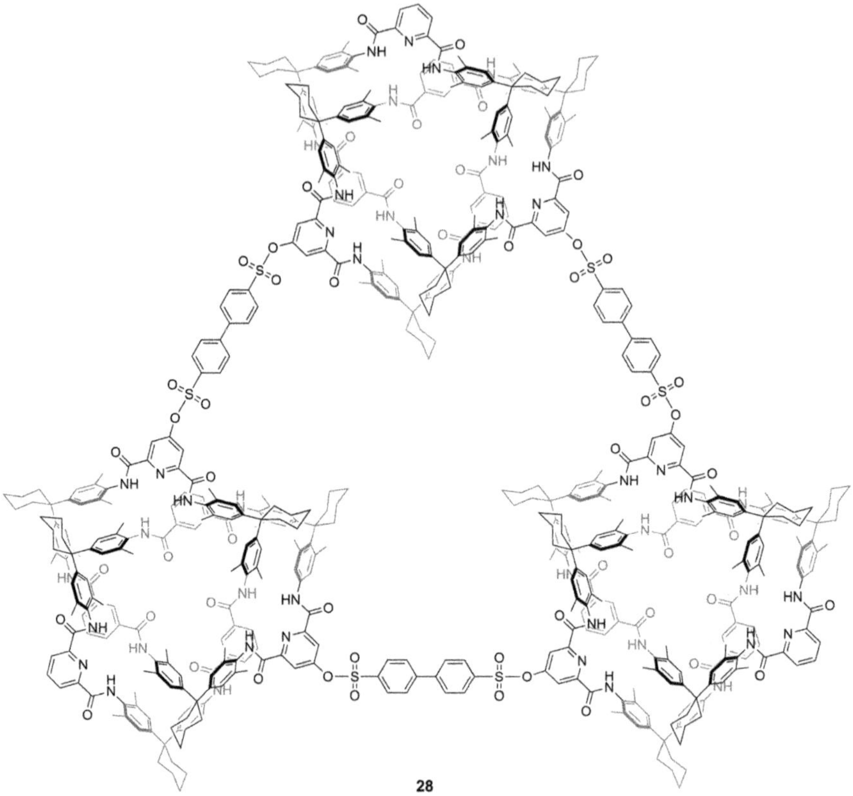

28

Figure 19.23 A covalent trimer of trefoil knots. The isomer shown (**28**) is chiral and made from homochiral (Δ) trefoil knot subunits.

central phenanthroline pairs were connected by a heptaethyleneglycol tether and those of the distal phenanthroline pairs were substituted with a propargyl function afforded a mixture of the dinuclear copper(I) complexes of the molecular "granny" (**29a**, Figure 19.24a) and "square" (**29b**, Figure 19.24b) knots [79]. $[Cu_4(\mathbf{29a})]^{4+}$ was obtained by coupling of the homochiral open knot dicopper complex, whereas $[Cu_4(\mathbf{29b})]^{4+}$ was obtained by coupling of open knot dicopper complexes of opposite chiralities and is therefore achiral.

More recently, a very efficient preparation of another topologically chiral composite trefoil knot was reported by Leigh et al. It used the single metal template-based approach, which relied on the D_3 presentation of the trefoil knot. All three stereoisomers were obtained stereoselectively via a two-step procedure. Figure 19.25 represents the structure of the dinuclear Lu(III) complex ($[Lu_2(\mathbf{30})]^{6+}$) of the enantiomer (Λ,Λ) of the composite trefoil knot **30** [80].

We shall conclude this section with the highest-order composite trefoil knot, which was synthesized in the group of Leigh, using the circular helicate strategy that had earlier allowed these authors to obtain a six-crossing [2]catenane [81]. This composite trefoil knot shown in Figure 19.26 as its hexanuclear Fe^{2+} complex

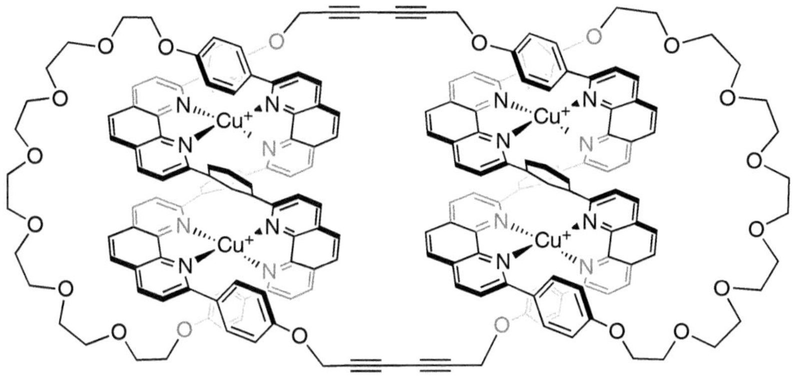

(a) $[Cu_4(\mathbf{29a})]^{4+}$

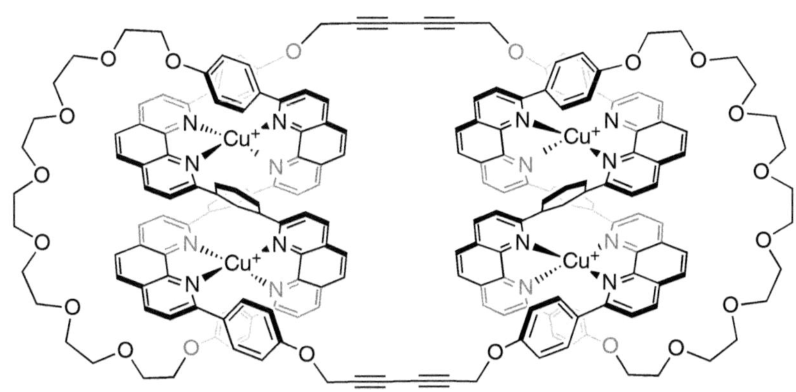

(b) $[Cu_4(\mathbf{29b})]^{4+}$

Figure 19.24 Copper(I) complexes of composite knots made by combining two double helical trefoil knot precursor complexes: (a) complex of the **29a** made from homochiral (Λ) open knot complex subunits and (b) complex of the knot **29b** made from heterochiral open knot complex subunits.

$[Fe_6(\mathbf{31})]^{12+}$ is made by fusing together three open trefoil knot fragments [82]. The principle of the construction is shown in Figure 19.17. The open knot fragments are formed around three Fe^{2+} cations. The three other Fe^{2+} cations interconnect the open knot fragments. Once the circular hexanuclear helicate is assembled, cyclization by alkene metathesis closes the two peripheral lobes of each knot fragment. It is interesting to compare the structure of the hexanuclear complex of this composite knot with that of the metalated six-crossing [2]catenane $[Fe_6(\mathbf{32})]^{12+}$. The outcome (i.e. the topology of the product) of the final cyclization step depended on the length and the geometry of the spacer between the peripheral bipyridine chelate subunits and the alkene functions. The desired composite knot was obtained using a longer spacer and an appropriate U turn.

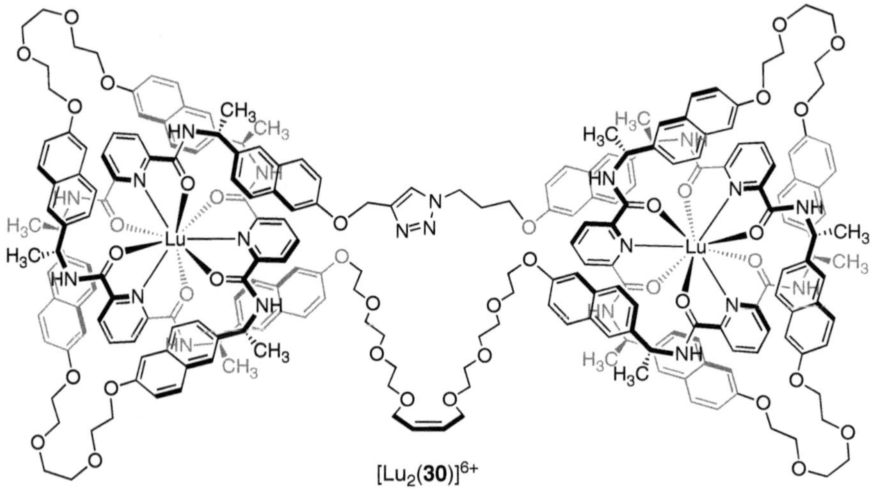

$[Lu_2(\mathbf{30})]^{6+}$

Figure 19.25 The dinuclear lutetium complex $[Lu_2(\mathbf{30})]^{6+}$ of the composite knot (**30**) made from homochiral (Λ) trefoil knot subunits.

19.7 Conclusions

In this essay, we have given few elements of chemical topology and defined the topological chirality of catenanes and knots, as well as the inherent chirality of rotaxanes (which obeys the same conditions as the so-called mechanical chirality). We have illustrated and discussed these particular chirality forms with selected examples of the literature. Finally, we have shown how several of these compounds or fragments of these compounds were used in the construction of more topologically complex molecules, that is, how they could be considered as chiral building blocks of their own. The combination of topologically chiral building blocks has led to the concept of topological diastereomerism, which has been compared with the classical diastereomerism based on centers of chirality. The combination of chiral building blocks based on interlocked, knotted, and threaded rings has been done in two ways: either "fusion" of fragments of the building blocks or covalent bridging of the building blocks, which thus behave independently. Examples of the latter are molecules incorporating several trefoil knot subunits while examples of the former are composite knots made by linking together open trefoil knot fragments and [3]rotaxanes incorporating oriented rings.

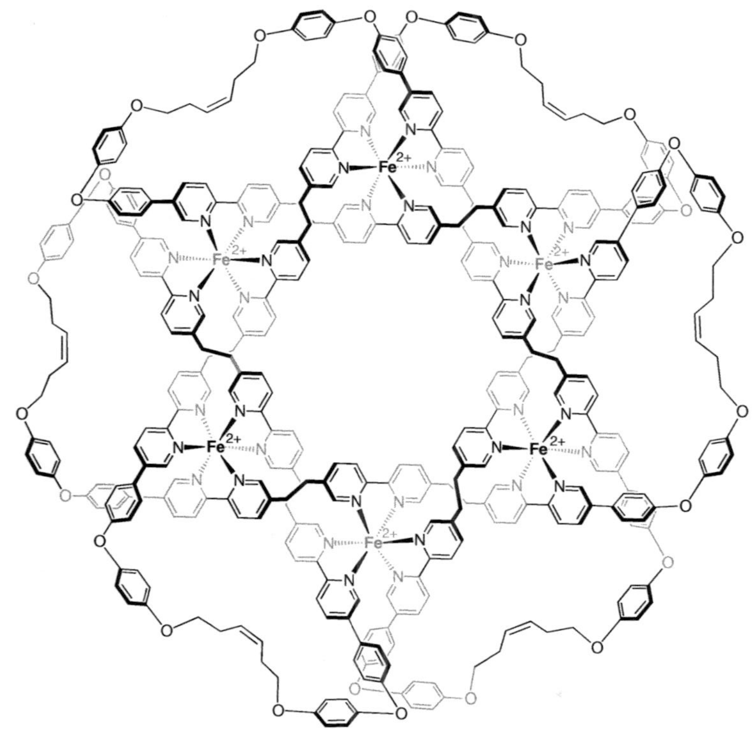

(a) $[Fe_6(\mathbf{31})]^{6+}$

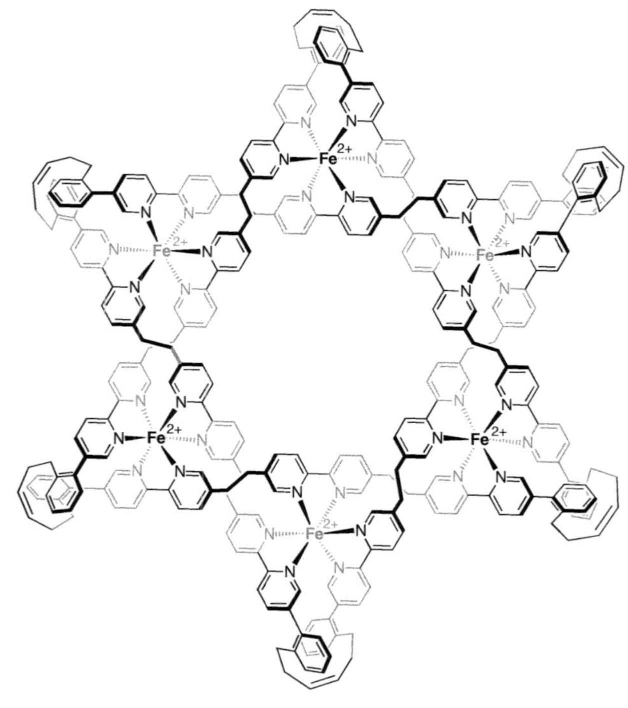

(b) $[Fe_6(\mathbf{32})]^{6+}$

Figure 19.26 Comparison of the structures of (a) the Fe^{2+} complex of the composite trefoil knot ($[Fe_6(\mathbf{31})]^{12+}$) made by fusing three open trefoil knot fragments and (b) the Fe^{2+} complex of the six-crossing [2]catenane ($[Fe_6(\mathbf{32})]^{12+}$) (in both cases one enantiomer is shown).

References

1 O'Loane, J.K. (1980). *Chem. Rev.* 80: 41–61.
2 Frisch, H.L. and Wasserman, E. (1961). *J. Am. Chem. Soc.* 83: 3789–3795.
3 Walba, D.M. (1985). *Tetrahedron* 41: 3161–3212.
4 Schill, G. (1971). *Catenanes, Rotaxanes and Knots*. New York: Academic Press.
5 Sauvage, J.-P. (2017). *Angew. Chem. Int. Ed.* 56: 11080–11093.
6 Stoddart, J.F. (2017). *Angew. Chem. Int. Ed.* 56: 11094–11125.
7 Forgan, R.S., Sauvage, J.-P., and Stoddart, J.F. (2011). *Chem. Rev.* 111: 5434–5464.
8 Beves, J.E., Blight, B.A., Campbell, C.J. et al. (2011). *Angew. Chem. Int. Ed.* 50: 9260–9327.
9 Neal, E.A. and Goldup, S.M. (2014). *Chem. Commun.* 50: 5128–5142.
10 Fielden, S.D.P., Leigh, D.A., and Woltering, S.L. (2017). *Angew. Chem. Int. Ed.* 56: 11166–11194.
11 Evans, N.H. (2018). *Chem. Eur. J.* 24: 3101–3112.
12 Jamieson, E.M.G., Modicom, F., and Goldup, S.M. (2018). *Chem. Soc. Rev.* 47: 5266–5311.
13 Chambron, J.-C., Dietrich-Buchecker, C., and Sauvage, J.-P. (1993). *Top. Curr. Chem.*, (guest ed. E. Weber), Springer-Verlag 165: 131–162.
14 Liang, C. and Mislow, K. (1994). *J. Math. Chem.* 15: 245–260.
15 Mislow, K. (1996). *Croat. Chem. Acta* 69: 485–511.
16 Dietrich-Buchecker, C.O., Sauvage, J.-P., and Kintzinger, J.-P. (1983). *Tetrahedron Lett.* 24: 5095–5098.
17 Feigel, M., Ladberg, R., Engels, S. et al. (2006). *Angew. Chem. Int. Ed.* 45: 5698–5702.
18 Huang, S.-L., Hor, T.S.A., and Jin, G.-X. (2017). *Coord. Chem. Rev.* 333: 1–26.
19 Ponnuswamy, N., Cougnon, F.B.L., Clough, J.M. et al. (2012). *Science* 338: 783–785.
20 Ponnuswamy, N., Cougnon, F.B.L., Dan Pantoş, G., and Sanders, J.K.M. (2014). *J. Am. Chem. Soc.* 136: 8243–8251.
21 Cougnon, F.B.L., Caprice, K., Pupier, M. et al. (2018). *J. Am. Chem. Soc.* 140: 12442–12450.
22 Feng, H.-J., Gao, W.-X., Lin, Y.-J., and Jin, G.-X. (2019). *Chem. Eur. J.* 25: 15687–15693.
23 Dang, L.-L., Sun, Z.-B., Shan, W.-L. et al. (2019). *Nat. Commun.* 10: 2057.
24 Singh, J., Kim, D.H., Kim, E.-H. et al. (2020). *J. Am. Chem. Soc.* 142: 9327–9336.
25 Dang, L.-L., Feng, H.-J., Lin, Y.-J., and Jin, G.-X. (2020). *J. Am. Chem. Soc.* 142: 18946–18954.
26 Chambron, J.-C., Heitz, V., and Sauvage, J.-P. (1992). *J. Chem. Soc., Chem. Commun.* 1131–1133.
27 Burns, C.J. and Stoddart, J.F. (2017). *The Nature of the Mechanical Bond. From Molecules to Machines*. Hoboken, NJ: Wiley.
28 Mitchell, D.K. and Sauvage, J.-P. (1988). *Angew. Chem. Int. Ed. Engl.* 27: 930–931.
29 Chambron, J.-C., Mitchell, D.K., and Sauvage, J.-P. (1992). *J. Am. Chem. Soc.* 114: 4625–4631.

30 Kaida, Y., Okamoto, Y., Chambron, J.-C. et al. (1993). *Tetrahedron Lett.* 34: 1019–1022.

31 McArdle, C.P., Vittal, J.J., and Puddephatt, R.J. (2000). *Angew. Chem. Int. Ed.* 39: 3819–3822.

32 Pentecost, C.D., Chichak, K.S., Peters, A.J. et al. (2007). *Angew. Chem. Int. Ed.* 46: 218–222.

33 Schouwey, C., Holstein, J.J., Scopelliti, R. et al. (2014). *Angew. Chem. Int. Ed.* 53: 11261–11265.

34 Song, Y.H., Singh, N., Jung, J. et al. (2016). *Angew. Chem. Int. Ed.* 55: 2007–2011.

35 Cui, Z., Lu, Y., Gao, X. et al. (2020). *J. Am. Chem. Soc.* 142: 13667–13671.

36 Ottens-Hildebrand, S., Schmidt, T., Harren, J., and Vögtle, F. (1995). *Liebigs Ann.* 1855–1860.

37 Denis, M., Lewis, J.E.M., Modicom, F., and Goldup, S.M. (2019). *Chem* 5: 1512–1520.

38 Dietrich-Buchecker, C.O., Edel, A., Kintzinger, J.-P., and Sauvage, J.-P. (1987). *Tetrahedron* 43: 333–344.

39 Dietrich-Buchecker, C.O. and Sauvage, J.-P. (1987). *Chem. Rev.* 87: 795–810.

40 Nierengarten, J.-F., Dietrich-Buchecker, C.O., and Sauvage, J.-P. (1994). *J. Am. Chem. Soc.* 116: 375–376.

41 Dietrich-Buchecker, C.O., Sauvage, J.-P., De Cian, A., and Fischer, J. (1994). *J. Chem. Soc., Chem. Commun.* 2231–2232.

42 Dietrich-Buchecker, C. and Sauvage, J.-P. (1999). *Chem. Commun.* 615–616.

43 Beves, J.E., Danon, J.J., Leigh, D.A. et al. (2015). *Angew. Chem. Int. Ed.* 54: 7555–7559.

44 Chambron, J.-C. and Sauvage, J.-P. (2013). *New J. Chem.* 37: 49–57.

45 Nakazono, K. and Takata, T. (2020). *Symmetry* 12: 144.

46 Dalla-Cort, A., Mandolini, L., Pasquini, C., and Schaffino, L. (2004). *New J. Chem.* 28: 1198–1199.

47 Ogino, H. (1981). *J. Am. Chem. Soc.* 103: 1303–1304.

48 Ogino, H. and Ohata, K. (1984). *Inorg. Chem.* 23: 3312–3316.

49 Jäger, R., Händel, M., Harren, J. et al. (1996). *Liebigs Ann.* 1201–1207.

50 Makita, Y., Kihara, N., Nakakoji, N. et al. (2007). *Chem. Lett.* 36: 162–163.

51 Kameta, N., Hiratani, K., and Nagawa, Y. (2004). *Chem. Commun.* 466–467.

52 Kameta, N., Nagawa, Y., Karikomi, M., and Hiratani, K. (2006). *Chem. Commun.* 3714–3716.

53 Bordoli, R.J. and Goldup, S.M. (2014). *J. Am. Chem. Soc.* 136: 4817–4820.

54 Cakmak, Y., Erbas-Cakmak, S., and Leigh, D.A. (2016). *J. Am. Chem. Soc.* 138: 1749–1751.

55 Chambron, J.-C. and Keene, R. (2017). *Chirality in Supramolecular Assemblies – Causes and Consequences* (ed. F.R. Keene). Chichester: Wiley.

56 Dietrich-Buchecker, C.O., Guilhem, J., Pascard, C., and Sauvage, J.-P. (1990). *Angew. Chem. Int. Ed. Engl.* 29: 1154–1156.

57 Prakasam, T., Lusi, M., Elhabiri, M. et al. (2013). *Angew. Chem. Int. Ed.* 52: 9956–9960.

58 Hasenknopf, B., Lehn, J.-M., Kneisel, B.O. et al. (1996). *Angew. Chem. Int. Ed. Engl.* 35: 1838–1840.

59 Ayme, J.-F., Beves, J.E., Leigh, D.A. et al. (2012). *Nat. Chem.* 4: 15–20.

60 Marcos, V., Stephens, A.J., Jaramillo-Garcia, J. et al. (2016). *Science* 352: 1555–1559.

61 Zhang, L., August, D.P., Zhong, J. et al. (2018). *J. Am. Chem. Soc.* 140: 4982–4985.

62 Zhong, J., Zhang, L., August, D.P. et al. (2019). *J. Am. Chem. Soc.* 141: 14249–14256.

63 Danon, J.J., Krüger, A., Leigh, D.A. et al. (2017). *Science* 355: 159–162.

64 Dietrich-Buchecker, C., Rapenne, G., and Sauvage, J.-P. (1997). *Chem. Commun.* 2053–2054.

65 Sokolov, V.I. (1973). *Russ. Chem. Rev.* 42: 452–463.

66 Guo, J., Mayers, P.C., Breault, G.A., and Hunter, C.A. (2010). *Nat. Chem.* 2: 218–220.

67 Gil-Ramírez, G., Hoekman, S., Kitching, M.O. et al. (2016). *J. Am. Chem. Soc.* 138: 13159–13162.

68 Ayme, J.-F., Gil-Ramírez, G., Leigh, D.A. et al. (2014). *J. Am. Chem. Soc.* 136: 13142–13145.

69 Barran, P.E., Cole, H.L., Goldup, S.M. et al. (2011). *Angew. Chem. Int. Ed.* 50: 12280–12284.

70 Safarowsky, O., Nieger, M., Fröhlich, R., and Vögtle, F. (2000). *Angew. Chem. Int. Ed.* 39: 1616–1618.

71 Brüggemann, J., Bitter, S., Müller, S. et al. (2007). *Angew. Chem. Int. Ed.* 46: 254–259.

72 Mohry, A., Vögtle, F., Nieger, M., and Hupfer, H. (2000). *Chirality* 12: 76–83.

73 Schmieder, R., Hübner, G., Seel, C., and Vögtle, F. (1999). *Angew. Chem. Int. Ed.* 38: 3528–3530.

74 Kishan, M.R., Parham, A., Schelhase, F. et al. (2006). *Angew. Chem. Int. Ed.* 45: 7296–7299.

75 Lee, S., Chen, C.-H., and Flood, A.H. (2013). *Nat. Chem.* 5: 704–710.

76 Lukin, O., Kubota, T., Okamoto, Y. et al. (2004). *Chem. Eur. J.* 10: 2804–2810.

77 Lukin, O. and Vögtle, F. (2005). *Angew. Chem. Int. Ed.* 44: 1456–1477.

78 Lukin, O., Kubota, T., Okamoto, Y. et al. (2003). *Angew. Chem. Int. Ed.* 42: 4542–4545.

79 Carina, R.F., Dietrich-Buchecker, C., and Sauvage, J.-P. (1996). *J. Am. Chem. Soc.* 118: 9110–9116.

80 Leigh, D.A., Pirvu, L., and Schaufelberger, F. (2019). *J. Am. Chem. Soc.* 141: 6054–6059.

81 Zhang, L., Stephens, A.J., Nussbaumer, A.L. et al. (2018). *Nat. Chem.* 10: 1083–1088.

82 Leigh, D.A., Pritchard, R.G., and Stephens, A.J. (2014). *Nat. Chem.* 6: 978–982.

Index

Chiral Building Blocks in Asymmetric Synthesis: Synthesis and Applications, First Edition.
Edited by Elżbieta Wojaczyńska and Jacek Wojaczyński.

b

c

e

p

q